Withdrawn
University of Waterloo

PLANTHOPPERS

Their Ecology and Management

Planthoppers: Their Ecology and Management

Robert F. Denno
Department of Entomology
University of Maryland
College Park, MD 20742

T. John Perfect
Natural Resources Institute
Central Avenue
Chatham Maritime
Chatham, Kent ME4 4TB
United Kingdom

PLANTHOPPERS

Their Ecology and Management

Edited by

Robert F. Denno & T. John Perfect

Withdrawn
University of Waterloo

CHAPMAN & HALL
New York • London

First published in 1994 by

Chapman & Hall
One Penn Plaza
New York, NY 10119

Published in Great Britain by

Chapman & Hall
2-6 Boundary Row
London SE1 8HN

© 1994 Chapman & Hall, Inc.

Printed in the United States of America on acid-free paper.

All rights reserved. No part of this book may be reprinted or reproduced or utilized in any form or by any electronic, mechanical, or other means, now known or hereafter invented, including photocopying and recording, or by an information storage or retrieval system, without permission in writing from the publishers.

Library of Congress Cataloging in Publication Data

Planthoppers : their ecology & management / Robert F. Denno & T. John
Perfect, editors.
p. cm.
Includes bibliographical references (p.).
ISBN 0-412-02341-5
1. Planthoppers—Ecology. 2. Planthoppers—Management. I. Denno,
Robert F. II. Perfect, T. John
SB945.P644P58 1993
632'.752—dc20 93-2764
CIP

ISBN 0-412-02341-5

Contents

Contributors

Betty Benrey: *Centro de Ecología, Universidad Nacional Autónoma de México, Apartado Postal 70-275, Ciudad Universitaria, 04510, MEXICO, D.F.*

Jiaan Cheng: *Department of Plant Protection, Zhejiang Agricultural University, Hangzhou, Zhejiang, PEOPLE'S REPUBLIC OF CHINA.*

Michael F. Claridge: *School of Pure and Applied Biology, University of Wales, P.O. Box 915, Cardiff CF1 3TL, UK.*

Anthea G. Cook: *Natural Resources Institute, Central Avenue, Chatham Maritime, Chatham, Kent ME4 4TB, UK.*

James T. Cronin: *Bodega Marine Laboratory, University of California, P.O. Box 247, Bodega Bay, California 94923, USA.*

Robert F. Denno: *Department of Entomology, University of Maryland, College Park, Maryland 20742, USA.*

Hartmut G. Döbel: *Department of Entomology, University of Maryland, College Park, Maryland 20742, USA.*

Kevin D. Gallagher: *FAO Intercountry IPM Programme, P.O. Box 1864, Manila, PHILIPPINES.*

J. Daniel Hare: *Department of Entomology, University of California, Riverside, California 92521, USA.*

Elvis A. Heinrichs: *West Africa Rice Development Association, Bouaka, Côte d'Ivoire, WEST AFRICA.*

Johnson Holt: *Natural Resources Institute, Central Avenue, Chatham Maritime, Chatham, Kent ME4 4TB, UK.*

Peter E. Kenmore: *FAO Intercountry IPM Programme, P.O. Box 1864, Manila, PHILIPPINES.*

Ryoiti Kisimoto: *Entomology Laboratory, Department of Agriculture, Mie University, 1514 Kamihama, Tsu, Mie 514, JAPAN.*

William O. Lamp: *Department of Entomology, University of Maryland, College Park, Maryland 20742, USA.*

Patricia C. Matteson: *Department of Entomology, Iowa State University, Ames, Iowa 50011, USA.*

Charles Mitter: *Department of Entomology, University of Maryland, College Park, Maryland 20742, USA.*

Lowell R. Nault: *Department of Entomology, Ohio Agricultural Research and Development Center, The Ohio State University, Wooster, Ohio 44691, USA.*

Geoff A. Norton: *Cooperative Research Centre for Tropical Pest Management, University of Queensland, Brisbane, Queensland 4072, AUSTRALIA.*

James R. Ott: *Department of Entomology, University of Maryland, College Park, Maryland 20742, USA.*

T. John Perfect: *Natural Resources Institute, Central Avenue, Chatham Maritime, Chatham, Kent ME4 4TB, UK.*

George K. Roderick: *Department of Entomology, University of Maryland, College Park, Maryland 20742, USA.*

Jane Rosenberg: *Natural Resources Institute, Central Avenue, Chatham Maritime, Chatham, Kent ME4 4TB, UK.*

Kazushige Sōgawa: *Laboratory of Insect Pest Forecasting, Kyushu National Agricultural Experiment Station, Ministry of Agriculture, Forestry and Fisheries, Izumi 496, Chi Kugo, Fukuoka 833, JAPAN.*

Peter D. Stiling: *Department of Biology, University of South Florida, Tampa, Florida 32620, USA.*

Donald R. Strong: *Bodega Marine Laboratory, University of California, P.O. Box 247, Bodega Bay, California 94923, USA.*

Peter W. F. de Vrijer: *Laboratory of Entomology, Agricultural University of Wageningen, P.O. Box 8301, 6700 EH, Wageningen, NETHERLANDS.*

Michael R. Wilson: *National Museum of Wales, Cathys Park, Cardif CF1 3NP, UK.*

Stephen W. Wilson: *Department of Biology, Central Missouri State University, Warrensburg, Missouri 64093, USA.*

Introduction: Planthoppers as Models for Ecological Study and Effective Pest Management

Robert F. Denno and T. John Perfect

Planthoppers constitute a large group of phytophagous insects in the order Homoptera. Currently, systematists recognize 19 families of planthoppers which collectively form the superfamily Fulgoroidea and derive their vernacular name from their jumping ability. Of this diverse array of families, one, the Delphacidae, albeit small compared to other sap-feeding herbivores such as aphids, has received by far the most attention. Of the approximately 2000 described species of delphacids, most are small in size (2–4 mm in length), dull in color, cryptic in habit, and go unnoticed by most biologists. It is this family of mostly grass-feeding species which has engaged our efforts for this volume.

Over the last century, only a small handful of renegade entomologists have studied delphacids in natural habitats and most of these workers have been taxonomists. Nevertheless, the utility of planthoppers as tools for the pursuit of ecological, evolutionary, and behavioral issues was recognized by a few. Past ecological studies involving planthoppers in nonagricultural habitats have addressed basic issues in mating behavior, communication, and species formation (Booij 1982a; de Vrijer 1984; Drosopoulos 1985; den Bieman 1987d), population dynamics (Denno et al. 1980; Prestidge and McNeill 1982, 1983), natural enemy/planthopper interactions (Rothschild 1966; Waloff 1980; Stiling and Strong 1982a; Waloff and Jervis 1987), and community organization (Waloff 1979; Denno 1980), but until recently such studies have been rather few and far between. In the last several years, however, an increasing number of studies has targeted delphacid planthoppers as model organisms to test ecological theory (Denno et al. 1989;

Cronin and Strong 1990a; Strong et al. 1990; Denno et al. 1991; Stiling et al. 1991b; Denno and Roderick 1992; de Winter 1992).

One key feature of the Delphacidae which has made them so attractive for ecological study has been wing dimorphism; both winged (macropterous) and flightless (brachypterous) adults with reduced wings can occur in the same population (Waloff 1980). The ease with which migratory and sedentary individuals can be identified has, in particular, facilitated studies of population dynamics and life history evolution (Kisimoto 1965; Cook and Perfect 1985b, 1989; Denno et al. 1991).

Paralleling the study of delphacid planthoppers in natural systems has been a much more intense effort to understand factors that contribute to their status as agricultural pests. Delphacids, due to their predilection for plants in the grass family, include some of the most devastating pests of major agricultural crops throughout the world. Contained in their feeding repertoire are staples such as rice, corn, sugarcane, and cereal crops on which they can inflict serious losses in production (Denno and Roderick 1990). Planthoppers damage plants directly by feeding in the phloem and by ovipositing in plant tissues; they also vector a variety of plant pathogens and are thus responsible for additional yield reduction (Sogawa 1982; Nault and Ammar 1989). The planthopper complex in Asia has ranked as a major threat to temperate rice production for centuries (Dyck and Thomas 1979). In the last two decades planthoppers have become a significant obstacle to the success of the Green Revolution in tropical Southeast Asia where losses are estimated in the hundreds of millions of dollars (Herdt 1987; Kenmore 1987). Outbreaks have occurred in response to new agronomic practices, to the breakdown in resistance of modern high-yielding varieties, and, in particular, to the widespread misuse of insecticides which killed the natural enemies that normally play such a key role in suppressing planthopper populations in tropical rice (Heinrichs et al. 1982a; Reissig et al. 1982a; Kenmore et al. 1984). The promiscuous use of pesticides also promoted the rapid development of resistance in several species of rice planthoppers (Nagata and Mochida 1984; Endo 1988a, 1988b). The result was a classic instance of insecticide-induced pest resurgence with frequent outbreaks of planthoppers and extensive crop loss (Heinrichs and Mochida 1984; Kenmore et al. 1987; Kenmore 1991). This critical dilemma stimulated research for the development of more ecologically sound management alternatives for rice planthoppers. Currently, regional management of planthoppers is being attempted by deploying resistant varieties, using insecticides judiciously, and conserving natural enemies (Lim and Heong 1984; Kenmore 1991). However, the balance is precarious and planthoppers continue to pose a widespread threat to rice.

Similar scenarios have occurred for planthoppers in other agroecosystems. For example, around the turn of the century, the sugarcane planthopper was single-handedly responsible for the demise of the Hawaiian sugar industry; recovery was impossible until control was achieved primarily through the introduction of an invertebrate predator (Swezey 1936).

Because of their primary pest status, a few species of planthoppers [*Nilaparvata lugens* Stål and *Sogatella furcifera* Horvath on rice, *Perkinsiella saccharicida* Kirkaldy and *Saccharosydne saccharivora* (Westwood) on sugarcane, *Peregrinus maidis* (Ashmead) on corn and sorghum, and *Javesella pellucida* (Fabricius) on cereals] have been the subject of continuous research attention for many years; this has given rise to an impressive database that includes nutritional physiology (Sogawa 1982), reproductive biology and mating systems (Ichikawa 1979; Claridge et al. 1985a; Zhang and Chen 1987), population dynamics and migration (Kuno and Hokyo 1970; Metcalfe 1972; Napompeth 1973; Kisimoto 1976; Kuno 1979; Cook and Perfect 1985b; Iwanaga et al. 1987; Cheng et al. 1989), host plant resistance (Claridge and Den Hollander 1982; Khush 1984; Romena et al. 1986), and natural enemy interactions (Raatikainen 1967; Kenmore et al. 1984; Nakasuji and Dyck 1984; Ooi 1988). The information has been generated by diverse groups of workers and often with a narrow species, geographic, and/or crop focus. The preeminent work on host plant nutrition, varietal resistance, and population dynamics has been conducted with rice planthoppers and much of the literature is published in Japanese and Chinese. Outstanding research on mating behavior, acoustical communication, and speciation has been conducted on the planthopper pests of cereal crops (de Vrijer 1984, 1986). By far the most convincing examples of classical biological control exist for planthoppers on sugarcane and taro (Swezey 1936; Matsumoto and Nishida 1966). Elegant research on plant/pathogen/planthopper interactions has been conducted in the corn system (Nault 1985). Furthermore, model management programs have been developed in rice with planthoppers as focal pests (Kenmore 1991).

To say the least, there exists an extremely scattered agricultural literature on planthoppers that has never been synthesized from an ecological perspective. Likewise, many basic scientists who have used planthoppers and their relatives to pursue theoretical issues in biology have remained largely unaware of the extensive work on the population biology of pest planthoppers that has been published in the Asian literature. There is, thus, an identifiable and important need to integrate the diversity of information on planthoppers, not only across natural and managed systems but also in a theoretical context. These, in brief, are the considerations that led us to compile the present volume on delphacid planthoppers within the framework of contemporary ecology.

Inspiration for this current work stems from the many interactions and discussions with colleagues at international congresses, crop protection symposia, and numerous ecological and entomological gatherings. Despite the long, and at times frustrating, gestation period, we believe the nature of the contributions herein combine to form a diverse and stimulating publication that will be of general value to both ecologists studying plant/herbivore/enemy interactions and pest managers striving to control agricultural pests.

The volume is structured in six parts and is based on what we hope readers will find to be a logical organization which flows from planthopper interactions with their host plants through life history evolution and behavior to population dynamics, interactions with natural enemies, and community structure. Part One begins with a consideration of the host plant relationships of planthoppers and the phylogenetic and ecological factors including plant chemistry and habitat structure which influence host plant selection, use, and dietary specialization. The life history styles of planthoppers and the ecological forces which shape them are addressed in Part Two as are reproductive behavior, acoustic communication, and mating systems. Assembled in Part Three are discussions of population dynamics and long-distance migration. Part Four examines the interactions of planthoppers with predators, parasitoids, and vectored plant viruses and considers factors which structure planthopper communities.

The focus of the volume then shifts to a treatment of those ecological and genetic factors that promote planthoppers as major agricultural pests. Part Five focuses on ecological approaches to effective planthopper pest management. Biological control successes are examined as are the factors which optimize natural enemy effectiveness. The genetics of host plant adaptation, varietal resistance and breakdown, and the development of insecticide resistance are considered in relation to the deployment of new crop varieties, natural enemy conservation, and the effective use of insecticides. In Part Six, these control considerations are integrated in a systems context for applied entomologists, and both future management options for planthoppers and how they might be realized are explored in the framework of Asian rice pest management.

PART ONE

Host Plant Relationships

Most if not all of the activities of delphacid planthoppers take place on or in very close proximity to their host plants. For example, all species feed, mate, and oviposit on the host, and overwintering/ diapause takes place either as embedded eggs in host plant tissues or as nymphs or adults nestled into litter near the base of the plant. Thus, most species, particularly brachypterous ones, are intimately associated with their host plants and it is only during bouts of dispersal that individuals relinquish plant contact. Consequently, the host plant, its taxonomy, chemistry, morphology, architecture, and persistence plays a central role in the lives of these insects and it is this focus which forms the theme for the initial section of this volume.

In Chapter 1, Wilson, Mitter, Denno, and Wilson examine the host plant relationships of delphacids and their planthopper relatives from an evolutionary perspective. In addition to determining which taxa of host plants are attacked, evidence for dietary specialization is provided. Host information is integrated with estimates of host plant and planthopper phylogeny and the fossil record to examine the possibility of parallel phylogenesis. The phylogenetic constraints on feeding location on the host plant (concealed or exposed above ground) are examined also. The radiation of insular delphacid species on dicots is contrasted with the usual habit of monocot feeding for continental species.

Cook and Denno (Chapter 2) examine the feeding behavior and niche of planthoppers and how they are affected by phloem physiology. The role of host plant chemistry in population biology is addressed with an emphasis on host plant selection, performance, and population dynamics. Plant defense and factors influencing host plant specificity are considered with special reference to the characteristics of crop varieties resistant to delphacid attack.

The effects of habitat structure and host plant architecture on the abundance and species richness is examined in Chapter 3 (Denno). Species–area relationships are assessed at both geographic and plant patch scales. The effects of patch size, plant density and plant diversity on planthopper abundance are explored also. The causal mechanisms for planthopper response to vegetation texture and persistence are discussed in the context of planthopper life history traits such as dispersal ability and overwintering stage which influence colonization and persistence.

1

Evolutionary Patterns of Host Plant Use by Delphacid Planthoppers and Their Relatives

Stephen W. Wilson, Charles Mitter,
Robert F. Denno and Michael R. Wilson

Introduction

Planthoppers (Homoptera: Fulgoroidea) are found on every continent except Antarctica and in all major biomes, including tropical rainforests, deserts, grasslands, and the arctic tundra (O'Brien and Wilson 1985). The more than 9000 described species are divided into 19 families (O'Brien and Wilson 1985; Wheeler and Wilson 1987). All species of Fulgoroidea are phytophagous, sucking fluids from leaves, stems, roots, or fungal hyphae. There are species which feed on woody plants (both angiosperms and gymnosperms), herbs, ferns, and even fungi (O'Brien and Wilson 1985).

The largest and most studied family of planthoppers is the Delphacidae which includes more than 2000 species in approximately 300 genera and 6 subfamilies (Asche 1985, 1990). Most continental delphacids are associated with monocots, particularly grasses and sedges (Wilson and O'Brien 1987; Denno and Roderick 1990), whereas some species, especially those on oceanic archipelagos, have radiated on dicots as well (Giffard 1922; Zimmerman 1948). Their predilection for monocots, phloem-feeding habit, oviposition behavior and ability to transmit pathogens have contributed to the severe pest status of several delphacid species on rice, corn, sugarcane, and several cereal crops (Wilson and O'Brien 1987).

Delphacids are intimately associated with their host plants and use them for feeding, mating, oviposition, and protection during winter and from their natural enemies (Claridge 1985a; Denno and Roderick 1990). As might be expected for delphacids from this close association, host plant nutrition, and allelochemistry have figured prominently in studies of host plant selection (Sogawa 1982; Woodhead and Padgham 1988), performance (Mitsuhashi 1979; Denno et al. 1986; Metcalfe 1970), population dynamics including dispersal (McNeill and Prestidge 1982; Denno and Roderick 1990), and community structure (McNeill and Prestidge 1982; Waloff 1980). Moreover, other aspects of the host plant, such as persistence and growth form, play an important role in shaping the life history strategies of planthoppers (Denno and Roderick 1990, Denno et al. 1991). Thus, elucidating the causes of current patterns of host plant use in the Delphacidae is central to understanding many aspects of their biology. Toward that end, we examine planthopper/host plant relationships from a phylogenetic and evolutionary perspective.

Synopsis of Issues

A recurring question concerning the evolution of host plant use by insects has been the degree to which it is genetically "constrained" (Mitter and Farrell 1991). If the genetic variation required for recognition and use of novel hosts were readily available, host preference might be locally optimized by natural selection with respect to intrinsic plant traits such as phenology or chemistry and to ecological variables such as plant and natural enemy abundance. If this were the case, there should be little phylogenetic pattern to host plant associations, which would be best understood as adaptation to local environments. The genetic variation for host use traits commonly observed in phytophage populations (Futuyma and Peterson 1985), and the rapid accumulation of herbivore species on some introduced plants (Strong et al. 1984), suggest that host use evolution might be relatively unconstrained by genetic barriers.

If, in contrast, adoption of a novel host required simultaneous change in a number of genetically independent traits, genotypes capable of such a shift might rarely occur. Such genetic "constraint" could slow or channel the evolution of host associations and impede local optimization of host preference (Mitter and Farrell 1991). Most species might be expected to feed on hosts similar or identical to those of their near relatives elsewhere, with local host use only fully interpretable in a phylogenetic context (Farrell and Mitter, in press).

The association of many insect genera or higher-ranking taxa with particular taxa of host plants (Ehrlich and Raven 1964) is consistent with the existence of strong genetic limits on host use. If major host shifting were rare, the association of particular herbivore and plant lineages might persist over geological time and through their subsequent diversification. One result of such "parallel phylogenesis" is that the present-day distribution of phytophage species across host taxa might reflect simply the age or phylogenetic divergence sequence of the associated taxa (Zwölfer 1978). Second, lineages which evolve while associated could have the opportunity to strongly influence each other's evolution. Such "coevolution," in the broad sense, might take several forms. For example, Ehrlich and Raven (1964) envisioned an "arms race" between plant defense and insect counteradaptation, with each innovation permitting an episode of radiation.

A major expectation, if a given insect and plant lineage have diversified significantly while in association, is that phylogenetic relationships among the host taxa should correspond in some fashion to those among their respective herbivores. If insect species were so host specific that each plant species were attacked by just the descendants of the herbivores of its immediate ancestors, there should be an exact match between insect and plant phylogenies, except where one lineage has speciated while its partner has not (Mitter and Brooks 1983). Such strict "parallel phylogenesis" may occur between some vertically transmitted endosymbionts and their hosts (Munson et al. 1991), but seems unlikely for planthoppers, in which the dispersive adults have the opportunity to select alternative hosts. However, substantial cladogram correspondence between synchronously evolving insect and plant lineages is also expected if insects colonize (transfer or shift to) host plant species other than the ones on which they developed, provided that these hosts are related (Farrell and Mitter 1990). Ehrlich and Raven's (1964) model, for example, implies periodic radiation by plants that evolve temporary immunity from herbivores, with subsequent colonization most likely to come from herbivores on related plants bearing the antecedent defense. Even if newly available hosts were colonized at random, cladistically basal (and hence older) herbivore taxa should tend to occur on correspondingly primitive plant taxa, unless early host associations were rapidly obscured through extinction or phyletic evolution of host choice.

Although concordance of phylogenies suggests diversification in concert, it could also result from herbivore colonization and speciation entirely subsequent to plant diversification if the plant similarities (in secondary chemistry or other traits) constraining host transfers were correlated with phylogenetic relatedness. To reject such "host tracking" in favor of parallel diversification, independent evidence is needed from fossils, distribution patterns, or "molecular clocks" that plant lineages are comparable in age to their associated insect lineages. Note that parallel phylogenesis itself does not necessarily imply coevolution, as the associated lineages might have simply responded independently to a shared history of geographic vicariance. However, the existence or apparent absence of parallel phylogenesis is an important indicator of the potential influence of evolutionary history, including coevolution, on present-day host use patterns. A review of the sparse evidence to date suggests that although closely parallel phylogenesis is rare, partial concordance of insect and host plant cladograms is widespread (Mitter and Farrell 1991). Close examination of many additional herbivore groups is needed to determine the overall influence of parallel phylogenesis on patterns of host use by phytophagous insects.

Distinct from the issue of which host plant taxa are attacked, much attention has also focused on how many host taxa are used (Futuyma 1983; Jermy 1984; Thompson 1988b; Jaenike 1990). Many selective factors have been invoked to explain the predominance of host specificity in phytophagous insect species, including plant toxins, repellents, and attractants (Dethier 1954; Levins and MacArthur 1969; Futuyma 1983), host plant abundance (Hsiao 1978; Futuyma 1983), the extent to which herbivores mate on their hosts (Jaenike 1990), and natural enemy attack rates (Price et al. 1980; Bernays and Graham 1988; Denno et al. 1990). It is becoming more apparent, however, that host plant range itself, like affiliation with a particular host taxon, may be phylogenetically conserved. Thus, host plant range in any given species should also be studied in a phylogenetic context. However, such studies on the phylogenetic patterns of host range are rare for herbivorous insects (but see Jermy 1984; Bernays and Chapman 1987; Mitter and Farrell 1991).

In this chapter, we synthesize the scattered data on delphacid host plants in order to establish broad patterns of host use including the degree of feeding specialization. We use this information, in conjunction with estimates of delphacid and host plant phylogenetic relationships and geological age, to explore the degree of phylogenetic conservatism of host use exhibited by the delphacids and to examine the possibility of parallel phylogenesis. To place delphacid patterns in a broader context, we provide first a review of phylogeny and host use in the Fulgoroidea as a whole. Our analysis was limited by the lack of delphacid phylogenies below the tribal level, by the uncertainty about phylogenetic position for many host plant groups, and by the relative dearth of information on the feeding habits for tropical as compared to temperate planthoppers and for most fulgoroid groups other than delphacids. Nonetheless, we are able to provide at least provisional evidence on the issues raised above and demonstrate statistically several phylogenetic trends in planthopper host use.

Phylogenetic Relationships Among Fulgoroid Families

There is ample evidence that members of the Fulgoroidea (= Fulgoromorpha) form a monophyletic group (Muir 1923, 1930a; Evans 1963, 1964;

Hennig 1981; Bourgoin 1986; Asche 1987; Emeljanov 1987, 1991). The extant families include the Tettigometridae, Delphacidae, Cixiidae, Kinnaridae, Meenoplidae, Derbidae, Achilidae, Achilixiidae, Dictyopharidae, Fulgoridae, Issidae including Acanaloniidae, Nogodinidae, Ricaniidae, Flatidae, Hypochthonellidae, Tropiduchidae, Lophopidae, Eurybrachidae, and Gengidae. The fulgoroids together with their apparent sister group the Cicadomorpha (Cicadoidea, Cercopoidea, and Cicadelloidea) make up the Auchenorrhyncha, which, in turn, is probably the sister group to the Heteroptera + Coleorrhyncha [Carver et al. 1991; but see Emeljanov (1987) and Hamilton (1981) for alternative views].

A preliminary cladistic assessment of phylogenetic relationships among the fulgoroid families (Fig. 1.1A) was provided by Asche (1987), who built on the work of Muir (1923, 1930a). Asche's hypothesis was subsequently criticized and expanded on by Emeljanov (1987, 1991), whose arrangement is depicted in Figure 1B. Although an exhaustive, quantitative cladistic study has yet to be conducted, synthesis of these largely concordant views provides a reasonable working hypothesis for our subsequent discussion of host plant relationships.

At least some character evidence has been advanced for the monophyly of most fulgoroid families, although a phylogenetic definition is still lacking for some, such as the Kinnaridae and Issidae (Asche 1987; Emeljanov 1991). Like Muir (1923), both Asche (1987) and Emeljanov (1991) regard the Tettigometridae as the most primitive family of Fulgoroidea, although Bourgoin (1985) offered some evidence for affinity with the higher fulgoroids. There is also agreement that cixiids and delphacids are basal to the remaining, more advanced families (see also Muir 1930a); the only disagreement concerns the exact placement of these two families. Emeljanov (1987, 1991) disputes the hypothesis of a delphacid–cixiid sister group relationship, offering four characters to support grouping of the cixiids with the more advanced families; this scheme requires parallel origin or loss of the fusion of the median gonapophyses of the ovipositor (Asche 1987). Both authors regard the Kinnaridae plus Meenoplidae as a monophyletic group, a relationship recognized by Muir (1930a); Emeljanov (1991) further postulates these to be sister group to the remaining, higher fulgoroids, an arrangement consistent with the sequence of families in the classification of Muir (1930a). Asche (1987) and Emeljanov (1991) agree with Muir (1930a) in grouping the Dictyopharidae with the Fulgoridae, and agree that all seven families named so far plus the Derbidae and Achilidae are excluded from an advanced clade consisting of the Issidae, Flatidae, Tropiduchidae, and their relatives; Muir (1930a) provides a very similar ordering of families.

In summary, the phylogeny (Fig. 1B) which we provisionally adopt for

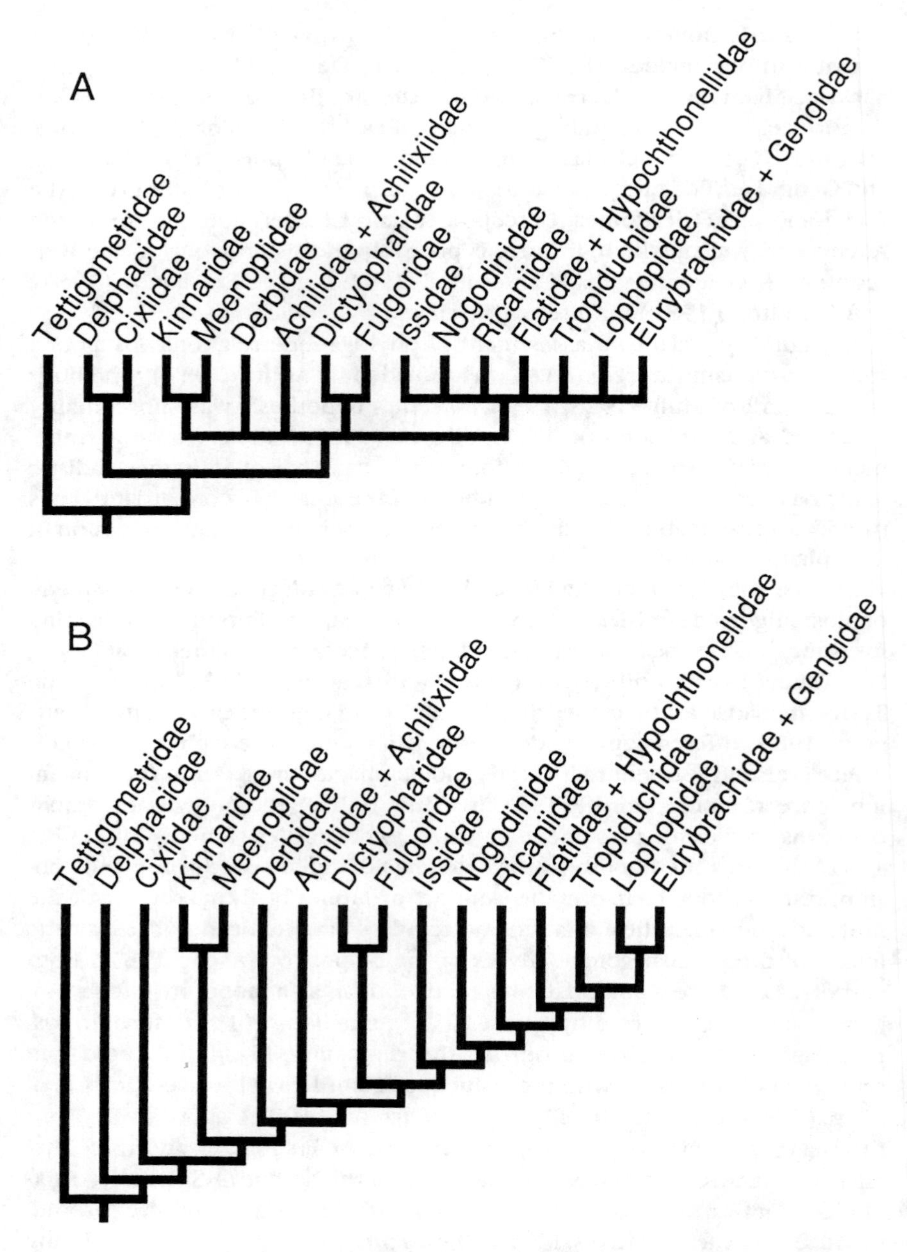

Figure 1.1. Hypotheses for phylogenetic relationships among fulgoroid families. (**A**) Phylogeny inferred from text and figures in Asche (1987); (**B**) phylogeny from Emeljanov (1991).

subsequent discussion of feeding habit evolution is consistent with, and at many nodes supported by, the main previous hypotheses, save for the precise placement of Cixiidae.

Fossil Record of the Fulgoroidea

Undoubted members of the Hemiptera *sensu lato* are first known from the Permian (Wootton 1981; Hennig 1981; Hamilton 1990). The Permian forms probably include true Auchenorrhyncha (but see Hamilton 1990), but assignments to modern subgroups are uncertain (Wootton 1981; Hennig 1981; Hamilton 1990). The time of first appearance for the Fulgoroidea has been particularly controversial. Some Permian fossils have been interpreted as possible fulgoromorphs (Evans 1963, 1964; Hennig 1981; Wootton 1981), and several extant families of Fulgoroidea have been reported from the Triassic (Muir 1923; Metcalf and Wade 1966; Hennig 1981). However, most of these fossils consist only of forewings (Evans 1964), which provide few characters useful for defining them as Fulgoroidea, let alone as members of extant families (Evans 1963, 1964; Hennig 1981; Emeljanov 1987). Numerous fossils of Fulgoriidae are known from the Jurassic, but assignment of this family to the Fulgoroidea has now been convincingly rejected (Hamilton 1990; Emeljanov 1987). Although two other Jurassic fossils may be fulgorids (Hamilton 1990), the most definitive early fulgoroid remains appear in the Lower Cretaceous, with an especially rich deposit in Brazil (Hamilton 1990). This fauna appears to include representatives of the modern family Achilidae and possibly the Cixiidae (see also Fennah 1987), as well as a profusion of species which may belong to the stem groups of several modern clades (Hamilton 1990). Dictyopharidae are known from the Upper Cretaceous, but definite fossils of other fulgoroid families, including the Delphacidae and higher families such as the Fulgoridae, Flatidae, Issidae, and Ricaniidae, appear first in the Tertiary, with most records from the Eocene, Oligocene, and Miocene (Metcalf and Wade 1966; Fennah 1968; Hamilton 1990).

In conjunction with Emeljanov's phylogenetic estimate (Fig. 1B), these data suggest the following conclusions. Although surely incomplete, the fossil record generally supports the hypothesized phylogeny in that families represented by the oldest fossils (Achilidae and Cixiidae) are relatively basal in the phylogeny. Given that the Achilidae arose by the Lower Cretaceous, their phylogenetic position implies that the cladistically more basal lineages (Tettigometridae, Delphacidae, Cixiidae, Kinnaridae/Meenoplidae, and Derbidae) must have arisen still earlier, even though Mesozoic fossils of these groups are largely lacking. Divergence of the Fulgoromorpha from the Cicadomorpha must have occurred even earlier. Among the more advanced fulgoroids, the Fulgoridae must be as old as their sister group, the Dictyopharidae, which occurred in the Upper Cretaceous; it is possible that the remaining higher families did not differentiate until the Tertiary (Hamilton 1990). Thus, although the evolution of the Fulgoroidea as a whole greatly overlaps the radiation of the angiosperms, the divergences among and perhaps within the older fulgoroid families including the Delphacidae may have preceded the appearance of angiosperms in the Lower Cretaceous (130 million years ago) and their proliferation in the Upper Cretaceous (Cronquist 1981).

Host Plant Relationships of the Fulgoroidea

Appendices 1 and 2 are the result of our effort to compile as complete a list of host plant records for the Fulgoroidea as possible. The data were assembled by examining the taxonomic, ecological, and agricultural literature, in addition to faunal surveys. Spurious records can result when adults are collected from nonhost vegetation, especially by sweep net, in the absence of information on feeding or presence of nymphs. To minimize such false records, we excluded reports for species (1) collected by general sweeping, (2) stated by the author to have been observed resting but not feeding on the plant, (3) taken in surveys of crops of which they are not known to be pests (e.g., Ballou 1936; Bruner et al. 1945; Kramer 1978; Maes and O'Brien 1988), or (4) whose taxonomic identities were questionable (Swezey 1904).

Our subsequent discussion of evolutionary patterns in host use is based on the following synopsis of host associations for each fulgoroid family, in conjunction with the data summaries in Tables 1.1–1.4. A "record" within any table is the occurrence of one planthopper species on one of the plant taxa specified in that table. Multiple observations of a species on the same plant taxon were not considered separate records. The number of records contributed by any individual species is the number of plant taxonomic categories on which it occurs. Table 1.1 gives the percentage of host records for each planthopper family from the Pteridophyta, Gymnospermae, Monocotyledoneae, and Dicotyledoneae. Table 1.2 summarizes the proportion of just monocot records that occur on each monocot family. Table 1.3 gives the percentage of dicot records occurring on each of the dicot subclasses recognized by Cronquist (1981).

The degree of host plant specialization for each family (Table 1.4) was assessed by sorting species into one of three categories: 1) **monophagous** (feeding primarily on one plant species or plant species in the same genus),

Table 1.1. Percentage of planthopper species records on the Pteridophyta, Gymnospermae, Monocotyledoneae, and Dicotyledoneae for the families of Fulgoroidea. Polyphagous species were scored on all of their host plant taxa. Summarized from host plant records in Appendices 1 and 2.

Planthopper Family	Pter. (%)	Gym. (%)	Monocot. (%)	Dicot. (%)	No. Records
Tettigometridae	0	0	33	67	6
Delphacidae[a]	3	0	65	32	482
Delphacidae[b]	7	0	6	87	150
Delphacidae[c]	1	0	92	7	332
Cixiidae	5	7	30	58	174
Kinnaridae	15	8	8	69	13
Meenoplidae	0	0	80	20	5
Derbidae[d]	4	0	54	42	81
Achilidae[d]	0	34	6	60	50
Dictyopharidae	4	2	14	80	44
Fulgoridae	0	3	16	81	31
Issidae	0	12	17	71	104
Nogodinidae	0	0	0	100	4
Ricaniidae	5	0	25	70	20
Flatidae	0	0	16	84	69
Hypochthonellidae	0	0	50	50	2
Tropiduchidae	5	0	40	55	42
Lophopidae	0	0	71	29	7
Eurybrachidae	0	0	0	100	2
Total	3	4	42	51	1143

[a]All records.
[b]Oceanic island (Hawaiian Archipelago, Polynesia, S. Pacific) records (Fennah 1957, 1958, 1959a, 1962a, 1964, 1976; Zimmerman 1948) only.
[c]Nonisland records only.
[d]Nymphs mycophagous.

2) **oligophagous** (feeding on several genera in the same plant family), and 3) **polyphagous** (feeding on two or more plant families). Patterns of host plant specialization may change as our knowledge of planthopper host plants becomes more sophisticated. For example, many tropical species have been observed or sampled only once on a host taxon. As a consequence, we would categorize such a species as monophagous when indeed it may be polyphagous. Thus, our estimates of diet breadth should be taken in the context of the rather incomplete state of knowledge concerning the host plant relationships of planthoppers, particularly tropical taxa.

The following is a brief discussion of the host plant affiliations, degree of host plant specialization and feeding habits (subterranean or above ground)

Table 1.2. Percentage of planthopper species records on the families of Monocotyledoneae for the families of Fulgoroidea. Only monocot-feeding species considered. Polyphagous species were scored on all of their host plant taxa. Summarized from host plant records in Appendices 1 and 2.

	Family of Monocotyledoneae														
Planthopper Family	AC (%)	AG (%)	AR (%)	CM (%)	CY (%)	HE (%)	JU (%)	LI (%)	NY (%)	PA (%)	PO (%)	PN (%)	TY (%)	ZI (%)	No. Records
Tettigometridae								50			50				2
Delphacidae	1[a]		2[a]	<1	18[a]	<1[a]	5[a]	1[b]	<1[a]	2[a,b]	69[a]	<1[a]	1		334
Cixiidae		24[a]	25	2	3		10	2[b]		3[b]	31[a]				59
Kinnaridae		100[a]													1
Meenoplidae	25[a]			12	38						25				8
Derbidae	16		56		2	2				2[b]	18			4	45
Achilidae			33								67				3
Dictyopharidae		50[a]									50				6
Fulgoridae			40								60[a]				5
Issidae			11[a]	5	11			5			68[a]				18
Ricaniidae	20		20							40[b]	20				5
Flatidae			27[a]							18[b]	55[a]				11
Hypochthonellidae											100				1
Tropoduchidae			38[a]		6[a]					6	50[a]				18
Lophopidae			60[a]								40[a]				5
Total	2	4	12	1	13	<1	5	1	<1	3	57	<1	1	<1	520

Note: AC (Araceae); AG (Agavaceae); AR (Arecaceae); CM (Commelinaceae); CY (Cyperaceae); HE (Heliconiaceae); JU (Juncaceae); LI (Liliaceae); NY (Nymphaceae); PA (Pandanaceae); PO (Poaceae); PN (Pontederiaceae); TY (Typhaceae); ZI (Zingiberaceae).

[a]Nymphs observed or sampled in some cases.

[b]Most records from Oceanic Islands (Hawaiian Archipelago, Polynesia, S Pacific) (Fennah 1957, 1958, 1959a, 1962a, 1964, 1976; Zimmerman 1948).

for species in each of the fulgoroid families. We were unable to locate any host plant records for planthoppers in the Achilixiidae and Gengidae.

Tettigometridae

Most species of tettigometrids occur in the Ethiopian tropics and Palaearctic region and are associated primarily with dicots (75% of records) in the three subclasses Magnoliidae, Rosidae, and Asteridae, although monocot feeding (Liliaceae and Poaceae) has been reported (Weaving 1980; Gunthart 1987; Wilson and O'Brien 1987). Most species (62%) appear to be monophagous, with the remainder polyphagous [e.g., *Hilda patruelis* Stål (Weaving 1980)].

Table 1.3. Percentage of planthopper species records for the families of Fulgoroidea occurring on the subclasses of Dictoyledoneae listed from primitive to advanced. Summarized from host plant records in Appendices 1 and 2.

Planthopper Family	Magnoliidae (%)	Hamamelidae (%)	Caryophillidae (%)	Dilleniidae (%)	Rosidae (%)	Asteridae (%)	No. Records
Tettigometridae	40				40	20	5
Delphacidae[a]	2	5	6	10	32	45	191
Delphacidae[b]	2	5	3	11	36	43	167
Delphacidae[c]			25	4	8	63	24
Cixiidae		11	16	21	24	28	114
Kinnaridae				12		88	8
Meenoplidae					50	50	2
Derbidae		26		26	33	15	46
Achilidae	4	29		26	24	17	46
Dictyopharidae		5	15	8	13	59	39
Fulgoridae		11		17	55	17	29
Issidae	5	7	10	14	34	30	83
Nogodinidae				33	33	33	3
Ricaniidae		13		31	31	25	16
Flatidae	5	9	6	14	34	32	65
Hypochthonellidae						100	1
Tropiduchidae	4	12	12	16	32	24	25
Lophopidae					67	33	3
Eurybrachidae					100		2

[a]All records.
[b]Oceanic island (Hawaiian Archipelago, Polynesia, S. Pacific) records (Fennah 1957, 1958, 1962a, 1964, 1976; Zimmerman 1948) only.
[c]Nonisland records only.

Table 1.4. Diet breadth of the nymphs and adults of planthopper species in the families of Fulgoroidea. The percentage of (M) monophagous (feeding primarily on one plant species or plant species in the same genus), (O) oligophagous (feeding on several genera in the same plant family), and (P) polyphagous planthopper species (feeding on 2 or more plant families) and the number of species records for each family is shown. Summarized from host plant records in Appendices 1 and 2.

	Nymphs				Adults			
Planthopper Family	M (%)	O (%)	P (%)	No. Rec.	M (%)	O (%)	P (%)	No. Rec.
Tettigometridae	—	—	—		62	0	38	8
Delphacidae[a]	—	—	—		74	14	12	472[b]
Delphacidae[c]	—	—	—		73	4	23	158[b]
Delphacidae[d]	—	—	—		74	19	7	314[b]
Cixiidae	70	15	15	13	67	13	20	158
Kinnaridae	100	0	0	1	64	9	27	11
Meenoplidae	100	0	0	1	14	0	86	7
Derbidae		Mycophagous			63	16	21	75
Achilidae		Mycophagous			37	8	55	38
Dictyopharidae	100	0	0	12	84	5	11	43
Fulgoridae	75	0	25	4	62	9	29	21
Issidae	50	23	27	22	64	9	27	96
Nogodinidae	100	0	0	2	100	0	0	4
Ricaniidae	50	0	50	2	47	6	47	17
Flatidae	58	21	21	14	48	6	46	70
Hypochthonellidae	0	0	100	1	0	0	100	1
Tropiduchidae	36	18	46	11	62	8	30	37
Lophopidae	67	33	0	3	43	43	14	7
Eurybrachidae	100	0	0	1	100	0	0	2

[a]All records.
[b]Records of diet breadth for adult delphacids approximates pattern of host use by nymphs.
[c]Oceanic island (Hawaiian Archipelago, Polynesia, S. Pacific) records (Fennah 1957, 1958, 1959a, 1962a, 1964, 1976; Zimmerman 1948) only.
[d]Nonisland records only.

Tettigometrid nymphs typically live on plant roots, whereas adults are found above ground on the foliage of their hosts (Emeljanov 1987). However, both the nymphs and adults of *H. patruelis* feed on peanut leaves during the early morning and move to subterranean chambers surrounding the roots as the day becomes warmer (Weaving 1980).

Delphacidae

The Delphacidae are cosmopolitan, occurring in both the Old and New World from the tropics to arctic latitudes (O'Brien and Wilson 1985). The nymphs and adults of delphacids occur together on the above-ground parts of their host plants, although many species inhabit the basal portion of their host and are associated with the crown (Denno 1980; Cook and Denno Chapter 2). Delphacids are relatively host-specific, and most mainland species (92% of records) attack monocots; dicot feeding dominates (82% of records) only on oceanic islands (Zimmerman 1948). Delphacid host associations are discussed in detail in a later section.

Cixiidae

The Cixiidae occur at both temperate and tropical latitudes in the Old and New World (Kramer 1983). Nymphs are subterranean, feeding on plant roots and perhaps fungi (O'Brien and Wilson 1985; Emeljanov 1987; Hoch and Howarth 1989a, 1989b). For nymphs, 62% of the records are from grasses and are represented by *Tachycixius pilosus* (Olivier) (Le Quesne 1960), *Oliarus felis* Kirkaldy (Hacker 1925a), *Pentastiridius pachyceps* (Matsumura) (Tsaur et al. 1988a), and *Myndus crudus* (Van Duzee) (Tsai and Kirsch 1978). Nymphs have also been reported feeding on the roots of ferns (Swezey 1906; Zimmerman 1948), gymnosperms (Pinaceae, Sheppard et al. 1979), other monocots (Agavaceae, Cumber 1952; Arecaceae, Sheppard et al. 1979),) and dicots in the Asteraceae, Chenopodiaceae, Convolvulaceae (Brĉak 1979), and Tamaricaceae (Hopkins and Carruth 1954).

As adults, cixiids occur above ground and most host records (58%) are from woody dicots, spanning five subclasses (Le Quesne 1960; Reinert 1980; Kramer 1983). Some cixiid adults are reported from ferns (*Iolania* spp., Zimmerman 1948), gymnosperms (*Ankistrus pini* Tsaur and Hsu and several *Cixius* spp. on Pinaceae; Kramer 1981 and Tsaur et al. 1988b) and monocots. Within the Monocotyledoneae, most records are from the Poaceae (some species of *Bothriocera*, *Kirbyana*, *Myndus*, *Oecleus*, and *Oliarus*), Arecaceae (species of *Myndus*, *Nymphocixia*, *Oecleus*, and *Oliarus*), and Agavaceae (some species of *Myndus*, *Oecleus*, and *Platycixius*).

Most species (70% of nymphal and 67% of adult records) have been reported from only a single host genus. However, within cixiid genera, there are no clear patterns of association with particular host plant taxa, raising the possibility that the adults are less host-specific than these figures would indicate. For example, within the genera *Bothriocera* and *Oecleus*, there are some species associated with monocots and others with dicots, and in the genera *Cixius*, *Myndus*, and *Oliarus*, there are several species associated with gymnosperms and others with angiosperms (Appendix 1).

Kinnaridae

The Kinnaridae is a primarily tropical, mostly Neotropical and Oriental family (O'Brien and Wilson 1985). Nymphs are subterranean, feeding on roots. For example, nymphs of *Quilessa grenadana* Fennah and two species of *Oeclidius* have been found feeding on roots (Fennah 1948, 1980). Adults are generally collected from the upper portions of plants, but the adults of *Oeclidius* have also been found on the roots of their hosts (Fennah 1980). Most kinnarid species (69%) are associated with dicots in the Asteridae and Dilleniidae (Fennah 1948; Goeden and Ricker 1989). The remaining records are from ferns, gymnosperms in the Ephedraceae, and monocots in the Agavaceae. As adults, most kinnarid species (64%) have been reported on a single host plant genus.

Meenoplidae

The Meenoplidae is an Old World family with many tropical species (O'Brien and Wilson 1985). Few studies, however, have provided details of their life histories. The limited information which exists suggests that adults and nymphs occur above ground on their hosts (Tsaur 1989a). Most species (80% of records) are associated with monocots, particularly the Cyperaceae, Poaceae, Araceae, and Commelinaceae, and the remainder (20%) with dicots in the Rosidae and Asteridae (Synave 1961; Tsaur et al. 1966).

As adults, meenoplid species (86%) may be among the most polyphagous in the Fulgoroidea. For example, species of *Nisia* feed across several families of monocots (Tsaur et al. 1986; Wilson and Claridge 1991). However, some apparently polyphagous species may, in fact, represent clusters of sibling species with more limited host plant ranges (Wilson and Claridge 1991). Thus, meenoplids may be more specialized in their use of host plants than the present records indicate.

Derbidae

Derbids occur both in the tropical and temperate regions. Nymphs are thought to be obligate fungal feeders (O'Brien and Wilson 1985; Emeljanov 1987). Nymphs have been collected under the bark of living (*Apache degeerii* Kirby; Wilson 1982b) and dead trees (*Otiocerus coquebertii* Kirby; Felt 1916), from palm stumps (*Heronax maculipennis* Melichar; Muir 1917) and from decaying logs (*Sikiana harti* Metcalf; Willis 1982), whereas adults occur above ground on their hosts (O'Brien and Wilson 1985).

Adult derbids are associated primarily with monocots (54% of records) and woody dicots (42%), and a few species (4%) occur on ferns (Gagne 1972; Flynn and Kramer 1983). Of the monocot records, 56% were on the Arecaceae, 18% on Poaceae, and 16% on Araceae (Schumacher 1920; Howard and Mead 1980; Wilson and O'Brien 1987). Dicotyledonous hosts range across four subclasses.

Adults of most derbid species (63%) have been reported on only one host plant genus. Some genera appear to be restricted to a single plant family (*Neocenchrea* and genera in the Sikaianini to Arecaceae, *Vekunta* to the Poaceae, and *Malenia* to Salicaceae); however, others are not so constrained. For example, there are records of *Cedusa* on ferns, monocots, and dicots, and records of *Omolicna* on monocots and dicots.

Achilidae

Like the Derbidae, achilids are thought to feed on decaying vegetation or fungi as nymphs and the sap of higher plants as adults (O'Brien 1971; Emeljanov 1987; Wilson 1989). Nymphs are typically collected under the bark of decaying logs (Downes 1927; Hepburn 1967; Chen et al. 1989), whereas adults occur above ground on their hosts (O'Brien 1971). Nymphs of *Synecdoche nemoralis* (Van Duzee) were reared on fungi (O'Brien 1971) and adults of the myconine *Epiptera slossonae* (Van Duzee) were collected in rotten wood or under stumps. Adults of this temperate family are associated primarily with dicots (60% of records, spanning five subclasses). However, achilids are more strongly associated with gymnosperms (Pinaceae and Cupressaceae; 34% of records) than any other fulgoroid family. The few monocot associations (6% of records) are with Poaceae and Arecaceae. As adults, a majority of species (55%) are polyphagous. For example, species of *Catonia* have been taken on both gymnosperms and angiosperms. However many achilids (45%) are known from a single plant family (e.g., several species of *Junipertha* on Cupressaceae, and several species of *Synecdoche* from Ericaceae; O'Brien 1971).

Dictyopharidae

This family occurs in both temperate and tropical regions of the Old and New World (O'Brien and Wilson 1985; Wilson and O'Brien 1987). Both nymphs and adults occur above ground on their hosts (Ball 1930; Goeden and Ricker 1975). Most species (80% of records) are associated with dicots in the Hamamelidae, Caryophyllidae, Dilleniidae, Rosidae, and Asteridae. Of the remaining species records, 14, 4, and 2% are from monocots (Poaceae and Agavaceae), ferns, and gymnosperms (Ephedraceae) respectively.

Most dictyopharid species (84%) are monophagous, more so than species in most of the other families of Fulgoroidea. For instance, some species such as *Phylloscelis pallescens* Germar and *P. rubra* Ball are monophagous on dicots as are all known species of *Scolops*. A few species (11%) are polyphagous on dicots (e.g., *Dictyophara europeae* L. and *Nersia florens* Stål).

Fulgoridae

Both the nymphs and adults of this primarily tropical family occur above ground on their host plants (Kershaw and Kirkaldy 1910; Chu 1931; Wilson and O'Brien 1987; Wilson and Wheeler 1992). Most fulgorid species (81%) are affiliated with dicots in the Hamamelidae, Caryophyllidae, Dilleniidae, Rosidae, and Asteridae. Some species (16%) are associated with monocots in the Poaceae and Arecaceae and one species has been reported feeding on gymnosperms (Pinaceae). Nymphs are rarely collected with adults and the nymphs of some species may feed on hosts not used by adults. For example, nymphs of *Rhabdocephala brunnea* Van Duzee were found on a grass, a host apparently not fed upon by the adults (Wilson and Wheeler 1992). Many fulgorid species (62%) have been reported on a single host and 29% are polyphagous.

Issidae

The Issidae occur in both tropical and temperate regions in the Old and New World (O'Brien and Wilson 1985). Issids, in general, feed as nymphs and adults on the above-ground parts of their host plants (Denno 1980; Wheeler and Hoebeke 1982; Wilson 1987; Wheeler and Wilson 1988; Goeden and Ricker 1989). Most issid host plant records (71%) are from dicots, including all subclasses, with some on monocots (17%) and a few on gymnosperms (12%). Of the monocot records, most are on the Poaceae (68%) followed by the Arecaceae (11%) and Cyperaceae (11%). More issids are monophagous (64%) than polyphagous (27%).

Host plant associations differ among the subfamilies of Issidae. As both nymphs and adults, acanaloniines are polyphagous on a wide variety of mainly woody dicots (Wheeler and Hoebeke 1982, Wilson and McPherson 1980). Most Issinae feed on woody dicots with a few species feeding on gymnosperms. For example, *Thionia simplex* (Germar) is polyphagous on dicots (Wheeler and Wilson 1988), *Thionia elliptica* Germar is associated with oaks (Wheeler and Wilson 1987), *Thionia bullata* (Say) feeds as adults and nymphs on three species of pines (Wheeler and Wilson 1988), and *Thionia producta* Van Duzee is found on junipers (Doering 1938). Some issines deposit mud egg cases on a variety of woody plants on which they apparently do not feed (Caldwell and DeLong 1948; Doering 1958, Schlinger 1958). All host records of Tonginae are from dicots (Fennah 1958; Ward et al. 1977). Bladinines were transferred from Issidae to Nogodinidae by Fennah (1984) and then back to Issidae by Emeljanov (1991). *Bladina* occurs primarily on monocots, whereas *Dictyssa* is primarily on dicots (Fennah 1945, 1952; Doering 1940). All species of Caliscelinae feed on monocots, especially grasses, sedges, and palms, and most species restrict their feeding to plants in the same family (Ball 1935a, 1935b; Denno 1980).

Nogodinidae

Very little is known about the biology of this primarily tropical family. Nogodinids apparently feed above ground as nymphs and adults. All known nogodinids are monophagous on woody dicots (Fennah 1978; Dlabola 1981; Carver et al. 1991).

Ricaniidae

Ricaniids are a primarily tropical Old World group (O'Brien and Wilson 1985). Both nymphs and adults occur above ground on their host plants (Cumber 1966). Host associations are primarily with woody dicots (70% of records) in the Dilleniidae, Rosidae, and Asteridae, but some records (25%) are from monocots (Poaceae, Pandanaceae, and Araceae) and ferns (5%). Polyphagy is common (47% of species) (Lee and Kwon 1977; Williams and Fennah 1980). For example, *Scolypopa australis* Walker feeds on ferns, clubmosses, and angiosperms (Cumber 1966), and *Ricania taeniata* Stål feeds on both monocots and dicots (Lee and Kwon 1977).

Flatidae

The Flatidae is a large, mostly pantropical family (O'Brien and Wilson 1985). Both the nymphs and adults occur on the above-ground parts of their host plants (Metcalf and Bruner 1948; Wilson and McPherson 1980; Goeden and Ricker 1986a). Most species (84% of records) feed on woody dicots, and all subclasses are used (Fennah 1941, 1942b). The remaining records are from monocots, mostly in the Poaceae and Arecaceae (Howard and Mead 1980, Wilson and Tsai 1984).

Many flatid species (46%) are polyphagous (Hoffman 1935; Fletcher 1985). For example, the polyphages *Anormenis chloris* (Melichar), *Metcalfa pruinosa* (Say), and *Ormenoides venusta* (Melichar) have been recorded from 41, 85, and 4 species of plants, respectively (Wilson and

McPherson 1980). Nymphs of some species may have narrower host ranges than adults; for example, adults of *Ormenaria rufifascia* (Walker) fed on 15 species of palms and 3 species of dicots, but nymphs developed only on 2 species of palms (Howard and Mead 1980, Wilson and Tsai 1984).

Hypochthonellidae

This monotypic family is known only from the Ethiopian region (O'Brien and Wilson 1985). Nymphs and adults of *Hypochthonella caeca* China and Fennah are subterranean, feeding on the roots of plants in the Fabaceae, Solanaceae, and Poaceae (China and Fennah 1952).

Tropiduchidae

The tropiduchids are primarily pantropical (O'Brien and Wilson 1985; Wilson and O'Brien 1987). Both nymphs and adults feed above ground on their hosts (Caldwell and Martorell 1950; Wilson 1986, 1987; Yang et al. 1989). The majority of species (55% of records) feed on woody dicots, spanning all subclasses (Carnegie 1980; Wilson and Wheeler 1984; Wilson and Hilburn 1991). Monocot feeding (40% of records) occurs primarily on the Poaceae and Arecaceae (species in the Catulliini, Tambiniini, and Trypetimorphini) (Asche and Wilson 1989a; Yang et al. 1989). *Ommatissus lybicus* Bergevin is a pest of date palms (Asche and Wilson 1989a). As adults, most tropiduchids (62%) have been taken on a single host genus, whereas 30% are polyphagous (Carnegie 1980; Wilson and Wheeler 1984; Wilson and Hilburn 1991).

Lophopidae

Lophopids occur largely in the Old World tropics (Wilson and O'Brien 1987). Adults and nymphs occur above ground on their hosts (Rahman and Nath 1940; Smith 1980; Wilson 1987). Although there are few data, most lophopid records (71%) are from monocots (Arecaceae and Poaceae), with the remainder from dicots in the Rosidae and Asteridae (Rahman and Nath 1940; Ghauri 1966; Woodward et al. 1970; Smith 1980; Wilson 1987). Most lophopid species are either monophagous (43%) or oligophagous (43%) as adults; nymphs are even more specialized (67% monophagous; Rahman and Nath 1940; Smith 1980; Wilson 1987).

Eurybrachidae

Very little is known about the biology of this small Australasian family (O'Brien and Wilson 1985). The only species for which any detailed biological information is available, *Platybrachys leucostigma* Walker, has been found above ground as both nymphs and adults on *Eucalyptus maculatus* (Hacker 1925b).

Host Plant Use Patterns in the Fulgoroidea: Parallel Phylogenesis?

The ages of origin of the fulgoroid families appear to span at least those of the angiosperms. It is, therefore, of interest to ask whether the present distribution of fulgoroid families across plant taxa in some way reflects that shared history. This inquiry faces several difficulties. First, although we have attempted to eliminate incidental host records, plant associations of fulgoroids apart from delphacids are known with little precision; this may be reflected by the apparent high proportion of polyphagous species. Second, the more basal families tend to be subterranean feeders as immatures, some using only fungi. Plants used by the free-living adults, which constitute the great majority of records, may not be the "definitive" hosts to which long-term fidelity is most likely; the nymphal host range is known to be narrower than that of adults. Third, fulgoroid families show great overlap in their use of host taxa of subclass and higher rank. Thus, for an explicit comparison to plant phylogeny, it may be desirable to examine fulgoroid phylogeny and patterns of host use at a finer taxonomic scale, or at least to identify the ancestral habit within each family. Except in the Delphacidae, however, phylogenetic relationships within fulgoroid families are little understood. Finally, higher-level relationships within the angiosperms remain poorly known (Dahlgren and Bremer 1985; Olmstead et al. 1992).

Given the obstacles to an explicit comparison of planthopper and host plant phylogenies, we take a heuristic, statistical approach following Zwölfer (1978). Based on our summary of the phylogenetic and fossil evidence, fulgoroids can be objectively divided into a set of relatively ancient, primitive lineages and a set of more derived, recent lineages. The "primitive" grouping includes the families Tettigometridae, Delphacidae, Cixiidae, Derbidae, and Achilidae, plus the lineage Meenoplidae + Kinnaridae, all of which we infer (sometimes indirectly) to have arisen by the Lower Cretaceous. For the remaining 12 derived families there is no evidence for appearance before the late Cretaceous, and some may date only from the Tertiary (Hamilton 1990). In the analyses below, we test for quantitative phylogenetic trends in fulgoroid host use by contrasting mean percentage of associations of various kinds (e.g., angiosperms versus nonangiosperms) between these two fulgoroid groupings, using individual families or lineages as quasi-independent replicates. For reasons discussed below, oceanic island records for delphacids were excluded from this analysis. Statistical significance was evaluated by the Mann–Whitney U test, which was performed both with all families included and with those providing fewer than five records excluded.

Suppose, as seems probable from the fossil evidence, that the primitive families are older than the angiosperms, whereas the younger families diverged contemporaneously with the angiosperms. If the fulgoroids diversified in tandem with the vascular plants over this period, we might expect the earlier-diverging families to show an elevated proportion of records from plant groups that became distinct before the angiosperms did, such as ferns and most groups of gymnosperms. The mean fraction of nonangiosperm records (Pteridophyta and Gymnospermae) is, in fact, somewhat higher for the basal lineages (11% versus 3%); however, the difference is not quite significant if planthopper families with less than five records are excluded from the analysis ($P<0.1$, $P>0.05$, one-tailed test).

If the fulgoroids were still older, such that all the families predate the angiosperms, the tendency for nonangiosperm habits to be cladistically basal might be more apparent within than among families. Lack of phylogenies for all families but the Delphacidae prevents a test of this hypothesis; however, several authors have opined that nonangiosperm feeders are derived from angiosperm feeders (Emeljanov 1987; Hamilton 1990). In any case, nonangiosperm associations constitute a small fraction of records even for primitive fulgoroids. Either planthoppers did not actually predate and feed on older relatives of the angiosperms or the host associations of present-day species have largely departed from the original habits.

Finally, suppose that the fulgoroids, including the primitive families, are equal in age to, and have diversified with their predominant hosts, the angiosperms. Then we might expect the older fulgoroid families to show a greater proportion of records on the earliest-diverging angiosperms. Phylogenetic relationships among the host plant families of fulgoroids are poorly resolved (Dahlgren and Bremer 1985; Olmstead et al. 1992), but following Zwölfer (1978) we can make one reasonably secure broad contrast: Families in the basal subclasses Magnoliidae and Hamamelidae are mostly older than those in the recent, most diverse subclass, the Asteridae. There is a tendency for the basal compared to the advanced fulgoroids to be more strongly associated with the Magnoliidae and Hamamelidae (18% versus 7%; Table 1.3), but the difference is not significant ($P\gg0.10$). Moreover, the basal fulgoroid lineages are recorded slightly more often from Asteridae than are the more advanced lineages.

Perhaps the strongest trend is that the primitive fulgoroid families tend to be associated more with monocots than with dicots, whereas the reverse is true of the more recent families (Table 1.1). The difference in mean fraction of records on dicots (48% versus 72% for primitive and advanced fulgoroids, respectively) is significant ($P<0.05$, two-tailed test), although that for monocot records (48% versus 25%) is not ($P>0.10$).

This pattern, if real, would be difficult to interpret as parallel phylogenesis. The monocots are neither cladistically basal to nor more primitive in features than the dicots; rather they appear to be sister group to, and to have diversified contemporaneously with, the bulk of the dicots (Dahlgren and Bremer 1985; Duvall et al. in press). Moreover, the families representing the major portion of planthopper monocot records, the Poaceae, Cyperaceae and Arecaceae, are relatively derived within the monocots (Dahlgren and Bremer 1985; Duvall et al. in press).

Thus, as far as we can judge from adult records, the distribution of fulgoroid families across host taxa is largely modern and not relictual: old versus young planthopper lineages are associated very weakly at best with correspondingly old versus young plant taxa.

Host Plant Use Patterns in the Fulgoroidea: Subterranean to Above-ground Life

Although there is little evidence for parallel phylogenesis of fulgoroid families with their hosts, there are marked phylogenetic progressions in the location and type of plant tissue fed on by the fulgoroid families (Table 1.5). In five (Tettigometridae, Cixiidae, Kinnaridae, Derbidae, and Achilidae) of the seven primitive families, the nymphs, and sometimes the adults as well, feed in concealment either underground or under bark. Of the two exceptions, Meenoplidae and Delphacidae, the latter, while feeding above ground, are often located near the base of the plant (Cook and Denno Chapter 2). In contrast, among the more advanced families, only the Hypochthonellidae appear to be concealed (subterranean) feeders. These proportions are significantly different ($X^2 = 6.1, P<0.01$). Subterranean feeding is probably ancestral for fulgoroids and for Auchenorrhyncha as a whole, as it is also widespread in the Cicadomorpha (Carver et al. 1991). However, Emeljanov (1987) doubts this conclusion and hints at independent acquisition of subterranean feeding in the Tettigometridae and Cixiidae.

A correlated trend among fulgoroids is that, although exposed feeders (including the adults of most primitive families) tap the phloem, concealed feeders feed on only the relatively undifferentiated cells of either fungi (nymphs of Achilidae, Derbidae, and some Cixiidae) or of root parenchyma (most Cixiidae) (Hamilton 1990). Feeding on undifferentiated cells, which also occurs in the moss-feeding Coleorrhyncha, could be primitive for Hemiptera *sensu lato* (Hamilton 1990), particularly because the probable

Table 1.5. Feeding locations[a] on the host plant of the nymphs and adults of the families of Fulgoroidea.

Planthopper Family	Nymphs	Adults
Etttigometridae	Subterranean (roots)	Above ground
Delphacidae	Above ground (some sp. near surface)	Above ground (some sp. near surface)
Cixiidae	Subterranean (roots)	Above ground
Kinnaridae	Subterranean (roots)	Subterranean/ above ground
Meenoplidae	Above ground	Above ground
Derbidae	Under bark and logs (mycophagous)	Above ground
Achilidae	Under bark and logs (mycophagous)	Above ground
Dictyopharidae	Above ground	Above ground
Fulgoridae	Above ground	Above ground
Issidae	Above ground	Above ground
Nogodinidae	Above ground	Above ground
Ricaniidae	Above ground	Above ground
Flatidae	Above ground	Above ground
Hypochthonellidae	Subterranean (roots)	Subterranean (roots)
Tropiduchidae	Above ground	Above ground
Lophopidae	Above ground	Above ground
Eurybrachidae	Above ground	Above ground

[a]Information on feeding locations extracted from: Kershaw and Kirkaldy (1910); Muir (1917); Hacker (1925a, 1925b); Ball (1930); Hoffman (1935); Rahman and Nath (1940); Metcalf and Bruner (1948); Cumber (1966); Ghauri (1966); China and Fennah (1952); O'Brien (1971); Goeden and Ricker (1975, 1986a, 1989); Fennah (1978, 1980); Denno (1980); Reinert (1980); Weaving (1980); Williams and Fennah (1980); Wilson and McPherson (1980); Dlabola (1981); Wheeler and Hoebeke (1982); Willis (1982); Wilson (1982, 1987, 1989); Kramer (1983); Wilson and Wheeler (1984); O'Brien and Wilson (1985); Emeljanov (1987); Gunthart (1987); Wheeler and Wilson (1987, 1988); Chen et al. (1989); Hoch and Howarth (1989a); Tsaur (1989a); Yang et al. (1989); Wilson and Wheeler (1992).

sister order to the Hemiptera (Homoptera + Heteroptera), the Thysanoptera, consists primarily of cell piercers feeding on fungi and higher plants (Mound et al. 1980). Moreover, the earliest fossil Homoptera appear to have had small cibarial chambers, judging from the small overlying frons (Hamilton 1990). Among extant groups, reduction or absence of the filter chamber is typically associated with feeding on nonvascular plant tissues (Emeljanov 1987). In contrast, Emeljanov (1987) postulates a hemipteran ancestor feeding on the phloem of arboreal gymnosperms.

These observations on feeding location suggest an explanation for the possible phylogenetic trend in the degree of association with monocots.

Ancestral fulgoroids probably fed as nymphs on roots or fungi in concealed habitats such as in the soil or under bark or logs. These nymphal habitats were probably moist, like those of most extant primitive fulgoroids including the Delphacidae (Ossiannilsson 1978; Denno and Roderick 1990). This habitat association may have predisposed early fulgoroids to the colonization of monocots, which were probably primitively aquatic (Cronquist 1981).

Finally, the trend toward above-ground feeding on woody dicots, among successively more advanced fulgoroids, may be associated with a progression in ovipositor morphology (Table 1.6). The sword-shaped orthopteroid ovipositor seen in the Delphacidae and Cixiidae, in which a piercing–sawing structure is formed from the fusion of the median gonapophyses of the ninth abdominal segment (Asche 1987; Emeljanov 1987), is probably ancestral for the superfamily. Ancestral fulgoroids may have used this structure to bury their eggs in soil or plant debris, as some Cixiidae do today (Zimmerman 1948; Kramer 1981). Such ovipositors may also have allowed ancestral fulgoroids to oviposit into the above-ground, soft plant tissues of monocots, a behavior exhibited by extant Delphacidae and some Cixiidae (Asche 1987).

Higher planthoppers have a "fulgoroid ovipositor," in which a piercing, excavating, or raking structure is formed from the gonapophyses of the eighth segment; the gonapophyses of the ninth segment may be paired and lobate or reduced (Muir 1923; Stephan 1975; Asche 1987). The families Derbidae, Meenoplidae, Kinnaridae, Achilidae, Dictyopharidae, and Fulgoridae possess a more primitive, "egg deposition" version of this ovipositor; they cover or cement the eggs rather than inserting them. The ovipositor is used to excavate soil or loose material, to cover eggs with wax or exogenous material, or to glue eggs to the substrate (Asche 1987; Table 1.6). The most advanced ovipositor type, found in the Issidae (some), Nogodinidae, Ricaniidae, Flatidae, Hypochthonellidae, Tropiduchidae, and Gengidae is a piercing–excavating ovipositor in which the gonapophyses of the eighth abdominal segment function in excavating cavities for egg deposition in tough woody tissues.

This phylogenetic progression may be related to the greater difficulty of ovipositing into woody plants with tough tissues and to selection for protection from the increased risk of mortality from natural enemies or desiccation for eggs positioned on above-ground woody tissues.

Phylogeny of the Delphacidae

The higher taxa of Delphacidae have recently been the focus of phylogenetic analysis by Asche (1985, 1990). The following discussion and cla-

Table 1.6 Structure of the ovipositor in the families of Fulgoroidea.

Planthopper Family	Ovipositor Structure[a]	Reference
Tettigometridae	Reduced[b]	Muir (1923), Asche (1987), Emeljanov (1987)
Delphacidae	Orthopteroid (piercing–sawing)[c]	Asche (1987), Emeljanov (1987)
Cixiidae	Orthopteroid (piercing–sawing)[c] or reduced[d]	Asche (1987), Emeljanov (1987)
Kinnaridae	Fulgoroid (egg deposition)[e]	O'Brien and Wilson (1985)
Meenoplidae	Fulgoroid (egg deposition)[e] or reduced[f]	O'Brien and Wilson (1985), Tsaur et al. (1986)
Derbidae	Fulgoroid (egg deposition)[e]	Muir (1923)
Achilidae	Fulgoroid (egg deposition)[f]	Muir (1923), Fennah (1950)
Dictyopharidae	Fulgoroid (egg deposition)[e]	Emeljanov (1987), Wilson and McPherson (1981b)
Fulgoridae	Fulgoroid (egg deposition)[e]	Emeljanov (1987)
Issidae	Fulgoroid (piercing–excavating)[g] or reduced[h]	Muir (1923), Stephan (1975)
Nogodinidae	Fulgoroid (piercing–excavating)[g]	Stephan (1975)
Ricaniidae	Fulgoroid (piercing–excavating)[g]	O'Brien and Wilson (1985), Stephan (1975)
Flatidae	Fulgoroid (piercing–excavating)[g] or reduced[h]	Stephan (1975), Muir (1923)
Hypochthonellidae	Fulgoroid (piercing–excavating)[g]	Stephan (1975)
Tropiduchidae	Fulgoroid (piercing–excavating)[g]	Stephan (1975), Asche (1987)
Lophopidae	Reduced[i]	Muir (1923), Stephan (1975)
Eurybrachidae	Reduced[i]	Muir (1923), Stephan (1975)
Gengidae	Fulgoroid (piercing–excavating)[j]	Fennah (1949b)

[a]Ovipositor formed of the gonapophyses of the eighth (**g8**) abdominal segment (= 1st valvula), the lateral gonapophyses of the ninth (**lg9**) abdominal segment (= 3rd valvula), and the median gonapophyses of the ninth (**mg9**) abdominal segment (= 2nd valvula).

[b]**g8** and **lg9** reduced to lobes or absent; assumed to be an apomorphic reduction.

[c]**mg9** fused and serves as a piercing structure.

[d]Gonapophyses reduced, platelike wax producing structure present.

[e]**g8** expanded, often with teeth or lobes, can be variably reduced; used for egg deposition by raking or sweeping substrate and placing eggs therein. **mg9** not fused, generally greatly reduced.

[f]**g8** and **mg9** greatly reduced.

[g]**g8** serve as piercing–excavating structure.

[h]**g8** greatly reduced.

[i]**g8** lost or fused to valvifer on 8th abdominal segment, **lg9** greatly expanded.

[j]Inferred from description and illustration in Fennah (1949b).

dogram are summarized from Asche's (1985, 1990) examination of more than half of the world species using 52 characters (Fig. 1.2). The presence of the moveable hind tibial spur, a sclerotized diaphragm on the male pygofer, pincerlike styles, and reduction of the frontal ocellus support monophyly of the family Delphacidae. Six subfamilies and five tribes are currently recognized (Asche 1985, 1990). In phylogenetic sequence they are as follows: Asiracinae (176 spp.) including the Ugyopini (148 spp.) and Asiracini (28 spp.); Vizcayinae (5 spp.); Kelisiinae (44 spp.); Stenocraninae (64 spp.); Plesiodelphacinae (7 spp.); and Delphacinae (1221 spp.) including the Delphacini (1090 spp.), Tropidocephalini (122 spp.), and Saccharosydnini (9 spp.).

At present, the group Asiracinae is maintained only for convenience as the taxa within this subfamily are united by symplesiomorphies. Two tribes, the Ugyopini and Asiracini, have been recognized. The Ugyopini has been shown to represent a monophyletic taxon based on the presence of a quadrangular metatibial spur bearing four rows of bristles and proximal loca-

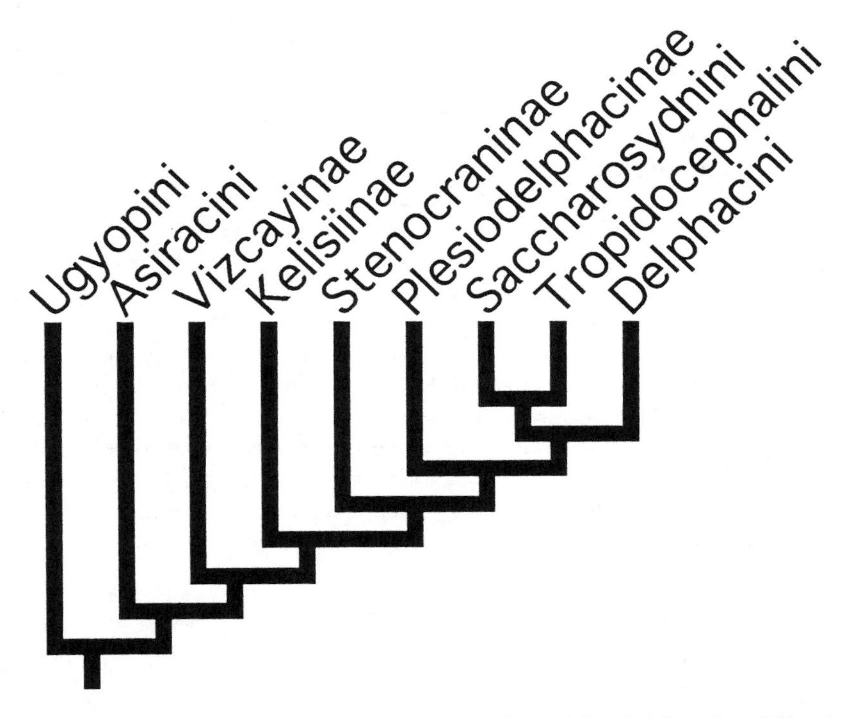

Figure 1.2. Phylogeny for the subfamilies and tribes of the Delphacidae, following Asche (1985, 1990).

tion of the middle spine on the first metatarsomere. The Asiracini appears to be a paraphyletic assemblage (see Asche 1990).

The remaining non-asiracine delphacids form a monophyletic assemblage based on nine synapomorphies, including structures associated with the sexually dimorphic drumming organ, teeth on the metatibial spur, spination on the hind legs, and reduction of the distal portion of the aedeagus. The Vizcayinae have retained some asiracine features and are characterized by four autapomorphies involving shape of the male drumming organ, apex of the vertex, spination of the basal metatarsomere, and form of the antennal segments. The remaining delphacids form a monophyletic group based on five apomorphies including sclerotization of a portion of the aedeagus, presence of a central plate on the male drumming organ, venation of the hind wing, number of nymphal pits on the head, and number and arrangement of antennal sensory fields.

The taxa forming Kelisiinae are united by the presence in males of a subanal appendage and fusion of parts of the aedeagus. The remaining, "higher" delphacids form a monophyletic group based on the presence of a ringlike opening on the egg chorion. Synapomorphies uniting Stenocraninae are the ditrysic female genitalia and a basal process arising from the aedeagus. The rest of the higher delphacids are united by the presence of elongate, dorsally-directed processes in the male drumming organ. Plesiodelphacinae are characterized by carination of the vertex, partial fusion of M and Cu veins on the hind wing, and a ugyopinelike first metatarsomere.

The monophyly of the Delphacinae is supported by several features, the most important of which is the thin-walled central part of the aedeagus. This subfamily is subdivided into three tribes, the Tropidocephalini, Saccharosydnini, and Delphacini. Synapomorphies uniting the Tropidocephalini include asymmetry of the base of the aedeagus. The Saccharosydnini are characterized by an elongate aedeagus contained within a large phragma bag. These two tribes are considered sister taxa united by reduction of the anal tube and a close connection between it and the aedeagus. Monophyly of the Delphacini is supported by the presence of an aedeagal suspensorium, loss of symbiont x, and the presence of symbionts H and f. Another possible delphacine synapomorphy, although examined in relatively few taxa, is an oviduct gland in the female, lacking only in three genera.

Host Plant Relationships within the Delphacidae

Most delphacid species (65%) feed and develop on monocots, with 32% of records from dicots and only 3% from ferns (Tables 1.1 and 1.7). Of the monocot records (Tables 1.2 and 1.8), most are on the Poaceae (69%), Cyperaceae (18%), and Juncaceae (5%).

Table 1.7. Percentage of planthopper species records on the Pteridophyta, Gymnospermae, Monocotyledoneae, and Dicotyledoneae for the subfamilies and tribes of Delphacidae. Polyphagous species were scored on all of their host plant taxa. Summarized from host plant records in Appendix 2.

Delphacid Taxon	Pter. (%)	Gym. (%)	Monocot. (%)	Dicot. (%)	No. Records
Asiracinae	25	0	25	50	12
Ugyopini	25	0	12	63	8
Asiracini	25	0	50	25	4
Kelisiinae	0	0	100	0	19
Stenocraninae	0	0	100	0	13
Plesiodelphacinae	0	0	100	0	2
Delphacinae[a]	3	0	63	34	437
Tropidocephalini	0	0	100	0	42
Saccharosyndnini	0	0	100	0	6
Delphacini	3	0	58	39	389
Delphacinae[b]	1	0	93	6	286
Delphacini[b]	7	0	5	88	151
Delphacini[c]	1	0	92	7	238

[a]All records.
[b]Oceanic Islands (Hawaiian Archipelago, Polynesia, S. Pacific) records (Fennah 1957, 1958, 1959a, 1962a, 1964, 1976; Zimmerman 1948) only.
[c]Nonisland records only.

A preponderance of the dicot records are from tropical islands in the Hawaiian archipelago and Polynesia (e.g., Fennah 1957, 1958, 1959a, 1962a, 1964, 1976; Zimmerman 1948), mostly in the Delphacinae. Insular host plant records include woody species in all subclasses of dicots (Cronquist 1981). If these island records are removed, the proportion of monocot hosts for the Delphacidae rises to approximately 92%, dicots fall to 7%, and ferns constitute about 1% (Table 1.1). Denno and Roderick (1990) suggested that the prevalence of dicot feeders on the Hawaiian archipelago might be unique to islands rather than typical of the tropics as a whole. Comparison of Zimmerman's (1948) records from the Hawaiian Islands with those of Fennah from other tropical islands indicates that tropical islands do have an unusually high proportion of dicot-feeding species (87%), whereas monocots dominate the limited host records of delphacids from the Neotropical mainland (Muir 1926; Table 1.1). Therefore, our discussion on host use patterns takes into account this evolutionary difference of mainland clades from those on islands, which are likely to be much more recent. We discuss possible causes for the radiation of island delphacids on dicots in a later section.

Table 1.8. Percentage of planthopper species records on the families of Monocotyledoneae for the subfamilies and tribes of Delphacidae. Only monocot-feeding species considered. Polyphagous species were scored on all of their host plant taxa. Summarized from host plant records in Appendix 2.

Delphacid Taxon	Family of Monocotyledoneae												Total Records
	AC (%)	AR (%)	CM (%)	CY (%)	HE (%)	JU (%)	LI (%)	NY (%)	PA (%)	PO (%)	PN (%)	TY (%)	
Asiracinae		33		67									3
Ugyopini		100											1
Asiracini				100									2
Kelisiinae				83		17							23
Stenocraninae				38						62			13
Plesiodelphacinae					100								2
Delphacinae	1	2[a]	<1	11		4	1[a]	<1	2[a]	76	<1	4	293
Tropidocephalini										100			42
Saccharosyndnini										100			6
Delphacini	1	2[a]	<1	13		5	1[a]	<1	3[a]	71	<1	2	245

Note: AC (Araceae); AR (Arecaceae); CM (Commelinaceae); CY (Cyperaceae); HE (Heliconiaceae); JU (Juncaceae); LI (Liliaceae); NY (Nymphaceae); PA (Pandanaceae); PO (Poaceae); PN (Pontederiaceae); TY (Typhaceae).

[a]Most records from Oceanic Islands (Hawaiian Archipelago, Polynesia, S. Pacific) (Fennah 1957, 1958, 1959a, 1962a, 1964, 1976; Zimmerman 1948).

Host use patterns of the major delphacid clades can be summarized, in phylogenetic sequence, as follows (see Fig. 1.2 and Tables 1.7 and 1.8). The small tribe Ugyopini, the most primitive delphacids, appear to feed mostly on woody dicots (63% of records) and ferns (25%); of the monocots, only Arecaceae have been recorded as hosts (Table 1.8). However, because many of the records for the Ugyopini are from islands (Appendix 2), the predominance of dicot records may reflect insular radiation, possibly obscuring the ancestral host associations in this tribe.

The four records for the small tribe Asiracini include dicots, ferns, and monocots (Cyperaceae). Host plants of the Vizcayinae are unknown. The Kelisiinae feed only on monocots and appear restricted to the Cyperaceae (83% of records) and Juncaceae (17%), and the Stenocraninae occur only on the Poaceae (62%) and Cyperaceae (38%). For the Plesiodelphacinae there are so far only two records, both from the monocot genus *Heliconia* (Heliconiaceae; Muir 1926; Fennah 1959a; Asche, personal communication).

Within the Delphacinae, the Tropidocephalini and Saccharosydnini feed exclusively on Poaceae (Table 1.8). The Delphacini also show very strong affiliation with the Poaceae and Cyperaceae (71% and 13% of monocot records, respectively). Nonetheless, it is this tribe which has radiated on all subclasses of dicots on the Hawaiian archipelago and Polynesia (Fennah 1957, 1958, 1959a, 1962a, 1964, 1976; Zimmerman 1948; Table 1.3), probably within the last 6 million years based on the age of the large Hawaiian Islands (Howarth 1990). On the temperate mainland, dicot feeding in the Delphacini is most prominent in *Pissonotus* and *Stobaera*, both occurring mostly on the Asteraceae.

Host Plant Use Patterns in the Delphacidae: Parallel Phylogenesis?

Host associations in delphacids show marked evolutionary conservatism. Except on islands, most genera and some higher taxa are restricted to a single or a few related plant families, and delphacids, as a whole, primarily attack just three monocot families (Poaceae, Cyperaceae, and Juncaceae). To what extent might this fidelity have led to parallel diversification, as evidenced by concordant plant and delphacid relationships? We examine this question at four plant taxonomic levels (Tracheophyta [vascular plants], Angiospermae, Monocotyledoneae, and Poaceae) over which such concordance might be expected.

Because the Delphacidae may date to at least the Lower Cretaceous, they could have begun diversifying on nonflowering plants and evolved subsequently with the angiosperms, a possibility explored previously for the fulgoroids as a whole. The only observations that might represent such ancient relictual associations are the occurrence of two species of Ugyopini (one Polynesian) and one of Asiracini, the two oldest clades, on ferns (Table 1.7; Appendix 2). In more advanced delphacids, fern feeding is known only for nine species, mostly Hawaiian, in the huge tribe Delphacini, two species of which are also recorded on *Equisetum*. No other nonisland delphacids feed on any nonflowering plant or on the relatively primitive dicot subclasses Magnoliidae or Hamamelidae. Even within the Ugyopini and Asiracini, it is not known if the fern feeders are cladistically basal. Thus, present-day host use by the Delphacidae shows little sign of any shared history with the major lineages of the Tracheophyta or of the Angiospermae.

Most delphacids feed on, and are restricted to habitats dominated by, herbaceous monocots (Table 1.8). Thus, it might be more plausible that delphacids diversified with the monocot families rather than with seed or vascular plants as a whole. A recent DNA-sequence-based estimate of relationships among the major monocot-host families for each delphacid clade is shown in Figure 1.3 (Duvall et al. in press); this hypothesis largely agrees

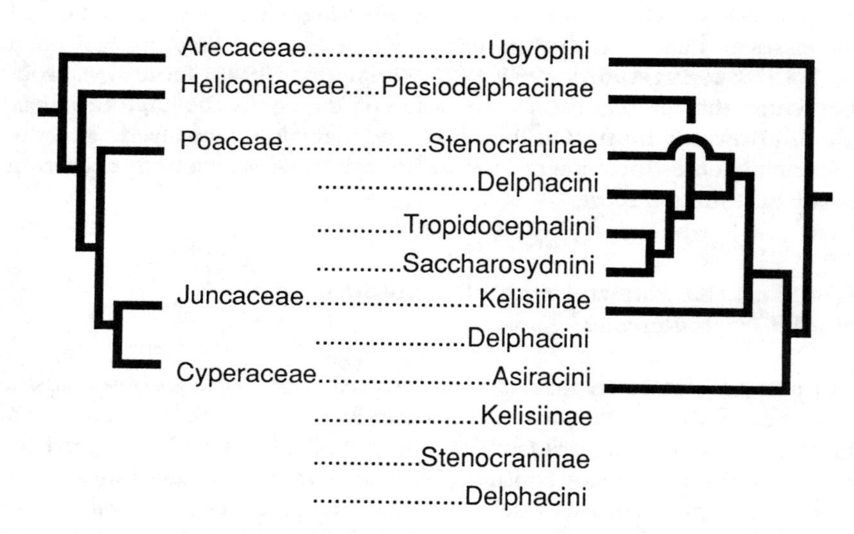

Figure 1.3. Phylogenetic relationships among monocot families serving as major hosts to the Delphacidae, following Duvall et al. (in press). Across from and below each plant family are given the delphacid taxa for which that family constitutes 5% or more of host records. The delphacid phylogeny (right-hand side of figure) is identical to that in Figure 1.2. The host plant associations and phylogenies in this figure form the basis for the cladogram concordance analysis in the text.

with a prior assessment based on plant morphology (Dahlgren and Bremer 1985). A strong point of possible concordance with delphacid phylogeny, reproduced on the right-hand side of Figure 1.3, is the occurrence of an island ugyopine species (*Ugyops osborni*) on the Arecaceae (Caldwell and Martorell 1950). However, several points of cladogram disagreement are evident as well. For example, the plant cladogram would predict that the Plesiodelphacinae, restricted to Heliconiaceae, should branch off before the Asiracini, in which both monocot records are from Cyperaceae, rather than the reverse as is evident from delphacid phylogeny (Fig. 1.2).

A widely used measure of overall cladogram agreement is the "component analysis" of Nelson and Platnick (1981), which counts the number of unseen speciation and extinction events required to account for present-day host associations if the insects had diversified strictly in tandem with the hosts. We used the COMPONENT package of Page (1990; FIT command under assumption 1) to enumerate these "items of error." We excluded the small scattering of records for Delphacini on several other monocot families which are mostly from islands and seem clearly secondary (see Table 1.8).

The fit of the monocot phylogeny to the delphacid phylogeny (Fig. 1.3) is closer than expected for host phylogenies generated under several random models (Page 1990; "items of error" = 26, $P<0.05$). Thus, although there is some suggestion that delphacids have evolved in association with monocots, this inference is weak. It is questionable whether the single ugyopine record on palms represents a retained ancestral association. This species, like other Pacific island endemics, may have undergone a relatively recent host shift; moreover, it appears to feed on other trees in addition to palms and may be polyphagous (Caldwell and Martorell 1950). Cladogram congruence is lacking if the ugyopine record is deleted ("items of error" = 26, $P = 0.20$). Similarly, if attention is restricted to associations with Poaceae, Cyperaceae, and Juncaceae, the hosts of nearly all mainland delphacids, all three possible host phylogenies fit the delphacid phylogeny equally well.

Another potential phylogenetic trend in association of delphacids with monocot families is a gradual shift from concentration on the Cyperaceae, seen in the relatively basal Asiracini and Kelisiinae, through association with both Cyperaceae and grasses, as in the Stenocraninae and Delphacinae, to strict association with Poaceae, seen in the Tropidocephalini and Saccharosydnini (Table 1.8). This progression is unlikely to reflect parallel phylogenesis as the Cyperaceae cannot be regarded as basal to the Poaceae. However, relationships of these delphacid taxa with different subfamilies of grasses, leaving aside other host associations, seem a more likely possibility for parallel delphacid/host phylogenesis. Of the delphacine tribes, the cladistically basal Delphacini are associated most often with the subfamily Pooideae (78% of records). The Tropidocephalini are found mostly on bambusoid grasses (78% of records), and their sister taxon, the Saccharosydnini, occur primarily on the Panicoideae (64%); the remainder of species are found on the Bambusoideae. The branching order of grass subfamilies implied by that of their delphacine herbivores conflicts with the view best supported by recent molecular studies, namely, that bambusoids are basal with respect to pooids and panicoids (Doebley et al. 1990; Duvall et al. in press; Fig. 1.4). However, parallel phylogenesis cannot be ruled out, as grass subfamily relationships are not conclusively settled (Doebley et al. 1990; Kellogg and Campbell 1987).

An alternative scenario of evolution of the Delphacinae with the grasses is suggested by Asche's (1985) view that the Delphacini are morphologically advanced in comparison to their sister tribes. The present-day association of the morphologically primitive Tropidocephalini with (possibly primitive) bambusoid grasses might reflect retention of primitive feeding habits. The morphologically advanced Delphacini, in contrast, might have subsequently shifted to the (possibly more derived) pooid grasses.

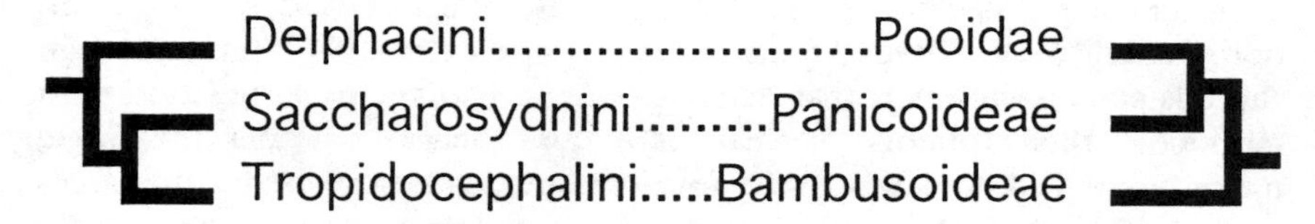

Figure 1.4. Comparison of branching relationships among delphacid tribes with relationships among the grass subfamilies used predominantly by each. Tree for grasses is the one best supported by the analyses of Doebley et al. (1990).

Host Plant Use Patterns in the Delphacidae: Plant Taxonomy Versus Habitat

While delphacid host use patterns seem little attributable to parallel diversification, they do exhibit an evolutionary conservatism comparable to that found in other phytophagous insects. Mapping of host associations onto published species-level cladograms for a variety of phytophages (Mitter and Farrell 1991) suggested that, on average, a major host shift (e.g., to a new plant family) accompanies about 20% of herbivore speciation events. Species-level cladograms are not available for any delphacids, but we can make a minimum estimate of host shift frequency by examining the habits of congeners. The minimum number of speciation events in the history of a genus is one less than the number of species, and the minimum number of host plant-family shifts is one less than the number of different host family associations among those species. Polyphagous species, which complicate this calculation somewhat, were taken as evidence for a host family shift only if they were recorded from at least one family that was not fed on by congeners.

Considering just mainland delphacids with known hosts (Appendix 2) and summing over genera containing at least two species (so that a host shift is in principle detectable), we find a minimum of 41 host family shifts out of 199 speciation events, a frequency of 0.21 per speciation event. Thus, delphacid species appear to retain the host-taxon association of their immediate ancestor, to a degree (~80%) approximately typical for phytophagous insects. The actual frequency of shifts in delphacids may be larger than this minimum estimate if the same host has been independently acquired by two or more congeneric species. On the other hand, the plant-family criterion for host shift may underrepresent the conservatism of delphacid habits, because the records are concentrated on just a few closely related families.

Conservatism of host plant-taxon affiliation, in delphacids as in other phytophages, is probably traceable in large part to host chemistry, though exactly how is still debated (Jaenike 1990). Host chemistry may pose a "constraint" on diet evolution, in the sense that genotypes required for a shift to a chemically novel host probably occur less frequently than those permitting a shift to a chemically similar host. Direct evidence for such genetic limitation, however, has rarely been sought (Futuyma and McCafferty 1990).

Exceptions to the general pattern of host-taxon conservatism in delphacids and other phytophages suggest that under some circumstances, genetic constraints imposed by host plant chemistry are somehow less stringent or that other kinds of constraint such as adaptation to extreme environments become more important. However, such alternative phylogenetic patterns are difficult to define and quantify. One factor suggested by some host use patterns is host proximity, hence probable frequency of encounter by host-seeking adults (Mitter and Farrell 1991). The relative importance of host proximity may be reinforced by the abundance of the alternative host and by specialization to a physically extreme environment. A possible example of lability mediated by habitat specialization and host proximity is the genus *Megamelus*, which feeds on a variety of aquatic hosts in the taxonomically distant Pontederiaceae and Nymphaeaceae, in addition to Cyperaceae, Juncaceae, and Poaceae. Similar habitat constraint is suggested by transfers among distantly related hosts in some other fulgoroid groups. For example, species in *Oecleus* (Cixiidae) and *Scolops* (Dictyopharidae) exploit xerophytic plants in distantly related families such as Chenopodiaceae, Asteraceae, and others which commonly co-occur in desert and prairie habitats (Ball 1930; Kramer 1977).

In delphacids and other fulgoroids, it will be difficult to separate the influence of chemical or taxonomic relatedness from that of ecological proximity because the main host families (Poaceae, Cyperaceae, and Juncaceae) are closely related, chemically similar, and occurr in similar, moist habitats (Cronquist 1981). Much further work will also be needed to assess the relative importance of other possible influences on host shift patterns of planthoppers such as differences in plant abundance (Nielson and Don 1974; Wasserman and Futuyma 1981; Futuyma 1983; Thompson 1988a), suitability as a mate location site (Claridge 1985a; Jaenike 1990), and resident natural enemies (Price et al. 1980).

The most strikingly consistent departure from host-taxon conservatism is found on oceanic islands. Exaggerated niche divergence in island versus continental radiations has been widely accepted but not often quantitatively demonstrated (Schluter 1988). The eight delphacid genera endemic to Pacific islands (primarily the Hawaiian chain) show significantly greater rates of host plant family shift than the remaining, mostly mainland genera, both in overall rate (53 shifts/134 speciation events versus 41/199; X^2 = 14.1, $P<0.01$) and in generic mean (0.43 versus 0.20, Mann–Whitney U-statistic = 291.5, $P<0.01$). This is due primarily to the much greater fraction of mainland genera which are restricted to a single host family (31/52 versus 2/8). The oceanic Pacific species also include a significantly larger proportion of polyphages (23% versus 7% for mainland species; X^2 = 10.0, $P<0.01$) and are recorded from a total of 62 families, mostly dicots, whereas the remaining genera (mostly mainland) use only 20 families, nearly all monocots.

Although it is not certain that the island genera are all phylogenetically independent, delphacids clearly exhibit exaggerated divergence in host use on Pacific islands, analogous to such examples as the greater diversity in beak dimensions, correlated with seed selection, among island versus continental species of granivorous finches (Schluter 1988). Both genetic and ecological causes may underlie such island radiations (reviewed in Schluter 1988). The typical conservatism of ecological traits such as host plant use might reflect the resistance to change of underlying coadapted gene complexes, in conjunction with the stabilizing effect of gene flow among populations (Mayr 1963). The "genetic revolutions" resulting from repeated "founder events," coupled with relative isolation from gene flow, might then greatly facilitate divergence in island populations (Carson and Templeton 1984).

Ecological hypotheses for island radiation have emphasized both resource availability and "release" from enemies and/or competitors (Schluter 1988). Departure of island colonists from the ancestral monocot niche may be initially mandated by the unique island environment. Thus, limited availability on Pacific islands of habitats dominated by herbaceous monocots may explain in part the predominant shift of endemic delphacids to woody, often arboreal dicots. Subsequent repeated shifts among host families might similarly arise by isolation of founder populations in floristically different local habitats on the same or nearby islands. It is also possible that island dicots are less chemically distinct than their mainland relatives, diminishing the barriers to host transfer; several examples of such loss in defense are known in island plant species (Carlquist 1970).

Absence of continental competitors has been a favored explanation for some island radiations (Schluter 1988), but the degree to which phytophagous insect niches are molded by interspecific competition is very uncertain (Lawton and Strong 1981; Strong et al. 1984). Natural enemies, including both predators and parasitoids, have been accorded a larger role in the population dynamics and community structure of phytophagous insects (Price et al. 1980; Bernays and Graham 1988). It is possible that delphacid host use radiation on islands is facilitated by a paucity of native invertebrate predators; for example, the Hawaiian Islands completely lack native species of ants (Howarth 1990). Closer inquiry into insular radiations may help illuminate the causes of host specificity and conservatism in mainland delphacids.

Host Range in the Delphacidae and Other Fulgoroids

The Delphacidae show a strong tendency toward monophagy. Of species with recorded habits, 74% are reported from a single host plant genus; only 14% are oligophagous and 12% polyphagous (Table 1.9). Even though the delphacids on tropical islands have radically different host plant affinities from mainland species (dicots versus monocots) and a higher percentage of polyphages, they show the same prevalence of monophagy (73% of island species versus 74% of mainland species; Table 1.4).

Narrow host range is characteristic of the species in all delphacid subfamilies and tribes (Table 1.9), from primitive (e.g., Asiracinae) to advanced (e.g., Delphacinae), and is probably the ancestral condition. Examples include the asiracine genus *Pentagramma*, all species of which are restricted to *Scirpus* (Cyperaceae), the kelisiine *Anakelisia*, reported only on *Carex*, and the plesiodelphacine genus *Burnilia*, known only from *Heliconia*. In the Delphacini, *Conomelus* feeds only on *Juncus*, *Nilaparvata* feeds on *Oryza* and the closely related *Leersia*, and *Tarophagus* is restricted to

Table 1.9. Diet breadth of planthopper species in the subfamilies and tribes of Delphacidae. The percentage of (M) monophagous (feeding primarily on one plant species or plant species in the same genus), (O) oligophagous (feeding on several genera in the same plant family), and (P) polyphagous planthopper species (feeding on two or more plant families), and the number of species records for each family is shown. Summarized from host plant records in Appendix 2.

Delphacid Taxon	M (%)	O (%)	P (%)	No. Records
Asiracinae	83	0	17	12
Ugyopini	89	0	11	9
Asiracini	67	0	33	3
Kelisiinae	68	11	21	19
Stenocraninae	77	23	0	13
Plesiodelphacinae	100	0	0	2
Delphacinae	74	14	12	426
Tropidocephalini	83	17	0	42
Saccharosydnini	83	17	0	6
Delphacini[a]	72	14	14	378
Delphacini[b]	73	4	23	158
Delphacini[c]	72	21	7	220

[a]All records.
[b]Oceanic island (Hawaiian Archepelago, Polynesia, S. Pacific) records (from Fennah 1957, 1958, 1959a, 1962a, 1964, 1976; Zimmerman 1948) only.
[c]Nonisland records only.

Colocasia. Detailed ecological studies on several taxa further corroborate a monophagous feeding habit for most delphacids. For example, most species of *Javesella* (*J. pellucida* excepted, Mochida and Kisimoto 1971; Ossiannilsson 1978; de Vrijer 1981), *Muellerianella* (Drosopoulos 1977; Asche 1982b; Booij 1982a), *Nilaparvata* (Claridge et al. 1988; Yang 1989), *Prokelisia* (Denno et al. 1987; S. Wilson unpublished data), and *Ribautodelphax* (den Bieman 1987a, 1987b) have been conclusively shown to be either monophagous or to feed on related plant genera in the same family.

All cases of "polyphagy" in the Kelisiinae (*Anakelisia* and *Kelisia*) and some in the Delphacini consist of species feeding on the closely related Cyperaceae and Juncaceae. Similarly, all instances of "polyphagy" in the Stenocraninae (*Stenocranus* spp.) and some in the Delphacini (e.g., *Javesella pellucida*) represent feeding on Cyperaceae and Poaceae. Thus, our somewhat arbitrary criterion for polyphagy, feeding on two or more plant families, results in the underrepresentation of the degree of feeding specialization in Delphacidae.

Monophagy is also the predominant condition in most other families of Fulgoroidea, including basal families such as the Tettigometridae, Kinnaridae, and nymphs of the Cixiidae (Table 1.4). The percentages of adult monophagy in Table 1.4 probably underestimate the degree of host specificity because in planthopper families for which early stages are known (e.g., Cixiidae, Meenoplidae, Kinnaridae, Dictyopharidae, Fulgoridae, and Flatidae), nymphs develop on a narrower range of hosts than that on which adults feed (Table 1.4). Monophagy may well represent the ancestral condition for the superfamily, with the exceptional degree of polyphagy seen in a few families (e.g., Ricaniidae, Flatidae, Lophopidae, and adult meenoplids and achilids) representing a derived condition.

Summary

1. The family Delphacidae belongs to a monophyletic assemblage of 19 families constituting the Fulgoroidea. A provisional phylogeny shows delphacids to be among the oldest fulgoroid lineages. The sparse fossil record, interpreted according to this phylogeny, suggests that a set of basal lineages including the Tettigometridae, Delphacidae, Cixiidae, Derbidae, Meenoplidae/Kinnaridae, and Achilidae were probably extant by the Lower Cretaceous. The remaining, more advanced families form a monophyletic group and are first represented by fossils from the Upper Cretaceous and Tertiary.
2. At least some host plant associations are known for nearly all fulgoroid families, but by far the most complete and accurate information available is for the Delphacidae. Fulgoroid associations span all major groups of vascular plants, including all the subclasses of angiosperms. Evidence was sought for parallel diversification of the Fulgoroidea with their host plant taxa by contrasting basal and advanced families for the fraction of adult records on angiosperms versus nonangiosperms, and on primitive versus advanced angiosperms. No significant trends were found, though the basal families are associated somewhat more strongly with both nonangiosperms and basal angiosperms. There was a slight tendency, not interpretable as parallel phylogenesis, for primitive families to feed more on monocots, whereas advanced families favor dicots.
3. In contrast to the lack of phylogenetic trends in host-taxon association, at least for adult fulgoroids, there were clear phylogenetic patterns in feeding location. There was a significant tendency for the primitive families to feed underground or under bark, on roots or fungi, which may have predisposed fulgoroids to preference for moist habitats, and hence to association with monocots. The trend toward exposed feeding

on woody dicots among more advanced fulgoroids appears correlated with evolution of an ovipositor capable of excavating cavities for egg deposition in woody tissues.

4. The Delphacidae are recorded from a great range of vascular plant taxa, but comparisons of host phylogeny with a morphology-based phylogeny for delphacid tribes/subfamilies yields no evidence for parallel phylogenesis with either vascular plants or angiosperms as a whole. Delphacids feed primarily on monocots, and the relationships among just their monocot host families show significant concordance to the ordering of those hosts on the delphacid cladogram. However, this pattern results primarily from the occurrence on the Arecaceae of a single island-endemic and possibly polyphagous species in the most basal tribe (Ugyopini); it is, thus, weak evidence for parallel diversification. Delphacids are concentrated on two closely related and phytochemically similar families, the Cyperaceae and Poaceae. There is a tendency for primitive lineages (Asiracini and Kelisiinae) to feed on Cyperaceae, whereas two more advanced subfamilies (Stenocraninae and Delphacinae) feed mostly on Poaceae. Each of the three tribes of Delphacinae is associated most strongly with a different grass subfamily. Branching relationships among the three tribes of Delphacinae conflict with the view of grass subfamily phylogeny best supported by recent molecular studies; however, grass relationships are still debated, and parallel phylogenesis of delphacids with grasses cannot be ruled out.
5. Although delphacids show little sign of parallel diversification with their host plants, their host-taxon associations appear about as evolutionarily conservative as those of other phytophagous insects; up to 80% of species feed within the same host family as their nearest relatives. Departures from this pattern suggest that host shifts are sometimes more constrained (or facilitated) by habitat fidelity than by intrinsic host similarity as reflected in plant taxonomy.
6. The large delphacid fauna endemic to Pacific islands, mostly Hawaiian, shows significantly elevated rates of shifting to new host families and frequency of polyphagous species, and in contrast to mainland groups occurs mainly on dicots. One of several plausible explanations for this island radiation is reduced pressure from predators.
7. Delphacids are strongly host specific, with 74% of species restricted to a single host plant genus and often to a single species, and only 12% recorded from more than one plant family. The predominance of monophagy holds for both monocot and dicot feeders, and both primitive and advanced tribes/subfamilies, suggesting that host specialization represents the ancestral state in the Delphacidae. Monophagy is

prevalent in the other primitive planthopper families and may be ancestral for the Fulgoroidea. There is a slight tendency toward increased polyphagy in the advanced fulgoroid families (e.g., Ricaniidae, Flatidae, and Lophopidae).

8. Further understanding of planthopper/host plant relationships will require, above all, an increased knowledge of planthopper phylogeny and of the biology of immature stages and tropical species.

ACKNOWLEDGMENTS

We thank M. Asche and H. Hoch for helpful discussions, C. Humphries for comments on an earlier draft of this chapter, and D. Castaner, A. F. Emeljanov, K. G. A. Hamilton, R. Kisimoto, and A. G. Wheeler, Jr. for a number of host plant records and literature citations. This work was supported in part by USDA-NRICGP Grant 90-37250-5482 to C. M. This is Contribution Number 8599 of the Maryland Agricultural Experiment Station, Department of Entomology.

Appendix 1. Recorded host plants* of the Fulgoroidea excluding the Delphacidae. Host plant records are for adult planthoppers unless marked (N) indicating that nymphs were observed, sampled or collected.

Planthopper taxon	Host Plant: Genus and Species	Host Plant: Family	Reference
TETTIGOMETRIDAE			
Egropinae			
Egropa breviceps (Stål)	*Annona squamosa*	Annonaceae	Baker (1915)
Hildinae			
Hilda patruelis Stål	Polyphygous		Weaving (1980)
Hilda undata (Walker)	*Anacardium occidentale*	Anacardiaceae	Akingbohungbe (1982)
	Arachis hypogaea	Fabaceae	
	Helianthus	Asteraceae	
Megahilda rhodesiana Fennah	*Aloe excelsa*	Liliaceae	Fennah (1959b)
Tettigometrinae			
Phalix titan Fennah	*Acacia decurrens* var. *mollis*	Fabaceae	Ghauri (1964)
Tettigometra hexaspina Kolenati	*Papaver somniferum*	Papaveraceae	Weaving (1980)
Tettigometra impressopunctata Duf.	Polyphagous		Gunthart (1987)
Tettigometra leucophaea (Preysster)	*Secale cereale*	Poaceae	Wilson and O'Brien (1987)
CIXIIDAE			
Bothriocerinae			
Achaemenes synavei Williams	*Dodonaea viscosa*	Sapindaceae	Williams (1975)
Bothriocera cognita Caldwell	*Spartina*	Poaceae	Kramer (1983)
	Juncus	Juncaceae	
	Cephalanthus occidentalis	Rubiaceae	
	Quercus	Fagaceae	
Bothriocera drakei Metcalf	Pteridophyta	?	Kramer (1983)
Bothriocera eborea Fennah	*Coccoloba uvifera*	Polygonaceae	Fennah (1949a)

Bothriocera maculata Caldwell	*Spartina patens* (N)	Poaceae	Denno (unpublished data)
	Spartina cynosuroides	Poaceae	Kramer (1983)
	Spartina patens	Poaceae	
	Juncus roemerianus	Juncaceae	
	Eupatorium capillifolium	Asteraceae	
Bothriocera signoreti Stål	*Cyperus papyrus*	Cyperaceae	Myers (1929)
	Commelina nudiflora	Commelinaceae	
	Two spp. of ferns		
	Polygonum acre	Polygonaceae	
Bothriocera transversa Caldwell	*Chrysobalanus icaco*	Chrysobalanaceae	Kramer (1983)
	Cestrum nocturnum	Solanaceae	
	Cocos nucifera	Arecaceae	
	Borrichia arborescens	Asteraceae	
Bothriocera venosa Fowler	*Coffea*	Rubiaceae	Dozier (1931)
Bothriocera sp.	*Veitchia merrillii*	Pinaceae	Howard and Mead (1980)
Brixia belouvensis bipunctata Synave	*Quivisia mauritiana*	Meliaceae	Williams (1975)
Brixia discolor Williams	*Murraya paniculata*	Rutaceae	Williams (1975)
Brixia fulgida Van Duzee	Pimento bush	?	Van Duzee (1907)
Brixia mauritii Synave	*Coffea macrocarpa*	Rubiaceae	Williams (1975)
	Myonima multiflora	Rubiaceae	
Brixia vaughani Williams	*Coffea vaughani*	Rubiaceae	Williams (1975)
Brixia venulosa Williams	*Embelia*	Myrsinaceae	Williams (1975)
Nymphocixia caribbea Fennah	*Cocos nucifera*	Arecaceae	Howard et al. (1981)
Nymphocixia unipunctata Van Duzee	*Corypha elata*	Arecaceae	Howard and Mead (1980)
	Rhizophora mangle	Rhizophoraceae	Kramer (1983)
	Laguncularia racemosa	Combretaceae	
	Avicennia nitida	Avicenniaceae	
Cixiinae			
Euryphlepsia cocos Muir	*Cocos nucifera*	Arecaceae	Muir (1924a)
Euryphlepsia pallescens (Metcalf)	*Hibiscus tiliaceus*	Malvaceae	Fennah (1956)
Hemitropis limonii Emeljanov	*Limonium suffruticosum*	Plumbaginaceae	Emeljanov (1964)
	Limonium gmelini	Plumbaginaceae	

(*continued*)

Appendix 1. (*Continued*)

Planthopper taxon	Host Plant		Reference
	Genus and Species	Family	
Hemitropis seticulosa (Lethierry)	*Tamarix*	Tamaricaceae	Linnavuori (1964)
Hemitropis tamaricus (Puton & Leth.)	*Tamarix*	Tamaricaceae	Dlabola (1981)
Hemitropis tatianae Emeljanov	*Tamarix*	Tamaricaceae	Emeljanov (1964)
Hemitropis sp.	*Tamarix gallica*	Tamaricaceae	Hopkins and Carruth (1954)
Kirbyana pagana Melichar	*Saccharum officinarum*	Poaceae	Schumacher (1920)
Kuvera ligustri Matsumura	*Ligustrum*	Oleaceae	Lee and Kwon (1977)
Microledrida arida Caldwell	*Cordia nitida*	Ehretiaceae	Caldwell and Martorell (1950)
Microledrida fuscata Van Duzee	*Heteromeles*	Rosaceae	Kramer (1983)
Cixiini			
Cixius adornatus iranicus Dlabola	*Pterocarya*	Juglandaceae	Dlabola (1979b)
Cixius alpestris Wagner	*Pinus montana*	Pinaceae	Gunthart (1987)
Cixius angustatus Caldwell	*Prunus*	Rosaceae	Kramer (1981)
Cixius balli Kramer	*Pinus*	Pinaceae	Kramer (1981)
Cixius basalis Van Duzee	*Abies balsamea*	Pinaceae	Beirne (1950)
	Picea mariana	Pinaceae	
Cixius cinctifrons Fitch	*Carya*	Juglandaceae	Packard (1890)
Cixius coloepium Fitch	*Gaylussacia*	Ericaceae	Van Duzee (1905)
Cixius cultus Ball	Polyphagous	Dicots	Kramer (1981)
Cixius cunicularis (L.)	Polyphagous on woody plants		Gunthart (1987)
Cixius dubius Wagner	*Prunus spinosa*	Rosaceae	Gunthart (1990)
	Quercus pubescens	Fagaceae	
	Trees and bushes		
Cixius ephratus Ball	*Salsola kali*	Chenopodiaceae	Kramer (1981)
Cixius haupti Dlabola	*Pinus*	Pinaceae	Gunthart (1984)
Cixius misellus Van Duzee	*Abies balsamea*	Pinaceae	Beirne (1950)
	Picea glauca	Pinaceae	

Cixius nervosus (L.)	Polyphagous on woody plants		Gunthart (1987)
Cixius pallipes Fieber	*Tamarix*	Tamaricaceae	Gunthart (1990)
Cixius pini Fitch	*Gaylussacia*	Ericaceae	Procter (1938)
	Pinus, Picea, Abies	Pinaceae	Kramer (1981)
Cixius pinicola Dufour	*Pinus*	Pinaceae	Synave (1951)
Cixius praecox Van Duzee	*Salix, Populus*	Salicaceae	Kramer (1981)
Cixius setinervis Stål	*Cliffortia scrabilifera*	Rosaceae	Van Stalle (1988)
Cixius simplex (Herrich-Schaffer)	Polyphagous on woody plants		Gunthart (1987)
Cixius sp.	*Atriplex canescens*	Chenopodiaceae	Stroud (1950)
Tachycixius pilosus (Olivier)	*Prunus spinosa*	Rosaceae	Gunthart (1987)
	Grass (N)	Poaceae	Le Quesne (1960)
	Picea	Pinaceae	Gunthart (1990)
	Juniperus	Cupressaceae	
	Quercus pubescens	Fagaceae	
	Crataegus monogyna	Rosaceae	
	Trees and bushes		
Oecleini			
Myndus beameri Ball	*Agave americana*	Agavaceae	Kramer (1979)
	Yucca	Agavaceae	
Myndus catalinus Ball	*Yucca*	Agavaceae	Kramer (1979)
Myndus crudus Van Duzee	*Phoenix canariensis*	Arecaceae	Meyerdirk and Hart (1982)
	Cocos nucifera	Arecaceae	Tsai et al. (1976)
	Grasses (N)	Poaceae	Tsai and Kirsch (1978)
	Stenotaphrum secundatum (N)	Poaceae	Reinert (1980)
	Paspalum notatum (N)	Poaceae	
	Cynodon dactylon (N)	Poaceae	
Myndus dibaphus Fennah	*Pandanus tectorius*	Pandanaceae	Fennah (1956)
Myndus enotatus Van Duzee	Grass	Poaceae	Osborn (1926)
	Juncus	Juncaceae	Kramer (1979)
	"Marsh grasses"	Poaceae	
Myndus gabrielensis (Flock)	*Yucca whipplei*	Agavaceae	Kramer (1979)
	Sedges	Cyperaceae	

(*continued*)

Appendix 1. (*Continued*)

Planthopper taxon	Host Plant		Reference
	Genus and Species	Family	
Myndus glyphis Kramer	*Populus*	Salicaceae	Kramer (1979)
Myndus irreptor Fennah	*Pandanus*	Pandanaceae	Fennah (1956)
Myndus lophion Kramer	*Pinus*	Pinaceae	Kramer (1979)
Myndus mojavensis Ball	*Yucca brevifolia*	Agavaceae	Kramer (1979)
Myndus musivus Germar	*Salix*	Salicaceae	Villiers (1977)
Myndus nigrifrons Ball	*Nolina*	Agavaceae	Ball (1937)
Myndus nolinus Ball	*Nolina microcarpa*	Agavaceae	Kramer (1979)
	Triodia pulchella	Poaceae	
Myndus occidentalis Van Duzee	*Juncus*	Juncaceae	Kramer (1979)
	Washingtonia filifera	Arecaceae	
Myndus radicis Osborn	*Impatiens* (N)	Balsaminaceae	Osborn (1903)
	Nettles (N)	Urticaceae	
	Grasses (N)	Poaceae	
Myndus rubidus Ball	Palmetto	Arecaceae	Kramer (1979)
Myndus simplicatus (Caldwell)	*Panicum barbinode*	Poaceae	Kramer (1979)
Myndus slossonae Ball	*Juncus*	Juncaceae	Dozier (1926)
	"Grasses"	Poaceae	Kramer (1979)
Myndus taffini Bonfils	*Cocos nucifera*	Arecaceae	Bonfils (1982); Julia (1982)
Myndus tekmar Kramer	*Agave*	Agavaceae	Kramer (1979)
Myndus texensis Kramer	*Yucca thompsoniana*	Agavaceae	Kramer (1979)
Myndus yuccandus Ball	*Yucca*	Agavaceae	Kramer (1979)
Oecleus arnellus B & K	*Chrysothamnus speciosus*	Asteraceae	Kramer (1977)
Oecleus balli Kramer	*Atriplex* (two sp.)	Chenopodiaceae	Kramer (1977)
	Pluchea sericea	Asteraceae	
Oecleus borealis Van Duzee	*Chilopsis linearis*	Bignoniaceae	Kramer (1977)
Oecleus campestris Ball	*Artemesia filifolia*	Asteraceae	Kramer (1977)
Oecleus centronus B & K	*Salicornia*	Chenopodiaceae	Kramer (1977)

Oecleus cucullus Kramer	*Trixis californica*	Asteraceae	Goeden and Ricker (1989)
	Artemesia	Asteraceae	Kramer (1977)
	Chilopsis linearis	Bignoniaceae	
Oecleus fulvidorsum Ball	*Atriplex canescens*	Chenopodiaceae	Kramer (1977)
Oecleus glochin Kramer	*Chrysothamnus, Hymenoclea*	Asteraceae	Kramer (1977)
Oecleus lineatus Ball	*Lygodesmia spinosa*	Asteraceae	Kramer (1977)
Oecleus lyra Kramer	*Dasylirion*	Agavaceae	Kramer (1977)
Oecleus martharum Kramer	*Atriplex* (three sp.)	Chenopodiaceae	Kramer (1977)
Oecleus monilipennis Van Duzee	*Atriplex*	Chenopodiaceae	Van Duzee (1923)
Oecleus natatorius Ball	Grasses	Poaceae	Ball (1937)
Oecleus nolinus B & K	*Nolina, Yucca*	Agavaceae	Kramer (1977)
Oecleus obtusus Ball	*Pluchea*	Asteraceae	Kramer (1977)
Oecleus perpictus Van Duzee	*Muhlenbergia porteri*	Poaceae	Kramer (1977)
Oecleus pigmy B & K	*Atriplex* (three sp.)	Chenopodiaceae	Kramer (1977)
Oecleus piperatus B & K	*Atriplex linearis*	Chenopodiaceae	Kramer (1977)
Oecleus planus B & K	*Chrysothamnus panniculatus*	Asteraceae	Kramer (1977)
Oecleus productus Metcalf	Grasses	Poaceae	Kramer (1977)
	Composites	Asteraceae	
Oecleus sagittanus B & K	*Pluchea sericea*	Asteraceae	Kramer (1977)
Oecleus snowi Ball	*Baccharis glutinosa*	Asteraceae	Kramer (1977)
Oecleus subreflexus Van Duzee	*Atriplex lentiformis*	Chenopodiaceae	Kramer (1977)
	Pluchea, Franseria, Chrysoth.	Asteraceae	
Oecleus texanus Ball	*Yucca glauca*	Agavaceae	Ball (1937)
Oecleus sp.	*Cocos nucifera*	Arecaceae	Howard et al. (1981)
Oecleus sp.	*Tamarix gallica* (N)	Tamaricaceae	Hopkins and Carruth (1954)
Pentastirini			
Helenolius dividens (Walker)	*Aster glutinosa*	Asteraceae	Van Stalle (1986c)
	Commidendron rugosum	Asteraceae	
Helenolius insulicola Van Stalle	*Frankenia postulatifiola*	Frankeniaceae	Van Stalle (1986c)
	Commidendron rugosum	Asteraceae	

(*continued*)

Appendix 1. (*Continued*)

Planthopper taxon	Host Plant		Reference
	Genus and Species	Family	
Hyalesthes angustulus Horvath	Polyphagous		Hoch and Remane (1985)
Hyalesthes flavipennis Horvath	Polyphagous		Hoch and Remane (1985)
Hyalesthes obsoletus Signoret	*Betula*	Betulaceae	Villiers (1977)
	Convolvulus arvensis L. (N)	Convolvulaceae	Brĉak (1979)
	Lepidium (N)	Brassicaceae	
	Chenopodium (N)	Chenopodiaceae	
	Amaranthus (N)	Amaranthaceae	
	Solanum (N)	Solanaceae	
	Cirsium (N)	Asteraceae	
Mnemosyne consoleae Van Stalle	*Consolea moniliformis*	Cactaceae	Van Stalle (1987b)
Norialsus elytropappi Van Stalle	*Elytropappus rhinocerotis*	Asteraceae	Van Stalle (1986b)
Norialsus salsolarum Van Stalle	*Salsola*	Chenopodiaceae	Van Stalle (1986b)
Oliarus acaciae Kirkaldy	*Acacia koa*	Fabaceae	Zimmerman (1948)
Oliarus acicus Caldwell	*Phoenix canariensis*	Arecaceae	Meyerdirk and Hart (1982)
	Prosopis	Fabaceae	Ward et al. (1977)
Oliarus annandalei cocosivora Muir	*Cocos nucifera*	Arecaceae	Wilson and O'Brien (1987)
Oliarus artemisiae Matsumura	*Artemisia*	Asteraceae	Lee and Kwon (1977)
Oliarus aridus Ball	*Phoenix canariensis*	Arecaceae	Meyerdirk and Hart (1982)
	Prosopis	Fabaceae	Ward et al. (1977)
Oliarus atkinsoni Myers	*Phormium tenax* (N)	Agavaceaae	Cumber (1952)
Oliarus barajus Dlabola	*Tamarix*	Tamaricaceae	Dlabola (1981)
Oliarus beieri (Wagner)	*Alnus*	Betulaceae	Gunthart (1987)
	Myricaria	Tamaricaceae	
	Salix	Salicaceae	
Oliarus borinquensis Caldwell	*Rhizophora mangle*	Rhizophoraceae	Caldwell and Martorell (1950)
Oliarus cocosivora Muir	*Cocos nucifera*	Arecaceae	Howard and Mead (1980)

Oliarus complectus Ball	*Cocos nucifera*	Arecaceae	Howard et al. (1981)
	Carica papaya	Caricaceae	Martorell and Adsuar (1952)
	Solanum melongena	Solanaceae	Caldwell & Martorell (1950)
	Castilla elastica	Moraceae	
	Volkameria aculeata	Verbenaceae	
	Batis maritima	Batidaceae	
	Swept from many plants		
	Saccharum (N)	Poaceae	Sein (1932, 1933)
	Panicum purpuracscens (N)		
Oliarus ecologus Caldwell	*Chamaedaphne*	Ericaceae	Beirne (1950)
Oliarus discrepans Giffard	*Gossypium tomentosum*	Malvaceae	Zimmerman (1948)
Oliarus euphorbiae Giffard	*Euphorbia*	Euphorbiaceae	Zimmerman (1948)
Oliarus felis Kirkaldy	*Sporobolus virginicus* (N)	Poaceae	Hacker (1925a)
Oliarus fici Van Stalle	*Ficus thorningii*	Moraceae	Van Stalle (1987a)
Oliarus filicola Kirkaldy	*Cibotium*	Dicksoniaceae	Zimmerman (1948)
Oliarus frontalis Melichar	*Acacia*	Fabaceae	Linnavuori (1964)
Oliarus haleakalae Kirkaldy	*Cibotium chamissoi*	Dicksoniaceae	Zimmerman (1948)
	Cyrtandra	Gesneriaceae	
Oliarus halehaku Giffard	*Cibotium*	Dicksoniaceae	Zimmerman (1948)
	Sadleria	Blechnaceae	
	Pipturus	Urticaeae	
Oliarus hodgarti Distant	*Xanthium strumarium*	Asteraceae	Hilgendorf and Goeden (1982)
Oliarus hottentottus (Stål)	*Diospyros mepiliformis*	Ebenaceae	Van Stalle (1987a)
Oliarus immaculatus Giffard	Ferns		Zimmerman (1948)
Oliarus kahavalu Kirkaldy	*Metrosideros*	Myrtaceae	Zimmerman (1948)
Oliarus kanakanus Kirkaldy	*Metrosideros*	Myrtaceae	Zimmerman (1948)
Oliarus kaonohi Kirkaldy	*Broussaisia*	Hydrangiaceae	Zimmerman (1948)
	Rotting tree fern fronds		
Oliarus kasachstanicus Emeljanov	*Atriplex cana*	Chenopodiaceae	Emeljanov (1964)
Oliarus koae Giffard	*Acacia koa*	Fabaceae	Zimmerman (1948)
Oliarus koanoa Kirkaldy	Tree fern roots (N)		Swezey (1906)

(*continued*)

Appendix 1. (*Continued*)

Planthopper taxon	Host Plant		Reference
	Genus and Species	Family	
	Maba sandwicensis	Ebenaceae	Zimmerman (1948)
Oliarus koele Giffard	Ferns		Zimmerman (1948)
Oliarus lacotei Dlabola	*Thymus vulgaris*	Lamiaceae	Dlabola (1970)
Oliarus littoralis Ball	*Distichlis, Sporobolus*	Poaceae	Mead and Kramer (1982)
Oliarus lupialensis Synave	*Cocos nucifera*	Arecaceae	Van Stalle (1987a)
Oliarus maidis Fennah	*Zea mays*	Poaceae	Fennah (1945)
Oliarus major (Kirschbaum)	*Zygophyllum*	Zygophyllaceae	Dlabola (1981)
Oliarus mori Matsumura	*Morus alba*	Moraceae	Tsaur et al. (1988a)
Oliarus myoporicola Giffard	*Myoporum sandwicense*	Myoporaceae	Zimmerman (1948)
Oliarus oleae Van Stalle	*Olea africana*	Oleaceae	Van Stalle (1987a)
Oliarus olympus Giffard	*Metrosideros*	Myrtaceae	Zimmerman (1948)
Oliarus opuna Kirkaldy	*Astelia*	Liliaceae	Zimmerman (1948)
	Dubautia	Asteraceae	
	Nephrolepis exaltata	Oleandraceae	
Oliarus orithyia Fennah	*Salicornia*	Chenopodiaceae	Van Stalle (1987a)
Oliarus oryzae Matsumura	*Saccharum offinarum*	Poaceae	Tsaur et al. (1988a)
Oliarus pele Kirkaldy	Tree ferns		Zimmerman (1948)
Oliarus pinicolus Osborn	*Pinus*	Pinaceae	Osborn (1926)
Oliarus sonoitus Ball	*Prosopis*	Fabaceae	Ward et al. (1977)
Oliarus texanus Metcalf	*Prosopis*	Fabaceae	Ward et al. (1977)
Oliarus vicarius (Walker)	Saw palmetto	Arecaceae	Thompson et al. (1979)
	Pine	Pinaceae	
Oliarus zyxus Caldwell	*Prosopis*	Fabaceae	Ward et al. (1977)
Pentastiridius apicalis (Uhler)	*Oryza sativa*	Poaceae	Wilson and O'Brien (1987)
Pentastiridius iphis (Linnavuori)	*Ludia*	Flacourtiaceae	Van Stalle (1986a)
Pentastiridius pachyceps (Matsum.)	*Cynodon dactylon* (N)	Poaceae	Tsaur et al. (1988a)
Pseudoliarus circularis Dlabola	*Tamarix*	Tamaricaceae	Dlabola (1981)

Subfamily Undetermined			
Ankistrus pini Tsaur & Hsu	*Pinus*	Pinaceae	Tsaur et al. (1988b)
Pintalia delicata (Fowler)	*Saggittaria latifolia*	Alismataceae	Kramer (1983)
Platycixius calvus Van Duzee	*Yucca*	Agavaceae	Kramer (1983)
Solonaima sp.	"Tree roots" (N) *Ficus*?	Moracerae?	Hoch and Howarth (1989b)
Undarana sp.	"Tree roots" (N)	Myrtaceae?	Hoch and Howarth (1989a)
KINNARIDAE			
Kinnarinae			
Adolendini			
Adolenda ephedrina Emeljanov	*Ephedra equisetina*	Ephedraceae	Emeljanov (1985)
Emeljanopleromini			
Kinnacana clara Remane	*Bystropogon*	Lamiaceae	Remane (1985b)
	Micromeria	Lamiaceae	
	Lavandula	Lamiaceae	
Kinnoccia chromata Remane	*Lavandula*	Lamiaceae	Remane (1985b)
	Argyranthemum	Asteraceae	
	Schizogyne sericea	Asteraceae	
Oeclidius brickellus Ball	*Trixis californica*	Asteraceae	Goeden and Ricker (1989)
Oeclidius sp.	*Hymenoclea salsola*	Asteraceae	Goeden and Ricker (1986a)
Southia nemoralis (Fennah)	*Theobroma cacao*	Sterculiaceae	Fennah (1945)
Prosotropinae			
Oreopenes luteifacies Ramos	*Petitia domingensis*	Verbenaceae	Ramos (1957)
Prosoptropis decorata Uhler	*Tabernaemontana*	Apocynaceae	Fennah (1942a)
Prosoptropis trinervosa Fennah	*Tabernaemontana*	Apocynaceae	Fennah (1942a)
Quilessa lutea Fennah	*Cyathea*	Cyatheaceae	Fennah (1942a)
Quilessa gladiolata Fennah	Ferns, shrubs		Fennah (1942a)
Quilessa grenadana Fennah	*Sanseveria* roots (N)	Agavaceae	Fennah (1948)
Quilessa maculata Fennah	Ferns, shrubs		Fennah (1942a)

(*continued*)

Appendix 1. (*Continued*)

Planthopper taxon	Host Plant		Reference
	Genus and Species	Family	
MEENOPLIDAE			
Kermesiinae			
Eponisia albovittata Fennah	On seven herbaceous plants swept from many others		Synave (1961)
Eponisia brunnescens Synave	On four herbaceous plants swept from many others		Synave (1961)
Kermesia plurimaculata Synave	*Vitex doniana*	Verbenaceae	Synave (1961)
	Eriosema psoraleoides	Fabaceae	
	Stereospermum kunthianum	Bignoniaceae	
Nisia australiensis Woodward	*Cyperus rotundus*	Cyperaceae	Tsaur et al. (1986)
	Commelina diffusa	Commelinaceae	
Nisia carolinensis Fennah	*Cyperus rotundus*	Cyperaceae	Tsaur et al. (1986)
	Oryza sativa	Poaceae	
Nisia nervosa (Motschulsky)	Sedges	Cyperaceae	Wilson and Claridge (1991)
	Colocasia esculenta	Araceae	Mitchell and Maddison (1983)
	Saccharum offinarum	Poaceae	Schumacher (1920)
	Oryza sativa	Poaceae	
Nisia serrata Tsaur	*Alocasia cucullata* (N)	Araceae	Tsaur (1989a)
DERBIDAE			
Cenchreini			
Cedusa aburiensis Muir	*Elaeis*	Arecaceae	Howard and Mead (1980)
	Raphia vinifera	Arecaceae	
Cedusa californica (Van Duzee)	*Salix*	Salicaceae	Van Duzee (1914)
Cedusa caribbensis Caldwell	*Pennisetum purpureum*	Poaceae	Caldwell and Martorell (1950)

Cedusa cedusa McAtee	*Carex*	Cyperaceae	Ball (1928)
	Osmunda	Osmundiaceae	
Cedusa chuluota Ball	*Woodwardia*	Blechnaceae	Ball (1928)
Cedusa incisa Metcalf	*Betula*	Betulaceae	Ball (1928)
	Andropogon	Poaceae	
Cedusa kedusa McAtee	*Platanus occidentalis*	Platanaceae	Ball (1928)
Cedusa mallochi McAtee	*Rosa*	Rosaceae	Wray (1950)
Cedusa nigripes Muir	Palms	Arecaceae	Muir (1930b)
Cedusa vulgaris (Fitch)	*Crataegus*	Rosacee	Wellhouse (1920)
Cedusa wolcotti Muir	Palms	Arecaceae	Howard and Mead (1980)
Lamenia caliginea (Stål)	*Colocasia esculenta*	Araceae	Mitchell and Maddison (1983)
Lamenia caliginea charon Fennah	*Messerschmidia*	Boraginaceae	Fennah (1956)
Lamenia epiensis Muir	*Cocos nucifera*	Arecaceae	Bonfils (1982)
Malenia licea Dlabola	*Salix*	Salicaceae	Dlabola (1979b)
Malenia mesasiatica Dubovsky	*Salix*	Salicaceae	Anufriev (1968)
Malenia sarmatic Anufriev	*Salix*	Salicaceae	Anufriev (1968)
Malenia turanica Anufriev	*Salix*	Salicaceae	Anufriev (1968)
Malenia turkestanica Dubovsky	*Salix*	Salicaceae	Anufriev (1968)
Malenia ussurica Anufriev	*Populus*	Salicaceae	Anufriev (1968)
Neocenchrea sp.	*Cocos nucifera*	Arecaceae	Howard et al. (1981)
	Sabal umbraculifera	Arecaceae	
	Veitchia merrillii	Arecaceae	
Nesorhamma chalcas chalcas Fennah	*Pandanus*	Pandanaceae	Fennah (1956)
Omolicna mcateei (Dozier)	*Physalis*	Fabaceae	Wray (1950)
Omolicna puertana Caldwell	*Carica papaya*	Caricaceae	Martorell and Adsuar (1952)
	Solanum melongena	Solanaceae	Caldwell and Martorell (1950)
	Melia azedarach	Meliaceae	
	Swept from many plants		
Omolicna sp.	*Pritchardia thurstonii*	Arecaceae	Howard and Mead (1980)
	Cocos nucifera	Arecaceae	Howard et al. (1981)
	Roystonea hispaniolana	Arecaceae	
	Sabal umbraculifera	Arecaceae	
	Veitchia merrillii	Arecaceae	

(*continued*)

Appendix 1. (*Continued*)

Planthopper taxon	Host Plant		Reference
	Genus and Species	Family	
Paraphenice mbaleensis Synave	*Theobroma cacao*	Sterculiaceae	Synave (1973)
Patara albida Westwood	*Inga vera*	Fabaceae	Caldwell and Martorell (1950)
Patara eloeidis Muir	*Elaeis guineensis*	Arecaceae	Muir (1930b)
Patara hargreavesi Muir	*Elaeis guineensis*	Arecaceae	Muir (1930b)
Phaciocephalus carolinensis Metcalf	*Wedelia biflora*	Asteraceae	Fennah (1956)
Phaciocephalus onoi Fennah	*Wedelia biflora*	Asteraceae	Fennah (1956)
	Artocarpus altilis	Moraceae	
Phaciocephalus phaedra Fennah	*Alocasia*	Araceae	Fennah (1956)
Phaciocephalus sigaleon Fennah	*Alpinia*	Zingiberaceae	Fennah (1967b)
Phenice stellulata (Boh.)	*Elaeis guineensis*	Arecaceae	Muir (1928)
Vekunta nigrolineata Muir	*Saccharum offinarum*	Poaceae	Schumacher (1920)
Vekunta stigmata Matsumura	*Saccharum offinarum*	Poaceae	Schumacher (1920)
Derbini			
Dysimia maculata Muir	*Sabal*	Arecaceae	Ball (1928)
	Coccothrinax gracilis	Arecaceae	Dozier (1931)
	Inga vera	Fabaceae	Muir (1924b)
	Inga laurina	Fabaceae	
Mysidia mississippiensis Dozier	*Acer*	Aceraceae	Ball (1928)
	Sabal palmetto	Arecaceae	
Otiocerini			
Anotia fitchi (Van Duzee)	*Saccharum officinarum*	Poaceae	Osborn (1926)
	Spartina pectinata	Poaceae	Holder (1990)
	Carya	Juglandaceae	Van Duzee (1894)
Anotia uhleri (Van Duzee)	*Acer*	Aceraceae	Van Duzee (1894)
Apache degeerii (Kirby)	*Fagus*	Fagaceae	Van Duzee (1889)
	Quercus (N under bark)	Fagaceae	Wilson (1982b)
	Acer	Aceraceae	
	Carya	Juglandaceae	

Kamendaka pulchra (Muir)	*Cocos nucifera*	Arecaceae	Muir (1917)
Kamendaka saccharivora (Matsum.)	*Saccharum offinarum*	Poaceae	Schumacher (1920)
Kamendaka spio Fennah	*Heliconia*	Heliconiaceae	Fennah (1967b)
	Alpinia	Zingiberaceae	
	Colocasia esculenta	Araceae	Mitchell and Maddison (1983)
Muiralyricen ruber Metcalf	*Icacorea* (*Ardisia*)	Myrsinaceae	Metcalf (1946b)
Mysidioides tagalica Muir	*Asplenium nidus*	Aspleniaceae	Muir (1917)
Nesokaba lineata Muir	*Cocos nucifera*	Arecaceae	Muir (1917)
Nesokaba phillipina Muir	*Cocos nucifera*	Arecaceae	Muir (1917)
Otiocerus coquebertii Kirby	*Acer*	Aceraceae	Van Duzee (1889)
	Fagus	Fagaceae	
Otiocerus stollii Kirby	*Quercus*	Fagaceae	Van Duzee (1889)
Otiocerus wolfi Kirby	*Acer*	Aceraceae	Ball (1928)
	Fagus	Fagaceae	Van Duzee (1894)
Paralyricen tephrias Fennah	*Fitchia speciosa*	Asteraceae	Fennah (1958)
Pyrrhoneura maculata Muir	*Cocos nucifera*	Arecaceae	Muir (1917)
Pyrrhoneura saccharicida Kirkaldy	*Colocasia esculenta*	Araceae	Mitchell and Maddison (1983)
Shellenius balli McAtee	*Acer*	Aceraceae	Ball (1928)
	Carpinus caroliniana	Betulaceae	
	Sabal palmetto	Arecaceae	
	Fraxinus	Oleaceae	Wilson (unpublished data)
Shellenius schellenbergi	*Acer*	Aceraceae	Ball (1928)
	Carpinus caroliniana	Betulaceae	
	Sabal palmetto	Arecaceae	
	Fraxinus	Oleaceae	Wilson (unpublished data)
Swezeyia lyricen Kirkaldy	*Colocasia esculenta*	Araceae	Mitchell and Maddison (1983)
Swezeyia polyxo Fennah	*Artocarpus altilis*	Moraceae	Fennah (1956)
	Mangifera	Anacardiaceae	
Rhotanini			
Levu africanum Muir	*Theobroma cacao*	Sterculiaceae	Synave (1973)

(*continued*)

Appendix 1. (*Continued*)

Planthopper taxon	Host Plant		Reference
	Genus and Species	Family	
Levu pallescens pagana Fennah	*Alocasia*	Araceae	Fennah (1956)
Levu pallescens baedulus Fennah	*Cyrtospermum*	Araceae	Fennah (1956)
Sikaianini			
Distantinia nigrocacuminis Muir	Palms	Arecaceae	Muir (1917)
Leomelicharia delicata Muir	Palms	Arecaceae	Muir (1917)
Leomelicharia nigrovittata Muir	Palms	Arecaceae	Muir (1917)
Mula resonans Ball	Sabal palmetto	Arecaceae	Ball (1928)
Sikaiana makii Muir	Palms	Arecaceae	Muir (1917)
Sikaiana vitriceps Muir	Palms	Arecaceae	Muir (1917)
Zoraidini			
Diostrombus alcmena Fennah	*Theobroma cacao*	Sterculiaceae	Synave (1973)
Diostrombus cocos Muir	*Cocos nucifera*	Arecaceae	Muir (1928)
Diostrombus politus Uhler	Polyphagous		Lee and Kwon (1977)
Lydda elaeidis Muir	*Elaeis guineensis*	Arecaceae	Muir (1928)
Pamendanga matsumurai Muir	*Carpinus cordata*	Betulaceae	Anufriev (1968)
	Acer	Aceraceae	Anufriev (1968)
Proutista lurida Muiry	*Musa*	Musaceae	Van Stalle (1986d)
Proutista moesta (Westwood)	*Saccharum officinarum*	Poaceae	Schumacher (1920)
	Zea mays	Poaceae	Wilson and O'Brien (1987)
Proutista sacchari Van Stalle	*Saccharum officinarum*	Poaceae	Van Stalle (1986d)
ACHILIDAE			
Myconini			
Epiptera woodworthi (Van Duzee)	*Pinus jeffreyi*	Pinaceae	Van Duzee (1916)
	Juniperus (N)	Cupressaceae	

Plectoderini			
Ballomarius mongaensis Synave	*Ixora radiata*	Rubiaceae	Synave (1962b)
	Bridelia micrantha	Euphorbiaceae	
	Setaria sphacelata	Poaceae	
	Sporobolus pyramidalis	Poaceae	
Catonia arbutina Ball	*Pinus cembroides*	Pinaceae	O'Brien (1971)
Catonia bicinctura Van Duzee	*Pinus*	Pinaceae	O'Brien (1971)
	Callicarpa americana	Verbenaceae	
Catonia cinerea Osborn	*Montezuma speciosissima*	Bombacaceae	Caldwell and Martorell (1950)
	Inga vera	Fabaceae	
	Piper aduncum	Piperaceae	
	Swept from many plants		
Catonia cinctifrons (Fitch)	*Pinus clausa*	Pinaceae	O'Brien (1971)
	Pinus	Pinaceae	
	Quercus	Fagaceae	
	Carya	Juglandaceae	
Catonia lunata Metcalf	*Pinus*	Pinaceae	O'Brien (1971)
	Quercus	Fagaceae	
	Vaccinium macrocarpon	Ericaceae	
Catonia nava (Say)	*Cornus*	Cornaceae	O'Brien (1971)
	Platanus	Platanaceae	
	Acer	Aceraceae	
Catonia picta Van Duzee	*Pinus*	Pinaceae	O'Brien (1971)
Catonia pini Metcalf	*Baptisia tinctoria*	Fabaceae	O'Brien (1971)
Catonia pumila Van Duzee	*Pinus*	Pinaceae	O'Brien (1971)
	Quercus	Fagaceae	
	Carya	Juglandaceae	
Cnidus naevius Jacobi	*Erythrophleum guineensis*	Fabaceae	Synave (1962b)
	Sorghum	Poaceae	
	Swept from others		
Cnidus pallidus Synave	*Piliostigma thonningii*	Asclepiadaceae	Synave (1962b)
	Swept from many others		

(*continued*)

Appendix 1. (*Continued*)

Planthopper taxon	Host Plant		Reference
	Genus and Species	Family	
Juniperthia indella (Ball)	*Juniperus occidentalis*	Cupressaceae	O'Brien (1971)
	Juniperus california	Cupressaceae	
	Libocedrus decurrens	Cupressaceae	
	Arctostaphylos tomentosa	Ericaceae	
	Baccharis	Asteraceae	
	Salix	Salicaceae	
Juniperthia majuscula (Van Duzee)	*Juniperus deppeana*	Cupressaceae	O'Brien (1971)
Juniperthia producta (Van Duzee)	*Juniperus californica*	Cupressaceae	O'Brien (1971)
	Libocedrus decurrens	Cupressaceae	
Juniperthia succinea (Van Duzee)	*Juniperus californica*	Cupressaceae	O'Brien (1971)
	Libocedrus decurrens	Cupressaceae	
Juniperthia unimaculata O'Brien	*Juniperus californica*	Cupressaceae	O'Brien (1971)
Leptarciella saegeri Synave	*Stereospermum kunthianum*	Bignoniaceae	Synave (1962b)
	Nauclea latifolia	Rubiaceae	
	Canthium	Rubiaceae	
	Irvingia smithii	Araliaceae	
	Erythrophleum guineensis	Fabaceae	
	Swept from many others		
Momar fumidus (Ball)	*Platanus wrightii*	Platanaceae	O'Brien (1971)
Momar maculifrons (Van Duzee)	*Platanus wrightii*	Platanaceae	O'Brien (1971)
	Quercus	Fagaceae	
	Vitis	Vitaceae	
Synecdoche albicosta (Van Duzee)	*Arctostaphylos*	Ericaceae	O'Brien (1971)
Synecdoche autumnalis O'Brien	*Pinus lambertiana*	Pinaceae	O'Brien (1971)
	Quercus chrysolepis	Fagaceae	
Synecdoche cara (Van Duzee)	*Platanus*	Platanaceae	O'Brien (1971)
	Libocedrus decurrens	Cupressaceae	
	Alnus rhombifolia	Betulaceae	
	Chrysopsis villosa	Asteraceae	

Synecdoche clara (Van Duzee)	*Baccharis*	Asteraceae	O'Brien (1971)
Synecdoche constellata (Ball)	*Pinus sabiniana*	Pinaceae	O'Brien (1971)
	Cercocarpus betuloides	Rosaceae	
	Pseudotsuga menziesii	Pinaceae	
Synecdoche dimidiata (Van Duzee)	*Pinus*	Pinaceae	O'Brien (1971)
	Fagus	Fagaceae	
Synecdoche fusca (Van Duzee)	*Arctostaphylos bicolor*	Ericaceae	O'Brien (1971)
	Arbutus menziesii	Ericaceae	
	Quercus	Fagaceae	
Synecdoche grisea (Van Duzee)	*Tilia*	Tiliaceae	O'Brien (1971)
Synecdoche helenae (Van Duzee)	*Washingtonia filifera*	Arecaceae	O'Brien (1971)
Synecdoche impunctata (Fitch)	*Quercus*	Fagaceae	O'Brien (1971)
	Prunus	Rosaceae	
Synecdoche irrorata (Van Duzee)	*Arctostaphylos*	Ericaceae	O'Brien (1971)
Syncecdoche necopina (Van Duzee)	*Arctostaphylos pringlei*	Ericaceae	O'Brien (1971)
Synecdoche nemoralis (Van Duzee)	Fungi (N)		O'Brien (1971)
	Pinus muricata	Pinaceae	
	Pinus sabiniana	Pinaceae	
	Pinus murrayana	Pinaceae	
	Pinus muricata	Pinaceae	
	Pinus radiata	Pinaceae	
	Tsuga	Pinaceae	
Synecdoche nervata (Van Duzee)	*Arctostaphylos glauca*	Ericaceae	O'Brien (1971)
	Cercocarpus ledifolius	Rosaceae	
	Alnus rhombifolia	Betulaceae	
	Pinus lambertiana	Pinaceae	
	Mint	Lamiaceae	
Synecdoche ocellata O'Brien	*Umbellularia californica*	Lauraceae	O'Brien (1971)
Synecdoche pseudonervata O'Brien	*Arctostaphylos glandulosa*	Ericaceae	O'Brien (1971)

(*continued*)

Appendix 1. (*Continued*)

	Host Plant		
Planthopper taxon	**Genus and Species**	**Family**	**Reference**
Synecdoche rubella (Van Duzee)	*Arctostaphylos glauca*	Ericaceae	O'Brien (1971)
	Arctostaphylos canescens	Ericaceae	
	Arctostaphylos montana	Ericaceae	
	Arctostaphylos tomentosa	Ericaceae	
	Arbutus menzesii	Ericaceae	
	Pasania	Fagaceae	
	Ceanothus cuneatus	Rhamnaceae	
	Quercus	Fagaceae	
DICTYOPHARIDAE			
Dictyopharinae			
Dictyopharini			
Dictyophara europea L.	Polyphyagous		Synave (1951)
Dictyophara nakanonis Matsumura	*Quercus*	Fagaceae	Lee and Kwon (1977)
Dictyophara patruelis Stål	*Saccharum offinarum*	Poaceae	Schumacher (1920)
	Oryza sativa	Poaceae	
Dictyophara sp.	*Xanthium strumarium*	Asteraceae	Hilgendorf and Goeden (1982)
Hastini			
Thanatodictya tillyardi Myers	*Pteridium esculentum*	Pteridiaceae	Myers (1923)
Nersiini			
Nersia florens Stål	*Rumex crispus* (N)	Polygonaceae	Wilson and McPherson (1981b)
	Eupatorium rugosum	Asteraceae	
Retiala viridis Fennah	*Coffea*	Rubiaceae	Fennah (1945)
	Casuarina	Casuarinaceae	
	Hibiscus	Malvaceae	

Orthopagini			
Orthopagus helios Melichar	*Saccharum offinarum*	Poaceae	Schumacher (1920)
Orthopagus lunulifer Uhler	*Pueraria*	Fabaceae	Lee and Kwon (1977)
	Mallotus	Euphorbiaceae	
Phylloscelini			
Phylloscelis pallescens Germar	*Pycnathemum tenuifolium* (N)	Lamiaceae	Wilson (unpublished data)
Phylloscelis rubra Ball	*Vaccinium macrocarpon* (N)	Ericaceae	Sirrine and Fulton (1914)
Scoloptini			
Scolops abnormis Ball	*Asclepias eriocarpa*	Asclepiadaceae	Isman et al. (1977)
Scolops graphicus Ball	*Gutierrezia californica* (N)	Asteraceae	Ball (1930)
Scolops grossus Uhler	*Melilotus*	Fabaceae	Strickland (1940)
Scolops luridus Breakey	*Artemesia*	Asteraceae	Breakey (1928)
Scolops osborni Ball	*Silphium laciniatum* (N)	Asteraceae	Beamer (1929)
Scolops pallidus Uhler	*Ambrosia confertiflora* (N)	Asteraceae	Goeden and Ricker (1975)
Scolops perdix Uhler	*Helianthus angustifolius*	Asteraceae	Ball (1930)
Scolops pungens (Germar)	*Ambrosia artemisifolia*	Asteraceae	Breakey (1928)
Scolops robustus Ball	*Ambrosia psilostachya*	Asteraceae	Ball (1930)
Scolops snowi Breakey	*Solidago trinervata* (N)	Asteraceae	Ball (1930)
Scolops stonei Breakey	*Stillingia angustifolia* (N)	Euphorbiaceae	Ball (1930)
Scolops sulcipes Say	*Convolvulus*	Convolvulaceae	Wirtner (1905)
Scolops uhleri Ball	*Dondia depressa*	Chenopodiaceae	Ball (1930)
Scolops uhleri marginatus Ball	*Dondia torreyana* (N)	Chenopodiaceae	Ball (1930)
Scolops viridis Ball	*Atriplex canescens* (N)	Chenopodiaceae	Ball (1930)
Taosini			
Taosa herbida (Walker)	*Coffea*	Rubiaceae	Fennah (1945)

(*continued*)

Appendix 1. (*Continued*)

Planthopper taxon	Host Plant		Reference
	Genus and Species	Family	
Orgeriinae			
Orgeriini			
Almanina			
Nymphorgerius balchanicus Emeljan.	*Ephedra equisetina*	Ephedraceae	Emeljanov (1979)
Nymphorgerius emaljanovi Dlabola	*Astragalus*	Fabaceae	Dlabola (1979b)
Nymphorgerius mullah Dlabola	*Astragalus*	Fabaceae	Dlabola (1979b)
Nymphorgerius prasinus Emeljanov	*Cirsium*	Asteraceae	Emeljanov (1981)
Nymphorgerius tryphema Emeljanov	*Seriphidium*	Asteraceae	Emeljanov (1981)
Orgeriina			
Acinaca lurida Ball & Hartzell	*Eriogonum fasciculatum*	Polygonaceae	Ball and Hartzell (1922)
Aridia compressa Ball	*Artemesia tridentata*	Asteraceae	Ball and Hartzell (1922)
Deserta obesa Ball	*Artemesia*	Asteraceae	Ball and Hartzell (1922)
Deserta obscura Ball	*Artemesia tridentata* (N)	Asteraceae	Ball and Hartzell (1922)
Deserta raptoria Ball	*Chrysathmnus*	Asteraceae	Ball (1937)
Loxophora dammersi Van Duzee	*Agave deserti*	Agavaceae	Comstock (1942)
Orgamara argentia Ball	*Yucca brevifolia*	Agavaceae	Ball (1937)
Orgerius foliatus Doering & Darby	*Atriplex canescens*	Chenopodiaceae	Stroud (1950)
Orgerius sp.	*Cirsium californicum* (N)	Asteraceae	Goeden and Ricker (1986b)
	Cirsium proteanum (N)	Asteraceae	
	Cirsium	Asteraceae	
Yucanda albida Ball	*Yucca baccata* (N)	Agavaceae	Ball and Hartzell (1922)
Subfamily Undetermined			
Doryphorina sobrina (Stål)	*Hibiscus*	Malvaceae	Fennah (1956)
	Helianthus	Asteraceae	
	Sida	Malvaceae	
Nesolyncides io Fennah	*Asplenium nidus* (N)	Aspleniaceae	Fennah (1958)
Raivuna sinica (Walker)	*Saccharum offinarum*	Poaceae	Schumacher (1920)
	Oryza sativa	Poaceae	

FULGORIDAE			
Amyclinae			
Rhabdocephala brunnea Van Duzee	*Muhlenbergia porteri* (N)	Poaceae	Wilson and Wheeler (1992)
Aphaeninae			
Lycorma delicatula (White)	*Ailanthus altissima* (N)	Simaroubaceae	Chu (1931)
	Melia azederach (N)	Meliaceae	
Enchophorinae			
Enchophora longirostris Distant	*Simarouba amara*	Simaroubaceae	Johnson and Foster (1986)
	Quararibea asterolepis	Bombacaceae	
	Trees		
Fulgorinae			
Diareusa conspersa Schmidt	*Poulsenia armata*	Moraceae	Johnson and Foster (1986)
	Ficus tonduzii	Moraceae	
Diareusa conspersa Schmidt	*Poulsenia armata*	Moraceae	Johnson and Foster (1986)
	Ficus tonduzii	Moraceae	
Diareusa imitatrix Ossiannilsson	*Cocos nucifera*	Arecaceae	Hogue et al. (1989)
Fulgora candelaria L.	*Nephelium longana*	Sapindaceae	Kershaw and Kirkaldy (1910)
	Mangifera indica	Anacardiaceae	
	Xanthium strumarium (N)	Asteraceae	
	Urena lobata (N)	Malvaceae	
	Citrus decumana (N)	Rutaceae	
Fulgora castresii Guerin Meneville	*Jacaranda acutifolia*	Bignoniaceae	Hogue et al. (1989)
Fulgora laternaria (L.)	*Hymenaea coubaril*	Fabaceae	Janzen and Hogue (1983)
	Hymenaea oblongifolia	Fabaceae	
	Simarouba amara	Simaroubaceae	
	Zanthoxylum	Rutaceae	

(*continued*)

Appendix 1. (*Continued*)

Planthopper taxon	Host Plant		Reference
	Genus and Species	Family	
Fulgora sp.	*Hymenaea coubaril*	Fabaceae	Hogue (1984)
	Vochysia tucanorum	Vochysiaceae	
	Lecythis	Lecythidaceae	
	Myroxylon balsamum	Fabaceae	
	Simaba versicolor	Simaroubaceae	
	Aspidosp. tambopatense	Apocynaceae	Hogue et al. (1989)
	Hura crepitans	Euphorbiaceae	
	Eugenia oerstediana	Myrtaceae	Johnson and Foster (1986)
Phrictus diadema (L.)	*Theobroma cacao*	Sterculiaceae	Kershaw and Kirkaldy (1910)
Phrictus quinquepartitus	*Terminalia oblonga*	Combretaceae	Johnson and Foster (1986)
Phenacinae			
Cerogenes auricoma (Burmeister)	*Quercus reticulata* (N)	Fagaceae	Hogue et al. (1989)
Poiocerinae			
Acmonia dichroa Germar	*Capparis*	Capparidaceae	O'Brien (unpublished data)
Cyrpoptus belfragei Stål	*Pinus*	Pinaceae	Osborn (1926)
Cyrpoptus metcalfi Ball	*Pluchea sericea*	Asteraceae	Ball (1933)
	Prosopis	Fabaceae	
Cyrpoptus reineckei Van Duzee	*Panicum repens*	Poaceae	Dozier (1926)
Cyrpoptus vanduzeei Ball	*Muhlenbergia porteri*	Poaceae	Ball (1933)
	Prosopis glandulosa	Fabaceae	Ward et al. (1977)
Hypaepa illuminata Distant	*Trichilia arborea*	Meliaceae	O'Brien (unpublished data)
Itzalana submaculata Schmidt	*Baccharis*	Asteraceae	Wilson and O'Brien (1986)
Lystra lanata L.	*Simarouba amara*	Simaroubaceae	Hogue (1984)
Poblicia fuliginosa Olivier	*Rhus*	Anacardiaceae	Dozier (1926)

Insect	Host plant	Family	Reference
ISSIDAE			
Acanaloniinae			
Acanalonia agilis (Melichar)	*Eugenia aeruginea*	Myrtaceae	Caldwell and Martorell (1950)
	Swept from many others		
Acanalonia bivittata (Say)	Polyphagous		Wilson and McPherson (1980)
Acanalonia bonducellae Fennah	*Caesalpinia bonducella*	Fabaceae	Fennah (1955)
Acanalonia clypeata Van Duzee	*Prosopis*	Fabaceae	Ward et al. (1977)
	Baccharis	Asteraceae	
	Atriplex canescens (N)	Chenopodiaceae	
	Salsola pestifer	Chenopodiaceae	Doering (1932)
Acanalonia conica (Say)	Polyphagous		Wilson and McPherson (1980)
Acanalonia badensis Caldwell	*Prosopis*	Fabaceae	Ward et al. (1977)
Acanalonia impressa Metcalf & Brun.	*Phoebe elongata*	Lauraceae	Metcalf and Bruner (1930)
Acanalonia invenusta Doering	*Prosopis*	Fabaceae	Ward et al. (1977)
Acanalonia laticosta Doering	*Prosopis*	Fabaceae	Ward et al. (1977)
	Helenium	Asteraceae	Doering (1932)
	Croton	Euphorbiaceae	
	Anthemis arvensis	Asteraceae	
	Chrysopsis	Asteraceae	
	Siderocarpus flexicaula	Fabaceae	
Acanalonia pumila Van Duzee	*Borrichia arborescens*	Asteraceae	Wheeler and Hoebeke (1982)
	Mallotinia gnaphalodes (N)	Bombacaceae	
	Batis maritima (N)	Batidaceae	
	Salicornia virginica (N)	Chenopodiaceae	
	Suaeda linearis (N)	Chenopodiaceae	
Acanalonia servillei Spinola	*Capparis comosa*	Capparidaceae	Dozier (1931)
Acanalonia theobromae Fennah	*Theobroma cacao*	Sterculiaceae	Fennah (1945)
	Flacourtia	Flacourtiaceae	
	Caesalpinia pulcherrima (N)	Fabaceae	
Acanalonia umbellicauda Fennah	*Theobroma cacao*	Sterculiaceae	Fennah (1945)
	Caesalpinia	Fabaceae	
	Asparagus	Liliaceae	

(*continued*)

Appendix 1. (*Continued*)

Planthopper taxon	Host Plant		Reference
	Genus and Species	Family	
Euthiscia tuberculata Van Duzee	*Lippia wrightii*	Verbenaceae	Ball (1935a)
Bladininae			
Bladina fuscana Stål	*Ananus comosus*	Bromeliaceae	Fennah (1945)
	Bromelia pinguin	Bromeliaceae	
	Rhoeo discolor	Commelianaceae	
Bladina fuscovenosa Stål	*Setaria poiretiana*	Poaceae	Fennah (1952)
Bladina molorchus Fennah	*Saccharum officinarum*	Poaceae	Kramer (1976)
Danepteryx artemesiae Kirkaldy	*Artemisia*	Asteraceae	Doering (1940)
Danepteryx manca Uhler	*Adenostoma* (N)	Rosaceae	Doering (1940)
	Artermisia (N)	Asteraceae	
Dictydea intermedia Uhler	*Prunus illicifolia*	Rosaceae	Doering (1940)
Dictydea texana O'Brien	*Ephedra aspersa*	Ephedraceae	O'Brien (1986b)
Dictyobia atra Van Duzee	*Cirsium proteanum*	Asteraceae	Goeden and Ricker (1986b)
Dictyobia permutata Uhler	*Trixis californica* (N)	Asteraceae	Goeden and Ricker (1989)
Dictyssa fenestrata Ball	*Simmondsia chinensis*	Buxaceae	Pinto and Frommer (1980)
Dictyssa leonilae O'Brien	*Juniperus californicus*	Cupressaceae	O'Brien (1986a)
Dictyssa marginepunctata (Melichar)	*Simmondsia chinensis*	Buxaceae	Pinto and Frommer (1980)
Dictyssa obliqua Ball	*Ambrosia chenopodiifolia*	Asteraceae	Goeden and Ricker (1976b)
Dictyssa schuhi O'Brien	*Purshia tridentata*	Rosaceae	O'Brien (1986b)
	Ceanothus velutina	Rhamnaceae	
Dictyssa transversa Van Duzee	*Artemisia*	Asteraceae	Doering (1936)
	Sideroxylon	Fabaceae	Van Duzee (1923)
	Zea mays	Poaceae	
Neaethus bicornis Doering	*Arctostaphylos manzanita*	Ericaceae	Doering (1941)
Neaethus consuetus Doering	*Arctostaphylos sensitiva*	Ericaceae	Doering (1941)
	Arctostaphylos tomentosa	Ericaceae	
Neaethus unicus Doering	*Arctostaphylos pechoensis*	Ericaceae	Doering (1941)
Osbornia arborea Ball	*Juniperus*	Cupressaceae	Doering (1940)

Caliscelinae			
Aphelonema decorata (Van Duzee)	Sedges	Cyperaceae	Doering (1941)
	Spartina alterniflora (N)	Poaceae	Denno (unpublished data)
Aphelonema scurrilis (Stål)	*Carex diriuscula*	Cyperaceae	Emeljanov (1977)
Aphelonema simplex (Uhler)	*Spartina patens* (N)	Poaceae	Denno (1980)
Asarcopus palmarum Horvath	*Caryota urens*	Arecaceae	O'Brien (1988)
	Phoenix dactylifera (N)	Arecaceae	Stickney et al. (1950)
	Phoenix roebelini	Arecaceae	
	Phoenix canariensis	Arecaceae	
	Washingtonia filifera	Arecaceae	
Bruchomorpha decorata Metcalf	*Chaetochloa grisebachii*	Poaceae	Ball (1935b)
Bruchomorpha rugosa Metcalf	Grasses	Poaceae	Ball (1935b)
Bruchomorpha triunata Ball	Grasses (N)	Poaceae	Ball (1935b)
Caliscelis bonellii (Latreille)	*Cynodon dactylon*	Poaceae	O'Brien (1967)
Papagona papoosa Ball	*Muhlenbergia porteri*	Poaceae	Ball (1935a)
Papagona succinea Ball	*Trodia mutica* (N)	Poaceae	Ball (1935a)
Issinae			
Agalmatium grylloides (Fabricius)	*Ficus*	Moraceae	Picard (1921)
Colpoptera brunneus Muir	*Piper aduncum*	Piperaceae	Caldwell and Martorell (1950)
	Swept from many others		
Colpoptera clerodendri Dozier	*Clerodendron fragrans* (N)	Verbenaceae	Dozier (1931)
Colpoptera cyatheae Fennah	*Cyathea* (N)	Cyatheaceae	Fennah (1955)
Colpoptera elevans Walker	*Cocos nucifera* (N)	Arecaceae	Wilson (1987)
	Coccoloba uvifera (N)	Polygonaceae	
Colpoptera maculifrons Dozier	*Volkameria aculeata*	Verbenaceae	Caldwell and Martorell (1950)
	Swept from many others		
Colpoptera rugosa Van Duzee	*Lantana*	Verbenaceae	Van Duzee (1907)
Colpoptera thyone Fennah	*Lantana* (N)	Verbenaceae	Fennah (1955)
Hysterodus taftanicus Dlabola	*Astragalus*	Fabaceae	Dlabola (1980)
	Artemisia	Asteraceae	Dlabola (1980)

(*continued*)

Appendix 1. (*Continued*)

Planthopper taxon	Host Plant		Reference
	Genus and Species	Family	
Hysteropterum bufo Van Duzee	*Lycium*	Solanaceae	Doering (1938)
Hysteropterum c. cornutum Melichar	*Artemisia tridentata*	Asteraceae	Doering (1938)
Hysteropt. cornutum utahnum Ball	*Artemisia cana*	Asteraceae	Doering (1938)
Hysteropterum fruticulinum Emeljan.	*Atriplex cana*	Chenopodiaceae	Emeljanov (1964)
	Suaeda physophora	Chenopodiaceae	
Hysteropterum sepulcralis Ball	*Flourensia cernua*	Asteraceae	Ball (1935a)
Hysteropterum severini Cald. & DeL.	*Avena sativa* (N)	Poaceae	Schlinger (1958)
Hysteropterum unum Ball	*Chrysothamnus nauseosus*	Asteraceae	Stroud (1950)
Issus cahipi Remane	*Pinus canariensis* (N)	Pinaceae	Remane (1985a)
	Pinus radiata (N)	Pinaceae	
Issus coleoptratus (Geoffroy)	Polyphagous on woody dicots		Le Quesne (1960)
Issus distinguendus Lindberg	*Euphorbia balsamifera*	Euphorbiaceae	Remane (1985a)
	Euphorbia obtusifolia	Euphorbiaceae	
Issus gratehigo Remane	*Euphorbia*	Euphorbiaceae	Remane (1985a)
	Hypericum	Hypericaceae	
Issus muscaeformis (Schrank)	Polyphagous woody plants (N)		Gunthart (1987)
Latematium cingulatum Dlabola	*Cupressus*	Cupressaceae	Dlabola (1983)
Latissus dilatatus (de Fourcroy)	*Quercus cerris*	Fagaceae	D'Urso et al. (1984)
Mycterodus hakkaricus Dlabola	*Quercus*	Fagaceae	Dlabola (1980)
Mycterodus izmiticus Dlabola	*Corylus avellana*	Corylaceae	Dlabola (1979b)
Mycterodus lodosicus Dlabola	*Pinus*	Pinaceae	Dlabola (1980)
Mycterodus pozanticus Kartal	*Verbascum*	Scrophulariaceae	Kartal (1983)
Mycterodus spinicordatus Dlabola	*Pinus eleagnifolia*	Pinaceae	Dlabola (1983)
	Arbutus	Ericaceae	
	Prunus communis	Rosaceae	
Neocolpoptera puertoricensis Dozier	*Vitex*	Vitaceae	Caldwell and Martorell (1950)
	Piper aduncum	Piperaceae	
	Swept from many others		

Phasmena cardinalis Emeljanov	*Atraphaxis*	Polygonaceae	Emeljanov (1979)
Phasmena ephedrae Emeljanov	*Ephedra przewalskii*	Ephedraceae	Emeljanov (1977)
Phasmena tardiviva Emeljanov	*Atraphaxis*	Polygonaceae	Emeljanov (1979)
Picumna maculata (Melichar)	*Pinus*	Pinaceae	Doering (1938)
Quadrastylum beysehiricum Dlabola	*Pyrus malus*	Rosaceae	Dlabola (1983)
Quadrastylum campanuliforma Dlab.	*Pinus eleagnifolia*	Pinaceae	Dlabola (1979a)
Quadrastylum conspurcatum (Spin.)	*Quercus*	Fagaceae	Dlabola (1980)
Sarima nigroclypeata Melichar	*Santalum album* (N)	Santalaceae	Chatterjee (1933)
	Polyphagous		Chatterjee (1933)
Thionia argo Fennah	*Coccoloba uvifera* (N)	Polygonaceae	Caldwell and Martorell (1950)
Thionia borinqueta Caldwell	*Coccoloba uvifera*	Polygonaceae	Caldwell and Martorell (1950)
Thionia clusiae Fennah	*Clusia* (N)	Clusiaceae	Fennah (1955)
Thionia bullata (Say)	*Pinus taeda* (N)	Pinaceae	Wheeler and Wilson (1988)
	Pinus echinata (N)	Pinaceae	
	Pinus virginiana (N)	Pinaceae	
Thionia elliptica Germar	*Quercus ilicifolia* (N)	Fagaceae	Wheeler & Wilson (1987)
	Quercus marilandica	Fagaceae	
Thionia producta Van Duzee	*Juniperus*	Cupressaceae	Doering (1938)
Thionia simplex (Germar)	Polyphagous on dicots (N)		Wheeler and Wilson (1988)
Thionia sp.	*Prosopis*	Fabaceae	Ward et al. (1977)
Tshurtshurnella bicornuta Dlabola	*Arbutus*	Ericaceae	Dlabola (1983)
	Tamarix	Tamaricaceae	
Tshurtshurnella diyarbakira Dlabola	*Punicum* (sic)	?	Dlabola (1980)
Tshurtshurnella extrema Dlabola	Grasses	Poaceae	Dlabola (1980)
Tshurtshurnella verbasci Dlabola	*Verbascum*	Scrophulariaceae	Dlabola (1983)
	Quercus	Fagaceae	
Tonginae			
Atylana astydamia Fennah	*Sophora*	Fabaceae	Fennah (1958)
Atylana palanto Fennah	*Metrosideros*	Myrtaceae	Fennah (1958)
Atylana parmula Fennah	*Glochidion* (N)	Euphorbiaceae	Fennah (1958)
Atylana vesontio Fennah	*Alyxia*	Apocynaceae	Fennah (1958)

(*continued*)

Appendix 1. (*Continued*)

Planthopper taxon	Host Plant		Reference
	Genus and Species	**Family**	
Atylana volumna Fennah	*Hernandia*	Hernandiaceae	Fennah (1958)
Tylana ustulata Uhler	*Prosopis juliflora*	Fabaceae	Ward et al. (1977)
Tylanira bifurca Ball	*Prosopis*	Fabaceae	Ward et al. (1977)
Subfamily Undetermined			
Acrestia quadracuta Dlabola	*Artemesia*	Asteraceae	Dlabola (1980)
Bubastia jatagana Dlabola	*Verbascum*	Scrophulariaceae	Dlabola (1980)
NOGODINIDAE			
Nogodinini			
Biolleyana costalis Fowler	*Theobroma cacao*	Sterculiaceae	Fennah (1945)
Epacriini			
Psiadiicola brevipennis Fennah	*Psiadia trinervia* (N)	Asteraceae	Fennah (1978)
Tribe Undetermined			
Hadjia nerii Dlabola	*Nerium* (N)	Apocynaceae	Dlabola (1981)
Nurunderia chrysopoides (Walker)	*Eucalyptus*	Myrtaceae	Carver et al. (1991)
RICANIIDAE			
Armacia clara clara (Stål)	*Pandanus*	Pandanaceae	Fennah (1956)
	Hibiscus tiliaceus	Malvaceae	
Armacia clara trukensis Fennah	*Cyrtospermum*	Araceae	Fennah (1956)
Armacia simaethis Fennah	*Gardenia*	Rubiaceae	Fennah (1956)
Euricania clara Kato	*Pueraria*	Fabaceae	Lee and Kwon (1977)
Euricania japonica Melichar	*Pueraria*	Fabaceae	Lee and Kwon (1977)
	Morus	Moraceae	
	Vicia	Fabaceae	

Pochazia fasciata (Fabricius)	*Cocos nucifera*	Arecaceae	Wilson (1987)
	Palm	Arecaceae	
Privesa laevifrons (Stål)	*Olea lancea*	Oleaceae	Williams and Fennah (1980)
Ricania fenestrata (Fabricius)	*Camellia sinensis*	Theaceae	Ghauri (1973)
	Santalum album	Santalaceae	
Ricania japonica Melichar	*Morus alba*	Moraceae	Schumacher (1920)
	Camellia sinensis	Theaceae	
	Cannabis sativa	Cannabidaceae	Takahashi (1919)
Ricania speculum (Walker)	*Camellia sinensis* (N)	Theaceae	Ghauri (1973)
Ricania taeniata Stål	*Pueraria*	Fabaceae	Lee and Kwon (1977)
	Oryza	Poaceae	
	Vicia	Fabaceae	
	Saccharum officinarum	Poaceae	Schumacher (1920)
Ricanoptera opaca Distant	*Camellia sinensis*	Theaceae	Hutson (1921)
Ricanoptera syrinx Fennah	*Wedelia biflora*	Asteraceae	Fennah (1956)
Scolypopa australis Walker	Polyphagous on clubmosses, ferns, and angiosperms (N)		Cumber (1966)
Tarundia cinctipennis Stål	*Pittosporum senacia*	Pittosporaceae	Williams and Fennah (1980)
	Eugenia tinifolia	Myrtaceae	
Tarundia servillei (Spinola)	*Gaetnera*	Rubiaceae	Williams and Fennah (1980)
	Erythrospermum mauritianum	Flacourtiaceae	
Tarundia straminea Muir	*Pandanus*	Pandanaceae	Williams and Williams (1988)
FLATIDAE			
Flatinae			
Flatini			
Byllisana brunnea Metcalf & Bruner	*Centrosema* (N)	Fabaceae	Metcalf and Bruner (1948)
	Miebomia supina (N)	Fabaceae	
Carthaemorpha balloui Metcalf & Brn.	*Amydalus persica*	Sterculiaceae	Metcalf and Bruner (1948)

(*continued*)

Appendix 1. (*Continued*)

Planthopper taxon	Host Plant		Reference
	Genus and Species	Family	
Dakshiana katharina Metcalf & Bruner	*Nectandra coriacea*	Lauraceae	Metcalf and Bruner (1948)
Flata ferrugata Fabricius	*Xanthium strumarium*	Asteraceae	Hilgendorf and Goeden (1982)
Lawana candida (Fabricius)	Polyphagous		Hoffman (1935)
Lawana conspersa (Walker)	Polyphagous		Hoffman (1935)
Lawana imitata (Melichar)	Polyphagous		Wilson and O'Brien (1987)
Mimophantia maritima Matsumura	*Miscanthus sinensis* (N)	Poaceae	Tsaur (1989b)
Salurnis formosana Jacobi	*Mussaenda parviflora* (N)	Rubiaceae	Tsaur (1989b)
Siphanta acuta (Walker)	*Eucalyptus*	Myrtaceae	Myers (1922)
Siphanta spp.	Polyphagous		Fletcher (1985)
Nephesini			
Anormenis chloris (Melichar)	Polyphagous		Wilson and McPherson (1980)
Anzora unicolor (Walker)	*Malus*	Rosaceae	Fletcher (1988)
	Pyrus	Rosaceae	
Cryptoflata vuattouxi Synave	*Cassonia barteri*	Araliaceae	Synave (1964)
	Bauhinia thonningii	Fabaceae	
Epormenis fuliginosa (Fennah)	*Coffea* (N)	Rubiaceae	Fennah (1945)
Epormenis unimaculata (Fennah)	*Theobroma cacao*	Sterculiaceae	Fennah (1945)
Flatormenis duplicata Caldwell	*Trema lamarkiana*	Ulmaceae	Caldwell and Martorell (1950)
	Inga	Fabaceae	
Flatormenis nefuscata Caldwell	*Canella winterana*	Lauraceae	Caldwell and Martorell (1950)
	Weeds and grasses		
Flatormenis pseudomarginata (Muir)	*Piper aduncum*	Piperaceae	Caldwell and Martorell (1950)
	Melastomes, shrubs	Melastomataceae	
	Weeds and bushes		

Flatormenis squamulosa (Fowler)	*Coffea* (N)	Rubiaceae	Fennah (1945)
Geisha distinctissima (Walker)	Polyphagous woody dicots		Lee and Kwon (1977)
Melicharia obtusanguloides Ghauri	*Camellia sinensis*	Theaceae	Ghauri (1973)
Melormenis antillarum (Kirkaldy)	Polyphagous		Caldwell and Martorell (1950)
Melormenis assymetrica Met. & Brun.	Polyphagous		Metcalf & Bruner (1948)
Melormenis basalis (Walker)	*Coccoloba uvifera*	Polygonaceae	Fennah (1965b)
	Acacia	Fabaceae	
	Polyphagous		Caldwell and Martorell (1950)
Metcalfa pruinosa (Say)	Polyphagous (N)		Wilson and McPherson (1980)
Metcalfa pruinosa cubana (Met.& Br.)	Polyphagous		Metcalf and Bruner (1948)
Ormenaria rufifascia (Walker)	*Latania lontaroides* (N)	Arecaceae	Wilson and Tsai (1984)
	Sabal palmetto (N)	Arecaceae	
Ormenis albigyna Campos	*Theobroma cacao*	Sterculiaceae	Wilson and O'Brien (1987)
Ormenis antoniae Melichar	*Coffea* (N)	Rubiaceae	Fennah (1945)
	Mangifera indica (N)	Anacardiaceae	
Ormenis coffeacola Dozier	*Coffea*	Rubiaceae	Dozier (1931)
Ormenis pygmaea Fabricius	*Coffea*	Rubiaceae	Van Zwaluwenburg (1917)
	Polyphagous		
Ormenis saucia Van Duzee	*Ambrosia acanthicarpa*	Asteraceae	Goeden and Ricker (1974a)
	Ambrosia dumosa (N)	Asteraceae	Goeden and Ricker (1976a)
	Ambrosia eriocentra	Asteraceae	
	Ambrisia psilostachya	Asteraceae	
	Hymenoclea salsola (N)	Asteraceae	Goeden and Ricker (1986a)
	Cirsium californicum	Asteraceae	Goeden and Ricker (1986b)
	Cirsium proteanum	Asteraceae	
	Cirsium	Asteraceae	
	Prosopis	Fabaceae	Ward et al. (1977)
Ormenis yumana Ball	*Tamarix gallica* (N)	Tamaricaceae	Hopkins and Carruth (1954)
	Prosopis	Fabaceae	Ward et al. (1977)
Ormenis sp.	*Bebbia juncea* (N)	Asteraceae	Goeden and Ricker (1989)

(*continued*)

Appendix 1. (*Continued*)

Planthopper taxon	Host Plant: Genus and Species	Host Plant: Family	Reference
Ormenoides venusta (Melichar)	Polyphagous		Wilson and McPherson (1980)
Petrusa marginata (Brunnich)	*Batis maritima*	Batidaceae	Caldwell and Martorell (1950)
	Lippia nodiflora	Verbenaceae	
	Beaten from many others		
Rhinophantia longiceps (Puton)	*Artemisia monosperma*	Asteraceae	Linnavuori (1964)
Trisephena anomala Medler	*Pipturus*	Urticaceae	Medler (1990)
Trisephena zestreya Medler	*Lantana camera*	Verbenaceae	Medler (1990)
Phantiini			
Mesophantia kanganica Dlabola	*Seidlitzia rosmarinus*	Chenopodiaceae	Krampl and Dlabola (1983)
Falcophantus acuminatus Fletcher	*Spinifex*	Poaceae	Fletcher (1988)
Falcophantus westcotti Fletcher	*Spinifex*	Poaceae	Fletcher (1988)
Phromniini			
Psenoflata brevis Van Duzee	*Cocos nucifera* (N)	Arecaceae	Wilson (1987)
Poekillopterini			
Poekilloptera phalaenoides (L.)	*Samanea* (N)	Fabaceae	Fennah (1945)
Selizini			
Barsac cocoa Fletcher	*Eucalyptus gamophylla*	Myrtaceae	Fletcher (1988)
Cyarda sp. nr. *acutissima* Metcalf	Polyphagous (N)		Wheeler and Hoebeke (1982)
Cyarda casuarinae Fennah	*Casuarina*	Casuarinaceae	Fennah (1965b)
Cyarda fuscifrons Metcalf & Bruner	*Casuarina*	Casuarinaceae	Metcalf and Bruner (1948)
Cyarda difformis Walker	*Cordia serrata* (N)	Ehretiaceae	Dozier (1931)
Cyarda haitiensis Metcalf & Bruner	*Pandanus*	Pandanaceae	Howard et al. (1981)
	Acacia lutea	Fabaceae	Metcalf and Bruner (1948)
Cyarda melichari Van Duzee	*Lantana camara*	Verbenaceae	Krauss (1953)
Cyarda salina (Dozier)	*Batis maritima*	Batidaceae	Caldwell and Martorell (1950)
Cyphopterum quartaui Linnavuori	*Suaeda vera*	Chenopodiaceae	Quartau (1975)
	Agropyron junceiforme	Poaceae	

Cyphopterum salvagensis Lindberg	*Suaeda vera*	Chenopodiaceae	Quartau (1975)
	Agropyron junceiforme	Poaceae	
Derisa pallida Fennah	*Tamarix*	Tamaricaceae	Linnavuori (1964)
Eubyloptera corticalis Fennah	*Lantana camara*	Verbenaceae	Fennah (1945)
	Cordia	Ehretiaceae	
	Bushes		
Jamella australiae Kirkaldy	*Pandanus*	Pandanaceae	Swezey (1906)
Ketumala thea Ghauri	*Camellia sinensis*	Theaceae	Ghauri (1971)
	Erythrina lithosperma	Fabaceae	
	Gliricidia sepium	Fabaceae	
Mistharnophantia caudata (Van Duz.)	*Helianthus*	Asteraceae	Van Duzee (1914)
Mistharnophantia sp.	*Prosopis*	Fabaceae	Ward et al. (1977)
Paradascalia edax (Van Duzee)	*Laccodesmia*	?	Van Duzee (1923)
	Prosopis	Fabaceae	
	Sideroxylon	Sapotaceae	
	Other plants		
Planodascalia sp.	*Sabal umbraculifera*	Arecaceae	Howard et al. (1981)
Seliza sp.	*Xanthium strumarium*	Asteraceae	Hilgendorf and Goeden (1982)
Flatoidinae			
Flatoides fecalfusca Caldwell	*Prosopis*	Fabaceae	Ward et al. (1977)
Flatoidinus caesalpiniae Fennah	*Acacia riparia*	Fabaceae	Fennah (1965b)
	Caesalpinia bundacella	Fabaceae	
	Tecoma leucoxylon	Bignoniaceae	
Flatoidinus cordiae Fennah	*Cordia cylindrostachya*	Ehretiaceae	Fennah (1945)
	Mangifera indica	Anacardiaceae	
	Various shrubs		
Flatoidinus fumatus (Melichar)	*Inga laurina*	Fabaceae	Caldwell and Martorell (1950)
	Inga vera	Fabaceae	
	Byrsonima spicata	Malpighiaceae	

(*continued*)

Appendix 1. (*Continued*)

Planthopper taxon	Host Plant: Genus and Species	Host Plant: Family	Reference
Flatoidinus fumatus angulatus Cald.	*Trema lamarkiana*	Ulmaceae	Caldwell and Martorell (1950)
Flatoidinus litoralis Fennah	*Acacia*	Fabaceae	Fennah (1965b)
Flatoidinus monae Fennah	*Jasminum multiflorum*	Oleaceae	Fennah (1965b)
Flatoidinus pallescens Met. & Brun.	*Lonchocarpus*	Fabaceae	Metcalf and Bruner (1948)
	Coffea arabica	Rubiaceae	
	Casuarina	Casuarinaceae	
HYPOCHTHONELLIDAE			
Hypochthonella caeca China & Fen.	*Nicotiana tabacum* (N)	Solanaceae	China and Fennah (1952)
	Zea mays	Poaceae	
	Arachis hypogaea	Fabaceae	
TROPIDUCHIDAE			
Alcestini			
Alcestis ingens Fennah	*Theobroma cacao*	Sterculiaceae	Fennah (1982a)
Alcestis vitrea Fennah	*Cecropia peltata*	Moraceae	Fennah (1982a)
Catulliini			
Catullia subtestacea Stål	*Digitaria*	Poaceae	Yang et al. (1989)
	Grasses (N)	Poaceae	
Numicia gauhati Wilson	*Oryza sativa*	Poaceae	Wilson (1984)
Numicia ghesquierei Lallemand	*Citrus*	Rutaceae	Lallemand (1938)
Numicia maculosa (Distant)	*Saccharum officinarum*	Poaceae	Fennah (1982b)
Numicia pusana Ghauri	*Saccharum officinarum*	Poaceae	Wilson (1984)
Numicia viridis Muir	Polyphagous on grasses and sedges (N)	Poaceae Cyperaceae	Carnegie (1980)
Cyphoceratopini			
Colgorma campestris Metcalf & Brun.	*Coffea arabica*	Rubiaceae	Metcalf and Bruner (1930)

Eporini			
Mesopora onukii Matsumura	*Citrus*	Rutaceae	Schumacher (1920)
Eutropistini			
Sakina boulardi Synave	*Triumphetta rhomboidea*	Tiliaceae	Synave (1978)
Remosini			
Neurotmeta sponsa (Guerin-Menev.)	*Psidium guajava*	Myrtaceae	Fennah (1982b)
	Coffea arabica	Rubiaceae	Metcalf and Bruner (1930)
Neurotmeta viridis (Walker)	*Croton humilis*	Euphorbiaceae	Caldwell and Martorell (1950)
	Suriana maritima (N)	Surianaceae	
	Swept from other plants		
	Coccoloba uvifera (N)	Polygonaceae	Fennah (1949a)
Remosa spinolae Guerin-Meneville	*Cestrum diurnum*	Solanaceae	Metcalf and Bruner (1930)
Tambiniini			
Athestia chariclo (Fennah)	*Chamaedorea* (N)	Arecaceae	Fennah (1974)
Kallitambinia australis Muir	*Aegiceras corniculatum* (N)	Myrsinaceae	Fletcher (1979)
	"Mangroves"	?	
Kallitaxila apicalis (Melichar)	*Cocos nucifera* (N)	Arecaceae	Wilson (1987)
Lanshu glochidionae Yang, Yang & W.	*Glochidion rubrum*	Euphorbiaceae	Yang et al. (1989)
	Woody plants		
	Miscanthus	Poaceae	
Ossoides lineatus Bierman	*Miscanthus* (N)	Poaceae	Yang et al. (1989)
Tambinia bizonata Matsumura	*Capillipedium kwashotensi*	Poaceae	Yang et al. (1989)
	Grasses (N)	Poaceae	
Tambinia guamensis Metcalf	*Hernandia*	Hernandiaceae	Metcalf (1946b)
	Piper guahamense	Piperaceae	
Tambinia theivora Fennah	*Camellia sinensis*	Theaceae	Fennah (1982b)
Tambinia verticalis Distant	*Cocos nucifera* (N)	Arecaceae	Wilson (1986)
	Coffea arabica (N)	Rubiaceae	
	Canthium	Rubiaceae	Wilson (1987)

(*continued*)

Appendix 1. (*Continued*)

Planthopper taxon	Host Plant: Genus and Species	Host Plant: Family	Reference
Tangiini			
Neotangia angustata (Uhler)	*Coffea*	Rubiaceae	Fennah (1982b)
Pelitropis rotulata Van Duzee	Polyphagous		Wilson and Wheeler (1984)
Tangella schaumi (Stål)	*Distictis lactiflora* (N)	Bignoniaceae	Caldwell and Martorell (1950)
	Coccoloba uvifera (N)	Polygonaceae	
	Melia azedarach (N)	Meliaceae	
	Swept from trees & shrubs		
Tangia breviceps (Metcalf & Bruner)	*Cocoloba uvifera*	Polygonaceae	O'Brien (1992)
Tangia litoralis (Fennah)	Polyphagous		Wilson and Hilburn (1991)
Tangia viridis (Walder)	*Citrus*	Rutaceae	Fennah (1982b)
Tropiduchini			
Leptovanua telamon Fennah	*Artocarpus altilis*	Moraceae	Fennah (1982b)
Swezeyaria viridana Metcalf	*Artocarpus altilis*	Moraceae	Fennah (1982b)
	Pandanus	Pandanaceae	Metcalf (1946b)
Tropiduchus biermani Bierman	*Mangifera indica*	Anacardiaceae	Fennah (1982b)
Trypetimorphini			
Ommatissus binotatus Fieber	*Chamaerops humilis*	Arecaceae	Asche and Wilson (1989a)
Ommatissus lofouensis Muir	Grasses	Poaceae	Asche and Wilson (1989a)
	Miscanthus	Poaceae	Yang et al. (1989)
Ommatissus lybicus Bergevin	*Phoenix dactylifera* (N)	Arecaceae	Asche and Wilson (1989a)
Ommatissus magribus Asche & Wils.	*Chamaerops humilis*	Arecaceae	Asche and Wilson (1989a)
Ommatisssus tumidulus Linnavuori	*Phoenix*	Arecaceae	Asche and Wilson (1989a)
Neommatissus basifuscus Kato	Pteridophyta	?	Yang et al. (1989)
Neommatissus formosanus Kato	*Cyathea lepifera*	Cyatheaceae	Yang et al. (1989)
	Pteridophyta	?	Yang et al. (1989)

LOPHOPIDAE			
Lophopinae			
Elasmoscelis similis Synave	*Canthium*	Rubiaceae	Synave (1962a)
	Lonchocarpus laxiflorus	Fabaceae	
	Kigelia aethiopica	Bignoniaceae	
	Swept from many other herbaceous plants		
Lophops saccharicida Kirkaldy	*Saccharum officinarum*	Poaceae	Woodward et al. (1970)
	Grasses	Poaceae	
Pyrilla perpusilla Walker	12 spp. grasses (N)	Poaceae	Rahman and Nath (1940)
Silvanana omani Metcalf	*Britoa acida*	Myrtaceae	Metcalf (1947)
Subfamily Undetermined			
Symplana major Fennah	Unspiked sandal	Arecaceae	Fennah (1962b)
Virgilia luzonensis Muir	*Cocos nucifera* (N)	Arecaceae	Wilson (1987)
Zophiuma lobulata Ghauri	*Cocos nucifera* (N)	Arecaceae	Smith (1980)
	Areca catechu	Arecaceae	Ghauri (1966)
EURYBRACHIDAE			
Eurybrachinae			
Eurybrachini			
Eurybrachys tomentosa (Fabricius)	*Santalum album*	Santalaceae	Wilson and O'Brien (1987)
Platybrachinae			
Platybrachini			
Platybrachys leucostigma Walker	*Eucalyptus maculatus* (N)	Myrtaceae	Hacker (1925b)

Note: Host plant records for the Fulgoroidea were compiled by examining the taxonomic (e.g., Beamer 1945; Metcalfe 1969; Kramer 1973; Hoch and Remane 1983; Wilson and Wheeler 1986), ecological (e.g., Mochida and Kisimoto 1971; Denno 1977; Tallamy and Denno 1979; Booij 1982a; den Bieman 1987a; Claridge et al. 1988), and agricultural literature (e.g., Wilson and O'Brien 1987) as well as faunal surveys (e.g., Zimmerman 1948; Mochida and Okada 1971, 1973; Ossiannilsson 1978) (also Appendix 2). The actual host plant on which planthopper feeding and development occurs can be difficult to assess because records are often based on sweep-net collections of adults. Spurious records result when adults are collected from nonhost vegetation in the absence of information on feeding or the presence of nymphs. Nevertheless, we made a sincere attempt to exclude spurious host records from our survey by omitting reports of species: (1) collected by general sweep netting, (2) observed simply resting and not feeding on the plant as noted by the author, (3) taken in surveys of particular crops on which the species in question are not known to be pests (e.g., Ballou 1936; Bruner et al. 1945, Kramer 1978, Maes and O'Brien 1988) and (4) whose taxonomic identities were questionable (e.g., Swezey 1904).

Appendix 2. Recorded host plants of delphacid planthoppers.

Planthopper Taxon	Host Plant		References
	Genus and Species	Family[a]	
	SUBFAMILY ASIRACINAE		
	Tribe Ugyopini		
Neopunana			
puertoricensis (Muir)	*Castilla elastica*	MR	Caldwell and Martorell (1950)
Ugyops			
caelatus (White)	*Leptospermum* sp.	MY	Fennah (1965a)
	Muehlenbeckia australis	PL	
	Coprosma rhamnoides	RU	
	Knightia excelsa	PR	
haliacmon Fennah	*Asplenium nidus*	AN	Fennah (1958)
kinbergi Stål	*Intsia bijuga*	FA	Metcalf (1946b)
osborni Metcalf	*Euterpe globosa*, trees	AR	Caldwell and Martorell (1950)
pelorus Fennah	*Muelenbeckia australis*	PL	Fennah (1965a)
samoaensis Muir	*Pemphis* sp.	LY	Metcalf (1946b)
tripunctatus (Kato)	Pteridophyta	PY	Yang and Yang (1986)
	Tribe Asiracini		
Asiraca			
clavicornis (Fabricius)	Polyphagous on dicots	DI	Asche (1982c)
Pentagramma			
longistylata Penner	*Scirpus americanus*	CY	Wilson and Wheeler (1986)
variegata Penner	*Scirpus validus*	CY	S. Wilson (unpublished data)
vittatus (Matsumura)	Pteridophyta	PY	Yang and Yang (1986)
	SUBFAMILY KELISIINAE		
Anakelisia			
fasciata (Kirschbaum)	*Carex* spp., *Carex riparia*	CY	Drosopoulos et al. (1983)
perspicillata (Boheman)	*Carex montana/humilis*	CY	Drosopoulos et al. (1983)

Kelisia			
brucki Fieber	*Juncus* sp.,	JU	Asche (1982a)
	Scirpus boloschoenus	CY	Drosopoulos et al. (1983)
confusa Linnavuori	*Carex riparia*	CY	Drosopoulos et al. (1983)
creticola Asche	*Juncus* sp.	JU	Asche (1982a)
	Scirpus boloschoenus	CY	Drosopoulos et al. (1983)
gargano Remane and Asche	*Carex montana/bumilis*	CY	Drosopoulos et al. (1983)
guttula (Germar)	*Carex flacca*	CY	den Bieman and Booij (1984)
guttulifera (Kirschbaum)	*Carex divulsa*	CY	Drosopoulos et al. (1983)
baupti Wagner	*Carex montana/bumilis*	CY	Drosopoulos et al. (1983)
benschii Horvath	*Carex* sp.	CY	Asche (1982b)
melanops Fieber	*Carex* sp.	CY	Drosopoulos et al. (1983)
monoceros Ribaut	*Carex verna*	CY	Gunthart (1987)
perrieri Ribaut	*Juncus* sp.	JU	Drosopoulos et al. (1983)
	Scirpus boloschoenus	CY	
preacox Haupt	*Carex* sp., *Cyperus* sp.	CY	Drosopoulos et al. (1983)
ribauti Wagner	*Carex elongata*	CY	den Bieman and Booij (1984)
riboceros Asche	*Carex* sp.	CY	Asche (1986)
sabulicola Wagner	*Carex arenaria*	CY	den Bieman and Booij (1984)
vittipennis Sahlberg	*Carex* sp., *Eriopborum* sp.	CY	Gunthart (1987)
yarkonensis Linnavuori	*Juncus* sp.	JU	Asche (1982b)
	Scirpus boloschoenus	CY	Drosopoulos et al. (1983)
SUBFAMILY STENOCRANINAE			
Stenocranus			
agamopsyche Kirkaldy	*Pbragmites communis*	PO	Yang (1989)
arundineus Metcalf	*Arundinaria* sp.	PO	Beamer (1946)
fuscovittatus (Stål)	*Carex* sp.	CY	Asche (1982b)
	Claudia mariscus	CY	Drosopoulos et al. (1983)
gialovus Asche & Hoch	*Phalaris aquatica*	PO	Asche and Hoch (1983)

(*continued*)

Appendix 2. (*Continued*)

Planthopper Taxon	Host Plant		References
	Genus and Species	Family[a]	
harimensis Matsumura	*Carex thunbergii*	CY	Kisimoto (personal communication)
lautus Van Duzee	*Carex lurida*	CY	Calvert and Wilson (1986)
major (Kirschbaum)	*Phalaris arundinacea*	PO	Ossiannilsson (1978)
matsumurai Metcalf	*Phragmites communis*	PO	Yang (1989)
minutus (Fabricius)	*Dactylis glomerata*	PO	Asche (1982b)
	Brachypodium pinnatum	PO	Cobben and Rozeboom (1983)
rufilinearis Kuoh	*Eleocharis tuberosa*	CY	Ding and Kuoh (1981)
	Cyperus rotundatus	CY	
similis Crawford	*Arundinaria tecta*	PO	Dozier (1922)
yasumatsui Ishihara	*Carex* sp.	CY	Lee and Kwon (1980)
Terauchiana			
nigripennis Kato	*Imperata cylindrica*	PO	Lee and Kwon (1980)
SUBFAMILY PLESIODELPHACINAE			
Burnilia			
heliconiae Muir	*Heliconia* sp.	HE	Muir (1926)
spinifera antillana Fennah	*Heliconia* sp.	HE	Fennah (1959a)
SUBFAMILY DELPHACINAE			
Tribe Tropidocephalini			
Arcofaciella			
verrucosa Fennah	*Bambusa oldhamii*	PO	Yang and Yang (1986)
	B. multiplex	PO	
Arcofacies			
fullawayi Muir	*Bambusa dolichoclada*	PO	Yang and Yang (1986)
	B. oldhamii/multiplex	PO	
huangi Ding and Hu	*Bambusa affinis*	PO	Ding and Hu (1982)

luodianensis Ding	Bamboo	PO	Ding (1982)
membranacea Yang and Yang	*Dendrocalamus latiflorus*	PO	Yang and Yang (1986)
taiwanensis (Muir)	*Dendrocalamus latiflorus*	PO	Yang and Yang (1986)
Belocera			
sinensis Muir	*Bambusa multiplex*	PO	Yang and Yang (1986)
Columbisoga			
chusqueae Muir	*Chusquea* sp.	PO	Muir (1926)
gynericola Muir	*Gynerium saccharoides*	PO	Muir (1926)
gynerii Muir	*Gynerium* sp.	PO	Muir (1926)
ornata Muir	*Gynerium* sp.	PO	Muir (1926)
Epeurysa			
abatana Asche	*Bambusa dolichoclada*	PO	Yang and Yang (1986)
	B. oldhamii	PO	
bengueta Asche	Bamboo	PO	Asche (1983)
infumata Huang & Ding	*Phyllostachys* sp.	PO	Yang and Yang (1986)
	Chimonobambusa naibunensis	PO	
maculata Yang & Yang	*Fargesia niitakayamensis*	PO	Yang and Yang (1986)
nawaii Matsumura	*Phyllostachys makinoi*	PO	Yang and Yang (1986)
	Chimonobambusa quadrangularis	PO	
remanei Asche	Bamboo	PO	Asche (1983)
sinobambusae Yang & Yang	*Sinobambusa kunishii*	PO	Yang and Yang (1986)
Jassidaeus			
lugubris (Signoret)	*Festuca* sp.	PO	Asche and Hoch (1982)
Macrocorupha			
gynerii Muir	*Gynerium sagittatum*	PO	Muir (1926)
Malaxa			
aurunca Yang and Yang	*Bambusa multiplex*	PO	Yang and Yang (1986)
bakeri Muir	*Bambusa stenosachya*	PO	Yang and Yang (1986)
	B. dolichoclada/oldhamii	PO	
fusca Yang and Yang	*Bambusa multiplex*	PO	Yang and Yang (1986)

(continued)

Appendix 2. (*Continued*)

Planthopper Taxon	Host Plant		References
	Genus and Species	Family[a]	
occidentalis Muir	*Gynerium* sp.	PO	Muir (1926)
Paranectopia			
lasaensis Ding and Tian	Poaceae	PO	Ding and Tian (1981)
Purohita			
cervina Distant	*Phyllostachys* sp.	PO	Yang and Yang (1986)
maculata Muir	*Dendrocalamus* sp.	PO	Yang and Yang (1986)
	Bambusa multiplex	PO	
nigripes Muir	*Bambusa vulgaris*	PO	Yang and Yang (1986)
picea Yang and Yang	*Dendrocalamus giganteus*	PO	Yang and Yang (1986)
	D. latiflorus	PO	
sinica Huang and Ding	*Dendrocalamus latiflorus*	PO	Yang and Yang (1986)
taiwanensis Muir	*Phyllostachys pubescens*	PO	Yang and Yang (1986)
	Bambusa arundinacea	PO	
	B. beechyana/multiplex	PO	
	B. oldhamii/stenostachya	PO	
	B. vulgaris, Dendrocalamus	PO	
	giganteus/latiflorus	PO	
Specinervures			
liquida Yang and Yang	*Bambusa edulis*	PO	Yang and Yang (1986)
Tropidocephala			
andropogonis Horvath	*Chrysopogon gryllus*	PO	Drosopoulos et al. (1983)
brunipennis Signoret	*Imperata cylindrica*	PO	Lee and Kwon (1977)
	Miscanthus sinensis	PO	
	Oryza sativa	PO	
dimidia Yang and Yang	*Imperata cylindrica*	PO	Yang and Yang (1986)
festiva (Distant)	*Imperata cylindrica*	PO	Yang and Yang (1986)
formosana Matsumura	*Miscanthus* spp.	PO	Yang and Yang (1986)

grata Yang and Yang	*Imperata cylindrica*	PO	Yang and Yang (1986)
	Miscanthus spp.	PO	
maculosa Matsumura	*Miscanthus* spp.	PO	Yang and Yang (1986)
saccharivorella Matsumura	*Miscanthus* spp.	PO	Yang and Yang (1986)
	Saccharum spp.	PO	
sinuosa Yang and Yang	*Imperata cylindrica*	PO	Yang and Yang (1986)
tuberipennis (Mulsant and Rey)	*Imperata cylindrica*	PO	Asche (1982c)
	Tribe Saccharosydnini		
Neomalaxa			
flava Muir	*Brachiaria adspersa*	PO	Wheeler (personal communication)
Saccharosydne			
ornatipennis Muir	*Paspalum intermedium*	PO	Muir (1926)
procerus (Matsumura)	*Zizania aquatica*	PO	Yang (1989)
	Z. caduciflora	PO	Ding et al. (1982)
	Z. latifolia	PO	Vilbaste (1968)
rostrifrons (Crawford)	*Paspalum virgatum*	PO	Metcalfe (1969)
saccharivora (Westwood)	*Andropogon bicornis*	PO	Metcalfe (1969)
	A. glomeratus	PO	
	Saccharum officinarum	PO	
	Sorghum sudanense	PO	
viridis Muir	*Oryza sativa*	PO	Muir (1926)
	Tribe Delphacini		
Acanthodelphax			
denticauda (Boheman)	*Deschampsia caespitosa*	PO	den Bieman and Booij (1984)
spinosus (Fieber)	*Nardus stricta*	PO	Gunthart (1987)
	Festuca cyllenica, F. spp.	PO	Drosopoulos et al. (1983)
Achorotile			
distincta Scudder sp.	*Calamagrostis rubescens*	PO	Scudder (1963)
	Carex sp.	CY	S. Wilson (unpublished data)
stylata Beamer	*Poa pratensis*	PO	Scudder (1963)

(*continued*)

Appendix 2. (*Continued*)

Planthopper Taxon	Host Plant		References
	Genus and Species	Family[a]	
Aloba			
artemisiae (Kirkaldy)	*Artemisia australis*	AS	Zimmerman (1948)
campylothecae Muir	*Campylotheca* sp.	AS	Zimmerman (1948)
dubautiae (Kirkaldy)	*Dubautia laxa*	AS	Zimmerman (1948)
	D. plantaginea	AS	
flavocollaris Muir	*Dubautia laxa*	AS	Zimmerman (1948)
	D. plantaginea	AS	
ipomoeae Kirkaldy	*Ipomoea batatas*	CO	Zimmerman (1948)
	I. bona-nox	CO	
	I. insularis	CO	
	I. pes-caprae	CO	
	I. pentaphylla	CO	
	I. tuberculata	CO	
kirkaldyi Muir	*Euphorbia hillebrandi*	EU	Zimmerman (1948)
majuma Fennah	*Cyrtandra* sp.	GE	Fennah (1958)
myoporicola Kirkaldy	*Myoporum sandwicense*	MP	Zimmerman (1948)
	Pelea volcanicola	RT	
plectranthi Muir	*Plectranthus parviflorus*	LA	Zimmerman (1948)
swezeyi Muir	*Bidens pilosa*	AS	Zimmerman (1948)
	Campylotheca macrocarpa	AS	
	Cheirodendron gaudichaudii	AL	
	Lipochaeta sp.	AS	
	Lythrum sp.	LY	
Ambarvalia			
pyrops Distant	*Stevensonia* sp.	AR	Fennah (1964)
Anchodelphax			
hagnon Fennah	*Pimelea* sp.	TH	Fennah (1965a)
olenus Fennah	*Muehlenbeckia australis*	PL	
	Lepidium oleraceum	BR	

Bakerella			
bidens Beamer	*Carex* sp.	CY	Beamer (1945)
cinerea Beamer sp.	*Eleocharis palustris*	CY	
	Juncus sp.	JU	S. Wilson (unpublished data)
Calligypona			
reyi (Fieber)	*Juncus* sp.	JU	Ossiannilsson (1978)
	Scirpus lacustris	CY	
	S. tabernaemontani	CY	
	Cyperus sp.	CY	Drosopoulos et al. (1983)
Changeondelphax			
velitchkovski (Melichar)	*Typha laxmanni*	TY	Kwon (1982)
Chloriona			
chinai Ossiannilsson	*Phragmites* sp.	PO	Ossiannilsson (1978)
clavata Dlabola	*Phragmites communis*	PO	Drosopoulos et al. (1983)
	Arundo donax	PO	
dorsata Edwards	*Phragmites* sp.	PO	den Bieman and Booij (1984)
flaveola Lindberg	*Phragmites communis*	PO	Asche (1982c)
	Arundo donax	PO	
glaucescens Fieber	*Phragmites communis*	PO	Drosopoulos et al. (1983)
	Arundo donax	PO	
ponticana Asche	*Phragmites communis*	PO	Drosopoulos et al. (1983)
	Arundo donax	PO	
smaragdula (Stål)	*Phragmites* sp.	PO	den Bieman and Booij (1984)
stenoptera (Flor)	*Phragmites* sp.	PO	Vilbaste (1974)
tateyamana Matsumura	*Phragmites communites*	PO	Yang (1989)
	Typha laxmanni	TY	Vilbaste (1968)
unicolor (Herrich-Schaffer)	*Phragmites communis*	PO	Asche (1982b)
	Arundo donax	PO	
vasconica Ribaut	*Phragmites communis*	PO	Drosopoulos et al. (1983)
	Arundo donax	PO	

(*continued*)

Appendix 2. (*Continued*)

Planthopper Taxon	Host Plant		References
	Genus and Species	Family[a]	
Chlorionidea			
bromi Emeljanov	*Bromus inermis*	PO	Emeljanov (1977)
flava Low	*Carex montana*	CY	Drosopoulos et al. (1983)
Conomelus			
anceps (Germar)	*Juncus effusus*	JU	Waloff and Solomon (1973)
odryssius Dlabola	*Juncus effusus/inflexus*	JU	Drosopoulos et al. (1983)
sagittifer Remane and Asche	*Juncus inflexus*	JU	Drosopoulos et al. (1983)
Coracodelphax			
obscurus Vilbaste	*Zoysia japonica*	PO	Kwon (1982)
Cormidius			
nigrifrons (Kusnezov)	*Agropyron pseudagropyrum*	PO	Emeljanov (1977)
Criomorphus			
albomarginatus Curtis	*Festuca rubra*	PO	den Bieman and Booij (1984)
	Holcus spp.	PO	Waloff and Solomon (1973)
borealis (Sahlberg)	*Calamagrostis canescens*	PO	den Bieman and Booij (1984)
moestus (Boheman)	*Calamagrostis canescens*	PO	Ossiannilsson (1978)
Delphacinus			
griceus Emeljanov	*Elymus* sp.	PO	Emeljanov (1977)
Delphacodes			
arcuata Beamer	*Manihot utilissima*	EU	Caldwell and Martorell (1950)
axonopi Fennah	*Axonopus compressa*	PO	Caldwell and Martorell (1950)
bellicosa Muir and Giffard	*Paspalum distichum*	PO	Wilson (1985)
capnodes (Scott)	*Carex riparia*	CY	Drosopoulos et al. (1983)
campestris (Van Duzee)	*Agropyron* sp.	PO	DuBose (1960)
	Agropyron cristatum	PO	S. Wilson (unpublished data)
	Calamavilfa longifolia	PO	
	Poa pratenensis	PO	

	Agrostis sp., *Festuca ovina*	PO	
	Stipa cornata	PO	
	Carex scoparia	CY	
cerberus Fennah	*Drimys confertifolia*	WI	Fennah (1957)
detecta (Van Duzee)	*Spartina patens*	PO	Raupp and Denno (1979)
idonea Beamer	*Panicum repens*	PO	Ballou et al. (1987)
laminalis (Van Duzee)	*Leersia hexandra*	PO	S. Wilson (unpublished data)
latidens Beamer	*Setaria texana*	PO	S. Wilson (unpublished data)
lutulenta (Van Duzee)	*Andropogon repens*	PO	S. Wilson (unpublished data)
	Poa pratensis	PO	
	Puccinellia nuttalliana	PO	
muirella Metcalf	*Corokia* sp.	CN	Fennah (1958)
	Metrosideros sp.	MY	
	Blechnum sp.	BL	
nigerrima Ishihara	*Murkania keisak*	CM	Kisimoto (personal communication)
nigrifacies Muir	*Paspalum notatum*	PO	Calvert et al. (1987a)
nigrigena Matsumura and Ishihara	*Poa sphondylodes*	PO	Lee and Kwon (1977)
parvula (Ball)	*Andropogon scoparius*	PO	S. Wilson (unpublished data)
penedetecta Beamer	*Spartina alterniflora*	PO	S. Wilson (unpublished data)
puella (Van Duzee)	*Galinsoga parviflora*	AS	Batra (1979)
	Panicum capillare	PO	S. Wilson (unpublished data)
rotundata Beamer	*Agrostis scabra*	PO	S. Wilson (unpublished data)
	Andropogon gerhardi	PO	
	Juncus marginatus	JU	S. Wilson (unpublished data)
schinias Asche and Remane	*Cyperus* sp.	CY	Drosopoulos et al. (1983)
	Phragmites sp.	PO	
stricklandi Metcalf	*Amaranthus elapious*	AM	Metcalf (1946a)
Delphacodoides			
anaxarchi (Muir)	*Hemarthia altissima*	PO	Drosopoulos et al. (1983)
Delphax			
armeniacus Anufriev	*Phragmites communis*	PO	Drosopoulos et al. (1983)
crassicornis (Panzer)	*Phragmites communis*	PO	Ossiannilsson (1978)
inermis Ribaut	*Phragmites communis*	PO	Asche (1982c)
	Arundo donax	PO	

(*continued*)

Appendix 2. (*Continued*)

Planthopper Taxon	Host Plant		References
	Genus and Species	Family[a]	
meridionalis (Haupt)	*Phragmites communis*	PO	Drosopoulos et al. (1983)
	Arundo donax	PO	
pulchellus (Curtis)	*Phragmites communis*	PO	Ossiannilsson (1978)
ribautianus Asche and Drosopoulos	*Phragmites communis*	PO	Drosopoulos et al. (1983)
	Arundo donax	PO	
Dicentropyx			
sublineata (Emeljanov)	*Elymus* sp.	PO	Emeljanov (1977)
Dicranotropis			
divergens Kirschbaum	*Nardus* sp.	PO	Gunthart (1987)
	Deschampsia flexuosa		
fumosa Matsumura	*Saccharum officinarum*	PO	Yang (1989)
hamata (Boheman)	Grasses	PO	Asche (1982b)
	Dactylis glomerata	PO	Gunthart (1987)
	Holcus sp.	PO	
Dictyophorodelphax			
mirabilis Swezey	*Euphorbia clusiaefolia*	EU	Zimmerman (1948)
	E. hillebrandi	EU	
predicta Bridwell	*Euphorbia hookeri*	EU	Zimmerman (1948)
swezeyi Bridwell	*Euphorbia celastroides*	EU	Zimmerman (1948)
usingeri Swezey	*Euphorbia* sp.	EU	Zimmerman (1948)
Ditropis			
pteridis (Spinola)	*Pteridium aquilinum*	PT	Asche (1982b)
Eoeurysa			
arundina Kuoh & Ding	*Arundo donax*	PO	Yang (1989)
flavocapitata Muir	*Saccharum officinarum*	PO	Yang (1989)
Euconomelus			
lepidus (Boheman)	*Juncus* spp.	JU	Drosopoulos et al (1983); Asche (1982b)
	Carex sp.	CY	

Euidelloides			
montana Muir	*Chusquea* sp.	PO	Muir (1926)
Euides			
alpina Wagner	*Phragmites communis*	PO	Lauterer (1983)
elegans (Muir)	*Guadua* sp.	PO	Muir (1926)
gerhardi (Metcalf)	*Scirpus americana*	CY	S. Wilson (unpublished data)
guaduae (Muir)	*Guadua* sp.	PO	Muir (1926)
speciosa (Boheman)	*Phragmites communis*	PO	Drosopoulos et al. (1983)
	Arundo donax	PO	
Euidopsis			
truncata Ribaut	*Imperata cylindrica*	PO	Drosopoulos et al. (1983)
Eurybregma			
nigrolineata Scott	*Dactylis glomerata*	PO	Drosopoulos et al. (1983)
pseudoagropyri Emeljanov	*Agropyron pseudoagropyrum*	PO	Emeljanov (1966)
Eurysa			
brunnea Melichar	*Secale montanum*	PO	Drosopoulos et al. (1983)
flavobrunnea Dlabola	*Helictotrichon convolutum*	PO	Drosopoulos et al. (1983)
fornasta Asche, Drosopoulos and Hoch	*Arrhenatherum elatius*	PO	Drosopoulos et al. (1983)
obesa Beamer	*Elymus* sp.	PO	S. Wilson (unpublished data)
rubripes (Matsumura)	*Dactylis glomerata*	PO	Drosopoulos et al. (1983)
Eurysula			
lurida (Fieber)	*Calamagrostis epigeios*	PO	Ossiannilsson (1978)
	C. canescens	PO	
Falcotoya			
miniscula (Horvath)	*Cynodon dactylon*, grasses	PO	Drosopoulos et al. (1983)
Flastena			
fumipennis (Fieber)	*Cyperus longus*	CY	Le Quesne (1983)
	Carex sp.	CY	Drosopoulos et al. (1983)

(*continued*)

Appendix 2. (*Continued*)

Planthopper Taxon	Host Plant		References
	Genus and Species	Family[a]	
Florodelphax			
leptosoma (Flor)	*Juncus* spp.	JU	Drosopoulos et al. (1983)
	Cyperus	CY	
mourikisi Drosopoulos	*Juncus acutus/maritimus*	JU	Drosopoulos et al. (1983)
Gravesteiniella			
boldi (Scott)	*Ammophila arenaria*	PO	Ossiannilsson (1978)
	Lasiagrostis splendens	PO	Vilbaste (1965)
mitjaevi Emeljanov	*Achnatherum splendens*	PO	Emeljanov (1982)
Halmyra			
aeluropodis (Emeljanov)	*Aeluropus littoralis*	PO	Drosopoulos et al. (1983)
Harmalia			
commelinae Yang	*Commelina diffusa*	CM	Yang (1989)
Horvathianella			
palliceps (Horvath)	*Chrysopogon gryllus*	PO	Drosopoulos et al. (1983)
Ilburnia			
dianae Fennah	Fern brakes	PY	Fennah (1976)
ignobilis White	*Dicksonia arborescens*	DK	Fennah (1976)
	Fern brakes	PY	
Ishiharodelphax			
matsuyamensis Ishihara	*Zoysia japonica/tenuifolia*	PO	Kwon (1982)
	Agrostis clavata	PO	
Iubsoda			
stigmatica (Melichar)	*Hyparrhenia hirta*	PO	Asche (1982c)
Javesella			
discolor (Boheman)	*Deschampsia flexuosa*	PO	de Vrijer (1981)
	Deschampsia cespitosa	PO	
	Poa nemoralis	PO	
	Avena sativa, Holcus mollis	PO	Wilson and O'Brien (1987)

dubia (Kirschbaum)	*Agrostis stolonifera*	PO	de Vrijer (1981)
	A. tenuis	PO	
	Arrhenatherum elatior	PO	
forcipata (Boheman)	*Poa pratensis/annua*	PO	de Vrijer (1981)
kilmani (Van Duzee)	*Equisetum* sp.	EQ	Strickland (1940)
obscurella (Boheman)	*Alopecurus geniculatus*	PO	de Vrijer (1981)
	Avena sativa	PO	Wilson and O'Brien (1987)
pellucida (Fabricius)	Polyphagous, grasses	PO	de Vrijer (1981)
	Festuca pratensis/rubra	PO	Gunthart (1987)
	Agrostis tenuis	PO	
	Dactylis glomerata	PO	
	Sieglingia decumbens	PO	
	Avena sativa	PO	Ossiannilsson (1978)
	Lolium perenne	PO	
	Carex rostrata/limosa	CY	DuBose (1960)
	Scirpus microcarpus	CY	S. Wilson (unpublished data)
salina (Haupt)	*Triglochin maritima*	JG	de Vrijer (1981)
	Juncus gerardi	JU	Ossiannilsson (1978)
simillima (Linnavuori)	*Eriophorum* sp., *Carex* sp.	CY	Ossiannilsson (1978)
stali (Metcalf)	*Equisetum* sp.	EQ	Ossiannilsson (1978)
Kakuna			
albipennis (Matsumura)	*Pennisetum alopecuroides*	PO	Yang (1989)
kuwayamai Matsumura	*Phragmites communis*	PO	Lee and Kwon (1977)
sapporonis (Matsum.)	*Typha laxmanni*	TY	Vilbaste (1968)
"*Kelisia*"[b]			
emoloa Muir	*Eragrostis variabilis*	PO	Zimmerman (1948)
eragrosticola Muir	*Eragrostis variabilis*	PO	Zimmerman (1948)
sporobolicola Kirkaldy	*Eragrostis atropiodes*	PO	Zimmerman (1948)
	Sporobolus virginicus	PO	
	Vincentia angustifolia	CY	
s. immaculata Kirkaldy	*Deschampsia australis*	PO	Zimmerman (1948)
	Vincentia angustifolia	CY	

(*continued*)

Appendix 2. (*Continued*)

Planthopper Taxon	Host Plant		References
	Genus and Species	Family[a]	
Kormus			
artemisiae Fieber	*Limonium gmelini*	PU	Emeljanov (personal communication)
Kusnezoviella			
antinoma Emeljanov	*Elymus* sp.	PO	Emeljanov (1977)
chalchica Emeljanov	*Hordeum* sp.	PO	Emeljanov (1977)
dimidatifrons (Kusnezov)	*Agropyron pseudagropyrum*	PO	Emeljanov (1977)
Laccocera			
obesa	*Thinopyrum ponticum*	PO	Spangler and MacMahon (1990)
sp.	*Stipa cornata*	PO	S. Wilson (unpublished data)
	Muhlenbergia richardsonis	PO	
	Puccinellia sp.	PO	
Laodelphax			
elegantulus (Boheman)	*Festuca rubra*	PO	Waloff and Solomon (1973)
	Deschampsia flexuosa	PO	Gunthart (1987)
striatellus (Fallen)	Polyphagous	PO	Wilson and O'Brien (1987)
	Oryza sativa	PO	
	Avena sativa	PO	
	Arrhenatherum elatius	PO	
	Triticum sp.	PO	
	Zea mays	PO	
	Hordeum vulgare	PO	
Leialoha			
cajeta Fennah	*Reynoldsia* sp.	AL	Fennah (1958)
	Rapanea sp.	MS	
	Weinmannia parviflora	CU	
	Metrosideros collina	MY	
	Alstonia sp.	AP	
	Cyrtandra sp.	GE	
	Cheirodendron sp.	AL	
hawaiiensis (Muir)	*Metrosideros* sp.	MY	Zimmerman (1948)

kauaiensis (Muir)	*Metrosideros* sp.	MY	Zimmerman (1948)
lanaiensis (Muir)	*Metrosideros* sp.	MY	Zimmerman (1948)
lehuae (Kirkaldy)	*Metrosideros* sp.	MY	Zimmerman (1948)
mauiensis (Muir)	*Metrosideros* sp.	MY	Zimmerman (1948)
naniicola (Kirkaldy)	*Metrosideros* sp.	MY	Zimmerman (1948)
oahuensis (Muir)	*Metrosideros* sp.	MY	Zimmerman (1948)
oceanides (Kirkaldy)	*Osmanthus sandwicensis*	OL	Zimmerman (1948)
ohiae (Kirkaldy)	*Metrosideros* sp.	MY	Zimmerman (1948)
scaevolae Muir	*Scaevola chamissoniana*	GO	Zimmerman (1948)
suttoniae Muir	*Myrsine sandwicensis*	MS	Zimmerman (1948)
Leptodelphax			
cyclops Haupt	*Saccharum* sp.	PO	Drosopoulos et al. (1983)
Liburniella			
ornata (Stål)	*Rhynchospora globularis*	CY	S. Wilson (unpublished data)
Maculidelphax			
maculipennis (Linnavuori)	*Imperata cylindrica*	PO	Drosopoulos et al. (1983)
Malaxodes			
farinosus Fennah	*Melinis minutiflora*	PO	Fennah (1967a)
Matutinus			
putoni (A. Costa)	*Typha latifolia/angustifolia*	TY	D'Urso and Guglielmino (1986)
Megadelphax			
sordidula (Stål)	*Avena sativa*	PO	Wilson and O'Brien (1987)
	Phleum pratense	PO	
	Triticum sp.	PO	
Megamelodes			
quadrimaculatus (Signoret)	Polyphagous	MO	Asche (1982b)
Megamelus			
davisi Van Duzee	*Nuphar advena*	NY	Wilson and McPherson (1981a)
lobatus Beamer	*Spartina patens*	PO	Raupp and Denno (1979)
metzaria Crawford	*Spartina pectinata*	PO	S. Wilson (unpublished data)
notula (Germar)	*Carex lasiocarpa*	CY	Vilbaste (1971)

(*continued*)

Appendix 2. (*Continued*)

Planthopper Taxon	Host Plant		References
	Genus and Species	Family[a]	
	Juncus sp.	JU	
	Carex riparia	CY	Drosopoulos et al. (1983)
palaetus (Van Duzee)	*Pontederia cordata*	PN	Wilson and McPherson (1979)
sp.	*Eleocharis* sp.	CY	S. Wilson (unpublished data)
	Distichilis stricta	PO	
sp.	*Spartina alterniflora*	PO	Denno 1977, S. Wilson (unpublished)
Metropis			
achnatheri Emeljanov	*Achnatherum splendens*	PO	Emeljanov (1977)
aris Asche, Drosop. and Hoch	*Festuca* sp.	PO	Drosopoulos et al. (1983)
inermis Wagner	*Festuca ovina*	PO	Gunthart (1987)
	Carex sp.	CY	
mayri Fieber	*Festuca cyllenica*	PO	Drosopoulos et al. (1983)
tolerans Emeljanov	*Festuca* spp.	PO	Emeljanov (1977)
Muellerianella			
brevipennis (Boheman)	*Deschampsia cespitosa*	PO	Booij (1982)
extrusa (Scott)	*Molinia caerulea*	PO	Asche (1982b)
fairmairi (Perris)	*Holcus lanatus/mollis*	PO	Asche (1982b)
relicta Logvinenko	*Luzula* sp.	JU	Drosopoulos (1977)
sp.	*Carex divulsa*	CY	Drosopoulos (1983)
	Setaria pumila	PO	
Muirodelphax			
atratus Vilbaste	*Zoysia japonica*	PO	Kwon (1982)
aubei (Perris)	*Ammophila arenaria*	PO	Drosopoulos et al. (1983)
	Elymus pycnanthus	PO	
matsuyamensis (Ishihara)	*Agrostis clavata*	PO	Lee and Kwon (1980)
	Zoysia japonica/tenuifolia	PO	
Neomegamelanus			
dorsalis (Metcalf)	*Spartina patens*	PO	Raupp and Denno (1979)

elongatus (Ball)	*Spartina patens*	PO	McDermott (1952)
elongatus reductus Caldwell	*Sporobolus virginicus*	PO	Caldwell and Martorell (1950)
Nesodryas			
antiope Fennah	*Cocos nucifera*	AR	Wilson (1987)
freycinetiae Kirkaldy	*Freycinetia arborea*	PA	Zimmerman (1948)
gigantea (Muir)	*Pritchardia* sp.	AR	Wilson (1987)
oenone Fennah	*Pandanus* sp.	PA	Fennah (1958)
swezeyi Zimmerman	*Pritchardia* sp.	AR	Wilson (1987)
Nesorestias			
filicola Kirkaldy	*Cibotium* sp.	DK	Zimmerman (1948)
	Elaphoglossum gorgonum	AN	
nimbata (Kirkaldy)	*Phegopteris* sp.	TL	Zimmerman (1948)
Nesorthia			
paronychiae Fennah	*Paronychia manicata*	CA	Fennah (1962a)
Nesosydne			
acastus Fennah	*Crossostylis biflora*	RH	Fennah (1958)
acuta (Muir)	*Cyrtandra mauiensis*	GE	Zimmerman (1948)
agenor Fennah	*Premna tahitensis*	VE	Fennah (1958)
	Morinda citrifolia	RU	
ahinahina (Muir)	*Argyroxiphium* sp.	AS	Zimmerman (1948)
aku (Muir)	*Cyanea tritomantha*	CP	Zimmerman (1948)
amaumau (Muir)	*Sadleria* sp.	BL	Zimmerman (1948)
anceps Muir	*Freycinetia* sp.	PA	Zimmerman (1948)
argyroxiphii Kirkaldy	*Argyroxiphium sandwicense*	AS	Zimmerman (1948)
asteliae Muir	*Astelia veratroides*	LI	Zimmerman (1948)
boehmeria (Muir)	*Boehmeria* sp.	UR	Zimmerman (1948)
bridwelli (Muir)	*Argyroxiphium virescens*	AS	Zimmerman (1948)
	A. sandwicense	AS	
	Dubautia sp.	AS	
calypso Fennah	*Drimys confertifolia*	WI	Fennah (1957)
	Gunnera masafuerae	GU	
campylothecae (Muir)	*Campylotheca* sp.	AS	Zimmerman (1948)

(*continued*)

Appendix 2. (*Continued*)

Planthopper Taxon	Host Plant		References
	Genus and Species	Family[a]	
chambersi Kirkaldy	*Raillardia ciliolata*	AS	Zimmerman (1948)
cheesmanae (Muir)	*Weinmannia* sp.	CU	Fennah (1958)
cleanthes Fennah	*Weinmannia* sp.	CU	Fennah (1958)
clitarchus Fennah	*Piper latifolium*	PI	Fennah (1958)
coprosmicola (Muir)	*Coprosma ernodiodes*	RU	Zimmerman (1948)
cyane Fennah	*Reynoldsia* sp.	AL	Fennah (1958)
	Weinmannia sp.	CU	
	Glochidion ramiflorum	EU	
	Loranthus sp.	LO	
	Cyrtandra sp.	GE	
cyathodis Kirkaldy	*Styphelia tameiameiae*	EP	Zimmerman (1948)
cyrtandrae Muir	*Cyrtandra* sp.	GE	Zimmerman (1948)
cyrtandricola Muir	*Cyrtandra* sp.	GE	Zimmerman (1948)
	Charpentiera obovata	AM	
dinomache Fennah	*Cyrtandra* sp.	GE	Fennah (1958)
	Sclerotheca sp.	CP	
	Vaccinium sp.	ER	
	Weinmannia sp.	CU	
dubautiae (Muir)	*Dubautia plantaginea*	AS	Zimmerman (1948)
eeke (Muir)	*Argyroxiphium* sp.	AS	Zimmerman (1948)
elatus Fennah	*Cyrtandra* sp.	GE	Fennah (1958)
	Bidens lantanoides	AS	
	Freycinetia sp.	PA	
fullawayi (Muir)	*Styphelia* sp.	EP	Zimmerman (1948)
geranii (Muir)	*Geranium arboreum*	GR	Zimmerman (1948)
giffardi Muir	*Cyrtandra grandiflora*	GE	Zimmerman (1948)
	Rollandia crispa	CP	

gigantea (Muir)	*Pritchardia* sp.	AR	Zimmerman (1948)
gouldiae Kirkaldy	*Cyrtandra grandiflora*	GE	Zimmerman (1948)
	Cyrtandra sp.	GE	
gunnerae Muir	*Gunnera petaloidea*	GU	Zimmerman (1948)
halia Kirkaldy	*Dubautia plantaginea*	AS	Zimmerman (1948)
	Freycinetia sp.	PA	
imbricola Kirkaldy	*Coprosma montana*	RU	Zimmerman (1948)
ipomoeicola Kirkaldy	*Antidesma* sp.	SG	Zimmerman (1948)
	Brassica sp.	BR	
	Cibotium sp.	DK	
	Cynodon dactylon	PO	
	Cyrtandra sp.	GE	
	Dolichos lablab	FA	
	Gouldia elongata	RU	
	Ipomoea batatas	CO	
	Ipomoea bona-nox	CO	
	Jussiaea villlosa	ON	
	Lythrum maritimum	LY	
	Mucuna gigantea	FA	
	Pipturus sp.	UR	
	Polygonum sp.	PL	
	Rumex sp.	PL	
	Sadleria sp.	BL	
	Solanum tuberosum	SO	
	Strongylodon lucidum	FA	
koae Kirkaldy	*Acacia koa*	FA	Zimmerman (1948)
koae-phyllodii Muir	*Acacia koa*	FA	Zimmerman (1948)
kokolau (Muir)	*Campylotheca* sp.	AS	Zimmerman (1948)
kuschei (Muir)	*Cyrtandra* sp.	GE	Zimmerman (1948)
lanaiensis (Muir)	*Styphelia* sp.	EP	Zimmerman (1948)
lanista Fennah	*Piper latifolium*	PI	Fennah (1958)
	Weinmannia sp.	CU	
	Sida sp.	MA	

(*continued*)

Appendix 2. (*Continued*)

Planthopper Taxon	Host Plant		References
	Genus and Species	Family[a]	
latona Fennah	*Weinmannia parviflora*	CU	Fennah (1958)
leahi (Kirkaldy)	*Lipochaeta calycosa*	AS	Zimmerman (1948)
linus Fennah	*Freycinetia* sp.	PA	Fennah (1958)
	Cyathea sp.	CT	
	Bidens lantanoides	AS	
	Hibiscus tiliaceus	MA	
	Metrosideros collina	MY	
lobeliae Muir	*Lobelia hypoleuca*	CP	Zimmerman (1948)
longipes (Muir)	*Cyrtandra mauiensis*	GE	Zimmerman (1948)
mamake (Muir)	*Pipturus* sp.	UR	Zimmerman (1948)
mauiensis (Muir)	*Campylotheca mauiensis*	AS	Zimmerman (1948)
	Lipochaeta integrifolia	AS	
	Raillardia menziesii	AS	
	Tetramolopium artemisia	AS	
melampus Fennah	*Weinmannia parviflora*	CU	Fennah (1958)
minos Fennah	*Gunnera masafuerae*	GU	Fennah (1957)
	Pernettya rigida	ER	
monticola Kirkaldy	*Coprosma montana*	RU	Zimmerman (1948)
montis-tantalus Muir	*Lobelia hypoleuca*	CP	Zimmerman (1948)
	Broussaisia arguta	HY	
naenae (Muir)	*Dubautia* sp.	AS	Zimmerman (1948)
	Raillardia sp.	AS	
neocyrtandrae (Muir)	*Gunnera mauiensis*	GU	Zimmerman (1948)
neoraillardiae (Muir)	*Lipochaeta subcordata*	AS	Zimmerman (1948)
neowailupensis (Muir)	*Coprosma longifolia*	RU	Zimmerman (1948)
nephrolepidis Kirkaldy	*Nephrolepis exaltata*	DV	Zimmerman (1948)
nesopele (Muir)	*Astelia veratroides*	LI	Zimmerman (1948)
nigrinervis (Muir)	*Styphelia* sp.	EP	Zimmerman (1948)

oahuensis Muir	*Charpentiera obovata*	AM	Zimmerman (1948)
olympica (Muir)	*Lobelia* sp.	CP	Zimmerman (1948)
oroanda Fennah	*Melochia velutina*	ST	Fennah (1958)
osborni Muir	*Raillardia* sp.	AS	Zimmerman (1948)
otus Fennah	*Fitchia* sp.	AS	Fennah (1958)
	Ferns	PY	
	Veronica sp.	SC	
painiu (Muir)	*Astelia veratroides*	LI	Zimmerman (1948)
phyllostegiae Muir	*Phyllostegia racemosa*	LA	Zimmerman (1948)
pilo (Muir)	*Coprosma ernodiodes*	RU	Zimmerman (1948)
pipturi Kirkaldy	*Pipturus* sp.	UR	Zimmerman (1948)
pseudorubescens Muir	*Acacia koa*	FA	Zimmerman (1948)
raillardiae Kirkaldy	*Raillardia scabra*	AS	Zimmerman (1948)
	R. ciliolata	AS	
	Rollandia sp.	CP	
raillardiicola (Muir)	*Raillardia menziesii*	AS	Zimmerman (1948)
	R. platyphyllum	AS	
rubescens (Kirkaldy)	*Acacia koa*	FA	Zimmerman (1948)
satyrion Fennah	*Coprosma* sp.	RU	Fennah (1958)
sharpi Muir	*Broussaisia* sp.	HY	Zimmerman (1948)
	Boehmeria stipularis	UR	
siderion Fennah	*Reynoldsia tahitensis*	AL	Fennah (1958)
sophonisba Fennah	*Fitchia* sp.	AS	Fennah (1958)
	Lautea sp.	CN	
	Asplenium nidus	AN	
sorix Fennah	*Metrosideros collina*	MY	Fennah (1958)
stenogynicola (Muir)	*Stenogyne kamehameha*	LA	Zimmerman (1948)
sulcata (Muir)	*Cyrtandra* sp.	GE	Zimmerman (1948)
tetramolopii (Muir)	*Tetramolopium humile*	AS	Zimmerman (1948)
timberlakei Muir	*Cyrtandra garnotiana*	GE	Zimmerman (1948)
	Cyanea truncata	AS	
ulehihi (Muir)	*Smilax sandwicensis*	SM	Zimmerman (1948)

(*continued*)

Appendix 2. (*Continued*)

Planthopper Taxon	Host Plant		References
	Genus and Species	Family[a]	
umbratica Kirkaldy	*Charpentiera obovata*	AM	Zimmerman (1948)
	Clermontia spp.	CP	
	Cyrtandra sp.	GE	
	Pipturus sp.	UR	
	Stenogyne sp.	LA	
	Urera sandwicensis	UR	
viridis (Muir)	*Phyllostegia* sp.	LA	Zimmerman (1948)
vulcan Fennah	*Rea micrantha*	AS	Fennah (1957)
waikamoiensis (Muir)	*Cyanea aculeatiflora*	CP	Zimmerman (1948)
	Pipturus sp.	UR	
wailupensis (Muir)	*Rollandia crispa*	CP	Zimmerman (1948)
Nesothoe			
antidesmae (Muir)	*Antidesma platyphyllum*	SG	Zimmerman (1948)
bobeae Kirkaldy	*Bobea* sp.	RU	Zimmerman (1948)
dodonaeae (Muir)	*Alphitonia* sp.	RA	Zimmerman (1948)
	Dodonaea sp.	SA	
	Myrsine sp.	MS	
dryope (Kirkaldy)	*Antidesma platyphyllum*	SG	Zimmerman (1948)
elaeocarpi (Kirkaldy)	*Cyrtandra paludosa*	GE	Zimmerman (1948)
	Elaeocarpus bifidus	EL	
	Scaevola mollis	GO	
eugeniae (Kirkaldy)	*Eugenia sandwicensis*	MY	Zimmerman (1948)
	Straussia kaduana	RU	
fletus Kirkaldy	*Antidesma platyphyllum*	SG	Zimmerman (1948)
	Myrsine sp.	MS	
giffardi (Kirkaldy)	*Cyrtandra grandiflora*	GE	Zimmerman (1948)
	Touchardia latifolia	UR	

gulicki (Muir)	*Euphorbia* sp.	EU	Zimmerman (1948)
	Metrosideros sp.	MY	
	Osmanthus sandwicensis	OL	
haa (Muir)	*Antidesma platyphyllum*	SG	Zimmerman (1948)
hula Kirkaldy	*Osmanthus sandwicensis*	OL	Zimmerman (1948)
	Pelea sp.	RT	
	Phyllostegia sp.	LA	
	Sideroxylon sp.	SP	
	Myrsine sp.	MS	
laka Kirkaldy	*Sida* sp.	MA	Zimmerman (1948)
maculata (Muir)	*Diospyros sandwicensis*	EB	Zimmerman (1948)
	D. hillebrandii	EB	
	Osmanthus sandwicensis	OL	
munroi (Muir)	*Dodonaea* sp.	SA	Zimmerman (1948)
perkinsi Kirkaldy	*Clermontia kakeana*	CP	Zimmerman (1948)
	Metrosideros sp.	MY	
	Myrsine sp.	MS	
piilani Kirkaldy	*Osmanthus sandwicensis*	OL	Zimmerman (1948)
pluvialis Kirkaldy	*Antidesma* sp.	SG	Zimmerman (1948)
semialba (Muir)	*Osmanthus sandwicensis*	OL	Zimmerman (1948)
seminigrofrons (Muir)	*Campylotheca* sp.	AS	Zimmerman (1948)
terryi Kirkaldy	*Osmanthus sandwicensis*	OL	Zimmerman (1948)
Nilaparvata			
bakeri (Muir)	*Leersia japanica*	PO	Yang (1989)
lugens (Stål)	*Oryza sativa*	PO	Claridge et al. (1988)
muiri China	*Leersia sayanuka*	PO	Mochida and Okada (1979)
sp.	*Leersia hexandra*	PO	Claridge et al. (1988)
Nothodelphax			
albocarinatus (Stål)	*Eriophorum* sp.	CY	Ossiannilsson (1978)
consimilis (Van Duzee)	*Scirpus microcarpus*	CY	DuBose (1960)
distinctus (Flor)	*Eriophorum vaginatum*	CY	Ossiannilsson (1978)
Nothorestias			
badia Muir	Ferns	PY	Zimmerman (1948)

(*continued*)

Appendix 2. (*Continued*)

Planthopper Taxon	Host Plant		
	Genus and Species	Family[a]	References
swezeyi Muir	*Aspidium* sp.	PY	Zimmerman (1948)
Numata			
muiri (Kirkaldy)	*Saccharum officinarum*	PO	Yang (1989)
Opiconsiva			
nigra Ding and Tian	*Paspalum distichum*	PO	Ding and Tian (1980)
paludum (Kirkaldy)	*Herpestis monnieria*	SC	Zimmerman (1948)
	Juncus sp.	JU	
	Sedge	CY	
Paracorbulo			
sirokata (Matsumura and Ishihara)	*Oryza sativa*	PO	Kwon (1982)
	Echinochloa crusgalli	PO	
	Phragmites communis	PO	
	Persicaria thunbergii	PL	Kwon (1982), Kisimoto (personal communication)
Paradelphacodes			
litoralis (Reuter)	*Carex rostrata*	CY	Anufriev (1980)
paludosa (Flor)	*Carex* sp.	CY	Asche (1982b)
Paraliburnia			
adela (Flor)	*Phalaris arundacea*	PO	Ossiannilsson (1978)
	Glyceria spp.	PO	
clypealis (Sahlberg)	*Calamagrostis canescens*	PO	den Bieman and Booij (1984)
	Eriophorum vaginatum	CY	
	E. angustifolium	CY	
	Molinia caerulea	PO	
	Carex spp.	CY	
Paraliburniella			
dalei (Scott)	*Agrostis tenuis*	PO	Waloff and Solomon (1973)
Parametopina			
yushaniae Yang	*Fargesia niitakayamensis*	PO	Yang (1989)

Peregrinus			
maidis (Ashmead)	*Zea mays*	PO	Wilson and O'Brien (1987)
	Sorghum halepense	PO	
Perkinsiella			
rivularis Linnavuori	*Saccharum* sp.	PO	Drosopoulos et al. (1983)
saccharicida Kirkaldy	*Saccharum officinarum*	PO	Wilson and O'Brien (1987)
sinensis Kirkaldy	*Saccharum officinarum*	PO	Yang (1989)
	Andropogon sorghum	PO	
	Phragmites communis	PO	
	Oryza sativa	PO	
thompsoni Muir	*Saccharum officinarum*	PO	Metcalf (1946b)
vastatrix (Breddin)	*Saccharum officinarum*	PO	Yang (1989)
Pissonotus			
albovenosus Osborn	*Lygodesmia grandiflora*	AS	Morgan and Beamer (1949)
aphidoides Van Duzee	*Castillija coccinea*	SC	Strickland (1940)
basalis Van Duzee	*Grindelia squarosa*	AS	Strickland (1940)
delicatus Van Duzee	*Grindelia* sp.	AS	Morgan and Beamer (1949)
	Happlopappus ciliatus	AS	
delicatus melanurus Morgan and Beamer	*Grindelia camporum*	AS	Morgan and Beamer (1949)
piceus (Van Duzee)	*Polygonum hydropiperoides*	PL	S. Wilson (unpublished data)
Prokelisia crocea (Van Duzee)	*Spartina pectinata*	PO	S. Wilson (unpublished data)
dolus Wilson	*Spartina alterniflora*	PO	Denno et al. (1987)
marginata Van Duzee	*Spartina alterniflora*	PO	Denno et al. (1987)
salina (Ball)	*Calamovilfa longifolia*	PO	Wilson 1982a, (unpublished data)
	Distichilis stricta	PO	
Pseudaraeopus			
bolivari (Melichar)	*Panicum turgidum*	PO	Linnavuori (1964)
lethierryi (Mulsant and Rey)	*Hyparrhenia hirta*	PO	Drosopoulos et al. (1983)
Pygospina			
aurantii (Crawford)	*Cyperus luzulae*	CY	Muir (1926)

(*continued*)

Appendix 2. (*Continued*)

Planthopper Taxon	Host Plant: Genus and Species	Family[a]	References
Remanodelphax			
cedroni Drosopoulos	*Dichanthium ischaemum*	PO	Drosopoulos et al. (1983)
Ribautodelphax			
albostriatus (Fieber)	*Poa pratensis*	PO	den Bieman (1987a)
	Agrostis sp.		Gunthart (1987)
	Dactylis sp.	PO	
	Bromus sp.	PO	
angulosus (Ribaut)	*Anthoxanthium odoratum*	PO	den Bieman (1987a)
collinus (Boheman)	*Agrostis capillaris*	PO	den Bieman (1987a)
falakron Asche, Drosopoulos and Hoch	*Festuca cyllenica*	PO	den Bieman (1987a)
fanari Asche, Drosopoulos and Hoch	*Elymus hispidus*	PO	den Bieman (1987b)
	E. pycnanthus	PO	
imitans (Ribaut)	*Festuca arundinacea fenas*	PO	den Bieman (1987b)
imitantoides Bieman	*Brachypodium phoenicoides*	PO	den Bieman (1987b)
kalonerensis Bieman	*Arrhenatherum elatius*	PO	den Bieman (1987b)
nogurae Bieman	*Carex nigra*	CY	den Bieman (1987b)
ochreata Vilbaste	*Setaria* sp.	PO	Vilbaste (1965)
pallens (Stål)	*Festuca ovina*	PO	den Bieman (1987a)
pungens (Ribaut)	*Brachypodium pinnatum*	PO	den Bieman (1987b)
	B. phoenicoides	PO	
	B. sylvaticum	PO	
ventouxianus Bieman	*Festuca rubra rubra*	PO	den Bieman (1987b)
vinealis Bieman	*Agrostis vinealis*	PO	den Bieman (1987b)
Sardia			
rostrata pluto (Kirkaldy)	*Cyperus rotundus*	CY	Beardsley (1990)
Sogata			
hakonensis (Matsum.)	*Miscanthus floridulus*	PO	Yang (1989)
hyalipennis (Matsumura)	*Miscanthus floridulus*	PO	Yang (1989)

mukwaensis Yang	*Miscanthus floridulus*	PO	Yang (1989)
nigrifrons (Muir)	*Miscanthus floridulus*	PO	Yang (1989)
Sogatella			
furcifera (Horvath)	*Oryza sativa*	PO	Asche and Wilson (1990)
kolophon (Kirkaldy)	*Panicum repens*	PO	Ballou et al. (1987)
molina (Fennah)	*Panicum purpurascens*	PO	Fennah (1963)
vibix (Haupt)	*Oryza sativa*	PO	Asche and Wilson (1990)
Stiroma			
affinis Fieber	*Dactylis glomerata*	PO	Drosopoulos et al. (1983)
bicarinata (Herrich-Schaffer)	*Deschampsia caespitosa*	PO	Drosopoulos et al. (1983)
Stiromella			
obliqua (Wagner)	*Phragmitetum* sp.	PO	Vilbaste (1974)
Stobaera			
affinis Van Duzee	*Ambrosia* sp.	AS	Kramer (1973)
bilobata Van Duzee	*Haplopappus squarrosus*	AS	Kramer (1973)
caldwelli Kramer	*Ambrosia* spp.	AS	Goeden and Ricker (1974a)
	Hymenoclea salsola	AS	Goeden and Ricker (1986a)
	Bebbia juncea	AS	Goeden and Ricker (1989)
	Trixis californica	AS	
concinna (Stål)	*Ambrosia* spp.	AS	Goeden and Ricker (1975)
			Calvert et al. (1987b)
	Parthenium hysterophorus	AS	McClay (1983)
giffardi Van Duzee	*Artemesia* sp.	AS	Kramer (1973)
muiri Kramer	*Ambrosia* spp.	AS	Goeden and Ricker (1974b)
pallida Osborn	*Baccharis halimifolia*	AS	Kramer (1973)
tricarinata (Say)	*Ambrosia* spp.	AS	Kramer (1973)
	Helianthus argophyllus	AS	
Struebingianella			
lugubrina (Boheman)	*Glyceria maxima*	PO	Ossiannilsson (1978)
	Glyceria aquatica	PO	Vilbaste (1971)
Sulix			
meridianalis (Muir)	*Scirpus frondosus*	CY	Fennah (1965a)

(continued)

Appendix 2. (*Continued*)

Planthopper Taxon	Host Plant		References
	Genus and Species	Family[a]	
tasmani (Muir)	*Muehlenbeckia* sp.	PO	Fennah (1965a)
vetranio Fennah	*Scirpus frondosus*	CY	Fennah (1965a)
Tagosodes			
cubanus (Crawford)	*Oryza sativa*	PO	Asche and Wilson (1990)
	Echinochloa sp.	PO	
	Cocos nucifera	AR	Wilson and O'Brien (1987)
orizicolus (Muir)	*Oryza sativa*	PO	Asche and Wilson (1990)
	Echinochloa sp.	PO	
Tarophagus			
colocasiae (Matsumura)	*Colocasia esculenta*	AC	Asche and Wilson (1989b)
persephone (Kirkaldy)	*Colocasia esculenta*	AC	Asche and Wilson (1989b)
proserpina (Kirkaldy	*Colocasia esculenta*	AC	Asche and Wilson (1989b)
Tertbron			
albovittata (Matsum.)	*Oryza sativa*	PO	Yang (1989)
Toya			
ibiturca Asche	*Cynodon dactylon*, grasses	PO	Drosopoulos et al. (1983)
obtusangula (Linnavuori)	*Cynodon dactylon*, grasses	PO	Drosopoulos et al. (1983)
propinqua (Fieber)	*Cynodon dactylon*, grasses	PO	Drosopoulos et al. (1983)
tuberculosa (Distant)[c]	*Panicum repens*	PO	Yang (1989)
	Cynodon dactylon, grasses	PO	Drosopoulos et al. (1983[e])

Tumidagena			
minuta McDermott	*Spartina patens*	PO	Raupp and Denno (1979)
Unkanodella			
ussuriensis Vilbaste	*Miscanthus sinensis*	PO	Kwon (1982)
Unkanodes			
excisa (Melichar)	*Elymus arenarius*	PO	Ossiannilsson (1978)
sapporana (Matsumura)	*Arundinella hirta*	PO	Lee and Kwon (1980)
	Imperata sp.	PO	
	Ischaemum anthephoroides	PO	
	Miscanthus sinensis	PO	

[a]AC —Araceae; AL—Araliaceae; AM—Amaranthaceae; AN—Aspleniaceae; AP—Apocynaceae; AR—Arecaceae; AS—Asteraceae; BL—Blechnaceae; BR—Brassicaceae; CA—Caryophyllaceae; CM—Commelinaceae; CN—Cornaceae; CO—Convolvulaceae; CP—Campanulaceae; CT—Cyatheaceae; CU—Cunoneaceae; CY—Cyperaceae; DI—Dicotyledonae; DK—Dicksoniaceae; DV—Davalliaceae; EB—Ebenaceae; EL—Elaeocarpaceae; EP—Epacridaceae; ER—Ericaceae; EQ—Equisetaceae; EU—Euphorbiaceae; FA—Fabaceae; GE—Gesnericaceae; GO—Goodeniaceae; GR—Geraniaceae; GU—Gunneraceae; HE—Heliconiaceae; HY—Hydrangeaceae; JG—Juncaginaceae; JU—Juncaceae; LA—Lamiaceae; LI—Liliaceae; LO—Loranthaceae; LY—Lythraceae; MA—Malvaceae; MO—Monocotyledonae; MP—Myoporaceae; MR—Moraceae; MS—Myrsinaceae; MY—Myrtaceae; NY—Nymphaeaceae; OL—Oleaceae; ON—Oleandraceae; PA—Pandanaceae; PI—Piperaceae; PL—Polygonaceae; PN—Pontederiaceae; PO—Poaceae; PR—Proteaceae, PT—Pteridiaceae, PU—Plumbaginaceae, PY—Pteridophyta, RA—Rhamnaceae, RH—Rhizophoraceae; RT—Rutaceae; RU—Rubiaceae; SA—Sapindaceae; SC—Scrophulariaceae; SG—Stilaginaceae; SM—Smilaceae; SO—Solancaecea; SP—Sapotaceae; ST—Sterculiaceae; TH—Thymelaeaceae; TL—Thylopteridaceae; TY—Typhaceae; UR—Urticaceae; VE—Verbenaceae; WI—Winteraceae.

[b]Species (and 1 subspecies) listed by Zimmerman (1948) in *Kelisia* are not members of the Kelisiinae and belong in Delphacini (Asche, personal communication).

[c]Listed as *Toya hispijimena* Asche, a synonym of *T. tuberculosa* (Distant) (Asche, personal communication).

2

Planthopper/Plant Interactions: Feeding Behavior, Plant Nutrition, Plant Defense, and Host Plant Specialization

Anthea G. Cook and Robert F. Denno

Introduction

Host plant allelochemistry and nutrition have figured prominently in understanding the feeding behavior, host plant range, population dynamics, and life history strategies of phytophagous insects (McNeill and Southwood 1978; Rosenthal and Janzen 1979; Denno and McClure 1983). For sap-feeding insects such as delphacids, host plant nutrition is thought to influence host plant selection, performance, and population growth and dynamics (Dixon 1970; Waloff and Solomon 1973; McNeill and Southwood 1978; Denno et al. 1980; Waloff 1980; Prestidge and McNeill 1982; Denno 1983, 1985a; Denno and Roderick 1990). However, far less information is available on the role of secondary plant chemistry in sap-feeder/plant interactions (Denno and Roderick 1990). Two factors have undoubtedly contributed to this historical emphasis on host plant nutrition. First, phloem-feeding planthoppers are thought to avoid many allelochemicals that are compartmentalized in nonvascular tissues (Waloff 1980; Sogawa 1982). Also, complex secondary chemicals such as glucosinolates, terpenoids, alkaloids, flavonoids, and phenolics are either absent or occur in low concentrations in grasses and related monocots, the most often recorded host plants of planthoppers (Culvenor 1970; Butler and Baily 1973; Harborne and Williams 1976; McNeill and Southwood 1978; van Etten and Tookey 1979; Waloff 1980; Prestidge and McNeill 1983). Nevertheless, some species of delphacids restrict their feeding to host plants rich in allelochemicals with known antiherbivore properties. For example, species in the North American genus *Stobaera* feed exclusively on Asteraceae

(Kramer 1973), a family of plants rich in sesquiterpenes and other allelochemicals (Mabry and Gill 1979). Furthermore, of the more than 135 species of Hawaiian delphacids, most feed on dicotyledonous trees and shrubs (Giffard 1922; Zimmerman 1948) with diverse secondary chemistries (Rosenthal and Janzen 1979). Improved understanding of the feeding physiology, host plant relationships, and population biology of dicot-feeding delphacids will undoubtedly require a greater emphasis on plant secondary chemistry.

In this chapter, we describe the feeding behavior and feeding niche of planthoppers and how they are influenced by plant physiology through the medium of phloem sap. Further, we provide evidence that host plant chemistry plays a major role in the population biology of planthoppers by dictating patterns of host plant selection, performance and fitness, and large-scale population dynamics including migration. Last, host plant defenses and dietary specialization are discussed with particular reference to the chemical properties of crop varieties resistant to planthopper attack.

Feeding Behavior

Planthoppers feed by inserting their stylets into the vascular tissue of plants and ingesting the sap (Sogawa 1977, 1982). Sustained ingestion occurs when planthoppers respond positively to feedback from a series of predominantly chemical cues experienced by the insect during the feeding process. A wide variety of peripheral sensory structures are associated with the stylets, labium, precibarium, and tarsi of planthoppers which provide information concerning the acceptability of the host plant. For a detailed discussion of the sensory physiology associated with feeding in planthoppers, see works of Sogawa (1977, 1982) and Backus (1985). Following **host plant selection**, the feeding behavior of planthoppers can be divided into three sequential and stereotyped behavioral phases, **plant surface exploration**, **stylet probing of plant tissues**, and **plant fluid ingestion** (Sogawa 1982; Backus 1985). In the sections that follow we (1) discuss factors influencing host plant colonization and the feeding behavior of planthoppers and how they are mediated by plant physiology, (2) provide evidence for phloem feeding, and (3) characterize the food and honeydew of planthoppers.

Host Plant Selection

For many phytophagous insects, visual and chemical cues play an important role in host plant selection (Harborne 1982; Dixon 1985). However, there is very little evidence to suggest such factors influence host plant location and colonization in planthoppers, even though most delphacid species are very host-specific (Denno and Roderick 1990). For example, despite the dependence of *N. lugens* on rice, the random fallout of migrants over plots of rice, other crops, and plowed nonirrigated fields in the Philippines was demonstrated by the similarity in abundance and periodicity of trapped macropters (Cook and Perfect 1985a). Also, there was no evidence to suggest that the visual stimulus of a rice crop elicited a landing response to migrating insects (Cook and Perfect 1985a). Although, in laboratory experiments, Saxena and Pathak (1979) showed that *N. lugens* was attracted to green color, high humidity, and the odor of plant extracts.

Host plant selection in planthoppers may conform to the scenario for aphids which also tend toward host specificity [Dixon 1985; see Kogan (1977) for a review of host selection strategies]. Aphids respond predominantly to visual cues showing a marked orientation to plant color, a cue which is a good indicator of the nutrient status of the host plant (Kennedy et al. 1961; Dixon 1985). Despite the distinctive odor of many plants, there is little unequivocal evidence that aphids use plant odor as the primary cue for host location (Pettersson 1970; Chapman et al. 1981; Dixon 1985). The accumulation of aphids on their host plants is due largely to reduced emigration from these plants (Kennedy et al. 1959; Dixon 1985). Departure rates are determined by the aphid's response to contact with the host during surface exploration, probing, and feeding (Dixon 1985). Thus, although host colonization is not random, it does not appear to be very actively directed by host odor. Host plant specificity results because insects selectively emigrate from nonhost vegetation, not because they fail to colonize it. Similarly, macropters of *Prokelisia marginata* emigrate at a higher rate from nutritionally inferior plants than they do from nitrogen-enriched ones (Roderick unpublished data).

Plant Surface Exploration

Upon arrival on the host plant, planthoppers explore the surface of a potential host plant by repeatedly tapping the tip of the labium against the plant surface (Sogawa 1973; Port 1978; Backus 1985). Electron microscopy has shown that the tip of the labium of *Nilaparvata lugens* bears several sensillae with both mechanoreceptor and chemoreceptor functions (Foster et al. 1983a). Sogawa (1982) suggests that labial tapping allows *N. lugens* to identify appropriate probing sites on the leaf surface. Specifically, labial tapping allows planthoppers to differentiate the smooth epidermis over the vascular bundles from the less suitable interveinal epidermis which is rough and covered by waxy scales. On susceptible rice varieties, 91% of stylet insertions occurred beside veins, whereas only 9% occurred between veins (Sogawa 1973). Similar studies on *N. lugens* showed that stylet insertion occurred over veins and between veins with similar frequency, though probes originating nearer to veins were more successful in locating vascular bundles (F.M. Kimmins unpublished data).

During surface tapping, information concerning the chemical nature of the plant surface waxes is probably obtained via the mouthpart chemoreceptors and a decision to probe is made. In *N. lugens*, small amounts of watery saliva are secreted during labial tapping and it seems likely that chemical components of the waxy surface dissolve in the saliva and are detected either by chemoreceptors on the labial tip or in the cibarial cavity if the saliva is taken up in the food canal (Foster et al. 1983b). Analytical studies by Cook and coworkers are in progress to identify the components of the surface wax which induce labial tapping and influence probing decisions. Preliminary results suggest that molecular size may be important because longer carbon chain compounds ($>C_{20}$) are associated with increased insect activity and surface exploration.

Stylet Probing and Evidence for Phloem Feeding

After location of a suitable probing site, the labium is firmly appressed to the plant surface and the stylets are inserted (Sogawa 1982; Backus 1985). During the course of stylet penetration, coagulable saliva (lipoprotein) is secreted, forming a tubular "stylet sheath" that bundles the stylets together during the probe and serves in their lubrication and directional control (Sogawa 1982; Backus 1985). Upon withdrawal of the stylets, the sheath remains in the plant tissue, leaving a record of the exact feeding site. The stylet sheaths of planthoppers may be unbranched or branched; the branched configuration of the stylet sheaths of *N. lugens* suggests that stylet insertion occurs repeatedly and changes directions through the initial point of entry (Sogawa 1982). Recent studies using electron microscopy have shown that probing is intracellular, causing considerable damage to plant cell organelles and cell death (Spiller 1990).

Early studies to identify the feeding site of planthoppers involved locating the terminal end of the stylet sheath within plant tissues using histological techniques (Sogawa 1982). Such studies have shown that planthoppers (*Laodelphax striatellus*, *Nilaparvata lugens*, *Peregrinus maidis*, and *Saccharosydne saccharivora*) feed predominantly in the phloem, although some stylet penetration occurs in other tissues as well (Metcalfe 1969; Sogawa 1973; Sonku and Sakurai 1973; Fisk et al. 1981). The anatomy of the precibarium of planthoppers is also consistent with the phloem as the primary tissue in which feeding occurs. The dilator muscles of planthoppers are small and the precibarium is less heavily sclerotized compared to those of the xylem-feeding Auchenorrhyncha (Backus 1985). The underdeveloped nature of precibarium and associated musculature suggests that they function to control inflow of the highly pressurized phloem sap via the precibarial valve rather than to pump against the negatively pressurized xylem sap (Backus 1985).

More recent attempts to identify specific feeding sites have utilized electronic recording systems. An electrical circuit is passed through a "wired" insect while it feeds. Characteristic waveform patterns produced during ingestion can be correlated with stylet location in specific tissues by immediately microsectioning leaves and locating the terminal end of the stylet sheath (McLean and Kinsey 1967; Tjallingii 1978). Once the ingestion waveforms from the various tissues are characterized, continuous monitoring will allow for the determination of the amount of time spent feeding in the different tissue types. By electronically monitoring the feeding of *Perkinsiella saccharicida* and *P. vitiensis* on sugarcane, Chang (1978) and Chang and Ota (1978) showed that the duration of phloem ingestion was significantly longer than ingestion from either xylem or parenchyma tissues.

In a detailed study of the probing behavior of *N. lugens* on rice, Kimmins (1989) indirectly correlated two waveform patterns with ingestion from phloem and xylem tissues by simultaneously recording honeydew production using a video technique. Honeydew produced during one of the ingestion patterns was basic, contained amino acids, and was thought to be derived from phloem. Honeydew produced during the second ingestion pattern was acidic, contained no amino acids and was probably derived from xylem. On susceptible rice varieties, ingestion from both phloem and xylem occurs (both waveforms are produced), but the duration of "phloem ingestion" is significantly longer.

Chemical Composition of Phloem Sap

New techniques being developed to collect pure samples of phloem sap have shed light on the chemical composition of the food source of planthoppers. Fukumorita and Chino (1982) collected rice phloem sap for analysis as it exuded from the laser-severed stylets of *N. lugens*. Sucrose was the only sugar present (17–25%, w/v) as were free amino acids (3–8%, w/v). Of the amino acids, asparagine, glutamate, serine, glutamine, threonine, and valine were dominant, and cystine and methionine were detectable. The inorganic ions potassium, phosphorus, sodium, sulphur, magnesium, calcium, and iron were identified and are listed in decreasing order of concentration (Fukumorita et al. 1983). Most of the phosphorus present occurred as orthophosphate. Hayashi and Chino (1986) used the same technique to collect wheat phloem for chemical analysis by laser-cutting the stylets of *L. striatellus*. They found that wheat sap also contained high concentrations of sucrose and that the major anion was potassium. Dominant amino acids were glutamic acid, aspartic acid, and serine. Sucrose was identified as the only sugar present in phloem in earlier studies where aphid stylectomy was used to collect phloem samples from *Ricinus* (Hall and Baker 1972), tobacco (Hocking 1980), and pea (Barlow and Randolph 1978).

Chemical Composition of Honeydew

The chemical composition of planthopper honeydew is also consistent with a phloem-feeding habit. During sustained ingestion, planthoppers produce copious quantities of honeydew (Sogawa 1977, 1982; Woodhead and Padgham 1988). The honeydew of *N. lugens* contains 2–5% carbohydrates. Glucose, fructose, sucrose, and oligosaccharides are present, and most of the carbohydrates occur as soluble polysaccharides (Sogawa 1982). Free amino acids (aspartic and glutamic acids) are also present as are amides, and there is a high concentration of potassium (Sogawa 1982). That the honeydew of planthoppers is a clear droplet rich in sucrose and potassium (common constituents of phloem sap) further attests to phloem feeding (Sogawa 1977, 1982).

Collectively, data on the feeding behavior and morphology of delphacids strongly suggests that they feed primarily in the phloem. Planthoppers probe on nonpreferred plants (e.g., nonhosts and resistant varieties) and insert their stylets into the phloem, but rejection occurs usually only after tasting the phloem sap (Sogawa and Pathak 1970; Sogawa 1982). In two cases (Fisk 1980; F.M. Kimmins unpublished data), varietal resistance was associated with the inability of planthoppers to locate the phloem during probing, suggesting that nonphloem factors may influence feeding decisions. Nonetheless, there is much evidence that the chemical composition of the host plant, particularly that of the phloem, influences probing and ingestion decisions and we address this issue now.

Plant Chemicals Influencing Probing and Ingestion

Host plant nutrition dramatically influences the feeding behavior of planthoppers. For example, *Nilaparvata lugens* probes more frequently, ingests much less phloem sap, and excretes much less honeydew when feeding on nitrogen-deficient plants (Sogawa 1970b, 1977, 1982; Cheng 1971). Tissues are probed and rejected, suggesting that ingestion is controlled by a suite of gustatory stimuli incurred either during the probing process or after sampling the phloem. A variety of probing stimulants isolated from rice have been identified for *N. lugens* using artificial diets. These include flavonoids, C-glucosylflavones (Sogawa 1976; Obata et al. 1981; Kim et al. 1985; Besson et al. 1985), and salicylic acid (Sekido and Sogawa 1976).

Although probing stimulants are not clearly nutritional in nature, many ingestion stimulants (= sucking stimulants of Sogawa) are. For example, certain amino acids, sucrose, and organic acids (succinic and malic) are potent ingestion stimulants in *N. lugens* (Sakai and Sogawa 1976). Of the amino acids tested, aspartic acid, glutamic acid, alanine, asparagine, and valine stimulated ingestion and the two dicarboxylic acids had a particularly strong effect (Sogawa 1982). Increases in these stimulatory amino acids in fertilized rice plants may account for decreased probing and increased ingestion rates (Sogawa 1982). Other chemicals have been identified which deter or inhibit ingestion in *N. lugens*. These include silicic acid (Yoshihara et al. 1979a) and oxalic acid (Yoshihara et al. 1979b, 1980). Silicic acid, though present in rice plants at concentrations which would inhibit ingestion, does not deter feeding on intact plants because it is not present in phloem sap; it may deter ingestion from the parenchyma where it is localized (Yoshihara et al. 1979a). Phenolic acids in sorghum appear to inhibit phloem location and subsequent ingestion in *Peregrinus maidis* (Fisk 1980).

Role of the Host Plant Nutrition in Planthopper Ecology

Compared to insects, plant tissues contain a relatively low proportion of protein (Southwood 1973; McNeill and Southwood 1978). Insects contain 7–14% nitrogen, whereas plants rarely reach concentrations of 7% nitrogen, and concentrations are generally much lower (2.1% average for nearly 400 species of plants) (Russell 1947; McNeill and Southwood 1978; Mattson 1980). Furthermore, for insects to be raised successfully on artificial diets, diets must contain 4–5% nitrogen (McNeill and Southwood 1978). These considerations have led ecologists to conclude that plant foliage is only marginally adequate for phytophagous insects and that nitrogen is scarce and perhaps limiting for many (Mattson and Addy 1975; McNeill and Southwood 1978; Mattson 1980; Strong et al. 1984; Denno 1985a). The problem may be compounded for females whose nitrogen demand is particularly high during periods of egg maturation (Waloff 1980; Prestidge and McNeill 1982). The negative effects of low plant nitrogen on individual fitness and population growth are widely documented for planthoppers and other families of Homoptera (Carrow and Betts 1973; McClure 1979,

1980a; Waloff 1980; Dixon 1985; Denno and Roderick 1990). Thus, it comes as no surprise that host plant nitrogen has been called on to explain the host plant interactions and population biology of planthoppers (Waloff 1980; Prestidge 1982a, 1982b; Prestidge and McNeill 1982, 1983; Denno 1983). In this section, we review the adverse effects of low plant nitrogen on the performance, fitness, and population growth of planthoppers and discuss several behavioral, morphological, and physiological adaptations that allow planthoppers to cope with low concentrations of ambient nitrogen.

Host Plant Nitrogen, Planthopper Performance, and Population Growth

In general, planthoppers survive better [*Nilaparvata lugens* on rice (Cheng 1971); *Peregrinus maidis* on sorghum (Fisk et al. 1981)], develop faster [*Prokelisia dolus* on cord grass (Denno unpublished data)], molt into larger adults [*Prokelisia marginata* on cord grass (Denno and McCloud 1985; Denno et al. 1986; Fig. 2.1A)], and are more fecund [*Saccharosydne saccharivora* on sugar cane (Metcalfe 1970; Fig. 2.1B); *N. lugens* on rice (Sogawa 1971a); *Dicranotropis hamata* on *Holcus* (Prestidge 1982a, 1982b)] if they develop on nitrogen-rich host plants. These increases in performance and fitness are associated with increased feeding rates on nitrogen-rich host plants (Sogawa 1982). Moreover, increased plant nitrogen tends to moderate the adverse effects of high population density on planthopper fitness (Denno et al. 1985, 1986). Following fertilizer application, host plant amino acids such as aspartic and glutamic acid [known feeding stimulants for planthoppers (Sogawa 1982)] increase dramatically in concentration (Prestidge and McNeill 1982; Denno unpublished data) and may be responsible for accelerated feeding.

In laboratory choice tests, planthoppers (*N. lugens* and *P. marginata*) select nitrogen-rich over nitrogen-poor plants on which to feed and oviposit (Sogawa 1970a; Denno 1985b). In the field, macropterous adults of *P. marginata* accumulate rapidly and emigrate less from nitrogen-rich host plants (Denno et al. 1980; Denno 1983; G. Roderick unpublished data; Fig. 2.2). Fertilization of the host grasses of planthoppers results in elevated levels of amino nitrogen, more robust plants with increased living space, and more favorable habitats (Nishida 1975; Andrzejewska 1976; Dyck et al. 1979; Prestidge 1982b; Denno and Roderick 1990). The combined effects of increased colonization and improved performance often result in rapid population growth and higher densities of planthoppers on host plants enriched with nitrogen by fertilization (Denno and Roderick 1990). Specific examples of planthoppers that show positive population increases to

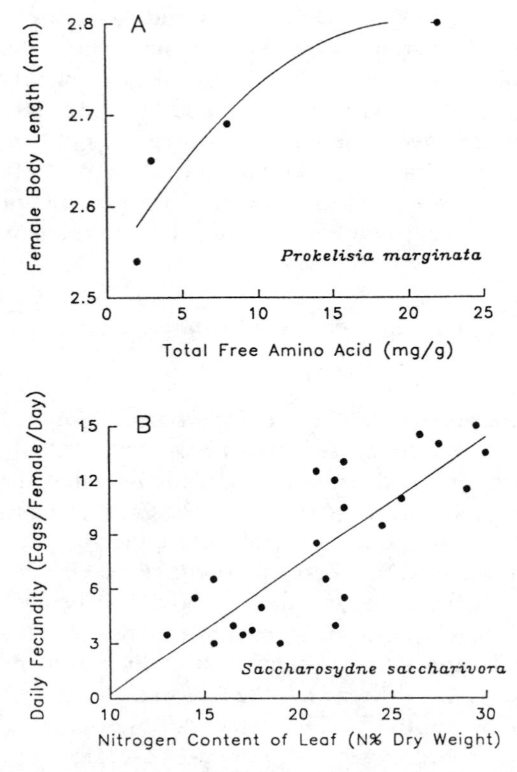

Figure 2.1. (**A**) Relationship between the body length (mm) of female *Prokelisia marginata* and the total free amino acid content (mg/g) of the host plant, *Spartina alterniflora*, on which they were raised [see Denno et al. (1986) for details on the experimental establishment of differing levels of host plant amino nitrogen by fertilization]. (**B**) Relationship between the daily fecundity (eggs/female/day) of *Saccharosydne saccharivora* and the nitrogen content of the leaves of (N% dry weight) of sugar cane; $Y = -6.82 + 0.71X$, $R^2 = 0.67$, $P<0.001$ (from Metcalfe 1970).

nitrogen fertilization under field conditions include *Dicranotropis hamata* (Prestidge 1982b), *Javesella pellucida* (Prestidge 1982b), *Nilaparvata lugens* (Cheng 1971; Kanno et al. 1977; Dyck et al. 1979; Heinrichs and Medrano 1985; Hosamani et al. 1986; Hu et al. 1986), *Prokelisia dolus* (Denno unpublished data), and *Prokelisia marginata* (Denno 1983, 1985b). Furthermore, nitrogen fertilization has been implicated as a cause for population outbreaks in pest planthoppers such as *N. lugens* (Dyck et al. 1979; Kenmore 1980).

Although delphacid planthoppers generally increase rapidly in abun-

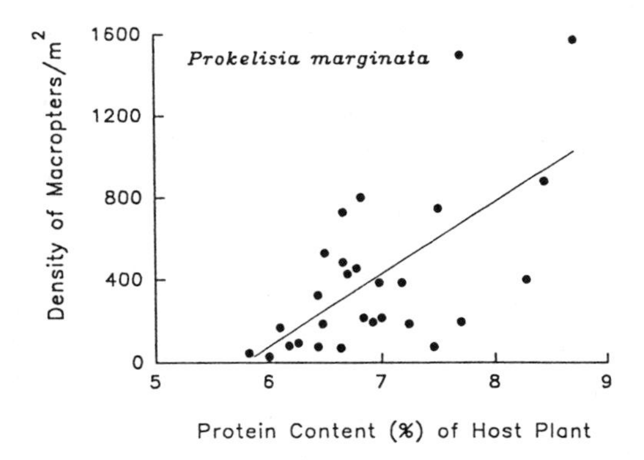

Figure 2.2. Relationship between the density of the macropters of *Prokelisia marginata* and the protein content (%) of the host plant (*Spartina alterniflora*) stand at Tuckerton, Ocean County, NJ; $Y = -2032.7 + 351.7X$, $R^2 = 0.41$, $P<0.001$ [replotted from Denno et al. (1980)].

dance on fertilized plants, not all species of Auchenorrhyncha do. In particular, many species of leafhoppers (Cicadellidae) show decreased abundance on fertilized plots of their host plants (Prestidge 1982b). Even some delphacid species (*P. marginata*) which can respond very strongly to nitrogen-enriched hosts (Denno 1983) do not always show dramatic population increases following fertilization (Vince et al. 1981; Silvanima and Strong 1991). In one case involving *P. marginata*, strong numerical responses of natural enemies (wolf spiders) can prevent the increase of planthopper populations on nitrogen-fertilized host plants (Denno unpublished data). In addition, several species of Auchenorrhyncha including the delphacid *D. hamata*, maximum nitrogen utilization efficiencies (positively correlated with daily fecundity) occur at different plant nitrogen levels and Prestidge (1982a) and Prestidge and McNeill (1982) suggest that each species may be associated with and seek out a particular concentration of plant nitrogen. Consequently, species of planthoppers may respond uniquely to nitrogen fertilization treatments and their response may be tempered by natural enemies.

Adaptations of Planthoppers to Low Host Plant Nitrogen

Because delphacid reproduction and development are so sensitive to levels of host plant nitrogen, it is not surprising to see a variety of adaptations that minimize the encounter with or buffer these insects against low host nitrogen (McNeill and Southwood 1978; Prestidge and McNeill 1982; Denno

1985a; Denno and Roderick 1990). McNeill and Southwood (1978) recognize six "strategies" (adaptations) which attest to the limiting nature of plant nitrogen for herbivorous insects (mostly sap-feeders). We have adopted their classification and provide the following evidence for these adaptations to low nitrogen in planthoppers:

1. **Increased feeding rate on nitrogen-deficient foliage.** Nymphs of *Dicranotropis hamata* show high ingestion rates and low nitrogen utilization efficiencies when leaf nitrogen is low; on nitrogen-rich plants, they show the opposite physiology (Hill 1976; Waloff 1980). Apparently, this mostly brachypterous species can partially compensate for low leaf nitrogen with increased feeding rate. Whether compensatory feeding and the ability to tolerate changing plant nitrogen are generally associated with brachypterous delphacids in persistent habitats remain to be seen (see Prestidge 1982a, 1982b; Denno et al. 1991). Nonetheless, more macropterous delphacids (*Javesella pellucida* and *Nilaparvata lugens*) which can escape deteriorating host plants show reduced feeding on nitrogen-deficient plants (Sogawa 1977, 1982; Prestidge 1982a).
2. **Selection of nitrogen-rich feeding sites on plants.** A survey of the feeding sites of grass-feeding species shows them to be located at the base of the plant near the crown or in the axil at the leaf–blade junction (Table 2.1). Such feeding sites are the locations of actively growing meristems where high concentrations of nitrogen occur (Langer 1979; Mattson 1980), but they may also provide protection from natural enemies (Denno and Roderick 1990). Similarly, dicot-feeding delphacids selectively feed on young leaves, buds, and succulent stems (Table 2.1), where nitrogen concentrations are high (Mattson 1980; Denno 1985a). Furthermore, *Dicranotropis hamata* (Hill 1976), *Laodelphax striatellus* (Noda 1986b), *Prokelisia marginata* (Denno et al. 1980), and *Sogatella furcifera* (Noda 1986b) all move from the leaves to the more nitrogen-rich inflorescences and seed heads of their host plants as they begin to flower.
3. **Synchronization of life cycle with periods of increased plant nitrogen.** Seasonal changes in the soluble nitrogen content of grasses follow a distinct pattern (McNeill and Southwood 1978; Waloff 1980; Prestidge and McNeill 1982, 1983; Denno unpublished data). In *Holcus lanatus*, *H. mollis*, *Lolium perenne*, and *Spartina alterniflora*, levels of free amino acids rapidly increase for a short period in spring as metabolites are translocated to the growing points. Subsequently, levels of soluble nitrogen decrease during summer and then rise again

Table 2.1. Feeding location of the nymphs (N) and adults (A) of delphacid planthoppers on their host plants.

Species	Host Plant	Plant Stratum	Plant Part	Reference
Grass and Sedge Feeders				
Delphacodes detecta	*Spartina patens*	Upper (N & A)	Convoluted leaf blade	Denno (1980)
Eoeurysa flavocapitata	*Saccharum officinarum*	Upper (N & A)	Young leaf sheath	Chatterjee and Choudhuri (1979)
Javesella pellucida	Polyphagous	Lower (N)	Stem	Kanervo et al. (1957); Raatikainen (1967)
Laodelphax striatellus	*Oryza sativa*	Upper (N & A)[a,b]	Upper leaf surface	Kisimoto (1956b); Noda (1986b)
Megamelus lobatus	*Spartina patens*	Lower (N & A)	Sheath and tillers	Denno (1980)
Megamelus metzaria	*Spartina pectinata*	Lower (N & A)	Sheath and tillers	Holder and Wilson (1992)
Neomegamelanus dorsalis	*Spartina patens*	Upper (N & A)	Convoluted leaf blade	Denno (1980)
Nilaparvata lugens	*Oryza sativa*	Lower (N & A)	Leaf sheath	Kisimoto (1956b); Sogawa (1982)
Peregrinus maidis	*Zea mays*	Lower (N & A)	Upper leaf surface, axil	Napompeth (1973)
Perkinsiella saccharicida	*Saccharum officinarum*	Lower (N & A)[a]	Upper leaf surface, axil	Williams (1957); Fennah (1969); Osborn (1974)
Perkinsiella vastatrix	*Saccharum officinarum*	Lower (N)	Upper leaf surface	Fennah (1969)
Prokelisia crocea	*Spartina pectinata*	Lower	Upper leaf surface, axil	Holder and Wilson (1992)
Prokelisia dolus	*Spartina alterniflora*	Mid-Lower (N & A)	Upper leaf surface, axil and basal 1/3	Denno and Cheng unpublished data
Prokelisia marginata	*Spartina alterniflora*	Mid-Upper (N & A)[b]	Upper leaf surface, axil and basal 1/3	Denno et al. (1980); Denno and Cheng (unpublished data)
Sacchar. saccharivora	*Saccharum officinarum*	Lower (N & A)[a]	Upper leaf surface, axil	Fennah (1969)
Sogatella furcifera	*Oryza sativa*	Upper (N & A)[a,b]	Upper leaf surface	Hinckley (1963); Noda (1986b)
Stenocranus lautus	*Carex lurida*	Lower (N)	Lower leaf surface, stems	Calvert and Wilson (1986)
Tagosodes orizicolus	*Oryza sativa*	Lower (N & A)[a]	Upper leaf surface, whorl	Everett (1969); King and Saunders (1984)

Toya propinqua	Polyphagous	Lower (N & A)	Sheaths (N & A), blades (N)	Raatikainen and Vasarainen (1990)
Tumidagena minuta	*Spartina patens*	Lower (N & A)	Sheaths and tillers	Denno (1980)
Broad-Leaved Monocot Feeders				
Megamelus davisi	*Nuphar lutum*	Lower (N & A)	Lower leaf surface Upper leaf surface, stems	S. Wilson (personal communication)
Megamelus palaetus	*Pontederia cordata*	Lower (N & A)	Lower leaf surface	S. Wilson (personal communication)
Tarophagus prosperpina	*Colocasia esculenta*	Lower (N & A)	Young leaves (axils)	Mitchell and Maddison (1983); Matsumoto and Nishida (1966)
Dicot Feeders				
Nesodryas freycinetiae	*Freycinetia arborea*	?	Young leaves, lower surface (N)	Zimmerman (1948)
Nesosydne gunnerae	*Gunnera petaloidea*	Lower (N & A)	Mature leaves (Midrib), Lower surface	Zimmerman (1948)
Nesosydne koae	*Acacia koa*	Upper (N & A)	Young leaves	Zimmerman (1948)
Stobaera concinna	*Ambrosia artemisiifolia*	Upper (N & A)	Veins of lower leaf surface (N_{1-3}) small stems (N_{4-5} & A)	Calvert et al. (1987b)
Stobaera pallida	*Baccharis halimifolia*	Upper (N & A)	Young leaves, buds	Denno (1978)
Stobaera tricarinata	*Ambrosia psilostachya*	Upper (N & A)	Veins of lower leaf surface (N_{1-2}) Stems (N_{3-5}); Upper leaf surface and succulent stems (A)	Reimer and Goeden (1982)

[a] Adults also feed in leaf whorls in the upper crown.
[b] Adults colonize seed heads.

at the end of the growing season when tillering and senescence occur. In particular, aspartic acid, glutamic acid, proline, and threonine show sharp and well-defined peaks in spring and fall (Prestidge and McNeill 1982).

Hill (1976) determined the nitrogen budgets for several leafhopper and planthopper species over the life cycle. He concluded that plant nitrogen levels were usually adequate to ensure nymphal development and adult survival during most of the year, but that high nitrogen demand during periods of adult reproduction (egg maturation) was most likely to result in nitrogen stress. Thus, selection should favor species which synchronize periods of reproduction with times when levels of host plant nitrogen are high (Prestidge 1982a, 1982b; Prestidge and McNeill 1983). In support of this prediction, *Stenocranus minutus* overwinters in the adult stage and females do not develop eggs until spring when they can take advantage of the nutrient flush (Prestidge and McNeill 1983). Certain multivoltine species (*Delphacodes detecta, Neomegamelanus dorsalis,* and *Prokelisia marginata*) also show synchrony in the spring occurrence of adults with peak nitrogen levels in their host grasses (Denno et al. 1981; Denno and Roderick 1990). Adult occurrence in *Dicranotropis hamata* is generally associated with the spring flush of nitrogen in *Holcus mollis*, but on the occasional year when adults are not well synchronized with peak plant nitrogen, reproduction is drastically reduced (Hill 1976; Waloff 1980). For some planthoppers, however, the advantages gained by synchronizing reproduction with periods of peak plant nutrition may be offset by selective pressures imposed by physical factors (spring freezes) and/or natural enemies (see Denno et al. 1981). Thus, the reproductive phenology of some planthopper species may be compromised by other factors and bouts of oviposition may not always correspond to flushes of plant nitrogen.

4. **Dispersal to more nutritious host plants.** Some species of planthoppers spatially synchronize their life histories with high host nitrogen by migrating to more nutritious plants. For example, macropters of *Prokelisia marginata* disperse annually from high to low marsh habitats where they encounter a more nitrogen-rich and robust growth form of the host grass, *Spartina alterniflora* (Denno and Grissell 1979; Denno 1983). By doing so, the offspring of the colonists emerge as large, very fecund adults which contribute to rapid population growth; adults remaining in the less nutritious *Spartina* meadows on the high marsh produce small progeny with reduced fecundity (Denno and McCloud 1985; Denno et al. 1985, 1986). The significant

relationship between the density of macropterous adults and the nitrogen content of *Spartina* across marsh habitats further suggests that *P. marginata* disperses and selects nutritious hosts on which to feed (Denno et al. 1980). Macropters of this species also selectively colonize salt-stressed *Spartina* plants which occur on the edges of mud flats and have elevated levels of free amino acids, particularly proline (Cavalieri and Huang 1981; Denno unpub. data).

Adult density of the highly polyphagous *Javesella pellucida* is also significantly related to the leaf nitrogen content of its hosts (Prestidge and McNeill 1983). In contrast, the adult densities of other planthopper species (*Ribautodelphax angulosa* and *Stenocranus minutus*) are not significantly correlated with the nitrogen content of their host plants (Prestidge and McNeill 1983). Prestidge and McNeill (1982) argue that responses to host plant nitrogen have selected for two discrepant life history styles in planthoppers and leafhoppers. First, there are species with highly specific nitrogen requirements which are very mobile (highly macropterous) and actively track specific nutritional requirements. *P. marginata* and *J. pellucida* are two highly mobile species whose behavior is suggestive of this pattern. Moreover, Prestidge (1982a) suggests that selection has promoted polyphagy in *J. pellucida* by favoring highly mobile individuals which switch hosts as they track a particular level of nitrogen. Second, there are species which are more tolerant of fluctuating levels of host plant nitrogen, but are closely tied to the phenology of their host plants (Prestidge and McNeill 1982). These species (e.g., *D. hamata*, *R. angulosa*, and *S. minutus*) tend to be much less mobile (brachypterous) and more monophagous in their feeding habits.

5. **Modification of plant nutritional physiology.** Heavy infestations of *N. lugens* on rice plants induce proteolysis and a 30-fold increase in the levels of several free amino acids (arginine, asparagine, lysine, proline, and tryptophan), whereas free sugars and moisture content decrease (Sogawa 1971b; Cagampang et al. 1974). During feeding, planthoppers leave salivary sheaths which clog the vascular tissue and interfere with translocation, resulting in the localized accumulation of free amino acids (Sogawa and Pathak 1970; Cagampang et al. 1974; Sogawa 1982). Similarly, feeding of *Prokelisia* planthoppers on *Spartina* results in significant increases in the concentration of free amino acids in the leaf blades, particularly proline and tryptophan (Bacheller 1990). Feeding on plants rich in amino nitrogen can result in higher fecundity for both *N. lugens* (Sogawa 1971a) and *P. marginata*. Thus, some species of planthoppers appear to modify plant nutritional phys-

iology to their own advantage by creating "nitrogen sinks," a phenomenon reported for aphids (Way and Cammell 1970). The aggregation behavior of many planthopper species during feeding (Williams 1957; Matsumoto and Nishida 1966; Fennah 1969; Metcalfe 1969; Sogawa 1982; King and Saunders 1984) may further facilitate the local accumulation of free amino acids.

However, the advantages of induced changes in plant nutritional physiology under high-density conditions are short-lived and are likely to benefit adults much more than nymphs. Even though there is an immediate increase in free amino acids on heavily infested plants, plants deteriorate rapidly under these conditions (Cagampang et al. 1974). Macropterous adults, especially females maturing eggs, stand to benefit by feeding on "nitrogen-induced" plants because they can take immediate advantage and then disperse to other plants. Nymphs, however, are relegated to the declining plant by virtue of their immobility and may suffer the consequences of a long development period on a rapidly deteriorating host. This argument may explain why adults of *P. marginata* select "stressed plants" high in amino nitrogen on which to feed, but prefer to oviposit on "healthy plants" (Denno and Cheng unpublished data).

6. **Nutrients from other sources.** Planthoppers possess yeast-like endosymbiotes (*Laodelphax striatellus*) dispersed throughout the fat body and rickettsia-like prokaryotic symbiotes (*Nilaparvata lugens*) occurring either extracellularly or in mycetocytes (Noda 1977; Chen et al. 1981a). Endosymbiotes are thought to provide nutrients otherwise rare or unavailable in host plants (Noda and Saito 1979a; Noda et al. 1979; Brooks 1985; Campbell 1989). The nutritional contribution of endosymbiotes to planthoppers has been assessed by the use of artificial diets and by direct comparison of the metabolism of insects treated with antibiotics, lysozyme, or heat to eliminate the endosymbiotes (Noda et al. 1979; Chen et al. 1981b; Campbell 1989). Treated planthoppers generally show adverse effects on nymphal development, molting, and fecundity, suggesting that endosymbiotes supply essential nutrients (Noda and Saito 1979b; Chen et al. 1981b). Sterols are strongly implicated as the essential nutrient synthesized by the yeast-like endosymbiotes in *Laodelphax striatellus* (Noda et al. 1979).

No data are available for planthoppers though studies on aphids suggest endosymbiotes of sap-feeding insects can provide their hosts with essential amino acids that may be deficient in their food source (Campbell 1989). For example, evidence suggests that tryptophan, methionine, and cysteine present are synthesized by endosymbiotes in some aphid species, and lipids (particularly sterols) can also be supplemented by endosymbiotes in Homoptera (reviewed by Campbell 1989).

Host Plant Resistance

Although there is an enormous effort to develop resistant crop varieties for use in planthopper pest management (Khush 1979; Claridge and Den Hollander 1982), there is very little information available on the actual mechanisms of resistance. It is known that planthoppers probe much more frequently, ingest much less and excrete very little honeydew on resistant varieties compared to susceptible ones (Sogawa 1977, 1982; Khan and Saxena 1984, 1985; Cook et al. 1987; Kimmins 1989). What little information exists on the chemical and physical factors involved in plant resistance to planthoppers comes from studies of varietal resistance in rice and sorghum (see Denno and Roderick 1990). There is evidence suggesting that the chemical mechanisms conferring resistance are diverse and may affect most aspects of planthopper feeding behavior such as plant surface exploration, probing or ingestion (Sogawa 1977, 1982).

For example, rejection of a resistant plant may occur during surface exploration prior to probing and ingestion. On a moderately resistant variety of rice (IR46), *N. lugens* did not probe after surface tapping, but frequently moved to other sites where surface exploration resumed (Cook et al. 1987). On susceptible rice varieties, the insects usually followed surface tapping by prolonged periods of inactivity during which probing and ingestion occurred. Woodhead and Padgham (1988) removed the wax layer from the leaf surface of a resistant rice variety (IR46) and applied it to a susceptible variety (IR22). The rejection response of *N. lugens* was simulated on the susceptible plants laminated with the surface waxes of the resistant variety. Both of these studies implicate a resistance factor associated with the surface wax which acts during surface exploration.

Although there is some evidence suggesting that the decision to accept or reject a host plant occurs prior to probing and ingestion, most research suggests that the major resistance factor to planthoppers is associated with phloem chemistry. For example, adults of *N. lugens* readily settle on some resistant varieties and effectively insert their stylets into the phloem, but rejection occurs only after ingesting the phloem sap (Sogawa and Pathak 1970; Sogawa 1982). Thus, on most resistant varieties there does not appear to be any physical or mechanical interference preventing stylet penetration of the phloem. For instance, a similar proportion of stylet sheaths of *N.*

lugens terminate in the phloem on susceptible and resistant varieties of rice, and the degree of stylet sheath branching is not related to varietal resistance (A. Cook and S. Woodhead unpublished data). However, the stylet sheaths of *N. lugens* fail to locate the vascular bundle and terminate in the air space within the leaf lumen on some wild rice varieties (F.M. Kimmins unpublished data), suggesting that varietal resistance may involve the absence of specific probing stimulants which facilitate phloem location.

In only a few cases have specific allelochemicals been implicated in varietal resistance to planthoppers. Oxalic acid which inhibits ingestion in *N. lugens* (Yoshihara et al. 1980) is twice as concentrated in resistant compared to susceptible varieties of rice (Yoshihara et al. 1979b). However, confirmation of the presence of free oxalic acid in the phloem sap is necessary before its causal role in varietal resistance can be rigorously established. Also, the low concentration of asparagine in the rice variety Mudgo has been implicated in its resistance to *N. lugens* (Sogawa and Pathak 1970). Varieties of sorghum with high levels of phenolic acids are resistant to attack from settling adults of *P. maidis* (Fisk 1980). High concentrations of phenolic acids in resistant varieties are correlated with a high frequency of branched stylet sheaths and the inability this planthopper to locate and ingest from the phloem (Fisk 1980).

Factors associated with the surface characteristics and/or phloem chemistry of resistant rice varieties increase the activity level and movement of planthoppers (Woodhead and Padgham 1988) and probably contribute to higher emigration rates and a redistribution of macropters in the field. However, because flight fuels in *N. lugens* build up only on acceptable hosts (Padgham 1983a, 1983b), the migratory ability of this species may be reduced by extended contact with resistant varieties. Any mechanism reducing dispersal and forcing contact with a resistant host genotype should accelerate adaptation (virulence). That small resident populations of *N. lugens* can occur on very resistant varieties such as IR36 (Padgham 1983a) may provide evidence for this phenomenon.

Plant Defense and Host Plant Specialization

Although some species of planthoppers feed on dicots and broad-leaved monocots, most species (74%) are monophagous on grasses and sedges (Waloff 1980; McClay 1983; Chu and Yang 1985; Claridge et al. 1985b; den Bieman 1987a; Tonkyn and Whitcomb 1987; Denno and Roderick 1990). Grasses have been categorized as apparent plants (McNaughton and Tarrants 1983) and as such are thought to employ "quantitative defenses" which reduce the digestibility of plant tissues and provide widespread protection from herbivores (Feeny 1976; Rhoades 1983). Silica and lignin, for example, can occur abundantly in many grasses and are thought to make tissues generally difficult to eat (Sasamoto 1961; Djamin and Pathak 1967; Pathak et al. 1971; Swain 1979; McNaughton and Tarrants 1983; Rhoades 1983).

For phytophagous insects with chewing mouthparts, this seems to be the case. For example, the mandibles of Lepidoptera larvae feeding on varieties of rice high in silica become severely worn and feeding efficiency is drastically reduced; feeding on low-silica varieties does not produce these adverse effects (Pathak et al. 1971). Similarly, the relative head mass of grass-feeding insects is larger than that for herbaceous feeders (Bernays 1986), and egg size in the Satyridae and Hesperiidae (Lepidoptera) is positively related to the leaf toughness of their host grass (Fukuda et al. 1984; Nakasuji 1987). Presumably, larvae hatching from large eggs have large heads and mandibles which retain their cutting and masticating capability until the mandibles are renewed at the next molt. Thus, large head and mandible size has apparently allowed some mandibulate herbivores to exploit grasses, but the price these species have paid for this dietary habit is reduced fecundity (Nakasuji 1987).

Although the life histories of some chewing insects have been adjusted for grass feeding, many mandibulate herbivores are excluded from grasses by leaf toughness. The domination of the herbivorous insect fauna of grasses by planthoppers and leafhoppers (Whittaker 1969a; Morris 1974; Denno 1980; Waloff 1980; Prestidge 1982b; Wilson and Claridge 1985) attests to the general effectiveness of silica as a defense against chewing, but not sap-feeding insects. Furthermore, most evidence from resistant crop varieties suggests that planthoppers are not excluded from feeding by any physical defense which prevents their stylets from penetrating the leaf tissue (Sogawa 1982). Nevertheless, trichomes on the leaf surfaces of dicots can act as a mechanical defense against leafhoppers and aphids by limiting access to preferred feeding and oviposition sites, by interfering with movement on the leaf surface, and by physical entrapment (Dixon 1985; Tingey 1985). Although some species of grasses posses trichomes (Anderson 1974), their small size and relatively low density compared to those on dicots is unlikely to pose much of a mechanical barrier for most planthoppers.

Allelochemicals present in plants are thought to play a major role in determining host specificity in phytophagous insects (reviewed in Futuyma 1983; Jaenike 1990). Thus, the relative scarcity of many complex allelochemicals in grasses (Culvenor 1970; Butler and Baily 1973; Harborne and Williams 1976; McNeill and Southwood 1978; Prestidge and McNeill 1983) poses a dilemma when attempting to explain monophagy in the

Delphacidae and is further compounded by the ability of phloem feeders to avoid allelochemicals that are compartmentalized in nonvascular tissues (Yoshihara et al. 1979a, 1980; Waloff 1980; Sogawa 1982). However, it may be that the diversity of allelochemicals present in grasses, though reduced, contributes to limited diet breadth. For example, cyanogenic glycosides (Woodhead and Bernays 1977; Conn 1979), alkaloids (Robinson 1979), phenolic acids (Fisk 1980), flavonoids (Sogawa 1976), organic acids (Yoshihara et al. 1979b, 1980; Sogawa 1982), and soluble silicic acid (Yoshihara et al. 1979a) occur in grasses and are known to influence feeding and/or host selection in planthoppers (Fisk et al. 1981; Sogawa 1982; Yoshihara et al. 1979a, 1979b, 1980). Furthermore, certain amino acids in grasses either stimulate or inhibit feeding in planthoppers (Sogawa 1972, 1982). Thus, grasses are not as devoid of allelochemicals as conventional wisdom might suggest.

In addition, there is some evidence that the phloem is not as deficient of allelochemicals as was once thought (Dixon 1985). For example, alkaloids, nonprotein amino acids, phenols, and cardenolides all occur in the phloem and are ingested by aphids (Eschrich 1970; Rothschild et al. 1970; Zeigler 1975; Wink et al. 1982), and the abundance of some phloem-feeding insects is inversely related to the concentration of secondary chemicals in their host plants (Argando et al. 1980; Wink et al. 1982).

It may be that monocot-feeding planthoppers are exposed to as unique plant chemistries as are dicot-feeding herbivores. As with other phytophagous insects, failure of a planthopper species to feed on a certain host plant taxon most likely results from the presence of feeding inhibitors or toxins (Yoshihara et al. 1979a, 1979b, 1980), the absence of feeding stimulants (Sakai and Sogawa 1976; Sogawa 1982), and/or inadequate plant nitrogen (Prestidge and McNeill 1982, 1983). For example, *N. lugens* is able to locate the phloem of *Oryza sativa* with its stylets, but is unable to do so on the closely related grass *Leersia hexandra*, suggesting the involvement of a specific feeding stimulant (F.M. Kimmins unpublished data).

Futuyma (1983) argues that given the ubiquity of broad-spectrum detoxifying mechanisms in phytophagous insects and the ease with which insects including planthoppers adapt to resistant crops and a variety of toxins (Claridge and Den Hollander 1982; Pathak and Heinrichs 1982; Heinrichs and Rapusas 1985), the selective reasons for dietary specialization seem less likely to lie in toxic barriers than with other ecological factors.

Examples of host range expansion and plant affiliation in the Delphacidae also suggest that toxins are not necessarily the major hurdle influencing feeding on novel hosts. There is some evidence that delphacids have increased their host range to include closely related plant taxa with presumably similar chemistries. For example, *Saccharosydne saccharivora* expanded from its native *Andropogon* host to include the introduced and closely related sugarcane cane (both are in the tribe Andropogoneae) in its feeding repertoire (Metcalfe 1969). Also, two sibling species of *Nilaparvata* occur on *Oryza sativa* and *Leersia hexandra* (Claridge et al. 1985b), two closely related grasses (Clayton and Renvoize 1986). There is evidence to the contrary as well. For example, the two salt marsh grasses, *Spartina patens* and *Distichlis spicata* are taxonomically removed [are placed in different tribes of the Chloridoideae (Clayton and Renvoize 1986), but share four species of planthoppers (Tallamy and Denno 1979; Denno et al. 1981)]. In addition, planthoppers in the genus *Megamelus* feed on host plant species in four families of monocots (Cyperaceae, Juncaceae, Poaceae, and Pontederiaceae) and the Nymphaceae, all of which are restricted to freshwater marsh habitats (Wilson et al. Chapter 1). These known instances of host range expansion and plant association are best characterized by planthoppers switching to structurally similar and abundant plants growing in a common habitat. Although close taxonomic relationship of host plants may enhance the chances for a host shift, distant taxonomic affiliation of hosts does not appear to preclude it altogether. Changes in the relative abundance of hosts in habitats to which planthoppers are restricted [many species are brachypterous (Denno et al. 1991)] may be the driving force behind host shifts.

Monophagy may evolve for a variety of reasons [reviewed in Futuyma (1983) and Jaenike (1990)]. Ovipositing on toxic plants (or on plants where performance is reduced) that cannot be distinguished from palatable species can select for host plant specialization (Levins and MacArthur 1969). Selection could favor genotypes that can survive on unsuitable hosts or it could favor genotypes capable of distinguishing unsuitable hosts (Futuyma 1983). If the latter were the case, there should be a correlation between oviposition preference and offspring performance across host plant species (Thompson 1988a). Although this correlation exists for some herbivorous insects (Thompson 1988a), it does not for others (Wiklund 1975; Futuyma 1983; Thompson 1988a; Jaenike 1990). That females of several species of *Ribautodelphax* prefer to feed and oviposit on hosts where nymphal performance is highest is suggestive of a positive correlation (den Bieman 1987a). Similarly, two sibling species of *Nilaparvata* preferred to oviposit on the host plant species on which nymphal survival was highest (Claridge et al. 1985b). One study with *N. lugens* on resistant and susceptible varieties of rice shows a correspondence between oviposition preference and nymphal performance (Choi et al. 1979), but in similar studies with *N. lugens*, *L. striatellus*, and *S. furcifera* no correlation was found (Song et al. 1974; Choi et al. 1979). A strong positive genetic correlation between preference and performance (implicated in *Ribautodel-*

phax) could foster monophagy and may preclude host shifting or polyphagy (Thompson 1988a). Nevertheless, other factors such as host plant abundance and natural enemies may influence the relationship between oviposition preference and offspring perfromance (Thompson 1988a; Denno et al. 1990; Jaenike 1990).

Feeding efficiency may be higher for monophagous insects because specialists would not incur the energy costs of tetoxifying a broad range of plant compounds (Dethier 1954). However, there is very mixed evidence for any energetic advantage resulting from host plant specialization in phytophagous insects (see Futuyma 1983). Host specificity in planthoppers and other herbivorous insects is more likely to be a consequence of behavioral responses to attractant and repellent chemicals rather than specialized physiological adaptation (see Dethier 1970; Futuyma 1983). The strong response of planthoppers to feeding and oviposition stimulants and repellents (Sogawa 1982) provides testimony for this notion. Futuyma (1983) suggests that specialization evolves first at the behavioral level, and once a population is specialized, selection will favor host-specific detoxification capability. That the potential host range of several ***Ribautodelphax*** species is wider than their actual host range (monophagous in most cases) is consistent with the scenario that host specificity results more from adult host-selection behavior than constraints of plant allelochemistry on planthopper performance (den Bieman 1987a).

Monophagy can also evolve in the absence of a choice if a particular host plant species is abundant (Hsiao 1978), as is the case for many of the grass hosts of planthoppers. For example, oviposition preferences for locally abundant hosts can evolve rapidly in aphids (Nielsen and Don 1974) and other phytophagous insects (Wasserman and Futuyma 1981). In addition, the low level of dispersal, characteristic of planthoppers on abundant hosts in persistent habitats, promotes contact with host plants (Denno et al. 1991). Under these conditions, female fecundity should not be compromized by the location of favorable hosts (Jaenike 1990), the evolution of a feeding and/or oviposition preference would accelerate and host plant specialization could result (Levins and MacArthur 1969).

For species that mate on their host plants, monophagy may facilitate mate finding by increasing the encounter rate with conspecifics (Jaenike 1990). Not only do planthoppers mate on their hosts, but mate location is dependent on acoustic signals transmitted through the plant substrate (Claridge 1985a). Furthermore, migratory planthoppers mate after dispersal and colonization of the new habitat (Denno and Roderick 1990). For these species, monophagy may be essential for mate location particularly in the absence of pheromone communication.

Host plant specialization also may provide some degree of "enemy-free space" (Price et al. 1980; Denno et al. 1990). Insects can benefit directly by sequestering plant toxins that provide protection from natural enemies (Rothschild 1973). Although some aphid species sequester toxins (Rothschild et al. 1970), the phenomenon is not reported for delphacids and is unlikely because no species are aposematic. Also, the taxonomic composition of the spider communities associated with two sympatric grasses (*Spartina patens* and *S. alterniflora*) is very different (Döbel et al. 1990). By specializing on *S. patens* (Denno 1980), several planthopper species avoid web-building spiders which are rare in *S. patens* but occur abundantly in *S. alterniflora*. Although natural enemies may shape the feeding and oviposition niches of planthoppers on a plant, there is little evidence to suggest that natural enemies are the primary factor contributing to host plant specialization in the Delphacidae. For example, most planthoppers on the Hawaiian Islands are monophagous (Zimmerman 1948) even though they have evolved in the absense of major predator groups such as native ant species (Howarth 1990).

A potentially powerful explanation of host plant specialization in insects is that adaptation to one host species reduces the potential for adaptation to another (Futuyma and Philippi 1987; Jaenike 1990). In planthoppers, however, virulence in "biotypes" of *N. lugens* is not readily lost on the original rice variety once it has been selected for on a novel variety (Claridge and Den Hollander 1982). Nevertheless, Jaenike (1990) favors genetically based trade-offs in offspring performance on different hosts as an explanation for monophagy in phytophagous insects and suggests that this hypothesis has been inadequately tested to date. The assumption of evolutionary constraints or trade-offs is not generally made with regard to behavioral adaptation (Gould 1991a). For example, a genetically based trade-off in adult host-selection behavior (e.g., sensitivity to a feeding stimulant in one plant species precludes the detection of a feeding stimulant in a different plant) could also promote monophagy, but only if the advantages of efficient host location are not offset by decreased offspring performance.

Clearly, our understanding of host plant specialization and host shifting in the delphacids, as for other phytophagous insects, is limited by an almost complete lack of data on how preference and performance are related genetically (Thompson 1988a) and if genetically based trade-offs exist in adult preference and/or offspring performance across hosts (Jaenike 1990). Specific attractants and feeding stimulants facilitate host location and recognition and act as proximate cues in the host plant selection process. Ultimately, host plant specialization is likely to result from ecological factors rather than toxic barriers. For planthoppers, important contributing factors probably are as follows: (1) genetically based trade-offs in offspring performance and/or adult oviposition preference on different hosts which promote adaptation to one host species while constraining adaptation to another, (2) the evolution of oviposition/feeding preferences under conditions of local host plant abundance, (3) selective pressures associated mate location, and (4) a genetic correlation between oviposition preference and performance of offspring across host plant taxa [implicated in ***Ribautodelphax***]. These four points should not be taken for certain, but should act as working hypotheses to guide future research on the host plant associations of planthoppers.

Summary

1. Host plant selection in planthoppers involves random settling of macropters and does not appear to be actively directed by plant odor, and host plant specificity results because adults accumulate on favorable plants and selectively emigrate from unsuitable hosts.
2. Planthoppers feed by inserting their stylets into the vascular tissue of plants and ingesting the sap. Feeding occurs primarily in the phloem as documented by plant histology, precibarial morphology, electronic monitoring of feeding behavior, and chemical composition of the honeydew.
3. The feeding behavior of planthoppers can be divided into three behavioral phases, **plant surface exploration**, **stylet probing of plant tissues**, and **plant fluid ingestion**. These behaviors are mediated by plant physiology (chemical stimulants and deterrents) and most evidence suggests that acceptance or rejection of a particular plant occurs only after sampling the phloem sap.
4. On nitrogen-rich host plants, planthoppers probe less frequently, ingest more phloem sap, and excrete more honeydew than on nitrogen-deficient plants. Certain amino acids are potent ingestion stimulants.
5. In the laboratory, planthoppers generally survive better, develop faster, molt into larger adults, and are more fecund if they develop on nitrogen-rich host plants. Increased plant nitrogen can moderate, to some degree, the adverse effects of high population density on planthopper performance.
6. In the field, the macropters of some species of planthoppers accumulate on nitrogen-rich plants. The combined effects of increased colonization and improved performance often result in rapid population growth and high densities of planthoppers on hosts enriched with nitrogen by fertilization. However, strong numerical responses by nat-

ural enemies can prevent planthopper population responses to increased plant nitrogen.

7. Several adaptations in planthoppers attest to the limiting nature of plant nitrogen and underscore its importance in planthopper population biology. These include: (1) increased feeding rate by some species on nitrogen-deficient plants, (2) selection of nitrogen-rich feeding sites (leaf axils and terminal meristems) on plants, (3) synchronization of adult reproduction with periods of increased plant nitrogen, (4) dispersal to more nutritious host plants, (5) induction of free amino acids during feeding which result in increased planthopper fitness, and (6) acquisition of nutrients from other sources such as endosymbiotes.
8. Varietal resistance in rice and sorghum to planthopper attack is conferred by a diversity of chemicals that affect plant surface exploration, probing, and ingestion. Most research suggests that the major resistance factor is associated with phloem chemistry (plants are rejected only after phloem penetration with the stylets) and likely involves the absence of specific feeding stimulants. Resistance is not achieved by any physical mechanism (feature of plant tissue) which might interfere with stylet penetration.
9. Most species of planthoppers feed on grasses and sedges, whereas some species use broad-leaved monocots and dicots as hosts. Communities of phytophagous insects on grasses are often dominated by sap-feeders such as planthoppers and leafhoppers. The high silica content of grasses provides an effective physical defense against attack from many mandibulate insects, but is largely ineffective against sap-feeders.
10. Seventy-four percent of the delphacid species are monophagous feeding on a single species or closely related group of plant species. Failure of a planthopper species to feed on a certain host plant taxon results from the presence of feeding inhibitors, the absence of feeding stimulants, and/or inadequate plant nitrogen. Specific attractants and feeding stimulants facilitate host location and recognition and act as proximate cues in the host plant selection process. Ultimately, host plant specialization is likely to result from ecological factors rather than toxic barriers. For planthoppers, these probably include: (1) genetically based trade-offs in offspring performance on different hosts which constrain adaptation to one host species while promoting adaptation to another, (2) the evolution of oviposition/feeding preferences under conditions of local host plant abundance, (3) selective pressures associated with mate location, and (4) a genetic correlation

between oviposition preference and performance of offspring across host plant taxa (some evidence in *Ribautodelphax*). It is important to realize, however, that very little data exists for planthoppers on the factors which influence host plant utilization. The four points outlined above should act as hypotheses to direct future research.

ACKNOWLEDGMENTS

We thank C. Mitter and T. J. Perfect for their comments on earlier drafts of this report. This research was supported in part by National Science Foundation Grants BSR-8206603 and BSR-8614561 to RFD. This is Contribution Number 8480 of the Maryland Agricultural Experiment Station, Department of Entomology.

3

Influence of Habitat Structure on the Abundance and Diversity of Planthoppers

Robert F. Denno

Introduction

Habitat structure can have a major influence on the organization of plant and animal communities (Bell et al. 1991). Specifically for herbivorous insects, several components of **vegetation structure** (**vegetation texture** and **architectural complexity**) are important contributors to variation in their abundance and diversity (Kareiva 1983; Denno and Roderick 1991). **Vegetation texture** includes several attributes: **plant density** (the distance between individuals of the same species), **patch size** (the geographical extent of the stand), and **vegetation diversity** (the frequency and identity of nonhost plant individuals (see Kareiva 1983; Denno and Roderick 1991). Also, individual host plants can vary in **architectural complexity**, which consists of two components: **plant size** (the spread of plant tissue through different positions in space) and **plant part diversity** (the number of different plant parts, both in form and persistence) (see Southwood et al. 1979; Lawton 1983; Denno and Roderick 1991). Together, these components of vegetation structure define the habitats of phytophagous insects. Variation in these host plant factors and interactions among them result in an array of habitat contingencies, each with different consequences for herbivore abundance and diversity.

Planthoppers are ideal insects for studying the population and community effects of habitat structure because many species are monophagous or feed only on a few related plant species, usually grasses (Sogawa 1982; Wilson and O'Brien 1987; Denno and Roderick 1990). Furthermore, their host plants range from architecturally simple to complex (Denno 1977) and often grow in pure stands (Denno 1980). Thus, the effects of patch size and architectural complexity can be studied without the confounding influence of mixed vegetation.

In this chapter, I review existing information on the texture and architectural complexity of host plants and how these factors affect the abundance and species richness of planthoppers (Delphacidae and Issidae) and other selected sap-feeding Homoptera (e.g., Cicadellidae and Aphididae). Moreover, I address how particular life history traits such as dispersal ability (percent macroptery) and overwintering style (protected versus exposed eggs or nymphs) influence colonization and persistence, two factors which influence the abundance and diversity of planthoppers on their host plants. Because habitat persistence and complexity as well as climatic severity can influence the evolution of planthopper life histories (e.g., dispersal ability) (Denno 1979; Denno et al. 1981, 1991), I review how these factors may indirectly affect the potential responses of planthoppers to changes in vegetation structure.

Responses of Herbivores to Vegetation Structure: An Overview

Patch size, or the area occupied by a herbivore's host, is an important consideration at both geographic and local scales (Cromartie 1975; Claridge and Wilson 1981; Rey 1981; Strong et al. 1984; Bach 1988a,b; Denno and Roderick 1991). Larger patches of host plants often support more species of herbivores for a variety of reasons, including increased habitat heterogeneity, increased encounter rate, and reduced probability of extinction (Williams 1964; Denno et al. 1981; Rey 1981; Kareiva 1983; Strong et al. 1984; Denno and Roderick 1991). The explanation for such species–area relationships lies with the observation that as patch size decreases, certain herbivores become rare and disappear altogether below a certain critical patch size (Kareiva 1982, 1983; MacGarvin 1982; Bach 1984; Strong et al. 1984). By defaunating "islands" of *Spartina alterniflora* in the Gulf of Mexico and monitoring recolonization, Rey (1981) was able to show that immigration rates increased and extinction rates declined with island area. Certain combinations of life history traits (e.g., high dispersal and concealed overwintering stages) will favor colonization and persistence on isolated resources (Denno et al. 1981; Simberloff 1981; Denno and Roderick 1991). Consequently, species with specific life history styles may be overrepresented or underrepresented on large or small patches of hosts. Nevertheless, the relationship between immigration and emigration rates dictate the potential abundance of a species in a particular patch of plants (Rey 1981; Kareiva 1983; Strong et al. 1984).

Some, but not all, herbivores show positive abundance–area relationships and the exceptions underscore the need to consider the unique responses of specific herbivores to changes in vegetation texture (Kareiva 1983). Several factors may influence an herbivore's response to increasing host patch size and alter any abundance–area relationship. These include (1) the diet breadth of the herbivore (Kareiva 1983), (2) the mobility of the herbivore (Denno et al. 1981; Kareiva 1982, 1983; Bach 1980), (3) the quality, density, and size of host plants as they vary with patch area (Farrell 1976; Bach 1980; Risch 1981), (4) the abundance, size, and allelochemistry of the background, nonhost vegetation (Tahvanainen and Root 1972; Rausher 1981; Kareiva 1983), and (5) the numbers of parasites, predators, and pathogens that vary with patch size and other components of vegetation texture (Root 1973; Speight and Lawton 1976; Kareiva 1983).

Another important feature of habitat structure that influences the species richness and abundance of herbivores on plants is the architectural complexity of the host plant (Moran 1980; Strong et al. 1984). Larger, more complex plants with a greater variety and persistence of above-ground parts provide more microhabitats, which result in a greater diversity and abundance of herbivores (Lawton 1978; Southwood 1978; Moran 1980). Strong et al. (1984) argue that trees provide a greater variety of niches for insects than herbs in at least three ways. First, microclimates are more diverse (Claridge and Reynolds 1972). Second, the degree of phenological change and age differ; for example, young trees and mature trees may support different species of herbivores (Niemela et al. 1980; Dixon 1985). Third, architecturally complex plants provide a greater variety of feeding and oviposition sites, hiding places from enemies, and overwintering sites than do structurally simple plants (Claridge and Wilson 1976; Denno et al. 1981; Price 1983).

Species–Area Relationships

Geographic-scale studies on the species–area relationships of delphacid planthoppers have not heretofore been conducted. However, using Giffard's (1922) account of the Hawaiian Delphacidae, I was able to examine the species–area relationship for planthoppers on the six largest Hawaiian islands. Data on the number of species per island, island area, and age were extracted from Giffard (1922), Armstrong (1983), and Kaneshiro (1983). Even though larger islands (Hawaii, Maui, and Oahu) supported more species on average than smaller ones (Kauai, Molokai, and Lanai), the species–

area relationship was not significant ($P>0.05$, Fig. 3.1A). In particular, the large island of Hawaii supported fewer species of planthoppers than expected. Hawaii is the youngest of the large islands (~0.5 million years) and has had less time to accumulate species compared to Kauai, which is ~6.0 million years old. When the number of species per island was corrected for island age (species richness/island age in millions of years), a significant species–area relationship emerged ($s = -110.3 + 43.2A$, $n=6$, $R^2=0.71$, $P=0.034$; Fig. 3.1B). Island age alone was not significantly related to species richness ($P>0.05$). Thus, the combined effects of island area and geologic age best explained the resident delphacid richness on an island. Large islands provide a greater variety of habitats and support a larger number of plant taxa (Wagner et al. 1990) on which host-specific delphacids (Giffard 1922; Zimmerman 1948) might diversify. Furthermore, in archipelagos characterized by rapid radiations of insect taxa (Howarth 1990), island age may prove to be a more important contributor to species richness than is currently thought [for a discussion of "area" and "geologic time" hypotheses, see Southwood (1961, 1962) and Strong (1979)].

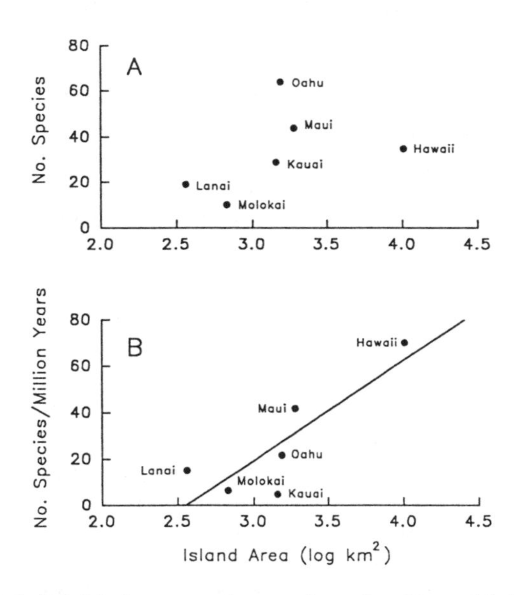

Figure 3.1. (**A**) Relationship between the number of resident delphacid species and island area (log km^2), based on data from the six largest islands in the Hawaiian Archipelago (Giffard 1922). The species–area relationship is not significant ($R^2 = 0.18$, $P>0.05$). (**B**) Relationship between the number of resident delphacid species corrected for island age (No. of species/island age in millions of years) and island area (log km^2) ($s = -110.3 + 43.2A$, $n=6$, $R^2=0.71$, $P=0.034$).

Claridge and Wilson (1976, 1981) examined the species–area relationship of mesophyll-feeding leafhoppers (Cicadellidae: Typhlocybinae) on British trees and shrubs. There was a significant positive relationship between the number of leafhopper species and the geographic range of the host plant [$\log_e(s+1) = -0.869 + 0.343 \log_e A$, $n=34$, $R^2 = 0.16$, $P<0.05$]. However, only 16% of the variation in leafhopper richness was attributable to host plant range. Taxonomic isolation of host plants was excluded as an important factor determining leafhopper richness, but the authors suggested that a large portion of the unexplained variation (84%) in the species–area regression was attributable to plant chemical factors (Claridge and Wilson 1981). The authors concluded that although species–area effects play a part in determining leafhopper richness on British trees, physicochemical and other factors were more important.

Sap-Feeder Abundance and Plant Patch Size

On a smaller spatial scale, Denno et al. (1981) studied the relationship between patch size of the salt marsh grass *Spartina patens* and the abundance of its sap-feeding Homoptera. *S. patens* is one of the dominant grasses of salt marshes along most of the Atlantic and Gulf Coasts of North America (Mobberley 1956). Variation in the elevational relief of the marsh surface results in an archipelago of usually pure *S. patens* patches ranging in size from just a few square meters to immense "islands" (> 20 ha) surrounded by another grass, *Spartina alterniflora*, that grows at a slightly lower elevation (Miller and Egler 1950; Blum 1968; Redfield 1972; Denno et al. 1981).

Six resident sap-feeding insects occur abundantly and develop on *S. patens* (*S. alterniflora* is not a host) in marshes from New England and the mid-Atlantic states south through Virginia. Of these, four are planthoppers, *Delphacodes detecta*, *Tumidagena minuta*, *Megamelus lobatus* (Delphacidae), and *Aphelonema simplex* (Issidae), and two are leafhoppers, *Amplicephalus simplex* and *Destria bisignata* (Cicadellidae).

Denno et al. (1981) sampled these sap-feeders on 15 "islands" of *S. patens* ranging in size from 8 m^2 to 30,000 m^2. All "islands" sampled were located within 25 m of large (> 1000 m^2) "source islands" that were included in a large archipelago of patches on a marsh near Tuckerton, New Jersey. Three species, *A. simplex*, *T. minuta*, and *D. detecta*, became less abundant as grass patch size decreased and *A. simplex* was absent altogether in the smallest patches (Fig. 3.2). Three other species, *M. lobatus*, *Am. simplex*, and *D. bisignata*, either were similarly abundant across the patch size gradient or populations decreased slightly on larger patches (Fig.

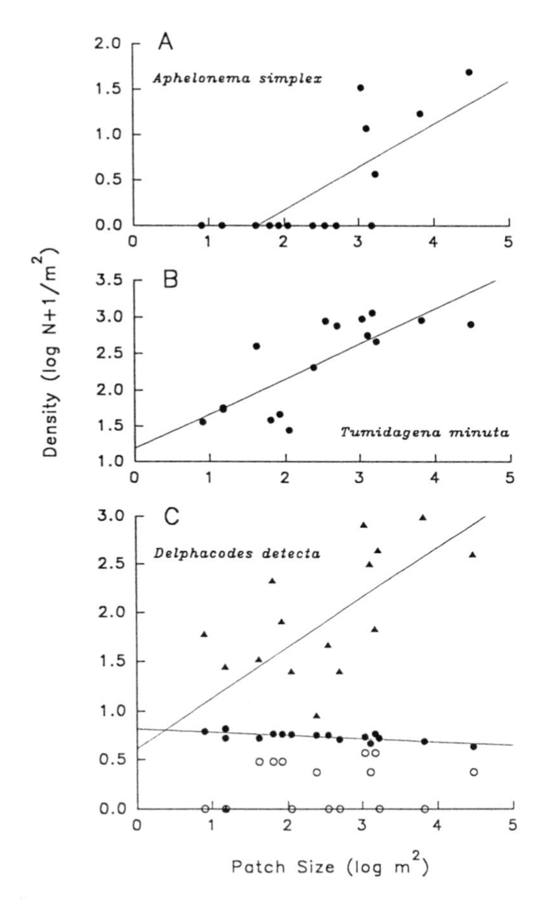

Figure 3.2. Relationship between the size of *Spartina patens* patches [log area (m^2)] and the density ($\log N + 1/m^2$) of *Aphelonema simplex* (**A**), *Tumidagena minuta* (**B**), and *Delphacodes detecta* (**C**). Density–area relationships during May (○), July (●), and September (▲) are *d*(●) = −0.781 + 0.475*A*, R^2 = 0.58, $P<0.01$ for *A. simplex*; *d*(●) = 1.184 + 0.482*A*, R^2 = 0.61, $P<0.01$ for *T. minuta*; *d*(●) = 0.817 − 0.034*A*, R^2 = 0.53, $P<0.01$; and *d*(▲) = 0.613 + 0.513*A*, R^2 = 0.43, $P<0.01$ for *D. detecta*.

3.3). These findings were confirmed to a great extent by another study where Raupp and Denno (1979) compared the densities of the same guild of sap-feeders between extremely large patches of *S. patens* (14–20 ha) and smaller ones (1–3 ha). The consequence of reducing patch size for this guild of sap-feeders is the selective elimination of certain species on the smallest islands, resulting in lower species richness in smaller patches.

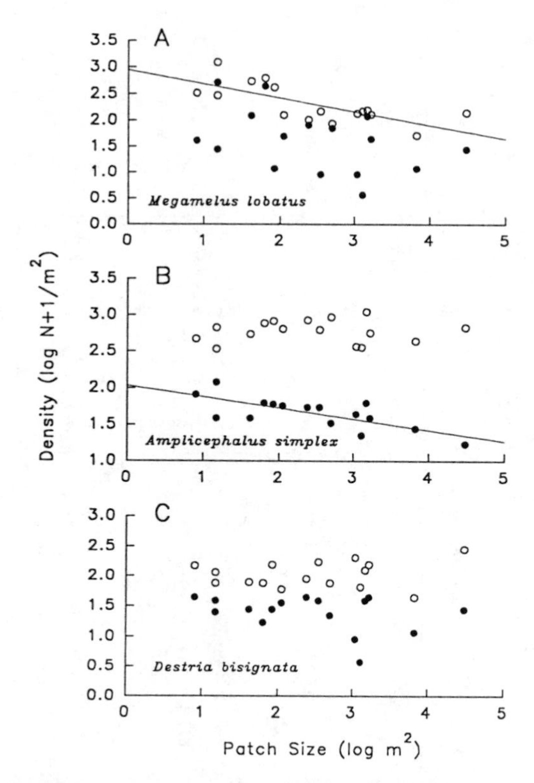

Figure 3.3. Relationship between the size of *Spartina patens* patches [log area (m^2)] and the density ($\log N + 1/m^2$) of *Megamelus lobatus* (**A**), *Amplicephalus simplex* (**B**), and *Destria bisignata* (**C**). Density–area relationships during May (○) and July (●) are d(○) = 2.944 − 0.259A, R^2 = 0.52, $P<0.01$ for *M. lobatus* and d(●) = 2.034 − 0.153A, R^2 = 0.53, $P<0.01$ for *Am. simplex*.

However, causal mechanisms producing abundance–patch area relationships are best elucidated by examining the life histories of the specific herbivores.

Abundance–Area Relationships and Sap-feeder Life Histories

Differences in the life history characteristics (e.g., dispersal ability and overwintering style) of the *S. patens* sap-feeders influence the ability of each species to colonize and remain on small compared to large patches of grass. The brachypterous (short-winged) morph in dimorphic populations of planthoppers is virtually "immobile" and unable to move more than a few meters (Raatikainen 1967; Denno et al. 1980). By contrast, long-winged macropters are capable of long-distance flight and are able to colonize distant patches, hundreds of kilometers away in some cases (Kisimoto 1976, 1979). By defaunating large plots of *S. patens* and measuring recolonization, Denno (1977) showed that fully winged insects like *Am. simplex* are excellent colonists. In contrast, species that produce mostly flightless brachypters (e.g., *A. simplex* and *T. minuta*) should be poor colonists (have low immigration rates) and should be underrepresented on small, isolated "islands" (Denno et al. 1981; Denno and Roderick 1991).

Overwintering style also affects the ability of sap-feeders to remain on small islands. For example, species that overwinter as exposed eggs or nymphs (*T. minuta* and *D. detecta*) should incur higher winter mortality than those which overwinter as eggs embedded in vegetation (*M. lobatus*, *Am. simplex*, and *D. bisignata*) (Denno et al. 1981). Extinction rates should be particularly high on small islands for exposed species because there is less protective thatch (dead litter layer) during the winter season (Denno et al. 1981) and washout rates due to tidal flooding are high (Raupp and Denno 1979). Furthermore, the overwintering success of sap-feeders with exposed stages (*T. minuta*) is very low in plots of grass where the protective thatch has been removed (Tallamy and Denno 1979).

Based on their life history traits, several predictions can be made concerning the success of specific sap-feeders on small and large patches of *S. patens*. Species with low immigration rates (those that are mostly brachypterous) and high extinction rates (those that overwinter as exposed eggs of nymphs) should be least likely to maintain populations on small islands. Sap-feeders with these characteristics are *A. simplex* and *T. minuta*. Populations of these planthoppers are composed of more than 95% brachypterous forms and all overwinter as exposed stages (Denno et al. 1981). As predicted, these species were either less abundant or absent on small patches (Figs. 3.2A and 3.2B).

Species with the opposite combination of life history traits (high colonization ability coupled with reduced winter extinction rate) should have the best chance for success on small patches. The only species with this combination of traits, *Am. simplex*, was similarly abundant on all sizes of *S. patens* patches during May and even showed a significant negative relationship between density and patch size during July (Fig. 3.3B).

Species with fair colonizing ability (10–20% macropterous forms in field populations) and exposed overwintering stages like *D. detecta*, or species with poor colonizing ability (> 95% brachypters) and protected overwintering stages like *M. lobatus* and *D. bisignata*, were more similarly abundant on small and large "islands," but for different reasons (Figs. 3.2C, 3.3A, and 3.3C). *D. detecta* was absent from the smallest "islands" during May, had colonized them by July, but populations continued to increase rapidly only on large islands which supported higher densities during September (Fig. 3.2C). Winter mortality is great on small islands, but macropterous adults colonize these sites early in the season and subsequently reproduce (Denno et al. 1981).

In contrast, *M. lobatus* (97% brachypters) and *D. bisignata* (100% of females brachypterous) have poor colonizing ability, but persist very well in all sizes of patches during winter because they embed their eggs in vegetation. Once they have colonized small islands, extinction rates for these species are apparently low as evidenced by their presence on the smallest of islands following winter in May prior to reproduction (Figs. 3.3A and 3.3C).

It is possible that other factors act selectively in small patches, resulting in reduced populations of certain sap-feeders. However, the nutritional quality (percent crude protein) and growth phenology of *S. patens* are the same along the patch size gradient (Denno et al. 1981). Furthermore, wolf spiders, the major predators of the sap-feeders (Döbel et al. 1990), are equally abundant in large and small patches (Denno et al. 1981).

However, the quantity of thatch is significantly greater in large compared to small patches of *S. patens* (Denno et al. 1981). Consequently, the quality of overwintering sites is lower in small patches and only those dispersive species which embed their eggs (e.g., *Am. simplex*) are consistently successful in these habitats. The life histories of most other sap-feeders in the *S. patens* system have been adjusted by varying degrees to maintain contact with large patches. The brachypterous adults and/or embedded eggs characteristic of most species are consistent with this hypothesis.

Overall, responses of sap-feeding herbivores to changes in patch size are mixed. For example, in the *S. patens* guild of sap-feeders, only two species (one-third of the total) showed strong density responses to patch size. Holder (1990) examined the effect of stand area on sap-feeder abundance and diversity in patches of *Spartina pectinata* ranging in size from $<50\ m^2$ to $400\ m^2$. There were significantly more species and individuals of some species on larger patches, but differences were not great. Specifically, the planthopper *Prokelisia crocea* was slightly more abundant on large stands of *S. pectinata*. The high mobility of this planthopper (98% macroptery) undoubtedly contributed to the similarity in its abundance across patches of hosts (Holder 1990).

Published data for planthoppers suggests that on a local scale, some species are not responsive to changes in patch size. For those which respond to patch size, densities are usually lower and only rarely higher on small host plant patches. Patterns of host patch size exploitation can be predicted to some extent by life history characteristics such as macroptery (%) and exposure of overwintering stages which reflect a species' ability to colonize and persist on small, isolated stands of host plants.

Host Plant Density

Close plant spacing often results in higher population densities of *Nilaparvata lugens* (Dyck et al. 1979; Oka 1979). For example, nymphs achieved significantly higher densities on 10 × 10cm plantings of rice (4 nymphs/tiller) than on 50 × 50cm plantings (<1 nymph/tiller) with plant size controlled (Dyck et al. 1979). Also, there was a positive relationship between nymphal density/hill and the number of tillers/hill ("plant size") (Dyck et al. 1979). It has been suggested that close plant spacing alters the microclimate of rice making conditions more humid and shaded and favorable for planthopper development (see Suenaga 1963; Pathak 1968; Kulshreshtha et al. 1974; Oka 1979). Furthermore, Nishida (1975) suggested that shading resulting from close plant spacing may be unfavorable for the development of the natural enemies of *N. lugens*.

Plant spacing also influences the movement of planthoppers. For example, in a mark-recapture study, adults of *Perkinsiella vitiensis* moved more within rows of sugar cane (plants touching) than between rows where plants were separated by 1.4m (Husain et al. 1967). On average, adults moved approximately 1 m/day, and the dispersal rate was higher (1–3 m/day) in fields of very young cane plants (8-week-old plants with low natural populations of planthoppers) than in fields of 40-week-old plants with high infestations of planthoppers (0.6–2 m/day).

Other studies with leafhoppers and aphids report that populations decline in abundance with increasing host plant density (Farrell 1976; Mayse 1978). However, as pointed out by Kareiva (1983), experimental manipulations of plant density often alter plant quality as evidenced by the fact that the nitrogen content of leaves was higher in sparse compared to dense plantings of collards (R. Root unpublished data). Host plant nutrition is known to alter patterns of colonization and persistence in *Prokelisia marginata*. For example, migratory forms selectively colonize and accumulate more rapidly on nutritionally superior host plants (Denno et al. 1980; Denno 1983), and emigration rates of macropters are lower on more nutritious plants (G. Roderick unpublished data). Also, few studies [Mayse (1978) excepted] take into account changes in plant size which could potentially confound the effects of plant density. Consequently, sap-feeder abundance and community structure should differ on plants varying in plant nutrition and size, but we do not yet know precisely how these factors interact with plant dispersion to influence rates of colonization and emigration and the abundance of planthoppers on their host plants. However, there is no reason to expect all species to respond similarly to changes in plant density given that planthopper species have different dispersal capabilities (Denno et al. 1991) and nutritional needs (Prestidge 1982a, 1982b; Prestidge and McNeill 1983).

Vegetation Diversity

Vegetation diversity or the presence of nonhost plants has the potential to influence the densities of planthoppers that accrue on their hosts. For example, it has been suggested that populations of *N. lugens* achieve higher densities in "weedy" rice fields (Cenda and Calora 1964; Fernando 1975). However, the mechanisms by which nonhost vegetation influences population size are poorly investigated for planthoppers in general, but for *N. lugens*, the increase is not due to the presence of alternate host plants because this species is monophagous (Wilson and Claridge 1985; Denno and Roderick 1990).

In contrast, densities (number/stem) of the leafhopper *Empoasca fabae* were significantly higher on pure stands of alfalfa than on stands in which alfalfa was grown in mixed plantings with grasses or broad-leafed weeds (Lamp et al. 1984; Oloumi-Sadeghi et al. 1987). Smith et al. (1992) suggest that volatile chemicals in nonhost grasses reduce the acceptability of alfalfa as a host for *E. fabae*. Similarly, several aphids colonize their host plants at significantly lower rates when those hosts are surrounded by weeds or a nonhost crop (Smith 1976; Horn 1981). Thus, background vegetation may influence the densities of herbivores on their hosts by altering herbivore movement, searching efficiency, and colonization (Tahvanainen and Root 1972, Rausher 1981), or by interfering chemically with host finding behavior (Altieri et al. 1977; Lamp et al. 1984; Smith et al. 1992).

However, Kareiva (1987) reported that outbreaks of aphids in patchy compared to continuous stands of goldenrod were not the result of any direct effect of habitat patchiness on aphid colonization, but rather the reduced ability of coccinellid predators to aggregate and control aphid populations in patchy configurations of hosts. Consequently, by changing the proportion of mortality attributable to natural enemies, vegetation texture may also influence herbivore populations.

Architectural Complexity and Species Richness

Studies of planthoppers and leafhoppers on salt marsh grasses provide insight into the effect of host plant architecture on the number of resident sap-feeding species. *Spartina patens* and *Distichlis spicata* are two complex-structured grasses with slender culms, narrow blades, and a dense persistent thatch (Blum 1968, Denno 1977, Tallamy and Denno 1979). In contrast, *Spartina alterniflora* is architecturally simple, consisting of stiff rosettes, wide blades, and a very loose thatch (Denno 1977, Tallamy and Denno 1979). The height, complex thatch, and general growth habit of *D. spicata* all contribute to a significantly higher foliage height diversity index compared to *S. alterniflora* (Tallamy and Denno 1979).

Grasses which produce a persistent thatch have an added component of structural diversity which thatchless species lack. Such grasses may provide additional feeding and oviposition sites, protection from physical stress, and refuge from enemies, characteristics which increase microhabitat diversity and should lead to an increase in the number of resident herbivores (Denno and Roderick 1991). Data from *S. patens* indicate that the various planthopper species specialize on different plant parts in the vertical strata of the grass system and can be categorized simply into upper and lower strata residents (Fig. 3.4; Denno 1980; Denno et al. 1981). Generally, vertical stratification appears to be a widespread phenomenon in Homoptera communities (Andrzejewska 1965; Halkka et al. 1977; Waloff 1979, 1980; Stiling 1980b; Morris 1981; McEvoy 1986; Denno and Roderick 1991).

Two lines of evidence suggest that structurally complex grasses including those with thatch provide greater microhabitat diversity and by doing so support a richer fauna of sap-feeders. First, direct positive correlations exist between sap-feeder species richness and various components of architectural complexity such as foliage height (Fig. 3.5A; Morris 1971, 1974), foliage height diversity (Murdoch et al. 1972; Tallamy and Denno 1979), and biomass (Fig. 3.5B; Denno 1977; Tallamy and Denno 1979). Grass height explained more of the variation in sap-feeder species richness than foliage height diversity or grass biomass, although all three components were significantly correlated with species richness (Tallamy and Denno 1979). Generally, complex-structured grasses such as *S. patens* and *D. spicata* support more species of planthoppers and leafhoppers (eight and six species, respectively) than do simple-structured grasses like *S. alterniflora* (five species) (Denno 1977; Denno and Roderick 1991). In particular, sap-feeders specializing in the lower stratum of *S. patens* are underrepresented on *S. alterniflora*, the grass with the least developed thatch (Denno and Roderick 1991).

More convincing evidence is provided by studies where grass architecture was experimentally simplified by dethatching (mowed and raked) and the subsequent changes in sap-feeder abundance measured. Dethatching *S. patens*, *S. alterniflora*, and *D. spicata* resulted in significant reductions in sap-feeder richness and diversity (Denno 1977; Tallamy and Denno 1979).

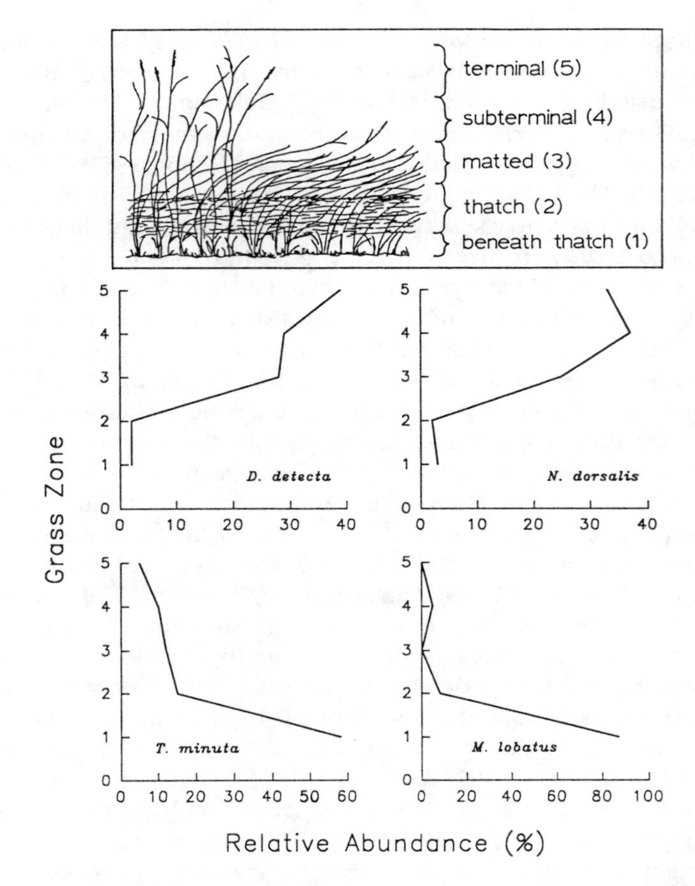

Figure 3.4. Microhabitat partitioning of *Spartina patens* by four species of delphacid planthoppers: *Delphacodes detecta*, *Neomegamelanus dorsalis*, *Tumidagena minuta*, and *Megamelus lobatus* on a salt marsh near Tuckerton, New Jersey. The relative abundance of each species in the terminal (5), subterminal (4), matted (3), thatch (2), and beneath thatch (1) zones of the grass is shown (see Denno 1980; Denno et al. 1981).

Species inhabiting the upper strata of the grasses (e.g., *Delphacodes detecta*) either increased in abundance on dethatched grass or were equally abundant there compared to unaltered vegetation (Fig. 3.6). However, sap-feeders inhabiting the lower strata (e.g., *Tumidagena minuta* and *Megamelus lobatus*) were adversely affected by thatch removal (Fig. 3.6; Denno 1977; Tallamy and Denno 1979). Dethatching apparently created new resources for upper strata residents, but destroyed microhabitats normally exploited by lower strata inhabitants (Denno 1977). Similarly, cutting and grazing of British grasslands resulted in significant population decreases in most planthoppers (e.g., *Criomorphus albomarginatus*, *Muellerianella fairmairei*, *Stenocranus minutus*), particularly when the treatment was applied just prior to the adult molt (Morris 1973, 1981; Morris and Lakhani 1979; Waloff 1980).

All evidence suggests that architecturally complex grasses provide a greater diversity of microhabitats and support larger assemblages of planthoppers and other sap-feeders than simple structured grasses (Denno and Roderick 1991). Similarly, more species of delphacids occur on the Hawaiian (only Oahu considered) trees *Metrosideros polymorpha* (six delphacids: *Leialoha lehuae, L. naniicola, L. oahuensis, L. ohiae, Nesothoë gulicki,* and *N. perkinsi*) and *Acacia koa* (three delphacids: *Nesosydne koae, N.*

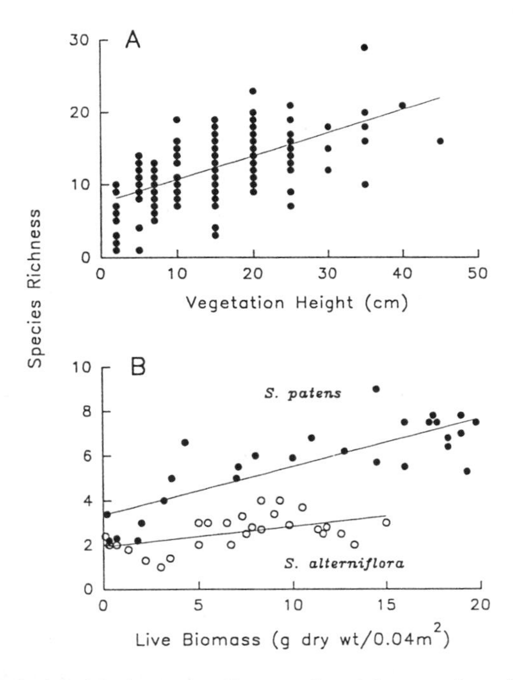

Figure 3.5. (**A**) Relationship between the species richness of sap-feeding Auchenorhyncha and the vegetation height (cm) of British grasslands ($y = 7.40 + 0.33x$, $R^2 = 0.36$, $P < 0.01$; Morris 1971). (**B**) Relationship between the species richness of sap-feeding Auchenorhyncha and the live biomass (g dry wt/0.04 m^2) of *Spartina patens* ($y = 3.41 + 0.21x$, $R^2 = 0.66$, $P < 0.01$) and *S. alterniflora* ($y = 1.94 + 0.10x$, $R^2 = 0.28$, $P < 0.01$) on a salt marsh near Tuckerton, New Jersey (Denno 1977).

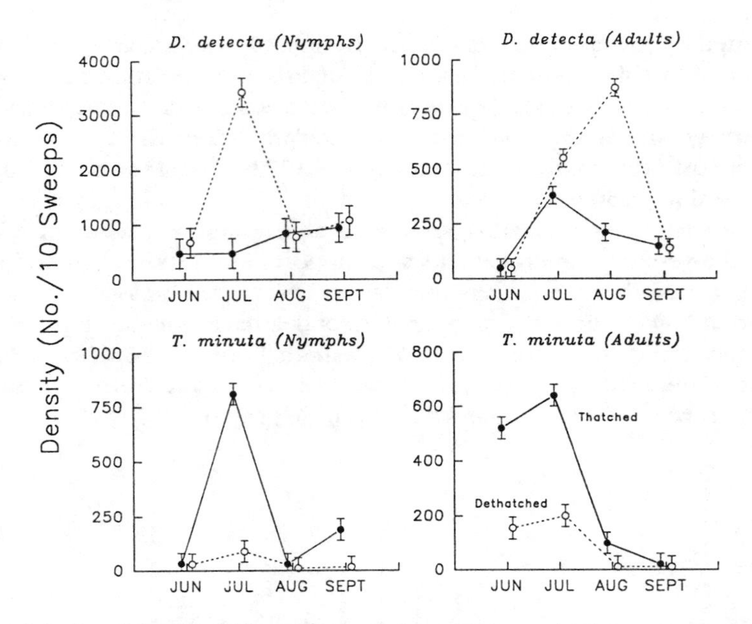

Figure 3.6. Densities (No./10 sweeps) of the nymphs and adults of *Delphacodes detecta* and *Tumidagena minuta* on control plots (thatch containing) and dethatched plots of *Spartina patens* at Manahawkin, New Jersey. Intervals surrounding means are Least Significant Intervals (LSIs). Nonoverlapping intervals denote significant differences among means at the 5% probability level (from Denno 1977).

koae-phyllodii, and *N. rubescens*) than are associated with small shrubs and herbs (one or occasionally two species) (Zimmerman 1948; Denno, Chapter 4).

Dixon (1985) also notes that more structurally complex trees are regularly infested with more species of aphids than herbaceous plants. Resource partitioning is common as evidenced by vertical stratification. However, there is very little evidence that niche diversification among sap-feeders results from competitive interactions for food (McClure and Price 1975, 1976; Waloff 1979; Denno et al. 1981; McEvoy 1986). More likely, stratification and niche specialization by sap-feeders results from specific nutritional or structural requirements provided locally by the host (McNeill and Southwood 1978; Denno et al. 1980; Waloff 1980; McNeill and Prestidge 1982; Prestidge and McNeill 1982; Sogawa 1982; McEvoy 1986), narrow tolerance of physical factors with restrictions to specific sites on the plant (Dixon and McKay 1970; Dixon 1985), or avoidance from parasites and predators in enemy-free sites (Raatikainen 1967; Denno and Roderick 1990, 1991).

Natural enemies of planthoppers also respond directly to vegetation texture. Within *Spartina alterniflora* habitats there is considerable variation in the quantity of thatch as well as the abundance of wolf spiders, the major predator of *Prokelisia* planthoppers (Döbel 1987; Döbel et al. 1990). Even though hunting spiders accumulate in habitats with more thatch, the thatch provides a protective refuge for planthoppers (Döbel 1987). Furthermore, the presence of thatch alters the functional response of *Pardosa* wolf spiders to increases in *Prokelisia* density from Type II to Type III (Döbel 1987), a change thought to stabilize predator/prey interactions (Murdoch and Stuart-Oaten 1975). These data underscore the complicated nature of interactions between sap-feeders, their natural enemies, and vegetation structure.

Host Plant Architecture and Sap-feeder Life History Traits

Reuter (1875) was the first to note that most wing-dimorphic species of planthoppers live on herbs or near the roots and that few dimorphic species occur on trees (summarized in Kirkaldy 1906). Subsequently, Denno (1979, Chapter 4) and Waloff (1983) have confirmed the observation that wing-polymorphic and brachypterous planthoppers, leafhoppers, and aphids are rare in arboreal habitats compared to low-structured vegetation. For example, most Hawaiian delphacids are host-specific, but as a group, feed on a wide variety of plant forms ranging from large trees to low-profile herbs (Giffard 1922; Zimmerman 1948). Macropterous species feed mostly on trees and large shrubs (7.1 ± 0.6 m in height, $n = 52$), whereas brachypterous and mostly brachypterous species (pooled for analysis) occur on significantly shorter vegetation, primarily small shrubs and herbs many of which contact the ground (1.4 ± 0.2 m, $n = 98$; see Fig. 4.7). A comparison of congeners in *Nesosydne* confirmed this pattern as well and allowed for an assessment of the effects of plant growth form on wing form independent of common ancestry (Denno, Chapter 4).

Several hypotheses may be invoked to explain the rarity of flightlessness on architecturally complex trees. First, planthoppers usually avoid predators by quickly leaping from their holds. In low-profile vegetation, an escaping planthopper or a dislodged one usually lands on or near a suitable host. However, if the host plant is a large tree and the planthopper is brachypterous, it may be virtually impossible to relocate suitable resources including mates (Denno 1979). Second, Waloff (1983) argues that retention of wings in arboreal habitats is essential to negotiate the three-dimensional nature of the environment. Because there are reproductive costs associated with the possession of wings in planthoppers, living in arboreal habitats may constrain the evolution of certain life history traits (early reproduction and high fecundity) which are characteristic of flightless female planthoppers living in low-profile vegetation (Denno et al. 1989).

Plant architecture also appears to influence other life history traits of sap-feeding insects. Although there are no data for planthoppers and leafhoppers, some data are available for aphids (Dixon 1985). Different species of aphids vary in body length from 0.7 to 7.0 mm, and those that feed on the deeply located phloem elements in the trunks of trees are larger than those which feed on the more accessible phloem elements of leaves. To reach phloem elements deep within the trunk, an aphid needs long stylets (Dixon 1985). Because there is a positive relation between stylet length and body size in aphid species, it is physically impossible for small aphid species to feed on tissues deep within a plant. Consequently, plant architecture (depth of phloem elements) places constraints on the body size of aphids (Dixon 1985).

In addition, there is some indication that larger species of aphids are more fecund than smaller species (Dixon 1985) and that larger species beat their wings slower and fly faster than smaller ones (Johnson 1969, Dixon 1985). Thus, in constraining body size, plant architecture may shape the entire life history of sap-feeder species from reproductive characters to dispersal capability (see Denno, Chapter 4).

Habitat Persistence and Disturbance

Habitat persistence dramatically influences planthopper abundance and community structure. For example, increasing the intensity and frequency of habitat disturbance (e.g., grazing, mowing, or burning) results in the decreased abundance of most sap-feeder species and adversely affects their community diversity (Andrzejewska 1965, 1978; Morris 1971, 1973, 1981; Hawkins et al. 1979). Furthermore, the timing of the disturbance relative to sap-feeder life history is an important consideration. In general, disturbance at times when sap-feeders were in the adult stage resulted in the greatest decreases in population size (Morris 1981). Reasons for the reduced abundance and diversity of Auchenorrhyncha following disturbance include simplification of plant architecture with reductions in niche diversity (Morris 1974; Denno 1977; Morris and Lakhani 1979; Tallamy and Denno 1979), and the selective elimination of species with poor dispersal ability (Andrzejewska 1962; Denno 1977; Waloff 1980; Denno and Roderick 1990).

Host plant persistence also influences the life histories and, in particular, the dispersal capability of planthoppers (Southwood 1962, 1977a; Denno and Grissell 1979; Denno 1978, 1979, 1983, 1985b; Denno et al. 1991). Levels of migration (percent macroptery) in field populations of planthoppers decreased significantly as the persistence of their habitats increased (Denno et al. 1991). For this analysis, habitats ranged from ephemeral agricultural crops (< 1 year) to very persistent marsh and bog habitats up to 12,000 years old. For example, of 15 species surveyed on persistent, low-profile, salt marsh vegetation, an average of 96% of adults were brachypterous (Denno 1978). Similarly, for five species sampled on persistent freshwater vegetation, 98% of adults were brachypterous. By contrast, an average of 70% of the adults were macropterous for five species of planthoppers inhabiting temporary host plants like agricultural crops and mowed lawns (Denno 1978). In addition, males are more macropterous than females for species inhabiting temporary habitats, a characteristic which facilitates mate location. For species in permanent habitats where mates are apparently more available and the wing-form ratios of the sexes are similar (Denno et al. 1991). These findings are consistent with those of Brown (1982, 1986) who found that early successional vegetation was exploited effectively by macropterous species, and that wing-polymorphism was most prevalent in species occurring during the middle stages of plant succession when long-lived grasses were common.

Habitat persistence appears to influence the migratory capability of planthoppers in two ways: (1) by selecting for habitat escape and (2) by dictating the availability of mates (Denno et al. 1991). In persistent habitats, wings are less necessary for habitat escape and they are rarely required for mate location. Because wings impose a reproductive penalty (Denno et al. 1989), flightlessness is widespread. For colonizing species which occur in temporary habitats, wings are favored in males to locate females at low densities and are favored in both sexes at high densities for reasons of habitat escape (Denno et al. 1991).

Summary and Conclusions

Planthopper and leafhopper communities on low-profile vegetation have proven to be ideal systems for investigating the consequences of vegetation texture on species abundance and diversity. Host relationships are sufficiently simple, the players are not so numerous so as to intimidate the investigator, and plant, sap-feeders, and enemies can be experimentally manipulated in the field (Denno 1977; Tallamy and Denno 1979; Strong and Stiling 1983; Döbel 1987; Roderick 1987). Consequently, sap-feeder communities provide the potential for identifying the relative importance of factors that structure communities, for example, in separating the effects of plant architecture, plant quality (especially nutrition), and natural enemies. Studies to date suggest that habitat and vegetation structure influence the diversity, abundance, and life history styles of delphacid planthoppers and other sap-feeding Auchenorrhyncha in the following ways:

1. For planthoppers on the six largest Hawaiian islands, there is a significant species–area relationship when the number of planthopper species per island is corrected for island age (species richness/island age in millions of years). The number of planthopper species/island is not significantly related to either island area or age alone.
2. Limited evidence suggests that the richness of leafhopper (Typhlocybinae) species is positively related to the distributional area of the host plant. Such geographic-scale data are not available for planthoppers.
3. On a smaller spatial scale, some species of Auchenorrhyncha (Delphacidae, Issidae, and Cicadellidae), but not all, are responsive to changes in the patch size of their host plants. For those species which respond to patch size, densities are usually lower and rarely higher on small patches.
4. Species with low immigration rates (those that are mostly brachypterous) and high extinction rates (in part, those that overwinter in an exposed stage) are least likely to maintain populations on small patches of host plants.
5. Population densities of *Nilaparvata lugens* are often higher in dense compared to sparse stands of rice. In contrast, aphids and leafhoppers often achieve higher densities in sparse plantings of their host plants. Mechanisms for these patterns are poorly understood.
6. Most aphids and leafhoppers reach higher densities on their host plants when hosts are growing in pure stands than when hosts are grown in mixed plantings with other nonhost vegetation. This density pattern results because background vegetation either alters herbivore movement, chemically interferes with host plant location, or influences the action of natural enemies.
7. Architecturally complex host plants provide a greater variety of microhabitats and support richer assemblages of Auchenorrhyncha than simple structured hosts.
8. Plant architecture constrains planthopper life histories and dispersal ability. Flightlessness (brachyptery) and wing polymorphism are rare in planthopper taxa which inhabit high-profile vegetation (trees) compared to low-profile vegetation (grasses).

9. Increasing the intensity and frequency of habitat disturbance (e.g., grazing, mowing, or burning) decreases the abundance of many Auchenorrhyncha species and reduces community diversity.
10. Host plant persistence influences the life histories and the dispersal capability of planthoppers. Levels of migration (percent macroptery) in field populations of planthoppers are inversely related to the persistence of their habitats.
11. Vegetation texture may influence sap-feeder abundance indirectly through natural enemies by altering predator behavior and/or foraging space or by influencing the availability of refugia for prey.

ACKNOWLEDGMENTS

This research was supported by National Science Foundation Grants BSR-8206603 and BSR-8614651. This is Contribution Number 8481 of the Maryland Agricultural Experiment Station, Department of Entomology.

PART TWO

Life History Patterns, Reproductive Biology, and Speciation

One feature of delphacid planthoppers that has captured the attention of investigators for over a century now is their wing dimorphism; both flight-capable forms with fully developed wings (macropters) and flightless forms with reduced wings (brachypters) can occur in the same population. Organisms displaying such dispersal polymorphisms have provided ecologists with ideal systems to study life history evolution and in particular the costs, benefits, and population consequences of flight and the habitat characteristics promoting it and other associated traits.

In Chapter 4, Denno examines delphacid life histories and the factors which shape and constrain them within the broad context of flight dimorphism. Phenotypic trade-offs between dispersal and reproduction, body size evolution and the integration of life history traits are discussed. In particular, the effects of habitat persistence and structure on dispersal characters and associated traits are featured. Following a debate on migration, traits (overwintering stage, voltinism, and diapause) which synchronize reproduction with favorable resources in time are considered. Last, planthopper life histories are evaluated in the framework of existing life history theory.

From the ecology and evolution of life histories, the tone of this part turns to behavior with a specific focus on the reproductive biology and mating system of delphacid planthoppers. Acoustic communication through substrate vibration mediates almost every aspect of mating behavior from mate location to courtship and apparently mate choice. Claridge and de Vrijer (Chapter 5) introduce the mating system of planthoppers with a detailed discussion of acoustic signaling, mate location, mate choice, and male–male interactions. The role of acoustic signals is then developed in the context of mate recognition, reproductive isolation, and species formation.

Ott (Chapter 6) continues the discussion by establishing an ecological framework to guide future research on the structure and determinants of delphacid mating systems. Emphasis is placed on the quantification of reproductive success, the measurement of intrasexual and intersexual selection, and establishing the relationship between variation in reproductive success and variation in behavior (e.g., courtship and mobility), morphology (body size and wing form), and ecology (population density and dispersion).

4

Life History Variation in Planthoppers

Robert F. Denno

Introduction

Life history theory seeks to explain variation in reproductive traits such as fecundity, age to first reproduction, and egg size as well as traits which synchronize reproduction with favorable resources in space (migration) and time (diapause and voltinism) (Stearns 1976, 1977, 1980; Denno and Dingle 1981). Individuals are bound by natural selection to maximize the number of progeny they leave in subsequent generations (Stearns 1976). However, the combinations of characteristics (life history strategies) that result in maximum offspring survival are tremendously diverse among the insects and for planthoppers as well (Denno and Dingle 1981; Denno et al. 1991). For example, in planthoppers, *Toya propinqua* deposits many (lifetime fecundity > 1000) small eggs (0.66 mm in length) (Raatikainen and Vasarainen 1990), whereas *Perkinsiella saccharicida* deposits fewer (300), but larger (1.2 mm) eggs (Williams 1957; Fennah 1969). The particular combination of traits that evolves will depend not only on selection but also on the phenotypic trade-offs and costs between traits, the genetic variance and covariance among traits, and phylogenetic constraints (Denno and Dingle 1981; Denno and Roderick 1990).

Three different though nonexclusive hypotheses exist which attempt to associate specific environmental conditions with the occurrence of particular combinations of traits (Stearns 1976, 1977; Tallamy and Denno 1981). These are the *r*- and *K*-selection models of MacArthur and Wilson (1967), the bet-hedging hypothesis formulated by Stearns (1976), and the balanced-mortality hypothesis formalized by Price (1974). *K*-selection is said to occur in resource-limited environments that favor traits (e.g., few, large offspring and long life) which improve competitive ability and defense against natural enemies. Alternatively, *r*-selection favors traits that maximize reproductive rate (high fecundity and early age of first reproduction) in harsh, unstable environments in which density-independent conditions prevail. The bet-hedging hypothesis emphasizes the consequences of fluctuating mortality schedules (Stearns 1976). For instance, when juvenile mortality fluctuates due to unpredictable environmental conditions, selection favors the same set of traits predicted by *K*-selection in stable habitats (e.g., delayed reproduction, small reproductive effort)(see Stearns 1976; Tallamy and Denno 1981). In the instance of unpredictable adult mortality in stable and fluctuating environments, the set of traits predicted by the bet-hedging and *r-K* hypotheses are the same (Schaffer 1974; Tallamy and Denno 1981). The balanced-mortality hypothesis purports that selection will balance high mortality with high fecundity in a hostile environment (Price 1973, 1974).

Regardless of the hypothesis, each assumes that particular combinations of traits will coevolve under certain selection regimes (Tallamy and Denno 1981). The literature provides many examples of trait combinations which do not conform to the predictions of life history theory (reviewed in Stearns 1977). On the one hand, these exceptions point to the weakness inherent in ambiguous or flawed theory (see Stearns 1977; Roff 1992), but, on the other hand, they indicate that fitness may be maximized differently (alternative combinations of traits) for species in the same environment due possibly to differences in trade-offs and correlations between traits and/or design constraints associated with phylogeny (Stearns 1977; Tallamy and Denno 1981; Roff 1990).

Despite the shortcomings of life history theory (see Stearns 1977), most investigators agree that the most rigorous analysis of life history variation is provided through experiment or by comparisons among closely related species, among allopatric populations of the same species, or between the morphs of a species polymorphic for a particular trait (Stearns 1977; Price 1984; Denno et al. 1989). Such comparisons help assure that traits have diverged as a result of differential selection under dissimilar ecological conditions rather than by differences in constraint imposed by phylogeny (see Denno et al. 1991).

For several reasons, delphacid planthoppers make ideal organisms for testing life history hypotheses and investigating patterns of reproduction. First, many species of delphacids are wing dimorphic with both macropterous and brachypterous adults (Ossiannilsson 1978; Waloff 1980; Denno and Roderick 1990). Because reproductive effort and dispersal capability are so intimately related both physiologically and ecologically in insects (Dingle 1972, 1974, 1985; Solbreck 1978; Denno and Dingle 1981; Roff 1986a; Fairbairn and Desranleau 1987), an examination of the integration of traits within wing forms and the conditions which favor a particular wing form can provide powerful insight into the evolution of insect life history strategies (see Denno and Roderick 1990). For example, a comparison of life history traits between wing forms of the same species allows for a critical assessment of trade-offs between reproduction and dispersal capability (Denno et al. 1989).

Second, delphacids inhabit a diversity of habitats ranging from very temporary (fluctuating) to extremely persistent (stable) (Denno 1978; Denno et al. 1991). Thus, predictions concerning life history evolution and habitat stability can be critically tested. Third, many delphacid lineages are sufficiently rich in species so as to allow for interspecific comparisons of life history traits independent of phylogenetic constraint (see Denno et al. 1991). Last, planthoppers are easily raised in the laboratory (Kisimoto 1965; Denno et al. 1985, 1986; Roderick 1987), permitting an experimental analysis of life history variation.

Despite the characteristics of planthoppers which serve to facilitate life history analysis, there is much yet to be learned concerning the forces which shape and constrain the evolution of delphacid life histories. Particularly evident is the need for an experimental approach to life history investigation (in addition to the comparative one) and a better understanding of the genetic basis for life history variation (see Stearns 1976, 1977; Denno and Dingle 1981).

This chapter (1) reviews dispersal capability and the degree of wing dimorphism in the Delphacidae, (2) discusses the genetic and environmental control of wing form, (3) provides evidence for a phenotypic trade-off between flight and reproduction and compares differences in the integration of traits between wing forms, (4) discusses the conditions under which each wing form is successful in the context of trait combinations, (5) examines body size evolution, (6) explores the influence of habitat persistence and structure (host plant architecture) on dispersal characters and associated traits, (7) discusses traits (overwintering stage, voltinism and diapause) which synchronize reproduction temporally with favorable resources, (8) considers spatial patterns of oviposition, egg batch size and egg size, and constraining factors, (9) reviews environmental effects on fitness components, (10) provides evidence for a coordinated life history strategy, and (11) evaluates planthopper life histories in the context of classical theoretical predictions.

Wing Dimorphism

Dispersal polymorphism is a widespread phenomenon among the insects but is particularly evident in the Homoptera (Rose 1972; Harrison 1980;

Dixon 1985; Denno et al. 1989). In delphacid planthoppers, the dimorphism is extreme and involves fully winged individuals capable of flight (macropters) and flightless adults with their wings reduced to varying degrees (brachypters) (Kirkaldy 1906; Okada 1977; Ossiannilsson 1978; Denno et al. 1985; Fig. 4.1). The brachypters of many delphacids (e.g., *Perkinsiella saccharicida*) have reduced mesothoracic wings which barely extend to the middle of the abdomen and metathoracic wings which are

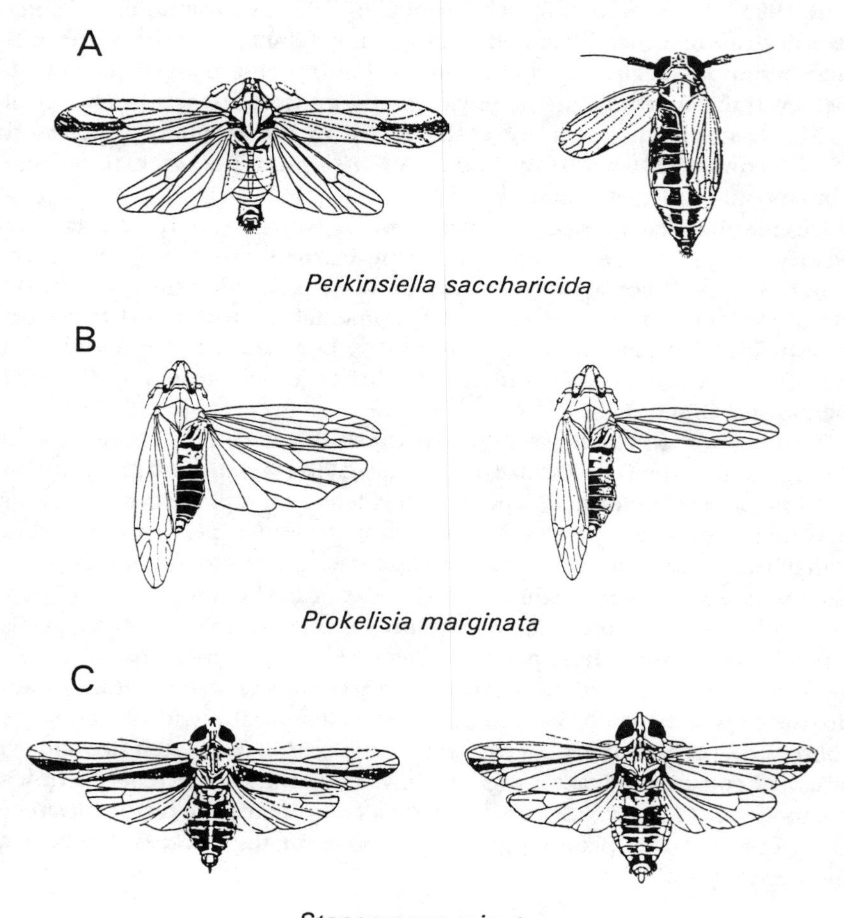

Figure 4.1. (**A**) Macropter (left) and brachypter (right) of *Perkinsiella saccharicida* (from Kirkaldy 1906). (**B**) Macropter (left) and brachypter (right) of *Prokelisia marginata* (from Denno et al. 1985). (**C**) Macropter (left) and "submacropter" (right) of *Stenocranus minutus* (from May 1975).

vestigial (Fig. 4.1A). In other species (e.g., *Prokelisia marginata*), the mesothoracic wings of the flightless brachypter extend to the end of the abdomen but the metathoracic wings remain vestigial (Fig. 4.1B). In a few dimorphic species such as *Stenocranus minutus*, there is a very mobile macropterous morph and a "submacropterous or koeliopterous" form with limited flight capability due to partially reduced mesothoracic and metathoracic wings (May 1975; Waloff 1980; Fig. 4.1C). For planthoppers there is no evidence that flightlessness results from wing muscle histolysis (J. Perfect personal communication), a phenomenon known to occur in other wing-dimorphic insects (Borden and Slater 1969; Dingle 1985).

Brachypters and submacropters move mainly by walking or hopping and their active home range is very limited (<10 m) (Raatikainen 1967; May 1975; Mochida and Okada 1979; Denno et al. 1980). Macropters are capable of long-distance migration by flight, but there is considerable interspecific variation. For example, the macropters of *Sogatella furcifera* and *Nilaparvata lugens* migrate transoceanically from China to Japan (600–1000 km) (Kisimoto 1971, 1976, 1979; Rosenberg and Magor 1983a), but the macropters of most other species disperse much shorter distances (1–3 km) (Raatikainen 1972; Schultz and Meijer 1978; Denno and Grissell 1979; Taylor 1985; Antolin and Strong 1987). Kisimoto (1979, 1981) attributes the exceptional migratory and colonizing ability of the macropters of *S. furcifera* to low wing loading (♂ = 0.0807 mg/mm^2, ♀ = 0.0932 mg/mm^2) and a high proportion of long-duration fliers in the population.

Although many species of delphacids exhibit wing dimorphism [89% of 75 Fennoscandian species in 39 genera are wing-dimorphic (Ossiannilsson 1978) and 72% of 93 Greek species in 45 genera are wing-dimorphic (Drosopoulos 1982, Drosopoulos et al. 1983)], others are monomorphic, being entirely brachypterous (e.g., *Javesella simillima*, Ossiannilsson 1978) or macropterous (e.g., *Saccharosydne saccharivora*, Metcalfe 1969). Field populations of dimorphic delphacids contain both wing forms, but the proportion of each morph can vary tremendously among different species (Denno et al. 1981, 1991; Waloff 1980; Denno and Roderick 1990), among allopatric populations of the same species (Denno and Grissell 1979; Iwanaga et al. 1987; Denno et al. 1991), throughout the season (Waloff 1973; Denno 1976; Cook and Perfect 1985b), and between the sexes (Kisimoto 1965; Denno et al. 1985, 1991; Roderick 1987).

Life history theory assumes that traits such as dispersal capability have a genetic basis, at least in part (Stearns 1976, 1977). Wing form in planthoppers is determined by a developmental switch that responds to environmental cues (Kisimoto 1965; Mochida 1973; Cook and Perfect 1985b; Denno et al. 1985, 1989, 1991; Iwanaga et al 1985; Denno and Roderick 1990). For wing-dimorphic insects, in general, it is thought that the developmental switch is controlled by the level of a hormone (Nijhout and Wheeler 1982; Roff 1986a; Zera and Tiebel 1989; Denno et al. 1991). In the brown planthopper, *N. lugens*, there is strong evidence for the involvement of juvenile hormone in wing-form determination (Iwanaga and Tojo 1986). However, the sensitivity of the switch is heritable and under polygenic control in *Javesella pellucida*, *Nilaparvata lugens*, *Laodelphax striatellus*, and *Prokelisia marginata* (Ammar 1973; Mochida 1973; Mahmud 1980; Roderick 1987; Mori and Nakasuji 1990). Using midparent-offspring regression, Roderick (1987) estimated the heritability of wing form in *P. marginata* and found it to be high, particularly for males (0.42). Similarly, Mori and Nakasuji (1990) reported a realized heritibility of 0.36 for macroptery in *L. striatellus*. Thus, selection for increased or decreased macroptery should result in rapid change (Roff 1990), a prediction that was verified by artificial selection experiments with *P. marginata* (Roderick 1987).

Various environmental cues such as crowding, host plant quality, temperature, and photoperiod trigger the developmental switch and influence wing form (Johno 1963; Kisimoto 1965; Mochida 1973; Drosopoulos 1977; Saxena et al. 1981; Denno et al. 1985; Iwanaga and Tojo 1986; Iwanaga et al. 1987). Of the factors known to affect wing form, population density is by far the most influential (Johno 1963; Kisimoto 1965; Mori and Kiritani 1971; Takagi 1972; Fisk et al. 1981; Cook and Perfect 1985b; Denno et al. 1985; Iwanaga and Tojo 1986; Denno and Roderick 1990). In most delphacids, the production of macropters is generally density-dependent and is often escalated on nutritionally inadequate (nitrogen deprived, aging, senescing, or wilting) host plants (Kisimoto 1965; May 1975; Saxena et al. 1981; Zhang 1983; Cook and Perfect 1985b; Denno et al. 1985, 1991; Iwanaga and Tojo 1988). However, the threshold density that triggers the production of macropters varies within and among species (Denno et al. 1991). Unlike flightless brachypters, macropters are able to escape local fitness-reducing conditions on deteriorating plants and colonize distant habitats (Denno et al. 1985, 1986; Denno and Roderick 1990).

Trade-offs Between Reproduction and Dispersal

The mere occurrence of flight dimorphism suggests that there are fitness costs associated with the ability to fly (Roff 1984). Energy used to construct wings and wing muscles and fuel the flight process may not be available for reproductive investment (Denno et al. 1989). Reproductive costs associated with dispersal capability become evident when traits are compared between the two wing forms of planthoppers (Denno and Rod-

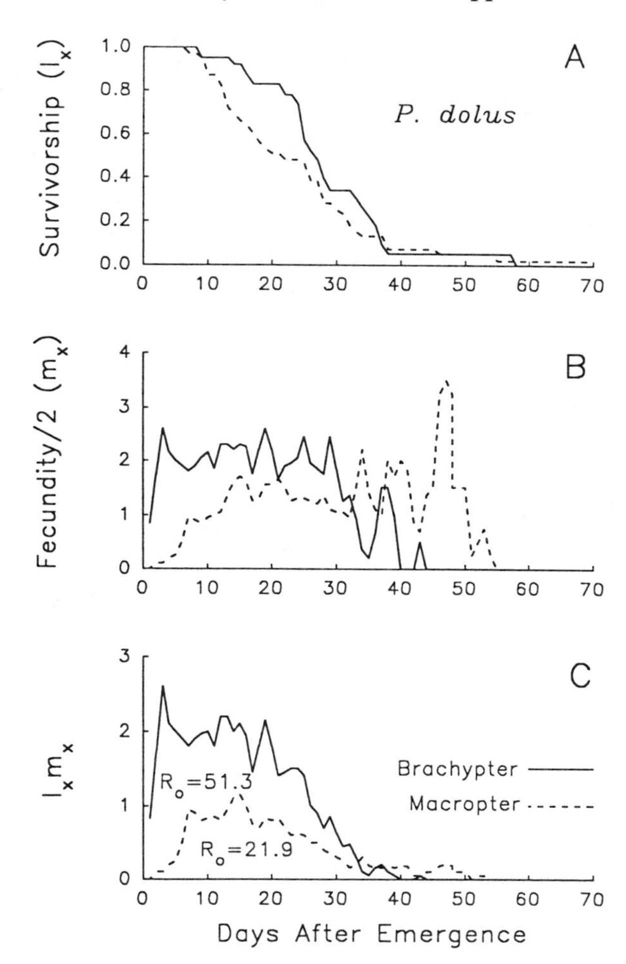

Figure 4.2. (A) Survivorship, (B) fecundity, and (C) reproductive function schedules for the macropterous and brachypterous females of *Prokelisia dolus*. Compared to macropters, brachypters survive longer, reproduce earlier in life, are more fecund and have a much greater net replacement rate ($R_o = \Sigma l_x m_x$) (from Denno et al. 1989).

erick 1990). For example, the macropters of *Laodelphax striatellus*, *Nilaparvata lugens*, *Prokelisia dolus*, and *Sogatella furcifera* have proportionally larger thoraces and exhibit a greater investment in flight apparatus than brachypters (Kisimoto 1965; Denno et al. 1989). However, macropterous females of these species and most other planthopper species as well are

Table 4.1. Comparison of the nymphal development time (days), age to first reproduction (days), total lifetime fecundity, and longevity (days) between the brachypterous (B) and macropterous (M) females of delphacid planthoppers. Extracted from Denno et al. (1989).

Species	Nymphal Development		Days to First Reproduction		Fecundity		Longevity		Reference
	B,	M	B,	M	B,	M	B,	M	
Javesella pellucida	22.2,	22.0 ns					23.8,	25.2 ns	Ammar (1973)
	29.2,	32.2 ns	5.4,	7.1**	156,	156 ns	20.8,	21.4 ns	Mochida (1973)
Laodelphax striatellus	14.9,	16.6							Kisimoto (1956c)
					158,	119			Tsai et al. (1964)
			3.0,	5.0	555,	570	25.3,	28.3	Kisimoto (1965)
					580,	351[a]	24.9,	22.1[a,e]	Kisimoto (1965)
					106,	61			Nasu (1969)
	23.6,	23.3			88,	110	17.4,	17.7	Mitsuhashi and Koyama (1974)
Muellerianella brevipennis			7.0,	10.0**	161,	155 ns	51.6,	44.5 ns	Drosopoulos (1977)
Muellerianella fairmairei			5.6,	9.4**	357,	331 ns	45.1,	46.3 ns	Drosopoulos (1977)
Nilaparvata lugens					202,	196			Fukuda (1934)
	18.0,	20.3							Kisimoto (1956c)
					406,	236			Kuwahara et al. (1956)
					598,	543			Kuwahara et al. (1956)
					164,	148			Mochida (1964)
					108,	140			Mochida (1964)
	14.7,	15.4	2.7,	6.0	599,	543[e]	26.1,	30.7	Kisimoto (1965)
					486,	422[f]			Manjunath (1977)
	14.3,	14.1	1.5,	6.2	300,	249	22.5,	24.7	Mochida (1970)
Nilaparvata lugens									Kenmore (1980)
24°C					258,	85			
29°C					340,	213			
Field cage					363,	133			

Nilaparvata lugens					Oh (1979)
First mating			289, 294 ns		
Second mating			141, 135 ns		
Third mating			74, 52		
Prokelisia marginata	37.9, 37.4 ns	7.3, 8.5*	81, 77 ns		Denno (1985b)
	27.5, 24.9 ns	16.7, 18.0*	67, 67 ns	16.3, 10.7**	Roderick (1987)
Prokelisia dolus	26.4, 25.1 ns	2.0, 8.1**	103, 44**	27.6, 21.7*	Denno et al. (1989)
Sogatella furcifera	12.9, 14.3				Kisimoto (1956c)
		2.9, 3.7	484, 385[e]	22.9, 24.4	Kisimoto (1965)
Stenocarnus minutus		225, 233	59, 33[b]	282, 273[c]	May (1975)
Consensus[†]	B = M (P = 1.000)	B < M (P = 0.001)	B > M (P = 0.001)	B = M (P = 0.582)	

[a]Emerged from diapausing nymphs.
[b]Mean maximum no. eggs/day.
[c]Females only.
[d]Days to median reproduction.
[e]Reanalyzed by Roff (1986a) and found significant ($P < 0.05$).
[f]Reanalyzed by Roff (1986a) and found not significant ($P > 0.05$).
[g]Days to first oviposition includes egg development period.
Note: $^{**}P < 0.01$, $^{*}P < 0.05$, ns = not significant.
[†]Consensus decisions determined by two-tailed sign test (Sokal and Rohlf 1981).

generally less fecund and reproduce later in life (preoviposition period is significantly longer) than brachypterous females (reviewed in Denno et al. 1989; Fig.4.2; Table 4.1). In macropterous females, the reproductive delay is associated with protracted oogenesis and deferred weight gain, sexual maturity, and mate receptivity (Kisimoto 1965; Mochida 1973; Drosopoulos 1977; Heady and Denno 1991). The reproductive delay and reduced fecundity of the macropterous forms of planthoppers support the notion that flight capability is costly (there are developmental constraints) and that phenotypic trade-offs between flight and reproduction exist (Denno et al. 1989). Furthermore, under field conditions where mortality is high and longevity is much reduced, the difference in realized fecundity between the two wing forms is further exaggerated with the brachypter having the greater advantage (Denno et al. 1989). Consistent with this assertion is the observation that outbreaks of pest planthoppers (*N. lugens* and *Perkinsiella saccharicida*) often arise from local areas of brachypter abundance (Zimmerman 1948; Kisimoto 1965).

Other evidence for costs associated with flight capability also exist. For example, in *Prokelisia dolus*, *P. marginata*, and *Stenocranus minutus*, brachypters survive longer than macropters in the laboratory (May 1975; Roderick 1987; Denno et al. 1989); however, this difference in longevity does not occur in all species (Table 4.1). Increased survival contributes to higher total fecundity because females are iteroparous and produce eggs throughout their entire lives (Denno et al. 1989). Consequently, there is a positive relationship between longevity and fecundity in both wing forms of many planthopper species [*Laodelphax striatellus*, *Nilaparvata lugens*, and *Sogatella furcifera* (Kisimoto 1965); *Javesella pellucida* (Mochida 1973); *Muellerianella brevipennis* and *M. fairmairei* (Drosopoulos 1977); *S. minutus* (Waloff 1980); *P. dolus* (Denno et al. 1989); Fig. 4.3]. Analysis of covariance with longevity as the covariate confirms that brachypterous females of *P. dolus* the same age as macropterous females are proportionally more fecund (Denno et al. 1989; Fig. 4.3). With longevity controlled, the elevated fecundity of the brachypter is also evident for *N. lugens*, *L. striatellus*, and *S. furcifera* (replotted from Kisimoto (1965); Fig. 4.3).

Trade-offs between flight and reproduction in male planthoppers are less evident compared to females. For instance, there is no evidence for a delay in gonadal development in macropterous males of *Javesella pellucida* as there is in females (Mochida 1973). However, brachypterous males of *Nilaparvata lugens* are more successful than macropterous males in obtaining mates under certain conditions (Ichikawa 1982). Also, brachypterous males of *Prokelisia marginata* outlive macropters in the absence of flight (Roderick 1987). These data suggest that there is a reproductive cost associated with flight capability in males. However, the reproductive pen-

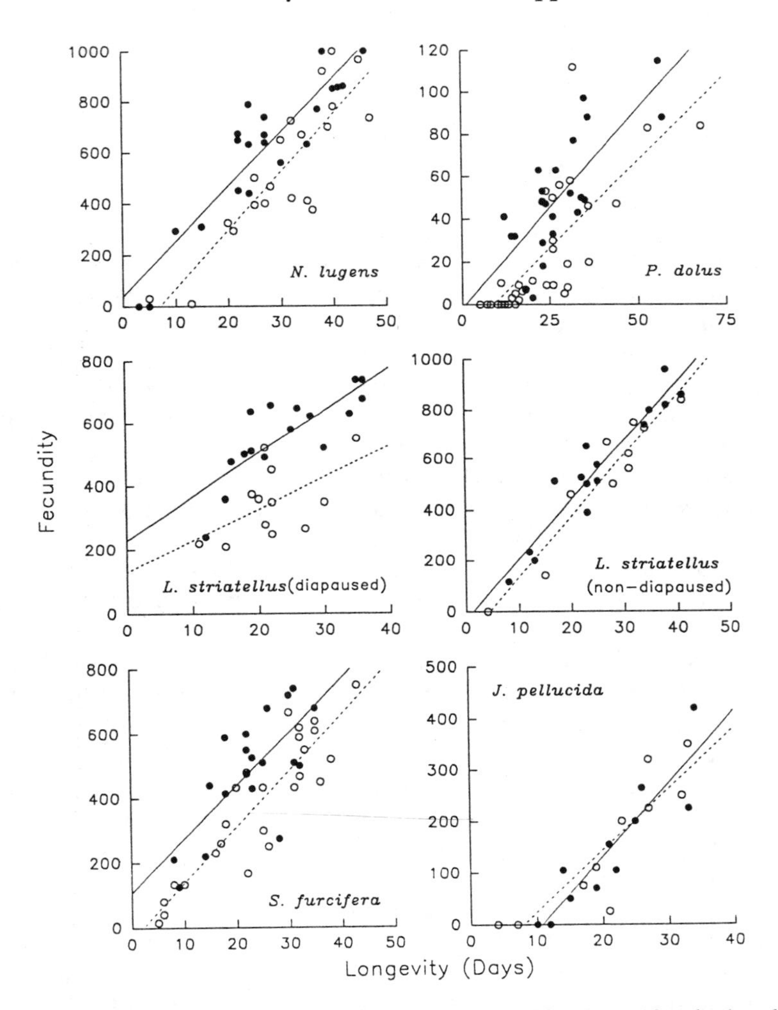

Figure 4.3. Relationship between lifetime fecundity and longevity for the brachypters (●) and macropters (○) of *Nilaparvata lugens*, *Prokelisia dolus*, *Laodelphax striatellus* (both diapaused and nondiapaused individuals), *Sogatella furcifera*, and *Javesella pellucida*. There is a significant relationship between fecundity and longevity in both macropterous and brachypterous females of all species shown ($P<0.01$; Kisimoto 1965, Mochida 1973, Denno et al. 1989). Data also show that brachypters are more fecund than macropters for *P. dolus* ($P<0.01$, ANCOVA with longevity as covariate, Denno et al. 1989), *N. lugens*, diapaused *L. striatellus*, and *S. furcifera* [$P=<0.001$, fecundity data from Kisimoto (1965) reanalyzed using ANCOVA with longevity as covariate].

alty for flight may be less in males than females due to their smaller gamete size.

Wing-form Success and the Integration of Life History Traits

As long as conditions remain suitable for development in the local habitat, brachypters should be the favored morph (Denno 1983, Denno et al. 1985). Because there is no reproductive delay, the offspring of brachypters should have a "head start" on development. Consequently, a brachypterous female should leave more offspring which grow to a larger size prior to the onset of winter (Denno et al. 1989). Also, brachypters compared to macropters should be better able to balance high offspring mortality (see Waloff 1980; Denno and Roderick 1990) with increased fecundity (Denno et al. 1989). Furthermore, brachypters are better able to spread the risks of reproduction in time due to their extended oviposition period, which results from the combination of early age to first reproduction and increased survivorship (Denno et al. 1989). Last, brachypters appear to reproduce more successfully following periods of stress and exposure to poor food quality. For example, following diapause, brachypters of *Laodelphax striatellus* are much more fecund than macropters, but this difference is minimized in nondiapaused females which have fed continuously as adults (Kisimoto 1965; Fig. 4.3). Also, the preoviposition period of the macropters of *Nilaparvata lugens* is delayed relatively more than that of the brachypters when they are fed on inferior quality food (regrown seedlings of rice); the reproductive delay of the macropter compared to the brachypter is minimized on high quality food (uncut rice seedlings) (Kisimoto 1965).

When the local habitat deteriorates critically, the fitness of the brachypterous morph plummets because of its inability to escape and colonize more favorable habitats elsewhere (Denno 1983). Under these conditions, the advantages of macroptery are realized. Yet, for a migrant to be successful it must effectively colonize the new habitat (Dingle 1978, 1985). For planthoppers, successful colonization involves arriving in the new habitat, locating mates (planthoppers mate after migration, see Denno et al. 1991), and locating favorable host plants for oviposition.

Arrival in the new habitat is facilitated by the ability of macropters to survive for several days in the complete absence of food, as occurs during migration (Kisimoto 1965). On arrival, colonists face the problems of small founding populations and conditions to which they may not be adapted (e.g., resistant plant genotypes) (Baker and Stebbins 1965; MacArthur and Wilson 1967; Safriel and Ritte 1980; Dingle 1985). These factors increase the risk of extinction and can be offset by efficient mate location, genetic variability, and high fecundity (Dingle 1985).

Planthoppers employ a "fly and call" strategy (Hunt and Nault 1991) for mate location whereby males fly among plants in search of virgin females (Ichikawa 1977). After settling on the host plant, males produce acoustical signals (calls) which are detected by females; females respond with return calls and, as duetting takes place, males move to mate with stationary females (Ichikawa 1976; Claridge 1985a). The "fly and call" strategy of mate location may be particularly advantageous under low-density conditions (Ichikawa 1977; Denno and Roderick 1990; Hunt and Nault 1991) as is often the case for colonizing planthoppers in new habitats (Kuno 1979).

Successful establishment in the new habitat is enhanced by high reproductive potential. The reproductive delay and reduced fecundity characteristic of the colonizing forms of planthoppers are not traits that contribute to a high intrinsic rate of increase (Denno et al. 1989). Within the constraints of wing form, however, selection may have acted to integrate traits in ways that maximize fecundity. For example, egg size may be traded off for egg number. The macropterous females of *Javesella pellucida* produce significantly smaller eggs (0.190 mm^3) than brachypterous females (0.203 mm^3) (Mochida 1973). Thus, although reproductive effort (egg size $\times$ total fecundity) is reduced for macropters (2.964) compared to brachypters (3.167), the total fecundity of the two wing forms is very similar (156 eggs/female) (Mochida 1973). Also, within the genus *Prokelisia*, there is much less difference between the fecundities of the macropter and brachypter of the migratory *P. marginata* than there is for the sedentary *P. dolus* (Table 4.1). A comparison of the egg sizes of the wing forms of these two species may elucidate differences in an egg size/fecundity trade-off between migratory and sedentary species. However, across genera there is no overall tendency for migratory species (>50% macroptery in field populations) to be more fecund than sedentary species (<50% macroptery in field populations) (Fig. 4.4A). In contrast, there is a tendency for the preoviposition period (6.0 $\pm$ 1.8 days, $x \pm$ SD) of the macropters of migratory species to be less than that (9.7 $\pm$ 1.0 days, $x \pm$ SD) of the macropters of sedentary species (Fig. 4.4B). These data suggest, but not in any rigorous way, that selection may have acted on the macropterous wing form of migratory species to integrate traits such that fecundity is enhanced.

Macropters also have the ability to selectively colonize and oviposit on nutrient-rich host plants where offspring performance and survival are enhanced (e.g., *Prokelisia marginata*; Denno and Grissell 1979; Denno 1983, 1985b; Denno et al. 1980, 1985, 1986, 1987). Thus, mobility provides macropters with an element of flexibility not possessed by brachypters

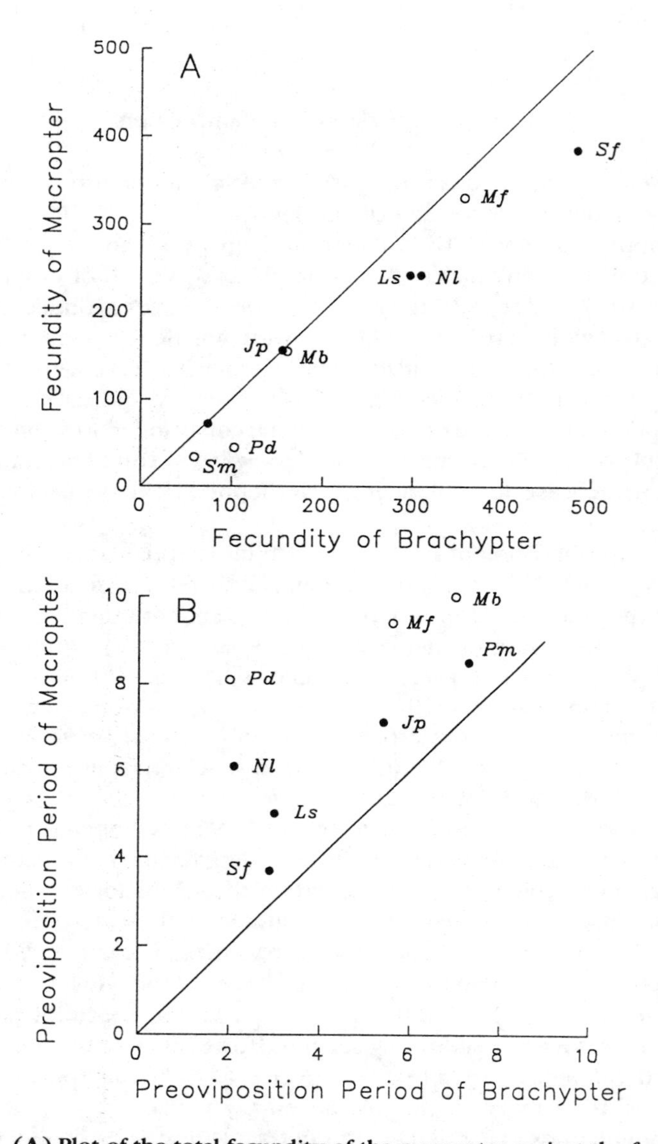

Figure 4.4. **(A)** Plot of the total fecundity of the macropter against the fecundity of the brachypter for five migratory (● = >50% macroptery in field populations) and four sedentary (○ = <50% macroptery in field populations) species of delphacids. Fecundities do not differ between the macropters or brachypters of migratory and sedentary species (*t*-test, $P>0.05$). **(B)** Plot of the preoviposition period (days) of the macropter against the preoviposition period of the brachypter for five migratory (● = >50% macroptery in field populations) and three sedentary (○ = <50% macroptery in field populations) species of delphacids. The preoviposition period for macropters of migratory species is significantly shorter (6.08±1.85 days) than that for the macropters of sedentary species (9.67±0.97 days) ($t = -2.63$, $P = 0.03$). The preoviposition periods for brachypters of migratory and sedentary species do not differ (*t*-test, $P>0.05$). Jp = *Javesella pellucida*, Ls = *Laodelphax striatellus*, Mb = *Muellerianella brevipennis*, Mf = *Muellerianella fairmairei*, Nl = *Nilaparvata lugens*, Pd = *Prokelisia dolus*, Pm = *P. marginata*, Sf = *Sogatella furcifera*, Sm = *Stenocranus minutus* (data from Table 4.1).

which better allows for the spread of reproductive risks in space (Waloff 1980; Denno and Roderick 1990).

A common "colonization syndrome" exists for several multivoltine planthoppers (*Nilaparvata lugens*, *Peregrinus maidis*, and *Sogatella furcifera*), which illustrates the adaptive nature of a polymorphic life history (Denno and Roderick 1990). Macropters colonize new habitats at low densities, but their offspring are mostly brachypterous (Kisimoto 1977; Dyck et al. 1979; Kuno 1979; Fisk et al. 1981). Even though the initial colonists have reduced fecundity, their brachypterous offspring have a much higher reproductive potential which contributes to rapid population growth and establishment.

Body Size, Rapid Development, and Migration

Body size, by virtue of correlations with fecundity, development time, age to first reproduction, and survival, reflects an integration of a large number of traits (Roff 1978, 1981, 1991; Calder 1984; Barbault 1988; Fairbairn 1992). For example, fecundity is positively related to body size in the planthopper, *Prokelisia marginata* (Denno and McCloud 1985). Furthermore, bioenergetic factors favor large size in migrating insects (Roff 1991). The distance an insect can fly increases with body mass which is proportional to the amount of fuel (Roff 1991), and evidence exists for a positive correlation between body size and dispersal capability for species of Hemiptera (Dingle et al. 1980; Fairbairn 1984), Diptera (Roff 1977; McLachlan 1983), Coleoptera (Davis 1986), and Lepidoptera (Roff 1991). Thus, body size is a useful metric for analyzing the collective evolution of life history traits including migration (Roff 1991).

Life history theory predicts that large body size should be favored in stable environments where predation and competition are more frequent (see Schoener and Janzen 1968; Stearns 1976; Blau 1981; Roff 1981). Small body size and rapid development are predicted by the *r-K* hypothesis in ephemeral habitats where mortality is density-independent, populations fluctuate violently, and frequent recolonization is necessary [see review by Stearns (1976)]. Yet, individuals of the waterstrider, *Limnoporus notabilis*, from ephemeral ponds were larger and developed more rapidly than individuals from permanent lake habitats (Fairbairn 1984). The large size of striders in pond habitats was related to dispersal ability because these habitats dry up and must be recolonized each spring. This example attests to the inadequacy of life history theory in predicting which combination of traits will evolve under a specific set of conditions.

It is generally agreed that migration allows insects to keep pace with changing (ephemeral) resources (Southwood 1962, 1977a; Denno et al.

1980; Dingle 1985; Denno et al. 1991; Roff 1991). If habitats have an uncertain future, selection should favor rapid development, but not at the expense of body size (Roff 1991). Thus, selection for migration should favor rapid growth, early eclosion, and large body size for energetic reasons (Fairbairn 1984). Nevertheless, rapid growth and large body size may be developmentally antagonistic and a trade-off may exist in some lineages precluding the maximization of both traits.

The hypothesis that the body size of migratory species should be larger than their nonmigratory counterparts (see Roff 1991) was tested by examining the body sizes of migratory and sedentary planthoppers. Because body size may be constrained by phylogeny, the assessment was restricted to congeneric comparisons. The migratory capability of each species was estimated from the wing-form composition of females in field populations. Species were sorted into three categories: **migratory species** in which macropters outnumbered brachypters in field populations (M>B), **sedentary species** in which at least 90% of females were brachypterous (B), and **intermediate species** in which brachypters outnumbered macropters and constituted between 50 and 90% of the population (B>M). Sufficient data were available for analysis on species in four genera: *Javesella*, *Nilaparvata*, *Prokelisia*, and *Sogatella*.

The mean body length of the migratory and sedentary species categories were compared across genera using a paired *t*-test. Migratory species were significantly larger than sedentary species in all four genera ($t = 7.276$, $P = 0.005$; Fig. 4.5). Moreover, the most endorsed long-distance migrators (*J. pellucida*, *N. lugens*, and *S. furcifera*; Raatikainen 1967; Kisimoto 1976, 1979, 1987) were the largest of their congeners. Even more convincing evidence that large body size is favored in migratory species comes from *P. dolus* and *P. marginata* for which there is variation in wing form (migratory capability) among populations (see Table 4.2). For both species, adults from the more migratory populations (New Jersey) were larger in body size than their sedentary counterparts (Gulf coast of Florida); the body size difference was particularly great in *P. dolus* (Fig. 4.5; ANOVA, $P < 0.05$; Denno and Roderick unpublished data). Moreover, population differences in body size persisted when these planthoppers were raised under the same conditions in the laboratory, suggesting that there is a genetic basis for body size variation.

Large body size was associated with short development time in the more macropterous New Jersey population of *Prokelisia dolus* (Denno and Roderick unpublished data). Individuals from New Jersey required 23 days for development from nymph (first instar) to adult, whereas those from Florida took 25 days (ANOVA, $P < 0.05$). These data are consistent with the results of Fairbairn (1984) in that large body size and rapid development are

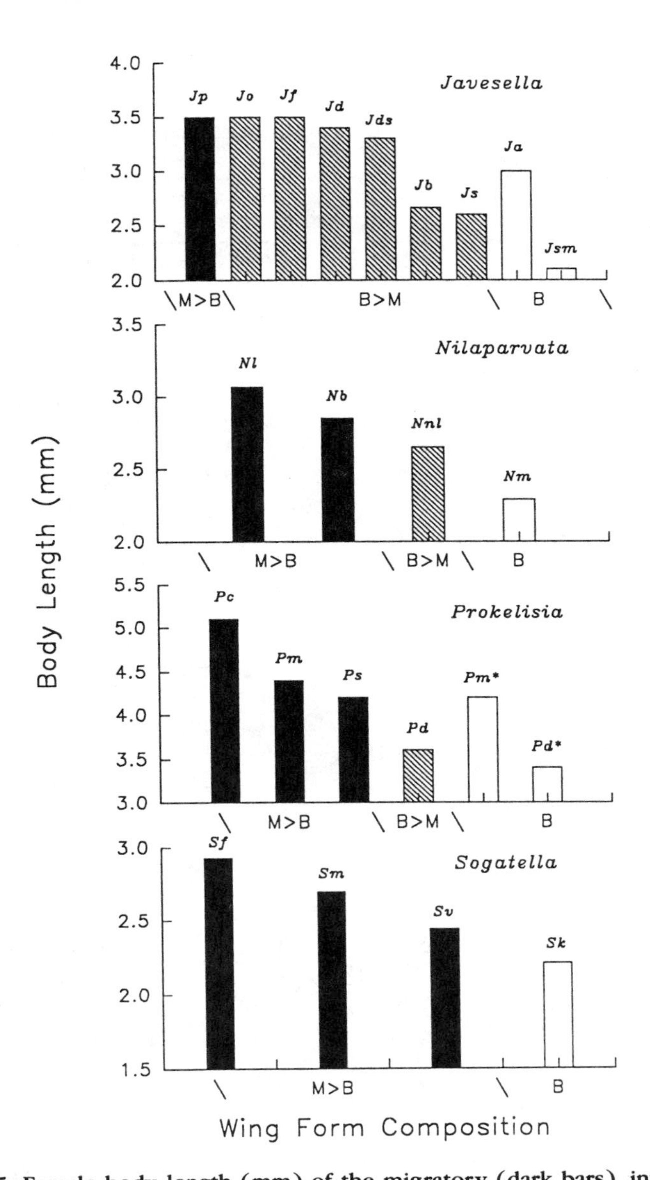

Figure 4.5. Female body length (mm) of the migratory (dark bars), intermediate (hatched bars), and sedentary species (open bars) of *Javesella*, *Nilaparvata*, *Prokelisia*, and *Sogatella*. Species are categorized on the basis of the wing-form composition of field populations: migratory = macropters outnumber brachypters, intermediate = brachypters constitute 50–90% of wing forms, sedentary = >90% of individuals brachypterous. Within a genus, migratory species are significantly larger than sedentary congeners (paired t-test, $P = 0.005$). Data for: *J. pellucida, J. obscurella, J. forcipata, J. dubia, J. discolor, J. bottnica, J. salina, J. alpina, and J. simillima* extracted from Ossiannilsson (1978); *N. lugens, N. bakeri, N. muiri,* and *N. nr. lugens* (*Leersia*) from specimens in University of Wales, Cardiff collection and Okada (1977); *S. furcifera, S. manetho, S. vibex*, and *S. krugeri* from specimens in British Museum collection; *P. crocea, P. marginata, P. salina,* and *P. dolus* from Wilson (1982a) and R.F. Denno collection. Sedentary (* = Gulf coast) and more migratory (Atlantic coast) populations of *P. marginata* and *P. dolus* are included.

Table 4.2. Geographic variation in the wing-form composition of delphacid planthopper populations.

Species	Geographic Location	Latitude	Macroptery (%)	Reference
Javesella pellucida	Iceland	65°N	Mostly Brachypters[F]	Lindroth et al. (1973)
	UK (England)	52°N	75[F]	Waloff (1980)
Megadelphax sordidulus	Finland	62°30′–65°N	85[F]	Raatikainen and Vasarainen (1976)
		61°–62°30′N	96[F]	
		59°30′–61°N	94[F]	
Nilaparvata lugens	Japan	33°N	70–80[L]	Nagata and Masuda (1980)
	Taiwan	24°N	70[L]	
	Thailand	15°N	59–70[L]	
	Philippines	15°N	10[L]	
Nilaparvata lugens	Japan	25–36°N	70–90 (variable)[H,14]	Iwanaga et al. (1987)
	Taiwan	24°N	40–95 (variable)[H,3]	
	China	24–31°N	25–60 (variable)[H,3]	
	Philippines	15°N	0[H,1]	
	Malaysia	6°N	0[H,1]	
	Indonesia	7°S	0[H,2]	
Prokelisia dolus	USA (NJ)	39°N	20–30[F]	Denno et al. (1987)
	USA (So. CA)	33–34°N	1–5[F]	Denno and Roderick (unpublished data)
	USA (FL, Gulf)	29°N	1–10[F]	Denno and Roderick (unpublished data)
Prokelisia marginata	USA (NJ)	39°N	85[F]	Denno et al. (1987)
	USA (No. CA)	37–38°N	40–70[F]	Roderick (1987); Denno and Roderick (unpublished data)
	USA (FL, Gulf)	29°N	10[F]	Denno and Grissell (1979); Denno and Roderick (unpublished data); McCoy and Rey (1981)

Note: F = macroptery (%) in field populations. L = macroptery (%) in the offspring of planthoppers cultured in the laboratory from field-collected parents. H = macroptery (%) in the female offspring of planthoppers raised at high experimental densities in laboratory lines; each line was initiated from parents field collected at a geographically distinct location. The number of different source populations within each country is indicated [see Iwanaga et al. (1987) for details].

associated with dispersal in unstable habitats. Surprisingly, traits were integrated differently for *P. marginata*. Large body size was associated with extended development (29 days) in the migratory population of this species; individuals from the sedentary Florida population were smaller and completed nymphal development in only 25 days (ANOVA, $P<0.05$; Denno and Roderick unpublished data).

Despite interspecific differences in the relationship between body size and development time, planthoppers provide strong evidence for the association between flight capability and large body size. It is essential to take into account the constraints of ancestry when exploring this relationship because there is no apparent relationship between dispersal ability and body size across planthopper genera (Kisimoto personal communication). Although large body size in migratory species may be favored for energetic reasons, it may also reflect selective pressures associated with increased fecundity in colonizing species. Examination of the difference in shape between morphs may elucidate this issue. If selection is on fecundity, then abdominal size should be dimorphic, but if selection is on the energetics of flight, then a thoracic dimorphism should be promoted (Fairbairn 1992).

Habitat Persistence and Migration

Habitat persistence figures prominently in the predictions of life history theory and in the evolution of insect migration (Southwood 1962, 1977a; Stearns 1976, 1977; Harrison 1980; Dingle 1985; Roff 1986b, 1990; Denno et al. 1991). Generally, migration is thought to be critical for the success of species which exploit temporary habitats (Southwood 1962, 1977a; Solbreck 1978; Dingle 1985; Roff 1986a, 1986b, 1986c, 1990; Denno et al. 1991). However, there are few rigorous assessments of the relationship between habitat age (persistence) and the dispersal ability of the inhabitants. One example is provided from the planthopper literature.

Denno et al. (1991) investigated the relationship between habitat persistence (maximum number of generations attainable during the existence of the habitat) and the wing-form composition of planthopper populations in the field. They found that levels of migration (percent macroptery) decreased significantly as the persistence of the habitat increased (Fig. 4.6A). Species inhabiting temporary habitats like agricultural crops were primarily macropterous, whereas species occurring in permanent habitats like bogs and marshlands were for the most part brachypterous. This analysis was restricted to species exploiting low-profile vegetation ($<$1–2 m in height) because host plant dimensionality (architecture) is known to influence the

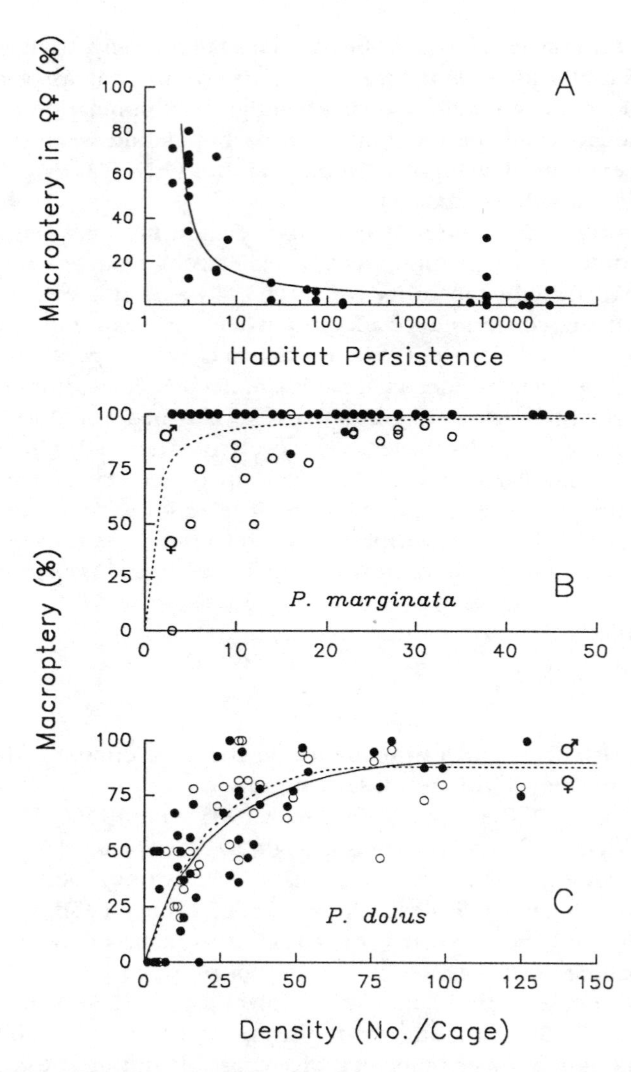

Figure 4.6. (**A**) Relationship between macroptery (%) in field populations of 35 species of delphacids and the persistence of their habitats (maximum number of generations attainable during the existence of the habitat). (**B**) Wing-form (percent macroptery) responses to rearing density in both sexes of *Prokelisia marginata* and (**C**) *P. dolus*. Replotted from Denno et al. (1991).

evolution of flight in insects (Denno 1978, 1979; Waloff 1983; Roff 1990; Denno and Roderick 1991).

Using phylogenetic contrasts between congeners, the same result was obtained suggesting that habitat persistence influences levels of migration independent of common ancestry (Denno et al. 1991). Also, there was a clear difference in the density–wing-form response between species in temporary (Fig. 4.6B) and persistent habitats (Fig. 4.6C). Generally, macropterous forms were triggered at a lower density for species (e.g., *Prokelisia marginata*) inhabiting temporary habitats compared to species in persistent habitats (e.g., *P. dolus*). However, males of species in temporary habitats were macropterous at both low and high rearing densities; macroptery was positively density dependent on the females of these species and on both sexes of species in permanent habitats (Denno et al. 1991). These results suggest that habitat persistence influences the dispersal capability and associated reproductive potential of planthoppers in two ways: (1) by selecting against habitat escape and (2) by dictating the availability of mates (Denno et al. 1991). In persistent habitats, wings are less necessary for habitat escape and they are rarely required for mate location. As a consequence and because wings impose a reproductive penalty, flightlessness is favored. In temporary habitats, wings are favored in males to locate females at low colonizing densities and are favored in both sexes for reasons of habitat escape.

Geographic Variation in the Wing-form Composition of Populations

Because rates of succession proceed more slowly at higher latitudes, Roff (1990) convincingly argues that the positive relationship between flightlessness and latitude for many insects results from clinal variation in habitat persistence. Data for some species of planthoppers are consistent with this hypothesis. For example, flightlessness (percent brachyptery) increases to the north in Finnish populations of *Megadelphax sordidulus* (Raatikainen and Vasarainen 1976; Table 4.2), and Sahlberg (1871) reports that an increase in brachyptery to the north is a general pattern for most dimorphic planthopper species in Europe. The higher levels of brachyptery in Icelandic compared to British populations of *Javesella pellucida* are probably also attributable to an increase in habitat persistence at high latitudes and not to any "oceanic island" phenomenon. For instance, Roff (1990) has rigorously demonstrated that contrary to "conventional wisdom," there is no tendency for oceanic islands to have higher frequencies of flightless insect species than comparable mainland sites when effects of latitude and altitude are removed.

Flightlessness in populations of two salt marsh-inhabiting planthoppers, *Prokelisia marginata* and *P. dolus*, does not increase with latitude; in fact, macroptery increases to the north. In both species, Atlantic coast populations (39°N) are much more macropterous than Gulf coast populations (Denno and Grissell 1979; McCoy and Rey 1981; Denno et al. 1987; Table 4.2). Nevertheless, the pattern reflects geographic differences in habitat persistence. Low marsh habitats where nymphs develop during summer do not persist through winter on the Atlantic coast, but do so on the Gulf coast where conditions are less harsh. Consequently, flight is favored in more northern populations because macropters can migrate from low to high marsh habitats prior to the onset of winter (Denno and Grissell 1979; Denno et al. 1985, 1987).

In some Gulf coast areas and in all Southern California locations, populations of *Prokelisia dolus* are almost completely brachypterous (>99%; Denno and Roderick unpublished data). In addition to the persistent nature of habitats in these areas (Denno and Grissell 1979), the small size of host plant patches (<10 m^2) and their isolation from one another may further intensify the advantages of flightlessness (McCoy and Rey 1981). Winged individuals are simply lost from very small, isolated patches placing a high selective premium on the evolution of flightlessness (Roff 1986a, 1990).

Geographic variation in the wing-form composition of populations of *Nilaparvata lugens* probably also results from differences in habitat persistence (Table 4.2). Japanese and Chinese populations are more macropterous than tropical populations in S.E. Asia (Nagata and Masuda 1980; Iwanaga et al. 1987). *N. lugens* does not overwinter successfully in Japan and Korea and populations are founded annually by migrants from lower latitudes (Kisimoto 1976, 1979). Thus, Japan and Korea can be viewed on a large scale as temporary habitats, which are colonized only by long-distance migrants. Although this example illustrates the selective elimination of flightless genes in northern Asia, it is essential to emphasize that no individuals survive the harsh winters in Japan. The wind-assisted migrations of *N. lugens* are probably an artifact of a dispersal "strategy" that is favored further to the south on mainland Asia (Denno and Roderick 1990).

For *Prokelisia marginata* and *Nilaparvata lugens*, geographic differences in the wing-form composition of populations have a genetic basis (Iwanaga et al. 1987; Denno et al. 1991). Significantly more adults of *P. marginata* emerged macropterous in the Atlantic compared to the Gulf coast population with host plant quality and rearing density controlled (Denno et al. 1991). Similarly, adults from temperate Asian populations of *N. lugens* emerged more macropterous than adults from tropical Asia

(Iwanaga et al. 1987). These data strongly suggest that the density-related sensitivity of the developmental switch which controls wing form differs among populations and is under genetic control (Denno et al. 1991).

Habitat Structure and Its Constraints on Life History Evolution

Host plant architecture (tree, shrub, or herb) or the dimensionality of the habitat (high- or low-profile vegetation) can have a dramatic influence on the evolution of flight in insects (Reuter 1875; Kirkaldy 1906; Marshall 1916; Denno 1978, 1979; Waloff 1983; Roff 1990; Denno and Roderick 1991). Two-dimensional habitats close to the ground can be reached by walking, but three-dimensional, arboreal habitats are difficult to negotiate without flight (Roff 1990). Brachypters which lose their holds (e.g., by jumping from predators) on large trees and shrubs may not be able to relocate easily (Denno 1978, 1979). Thus, selection should favor the retention of flight capability in arboreal species even though their habitats are persistent (Denno 1978; Roff 1990). The rarity of brachypterous and wing-dimorphic planthoppers in arboreal habitats has been noted before (Reuter 1875; Kirkaldy 1906; Denno 1978, 1979), but the habitat dimension hypothesis has not yet been rigorously tested for this group of insects.

The habitat dimensionality–flight capability hypothesis was tested by comparing the wing-form composition of delphacid species with the growth form of the host plant species on which they occur. Because most temperate delphacid species are monophagous on grasses and sedges and occupy low-profile vegetation (Denno et al. 1991), few comparisons with arboreal species are available. By contrast, most Hawaiian planthoppers, although monophagous, feed on a wide variety of host plants which vary in structure from herbs to very tall trees (Zimmerman 1948). Data on the wing-form composition (macropterous, brachypterous, or wing-dimorphic) of 122 species of delphacids (12 genera), their host plant records and the growth form (midpoint of height range in meters) of host plants were obtained from Giffard (1922), Zimmerman (1948), and Wagner et al. (1990) (see Table 4.3). Importantly, most of the host plants of native Hawaiian delphacids occur in habitats that are relatively stable (e.g., wet forests). Thus, the analysis was not confounded by major differences in the persistence of habitats. Fugitive species feeding on ephemeral agricultural crops (e.g., *Pekinsiella saccharicida*) and wing-dimorphic species were excluded from the analysis.

Macropterous species fed mostly on trees and large shrubs (7.1 ± 0.6 m, $\bar{x}\pm$SE, $n=52$), whereas brachypterous and mostly brachypterous species

Table 4.3. Wing form composition (macropterous, brachypterous, or wing-dimorphic) of Hawaiian Delphacid populations and the growth form of their host plants.

Planthopper Species	Host Plant Species[a]	Host Plant Family	Host Plant Growth Form[b]	Wing-Form Composition[c]
Leialoba hawaiiensis	*Metrosideros polymorpha*	Myrtaceae	Tree (7–10 m)	Macropterous
Leialoba kauaiensis	*Metrosideros polymorpha*	Myrtaceae	Tree (7–10 m)	Macropterous
Leialoba lanaiensis	*Metrosideros polymorpha*	Myrtaceae	Tree (7–10 m)	Macropterous
Leialoba lehuae	*Metrosideros polymorpha*	Myrtaceae	Tree (7–10 m)	Macropterous
Leialoba mauiensis	*Metrosideros polymorpha*	Myrtaceae	Tree (7–10 m)	Macropterous
Leialoba naniicola	*Metrosideros polymorpha*	Myrtaceae	Tree (7–10 m)	Macropterous
Leialoba oahuensis	*Metrosideros polymorpha*	Myrtaceae	Tree (7–10 m)	Macropterous
Leialoba oceanides	*Nestegis sandwicensis*	Oleaceae	Tree (8 m)	Macropterous
Leialoba ohiae	*Metrosideros polymorpha*	Myrtaceae	Tree (7–10 m)	Macropterous
Leialoba scaevolae	*Scaevola chamissoniana*	Goodeniacceae	Shrub/small tree (1.5–2.5 m)	Macropterous
Leialoba suttoniae	*Myrsine sandwicensis*	Mysinaceae	Small tree (1.5–4 m)	Macropterous
Nesothoë antidesmae	*Antidesma platyphyllum*	Euphorbiaceae	Tree (4–10 m)	Macropterous
Nesothoë bobeae	*Bobea*	Rubiaeae	Tree (10 m)	Macropterous
Nesothoë dodonaeae	*Alphitonia ponderosa*	Rhamnaceae	Tree (4–25 m)	Macropterous
	Dodonaea viscosa	Sapindaceae	Shrub/tree (2–8 m)	Macropterous
	Myrsine	Mysinaceae	Small tree (1.5–4 m)	Macropterous
Nesothoë dryope	*Antidesma platyphyllum*	Euphorbiaceae	Tree (4–10 m)	Macropterous
Neosthoë elaeocarpi	*Cyrtandra paludosa*	Gesneriaceae	Shrub (1–5 m)	Macropterous
	Elaeocarpus bifidus	Elaeocarpaceae	Tree (2.5–10 m)	Macropterous
	Scaevola mollis	Goodeniaceae	Large shrub (1–2.5 m)	Macropterous
Nesothoë eugeniae	*Psychotria kaduana*	Rubiaceae	Tree (2–4 m)	Macropterous
	Syzygium sandwicensis	Myrtaceae	Tree (3–25 m)	Macropterous
Nesothoë fletus	*Antidesma platyphyllum*	Euphorbiaceae	Tree (4–10 m)	Macropterous
	Myrsine	Mysinaceae	Small tree (1.5–4 m)	Macropterous

Nesothoë giffardi	*Cyrtandra grandiflora*	Gesneriaceae	Shrub (1–2 m)	Macropterous
	Rolandia crispa	Campanulaceae	Shrub (1–2 m)	Macropterous
	Touchardia latifolia	Urticaceae	Large shrub (1–3 m)	Macropterous
Nesothoë gulicki	*Metrosideros polymorpha*	Myrtaceae	Tree (7–10 m)	Macropterous
	Nestegis sandwicensis	Oleaceae	Tree (8 m)	Macropterous
Nesothoë haa	*Antidesma platyphyllum*	Euphorbiaceae	Tree (4–10 m)	Macropterous
Nesothoë hula	*Myrsine*	Mysinaceae	Small tree (1.5–4 m)	Macropterous
	Nestegis sandwicensis	Oleaceae	Tree (8 m)	Macropterous
	Pelea	Rutaceae	Tree (2–8 m)	Macropterous
	Pouteria sandwicensis	Sapotaceae	Tree (12 m)	Macropterous
Nesothoë laka	*Sida*	Malvaceae	Erect Shrub (0.5–1.5 m)	Macropterous
Nesothoë maculata	*Diospyros hillebrandii*	Ebenaceae	Tree (4–7 m)	Macropterous
	Diospyros sandwicensis	Ebenaceae	Tree (2–10 m)	Macropterous
	Nestegis sandwicensis	Oleaceae	Tree (8 m)	Macropterous
Nesothoë munroi	*Dodonaea viscosa*	Sapindaceae	Shrub/tree (2–8 m)	Macropterous
Nesothoë perkinsi	*Clermontia kakeana*	Campanulaceae	Shrub/tree (1–6 m)	Macropterous
	Metrosideros polymorpha	Myrtaceae	Tree (7–10 m)	Macropterous
	Myrsine	Mysinaceae	Small tree (1.5–4.0 m)	Macropterous
Nesothoë piilani	*Nestegis sandwicensis*	Oleaceae	Tree (8 m)	Macropterous
Nesothoë pluvialis	*Antidesma platyphyllum*	Euphorbiaceae	Tree (4–10 m)	Macropterous
Nesothoë semialba	*Nestegis sandwicensis*	Oleaceae	Tree (8 m)	Macropterous
Nesothoë seminigrofrons	*Bidens*	Asteraceae	Herb/shrub (0.1–4 m)	Macropterous
Nesothoë terryi	*Nestegis sandwicensis*	Oleaceae	Tree (8 m)	Macropterous
Nesodryas freycinetiae	*Freycinetia arborea*	Pandanaceae	Woody climber[V]	Macropterous
Nesodryas swezeyi	*Prichardia*	Arecaceae	Tree (4–25 m)	Macropterous
Aloha artemisiae	*Artemisia australis*	Asteraceae	Subshrub (0.2–0.5 m)[GC]	Brachypterous
Aloha campylothecae	*Bidens*	Asteraceae	Herb/shrub (0.1–4 m)	Brachypterous

(*continued*)

Table 4.3. (*Continued*)

Planthopper Species	Host Plant Species[a]	Host Plant Family	Host Plant Growth Form[b]	Wing-Form Composition[c]
Aloha dubautiae	*Dubautia laxa*	Asteraceae	Shrub (1–2 m)[GC]	Brachypterous
	Dubautia plantaginea	Asteraceae	Shrub (1–2 m)[GC]	Brachypterous
Aloha flavocollaris	*Dubautia plantaginea*	Asteraceae	Shrub (1–2 m)[GC]	Brachypterous
Aloha ipomoeae	*Ipomoea* (four species)	Convolvulaceae	Vines (procumbent)[V]	Wing-dimorphic[B]
Aloha kirkaldyi	*Chamaesyce multiformis*	Euphorbiaceae	Shrub (0.5–2 m)[GC]	Brachypterous
Aloha myoporicola	*Myoporum sandwicense*	Myoporaceae	Shrub (1–2 m)[GC]	Brachypterous
Aloha plectranthi	*Plectranthus parviflorus*	Lamiaceae	Perennial herb (0.1–0.5 m)	Brachypterous
Aloha swezeyi	*Bidens macrocarpa*	Asteraceae	Herb (0.5–2 m)	Wing-dimorphic[B]
	Bidens pilosa	Asteraceae	Annual herb (0.3–1.8 m)	Wing-dimorphic[B]
Nesorestias filicicola	*Cibotium*	Dicksoniaceae	Tree fern (2–7 m)	Brachypterous
	Elaphoglossum gorgonum	Lomariopsidaceae	Fern (1 m)[GC] (also epiphytic)	Brachypterous
Nesorestias nimbata	*Phegopteris*	Thelypteridiaceae	Fern (1 m)[GC]	Brachypterous
Nothorestias badia	Ferns	Thelypteridiaceae	?	Brachypterous
Nothorestias swezeyi	*Aspidium*	Aspidiaceae	Ferm (<1 m)[GC]	Brachypterous
Dictyophorodelphax mirabilis	*Chamaesyce clusiifolia*	Euphorbiaceae	Shrub (1 m)	Brachypterous
	Chamaesyce multiformis	Euphorbiaceae	Shrub (0.5–2 m)[GC]	Brachypterous
Dictyophorodelphax praedicta	*Chamaesyce multiformis*	Euphorbiaceae	Shrub (0.5–2 m)[GC]	Brachypterous
Dictyophorodelphax swezeyi	*Chamesyce celastroides*	Euphorbiaceae	Shrub (0.4–2 m)[GC]	Brachypterous

Nesosydne acuta	*Cyrtandra platyphylla*	Gesneriaceae	Shrub (1–2 m)	Brachypterous
Nesosydne ahinahina	*Argyroxiphium sandwicense*	Asteraceae	Rosette shrub (0.5–1.5 m)[GC]	Brachypterous
Nesosydne aku	*Cyanea tritomantha*	Campanulaceae	Palm-like shrub (1.5–2.5 m)	Brachypterous
Nesosydne amaumau	*Sadleria*	Blechnaceae	Fern (1–3 m)[GC]	Brachypterous
Nesosydne anceps	*Freycinetia arborea*	Pandanaceae	Woody climber[V]	Brachypterous
Nesosydne argyroxiphii	*Argyroxiphium sandwicense*	Asteraceae	Rosette shrub (0.5–1.5 m)[GC]	Brachypterous
Nesosydne asteliae	*Astelia menziesiana*	Liliaceae	Perennial herb (0.3–1.5 m)[GC]	Brachypterous
Nesosydne boehmeria	*Boehmeria grandis*	Urticaceae	Shrub (1–2 m)	Brachypterous
Nesosydne bridwelli	*Argyroxiphium sandwicense*	Asteraceae	Rosette shrub (0.5–1.5 m)[GC]	Brachypterous
	Argyroxiphium virescens	Asteraceae	Rosette shrub (0.5–1.5 m)[GC]	Brachypterous
Nesosydne campylothecae	*Bidens*	Asteraceae	Herb/shrub (0.1–4 m)	Brachypterous
Nesosydne chambersi	*Dubautia ciliolata*	Asteraceae	Shrub (1 m)[GC]	Brachypterous
Nesosydne coprosmocola	*Coprosma ernodeoides*	Rubiaceae	Prostate shrub (1 m)[GC]	Brachypterous
Nesosydne cyathodis	*Styphelia tameiameiae*	Epacridaceae	Spreading shrub (0.5–1.5 m)[GC]	Brachypterous
Nesosydne cyrtandrae	*Cyrtandra*	Gesneriaceae	Shrub (1–2 m)	Brachypterous
Nesosydne crytandricola	*Cyrtandra*	Gesneriaceae	Shrub (1–2 m)	Brachypterous
	Charpentiera obovata	Amaranthaceae	Shrub (1–2 m)	Brachypterous
Nesosydne dubautiae	*Dubautia plantaginea*	Asteraceae	Shrub (1–2 m)[GC]	Brachypterous
Nesosydne eeke	*Argyroxiphium*	Asteraceae	Rosette shrub (0.5–1.5 m)[GC]	Brachypterous
Nesosydne fullawayi	*Styphelia tameiameiae*	Epacridaceae	Spreading shrub (0.5–1.5 m)[GC]	Brachypterous
Nesosydne geranii	*Geranium arboreum*	Geraniaceae	Shrub (0.2–0.5 m)[GC]	Brachypterous
Nesosydne giffardi	*Cyrtandra grandiflora*	Gesneriaceae	Shrub (1–2 m)	Brachypterous
	Rolandia crispa	Campanulaceae	Shrub (1–2 m)	Brachypterous
Nesosydne gigantea	*Pritchardia*	Arecaceae	Tree (4–25 m)	Brachypterous
Nesosydne gouldiae	*Cyrtandra grandiflora*	Gesneriaceae	Shrub (1–2 m)	Brachypterous

(*continued*)

Table 4.3. (*Continued*)

Planthopper Species	Host Plant Species[a]	Host Plant Family	Host Plant Growth Form[b]	Wing-Form Composition[c]
Nesosydne gunnerae	*Gunnera petaloidea*	Gunneraceae	Large herb (1–2 m)	Brachypterous
Nesosydne halia	*Dubautia plantaginea*	Asteraceae	Shrub (1–2 m)[GC]	Brachypterous
Nesosydne imbricola	*Coprosma montana*	Rubiaceae	Shrub (1–2 m)[GC]	Brachypterous
Nesosydne ipomoeicola	*Ipomoea alba*	Convolvulaceae	Sprawling vine[GC]	Wing-dimorphic
	Mucuna gigantea	Fabaceae	Sprawling liana[V]	Wing-dimorphic
	Sadleria	Blechnaceae	Fern (1–3 m)[GC]	Wing-dimorphic
	Strongylodon ruber	Fabaceae	Climbing liana[V]	Wing-dimorphic
Nesosydne koae	*Acacia koa*	Fabaceae	Tree (5–35 m)	Wing-dimorphic
Nesosydne koae-phyllodii	*Acacia koa*	Fabaceae	Tree (5–35 m)	Macropterous
Nesosydne kokolau	*Bidens*	Asteraceae	Herb/shrub (0.1–4 m)	Brachypterous
Nesosydne kuschei	*Cyrtandra*	Gesneriaceae	Shrub (1–2 m)	Brachypterous
Nesosydne lanaiensis	*Styphelia tameiameiae*	Epacridaceae	Spreading shrub (0.5–1.5 m)[GC]	Brachypterous
Nesosydne leahi	*Lipochaeta lobata*	Asteraceae	Perennial herb (0.5–1.5 m)[GC]	Wing-dimorphic
Nesosydne lobeliae	*Lobelia hypoleuca*	Campanulaceae	Shrub (1–2 m)	Brachypterous
Nesosynde longipes	*Cyrtandra platyphylla*	Gesneriaceae	Shrub (1–2 m)	Brachypterous
Nesosydne mamake	*Pipturus*	Urticaceae	Shrub (1–2 m)	Brachypterous
Nesosydne mauiensis	*Bidens mauiensis*	Asteraceae	Decumbent herb (0.3 m)[GC]	Brachypterous
	Dubautia menziesii	Asteraceae	Shrub (1–2 m)[GC]	Brachypterous
	Lipochaeta integrifolia	Asteraceae	Spreading herb (0.2 m)[GC]	Brachypterous
	Tetramolopium	Asteraceae	Decumbent shrub (1 m)[GC]	Brachypterous

Nesosydne monticola	*Coprosma montana*	Rubiaceae	Shrub (1–2 m)[GC]	Brachypterous
Nesosydne montis-tantalus	*Broussaisia arguta*	Hydrangeaceae	Shrub seedlings (0.5 m)[GC]	Brachypterous
	Lobelia hypoleuca	Campanulaceae	Shrub (1–2 m)	Brachypterous
Nesosydne naenae	*Dubautia*	Asteraceae	Shrub (1–2 m)	Brachypterous
Nesosydne neocyrtandrae	*Gunnera petaloidea*	Gunneraceae	Large herb (1–2 m)	Brachypterous
Nesosydne neoraillardiae	*Lipochaeta subcordata*	Asteraceae	Perennial herb (0.5–2 m)	Brachypterous
Nesosydne neowailupensis	*Coprosma longifolia*	Rubiaceae	Shrub (1–2 m)	Brachypterous
Nesosydne nephrolepidis	*Nephrolepis exaltata*	Davalliaceae	Rhizomatous fern (0.5 m)[GC]	Brachypterous
Nesosydne nesopele	*Astelia menziesiana*	Liliaceae	Perennial herb (0.3–1.5 m)[GC]	Brachypterous
Nesosydne nigrinervis	*Styphelia tameiameiae*	Epacridaceae	Spreading shrub (0.5–1.5 m)[GC]	Brachypterous
Nesosydne oahuensis	*Charpentiera obovata*	Amaranthaceae	Shrub (1–2 m)	Brachypterous
Nesosydne olympica	*Lobelia*	Campanulaceae	Shrub (1–2 m)	Brachypterous
Nesosydne osborni	*Duabutia*	Asteraceae	Shrub (1–2 m)	Brachypterous
Nesosydne painiu	*Astelia menziesiana*	Liliaceae	Perennial herb (0.3–1.5 m)[GC]	Brachypterous
Nesosydne phyllostegiae	*Phyllostegia racemosa*	Lamiaceae	Climbing vine[V]	Wing-dimorphic
Nesosydne pilo	*Coprosma ernodeoides*	Rubiaceae	Prostrate shrub (1 m)[GC]	Wing-dimorphic
Nesosydne pipturi	*Pipturus*	Urticaceae	Shrub (1–2 m)	Brachypterous
Nesosydne pseudorubescens	*Acacia koa*	Fabaceae	Tree (5–35 m)	Macropterous
Nesosydne raillardiae	*Dubautia ciliolata*	Asteraceae	Shrub (1 m)[GC]	Wing-dimorphic[B]
	Dubautia scabra	Asteraceae	Shrub (1 m)	Wing-dimorphic[B]

(*continued*)

Table 4.3. (*Continued*)

Planthopper Species	Host Plant Species[a]	Host Plant Family	Host Plant Growth Form[b]	Wing-Form Composition[c]
Nesosydne raillardiicola	*Dubautia menziesii*	Asteraceae	Shrub (1–2 m)[GC]	Brachypterous
	Dubautia platyphylla	Asteraceae	Shrub (1–2 m)[GC]	Brachypterous
Nesosydne rubescens	*Acacia koa*	Fabaceae	Tree (5–35 m)	Macropterous
Nesosydne sharpi	*Boehmeria grandis*	Urticaceae	Shrub (1–2 m)	Brachypterous
	Broussaisia arguta	Hydrangeaceae	Shrub (2 m)	Brachypterous
Nesosydne stenogynicola	*Stenogyne kamehamehae*	Lamiaceae	Climbing vine[V]	Brachypterous
Nesosydne sulcata	*Cyrtandra*	Gesneriaceae	Shrub (1–2 m)	Brachypterous
Nesosydne tetramolopii	*Tetramolopium humile*	Asteraceae	Shrub (0.2 m)[GC]	Brachypterous
Nesosynde timberlakei	*Cyanea truncata*	Campanulaceae	Shrub (0.3–2 m)[GC]	Brachypterous
	Cyrtandra garnotiana	Gesneriaceae	Shrub (1–2 m)	Brachypterous
Nesosydne ulehihi	*Smilax melastomifolia*	Smilacaceae	Climbing liana[V]	Brachypterous
Nesosydne umbratica	*Charpentiera obovata*	Amaranthaceae	Shrub (1–2 m)	Wing-dimorphic
	Clermontia clermontioides	Campanulaceae	Shrub (1.5 m)	Wing-dimorphic
	Cyanea hamatiflora	Campanulaceae	Palm-like tree (3–8 m)	Wing-dimorphic
	Cyrtandra	Gesneriaceae	Shrub (1–2 m)	Wing-dimorphic
	Pipturus	Urticaceae	Shrub (1–2 m)	Wing-dimorphic
	Urera glabra	Urticaceae	Shrub (2 m)	Wing-dimorphic

Nesosydne viridis	*Phyllostegia*	Lamiaceae	Shrub (1–2 m)	Brachypterous
Nesosydne waikamoiensis	*Cyanea aculeatiflora*	Campanulaceae	Palm-like tree (2–7 m)	Brachypterous
	Pipturus	Urticaceae	Shrub (1–2 m)	Brachypterous
Nesosydne wailupensis	*Rolandia crispa*	Campanulaceae	Shrub (1–2 m)	Brachypterous
Perkinsiella saccharicida	*Saccharum officinarum*	Poaceae	Robust grass (2–3 m)^{GC}	Wing-dimorphic
Peregrinus maidis	*Zea mays*	Poaceae	Robust grass (1–2 m)	Wing-dimorphic
Kelisia emoloa	*Eragrostis variabilis*	Poaceae	Tufted grass (0.4–0.8 m)^{GC}	Brachypterous
Kelisia eragrosticola	*Eragrostis variabilis*	Poaceae	Tufted grass (0.4–0.8 m)^{GC}	Wing-dimorphic^{B}
Kelisia sporobolicola	*Eragrostis atropioides*	Poaceae	Tufted grass (0.1–0.2 m)^{GC}	Wing-dimorphic^{B}
	Sporobolus virginicus	Poaceae	Spreading grass (0.1–0.5 m)^{GC}	Wing-dimorphic^{B}
	Machaerina angustifolia	Cyperaceae	Sedge (0.5–1 m)^{GC}	Wing-dimorphic^{B}
Kelisia swezeyi	*Eragrostis variabilis*	Poaceae	Tufted grass (0.4–0.8 m)^{GC}	Wing-dimorphic^{B}
	Gahna	Cyperaceae	Tussock sedge (1.5 m)^{GC}	Wing-dimorphic^{B}
Tarophagus prosperpina	*Colocasia esculenta*	Araceae	Herb (0.5–1 m)^{GC}	Wing-dimorphic^{B}

Note: GC = Ground contact of host plant substantial by virtue of prostrate or decumbent stems, multiple stems, basal rosette, or rhizomatous growth. B = Wing-dimorphic, but mostly brachypterous. V = For analysis, the height of vines was estimated at 2 m and woody climbers and lianas at 3 m.

[a]Host plant records obtained from Giffard (1922) and Zimmerman (1948); nomenclature updated from Wagner et al. (1990).

[b]Growth form and height of host plants determined from Wagner et al. (1990).

[c]Wing form of planthoppers obtained from Giffard (1922) and Zimmerman (1948).

(pooled for analysis) occurred on significantly shorter vegetation, primarily small shrubs and herbs, many of which contact the ground (1.4 ± 0.2 m, $n=98$; Wilcoxon Rank Sum test, $Z=9.141$, $P=0.0001$; Fig. 4.7). A comparison of congeners in *Nesosydne* allowed for an assessment of the effects of plant growth form on wing form independent of common ancestry. Within the same lineage, macropterous species (e.g., *Nesosydne koae-phyllodii*) fed on large trees (20.0 ± 0.0 m, $\bar{x}\pm$SE, $n=3$) and brachypterous species (e.g., *N. argyroxiphii*) exploited mostly low-profile shrubs and herbs (1.6 ± 0.2 m, $n=72$; Wilcoxon Rank Sum test, $Z=3.066$, $P=0.002$). There were a few apparent exceptions to the pattern. For example, *Nesorestias filicicola* occurs on tree ferns (2–7 m), *Nesosydne gigantea* feeds on *Prichardia* palms (4–25 m), and *Nesosydne ulehihi* feeds on a climbing liana; all three of these species are unexpectantly brachypterous (Table 4.3). However, it was not possible to decern from the literature if these planthoppers oc-

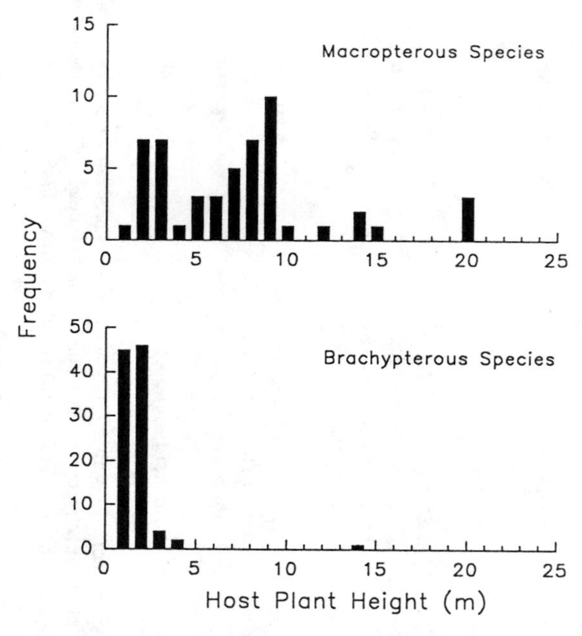

Figure 4.7. Frequency distributions of host plant heights (m) for 52 macropterous (top) and 98 species of brachypterous Hawaiian delphacids (bottom). Macropterous species feed mostly on trees and large shrubs (7.1 ± 0.6 m), whereas brachypterous species occur mostly on small shrubs and herbs (1.4 ± 0.2 m) (Wilcoxan Rank Sum test, $P=0.0001$). Information on wing-form composition of planthopper populations and on the height of mature host plants from Giffard (1922), Zimmerman (1948), and Wagner et al. (1990).

curred on mature host plants or if they were restricted to seedlings. Overall, the Hawaiian delphacid data provide strong support for the habitat dimensionality–flight capability hypothesis; macroptery is favored in arboreal habitats and flightlessness has evolved in persistent, low-profile vegetation.

Wing-dimorphic species, in which both wing forms were common, were few and occurred for the most part in either temporary agricultural habitats (*Perkinsiella saccharicida* and *Peregrinus maidis*) or were polyphagous (*Nesosydne ipomoeicola* and *N. umbratica*) (Table 4.3). It has been proposed for *Javesella pellucida* that wings are retained in this polyphagous species to facilitate the tracking of changing levels of plant nitrogen across host taxa (Prestidge 1982a; Prestidge and McNeill 1982, 1983; Denno and Roderick 1990). Perhaps this is the case for these dimorphic Hawaiian species as well.

The wing form of North American planthoppers in the genus *Stobaera* also suggests that macroptery prevails in arboreal habitats (Kramer 1973; Denno 1978; Table 4.4). *S. pallida*, a macropterous species, feeds exclusively on a large woody composite (1–3 m). Dimorphic species which are mostly macropterous (*S. concinna* and *S. tricarinata*) occur on erect shrubby composites (1 m+), and species which are primarily brachypterous (*S. bilobata* and *S. muiri*) feed on small, sprawling shrubby hosts ($\leqslant$1 m). One macropterous species, *S. caldwelli*, appears to retain its wings for reasons of habitat persistence; it is polyphagous within the Asteraceae and some of its host plants are weedy annuals.

Overwintering Stage, Voltinism, and Diapause

Most temperate and boreal species of delphacids overwinter as nymphs (73%), but some overwinter as eggs (17%) and a few as adults (10%) (Table 4.5). For the most part, congeners overwinter in the same stage. For example, species of *Anakelisia* and *Muellerianella* pass the winter as eggs, *Javesella*, *Prokelisia*, and *Ribauatodelphax* as nymphs, and *Stenocranus* as adults. One exception to this pattern is *Megamelus* in which the overwintering style varies across species with all three stages (egg, nymph, and adult) represented (Table 4.5). Overwintering as eggs embedded within plant tissue provides some degree of protection from cold winters. For example, in a guild of salt marsh-inhabiting species, those that overwinter as nymphs (e.g., *Delphacodes detecta* and *Tumidagena minuta*) show a strong negative relationship between winter survival and winter severity (Denno et al. 1981). For species overwintering as eggs (e.g., *Megamelus lobatus*), winter severity does not influence survival. In this same guild, those species that overwinter as nymphs become rarer at northern latitudes. This pattern is roughly supported for the species listed in Table 4.5;

Table 4.4. Wing-form composition (macropterous, brachypterous, or wing-dimorphic) of planthopper populations in the genus *Stobaera* and the growth form of their host plants.

Planthopper Species	Host Plant Species[a]	Host Plant Family	Host Plant Growth Form[b]	Wing-Form Composition[c]
Stobaera bilobata	*Haplopappus squarrosus*	Asteraceae	Shrub (3–10 dm)	Wing-dimorphic[B]
Stobaera caldwelli	*Franseria acanthicarpa*	Asteraceae	Weedy annual herb (1–7 dm)	Macropterous
	Franseria dumosa	Asteraceae	Shrub (2–6 dm)	Macropterous
	Franseria eriocentra	Asteraceae	Shrub (3–12 dm)	Macropterous
	Hymenoclea salsola	Asteraceae	Shrub (6–10 dm)	Macropterous
Stobaera concinna	*Ambrosia psilostachya*	Asteraceae	Erect shrub (5–12 dm)	Wing-dimorphic[M]
	Franseria confertiflora	Asteraceae	Erect shrub (3–12 dm)	Wing-dimorphic[M]
Stobaera muiri	*Franseria chamissonis*	Asteraceae	Sprawling shrub (1.5–3 dm)[GC]	Wing-dimorphic[B]
Stobaera pallida	*Baccharis halimifolia*	Asteraceae	Large shrub (1–3 m)	Macropterous
Stobaera tricarinata	*Ambrosia psilostachya*	Asteraceae	Erect shrub (5–12 dm)	Wing-dimorphic[M]

Note: GC = Ground contact of host plant substantial by virtue of prostrate stems. B = Wing-dimorphic, but mostly brachypterous. M = Wing-dimorphic, but mostly macropterous.

[a]Host plant records obtained from Kramer (1973); nomenclature according to Munz and Keck (1965).

[b]Growth form of host plants determined from Munz and Keck (1965).

[c]Wing form of planthoppers obtained from Kramer (1973).

67% of all delphacids occurring at latitudes of 55°N and higher overwinter as nymphs, whereas 82% of species living at latitudes <45°N pass the winter as nymphs. It has been hypothesized that by passing the winter as an active nymph, feeding can begin as soon as temperatures permit in the spring when levels of host plant nitrogen are at their highest (Denno et al. 1981; Prestidge and McNeil 1983).

The number of generations attained each year is highly variable and depends largely on the length of the growing season (Denno and Roderick 1990). In widespread species, there is a reduction in the number of generations at high latitudes. For instance, *Laodelphax striatellus* is bivoltine in Sweden (60°N), but produces eight generations annually in Israel (32°N) (Harpaz 1972; Ossiannilsson 1978; Fig. 4.8). Similarly, *Nilaparvata lugens* is trivoltine in Northern Japan (Kuno 1979), but can produce up to 12 generations a year in South East Asia (Dyck et al. 1979; Fig. 4.8). There is a significant negative relationship between the number of annual generations and degrees north latitude among all the species/populations listed in Table 4.5 (Fig. 4.8).

Diapause patterns also provide evidence for the flexible life histories of planthoppers (Denno and Roderick 1990). Northern populations of *Laodelphax striatellus* overwinter as diapausing nymphs, but southern populations never diapause and reproduce continuously (Harpaz 1972). Environmental cues inducing diapause in the nymphs of *L. striatellus* and *Javesella pellucida* and in the eggs of *Muellerianella* are short photoperiod and cold temperatures (Kisimoto 1958; Raatikainen 1967; Witsack 1971; Harpaz 1972; Drosopoulos 1977). In *Stenocranus munutus*, a univoltine species, adult reproductive diapause is induced by exposure to long-day conditions (18 hours light) during the nymphal stage; diapause is broken by exposing young adults to a short-day photoperiod (8 hours light) (Müller 1957). Nymphs developing under short-day conditions molt to significantly smaller (15%) adults (Müller 1957, 1958), which may be less fecund. There is little evidence for diapause capability in tropical species (e.g., *Nilaparvata lugens* and *Sogatella furcifera*), which most likely accounts for their inability to survive the cold winters at northern latitudes (Kisimoto 1976, 1987; Kuno 1979; Mochida and Okada 1979).

For planthoppers, it is not known if mixed diapause strategies exist within populations (e.g., some individuals entering diapause with others undergoing direct development) as is the case for other insects (Solbreck 1978; Vepsäläinen 1978; Istock 1981; Taylor and Karban 1986). Moreover, little is known of the costs associated with diapause. That the fecundity of the macropters, but not the brachypters of *Laodelphax striatellus* is reduced following diapause suggests that there is a wing-form-dependent reproductive penalty associated with diapause [fecundity of postdiapause macrop-

Table 4.5. Overwintering stage and number of generations produced per year in delphacid planthoppers occurring at different geographic locations and latitudes.

Species	Geographic Location	Latitude	Overwintering Stage	Number of Generations/Year	Reference
Anakelisia fasciata	Fennoscandia	59°N	Egg	1[a]	Müller (1957), Ossiannilsson (1978)
Anakelisia perspicillata	Fennoscandia	59°N	Egg	1[a]	Ossiannilsson (1978)
Calligypona reyi	Fennoscandia	59°N	Nymph	1[a]	Ossiannilsson (1978)
Chloriona glaucescens	Fennoscandia	59°N	Nymph	1[a]	Müller (1957), Ossiannilsson (1978)
Chloriona smaragdula	Fennoscandia	59°N	Nymph	1[a]	Müller (1957), Ossiannilsson (1978)
Conomelus anceps	UK (England)	52°N	Egg	2	Waloff (1980)
Criomorphus albomarginatus	Fennoscandia	59°N	Nymph	1[a]	Müller (1957), Ossiannilsson (1978)
Delphacinus mesomelas	Fennoscandia	59°N	Nymph	1[a]	Müller (1957), Ossiannilsson (1978)
Delphacodes capnodes	Fennoscandia	59°N	Adult	1[a]	Müller (1957), Ossiannilsson (1978)
Delphacodes detecta	USA (New Jersey)	39°N	Nymph	3	Denno et al. (1981)
Delphacodes venosus	Fennoscandia	59°N	Adult	1[a]	Ossiannilsson (1978)
Dicranotropis hamata	UK (England)	52°N	Nymph	2	Waloff (1980)
	Finland	61°N	Nymph	1	Raatikainen and Vasarainen (1964)
Ditropis pteridis	Fennoscandia	59°N	Nymph	1[a]	Müller (1957), Ossiannilsson (1978)
Eurybregma bielawskii	Greece[M]	39°N	Nymph	1	Drosopoulos et al. (1983)
Eurysa lineata	Greece[M]	39°N	Nymph	?	Drosopoulos (1982)
	Germany	52°N	Nymph	?	Müller (1957)
Eurysa rubripes	Greece[M]	39°N	Nymph	?	Drosopoulos (1982)
Eruysula lurida	Fennoscandia	59°N	Nymph	1[a]	Müller (1957), Ossiannilsson (1978)
Javesella discolor	Fennoscandia	59°N	Nymph	1	Ossiannilsson (1978)
Javesella dubia	UK (England)	52°N	Nymph	2	Waloff (1980)
	Fennoscandia	59°N	Nymph	1[a]	Ossiannilsson (1978)

Javesella forcipata	Fennoscandia	59°N	Nymph	1	Ossiannilsson (1978)
Javesella obscurella	Fennoscandia	59°N	Nymph	1[a]	Ossiannilsson (1978)
Javesella pellucida	Germany (Berlin)	52°N	Nymph	2	Raatikainen (1967)
	UK (England)	52°N	Nymph	2	Waloff (1980)
	Fennoscandia	59°N	Nymph (2–4)	1	Raatikainen (1967)
Kelisia ribauti	Feennoscandia	59°N	Egg	1[a]	Müller (1957), Ossiannilsson (1978)
Kelisia vittipennis	Fennoscandia	59°N	Egg	1[a]	Müller (1957), Ossiannilsson (1978)
Laodelphax elegantulus	UK (England)	52°N	Nymph	2	Waloff (1980)
Laodelphax striatellus	Israel	32°N	Continuous reproduction	8	Harpaz (1972)
	China	31°N	Nymph (3–5)	6–7	Harpaz (1972)
	China	33°N	Nymph (3–5)	6	Harpaz (1972)
	Japan	33°N	Nymph (3–5)	5–6	Harpaz (1972)
	China (North)	39°N	Nymph (3–5)	4–5	Tsai et al. (1964)
	USSR (Siberia)	55°N	Nymph (3–5)	3	Harpaz (1972)
	Sweden	59°N	Nymph	2	Ossiannilsson (1978)
Megadelphax sordidulus	Fennoscandia	59°N	Nymph	1	Müller (1957), Ossiannilsson (1978)
Megamelus davisi	USA (Illinois)	38°N	Nymph (5)	3	Wilson and McPherson (1981a)
Megamelus lobatus	USA (New Jersey)	39°N	Egg	2	Denno et al. (1981)
Megamelus notula	Fennoscandia	59°N	Adult	1[a]	Ossiannilsson (1978)
Muellerianella brevipennis	Netherlands	52°N	Egg	2	Drosopoulos (1977)
Muellerianella fairmairei	Netherlands	52°N	Egg	2	Drosopoulos (1977)
	Finland	62°N	Egg	1	Linnavuori (1952)
Muirodelphax aubei	Germany	52°N	Nymph	2	Ossiannilsson (1978)
Muriodelphax exiguus	UK (England)	52°N	Nymph	2	Waloff (1980)
Neomegamelanus dorsalis	USA (New Jersey)	39°N	Nymph	3	Denno et al. (1981)
Nilaparvata lugens	SE Asia	5°N	Continuous reproduction	12	Dyck et al. (1979)

(*continued*)

Table 4.5. (*Continued*)

Species	Geographic Location	Latitude	Overwintering Stage	Number of Generations/Year	Reference
	Taiwan	24°N	Continuous reproduction	8–11	Dyck et al. (1979)
	China (Central)	32°N		6–7	Dyck et al. (1979)
	Japan (South)	33°N	Fails to overwinter	5	Mochida and Okada (1979)
	Japan (North)	35°N	Fails to overwinter	3	Kuno (1979)
Nilaparvata muiri	Japan	33°N	Egg	3	Kisimoto (1981)
Paradelphacodes paludosa	Fennoscandia	59°N	Nymph	1[a]	Müller (1957), Ossiannilsson (1978)
Paraliburnia dalei	UK (England)	52°N	Nymph	2	Waloff (1980)
Perkinsiella saccharidica	USA (Hawaii)	20°N	Continuous reproduction	6	Fennah (1969)
Perkinsiella vastatrix	Philippines	15°N	Continuous reproduction	6	Fennah (1969)
Pissonotus piceus	USA (Missouri)	39°N	?	2	S. Wilson (personal communication)
Prokelisia crocea	USA (Missouri)	39°N	Nymph (2–3)	2	Holder (1990)
Prokelisia dolus	USA (New Jersey)	39°N	Nymph (3–5)	3	Denno et al. (1987)
Prokelisia marginata	USA (New Jersey)	39°N	Nymph (3–5)	3	Denno et al. (1987)
	USA (Florida)	29°N	Nymph	6	Denno (1983)
Ribautodelphax angulosus	Europe (Western)	50°N	Nymph (2–4)	2	den Bieman (1987a,c)
Ribautodelphax collinus	Europe (Western)	50°N	Nymph (2–4)	2	den Bieman (1987a,c)
Ribautodelphax imitantoides	Spain	40°N	Nymph (2–4)	3	den Bieman (1987a,c)
Ribautodelphax imitans	Greece	39°N	Nymph	?	Drosopoulos (1982)

Ribautodelphax pungens	Europe (Western)	50°N	Nymph (2–5)	2	den Bieman (1987a,c)
Ribautodelphax ventouxianus	France (Southeast)[M]	44°N	Nymph (2–3)	1	den Bieman (1987a,c)
Ribautodelphax vinealis	Netherlands	52°N	Nymph (2–3)	2	den Bieman (1987a,c)
Saccharosydne saccharivora	Jamaica	18°N	Continuous reproduction	7	Metcalfe (1969)
Sogatella nigeriensis	Egypt	30°N	Continuous reproduction	8	Ammar et al. (1980)
Sogatella vibex	Egypt	30°N	Continuous reproduction	8	Ammar (1977)
	Israel	32°N	Nymph (2–5)	?	Harpaz (1972)
Stenocranus lautus	USA (Missouri)	39°N	Egg	2	Calvert and Wilson (1986)
Stenocranus major	Fennoscandia	59°N	Adult	1[a]	Müller (1957), Ossiannilsson (1978)
Stenocranus minutus	UK (England)	52°N	Adult	1	Waloff (1980)
	Fennoscandia	59°N	Adult	1[a]	Müller (1957), Ossiannilsson (1978)
Stiroma affinis	Fennoscandia	59°N	Nymph	1[a]	Müller (1957), Ossiannilsson (1978)
Stobaera tricarinata	USA (California)[D]	34°N	Adult	2	Reimer and Goeden (1982)
Struebingianella lugubrina	Fennoscandia	59°N	Nymph	1[a]	Müller (1957), Ossiannilsson (1978)
Tagosodes orizicolus	Mexico	20°N	Continuous reproduction	4/crop	Everett (1969)
Toya propinqua	Greece	39°N	Nymph	?	Drosopoulos (1982)
Tumidagena minuta	USA (New Jersey)	39°N	Nymph	2	Denno et al. (1981)
Tryphodelphax distinctus	Fennoscandia	59°N	Nymph	1[a]	Müller (1957), Ossiannilsson (1978)
Xanthodelphax stamineus	Greece	39°N	Nymph	?	Drosopoulos (1982)
	Fennoscandia	59°N	Nymph	1[a]	Müller (1957), Ossiannilsson (1978)

[a]Based on the statement by Ossiannilsson (1978) that most Fennoscandian delphacids are univoltine.

Note: D = Desert location, M = Mountain location.

ters (348 ± 32 eggs/female) significantly less than that of nondiapaused macropters (526 ± 85, $\bar{x} \pm SE$; $t = 2.092$, $P < 0.05$); fecundity of postdiapause brachypters (565 ± 34 eggs/female) not significantly different from that of nondiapaused brachypters (559 ± 65; $t = 8.313$, $P > 0.05$); data reanalyzed from Kisimoto 1965].

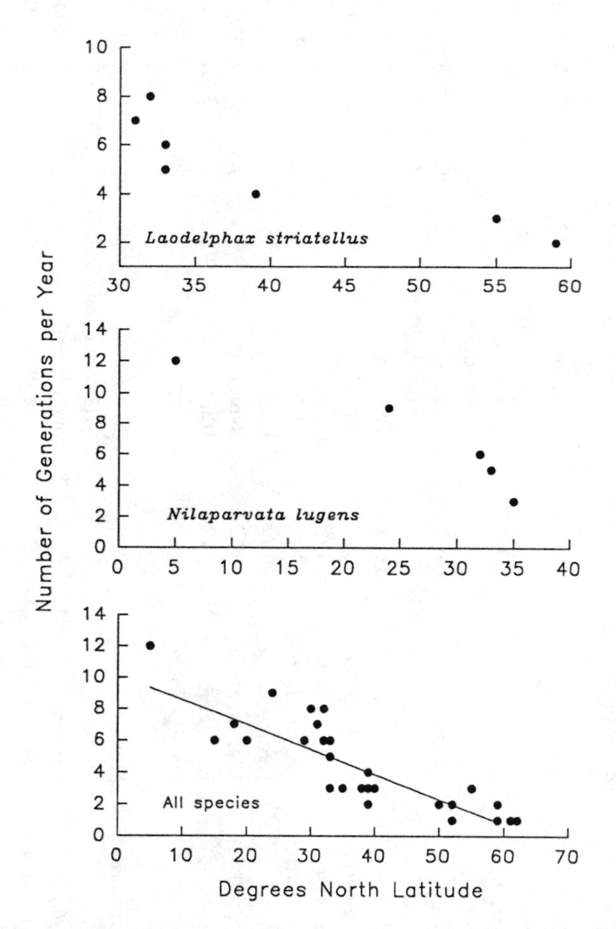

Figure 4.8. Relationship between the number of generations attained annually and latitude (°N) for *Laodelphax striatellus* (top), *Nilaparvata lugens* (middle), and all species/populations listed in Table 4.5 (bottom) ($R^2 = 0.80$, $P = 0.0001$, $y = -0.157x + 10.107$). Data from Harpaz (1972), Ossiannilsson (1978), Dyck et al. (1979), and Kuno (1979).

Oviposition Patterns, Batch Size, and Egg Size

Females of most planthoppers make a slit in plant tissue with their saw-like ovipositors and then insert their eggs, often in rows (Denno and Roderick 1990). Some species (e.g., *Prokelisia marginata*) simply slip their eggs between the ridges on the upper surface of the leaf blade (Denno et al. 1987). On grass hosts, egg rows are always oriented parallel to the vascular bundles of the plant tissue (Metcalfe 1969). Following oviposition, some species (e.g., *Perkinsiella saccharicida* and *Saccharosydne saccharivora*) cover their eggs with a fluffy wax or secretion which later solidifies into a protective covering (Williams 1957; Metcalfe 1969). For *S. saccharivora*, the deposition of one batch requires 15–20 minutes. Batches can be deposited side by side or they can be separated by several vascular bundles, and females can deposit up to three batches per day for 2–4 weeks (Metcalfe 1969).

Batch size or the number of eggs inserted together as a group varies substantially among planthopper species (Table 4.6). Most species deposit groups of two to eight eggs, but a few species deposit their eggs singly (e.g., *Dicranotropis muiri*) (Table 4.6). There is considerable interspecific variation in batch size and a general property of batch size distributions is that they are strongly skewed toward the smaller size classes (Mochida 1964; Kisimoto 1965; Raatikainen 1967; Drosopoulos 1977).

By depositing eggs in small groups, planthoppers may "spread their oviposition risks" in space, thereby increasing offspring survival (Prestidge 1982a). On the other hand, batch size may reflect physiological and morphological constraints. For example, in *Tagosodes orizicolus* batch size (7) corresponds to half the number of ovarioles in each ovary (McMillian 1963). Characteristics of the host plant also appear to have an important effect on batch size. Smaller batches of eggs are deposited in tougher substrates and in young tissues (e.g., seedlings) where there is less space (Mochida 1964; Kisimoto 1965; Drosopoulos 1977). For example, batch size in *Javesella pellucida* is inversely related to the thickness of the stem wall of oats (Raatikainen 1967; Table 4.6). Differences in tissue toughness may explain why batch size in *J. pellucida* varies among host plant species and varieties (Raatikainen 1967). Tissue toughness may also explain why *Stobaera tricarinata* deposits single eggs into the woody stems of its composite host (Reimer and Goeden 1982).

Beyond the constraints imposed by the host plant, the selective pressures driving batch size evolution in planthoppers remain unclear. However, natural enemies may play a role. For example, once it finds an oviposition scar, the parasitoid *Paranagrus optabilis* oviposits into every egg in the clutch of *Perkinsiella saccharicida* (Swezey 1936). Similarly, some species of

Table 4.6. Egg batch size and oviposition site in host plant for delphacid planthoppers.

Species	Geographic Location	Egg Batch Size Mean or Mode* (Range)	Oviposition Site	Reference
Dicranotropis hamata	Finland	8*	Stem	Raatikainen and Vasarainen (1964)
Dicranotropis muiri	Mauritius	1	Leaf blade, midrib	Williams (1957)
Javesella pellucida	UK (England)	2* (1–6)	Stem	Mochida (1973)
	Finland	17 (1–54)	Stem (wheat)	Raatikainen (1967)
		3 (1–12)	Leaf blade (wheat)	
		11 (1–83)	Stem (oats)	
		3 (1–35)	Leaf blade (oats)	
	Finland	15.9 ± 0.7	Stem wall thickness 0.2 mm	Raatikainen (1967)
		13.9 ± 1.1	Stem wall thickness 0.3 mm	
		7.6 ± 0.7	Stem wall thickness 0.4 mm	
		8.9 ± 1.0	Stem wall thickness 0.5 mm	
		8.6 ± 0.7	Stem wall thickness >0.5 mm	
Laodelphax striatellus	Japan	2* (1–15)	Leaf sheath	Kisimoto (1965)
	Israel	(2–5)	Leaf sheath	Harpaz (1972)
Megadelphax sordidulus	Finland	8*	Stem	Raatikainen (1960)
Muellerianella brevipennis	Netherlands	2* (1–6)	Leaf sheath	Drosopoulos (1977)
Muellerianella fairmairei	Netherlands	2* (1–5)	Leaf sheath	Drosopoulos (1977)

Nilaparvata lugens	Japan	4*	(1–15)	Lower stem	Mochida (1964)
	Japan	4*	(1–25)	Stem	Kisimoto (1965)
Peregrinus maidis	USA (Hawaii)	3	(1–4)	Leaf blade, midrib	Napompeth (1973)
	Mauritius		(3–5)	Leaf blade, midrib	Williams (1957)
Perkinsiella saccharicida	USA (Hawaii)		(1–12)	Leaf blade, midrib	Zimmerman (1948)
	Mauritius		(3–5)	Leaf blade, midrib	Williams (1957)
Perkinsiella vastatrix	Philippines		(1–4)	Leaf blade, midrib	Fennah (1969)
Prokelisia crocea	USA (Missouri)	1		Leaf blade, midrib	Holder and Wilson (1992)
Saccharosydne saccharivora	Venezuela		(4–12)	Leaf blade, midrib	Fennah (1969)
	Jamaica	7*	(5–8)	Leaf blade, midrib	Metcalfe (1969)
Sogatella furcifera	Japan	7*	(1–25)	Leaf blade, midrib	Kisimoto (1965)
	Japan		(4–8)	Leaf blade, midrib	Suenaga (1963)
Sogatella nigeriensis	Egypt	3	(1–4)	Leaf sheath, blade	Ammar et al. (1980)
Sogatella vibix	Egypt	3	(1–4)	Leaf sheath, blade, midrib	Ammar (1977)
Stobaera concinna	USA (Florida)		(3–5)	Stems, small	Calvert et al. (1987b)
Stobaera tricarinata	USA (California)	1		Stem, pith	Reimer and Goeden (1982)
Stenocranus lautus	USA (Missouri)	1		Stem	Calvert and Wilson (1986)
Tagosodes orizicolus	Mexico	7		Leaf blade, midrib	Everett (1969)
Toya propinqua	Turkey	3*	(1–7)	Leaf sheath, blade	Raatikainen and Vasarainen (1990)

Anagrus move and oviposit from one egg of *Nilaparvata lugens* to the next (M. Claridge personal communication). Such oviposition behavior by egg parasitoids should favor small, widely-spaced batches in planthoppers for reasons of risk spreading. In contrast, females of *Anagrus delicatus* deposits only a small fraction of their eggs in a clutch of *Prokelisia marginata* before dispersing (Cronin and Strong 1990a); such behavior should select for large clutch size in the host. Nevertheless, the energy demands associated with the drilling of plant tissues must be considered as well when attempting to elucidate the selective role of natural enemies in directing clutch size evolution in planthoppers.

Most delphacids prefer their developmental host(s) for oviposition, but *Muellerianella fairmairei* oviposits during its second generation on non-host vegetation where eggs successfully overwinter (Harpaz 1972; Drosopoulos 1977; Booij 1982a; Waloff 1980). Oviposition sites vary enormously among planthopper species (Table 4.6). Of the grass feeders, some prefer to deposit their eggs in stems or leaf sheaths (*Javesella pellucida*, *Laodelphax striatellus*, and *Nilaparvata lugens*), whereas others selectively oviposit in leaf blades (*Peregrinus maidis* and *Perkinsiella saccharicida*). Dicot feeders oviposit in the leaves (*Megamelus davisi*), young shoots (*Nesosydne koae*), or stems (*Stobaera tricarinata*) of their hosts (Zimmerman 1948; Reimer and Geoden 1982; Wilson and McPherson 1981a).

Several hypotheses have been proposed to explain the restriction of oviposition to certain plant parts. First, physical characteristics of certain plant tissues and plant species may preclude oviposition. For instance, *Javesella pellucida* deposits three times as many eggs in thin-walled oat stems (0.2 mm) as it does thick-walled stems (0.5 mm) in which oviposition is deterred (Raatikainen 1967). Second, interspecific competition may result in oviposition site divergence. Waloff (1980) suggested that competition for oviposition sites between *Dicranotropis hamata* and several leafhoppers is reduced by resource partitioning. However, experimental documentation for this claim is needed.

Choice of oviposition site may also result from selective pressures associated with the actions of natural enemies. For example, *Anagrus atomus* selectively oviposits in the eggs of *J. pellucida* deposited in leaf blades (53% of 1377 eggs parasitized) rather than those inserted in stems (0% of 17,189 eggs parasitized) (Raatikainen 1967). Similarly, females of the wasp *Panstenon oxylus* preferentially oviposit into eggs found in the thin-walled stems of wheat and oats; oviposition is prevented through the thick-walled stems because their ovipositors are too short to penetrate to the internodes where eggs are deposited. Also, larvae of *P. oxylus* forage internally in stems where they feed on the exposed ends of planthopper eggs. Eggs deposited in thick-walled stems do not protrude to the interior and consequently escape predation from larvae of *P. oxylus* (Raatikainen 1967). Thus, by preferentially ovipositing in stems, *J. pellucida* avoids parasitism from *A. atomus*, and oviposition in the thicker stems of cereals provides some space free from *P. oxylus*.

It is not known if the longer ovipositors of some species (*Sogatella furcifera* = 1.56 mm > *Laodelphax striatellus* = 1.05 mm; Kisimoto 1965) and wing forms (brachypters of *Nilaparvata lugens* and *S. furcifera* have proportionally longer ovipositors than macropters; Kisimoto 1965) translate into any advantage for penetrating tough plant tissues or placing eggs out of reach from natural enemies.

Egg size varies considerably among planthopper species, ranging from 0.65 mm in length for *Toya propinqua* (Raatikainen and Vasarainen 1990) to 1.15 mm in *Perkinsiella saccharicida* (Williams 1957). *T. propinqua* deposits smaller eggs and is more fecund than any other planthopper studied (Raatikainen and Vasarainen 1990). Thus, selection may have favored increased fecundity at the expense of egg size in this highly mobile, colonizing species. However, support for a trade-off between egg number and egg size is not provided by interspecific comparisons. For instance, the relationship between egg size and egg number is not significant across species (R^2 = 0.08, P>0.05, Fig. 4.9). This analysis is confounded by differences in ancestry and body size. What is needed is a comparison of egg size and fecundity among conspecifics and/or congeners with body size and

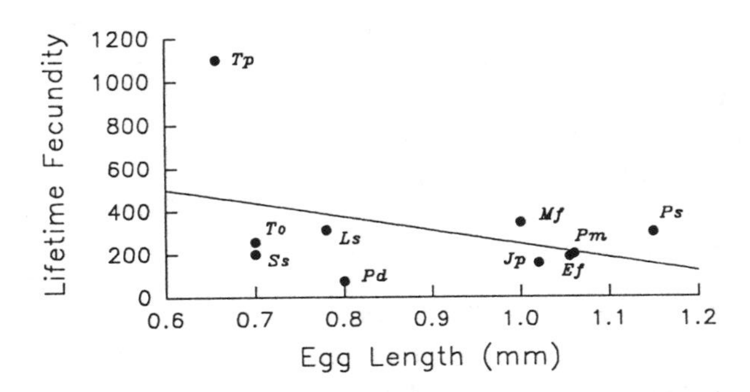

Figure 4.9. Relationship between egg number (lifetime fecundity) and egg length (mm) for 10 species of planthoppers: *Eoeurysa flavocapitata* (Chatterjee and Choudhuri 1979), *Javesella pellucida* (Harpaz 1972; Mochida 1973), *Laodelphax striatellus* (Harpaz 1972; Kisimoto 1965), *Muellerianella fairmairei* (Drosopoulos 1977), *Peregrinus maidis* (Fullaway 1918; Tsai and Wilson 1986), *Perkinsiella saccharicida* (Williams 1957; Fennah 1969), *Prokelisia dolus* (Denno et al. 1989, unpublished data), *Saccharosydne saccharicida* (Fennah 1969), *Tagosodes orizicolus* (McMillian 1963; King and Saunders 1984), and *Toya propinqua* (Raatikainen and Vasarainen 1990). R^2 = 0.08, P>0.05.

longevity controlled, but sufficient data are not yet available for this comparison. For insects in general, there is widespread evidence for a trade-off between egg size and number (Berrigan 1991).

Environmental Effects on Fitness Components

Population density influences the performance of many planthopper species. Under crowded conditions, intense intraspecific interactions often result in decreased survival (Kisimoto 1965; Fisk et al. 1981; Heong 1988; Denno and Roderick 1992), extended development (Kisimoto 1965; Drosopoulos 1977; Denno et al. 1986; Denno and Roderick 1992), and reduced fecundity due to competition for oviposition sites (Kenmore 1980; Heong 1988). However, certain species are more adversely affected under high-density conditions than others. For example, at the same high densities, most components of fitness are negatively affected in *Prokelisia marginata*, but not in *P. dolus* (Denno and Roderick 1992). In these two species as well as others, adverse effects of density on fitness are associated with mobility (Denno et al. Chapter 7). Selection appears to have favored high dispersal capability in species which are most sensitive to crowding. To a large extent, the negative effects of density on fitness are exacerbated on poor-quality host plants (Fisk et al. 1981; Denno et al. 1986).

Reductions in fitness may occur at very low densities as well. For instance, if population size is very small, as can be the case for colonizing species, mates may be limiting (Denno et al. 1991). Under such conditions for *Nilaparvata lugens*, single matings compared to multiple matings result in lower fecundity (Oh 1979). That the males of species inhabiting temporary habitats, such as most of our agricultural pests, emerge macropterous even at very low densities suggests that wings facilitate mate location (Denno et al. 1991; Fig. 4.6B).

Other environmental factors also affect the fitness and performance of delphacids. Under hot (30°C) compared to cool rearing temperatures (10°C), eggs and nymphs develop more rapidly, but survivorship is reduced (Suenaga 1963; Kisimoto 1965; Tsai and Wilson 1986). In both sexes of *Prokelisia marginata*, the body size of eclosing adults is inversely related to rearing temperature (Fig. 4.10), and body size is positively related to fecundity in *Prokelisia* planthoppers (Denno and McCloud 1985). The body length of *P. dolus* did not increase with a decrease in rearing temperature (Fig. 4.10). For tropical (*Nilaparvata lugens* and *Peregrinus maidis*) and temperate species (*Muellerianella fairmairei*, *P. dolus*, and *P. marginata*), egg and/or nymphal development is most rapid between 24 and 28°C, and at temperatures above and below these, development is

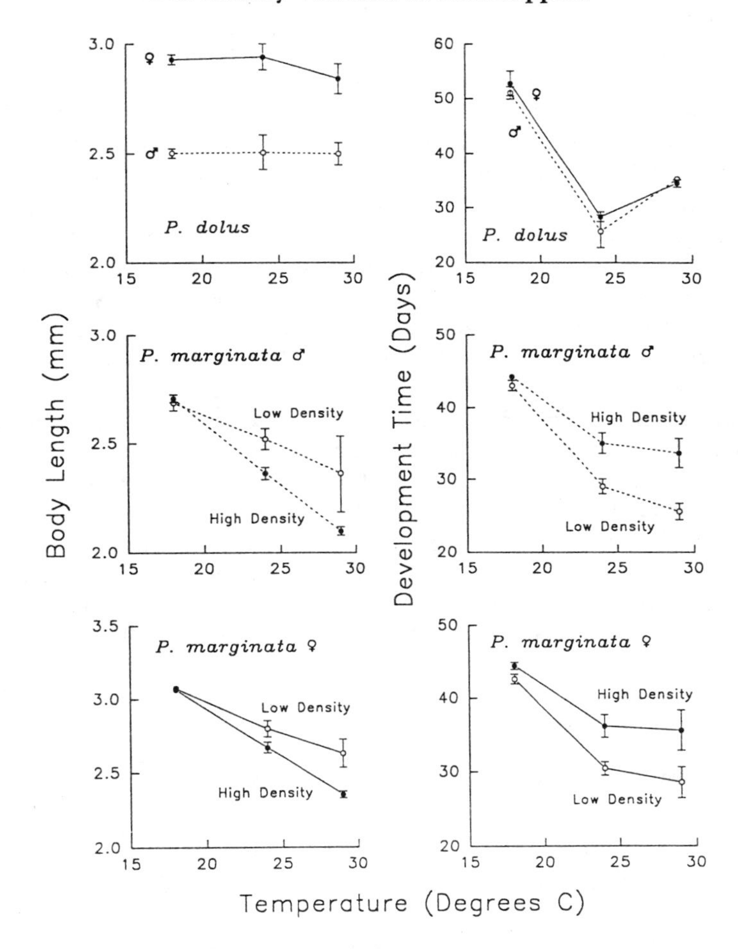

Figure 4.10. Influence of rearing temperature (18, 24, and 29°C) and density (5 and 25 nymphs/cage) on the body length (mm) and development time (days from nymph to adult) of the sexes of ***Prokelisia dolus*** and ***P. marginata***. There was no effect of temperature or density on the body length of ***P. dolus*** (P>0.05); temperature (P=0.001), but not density (P>0.05), did effect development rate in both sexes (ANOVA, $n = \sim 10$ for each temperature × density treatment). For both sexes of ***P. marginata***, there was a significant interactive effect of temperature and density on body length and development time (P<0.05, ANOVA, $n = \sim 10$ for each temperature × density treatment); thus, the adverse effects of density on body size and development were compounded at high rearing temperatures. See Denno et al. (1985) for cage design and rearing setup.

delayed (Suenaga 1963; Drosopoulos 1977; Tsai and Wilson 1986; Fig. 4.10). There is a significant interactive effect of temperature and population density on the development rate and body size of *P. marginata*; the adverse effects of density on these components of fitness are intensified at high temperatures (Fig. 4.10). In *N. lugens*, the preoviposition period of macropters but not brachypters is extended under cool temperatures, yet remains the same as that for brachypters at hot temperatures (Mochida and Okada 1979). Consequently, differences in age to first reproduction between the wing forms are maximized at cool temperatures. Also, for *N. lugens*, nymphal survival (98%) and egg hatch (92%) are maximized at 25°C (Mochida and Okada 1979; Cheng and Holt 1990), and peak fecundity occurs between 23 and 26°C (Ding et al. 1981). The simulated peak density of *N. lugens* was particularly sensitive to temperature changes during the height of the oviposition period (Cheng and Holt 1990). All factors considered, the population growth of *N. lugens* is maximized at temperatures of about 28°C (Mochida and Okada 1979).

Under low-humidity conditions (60%), *M. fairmairei* developed slower and survived less well than when humidity was high (100%) (Drosopoulos 1977). Even though data are limited, this result may explain why so many delphacid species occur in moist habitats (Denno 1978; Ossiannilsson 1978; Dyck et al. 1979; Booij 1982a).

Evidence for a Coordinated Life History Strategy

Life history theory predicts that particular combinations of traits will evolve under specific selective conditions (Stearns 1976, 1977, 1980). For example, in temporary habitats, selection should favor both high dispersal capability and high fecundity for reasons of colonization and establishment (Fairbairn 1984; Dingle 1985; Denno and Roderick 1990). However, tradeoffs (typical of delphacids) between dispersal (wing form) and reproductive potential appear to constrain the extent to which selection might maximize both traits (Roff 1984; Denno et al. 1989; Denno and Roderick 1990). Nevertheless, within the constraints of wing form, selection may have acted to integrate traits in ways that maximize fecundity. For example, egg size may be traded off for egg number and large body size may be favored.

Whether selection on one trait (dispersal) will influence another (fecundity) depends on the genetic correlation between them (Stearns 1980; Falconer 1981; Lande 1982; Fairbairn and Roff 1990; Roff 1990). In *Prokelisia marginata*, no significant genetic correlations were found between dispersal capability (wing form) and either survivorship or development time, which suggests that selection for dispersal will not influence these other traits (Roderick 1987). However, in females, there exists a significant negative correlation between body size, a correlate of fecundity, and development time (Roderick 1987). Across species within several genera and among populations of the same species, there is a positive association between dispersal ability (percent macroptery) and body size (Fig. 4.5), but within a species no data exists to support the contention that these two traits are genetically correlated. Furthermore, there are no data for planthoppers which suggest a genetic basis for correlations between wing form and reproductive characters. However, genetic correlations between wing form and reproductive traits exist for other wing-dimorphic insects (Roff 1990).

Thus, for planthoppers there is evidence for associations among traits [high dispersal capability and large body size (Fig. 4.5) and early age to first reproduction (Fig. 4.4B)], and the evolution of particular traits under certain circumstances [high dispersal capability in temporary (Fig. 4.6A) and arboreal habitats (Fig. 4.7)]. Nevertheless, few studies have attempted to elucidate a genetic basis for correlated traits. If we are ever able to understand why combinations of traits occur and the constraints on their association, a quantitative genetic approach is essential.

Planthopper life histories have been discussed in the context of an *r-K* continuum, whereby *r*-strategists in unpredictable environments are predicted to be highly mobile, very fecund, and reproduce at an early age (Kiritani 1979; Waloff 1980). An interspecific comparison of migratory and nonmigratory species provides weak evidence for an earlier age to first reproduction in migratory species, but fecundity was variable and not consistently higher for dispersive species (Fig. 4.4). The suggestion that selection has favored increased fecundity at the expense of egg size in the macropters of *J. pellucida* (Mochida 1973) is also consistent with the expectations of *r-K* theory for organisms in fluctuating environments (Stearns 1976). The bet-hedging hypothesis focuses on the relationship between fluctuations in juvenile and adult mortality (Stearns 1976; Tallamy and Denno 1981). Egg mortality is thought to be the key factor in the population dynamics of several delphacids (Waloff 1980; Waloff and Thompson 1980; Cook and Perfect 1989), and egg mortality is high (55-95%) in many species (Denno and Roderick 1990). Thus, juvenile mortality may fluctuate more than adult mortality in stable habitats, and selection should favor early and repeated reproduction (Murphy 1968; Stearns 1976). The preponderance of the brachypterous wing form in persistent habitats is consistent with this prediction.

Waloff (1980) has suggested that the collection of traits associated with the wing forms of planthoppers conforms to the predictions of life history theory, with macropters *r*-selected and brachypters *K*-selected. For this reason it is difficult to apply the predictions of life history at the specific level. Roff (1992) suggests that although the *r*- and *K*-selection model has been useful in formalizing the definition of fitness in density-regulated populations, attempts to transfer the concept to actual populations without regard to the complexities of life histories (e.g., correlations and trade-offs between traits) may be detrimental. Roff (1992) argues that the terms *r*- and *K*-selection should be interpreted strictly in terms of models of density dependence, and given the controversey surrounding this issue, it may be preferable to avoid their use altogether. Also, because the production of migratory forms in planthoppers is density-dependent and inversely correlated with population growth, the density-dependent suppression of populations is more characteristic of mobile species in temporary habitats than sedentary species in persistent habitats (Denno et al. 1991; Denno and Roderick 1992). Thus, density-dependent "regulation" appears to be characteristic of species in fluctuating habitats, a finding fundamentally at odds with the tenets of *r-K* theory.

Alternatively, a more productive approach to understanding life histories may be to examine variation in a particular trait, the ecological conditions which influence its expression, and correlations and trade-offs with other traits which constrain or promote its evolution (see Roff 1992). I have taken this tack with planthoppers by focusing on dispersal capability. In planthoppers, selection has resulted in the partitioning of dispersal and reproduction between two morphs, the ultimate in life history evolution (Denno and Roderick 1990). Macropterous individuals can locate new habitats, and brachypters contribute maximally to future generations (Dingle 1985). Thus, dispersal capability allows for spreading the risk of reproductive failure in space (Waloff 1980), and brachyptery may spread the risk of reproductive failure in time (Prestidge 1982a; Denno et al. 1989). Within a morph, selection may fine-tune the integration of traits (e.g., increasing egg number at the expense of egg size), and it is this aspect of life history variation in planthoppers which deserves future attention. Most certainly, however, the key to understanding the evolution of planthopper life histories is to focus on dispersal characters because reproductive and body size-related traits are so intimately linked to wing form.

SUMMARY

1. Delphacid planthoppers make ideal study organisms for testing life history hypotheses because they exhibit wing-dimorphism and because reproductive traits are integrated so differently in migratory

(macropterous) and flightless (brachypterous) forms. Thus, an examination of factors influencing wing form can provide powerful insight into the evolution of planthopper life histories.

2. Wing form in planthoppers is determined by a developmental switch that responds to environmental cues such as crowding and host plant quality. However, the sensitivity of the switch is heritable and under polygenic control.
3. Brachypterous females of most species have a short preoviposition period, reproduce at an earlier age, and are generally more fecund than their macropterous counterpart. The reproductive delay and reduced fecundity of macropters supports the notion that flight capability is costly and that phenotypic trade-offs between flight and reproduction exist.
4. Successful establishment in new habitats is enhanced not only by dispersal capability, but also by high reproductive potential. The reproductive penalty imposed on colonizing macropters may be offset in part by selection for small egg size (e.g., *Javesella pellucida*) and the ability to place offspring on nutritious host plants (e.g., *Prokelisia marginata*).
5. Within a genus, there is an association between large body size and flight capability; migratory species are larger than their sedentary congeners. This pattern most likely reflects selective pressures associated with bioenergetics because larger insects can fly further. However, large size in colonizing species may also be favored for reasons of increased fecundity (a correlate of body size). For species in the genus *Prokelisia*, there is no consistent association between large body size and rapid development.
6. Levels of migration (percent macroptery) decrease significantly as the persistence of the habitat increases. Species inhabiting temporary habitats (agricultural crops) are primarily macropterous, whereas those occurring in permanent habitats (bogs and marshlands) are mostly brachypterous. Also, males of many migratory species retain wings at low rearing densities, suggesting that habitat persistence influences dispersal capability by dictating the availability of mates as well as by selecting for habitat escape. Flightlessness is favored in permanent habitats for reasons of increased reproductive capability.
7. Within a species, geographic variation in wing-form composition is genetically based and is related to habitat persistence. Macropterous populations are associated with ephemeral habitats. Flightlessness increases to the north in some species where, at high latitudes, rates of

succession proceed slowly. However, this is not a general pattern for planthoppers because persistence increases to the south where winter destruction of habitats is less frequent.

8. Host plant architecture places constraints on the evolution of delphacid life histories. With habitat persistence controlled, macropterous species of Hawaiian delphacids feed mostly on trees and large shrubs, whereas brachypterous species occur mostly on low-profile vegetation such as small shrubs and herbs. Flightlessness is apparently disadvantageous in arboreal habitats because their negotiation is difficult without flight.
9. Most temperate and boreal species of delphacids overwinter as nymphs (73%), but some overwinter as eggs (17%) and a few as adults (10%). The number of generations attained each year depends largely on the length of the growing season and is negatively correlated with latitude. This pattern holds across species and among populations of widespread species. Most boreal species are univoltine, whereas tropical species produce as many as 12 generations annually. Diapause also varies geographically, with northern populations entering diapause and southern populations undergoing continuous reproduction. In *Laodelphax striatellus*, there is a wing-form-dependent reproductive penalty coupled with diapause; the fecundity of postdiapause macropters (not brachypters) is reduced. Mixed diapause strategies are poorly investigated in delphacids.
10. Most species deposit their eggs in small groups (2–8), and, by doing so, may "spread ovipositional risks" associated with natural enemies. However, physiological/morphological constraints (e.g., ovariole number) and host plant toughness may also influence egg batch size. Egg size varies considerably among planthopper species (0.65–1.15 mm in length), but there is no evidence for a trade-off between egg number and egg size using interspecific comparisons as there is between the wing forms of *J. pellucida*. A comparison among congeners may provide support for such a trade-off.
11. Fitness components are adversely affected in most species under crowded conditions and on nutritionally inferior host plants. Also, the optimal temperature for development of most delphacids is 24–28°C, and at temperatures above and below these, development is delayed. High temperatures in particular result in decreased survival, reduced body size, and they exacerbate the negative effects of crowding.
12. Life history theory predicts that particular combinations of traits will evolve under specific selective conditions. However, the particular combination of traits that evolves will depend on the genetic variance

and covariance among traits, phenotypic trade-offs and costs, phylogenetic constraints, and selection. Although there is substantial evidence for phenotypic trade-offs between traits (reproductive potential and flight capability), positive associations among traits (large body size and migration), and the influence of selection on a trait (association of macroptery with arboreal and ephemeral habitats), there is little information on the genetic correlation among life history traits for planthoppers. Further, there has been little emphasis on how traits may be integrated differently within the same wing form across species. These issues should form the focus for future research on the life histories of planthoppers.

ACKNOWLEDGMENTS

I thank M. Claridge, D. Fairbairn, and D. Roff for their comments on earlier drafts of this report. M. Claridge and M. Wilson provided me with the opportunity to study planthoppers in collections at the University of Wales and British Museum respectively; their hospitality is greatly appreciated. This research was supported by National Science Foundation Grants BSR-8206603 and BSR-8614561. This is Scientific Article Number A-6301, Contribution Number 8472 of the Maryland Agricultural Experiment Station, Department of Entomology.

5

Reproductive Behavior: The Role of Acoustic Signals in Species Recognition and Speciation

Michael F. Claridge and
Peter W.F. de Vrijer

Introduction

Biparental sexual reproduction is usual in planthoppers. In this they resemble their other relatives among the Auchenorrhyncha. Den Bieman and de Vrijer (1987) discovered the only known example of a truly parthenogenetic planthopper, but pseudogamy has been reported in a few species. The evolution of these forms of reproduction will be discussed further below. Biparental reproduction involves a complicated sequence of behavior that results in contact between receptive males and females of the same species. These behavior patterns, from long-range mate location to copulation and transfer of gametes, may best be included under the general title of mating behavior.

Acoustic signaling is central to the communication system known to be used by both sexes of planthoppers when courting and mate finding on their host plants. Such signals generally show very high species specificity (see Claridge 1985a, 1985b). Thus, here we discuss the nature of acoustic signaling systems and their possible roles in species recognition. Further we discuss what is known of possible modes of speciation in these insects which show great variation both in host plant specificity and in mating signals. Planthoppers, thus, provide ideal research material with which to test the validity of competing sympatric and allopatric models of speciation in specialist insect herbivores (Claridge 1988).

Mating Behavior

Few authors have attempted to describe the complete sequence of mating behavior in any planthopper. The most complete studies are provided by Booij (1982b), Drosopoulos (1985), Heady and Denno (1991), Ichikawa (1976, 1979) and McMillian (1963).

In all species, sexual receptivity is incomplete at the time of adult ecdysis. The duration of this period of sexual maturation varies between species, sex, and wing morph, from less than one day to eight or nine days. Usually, for any species, males mature earlier than females and brachypters earlier than macropters.

Close-range communication, that is, between insects on the same plant or on a group of plants the foliage of which is in physical contact, appears to be dominated by acoustic signaling (Claridge 1985a, 1985b). Longer-range communication has been little studied and may not exist. It seems probable that sexually receptive males move actively from individual plant to plant. On each, they probably alight and signal for a short time. If no response is received, then they will probably move on. No detailed field studies have yet been reported for planthoppers, but Hunt and Nault (1991) recently described such behavior in the leafhopper *Graminella nigrifrons*.

Acoustic Signaling and Mate Location

Ossiannilsson (1949) in a pioneering study first demonstrated that adult males of all major groups of Auchenorrhyncha are able to produce very faint acoustic signals, some of which he recorded for the first time. For some species, he also detected female calls. In addition, Ossiannilsson showed that males of all groups studied had a mechanism apparently homologous to the tymbal system of the well-known cicadas, though no striated tymbal was found in the Delphacidae.

Ossiannilsson was unable to be sure of the function of the quiet sounds produced by leafhoppers and planthoppers, but he suspected a relationship with courtship and mating (Ossiannilsson 1949, 1953). Strübing (1958a, 1985b) was the first to demonstrate the relationship of sound production in planthoppers with mate location. A major breakthrough in the study of the acoustic communication of planthoppers was the demonstration first by Ichikawa and Ishii (1974) that the signals are transmitted primarily through the substrate on which the insects are moving and, therefore, can be investigated by very simple techniques (see below). The vibrations are passed to the substrate through the legs of the signaling planthopper. The abdominal vibrations described by several authors are the visible signs of sound production, but do not involve percussion directly of the substrate.

The detailed physiological mechanism of sound production has not been fully elucidated and is a subject of some controversy (see Zhang and Chen 1987; Zhang et al. 1988; Mitomi et al. 1984). Whatever the detailed physiological mechanism, which will certainly differ between males and females, calls consist of patterned trains of damped pulses. The temporal patterns which may be very complex are amplitude modulated and generally show great species specificity (Claridge 1985a, 1985b).

Pair formation in planthoppers usually begins with the emission of what are generally termed calling signals by the male on an appropriate host plant. If this signal is received by a mature virgin female, she will normally respond with her own calling signal (Fig. 5.1). Usually this female response triggers further calling accompanied by intense locomotor activity by the male. This leads to an exchange or alternation of signals in which the female generally remains stationary. This sequence of alternating calling normally ends with the male making physical contact with the female.

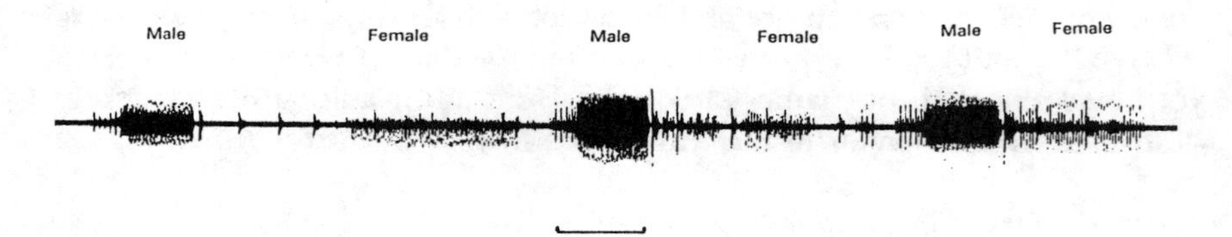

Figure 5.1. Oscillograms of an acoustic exchange between a male and a female *Nilaparvata lugens* (Stål) from rice in the Philippines. Scale line = 2.0 sec.

Courtship and Copulation

Once the male has made close contact with the female, courtship may be said to have begun. The exchange of acoustic signals continues during courtship and usually becomes more intense. At this stage, the male may also introduce new elements into his acoustic repertoire. In all species studied, the male takes up a similar orientation with respect to the female; that is, at the side and slightly to the rear of her, with both heading in the same direction. In some species, males show increased wing fluttering as courtship proceeds. Some males may also tap the female beneath her abdomen with front or middle legs (Claridge et al. 1984b). In *Prokelisia marginata*, males put one of their forelegs on top of the female's thorax (Heady and Denno 1991).

When making a copulation attempt, the male first lifts his wings and curves his abdomen so that the genital capsule points upward. He then attempts to copulate by moving his abdomen in this pose sideways under the female's abdomen. At this moment, the female has to lift her abdomen and hind legs to expose her genitalia and to allow the male to copulate. Various authors have described how a female may reject a male's copulation attempt by shaking her abdomen vigorously, and even by kicking movements with her hind legs (Booij 1982b; Claridge et al. 1984b, Drosopoulos 1985). Drosopoulos (1985) found that different species of *Muellerianella* differed markedly in the number of failed copulation attempts before the female finally allowed copulation. Thus, readiness to be courted does not automatically imply that a female is ready to mate.

The detailed mechanism of coupling of the male and female genitalia is little understood. The parameres and sclerotized hooks of the anal segment of the male are anchored under or behind special parts of the female's genital region, and, finally, the aedeagus is inserted into the female gonopore at the base of her ovipositor. The duration of copulation varies greatly between species, from about 1 sec in *Javesella pellucida* (de Vrijer unpublished) to sometimes more than 1 hour in *Muellerianella brevipennis* (Drosopoulos 1985). During copulation the orientation of the male toward the female may change from an angle of 45° to 120° or more. In some species, including *Javesella* and *Nilaparvata*, the males may continue to emit acoustic signals during copulation, albeit separated by larger time intervals, but the females are always silent.

The termination of copulation is usually initiated by the female. Sometimes coupling is so strong that the female has to use her hind legs to push the male away, which may require considerable force. After uncoupling, the female often may be observed to make a sort of pumping movement with her abdomen. Sometimes, the male is observed to curve his abdomen downward and to press his genital capsule against the substrate, possibly to restore his genital parts that might have become somewhat dislocated during a rather drastic uncoupling process.

After termination of a successful copulation, males may very soon start new calling behavior. However, recently mated females usually do not respond to male calling and reject further copulation attempts. Multiple mating in males seems to be much more common; for example, three to five copulations have been recorded in one day (Drosopoulos 1985, Oh 1979). Sometimes females were also observed to mate a second time, but this was only after a number of days had elapsed since the first mating (Drosopoulos 1985).

Male–Male Interactions

Male planthoppers have been reported to show competitive interactions during mating behavior, accompanied by typical acoustic signals. Ichikawa

(1982) observed increasing aggressive behavior between sexually active males of *Nilaparvata lugens* when kept in close contact. This aggressive behavior was expressed acoustically by two types of signals ("preaggressive" and "aggressive") and physically by "direct body attack." Other planthoppers also have been reported to emit similar types of alarm or distress signals (Booij 1982b; Drosopoulos 1985; Strübing and Hasse 1975).

Competitive behavior does not necessarily involve acoustic or physical aggression. Drosopoulos (1985) described the dominance of certain individual males of *Muellerianella*, which, in successive mating experiments, were repeatedly preferred over other ones by different females. In other species, males have been observed to display apparent group singing or chorusing, with no indication of aggression (Strübing 1959).

Experimental Studies on Acoustic Behavior

Studies on the reproductive behavior of planthoppers have focused necessarily on their acoustic behavior. Most studies have been related especially to species problems and evolutionary biology (see below). For a proper analysis of experimental recordings, an understanding of various problems that may affect the experimental results is essential. Apart from the recording technique itself, results may be influenced by the behavioral motivation of the insects, by modifications of the signal during transmission through the recording substrate, and by a variety of environmental variables.

Techniques of Recording and Analysis

Following the demonstration that acoustic communication of planthoppers involved transmission of vibratory signals through the host plant (already suggested by Ossiannilsson in 1949), it proved to be relatively easy to detect the signals using a simple gramophone cartridge and appropriate amplifier (Ichikawa and Ishii 1974; Strübing 1977; Claridge 1985a, 1985b). Technically more sophisticated, but more expensive and more difficult to apply, is a lightweight accelerometer in combination with a charge amplifier (Strübing 1977). Both types of vibration transducers have the same problem of changing the transmission properties of the substrate as soon as they are attached to it (Michelsen et al. 1982), although the effect of this problem can be reduced by suspending the transducer in a balanced position, just touching the substrate (de Vrijer 1984). The very specialized technique of laser vibrometry is the only completely loading-free recording method. However, recently a very elegant and simple technique which greatly reduces any loading effects was developed by Strübing and Rollenhagen (1988), based on the magnetodynamic principle.

Because the signals of planthoppers are amplitude modulated, simple oscillographic techniques are adequate for analysis. Indeed, it is unlikely that the loading problems of the simpler recording instruments discussed here materially affect the recorded signal in biologically meaningful ways. Apart from variations caused by the planthopper itself, signals detected are affected by the position of the insect on the recording substrate and by the filtering properties of the substrate, usually the host plant (Michelsen et al. 1982).

Behavioral Motivation

The acoustic signals emitted by planthoppers may vary in type and function, according to the particular behavior with which they are associated (see above). Which type of signal is produced in a given situation and with what intensity depends on the behavioral motivation of the insects. We have already seen that crowded conditions may lead to aggressive behavior and that females after mating lose, at least for some time, their readiness to respond. Other factors that might influence acoustic behavior are feeding condition, age, photoperiod, time of day, and previous acoustic experience.

The handling and manipulation of insects before and during recording requires special care and skill to cause as little disturbance as possible. During recording, the environmental conditions should be controlled as closely as possible. Temperature is known to have very strong effects on the rate and pitch of the signals (de Vrijer 1984). But light and humidity might also affect acoustic activity.

Classification and Terminology of Acoustic Signals

Acoustic signals emitted by planthoppers may be defined either in terms of their function (behavioral context) or their form. However, a functional definition of planthopper signals presupposes a good understanding of these functions and clear criteria to discriminate among them. Our knowledge of the behavioral functions of planthopper signals is largely based on inferences from their behavioral contexts, and relatively little on experimental data.

The most easily defined signal function is **calling** because it is associated with the initial stage of mating biology, in which individual males or females spontaneously start to emit signals to advertise their readiness to mate and to invite potential mates to respond. This function can be verified relatively easily by experiment. Prerecorded signals may be played back to test responses by members of the opposite sex. In planthoppers, calling is usually initiated by the male. To study calling signals, insects should be kept singly and isolated from acoustic stimulation by conspecifics, which could possibly lead to alteration in their behavioral motivation and, thus, signal type. For females, this procedure may give some practical difficulties because they do not often readily initiate spontaneous calling activities. Female calling behavior, however, can usually be evoked by playback stimulation with a prerecorded conspecific male calling signal.

In addition to **calling**, various other categories of acoustic signals are found in the literature, defined in terms of their different behavioral contexts. A frequently used term is **courtship sound**. The most essential distinction between **calling** and **courtship** is that the latter type of signal can only be displayed during acoustic interaction between two potential mates. Yet, both terms are sometimes used for the same signal type with the same function. This leads to confusion and misunderstanding and should be avoided so far as possible.

A second group of terms is used to describe the form or structural properties of acoustic signals. Although certain terms like pulse and chirp are widely used, various others have been created as convenient descriptors for particular signals of particular species. The possibility of confusion or misunderstanding, however, will not be serious as long as clear and sufficiently detailed oscillograms are presented to illustrate each signal.

Reproductive Isolation and Speciation

Reproductive isolation is the critical criterion for establishing the species status of populations of sexually reproducing organisms. Such isolation is maintained by species isolating mechanisms (Dobzhansky 1937). This is the essence of the classical biological concept of species, as developed particularly by Mayr (1942). Thus, to Dobzhansky, Mayr, and most succeeding authors, the phenomenon of speciation is seen as the evolution of reproductive isolation.

In recent years, Paterson (1985) has argued that species are primarily characterized by unique specific mate recognition systems. These serve to ensure the mating of conspecifics and only result secondarily in reproductive isolation. This concept Paterson terms the recognition concept in contrast to the classical biological, or isolation, concept of Mayr.

Whatever view may be taken in this discussion, biological species do have characteristic mate recognition systems which result in the reproductive isolation of species (Claridge 1988, 1990a). Thus, both concepts, in practice, produce similar descriptions of nature. Sympatric species have distinct mate recognition systems and are reproductively isolated from each other. It is then clear that the study of mate recognition systems will allow the precise identification of species. In addition, the process of speciation must involve primarily the evolutionary differentiation of mate recognition systems.

For planthoppers, it has been suggested that distinct acoustic calling signals are primarily responsible for maintaining reproductive isolation between related sympatric species [for reviews see Claridge (1985a, 1985b, 1988)]. Certainly, whenever groups of closely related sympatric species have been investigated, they have been found to be distinct in their acoustic signals, particularly those of males (Fig. 5.2). This strongly suggests that the

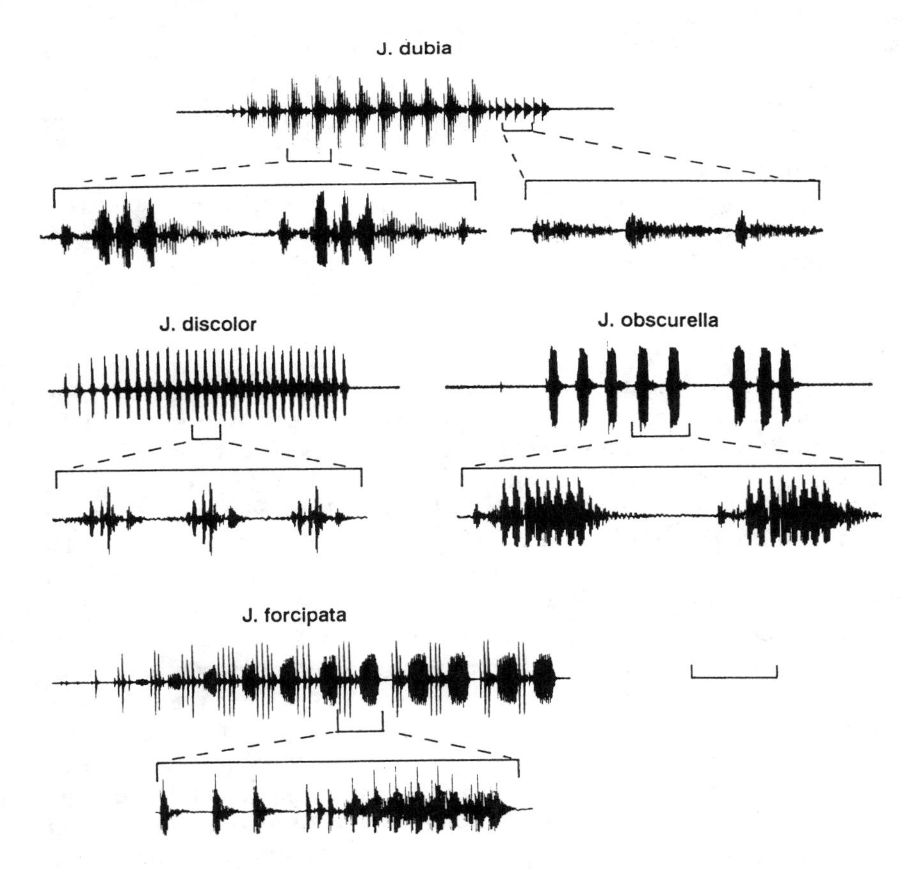

Figure 5.2. Slow and fast oscillograms of the male calls of *Javesella dubia*, *J. discolor*, *J. obscurella*, and *J. forcipata*. Scale line = 1.0 sec for slow and 0.1 sec. for fast oscillograms.

signals represent at least the major elements of their specific mate recognition systems.

For allopatric populations, the problems are considerable. It is impossible to demonstrate reproductive isolation between populations that do not have the opportunity of interbreeding in nature. Also, it is difficult to decide when differences in mate recognition signals between two populations are big enough to assign them to different species. Experimental hybridization may be helpful, but results have to be interpreted with caution. If hybridization attempts fail, it does not necessarily mean that this is due to an acoustic mating barrier, but playback experiments may help in determining this (see below). If hybridization is successful in the laboratory, this does not mean that it would occur under natural conditions. Indeed, there are several examples of distinct biological species of planthoppers that can be crossed easily in the laboratory, but which show no significant level of hybridization in the field, e.g., *Muellerianella* (Booij 1982c; Drosopoulos 1978), *Nilaparvata* (Claridge et al. 1985b, 1988), and *Ribautodelphax* (den Bieman 1988b).

Two particularly useful experimental approaches to determining the function of acoustic signals in mate recognition and species isolation of planthoppers are (1) mate choice tests and (2) playback experiments.

Mate Choice Tests

Mate choice experiments involve giving individuals of one sex (usually the female) a choice between mating with the opposite sex of its own population or with one from a test population. These experiments provide valuable data that are quantifiable and give good indications of field behavior.

Claridge et al. (1984b, 1985a) were able to show that populations of *Nilaparvata lugens* of different geographical origins could be crossed relatively easily, despite significant differences in the pulse repetition rate of the male calling signal. However, in mate choice experiments, females clearly preferred males of their own population to those from others. Similar results were reported for different species of *Muellerianella* (Drosopoulos 1985), and *Ribautodelphax* (den Bieman 1988b). In the same way, Claridge et al. (1985b, 1988) demonstrated strong assortative mating between two morphologically inseparable species of *Nilaparvata* from rice and *Leersia hexandra*, respectively. In this latter example, fully fertile hybrids were obtained after forced crossings, but there was no indication of hybridization in the field over a wide geographical area of Asia and Australia.

Although such experiments may give conclusive evidence on the species status of populations, they do not necessarily demonstrate the critical role of the acoustic signals. Another serious problem is that, the acoustic activity of planthoppers during experiments cannot be controlled. There may be various interactions taking place, both between males and between the female and the males. For example, it is quite possible that a heterospecific male does not recognize the female's response call and is, therefore, not stimulated to start searching behavior. When the proper female response call fails to appear, the male may soon stop calling, and, thus, may not even be detected by the female.

Playback Experiments

An elegant way to study the possible effects of particular signal types is by means of a playback technique, that is, by feeding back a prerecorded signal to the substrate on which a test planthopper is placed (Claridge et al. 1985a). Different functions of acoustic signals may be studied by choosing the appropriate experimental system.

Several authors have reported the successful use of the playback of calling songs of either sex to evoke response calling in the opposite sex. This clearly indicates that calling signals may function as effective mate recognition signals. If calling signals also have a function in acoustic isolation, then this might be reflected in a high level of nonresponse when played back to heterospecific females. Such experiments have been performed on various groups of planthoppers. Claridge et al. (1985b) demonstrated that females of the rice- and *Leersia*-feeding species of the *Nilaparvata lugens* group were able to differentiate between male calls of both populations. Females responded much sooner and more vigorously and called for longer periods in response to calls from their own species. In various combinations of closely related *Ribautodelphax* species, de Winter and Rollenhagen (1990) observed that females responded to heterospecific male calls at about 80% of the conspecific response level, irrespective of the level of cross-insemination that was found for the same combinations of species. In *Javesella*, where cross-insemination between species was extremely rare, considerable levels of female response could also be obtained to playback of heterospecific male calls (de Vrijer unpublished data). These results suggest that male calling signals alone for *Ribautodelphax* and *Javesella* are not sufficient to ensure reproductive isolation because the levels of female response to heterospecific male calls seem to be far too high. The results for *Nilaparvata* show a more complete isolating role for acoustic signals.

De Winter and Rollenhagen (1990) performed experiments with female calling signals on males. They found that, *Ribautodelphax* males do respond acoustically to heterospecific female calling signals, but do not start searching behavior. In choice tests, in which a conspecific and a heterospecific female signal were played back alternately, they found that, in various combinations of species, males approached the conspecific call significantly more often than that of the other species. It was concluded that at least part of the isolation between species was maintained by the ability of the males to discriminate between conspecific and heterospecific female calls. This is a remarkable conclusion for several reasons. First, it was unexpected because the relatively simple structure of the female signal did not seem to offer enough basis for effective species discrimination. Second, it is also in conflict with the theory that the sex with the greater parental investment, here the female, should be exerting the choice. It leaves us with the question of why male signals in planthoppers show so much more apparent differentiation than do female signals. In this context, it is significant that Claridge et al. (1985b, 1988) showed clear differentiation in female calls and male responses between the two sibling species currently known as *N. lugens*.

Evolutionary Differentiation of Acoustic Signals

The basis for all evolutionary change is genetic variation. Because only very little is known about the genetics of acoustic characters, the debate about their evolutionary change and the shaping forces involved has been largely speculative. In the view of Paterson (1978, 1985), each specific mate recognition system is subject to strong stabilizing selection, which keeps natural variation within narrow limits. It follows then that natural selection can operate only by very small steps because each selective change requires a change in the other partner to restore coadaptation of signal and receiver (Paterson 1985). According to this view, the main shaping force is provided by changing environmental conditions, to which the specific mate recognition system has to be adapted for optimal functioning.

However, the claims of Paterson seem to be in conflict with many reports of geographical variation in courtship and acoustic characters in various animal species. A well-studied example in planthoppers is the divergence in both male and female calling signals of *Nilaparvata lugens* (Claridge et al. 1985a, 1985b, 1988; Claridge 1990a), which seems to have originated in advance of any other major evolutionary change. Experimental tests revealed that various degrees of sexual isolation exist among populations, correlated with the degree of acoustic differentiation. In a related morphological species, *N. bakeri* (Muir), Claridge and Morgan (in press) have described major differences in male calling songs between allopatric populations from the Philippines, India, and Indonesia (Fig. 5.3). However, mate choice experiments showed random mating with no preference for

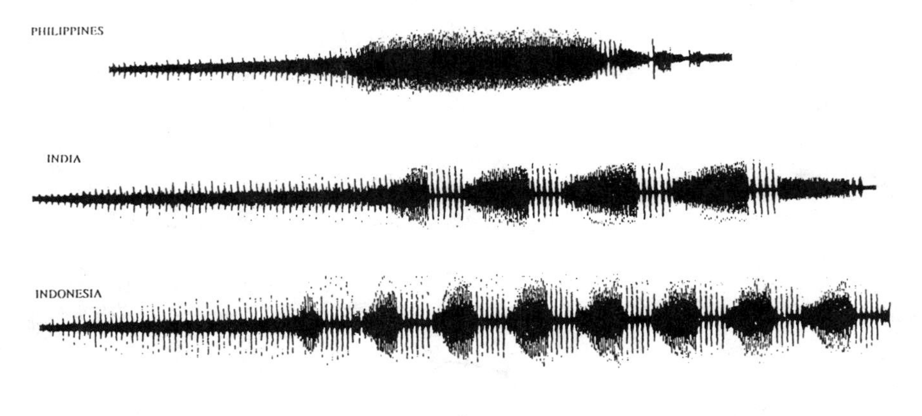

Figure 5.3. Oscillograms of typical individual male calls of *Nilaparvata bakeri* from the Philippines (Luzon), India (Orissa), and Indonesia (Bali) (after Claridge and Morgan in press).

homogametic pairing among these allopatric populations. Laboratory crossings produced fully fertile hybrids with intermediate calls. Thus, it is clear that distinct differences in male calls among allopatric populations cannot be taken necessarily to indicate biological species status.

The nature of the evolutionary forces that may be responsible for the geographical variation in planthopper calls is a major area for speculation. An interesting idea was advocated by West-Eberhardt (1983), who argued that mating signals are likely to be subject to selection caused by social competition for mates, that is, by sexual selection. According to this hypothesis, acoustic differences may be expected to develop among geographically isolated populations. How easily such changes may occur depends primarily on the amount of variation that exists within each population. In a very recent study, de Winter (1992) analyzed call variation for a population of *Ribautodelphax imitans* by subjecting it to artificial bidirectional selection on the pulse repetition frequency of the female call. Already after only five generations of selection, the divergent lines showed no overlap in their ranges of pulse repetition frequency (Fig. 5.4). Mate choice tests after 10 generations of selection revealed significant symmetrical assortative mating between males and females of the divergent lines. The divergent lines were experimentally crossed and the song types of the F1 and F2 generations were analyzed for the selected character, which was shown to be polygenic in nature, determined by at least six segregating genes. This is an exciting laboratory demonstration of the beginnings of

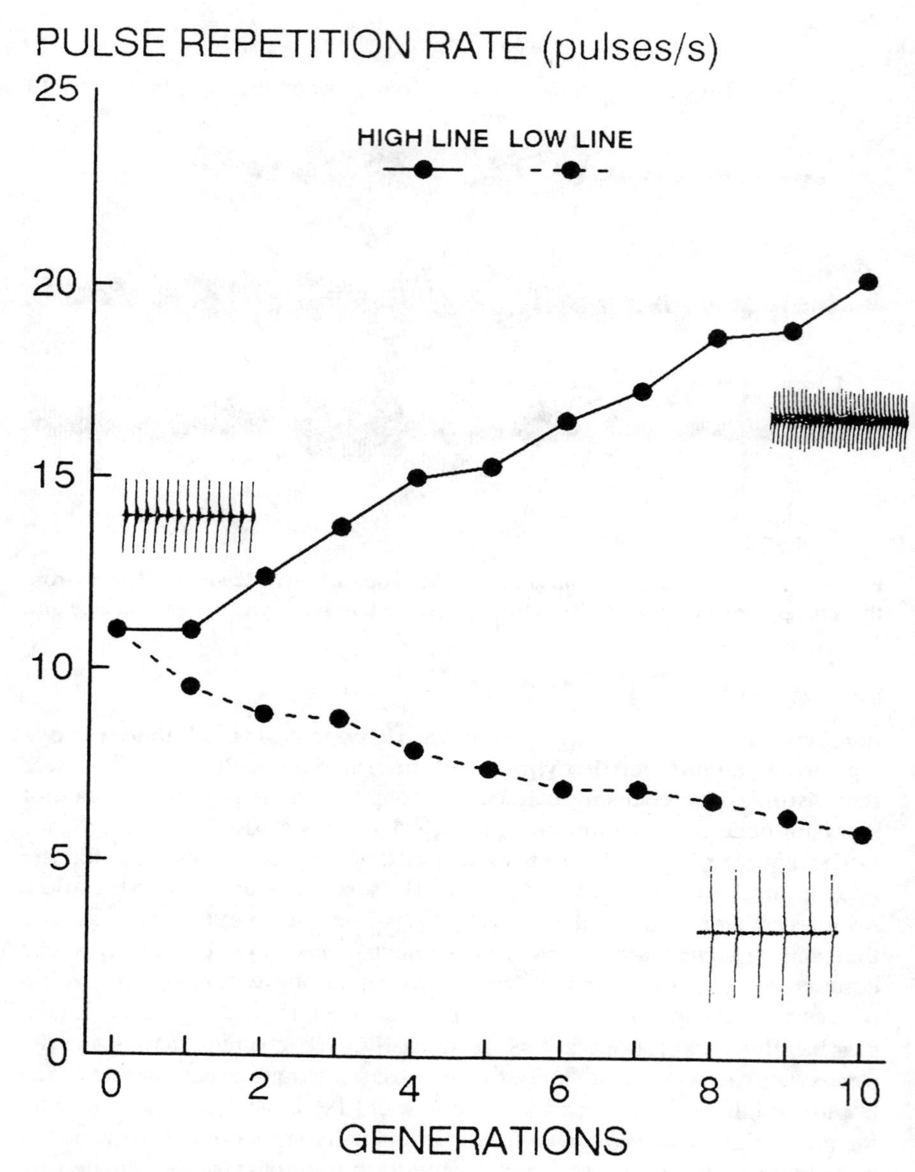

Figure 5.4. Results of bidirectional selection experiments on female calling signals of *Ribautodelphax pungens* (Ribaut). From the same parental population (generation 0), two lines were selected, a "High Line" for high Pulse Repetition Rate (PRR) and a "Low Line" for low PRR. PRR was measured in a specific and constant part of the calling signal. The graph shows the mean PRR values found in 10 successive generations of selection, based on 4 replicates per line and 40 females per replicate. At each generation of a replicate line, the 10 females having the most extreme PRR values were selected to start the next generation. After only five generations of selection, the ranges of variation of both lines no longer overlapped. Oscillograms of sample calls from the original population at generation 0 and of the high and low selected lines at generation 10 are shown on the graph. (Unpublished data by T. de Winter).

speciation and suggests that runaway sexual selection may, indeed, be a major evolutionary force for divergence in allopatric populations.

Reproductive Isolation and Host Plant Differentiation

Because the natural transmission channel for the vibratory signals of planthoppers is formed by their host plants, specific host plant relations are of direct relevance to the reproductive isolation of closely related species. Host plant differentiation not only results in greater spatial separation, but also in at least partial acoustic isolation. Both Booij (1982b) and Drosopoulos (1985) considered the different patterns of host plant utilization of *Muellerianella* species as the primary isolating factor. Certainly, in several groups of planthoppers, distinct patterns of host plant relations have been found among closely related species, for example, *Javesella* (de Vrijer 1981), *Muellerianella* (Drosopoulos 1985), *Ribautodelphax* (den Bieman 1987d), and *Nilaparvata* (Claridge and Morgan 1987). However, this of itself is not a strong argument for the importance of host plant differentiation in the process of speciation (Claridge 1988).

It is possible that the specific filtering properties of different host plants may have provided selective factors in shaping the acoustic signals produced by planthoppers feeding on them. Whether such a process might be strong enough to account for significant acoustic differentiation seems unlikely. In any case, for acoustic differentiation to take place, host plant differentiation is not an absolute prerequisite. This is well demonstrated by the geographically separated populations of *Nilaparvata lugens* (Claridge et al. 1985a) which all feed on rice. Also, a sibling species, attributed to *N. lugens*, and the related *N. bakeri* (Muir) both occur often together on *Leersia hexandra* (Claridge and Morgan 1987). Similarly, in North America, two sympatric *Prokelisia* species share the same host grass, *Spartina alterniflora*, and have structurally distinct calls (Heady and Denno, 1991). Another genus, currently under study in the Netherlands, that might be interesting in this respect is *Chloriona*, all species of which feed on the reeds *Phragmites communis* and *Arundo donax*.

Despite the frequently stated arguments for the sympatric speciation of many insect herbivores along host plant lines, the evidence is not strong for Auchenorrhyncha (Claridge 1988). Models of allopatric divergence and speciation are better supported by the limited data at present available on only a few groups of species.

Reproductive Isolation and Structural Differences in Genitalic Morphology

Traditionally, structural characters of the genitalia, especially those of males, have been considered to provide reliable features for the identification of planthopper species. Indeed, in most genera, at least small genitalic differences can be found, even among very closely related species. It is not surprising that the specific structure of copulatory organs have been widely thought to play a role in species isolation, the so-called lock and key principle (Eberhardt 1985). However, it is not likely that mechanical barriers are very effective in preventing interspecific mating, especially among closely related species because there are many examples where successful experimental cross-mating has been reported, despite genital differences among the species involved. Also, species recognition usually seems to occur before males make physical contact with females. An alternative explanation for such evolutionary differentiation may again lie in the pressures of sexual selection by female choice (West-Eberhardt 1983, Eberhardt 1985).

Pseudogamy and Parthenogenesis

In some planthoppers, reproductive biology is complicated by the phenomenon of triploid all-female populations living in association with normal populations of diploid males and females. This was first discovered by Drosopoulos (1976), who further found that the triploid females reproduce by pseudogamy or gynogenesis. This mode of reproduction implies that the triploid females depend on mating with males, but only to initiate the embryogenesis of their eggs. Fusion of gametes does not occur and the developing embryos receive maternal genes only. The resulting offspring are again exclusively female and triploid. Genetically, this type of reproduction is the same as parthenogenesis, but ecologically it requires coexistence with males of a related biparental species. However, for the males of the diploid population, mating with pseudogamous females means wasting gametes. For the diploid females, the presence of triploid females means competition for males and for sperm. Therefore, such triploid pseudogamous forms may be considered as reproductive parasites of the diploid population.

So far, pseudogamy has been found in two planthopper genera, *Muellerianella* (Drosopoulos 1976; Booij and Guldemond 1984) and *Ribautodelphax* (den Bieman 1987d, 1988a, 1988b, 1988c). However, because the phenomenon can be recognized only following careful cytological studies of female oocytes and by breeding experiments, it might prove to be more common when other genera are investigated, especially where female-biased sex ratios occur in the field.

Because pseudogamous females produce only female progeny, they have potentially a twofold reproductive advantage over diploid females, which also have to produce males (usually 50%). If this potential was fully realized, the number of pseudogamous females in mixed populations would increase rapidly compared to the diploid population. More and more pseudogamous females would have to mate with a relatively decreasing number of males. Eventually, the triploid females might become so numerous compared to the diploid females that some of the latter might not succeed in attracting males. Succeeding generations would yield less and less males, which finally would mate only with pseudogamous females, thereby causing the complete extinction of the population.

In field populations, the ratios of pseudogamous females to diploid females is relatively stable from year to year (Booij and Guldemond 1984, den Bieman 1988a), although considerable differences in ratio were found between populations. According to den Bieman (1988a) diploid males are able to discriminate between the two female forms, as demonstrated in mate preference tests. Such a form of assortative mating among males and females of the diploid population might be sufficient to counterbalance the reproductive advantage of the pseudogamous females.

Acoustic Behavior of Pseudogamous Forms

The most extensive studies on acoustic behavior of pseudogamous forms were carried out on ***Ribautodelphax pungens*** by den Bieman (1988a). Generally, triploid pseudogamous females appeared to produce the same type of signals as the females from the diploid populations with which they were coexisting. The same was found also in two other species, ***R. imitantoides*** (den Bieman 1988a) and ***Muellerianella fairmairei*** (Booij 1982c), which also are associated with pseudogamous forms. However, the variation in female calling signals within such mixed populations was much greater than in pure diploid populations of the same and other related species.

According to de Winter (personal communication), this large variation might well be explained by the effects of a constantly changing direction of selection on male mating preference. There will be selective advantage for males preferring female song types that have a higher chance of being produced by diploid females. The song types of diploid females, therefore, will tend to evolve away from the most frequently occurring song types among pseudogamous females. On the other hand, populations of pseudogamous females will be forced to track any evolutionary changes in the song type of the diploid females in order not to lose their chances of mating with males. This process would disturb completely the effect of stabilizing selection operating in unmixed diploid populations, where females with aberrant song types are assumed to have lower chances of successful mating.

Evolution of Pseudogamous Forms in Planthoppers

Pseudogamy is found in various groups of animals and is always associated with polyploidy (Stenseth et al. 1985). It may originate either spontaneously in a diploid population through autoploidy, or after hybridization with a closely related species (alloploidy). According to Drosopoulos (1976) and Booij and Guldemond (1984), pseudogamous forms in *Muellerianella* are probably of allopolyploid origin because they display a number of hybrid characters in morphology, host plant choice, and distribution. Drosopoulos (1978) even succeeded in synthesizing a triploid pseudogamous form experimentally, after hybridizing and backcrossing two related *Muellerianella* species. Den Bieman (1988a, 1988b), in contrast, considered the evidence in *Ribautodelphax* to be more in favor of an autopolyploid origin of pseudogamous forms because of the great similarity to the associated diploid population in host plant relations, acoustic behavior, and allozyme patterns. Extensive crossing and backcrossing experiments never resulted in the production of triploid forms.

In planthoppers, only one case of complete parthenogenetic reproduction has been found so far in the genus *Delphacodes* (den Bieman and de Vrijer 1987). The karyotype of the females appeared to be triploid. Although the parthenogenetic females are not dependent on males for their reproduction, they were acoustically active and responded to calling signals of *D. capnodes* males, and also readily mated with them. Because the parthenogenetic form behaved in many respects like the pseudogamous forms known from *Muellerianella* and *Ribautodelphax*, parthenogenesis might have evolved through an ancestral pseudogamous stage. From an ecological point of view, this would mean that the population has freed itself from obligate coexistence with a biparental species.

Summary

1. Acoustic signaling dominates the communication systems used by both male and female planthoppers during mate location and courtship on their host plants. In some species, it is also associated with competitive interactions between males.
2. Acoustic signals are transmitted as substrate vibrations, the natural substrate usually being the host plant. Communication only takes place between planthoppers sharing a common substrate, such as the same plant or touching plants.
3. Mate recognition and mate location are mediated by an exchange of calling signals between a generally stationary female and an active male. Both male and female calling signals usually show high species specificity.

4. Biological species are characterized by unique specific mate recognition systems (SMRS) that have the effect of producing reproductive isolation between them. In planthoppers, distinct acoustic signals have been shown to be an important component of the SMRS.
5. Mate choice tests indicate the degree of any assortative mating between planthopper populations. The contribution of acoustic signaling may be determined experimentally by observing responses to the playback of prerecorded calls.
6. In Paterson's (1985) viewpoint, the SMRS of a species should be stable throughout its range. Contrary to this view, the acoustic signals of planthoppers often show clear variation and differentiation among allopatric populations. The nature of the evolutionary processes responsible for such divergence are uncertain, but it is suggested that sexual selection may be important. Artificial selection experiments have shown that acoustic signals are capable of rapid evolutionary change.
7. In many groups of planthoppers, host plant differentiation is probably a major factor in species isolation. There is, however, little evidence to suggest that host plant shifts have been important in call divergence or that sympatric speciation has been a common mode of speciation among planthoppers.
8. Populations of some planthoppers include triploid parthenogenetic females in addition to normal diploid males and females. These triploid females are pseudogamous and must copulate with diploid males to reproduce, even though fusion of gametes does not occur.
9. Acoustic signals of triploid females resemble those of the diploids with which they are associated. However, variation in the signals of diploid females within such mixed populations is greater than in pure diploid ones. Interactions between males and triploid females may disturb the effect of stabilizing selection normally operating in pure diploid populations.
10. Planthoppers provide an ideal group in which to study the nature and pattern of variation in SMRSs and the possible modes of their evolutionary divergence.

ACKNOWLEDGMENTS

We are deeply indebted to our colleagues in Cardiff and Wageningen for their help and encouragement in the preparation of this chapter. We should particularly mention John Morgan in Cardiff and Ton de Winter in Wageningen. Sakis Drosopoulos, Athens, kindly read and made useful comments on an early draft of this chapter. We are especially grateful to Ton de Winter for allowing us to use his unpublished Fig. 5.4.

6

An Ecological Framework for the Study of Planthopper Mating Systems

James R. Ott

Introduction

Mating system studies seek to elucidate the selective regimes responsible for the evolution of observed reproductive strategies (Emlen and Oring 1977; Halliday 1978, 1983; Vehrencamp and Bradbury 1984; Davies 1991). Investigations of mating systems, therefore, attempt to (1) quantify variation in male and female reproductive success and evaluate the extent and form of sexual selection, (2) understand the role of the environment in shaping the mating system by examining spatial and temporal variation in ecological parameters thought to be important in structuring mating systems, (3) determine the adaptive significance of behaviors and morphologies involved in mate search, courtship, competition, and choice, (4) elucidate constraints acting on behaviors and morphologies associated with reproduction, and (5) explain among-species variation in reproductive strategies. Two central premises underlie this research. First, the analysis of species' mating systems in their current ecological context can be used to infer the environmental conditions and selection regimes responsible for the evolution of observed reproductive behaviors and morphologies. Second, the reproductive strategies of the sexes and species' mating systems reflect the outcome of selection to maximize reproductive success.

The analysis of insect mating systems has produced a wealth of detailed information on the operation of natural and sexual selection (Blum and Blum 1979; Thornhill and Alcock 1983; Endler 1986; Bradbury and Andersson 1987) and has contributed significantly to the growth of behavioral ecology (Krebs and Davies 1978, 1984, 1991). Whereas insect mating systems have received a great deal of both theoretical and empirical attention (Borgia 1979; Thornhill 1979; Thornhill and Alcock 1983), the ecology and evolution of planthopper (Homoptera: Delphacidae) mating systems have thus far been largely overlooked. Selected aspects of the mating biology of planthoppers have been examined from an evolutionary perspective (see, for example, Booij and Guldemond 1984; Drosopoulos 1985; den Bieman 1988a; Claridge 1990a; de Winter and Rollenhagen 1990) but, thus far, no investigations have provided a comprehensive analysis of the ecology and dynamics of the mating system for any planthopper species. For example, no studies of single species or populations have explicitly: (1) quantified variation in reproductive success of male and female planthoppers in field populations, (2) measured intrasexual or intersexual selection, (3) examined the relationship between variation in reproductive success and either variation in behavior (acoustic, courtship, or competitive) or morphology (body size or wing form in dimorphic species), (4) assessed the role of environmental variation as a determinant of variance in mating success, or (5) attempted to elucidate constraints acting on behavior and morphology associated with reproduction. Investigations in each of these areas will be necessary to understand more fully the ecology and evolution of delphacid mating systems.

In addition to providing an excellent opportunity to explore the relationships between variation in both reproductive success and ecology, morphology, and behavior, studies of delphacid mating systems will contribute to our general understanding of broader issues in planthopper biology. For example, the behavioral/ecological approach outlined herein will provide insight into three areas that traditionally have been central in applied and basic planthopper biology. These include: (1) the evolution of planthopper resistance to introduced cultivars (in which assortative mating plays an important role in the establishment and maintenance of local adaptation), (2) the transmission of plant pathogens (in which differential rates of movement and differences in the reproductive behaviors of the sexes and/or wing morphs may be important), and (3) the evolution of dispersal polymorphisms such as wing-dimorphism (in which differential reproductive success of alternative wing morphs under varying environmental conditions may play an important role in the evolution and maintenance of the dimorphism).

The paucity of research on planthopper mating systems has not been for lack of scientific interest in the Delphacidae. Because of their status as important agricultural pests and vectors of crop pathogens, planthoppers have been the focus of an impressive basic and applied research effort as evidenced by this volume, and others (International Rice Research Institute 1979; Nault and Rodriguez 1985), and a voluminous journal literature. This research, however, has been conducted principally at the population and community level (reviewed in Denno and Roderick 1990), whereas research on the breeding biology of individual species and species complexes has focused primarily on reproductive isolating mechanisms (Claridge and Reynolds 1973; Booij 1981, 1982b; Claridge 1985a, 1985b; Claridge et al. 1988), species recognition (Drosopoulos 1976, 1977, 1978, 1985; Claridge and Morgan 1987), and courtship behavior (McMillian 1963; Claridge et al. 1984b; Heady and Denno 1991).

Several characteristics of delphacid biology, including the monophagy of many species (Denno and Roderick 1990), dramatic temporal and spatial variation in population density (Kisimoto 1965; Kuno 1979; Denno et al. 1985b, 1986), wing dimorphism in many species, and shifts in body size and wing-form ratios as a function of host plant quality and population density (often with the sexes responding differently to density) (Kisimoto 1965; Saxena et al. 1981; Denno et al. 1991), all suggest that competition for mates and mating success may exhibit high levels of spatial and temporal variation. Moreover, these attributes of planthopper biology in conjunction with aspects of the reproductive biology and ecology of delphacids, for example, courtship signaling and behavior (McMillian 1963; Booij 1982b; Claridge 1985a, 1985b; Drosopoulos 1985), aggressive acoustic and physical male–male interactions (Ichikawa 1979, 1982), differences in the timing of reproductive maturation of the sexes (Heady and Denno 1991), and relationships between mating frequency, fecundity, and fertility (Oh 1979), shed light on the possible determinants of the structure and diversity of delphacid mating systems.

The goal of this chapter is to establish a framework to guide future studies of the structure and determinants of delphacid mating systems. To accomplish this goal, I document those features of planthopper reproductive biology and behavior likely to contribute to variation in individual reproductive success, examine ecological aspects of the environment likely to influence the structure of delphacid mating systems, and outline testable hypotheses to guide further study.

A Framework for the Study of Delphacid Mating Systems

Variation in Male and Female Reproductive Success

A species' mating system is defined on the basis of variation in the number of mates acquired by members of each sex (Trivers 1972; Emlen and Oring 1977; Wade and Arnold 1980; Thornhill and Alcock 1983). Variation results in different patterns of mating, that is, monogamy or polygyny for males or monogamy or polyandry for females (Thornhill and Alcock 1983).

Ecological classifications of mating systems (Emlen and Oring 1977; Bradbury and Vehrencamp 1977) which group species according to similarities in their reproductive ecologies highlight the role of the environment in shaping the mating system and focus attention on the determinants of variation in male and female mating success.

Determinants of Variation in Reproductive Success

Variation in male reproductive success typically exceeds that of females (Bateman 1948). This fundamental result is considered to arise ultimately from the basic dichotomy in the size of gametes produced by the two sexes. Because of their larger gamete size, female reproductive success is limited by the ability to produce offspring, whereas male success is limited by the ability to inseminate multiple mates (Bateman 1948; Trivers 1972). As a consequence, females are usually the limiting sex, and the form of the mating system is determined primarily by the degree of variation in male reproductive success (typically measured as the number of mates/male) (Thornhill and Alcock 1983). No quantitative data directly examine relative variation in reproductive success between the sexes in delphacids. In the absence of such data, no definitive statements can be made concerning the patterns of mating, the extent and form of sexual selection, or the spatial and temporal dynamics of delphacid mating systems under field conditions. Details of the breeding biology of delphacids (discussed herein) suggest that, qualitatively, variation in male mating success should exceed variation in females.

Several studies have examined variation in reproductive success among female delphacids and have shown that reproductive success is affected by (1) host plant quality (Metcalfe 1970; Denno and McCloud 1985; Denno et al. 1986), (2) population density and crowding (Kenmore 1980; Heong 1988), (3) body size (Denno et al. 1986; Roderick 1987), and (4) wing form (Roderick 1987; Denno et al. 1989). For male delphacids, similar data are anecdotal and incomplete (Takeda 1974; Drosopoulos 1985). Because the form of the mating system is determined by the degree of variation in male mating success, this chapter will primarily address potential sources of variation in male reproductive success.

Variation in male reproductive success can arise from both intrasexual and intersexual selection (Darwin 1871), processes that in planthoppers may be influenced by (1) adult density, (2) variation in resource quality, (3) adult spatial distribution, (4) operational sex ratios (Emlen 1976), (5) the intensity and frequency of male–male and male–female interactions, (6) wing form, (7) body size, and (8) characteristics of mate signaling such as call frequency, intensity, and pattern of amplitude modulation. The avail-

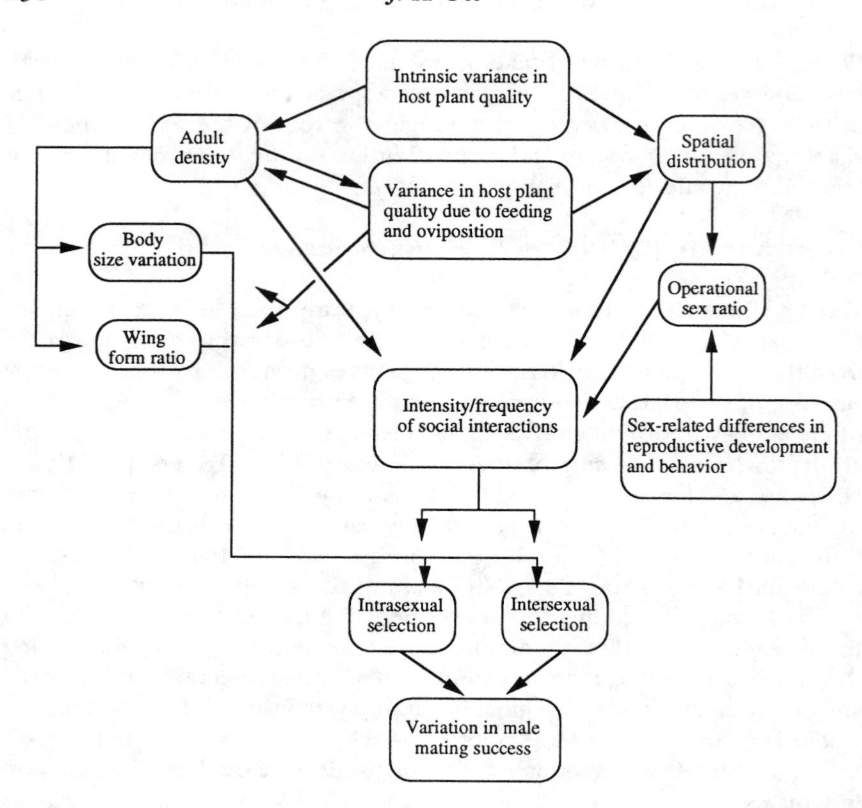

Figure 6.1. Determinants of variation in reproductive success of male delphacids. Variation in the reproductive success of male delphacids can arise from both intrasexual and intersexual selection. In planthoppers, both of these processes may be influenced by the ecological conditions experienced by adults. Both intrinsic variation in host plant quality and induced variation caused by feeding and oviposition influence the density and spatial distribution of adults. In turn, the density and spatial pattern of adults may, in part, control the frequency and intensity of social interactions within and between the sexes. In addition, adult density and spacing patterns may influence the operational sex ratio (the ratio of adult males to females at the time and location of breeding). Sex-related differences in the time required to attain sexual maturity (males develop faster than females) and sexual differences in postcopulatory behavior (once mated females are removed from the pool of potential breeders, whereas males may mate multiply) may also influence the operational sex ratio. The intensity and frequency of social interactions driven by adult density, spacing pattern, and operational sex ratio are important in determining the outcome of both intrasexual and intersexual selection and may influence, for example, the development of male–male aggressive behavior, the intensity of competition between males for mates, the frequency with which males acoustically signal to po-

able information on delphacid reproductive biology and ecology suggest that environmental variation and intrinsic features of planthopper biology may interact to produce variation in male reproductive success. These features and their possible interactions are presented in Figure 6.1 and are discussed each in turn.

The Ecological Template of Delphacid Mating Systems

Population Density

Delphacid populations typically experience extreme temporal and spatial variation in density throughout the course of their development on the host plant (Kisimoto 1965; Denno and Grissell 1979; Strong et al. 1990). Density fluctuations almost certainly influence all facets of the reproductive ecology of delphacids. For example, variation in planthopper density generates variation in host plant quality (Cagampang et al. 1974). Together, variation in resource quality and population density influence (1) the ratio of brachypterous to macropterous individuals produced in a population (Denno et al. 1985; Roderick 1987; Denno et al. 1991), hence impacting the encounter frequencies of sexually receptive and unreceptive premigratory virgins), (2) the body size of adults produced (Kisimoto 1965; Denno et al. 1986; Roderick 1987), (3) female fecundity (Kenmore 1980; Heong 1988), (4) the spatial distribution of potential mates and competitors (Denno and Roderick 1992), (5) the intensity and frequency of social interactions such as male–male aggression and establishment of dominance (Ichikawa 1979, 1982), and 6) rates of acoustic signaling by males (Ichikawa 1982). Because of the nearly ubiquitous influence of density on the reproductive ecology

---

tential mates, and mate choice by females. Intrinsic and induced variation in host plant quality also affect the body size and wing-form ratio of adults produced in populations. Variation in male body size and wing form may play an important role in intrasexual selection. Male body size and wing form as well as individual variation in mate attraction calls (call frequency, intensity, and structure), courtship displays (wing-flutter), and mating status of males (virgin as opposed to already mated) may serve as criteria for female choice and, hence, may be important in intersexual selection. The vagility of adult planthoppers within and among plants within habitats and the response of adults to both ecological and social stimuli determine the extent to which these stimuli effect the intensity and form of sexual selection.

of delphacids, the potential impact of density on mating system dynamics is incorporated, as appropriate, in the following sections.

Variation in Host Plant Quality and the Spatial Distribution of Adults

Within and between plant variation in the quality of feeding and oviposition sites can arise from intrinsic microsite variation within habitats (Denno et al. 1980; Denno 1983, 1985b), phenological variation within and between plants (Waloff 1980, Prestidge 1982a; Prestidge and McNeill 1982), and the feeding and oviposition actions of planthoppers (Cagampang et al. 1974; Sogawa 1982; Bacheller 1990). Data on the distribution of feeding and ovipositing females on their host plants suggest that females are sensitive to variation in feeding site quality (Denno et al. 1980; Waloff 1980; Sogawa 1982) and are selective with respect to sites for oviposition (Raatikainen 1967; Waloff 1980). This sensitivity may play an important role in the dynamics of mating systems by influencing the distribution of hoppers within and among plants and, hence, opportunities for interactions within and between the sexes.

Three predictions of how host plant variation (patchiness) influences the spatial distribution of adults, interactions within and between the sexes, and variation in male mating success follow from the ecological classification of mating systems of Emlen and Oring (1977) and await testing. First, the clumping of females and males increases with increasing variance in plant and patch quality. Second, the intensity of male–male interactions increases in response to clumping of females. Third, variation in host plant quality may lead to increased opportunities for female monopolization by individual males and/or female choice and, hence, may result in increased variation in male reproductive success.

Adult Vagility

Studies of planthopper movement have focused primarily on long-distance dispersal (Kisimoto 1976; Rosenberg and Magor 1983a; Riley et al. 1987; Loevinsohn 1991; Riley et al. 1991) and movement between adjacent habitats varying in resource quality (Denno and Grissell 1979; Prestidge and McNeill 1982; Denno 1983). However, movement within and among plants and within habitats, termed trivial movement by Kennedy (1961), is poorly investigated for planthoppers and is critical to the understanding of planthopper mating systems. The scale and frequency of such movement determine the extent to which planthoppers experience the spatial and temporal variation characterizing their environment, environmental grain, sensu

[Levins (1968)], and the rates and diversity of interactions with conspecific competitors and potential mates. In turn, the propensity of planthoppers to move in response to environmental variation and social stimuli, such as crowding, the distribution of potential mates, and male–male aggression, determines the influence of these factors on the dynamics of the mating system. In addition, information on small-scale movements also may be important for understanding the behavioral mechanisms that underlie the transmission of plant pathogens (Conti 1985; Nault and Ammar 1989) and may help to decipher the cues that trigger long-distance dispersal (Chen and Cheng 1980; Padgham et al. 1987).

The dispersal of adult planthoppers over short distances has been studied via mark and recapture techniques (Husain et al. 1967; Perfect et al. 1985, Roderick unpublished data). Although providing information on the general scale of local movement, these studies are of limited use in determining the movement activity of adults (frequency and timing of movements within and between plants) in response to (1) variation in host plant quality, (2) social interactions, (3) physiological state (mated versus unmated or premigratory versus postmigratory), and (4) sex differences. A combination of laboratory and field studies focusing on the movement of individually identified adults, the "focal samples" of Altmann (1974), are needed to quantify both the scale and frequency of trivial movements and to examine the response of planthoppers to environmental and social stimuli.

Four basic questions need to be addressed to understand the role of trivial movement in delphacid mating systems. First, do the sexes differ in patterns of trivial movement? The strategies employed by the sexes in locating mates may be influenced by basic differences in patterns of movement. For example, if receptive females are essentially stationary, then males must search among plants for females. Moreover, if female movements are dependent on host plant quality then significant variation in the distribution of females within and among plants is to be expected. Consequently, the benefits to males of locating clumped females would be high. Although abundant information is available on the movement of male planthoppers once acoustic exchange with females is initiated (Ichikawa and Ishii 1974; Ichikawa et al. 1975; Claridge 1985b; de Vrijer 1986; Heady and Denno 1991), little information is available on local movement patterns of males prior to acoustic contact with prospective mates.

Based on the relative abundance of male to female *Laodelphax striatellatus, Nilaparvata lugens*, and *Sogatella furcifera* in traps designed to capture planthoppers moving between plants in rice paddies, Ichikawa (1977) suggested a scenario for the movement of males. Ichikawa proposed that males move among plants in pursuit of receptive females, call briefly once on a plant, and, depending on the presence of females, either respond to female signals or move to another plant. This pattern of movement has been termed a "call–fly strategy" by Hunt and Nault (1991), who also suggested this pattern of movement based on the within and between plant movements of males and females of the leafhopper *Graminella nigrifrons* under laboratory conditions. The call–fly strategy forms a convenient hypothesis to guide research on the trivial movements of male planthoppers. Mark and release studies of cohorts of adult *Prokelisia marginata* (Roderick unpublished data) showed that males disperse over greater distances per/unit time than females. This suggests a higher rate of movement for males which is consistent with the expectations of the call–fly strategy. However, it is important to note for planthoppers that the call–fly hypothesis is not based on direct observations of movements of individuals but instead has been erected to account for indirect evidence of apparent sex-related differences in movement.

Second, does the relative vagility of planthoppers change in response to intrinsic and planthopper-generated variation in resource quality at the scale of within and between plants and patches? The propensity to move in response to variation in resource may influence the spatial and temporal distribution of hoppers within a habitat and, hence, the dynamics of social interactions. Ultimately, movement in response to resource variation may have direct effects on fitness components for both males and females. Because fecundity is related to host plant quality in females (Metcalfe 1970; Denno and McCloud 1985; Denno et al. 1986), movement to high-quality feeding and oviposition sites is an important component of reproductive success. For males, the propensity to move in response to variation in the availability of potential mates influences the encounter rate with receptive females. Dispersal studies of *Prokelisia marginata* (Roderick unpublished data) suggest that emigration from "high-quality" plants is reduced for both sexes.

Third, is there a relationship between wing form and vagility within a habitat? Determining whether mobility differences exist between the wing forms of dimorphic species is one piece of information necessary to understand the adaptive significance of changes in the relative abundance of wing forms throughout the growing season in field populations. For example, differential mobility of wing forms may be an important component of fitness differences (e.g., female encounter rate) between males, and the importance of this mobility difference may be a function of density. Also, wing-form-related differences in mobility may be involved in the establishment of dominance hierarchies among males on plants. Dominance interactions among males and the relationship between wing form and dominance are more fully explored as components of intrasexual selection below.

Fourth, what are the possible constraints on mobility within a habitat?

For example, what are the relationships between sex, wing form, mobility, and the risk of predation, parasitism, or displacement from the host plant? For a variety of insects, acoustic signaling to potential mates and searching for mates are associated with increased risk of predation or parasitism (Lloyd 1975; Soper et al. 1976; Thornhill 1978; Sakaluk and Belwood 1984; Gwynne 1987). Rates of acoustic signaling and mate search may, therefore, be constrained by predation risk (Trivers 1972).

For planthoppers, spiders appear to be major predators in salt marsh and rice ecosystems (Kobayashi 1975; Pfeiffer and Wiegert 1981; Vince et al. 1981; Kenmore et al. 1984; Döbel 1987). Because spiders can cue in on motion and are sensitive to substrate-borne vibrations (Stratton and Uetz 1981, 1983; Foelix 1982), differences in mobility between the sexes and/or wing forms in dimorphic delphacids may translate into differences in the risk of predation from spiders (Döbel and Denno unpublished data). The call–fly strategy proposed for male planthoppers and the postulated sedentary nature of females and brachypterous males (Ichikawa 1982) suggest three specific hypotheses relating sex, movement, and acoustic signaling to predation risk. First, females experience lower predation than males. Second, males searching for mates face greater predation risk compared to stationary, nonsignaling males. Third, if macropterous males move among plants at greater frequencies than brachypterous males, as suggested by Ichikawa (1982), then macropters may experience greater risks of predation. Increased rates of movement by macropterous males may stem from intrinsic differences between the wing forms or may reflect the outcome of male–male interactions leading to the displacement of macropterous males by brachypterous males (Ichikawa 1982). An additional possibility is that the mate searching pattern of males may change with female density and/or predator density. Although the mating behaviors of males have been examined in relation to female density and the presence of predators in other insect species (Parker 1970; Crowley et al. 1991; Travers and Sih 1991), these possibilities have not yet been explored for planthoppers.

Sexual Selection in the Delphacidae

Sexual selection results from unequal reproductive success (typically among males) and consists of two forms, intrasexual and intersexual selection which may or may not act simultaneously (Darwin 1871). Intrasexual selection is defined as variation in reproductive success resulting from competitive interactions among males that limit the access of some males to females. Intrasexual selection characterizes the reproductive ecology of many organisms. Intersexual selection (female choice) is defined as variation in male reproductive success resulting from the nonrandom selection by females of mating partners within a population. Although intersexual selection is considered to be both a common and important feature of mating systems, the mechanisms involved in intersexual selection remain controversial (Bradbury and Andersson 1987).

The Delphacidae have been poorly studied with respect to both intrasexual and intersexual selection. To date, no research has dealt explicitly with the question of whether sexual selection operates within delphacid populations. Techniques to analyze sexual selection in natural populations are presented in Wade (1979), Wade and Arnold (1980), Arnold and Wade (1984a, 1984b), and Endler (1986). Recent models of sexual selection are summarized in Borgia (1987), Kirkpatrick (1987), and Maynard Smith (1987). In the following sections, some features of the ecology, reproductive biology, and behavior of male and female delphacids possibly involved in intrasexual and/or intersexual selection are discussed.

The Operational Sex Ratio (OSR)

Although the sex ratio of adults in a population may be unity, differences in the developmental and reproductive biology of the sexes as well as environmental factors leading to variation in the spacing patterns of the sexes may alter the sex ratio of reproductively mature adults at the time and location of breeding (the operational sex ratio) (Emlen 1976). In both intrasexual and intersexual selection, skewed OSRs can influence variation in male reproductive success. The effect of the OSR on male mating success depends on the manner in which variation in mating success is determined (Partridge and Endler 1986). Male-biased OSRs can lead to increased male–male interactions which, in turn, can limit the access of some males to females and result in greater variance in mating success (Emlen and Oring 1977; Vehrencamp and Bradbury 1984). Among insects, male-biased OSRs are associated with highly skewed male mating success (Borgia 1980, 1981; Thornhill and Alcock 1983; Lawrence 1986).

In planthoppers, a number of differences in the reproductive biologies of the sexes may contribute toward a male-biased OSR. In *Prokelisia dolus*, *P. marginata* (Heady and Denno 1991), and *N. lugens* (Oh 1979), males on average attain sexual maturity faster than females. In these species, brachypterous and macropterous males develop at similar rates but sexual maturity is delayed in macropterous females. As a consequence, an abundance of males may be present when females become sexually receptive. The behavior of female planthoppers following mating may also contribute to OSRs exceeding 1:1. First, mated females may position themselves on plants differently from unmated females and in locations where males are less likely to encounter them, thereby increasing the OSR. For example, in the leafhopper *G. nigrifrons*, mated females move to the lower areas of their host plant, whereas unmated females spend more time on the upper areas of the plant (Hunt and Nault 1991). Males moving among plants tend to land on the lower third of the host plant and walk upward; as a consequence, the ratio of unmated to mated females increases as males work up the plant. For planthoppers, no data explicitly address the movements of the sexes on their host plants following mating or the spatial distribution of mated verses unmated adults. Everett (1969), however, reports that for the planthopper *Tagosodes orizicolus*, females and immature males usually are found on the lower portion of the plant. This anecdotal information hints at the possibility that sex, maturity, and mating status may influence the movements of planthoppers on their host plants. Second, the OSR may be male-biased because females typically cease responding to the attraction calls of males shortly after a single mating (Oh 1979; Claridge 1985a; Heady and Denno 1991). Finally, for species of planthoppers for which information is available [the *Muellerianella* complex (Drosopoulos 1985); *N. lugens*, (Oh 1979; Claridge 1985a); the *Ribautodelphax* complex (de Winter and Rollenhagen 1990)], females appear to copulate only once or recopulate only after a considerable time lag. Single mating by females is also observed in the leafhopper *G. nigrifrons* (Hunt and Nault 1991).

Features of the biology and behavior of males may also contribute to OSRs exceeding 1:1. In *Laodelphax striatellus* (Oyama 1972), *N. lugens* (Takeda 1974; Ichikawa 1979; Oh 1979), *Tagosodes orizicolus* (McMillian 1963), and in both the *Muellerianella* species complex (Drosopoulos 1985) and *Ribautodelphax* complex (de Winter and Rollenhagen 1990), males are capable of copulating frequently within a short time interval and can resume courtship immediately following mating. Hence, unlike females, males are neither temporarily or completely removed from the pool of potential breeders following mating.

Intrasexual Selection

Intrasexual selection is considered to be of primary importance in shaping male behaviors and morphologies associated with reproduction. A diverse array of competitive interactions among male insects can be involved in intrasexual selection including direct physical contests over potential mates or breeding sites, displacement of copulating males from females, establishment of dominance hierarchies, and defense of territories (Thornhill and Alcock 1983). Males may also compete indirectly via aggressive acoustic displays (Ichikawa 1982), physical displays, or by adopting variable mate search strategies (Borgia 1980; Thornhill and Alcock 1983). The intensity of competition among males for mating opportunities may be accentuated by the spatial and temporal distribution of females and resources important to females (Emlen and Oring 1977). Finally, density may also directly or indirectly effect the variance in male reproductive success (Wade 1980; Conner 1989).

In planthoppers, competitive interactions among males can involve aggressive calling and direct body contacts (Ichikawa 1982). Both behaviors can lead to the displacement of males from the location where the contest occurred, and aggressive signals may suppress the emission of attraction signals of conspecific males (Ichikawa 1982). The relationship between aggressive interactions and variation in male mating success, however, remains unexplored. To establish this relationship the context in which aggressive behavior occurs as well as the outcome of contests between males and the characteristics of "winning" males must be evaluated.

Development of Aggressive Behavior among Males

Aggressive interactions among male planthoppers have been studied most extensively by Ichikawa (1979, 1982). Ichikawa documented that male *N. lugens* cultured at high density produced aggressive calls structurally distinct from courtship calls (Booij 1982b; Claridge 1985b) or attraction calls (Claridge 1985a, 1985b) when placed on a rice stalk in the presence of another male. Males were observed to respond to the attraction signals of other males and engage in direct body contacts with signaling males. In addition, both the frequency of aggressive signaling and the occurrence of direct body contacts between paired males increased as a function of the density at which males had been maintained prior to the experiment. The emission of aggressive signals by one male of the pair also appeared to suppress the emission of acoustic signals by the other male.

Ichikawa (1982) proposed the following scenario for the development of aggressive behavior in male *N. lugens*: (1) aggressiveness is acquired through competitive communication during early adult life; (2) males emerging early on a plant expel those emerging later by frequent signal emission; and (3) brachypterous males (because of early emergence and flightlessness) become more aggressive than macropterous males due to continuous communication with other males on the plant.

Although Ichikawa's creative studies suggest the possibility of intriguing interactions between aggressive behavior, planthopper density, male wing form, and mating success, several important caveats concerning this work exist, and much remains to be learned concerning both the development and importance of aggressive behavior in male planthoppers. Ichikawa (1982) presented no direct evidence to suggest that brachypterous males were more aggressive than macropterous males (brachypterous and macropterous males were not paired together in the experiments), and the frequencies of aggressive calling and direct body contacts did not differ between the pairs of brachypterous and pairs of macropterous males that were examined. Furthermore, although the density at which males were housed prior to the experiments was related to subsequent aggressiveness; aggressive behavior between males was not evaluated as a function of density within the experimental arenas. Hence, no information bears directly on the effect of density on male aggressiveness. Also, although it was suggested that male wing form and aggressiveness were related to mating success, this was not evaluated. Clearly, more detailed research is needed on the development of aggressive behaviors and the relationship between these behaviors and mating success in the Delphacidae.

Three broad questions need to be addressed if we are to understand the relative importance of intrasexual selection in planthopper populations: (1) What aspects of male behavior and/or morphology (wing form, body size, call structure) are involved in male-male interactions? (2) How do ecological factors determine the frequency and intensity of male–male interactions? (3) Do male–male interactions contribute meaningfully to variation in mating success and, if so, how?

Intersexual Selection

Intersexual selection (female choice) involves the nonrandom selection of males for mates from a population of potential mates. Unequivocally demonstrating female choice requires documenting the rejection of some potential mates by females, the acceptance of others, and identification of the criteria by which choices are made. Female mate choice in planthoppers has been studied in the context of reproductive isolation between species (Claridge 1985a, 1985b, 1988; de Winter and Rollenhagen 1990), call differentiation between sympatric sibling species (Claridge 1985b, 1988; Heady and Denno 1991), and allopatric populations of the same species (Claridge et al. 1984b, 1985a). In these studies the criteria for demonstrating female choice appear to have been met. In general, female planthoppers preferentially mate with conspecific rather than heterospecific males in laboratory choice situations (Claridge 1985b; de Winter and Rollenhagen 1990; Heady and Denno 1991), and these decisions appear to be based, in part, on differences in the structure of attraction signals (Claridge et al. 1988; Heady and Denno 1991).

Inspection of the mating behavior of females and the behavioral and physical attributes of males suggests that the prerequisites for the occurrence of intersexual selection within planthopper populations may be present. However, to date, no studies have attempted to determine whether sexual selection via female choice occurs within populations or, if so, what features of male morphology and/or behavior influence female choice. Female reproductive behaviors suggesting mate choice and aspects of male behavior and morphology serving as potential criteria for female choice are described below.

Circumstantial Evidence for Female Choice: Female Rejection Behaviors

Females of a number of planthopper species including those in the *Muellerianella* complex (Booij 1982b; Drosopoulos 1985), *N. lugens* (Takeda 1974; Oh 1979; Claridge et al. 1984b), and *P. marginata* and *P. dolus* (Heady and Denno 1991) are known to physically reject copulation attempts from heterospecific and conspecific males. Rejection behaviors include abdomen shaking, kicking males away with the mid and/or hind legs, projection of the ovipositor, and walking away from males. What is missing from our knowledge of female rejection behavior in planthoppers, however, is information on the context in which rejection of conspecific males occurs and the relationship between rejection and male characteristics. Because receptivity in females is known to be related to sexual maturation (Heady and Denno 1991) and mating status (virgin or mated) (Oh 1979; Heady and Denno 1991), it is important to distinguish rejection of copulatory attempts related to male characteristics from rejection behavior related to the female's own developmental stage or mating status. This distinction could be made by controlling for female age and mating status and varying the characteristics of the males presented to females in future studies.

Possible Criteria for Female Choice

The ultimate basis of female choice is often not obvious because of correlations among male traits (Lande and Arnold 1983). Therefore, determining the importance of individual male traits in female choice will require experimental tests designed to isolate the influence of specific traits. Male delphacids differ in a number of attributes which females might use to base mate choice decisions. Notable aspects include (1) body size, (2) wing form, (3) courtship displays, (4) mating status (whether males have mated previously), and (5) characteristics of acoustic signaling (pulse type, repetition rate, duration, and intensity).

BODY SIZE

Body size is an important component of male mating success in many insect species (McCauley and Wade 1978; Borgia 1981; McLain 1982; Partridge et al. 1987; Conner 1989) and has been shown to be involved in both intersexual and intrasexual selection. In planthoppers, adult body size is sensitive to variation in host plant quality and planthopper density (Denno et al. 1985; Denno and Roderick 1992), and as a consequence in natural populations, the body size of males varies considerably. Although no information directly examines the relationships between body size and fitness for male planthoppers, the relationships between body size and components of fitness for males in other insects (Thornhill and Alcock 1983) suggest the hypothesis that large body size in males may be favored by intersexual selection.

WING FORM

Wing-dimorphism is common in both sexes of many delphacid species (Denno and Roderick 1990). In dimorphic species, with the exception of newly founded populations established by long-distance migrants (macropterous morphs), females have the option of mating with either brachypterous or macropterous males. Amazingly, no published information is available on the mating preference of female delphacids in relation to male wing form. This gap in our knowledge is important because the relationship between female mating preference and the reproductive success of brachypterous and macropterous males may be central to understanding seasonal shifts in wing-form ratios observed in both natural and agricultural populations of delphacids. Interestingly, Takeda (1974) presented data suggesting that, under highly structured laboratory conditions, males of both wing forms of *N. lugens* preferred to mate with brachypterous females. This result may, however, reflect the opportunity for mating rather than male preference in the strict sense because brachypterous females had a greater tendency to stay on the host plant than macropterous females and, hence, were more available.

MALE MATING STATUS

Whether or not a male has previously mated may be a criterion for mate choice when females rely primarily on a single mating to fertilize their complement of eggs, or when fecundity and fertility are dependent on the quantity of sperm and/or accessory fluids delivered by males during copulation (Lefevre and Jonsson 1962). Females of the brown planthopper, *N.*

lugens, mate when sexually mature, become unreceptive immediately following copulation, and then begin oviposition (Oh 1979). After ceasing to lay fertile eggs, some females may mate again after which both oviposition rate and fertility increase (Oh 1979). Remating several days after initial copulation has also been observed in female *Muellerianella* (Drosopoulus 1985). This behavior suggests the possibility that the output of fertile eggs may be limited by the availability of sperm and/or accessory fluid and that females may be under selection for the ability to choose males capable of fertilizing their entire complement of eggs. In laboratory studies of *Drosophila melanogaster*, females favored virgin males as mates and females that chose virgin males deposited more fertile eggs (Markow et al. 1978). Moreover, the decision to remate by females is influenced by the quantity of sperm in their storage organs (Gromko et al. 1984).

To date, no studies have investigated the relationship between female choice and male mating status. Similarly, no studies have examined the relationships between quantity of sperm delivered and the mating status, age, body size, or wing form of male delphacids. Both types of studies could help in determining whether intrinsic and age-related differences exist between males in their potential quality as mates and whether females are capable of assessing these differences.

COURTSHIP DISPLAYS AND FEMALE CHOICE

The courtship behavior of planthoppers consists of a sequential series of stereotypic behaviors that are produced once males are in the immediate vicinity of females (see Claridge and de Vrijer, Chapter 5). One of the most notable courtship behaviors exhibited by male planthoppers is a rapid raising and lowering of the wings known as wing fluttering. This behavior often accompanies acoustic signaling and only occurs when males are in the immediate vicinity of females. Wing fluttering is known in a number of planthopper genera, including *Laodelphax*, *Muellerianella*, *Nilaparvata*, *Prokelisia*, and *Sogatella* (McMillian 1963; Ichikawa 1976; Drosopoulos 1985; Heady and Denno 1991).

Ichikawa (1976) suggested that wing flutter is "an expression of extreme excitation of males" and concluded that wing flutter was not an essential sign stimulus because not all males in his study displayed the behavior during courtship. An alternative and testable hypothesis is that wing flutter is an explicit courtship display directed toward females that contains information concerning the quality of the male as a potential mate. The observation that wing flutter occurs in numerous planthopper genera and occurs in both macropterous and brachypterous males suggests that this behavior is an integral component of male courtship display. Similar behaviors exhibited by the males of other insect taxa are considered as courtship displays, and females may discriminate between males on the basis of variation in the vigor of the display (Thornhill and Alcock 1983). Studies that examine the relationship between wing flutter variation and mating success among male delphacids could provide evidence of female choice.

ATTRACTION AND COURTSHIP SIGNALS AND FEMALE CHOICE

Male mating signals are the most obvious feature of courtship behavior likely to be subject to sexual selection (West-Eberhardt 1983). In planthoppers, the structure of attraction and courtship signals produced by males exhibits greater complexity than female signals (Booij 1982b; Claridge 1985a, 1985b; Claridge et al. 1985a; de Vrijer 1986). On the basis of the female's preference for conspecific males in choice experiments involving conspecific and heterospecific males, female choice has been invoked as the mechanism responsible for the elaboration of male signals in *Prokelisia* (Heady and Denno 1991). This finding, while consistent with the process of female choice at the interspecific level, does not establish that female choice takes place for conspecific males in the same population nor does it establish the link between female choice and the evolution of the elaborate structure of the male call. Moreover, as described by Claridge and de Vrijer (Chapter 5), alternative explanations for male call complexity exist. To demonstrate that female choice acting *within* populations occurs and has resulted in the evolution of complex male signals, three conditions must be met: (1) sufficient among-male variation must exist in signal structure, (2) signal variation must be genetically based, and (3) female choice of males based on call variation must be documented.

Although between-population variation in delphacid signal structure has received a great deal of attention, within-population variation in signal structure has received only cursory study. Ichikawa et al. (1975) found that the structure of attraction calls did not differ between brachypterous and macropterous morphs for *N. lugens*. The same result was obtained by Heady and Denno (1991) for *Prokelisia* species. Importantly, Heady and Denno (1991) also found that among-male variation in call parameters (unrelated to wing form) exceeded within-individual variation for *P. dolus* and *P. marginata*. The relationship between call variation of males and female preference, however, has not been addressed in these species. At the other end of the spectrum, Drosopoulos (1985) found that females of *Muellerianella* spp. repeatedly preferred one particular male over other males in a test situation, but the criteria for choice were not determined. The results of Heady and Denno (1991) and Drosopoulos (1985) suggest that future investigations are likely to uncover both significant variation in call parameters among males within species and significant variation in reproductive success among males. The goal then will be to relate variation in male call parameters to variation in reproductive success due to female choice.

The unpublished selection studies of de Winter (see Claridge and de Vrijer, Chapter 5) on the pulse repetition rate of the attraction calls of female *Ribautodelphax imitans* suggests a genetic basis for this character. Moreover, these studies suggest that males may be able to discriminate among females varying in call structure within populations. Selection studies on other species or hybridization studies between species with well-characterized call structures would be useful in improving both our understanding of the genetic basis of call structure and ultimately the relationship between variation in call structure and mate choice.

Conclusions

A well-developed literature now exists on the ecology, population, and reproductive biology of delphacid planthoppers. Despite this broad base of study very little information is currently available concerning the determinants of variation in reproductive success (particularly for males) or on the structure of delphacid mating systems. The goal of this chapter has been to create an ecological framework to guide future studies of planthopper mating systems by pointing out current areas of deficiency and to outline testable hypotheses that address potential determinants of mating success and mating system structure. These hypotheses arise from consideration of the reproductive biology, behavior, and ecology of adult delphacids.

To develop more fully our understanding of the ecology and evolution of delphacid mating systems, additional studies are needed that will attempt to (1) quantify variation in male and female reproductive success in field populations, (2) measure intrasexual and intersexual selection, (3) examine the relationship between variation in reproductive success and variation in male courtship behavior, vagility, and morphology, (4) assess the role of spatial and temporal variation in host plant quality and population density as determinants of variance in mating success, and (5) elucidate constraints acting on male behaviors and morphologies associated with reproduction.

In addition to being of interest to behavioral ecologists and providing an excellent opportunity to explore the relationships between reproductive success and variation in ecology, morphology, and behavior, the study of delphacid mating systems will contribute to our broader understanding of the biology of planthoppers. The behavioral/ecological approach outlined herein will generate information providing insight into applied and basic areas of planthopper biology such as the evolution of resistance, the transmission of plant pathogens, and the evolution of wing-dimorphism. Each of these areas are important research foci and more complete information on the reproductive ecology and behavior of delphacids is likely to lead to advances in each field of study.

Summary

1. Despite our considerable knowledge of the ecology, population biology, and reproductive biology of the delphacidae, we currently know very little about delphacid reproductive ecology or mating systems under field conditions.
2. In particular, inadequate information exists on (1) both the magnitude and determinants of variation in male and female mating success in field populations including the roles of intrasexual and intersexual selection, (2) the significance of within-population variation in behaviors and morphologies such as acoustic signals, courtship displays, body size, and wing-dimorphism involved in mate search, courtship, competition, and choice, and (3) the relationship between spatial and temporal variation in ecological conditions such as host plant quality or planthopper population density, and the intensity of mate competition and variation in mating success. Studies that measure lifetime mating success or even variation in mating success over a segment of the adult life span would aid greatly in understanding the determinants of individual variation in mating success.
3. Planthopper mating systems may be shaped in part by ecological aspects of the environment such as spatial and temporal variation in population density and host plant quality which create the opportunity for both intrasexual and intersexual selection. For example, within- and between-host plant variation in quality arising from intrinsic differences between habitat patches, prior feeding and oviposition activities and phenological variation between plants may shape planthopper mating system dynamics by affecting (1) the distribution of adults within and among plants, (2) body size, (3) the intensity and frequency of interactions within and between the sexes, and (4) ultimately male mating success. More information is needed on the sensitivity of adult planthoppers to variation in host plant quality within and among plants in the habitat to understand fully the importance of host plant-induced variation in planthopper mating systems.
4. The scale and frequency with which planthoppers move within and among plants has received only cursory attention. Four questions con-

cerning trivial movement patterns must be addressed to understand how these small-scale movements influence delphacid mating systems. These are: (1) Do the sexes differ in patterns of movement? (2) Does vagility change in response to variation in resource quality? (3) Do the wing forms differ in patterns of trivial movement within a habitat? (4) Are there constraints on movement within a habitat?

5. Intrinsic features of planthopper biology such as sex-related differences in development and copulatory and postcopulatory behavior may also shape delphacid mating systems. For planthoppers, differences in the reproductive biology and behavior of the sexes suggest that male-biased operational sex ratios occur. Male-biased operational sex ratios are typically associated with increased variation in male mating success in other insects. No information, however, is currently available on operational sex ratios or their importance in natural populations of delphacids.
6. Intrasexual selection results from competition among males for mating opportunities and is considered to be a primary force in shaping behaviors and morphologies associated with reproduction. In planthoppers, males engage in both aggressive acoustic and physical interactions. However, the relationship between aggressive behavior and mate competition (intrasexual selection) has not been addressed for male planthoppers. Although it has been suggested that male wing form is related to aggressiveness and hence mating success, this relationship has not been tested rigorously.
7. Intersexual selection (the nonrandom selection of males as mating partners by females) is a potentially powerful force capable of shaping male behavior and morphology. For planthoppers, female choice has been studied only in the context of reproductive isolation between isolated populations and between species. Inspection of the mating behavior of females and variation in male acoustic behavior, courtship displays, and physical attributes, however, suggest that the prerequisites for female choice are present within populations.

ACKNOWLEDGMENTS

I thank G. Borgia, M. Claridge, R. Denno, M. Eubanks, R. Hunt, and G. Roderick for their comments on earlier drafts of this report. This work was supported in part by National Science Foundation Grant DEB-9209693 to R. F. Denno and J. R. Ott. This is Contribution Number 8529 of the Maryland Agricultural Experiment Station, Department of Entomology.

PART THREE

Population Ecology

Generally, the dynamics of planthopper populations can be characterized by wild fluctuations in both time and space. The high reproductive potential of the brachypterous wing form combined with the extraordinary dispersal ability of the macropter contribute to the effective colonization of new habitats followed by rapid population growth and emigration. Such a scenario is very characteristic for delphacids exploiting temporary habitats including most agricultural pests, but dramatic changes in population size are also characteristic of sedentary species in more permanent natural habitats. Many factors potentially contribute to population change in delphacids and these include intraspecific crowding (competition), host plant nutrition, natural enemies, and physical factors. Hotly contested in today's literature is whether or not population fluctuations are density dependent. Equally controversial is the frequency and strength of competition and its overall impact on the population dynamics of phytophagous insects. For these reasons, we have grouped together in Part Three those chapters in which the focus is primarily on the potential direct effects of density on population dynamics including migration. Whereas changing host plant chemistry and natural enemies most certainly contribute to planthopper population dynamics, the treatment of these factors has been dealt with elsewhere in this volume, host plants in Chapter 2, and natural enemies in Chapters 10–14.

The extent to which change in planthopper population size is density dependent is explored by Denno, Cheng, Roderick, and Perfect in Chapter 7. The effects of intraspecific crowding on various fitness components, dispersal capability (wing form), and spatial distribution are explored as are the contributions of these factors to population dynamics. Simulation models are also used to assess the importance of density-dependent variables (wing form, fecundity, and survivorship) on population growth. Evidence for interspecific density effects is reviewed, both in terms of traditional adverse effects on fitness components and by means of altered dispersal capability.

The brown Planthopper (*Nilaparvata lugens* Stål) is a devastating pest of rice from tropical to temperate Asia. Because it fails to successfully overwinter in north temperate Asia, its presence there is dependent on the spring/summer colonization by transoceanic migrants originating from mainland locations hundreds of kilometers to the southwest. In contrast, throughout tropical Asia where rice is grown continuously and climate is mild, the population dynamics of *N. lugens* is very different. In Chapter 8, Perfect and Cook compare regional differences in the population dynamics of this important rice pest and elucidate the underlying processes which contribute to the observed difference in its population biology between tropical and temperate Asia.

Kisimoto and Rosenberg (Chapter 9) extend our pursuit of the population dynamics of rice planthoppers with a detailed dissection and examination of long-distance migration. Factors contributing to the production of migrants, variation among individuals in their migratory ability, conditions eliciting mass migrations, and evidence for return migrations are all reviewed.

7

Density-related Effects on the Components of Fitness and Population Dynamics of Planthoppers

Robert F. Denno, Jiaan Cheng, George K. Roderick, and T. John Perfect

Introduction

The extent to which populations are "regulated" or suppressed by density-dependent factors continues to be the subject of extensive ecological debate (see Slade 1977; Strong 1984; Vickery and Nudds 1984; Hassell 1985a, 1986a; Dempster and Pollard 1986; den Boer 1988). Traditionally, discussions of density dependence have focused on its statistical detection (Morris 1959; Varley and Gradwell 1963; Eberhardt 1970; It 1972; Kuno 1973a; Bulmer 1975; Slade 1977; Vickery and Nudds 1984; Dempster and Pollard 1986; Hassell 1986a; Mountford 1988), its potential "stabilizing" role in population dynamics (Varley et al. 1973; Dempster 1983; Hassell 1985a; Dempster and Pollard 1986; den Boer 1986, 1988; Latto and Hassell 1987; Poethke and Kirchberg 1987), and its relative strength and frequency in natural populations (Murdoch 1970; Chesson 1978; Strong 1984, 1986). Despite the history of controversy, there currently appears to be widespread agreement that the dynamics of persisting populations result from both density-dependent and density-independent processes and that the variance associated with density relationships is an important consideration (Strong 1984, 1986; Hassell 1986a).

Hassell (1986a) suggests that feedback processes acting in a broadly density-dependent way (e.g., via competition for food, dispersal, and/or susceptibility to natural enemies) are necessary, at least sporadically, for the long-term persistence of populations. However, competition for food is generally considered to be too infrequent or weak to be an important determinant of population dynamics or the structure of most phytophagous insect communities (Lawton and Strong 1981; Price 1983; Strong et al. 1984). The consensus reflected in today's literature on herbivorous insects is that other factors such as natural enemies, fluctuating climate, or changing host plants (morphology, allelochemistry and nutrition) act to maintain densities below those where competition has frequent impact (Hairston et al. 1960; Lawton and Strong 1981; Lawton and Hassell 1983; Price 1983). Furthermore, if interspecific effects occur, they are usually much less common and intense than intraspecific ones (Strong et al. 1984).

Nevertheless, some insect taxa such as the sap-feeding Homoptera (e.g., planthoppers, leafhoppers, aphids, spittlebugs, and scale insects) are overrepresented in studies of phytophagous insects in which the effects of competition have been demonstrated (Lawton and Strong 1981; Strong et al. 1984; Karban 1986; Denno and Roderick 1992). For the Homoptera, interspecific interactions can result in adverse effects on various fitness components and population growth (Tamaki and Allen 1969; McClure and Price 1975; Addicott 1978; Whitham 1978, Karban 1986), intense territoriality and niche shifting (Whitham 1979; Salyk and Sullivan 1982), the reciprocal triggering of migratory forms (Denno and Roderick 1992), and even competitive exclusion (McClure 1980b, 1983).

Traditionally, density effects are assessed as reduced survival and/or fecundity, delayed development, or small-scale spatial displacement (Schoener 1983,1988; Abrams 1987). However, for wing-dimorphic insects such as planthoppers whose wing form is influenced by population density (Cook and Perfect 1985b; Denno and Roderick 1990; Denno et al. 1991), the potential exists for both intraspecific and interspecific effects on dispersal capability. If this were the case, the effects of competition could be extended to include life history traits such as migration which drive large-scale population dynamics (Denno and Roderick 1992).

In this chapter, we examine intraspecific and interspecific density effects on various components of fitness, dispersal capability, and population dynamics of Delphacid planthoppers. Only direct density effects and how these might be modified by such factors as the nutritional condition of the host plant are addressed. We do not discuss the influence of predators and parasites on planthopper population dynamics as this issue is addressed by other authors elsewhere in this volume. Specifically, we examine (1) how intraspecific crowding affects survivorship, development time, adult body size, fecundity, longevity, dispersal capability (wing form), and spatial distribution; (2) how host plant condition alters density responses; (3) if interspecific competition occurs between sympatric planthopper species; and (4) if direct density effects play an important role in the population dynamics of planthoppers in the field.

Intraspecific Density Effects

Fitness Components

Laboratory experiments show that the survivorship (first instar to adult) of many planthopper species is negatively related to rearing density (Fig. 7.1A). Adverse effects on survival have been reported for *Javesella pellucida* (Mochida 1973), *Peregrinus maidis* (Fisk et al. 1981), *Laodelphax striatellus*, *Nilaparvata lugens*, and *Sogatella furcifera* (Kisimoto 1956a,

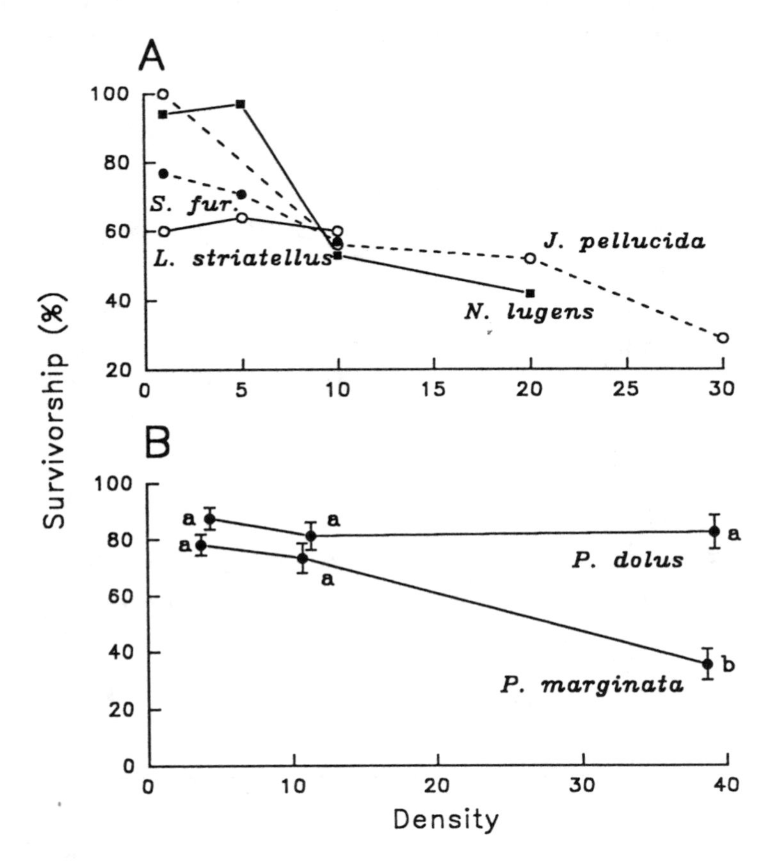

Figure 7.1. Relationship between nymphal survivorship (%) and rearing density in laboratory populations of (**A**) *Javesella pellucida* (from Mochida 1973), *Laodelphax striatellus*, *Nilaparvata lugens*, and *Sogatella furcifera* (from Kisimoto 1956a, 1956b, 1965), and (**B**) *Prokelisia dolus* and *Prokelisia marginata* (from Denno and Roderick 1992).

1956b, 1965). However, density effects on survival differ among species and even between congeners. For example, the survivorship of *Prokelisia marginata* is dramatically reduced under crowded conditions representative of those in the field, whereas the survivorship of *Prokelisia dolus* is unaffected at the same high rearing density (Fig. 7.1B; Denno and Roderick 1992). Similarly, the development time from first instar nymph to adult is extended for several planthopper species when they are raised under high population densities (Fig. 7.2; Kisimoto 1956a, 1956b, 1965; Drosopoulos 1977; Denno and Roderick 1992). Nevertheless, significant interspecific variation exists. Slow development at high densities is particularly evident for *P. marginata* and *N. lugens* (Figs. 7.2A,B). However, development time

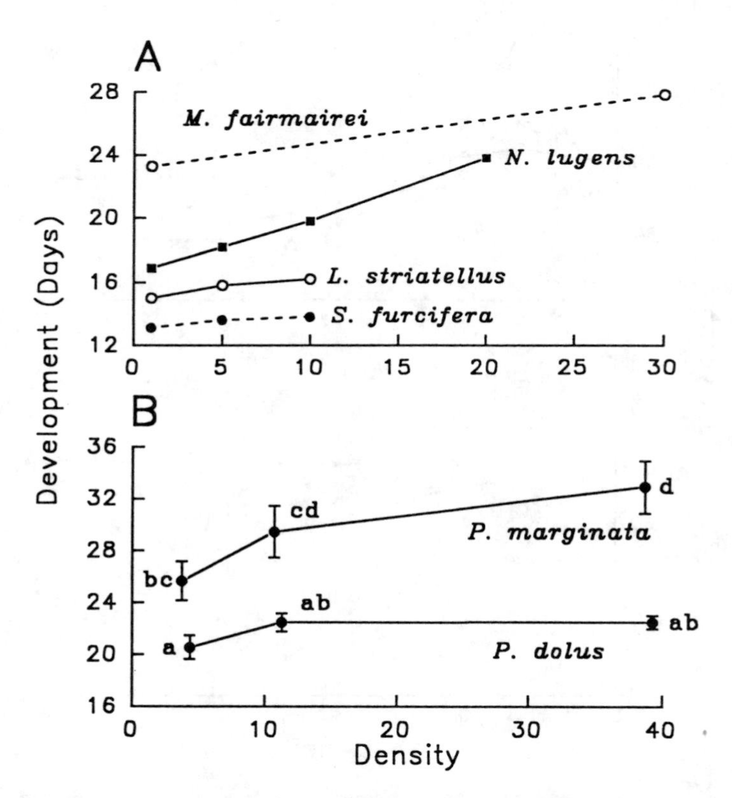

Figure 7.2. Relationship between development time to adult (days) and rearing density in laboratory populations of (**A**) *Muellerianella fairmairei* (from Drosopoulos 1977), *Laodelphax striatellus*, *Nilaparvata lugens*, and *Sogatella furcifera* (from Kisimoto 1956a, 1956b, 1965), and (**B**) *Prokelisia dolus* and *Prokelisia marginata* (from Denno and Roderick 1992).

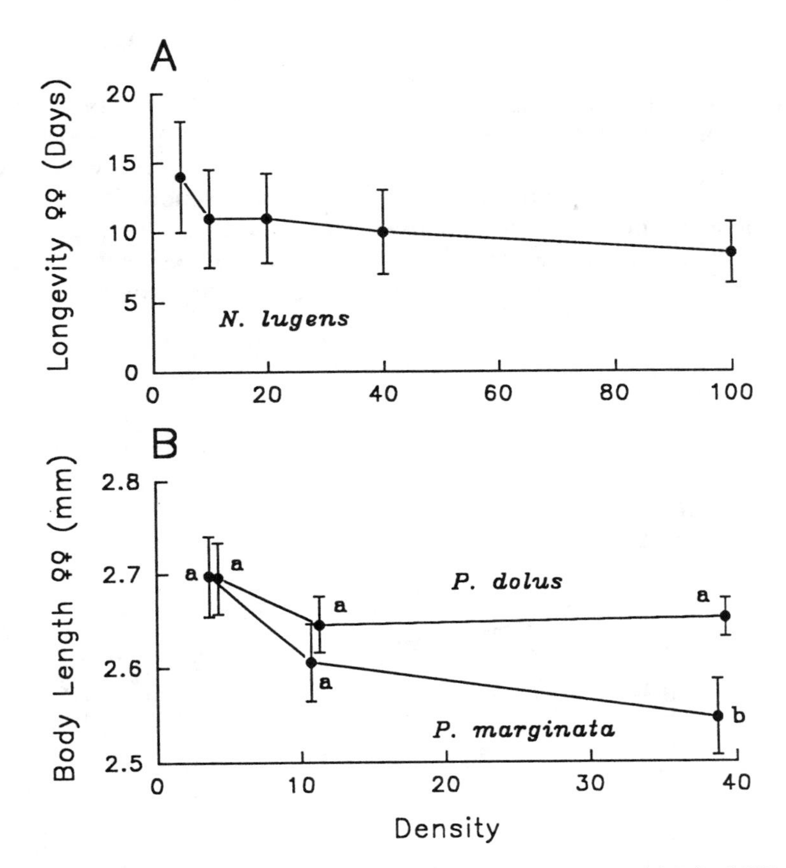

Figure 7.3. Influence of rearing density on (**A**) female longevity (days) of *Nilaparvata lugens* (from Heong 1988), and (**B**) female body length (mm) of *Prokelisia dolus* and *Prokelisia marginata* (from Denno and Roderick 1992) in laboratory populations.

is not extended under crowded conditions for *P. dolus* (Fig. 7.2B) or *J. pellucida* (Mochida 1973).

The adult longevity of *N. lugens* is reduced under high-density conditions (Fig. 7.3A; Heong 1988) as is the body length of *P. marginata* (Fig. 7.3B; Denno and Roderick 1992) which is positively correlated with fecundity (Denno and McCloud 1985). In contrast, the body length and related fecundity of *P. dolus* is not affected by crowding across the same range of experimental densities experienced by *P. marginata* (Fig. 7.3B).

Direct effects of population density on fecundity are also known for planthoppers. In a greenhouse study with *N. lugens*, both the number of egg batches/female and total fecundity/female were negatively related to female density (Figs. 7.4A,B; Heong 1988). Similarly, a significant negative relationship between the number of eggs/female and female density occurs in field populations of *N. lugens* (Fig. 7.5A; Kenmore 1980, Kenmore et al. 1984). By caging females at different densities over rice plants which were either never replaced or replaced every two days, Kenmore (1980) examined the mechanism for inverse density-dependent fecundity in *N. lugens*. If host plants were never replaced, strong density-dependent effects on fecundity were observed; on the other hand, fecundity was not reduced

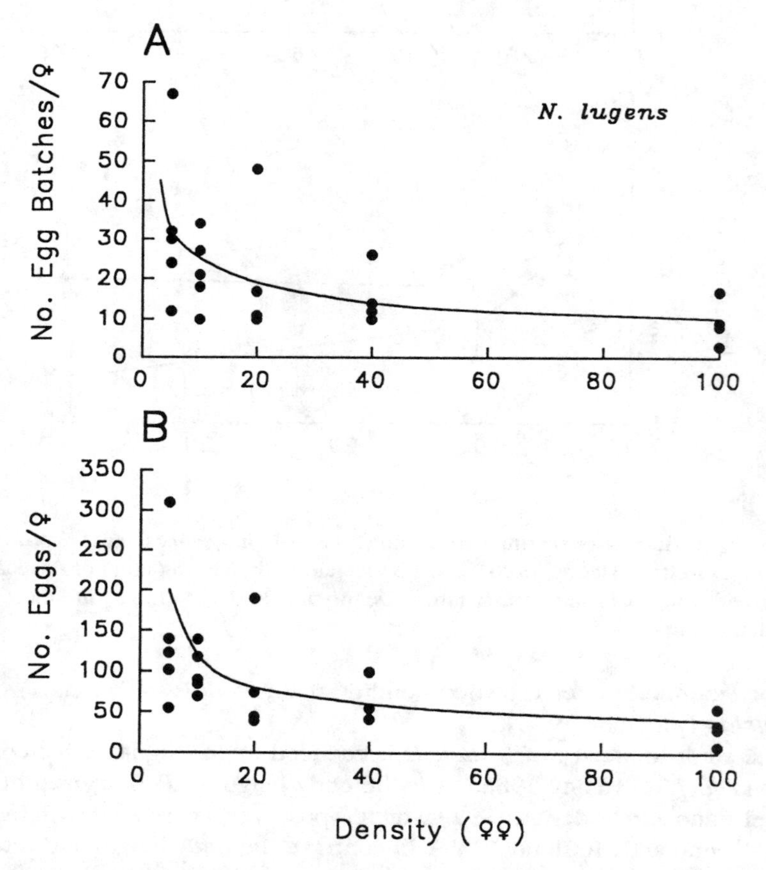

Figure 7.4. Influence of rearing density on (**A**) the number of egg batches/female and (**B**) total femal fecundity in greenhouse populations of *Nilaparvata lugens* (from Heong 1988).

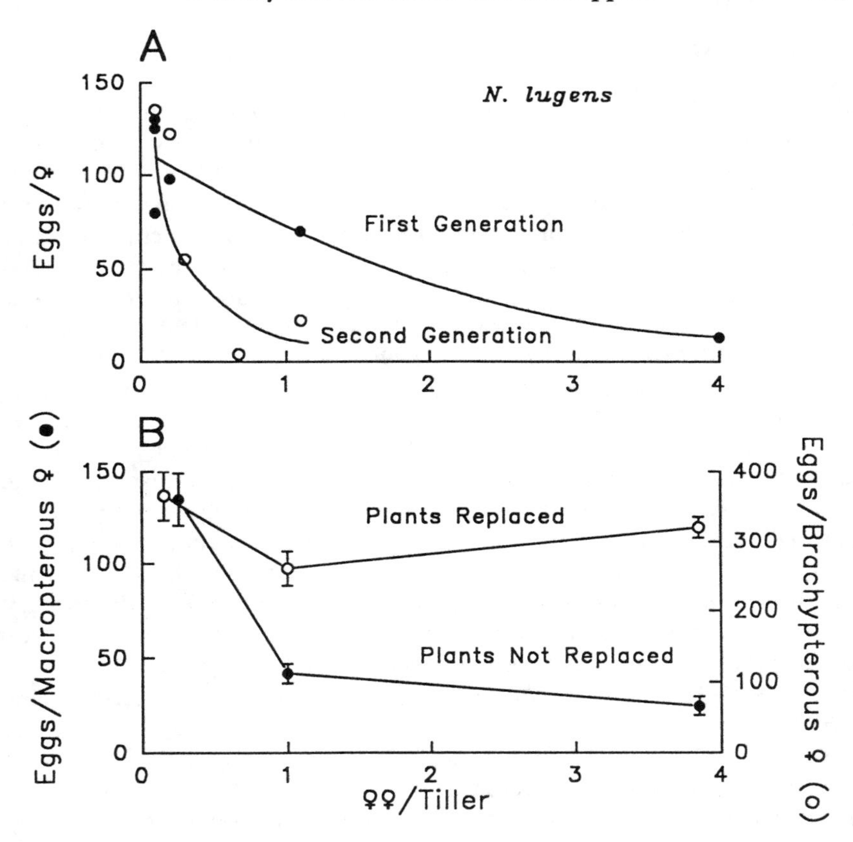

Figure 7.5. Influence of female density on the number of eggs/female in populations of *Nilaparvata lugens* (from Kenmore 1980). (**A**) Data from first (●) and second-generation (○) planthoppers in rice fields at Laguna Province, Philippines. (**B**) Data for macropterous females (●) on rice plants that were never replaced throughout the course of the study (oviposition sites limiting), and data for brachypterous females (○) on rice plants that were replaced every two days (oviposition sites and food not limiting).

under high-density conditions if rice plants were frequently replaced (Fig. 7.5B). Kenmore (1980) concluded that density-dependent effects on fecundity resulted from competition among females for oviposition sites, density-induced changes in host plant nutrition (see Cagampang et al. 1974; Bacheller 1990; Denno et al. unpublished data), or some combination of the two.

Dispersal Characters (Wing Form)

Many delphacid species exhibit wing-dimorphism with both adult forms (flightless brachypters with reduced wings and fully winged macropters) normally occurring in the same population (Waloff 1980; Denno et al. 1989; Denno and Roderick 1990; Denno et al. 1991). Wing form in planthoppers is determined by a developmental switch which responds to environmental cues (Kisimoto 1981; Cook and Perfect 1985b; Denno and Roderick 1990). The sensitivity of the switch is heritable and under polygenic control (Ammar 1973; Kisimoto 1981; Iwanaga et al. 1987; see Roff 1986a for review; Roderick 1987; Denno and Roderick 1990).

Of all the environmental factors known to affect wing form in planthoppers, population density is clearly the most influential for most species (Johno 1963; Kisimoto 1965; Mori and Kiritani 1971; Takagi 1972; Fisk et al. 1981; Cook and Perfect 1985b; Denno et al. 1985; Iwanaga and Tojo 1986; Denno and Roderick 1990). In most species, the production of migratory forms (macropters) is positively density-dependent (Fig. 7.6; Denno and Roderick 1990; Denno et al. 1991). However, there is extreme variation in the density–wing form response, and the threshold density that triggers the production of macropterous forms can differ between sexes and populations of the same species (Figs. 7.6A,B; Iwanaga et al. 1985, 1987; Denno et al. 1991; Denno and Roderick 1992), between congeners (Figs. 7.6C,D; Denno et al. 1991; Denno and Roderick 1992), and among less closely related species (Denno et al. 1991).

Generally, the threshold densities which result in the triggering of migratory forms are low for species which reside in temporary habitats (e.g., agricultural fields); triggering thresholds are high for planthoppers inhabiting persistent habitats (Denno et al. 1991). Also, a unique feature of species in temporary habitats (e.g., *N. lugens*) is a marked difference between the sexes in their wing-form responses to crowding; the production of macropterous females is density-dependent in these species, but males often emerge macropterous regardless of rearing density (Denno et al. 1991). In contrast, for species in permanent habitats (e.g., marshes and bogs), the wing-form responses of the sexes are similar and macroptery is positively density-dependent. The low triggering threshold for long-winged forms and the male bias toward macroptery are characteristics which facilitate habitat escape and mate location in temporary habitats (Denno et al. 1991).

The density-dependent triggering of winged forms affects reproductive capability as well as dispersal because macropterous females of most species are less fecund and reproduce at an older age than brachypters (Denno et al. 1989), The consequences of density-related dispersal and reproduction for planthopper population dynamics are substantial (see below).

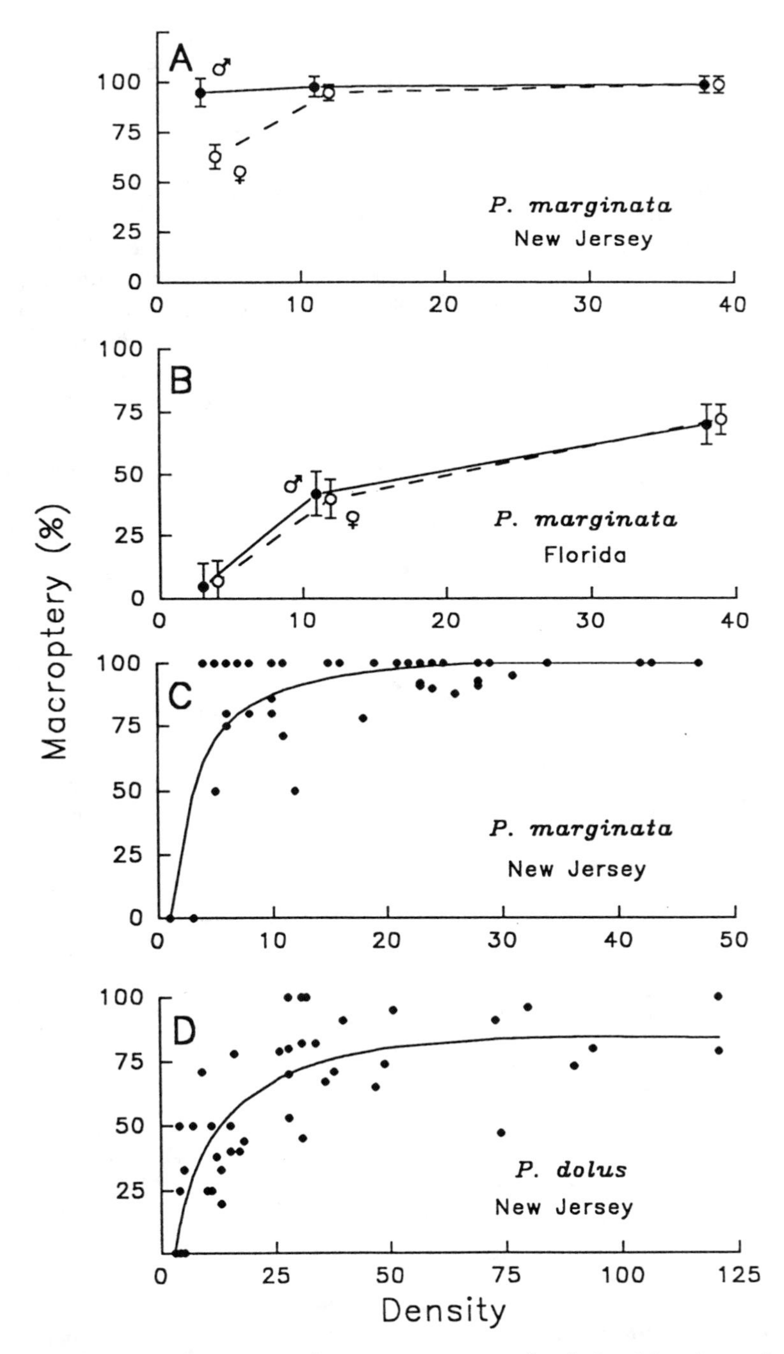

Figure 7.6. Density–wing form (percent macroptery) relationships determined in the laboratory for (**A**) Atlantic (Tuckerton, New Jersey) and (**B**) Gulf coast (Cedar Key, Florida) populations of *Prokelisia marginata*, and Atlantic coast (Tuckerton, New Jersey) populations of (**C**) *Prokelisia marginata* and (**D**) *Prokelisia dolus*; female macroptery (%) only in **C** and **D** (from Denno et al. 1991).

Niche Shifting

Some planthopper species change their spatial distribution on the host plant as population density increases. At a density of one individual/stem, most nymphs (especially the larger instars) of *N. lugens* occur within 3 cm of the base of rice plants; at a density of 10 nymphs/stem, the distribution is shifted up the stem (Fig. 7.7A; Kisimoto 1956a). Basal feeding sites are the locations of actively growing meristems where high concentrations of nitrogen occur (Langer 1979; Mattson 1980), but they may also provide protection from natural enemies (Denno and Roderick 1990). Thus, under high-

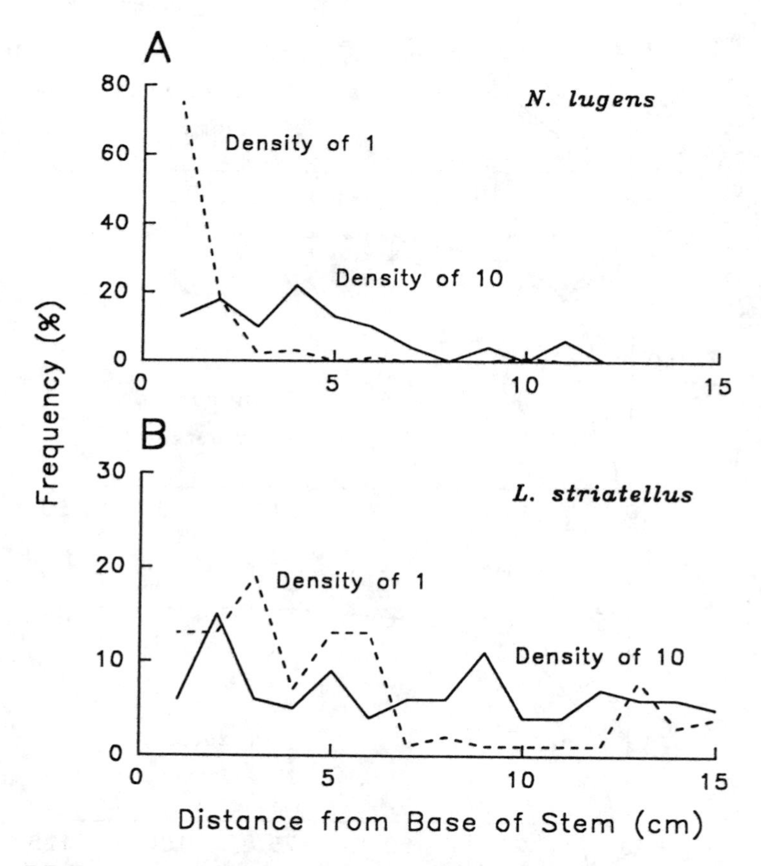

Figure 7.7. Frequency distributions of the nymphs of (**A**) *Nilaparvata lugens* and (**B**) *Laodelphax striatellus* along rice stems (distance from base in centimeters) at densities of 1 and 10 nymphs/stem (from Kisimoto 1956a). Only *N. lugens* exhibits density-related niche shifting.

density conditions, some individuals may be forced into suboptimal sites where food quality is inferior and risk of attack from natural enemies is increased. However, the potential fitness costs associated with niche shifting have not been explicitly demonstrated for any planthopper.

Not all species respond to density in the same way by niche shifting. For example and in contrast to *N. lugens*, the distribution of *L. striatellus* nymphs does not appear to change with an increase in density from 1 to 10 nymphs/stem (Fig. 7.7B; Kisimoto 1956a), and the fitness (survivorship and development) of this species is far less affected by density compared to *N. lugens* (Figs. 7.1 and 7.2).

Interactions between Population Density and Host Plant Quality

In general, density effects on various fitness components and the production of migratory forms are moderated on nutritious host plants. Fisk et al. (1981) raised nymphs of *P. maidis* at three densities (low, medium, and high) on two nutritional categories (high and low) of corn. Low-quality plants were established by subjecting them to previous feeding by planthoppers, and high-quality plants were kept planthopper-free prior to the experiment. On the high-density treatments, survivorship was least on poor-quality plants (Fig. 7.8A). Similarly, the body size of *P. marginata* females (a correlate of fecundity) is larger at high densities on nutritious (heavily fertilized) compared to inferior (lightly fertilized) host plants (Denno et al. 1986; Fig. 7.8B). The relationship between per capita fecundity and female density of *N. lugens* changes as host plants age from the first to the second generation and may be related to the nutritional condition of rice. Fewer eggs are produced by second-generation compared to first-generation females at the same population density (Fig. 7.5A; Kenmore 1980).

Interactions between population density and host plant quality are also evident for nymphal development time. In an elegant experiment, Kisimoto (1965) raised *N. lugens* at four different densities (1, 5, 10, and 20 nymphs/plant) on rice seedlings. However, seedlings were renewed at three different rates (once every 2, 4, and 6 days). Generally, nymphal development time was extended at high densities, but the density effect was dampened on planthoppers raised on frequently-replaced plants (Fig. 7.9A).

Kisimoto (1965) also determined the wing form of adult *N. lugens* emerging from a similar experiment. The frequent replacement of rice seedlings had a brachypterizing effect on both males and females at intermediate rearing densities (Fig. 7.9B). Similarly, the density-dependent pro-

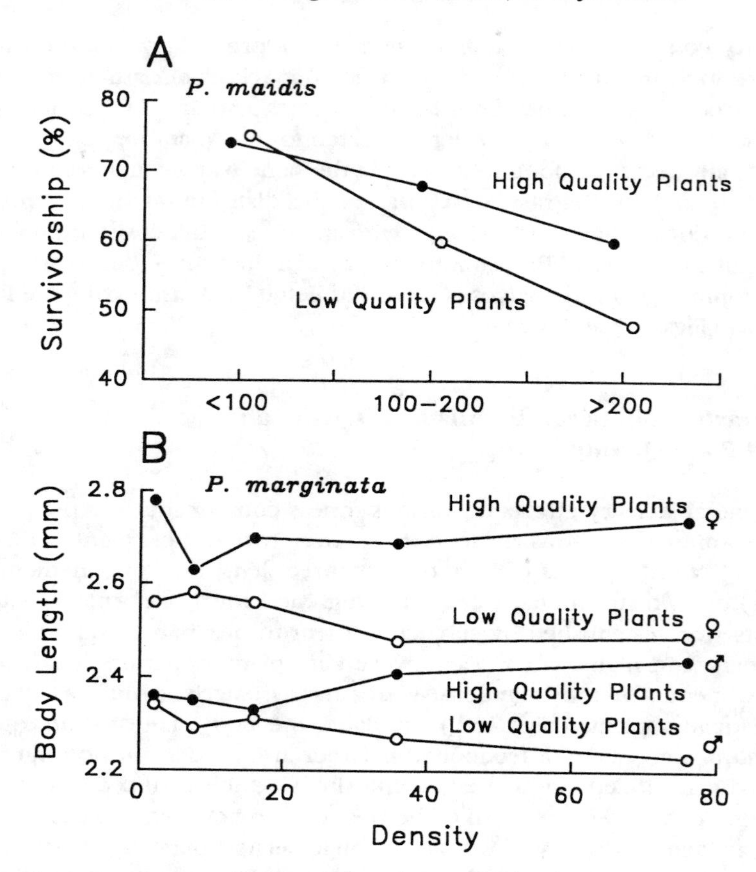

Figure 7.8. (**A**) Density-dependent survivorship of *Peregrinus maidis* reared on high-quality (no previous feeding by planthoppers) and low quality (previously fed upon by planthoppers) sorghum plants (from Fisk et al. 1981). (**B**) Density-related changes in the body length (mm) of *Prokelisia marginata* reared on high-quality (heavily fertilized) and low-quality (lightly fertilized) *Spartina* seedlings (from Denno et al. 1986).

duction of migratory forms in *P. marginata* and *P. maidis* was less pronounced on high-quality compared to low-quality host plants (Figs. 7.10A,B; Fisk et al. 1981; Denno et al. 1985).

Most available data suggest that density effects on wing form and various fitness components are mediated in part by planthopper-induced changes in host plant physiology which occur during feeding. For example, host plants heavily infested with planthoppers showed declines in soluble protein, chlorophyll, and moisture, but increases in the concentration of several free amino acids (Sogawa 1971a, 1982; Cagampang et al. 1974; Bacheller 1990; Denno et al. unpublished data). Presumably, when host plants are renewed, frequently the opportunity for insect-induced changes in plant nutrition is less as is any resulting effect of density on wing form or fitness.

Crowding may act directly on planthoppers in addition to the influence of induced plant physiology. Sogawa (1982) has shown that planthoppers probe more, move more, and contact other individuals more often on nitrogen-deficient plants. Similarly, aphids withdraw their stylets and wander about, making frequent contacts with other individuals on poor-quality host

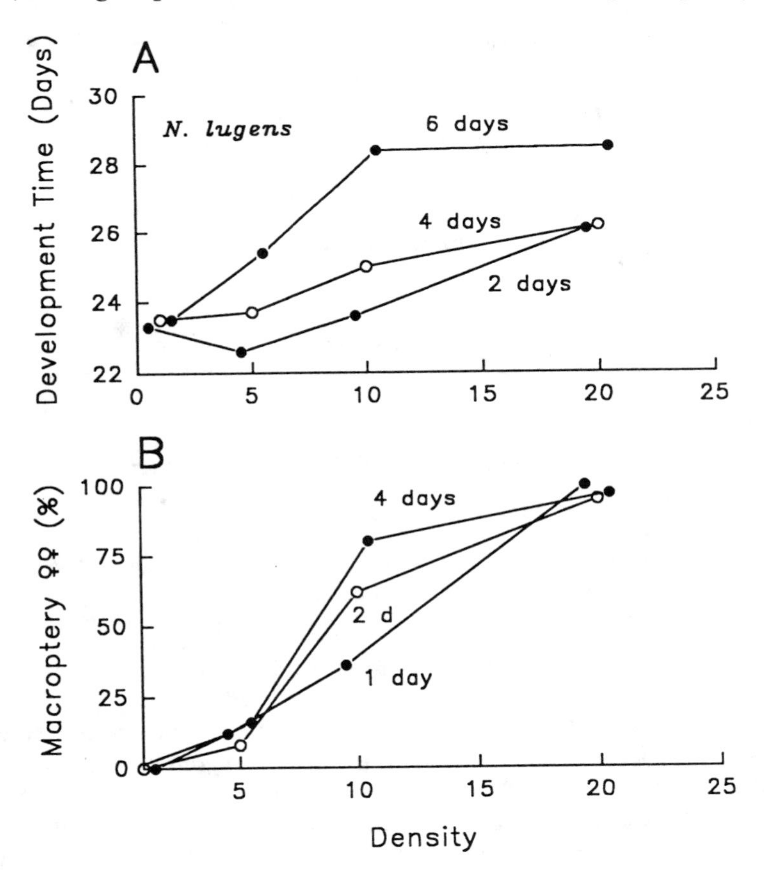

Figure 7.9. (**A**) Density-related development time and (**B**) wing-form response (percent macroptery) of both sexes of *Nilaparvata lugens* reared on rice plants that were renewed at 2-, 4-, and 6-day intervals (A) and 1-, 2-, and 4-day intervals (B) (from Kisimoto 1965).

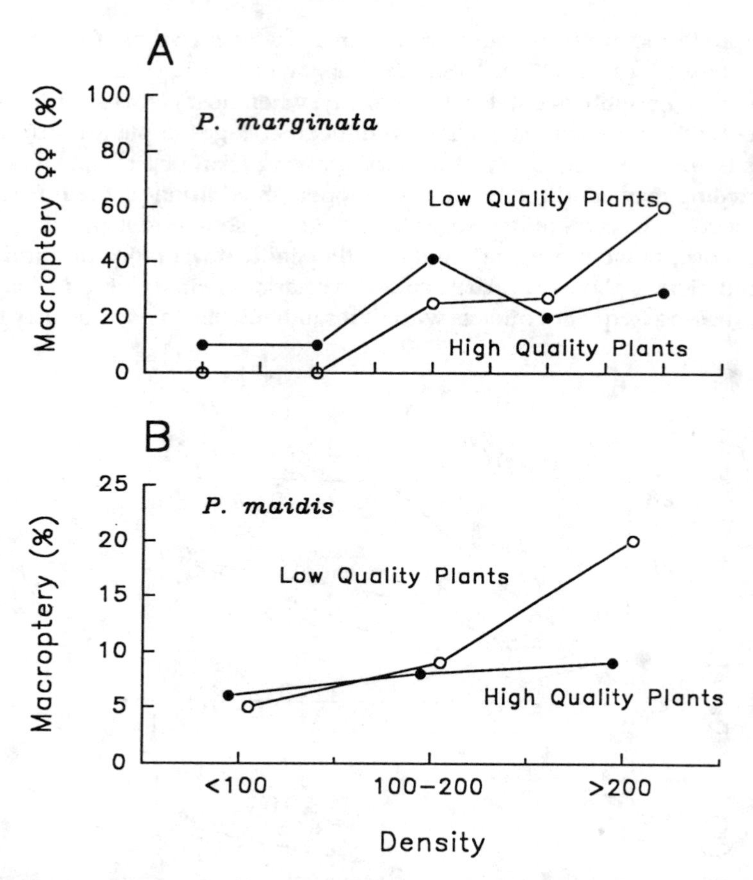

Figure 7.10. Density–wing form (percent macroptery) responses of *Prokelisia marginata* (**A**) and *Peregrinus maidis* (**B**) reared on high- and low-quality host plants (from Denno et al. 1985 and Fisk et al. 1981, respectively).

plants (Johnson and Birks 1960; Johnson 1965). Conversely, on favorable hosts, the locomotor response is reduced and tactile stimulation among aphids is minimized. Tactile stimulation alone is likely to be responsible for the production of alates in *Aphis craccivora* and *Megoura viciae* (Johnson 1965; Lees 1966, 1967). Thus, plant nutrition may affect wing-form and fitness responses both directly and indirectly by modifying crowding effects. Probable mechanisms include (1) the direct effects of plant quality on the developmental switch that determines wing form, (2) planthopper-induced changes in nutrition which influence the developmental switch, and (3) nutrition-mediated modifications in insect behavior which alter tactile stimulation and trigger the developmental switch.

Interspecific Interactions

Even though planthopper and leafhopper species partition their host plant resources in time and space (Waloff 1979, 1980; Denno 1980; Denno et al. 1981; Denno and Roderick 1991) and show inverse density relationships across habitats (Denno 1980), few studies have addressed interspecific interactions directly. Negative interspecific effects have recently been shown for two sympatric, salt marsh-inhabiting species, *Prokelisia marginata* and *P. dolus* (Denno and Roderick 1992).

Interspecific effects between the two *Prokelisia* species were assessed by rearing first instar nymphs to adults in pure and mixed cultures across a range of densities representative of those in the field. The effect of crowding on survivorship, development time, body size, and adult wing form were determined for both *Prokelisia* species. Though interspecific crowding had little effect on the survivorship, development time, and body size of either species, it did act as a powerful stimulus in triggering the production of migratory forms. In fact, in all cases, the effect of interspecific crowding on wing form was as significant as intraspecific crowding. By triggering macropterous forms, each species effectively influenced the dispersal capability of the other. This phenomenon was also validated in field populations (Denno and Roderick 1992). Furthermore, interspecific crowding also influences reproductive potential because macropterous females are less fecund and reproduce at an older age compared to brachypters (Denno et al. 1989).

The interspecific triggering of migrants may have important implications in population dynamics. If threshold densities that trigger the production of migrants differ between two species, as is the case in *Prokelisia* (Figs. 7.6C,D), one species can contribute to the selective emigration of the other from a shared habitat. These data suggest that the interhabitat migrations of *P. marginata* may reflect the combined effect of intraspecific crowding and interspecific interactions with *P. dolus* (Denno and Roderick 1992).

Evidence for Density Dependence in Field Populations

Although there is strong and widespread evidence from laboratory experiments with planthoppers for the negative effects of crowding on various fitness components and the production of less fecund migratory forms (Figs. 7.1–7.6), the question of the importance of density effects in field populations remains. Varley and Gradwell's (1963) technique for detecting density dependence in insect populations is a modification of Morris'

(1959) method. Slade (1977) reports Morris's logic: If populations increase or decrease independent of density, their change through time should be exponential. In the absence of density dependence, a plot of N_{t+1} (population size in the subsequent generation) against N_t (population size in the preceding generation) should produce a straight line with a slope (β) of 1. If density-dependent factors are important in determining a population's trajectory through time, then a plot of N_{t+1} against N_t should result in a slope of <1. Due to the bias of the regression coefficient as an estimator of β, Varley and Gradwell (1963) recommended regressing N_{t+1} against N_t and vice versa, and rejecting the hypothesis of $\beta = 1$ only when both calculated regression slopes differed from 1 on the same side of unity (see Slade 1977). Dempster (1975) concluded that Varley and Gradwell's (1963) technique was the best test available, but it has been criticized for a variety of reasons, two of which are its overconservatism and the serial correlation inherent in time series data (see Itô 1972; Kuno 1973a; Slade 1977; Vickery and Nudds 1984). Contemporary time series techniques do exist to analyze for serial correlations and detect for multiple lag effects and any form of delayed density dependence (see Box and Jenkins 1970; Draper and Smith 1981).

We chose Varley and Gradwell's technique (regression of N_{t+2} on N_t) to analyze for density dependence in populations *Nilaparvata lugens* and two *Prokelisia* species because it is straightforward and because our data are not amenable to time series analysis. *N. lugens* does not overwinter successfully in temperate Asia and spring populations are founded by immigrants from the south (Cheng et al. 1979; Kisimoto 1987). Similarly, the overwintering success of some temperate North American planthoppers (those which spend the winter as nymphs such as *Prokelisia*) is inversely related to the severity of the winter (Denno et al. 1981). For these reasons, we restricted our analysis to an assessment of density effects on within-growing season population growth. Specifically, we asked if there was evidence for the density-dependent suppression of population growth from the first to the third generation in trivoltine populations of these planthoppers. Our data is replicated in time (years) and space (plots), but because we assess density effects between two "time periods" (generations), it is not reasonable to test for serial correlations and the reciprocal regression of N_t on N_{t+2} recommended by Varley and Gradwell (1963) has no meaning. Also, it seems unlikely that any serial correlations at each spatially independent location would lead to an overall spurious pattern of density dependence. Furthermore, the biology of these species (high density-independent winter mortality) challenges the need for an assessment of serial correlation across years.

We examined population data of *Nilaparvata lugens* (taken from three rice fields during a 10-year period in China) for density dependence. By regressing peak population density (nymphs and adults) in the third generation ($\log_{10} N_{t+2}$) against the peak density of immigrating adults at the beginning of the season ($\log_{10} N_t$), we found clear evidence for density dependence (β for N_{t+2} against $N_t < 1, P < 0.01$; Fig. 7.11A). Similar analyses of Japanese and Philippine populations of *N. lugens* have found evidence both for (Kuno and Hokyo 1970; Kuno 1973b) and against (Kuno and Dyck 1985) density-dependent population growth. Using the same analysis, Napompeth (1973) concluded that the population change in *Peregrinus maidis* was strongly density-dependent and was influenced primarily by the adverse effects of density on the intrinsic rate of increase.

Several explanations for density dependence in populations of *N. lugens* exist. First, the survivorship of nymphs is reduced at high densities in lab cultures (Fig. 7.1) and there is weak evidence for density-dependent survivorship of nymphs in the field (Fig. 7.11B). Second, fecundity is reduced at high densities (Figs. 7.4 and 7.5), and third, the production of macropterous adults (emigrants) is strongly density-dependent in *N. lugens* (Fig. 7.12A). That wing form (%) is an important factor contributing to population growth in *N. lugens* is evidenced by the significant negative relationship between population growth rate (N_{t+2}/N_t) and macroptery (%) at generation $t+1$ (Fig. 7.12B). Thus, at high densities, populations may increase less because (1) fewer adults remain in the local habitat to reproduce (most macropters emigrate; Kuno 1973b; Cheng et al. 1989; Cheng and Holt 1990) and (2) the macropters which remain are inherently less fecund than brachypters (Denno et al. 1989).

Evidence also exists for density-dependent change in populations of *Prokelisia* planthoppers. For *P. marginata*, the regression slope (β) of peak population density (adults only) in the third generation ($\log_{10} N_{t+2}$) against the peak density of adults at the first generation ($\log_{10} N_t$) is <1 ($P<0.01$; Fig. 7.13A). There was no evidence for the density-dependent suppression of *P. dolus* populations; the slope of $\log_{10} N_{t+2}$ against $\log_{10} N_t$ was not significantly different from 1 (Fig. 7.13B). The results of these analyses are consistent with the population dynamics of these *Prokelisia* species. There was strong evidence for density-dependent change in populations of *P. marginata*, a very mobile species which undergoes annual interhabitat migrations (Denno and Grissell 1979; Denno et al. 1985, 1987, 1991). Mechanisms for population decrease under crowded conditions include the density-dependent triggering of emigrants (macropters) at low threshold densities (Fig. 7.6C), and the adverse effects of crowding on survival and development (Figs. 7.1B and 7.2B). In contrast, the lack of evidence for density-dependent change was found in populations of *P. dolus*, a much less mobile species (Denno et al. 1987) in which survival and

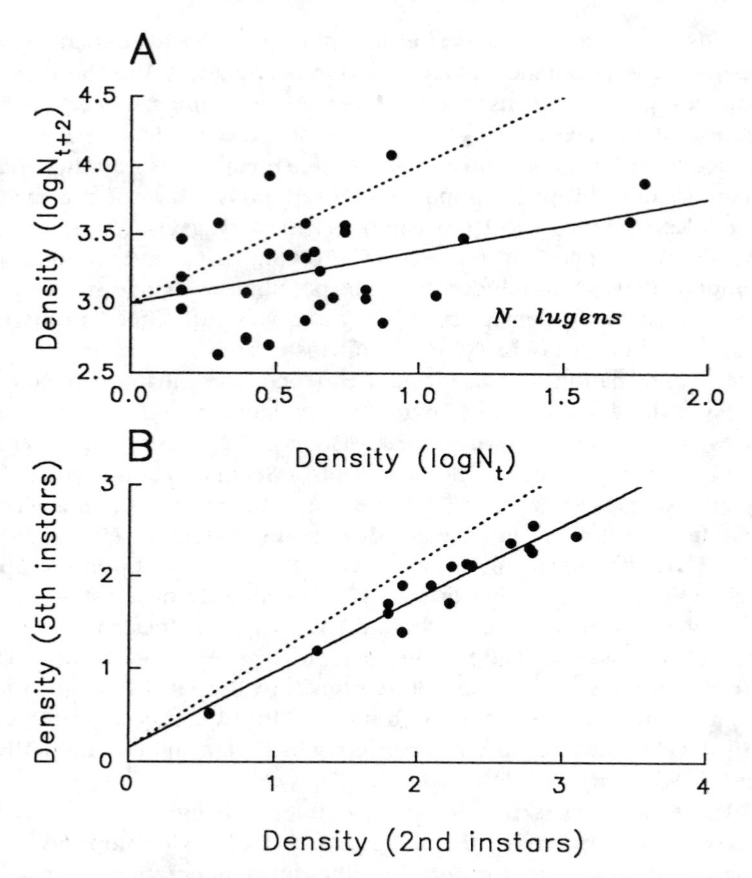

Figure 7.11. (**A**) Density dependence in Chinese (Jiaxing, Zhejiang Province) populations of *Nilaparvata lugens*. Relationship between peak population density (# nymphs + adults/100 hills) in the third generation ($\log_{10} N_{t+2}$) and the peak density of immigrating adults (#/100 hills) at the beginning of the season ($\log_{10} N_t$) is $y = 3.002 + 0.380x$; $\beta = 0.380$ is <1, $t_{30} = 3.780$, $P<0.01$. (**B**) Weak density dependence in Philippine (Los Baños) populations of *Nilaparvata lugens*. Relationship between peak population density of fifth instar nymphs ($\log_{10} N_5$/20 hills) and the peak density of second instar nymphs ($\log_{10} N_2$/20 hills) is $y = 0.156 + 0.803x$; $\beta = 0.803$ is <1, $t_{17} = 3.280$, $P<0.01$. Dotted lines have slopes of 1.0.

development are little affected by crowding (Fig. 7.1B and 7.2B) and macropters are produced at high threshold densities (Fig. 7.6D).

Further evidence for the importance of density-dependent processes in planthopper population dynamics comes from simulation models which predict peak population density. Peak density can be predicted with and without density modifications to the variables (e.g., survivorship and fecundity) included in the model, and the models can then be verified by regressing observed density in the field against predicted density (Cheng et al. 1989; Cheng and Holt 1990). If density effects are important in suppressing populations, then models without density effects included should predict peak populations larger than those observed in the field

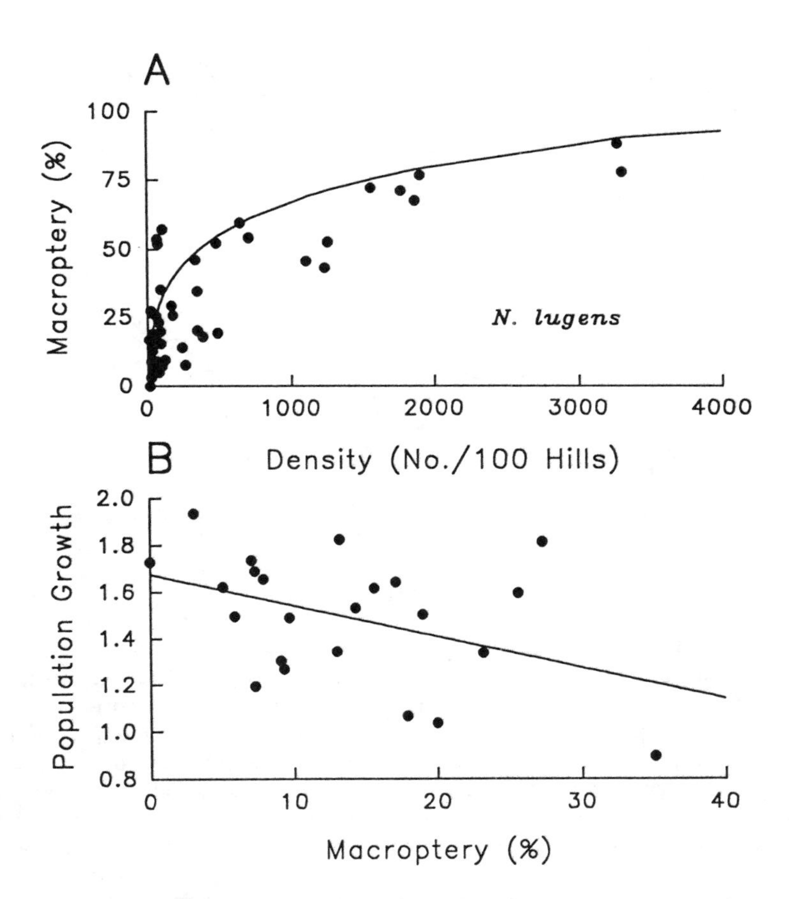

Figure 7.12. (**A**) Relationship between macroptery (%) and population density ($\log_{10} N_{nymphs}$/100 hills), and (**B**) relationship between population growth [$\log_{10}(N_{t+2}/N_t)$] and macroptery (%) in field populations of *Nilaparvata lugens* at Jiaxing, Zhejiang Province, China. N_t = the peak number of immigrating adults/100 hills, N_{t+2} = the peak number of nymphs + adults during the second generation; $y = 1.674 - 0.013x$, $R^2 = 0.18$, $P<0.05$.

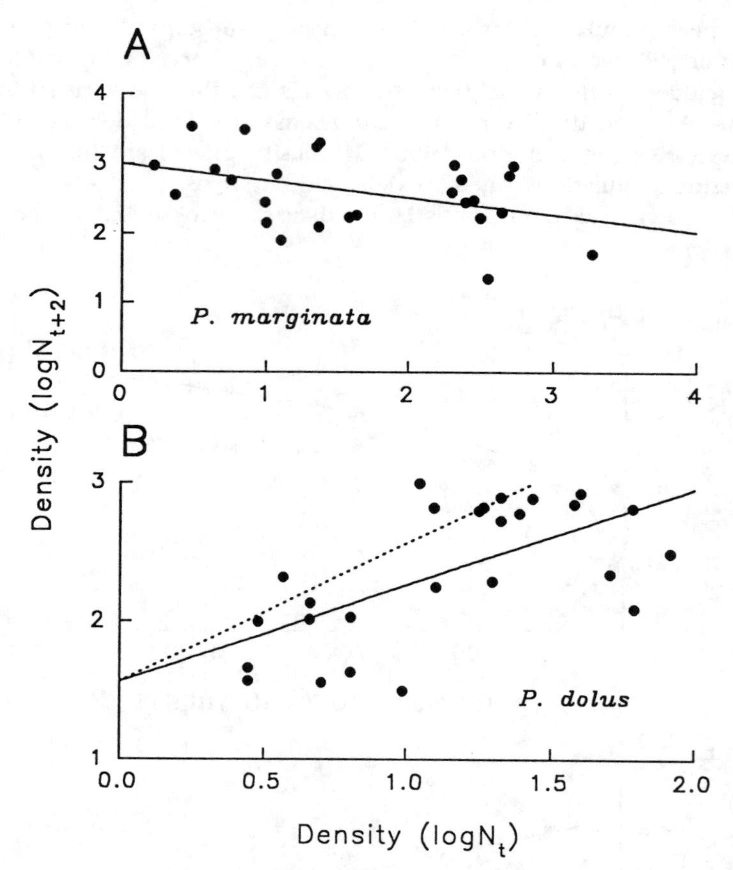

Figure 7.13. (**A**) Density dependence in field populations of *Prokelisia marginata* at Tuckerton, New Jersey, USA. Relationship between peak population density (# adults/m^2) in the third generation ($\log_{10} N_{t+2}$) and the peak density of adults (#/m^2) at the beginning of the season in the first generation ($\log_{10} N_t$) is $y = 2.997 - 0.242x$; $\beta = -0.242$ is <1, $t_{26} = 10.890$, $P<0.01$. (**B**) Lack of evidence for density dependence in field populations of *Prokelisia dolus* at Tuckerton, New Jersey, USA. Relationship between peak population density (# adults/m^2) in the third generation ($\log_{10} N_{t+2}$) and the peak density of adults (#/m^2) at the beginning of the season in the first generation ($\log_{10} N_t$) is $y = 1.561 + 0.700x$; $\beta = 0.700$ is ≈ 1, $t_{26} = 1.690$, $P>0.1$. Dotted line has a slope of 1.0.

($\beta<1$). Including density effects in the model should result in unity (observed = predicted density, $\beta = 1$). Indeed, this was the case for populations of *Nilaparvata lugens* in China. When survivorship, fecundity, wing form, and emigration rate were allowed to vary according to density (based on laboratory experiments in which density was considered), predicted peak density did not differ from observed density in the field (Fig. 7.14A;

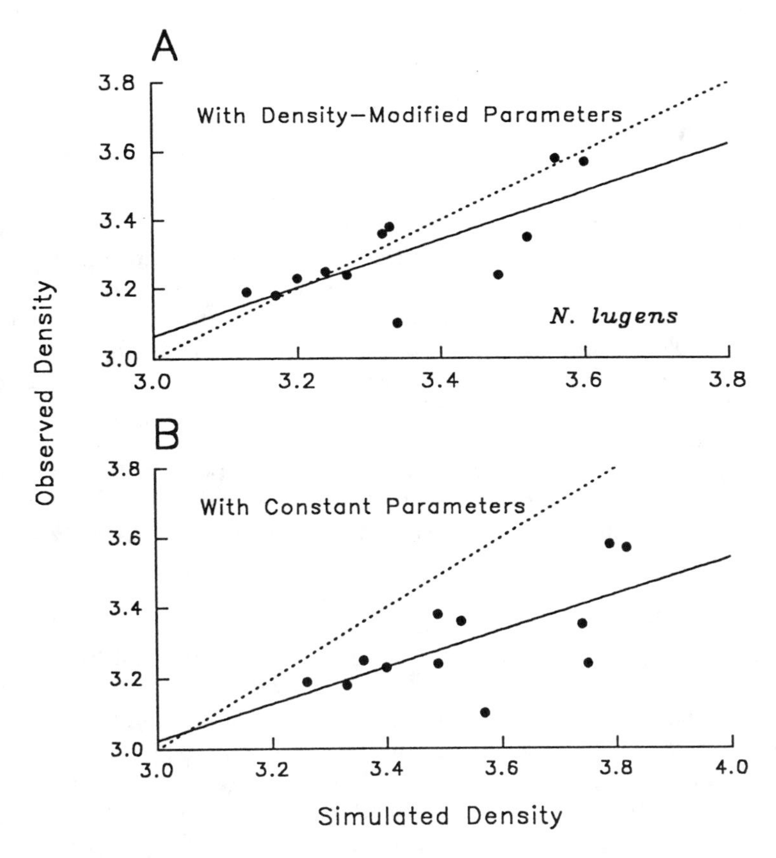

Figure 7.14. Relationship between observed population density (peak # of nymphs + adults/100 hills) and predicted population density (see Cheng et al. 1989 and Cheng and Holt 1990 for simulation model) in Chinese (Jiaxing, Zhejiang Province) populations of *Nilaparvata lugens*. (**A**) When survivorship, fecundity, wing form, and emigration rate were allowed to vary according to density (determined from laboratory experiments), the predicted density did not differ from the observed density ($y = 0.949 + 0.704x$; $\beta = 0.704$, $t_{12} = 1.473$, $P>0.1$. (**B**) If values for survivorship, fecundity, wing form, and emigration rate were held constant (laboratory maximums), predicted density was significantly higher than the observed density ($y = 1.473 + 0.517x$; $\beta = 0.517$, $t_{12} = 2.611$, $P<0.05$. Dotted lines have slopes of 1.0.

$\beta = 1, P > 0.1$; see Cheng et al. 1989 and Cheng and Holt 1990 for simulation model). In contrast, if values for survivorship, fecundity, wing form, and emigration rate were held constant (data from laboratory experiments conducted under low-density conditions), the predicted peak density ($\bar{x} = 6353$ planthoppers/100 hills) was significantly higher than the observed density ($\bar{x} = 3828$)(Fig. 7.14B; $\beta < 1, P < 0.05$).

Discussion

The general pattern of population development for many species of planthoppers is rapid population growth throughout the course of the growing season (Kisimoto 1965; Kuno 1979; Denno et al. 1981; Denno and Roderick 1990). Nevertheless, populations can fluctuate dramatically within and among seasons in the same habitat as well as spatially among habitat patches (Waloff 1980; Cook and Perfect 1985a, 1985b; Strong et al. 1990; Denno and Roderick 1990). The pressing question concerns the extent to which density dependence contributes to the fluctuation and "regulation" of planthopper populations.

In this chapter, we provided information in support of the density-dependent suppression in populations of some planthopper species. Strong density-dependent effects were detected in populations of *Nilaparvata lugens* and *Prokelisia marginata*. Specifically, the production of macropterous forms is strongly density-dependent (Figs. 7.6 and 7.12A), which has widespread effects on the growth (Fig. 7.12B) and the dispersion of planthopper populations (Kisimoto 1965; Cook and Perfect 1985a, 1985b; Denno and Roderick 1990). The density-related reduction in population growth at high densities (Figs. 7.11A and 7.13A) results from the combined effects of reduced survival and fecundity, extended development, and the production of macropterous forms which can emigrate from the current habitat. Thus, both temporal (reduced population growth) and spatial changes (increased potential for dispersal to new habitats) in planthopper populations are strongly influenced by density. In the temperate populations we studied, *Nilaparvata lugens* (30°N latitude) and *Prokelisia marginata* (39°N latitude), density effects were detected over the course of one growing or cropping season (spring to fall), during which time population growth is limited to three or four generations.

Density-independent factors are also implicated in the fluctuation of planthopper populations. For example, at temperate latitudes (39°N latitude), some species (e.g., *Delphacodes detecta*) can be reduced or eliminated from habitat patches following harsh winters, and their populations never attain the size at the end of the season that they do following mild winters (Denno et al. 1981). Generally, the overwintering success (early spring density) of such species is inversely related to winter severity (Denno et al. 1981). Consequently, for several temperate planthoppers, among-year variation in population size during spring can be influenced by winter kill (Denno et al. 1981), but within-year rates of population growth can be strongly density-dependent as shown herein.

The relative contribution of density-dependent and -independent factors to planthopper population dynamics may be influenced both by the habitats and life history characteristics of planthoppers. For example, species at high latitudes (Denno et al. 1981), in greatly perturbed habitats (Heinrichs 1979), with exposed life history stages (e.g., species which overwinter as nymphs rather than eggs embedded in vegetation; Denno et al. 1981) and with diapause limitations (Kisimoto 1981; Denno and Roderick 1990) are more likely subject to density-independent suppression. However, we provide evidence for density-dependent suppression in planthopper populations at temperate (Figs. 7.11A, 7.12, 7.13A) and tropical latitudes (Figs. 7.11B), and in temporary (Fig. 7.11A and 7.12) and permanent habitats (Figs. 7.13A). Nevertheless, it would be presumptuous to suggest that density-dependent suppression is characteristic of planthopper population dynamics in general. Density dependence must be documented on a case-by-case basis (species and populations), as evidenced by the large difference in its effect on population growth between two *Prokelisia* species which share the same habitats (Figs. 7.13A, B).

Several factors may explain why density effects are more frequent in the Homoptera compared to other phytophagous insect groups (see Lawton and Strong 1981; Strong et al. 1984; Karban 1986; Denno and Roderick 1992). First, many Homoptera are aggregative [delphacid planthoppers (Williams 1957; Fennah 1969; Metcalfe 1969; Sogawa 1982; King and Saunders 1984; Denno personal observation); treehoppers (Wood 1984); aphids (Way and Cammell 1970; Kidd 1976); whiteflies (Ekbom and Rumei 1990); and psyllids (Morgan 1984)]. Thus, contact among individuals is frequent and interactions for resources may be intensified.

Second, many Homopterans have very high reproductive potentials (Kuno 1979; Waloff 1980; Dixon 1985) and are migratory as well (Waloff 1973; Denno 1983; Taylor 1985; Kisimoto 1987; Perfect and Cook 1987). Consequently, rapid population growth may occur in newly colonized habitats under conditions of improved host plant quality and/or reduced natural enemies (Dixon 1985).

Third, many Homoptera feed on agricultural crops and ornamental plants (Wilson and Claridge 1985; Minks and Harrewijn 1987; Byrne et al. 1990). In these habitats, natural enemy/prey interactions can be destabilized from the selective kill of predators and/or parasites with pesticides (Heinrichs et al. 1982a; Reissig et al. 1982a), host plant nutrition can be elevated dramatically with the use of fertilizers (Metcalfe 1970; Dyck et al. 1979), high yielding crop varieties may be less resistant to pest attack (Dyck et al. 1979), and planting or management practices may facilitate crop colonization and pest persistence (Denno and Roderick 1991). All of these factors promote rapid population growth and high-density conditions in both managed and natural systems and may explain why density effects are often observed in the Homoptera.

Summary

1. Generally, intraspecific crowding adversely affects the survivorship, development time, and fecundity of planthoppers. However, density has far more effect on the fitness of some species (e.g., *N. lugens* and *P. marginata*) than others (e.g., *P. dolus*).
2. The production of migratory forms (macropters) is positively density-(intraspecific) dependent for most species. Nevertheless, the threshold density above which macropters are produced differs among species and between populations of the same species. Threshold densities are lower for species occurring in temporary habitats and higher for inhabitants of persistent habitats.
3. Niche shifting occurs in some species (e.g,. *N. lugens*) as populations increase. Crowding forces individuals into microhabitats which otherwise would not be occupied under low-density conditions.
4. The negative effects of crowding on various fitness components (survivorship and development time) and wing form (the production of less fecund macropters) are moderated by feeding on host plants of high nutritional quality.
5. There is little support for traditional interspecific competition in planthoppers. In the only case study, interspecific crowding between *P. marginata* and *P. dolus* does not result in reduced survival, development, or body size-mediated fecundity. However, interspecific crowding is as powerful a stimulus as intraspecific crowding in triggering migratory forms. Because macropters have reduced fecundity, interspecific crowding will influence reproductive potential as well as potential emigration rate.
6. Clear evidence exists for density dependence in populations of *N. lugens*, *P. maidis*, and *P. marginata*; populations increase less at high densities. At high densities, population growth is suppressed because

(1) the development and survival of nymphs are adversely affected, (2) fewer adults remain in the local habitat to reproduce (macropters are produced which can emigrate), and (3) the macropters which remain are inherently less fecund than brachypters.

7. Strong evidence for the density-dependent suppression of *P. marginata* populations in the field is corroborated by laboratory data showing density-related reductions in individual fitness. The lack of evidence for density dependence in field populations of *P. dolus* is consistent with laboratory data documenting a high tolerance for crowding.
8. The significant negative relationship between population growth rate and macroptery (%) in some field populations of *N. lugens* provides further evidence that a density-dependent factor (macroptery) contributes to population suppression at high densities.
9. Simulation models provide additional evidence for the importance of density-dependent processes in planthopper population dynamics. Predicted population densities of *N. lugens* do not differ from observed values in the field if model parameters (e.g., wing form, fecundity, and survivorship) are allowed to vary realistically according to density. Treating these parameters as constants results in the overestimation of population size.
10. The occurrence of density-dependent processes in planthopper population dynamics, and the overrepresentation of Homopterans in competition studies in which negative effects have been shown, may be related to two features of their life history. Many Homopterans form aggregations in which the frequency of contact among individuals and the probability of competition for food is high. Coupled with high reproductive potential, aggregative behavior may predispose these insects to density effects that adversely affect both individual fitness and population growth.

ACKNOWLEDGMENTS

We thank J. Cronin, J. Reeve, and M. McClure for their comments on earlier drafts of this report. This research was supported in part by National Science Foundation Grants BSR-8206603 and BSR-8614561 to RFD. This is Contribution Number 8479 of the Maryland Agricultural Experiment Station, Department of Entomology.

8

Rice Planthopper Population Dynamics: A Comparison between Temperate and Tropical Regions

T. John Perfect and Anthea G. Cook

Introduction

Situations in which insect habitats provide continuously favorable physical and biological conditions for population growth in both space and time are rare. For the vast majority of species, population continuity is achieved through one or both of two basic mechanisms: the ability to survive unfavourable conditions through diapause and sufficient mobility to permit the species to track spatial displacement of the requisite habitat conditions in time (Southwood 1962, 1977a; Dingle 1972; Denno 1983). The former is normally the realized solution under temperate conditions where low temperature is often the critical factor, favoring reduction of metabolic processes to a minimum for survival.

For some species whose range bridges tropical and temperate regions (e.g., *Nilaparvata lugens* Stål), the ability to migrate can be well developed (Kisimoto 1971). Also, migration can be characteristic of monophagous species which track spatial changes in the quality of their only host plant (e.g., *Prokelisia marginata* Van Duzee; Denno and Grissell 1979) or polyphagous species which exploit seasonal variation in nutrition across several host plant species (*Javesella pellucida* Fab.; Prestidge and McNeill 1982). Migration in the context of the present contribution is taken to mean major wind-assisted displacement involving the escape of the insect from its boundary layer and resulting in movement between isolated habitat patches, whereas dispersal relates to trivial flights which give rise to redistribution within such a patch (Johnson 1960). Delphacid planthoppers provide a good example of an insect group in which both migration and dispersal, as defined above, play an important part in population dynamics.

Migration is a prerequisite for the success of species in transient habitats (Denno et al. 1991). For such insects, given the low proportion of emigrants that will finally attain the required transfer to a new habitat, high fecundity and specialization of flight behavior to ensure escape from the boundary layer have contributed to their success. The evolution of wing-dimorphism, whose expression is sensitive to environmental conditions, has played an important role in the exploitation of temporary habitats (Roff 1986b).

Much of our knowledge of planthopper population dynamics derives from detailed studies of a relatively few species. Some of these are associated with climax or agricultural grassland communities (Raatikainen 1967; Denno 1976, 1983; Denno et al. 1980; Waloff 1980) but the majority are key pests of graminaceous crops. The most economically important are those associated with rice, preeminently the rice brown planthopper, *Nilaparvata lugens*, but also species of the genera *Sogatella* and *Laodelphax*; serious pests are also found on maize (*Peregrinus maidis* Kirkaldy) and sugarcane (*Perkinsiella* spp.) (O'Brien and Wilson 1985). The importance of most of these species derives both from the direct damage they inflict and their ability to transmit plant pathogens. The growing recognition that management of populations can represent the only sustainable option for pest control has focused attention on the need to understand population dynamics and led to a number of studies on individual pest species that have yielded a wealth of data.

Among the planthoppers, those species which attack both temperate and tropical rice have been subjected to particularly intense scrutiny over the last two decades. In this chapter, we attempt to highlight differences in population dynamics between such populations and to describe the underlying processes which contribute to the observed differences.

Wing-Dimorphism and Habitat Stability

Two adult morphs, long-winged macropters and short-winged brachypters, are produced by a number of delphacids in one or both sexes. Brachypters are normally flightless and effectively serve as a reproductive morph, being more cryptic in habit and more fecund than the dispersive macropters (Denno et al. 1989). Denno et al. (1991) examined the relationship between habitat stability and wing-morph composition as an indicator of species mobility across a variety of species occurring in temporary and persistent habitats. They found a highly significant inverse relationship between macroptery (%) and habitat persistence. Temporary habitats such as agricultural crops were dominated by macropterous species, whereas brachypterous species occurred in persistent habitats such as bogs and marshes.

The interhabitat migrations of *Prokelisia marginata* are associated with a shift in optimal habitat conditions between two forms of the host plant, *Spartina alterniflora* Loisel, at different times of the year (Denno and Grissell 1979). Survival of overwintering nymphs of *P. marginata* is better on the nutritionally inferior short form of the grass, which provided the only cover during the winter months. Population growth is, however, favored during the growing seasons on the more nutritious tall form of the host plant lining tidal creeks. Habitat separation is up to 100 m, sufficient to maintain a high level of macroptery (80%) within the population. In this system, optimal food resources are separated in space and time from overwintering sites and the retention of wing-dimorphism accommodates regular population shifts between the two habitats.

A similar pattern of interhabitat movement has been reported for *Javesella pellucida* (F.) by Raatikainen (1967) in Finland. This species overwinters in grassland dominated by *Phleum pratense* L. and migrates annually into spring-sown cereal crops. *Laodelphax striatellus* similarly exhibits wing-dimorphism and overwinters on the grasses surviving on the bunds of rice fields in southern Korea. Adults migrate into spring barley crops and then into flooded rice (Kim et al. 1987). This same pattern of immigrant macropters producing a predominantly brachypterous population which gives rise later in the season to mobile macropterous populations also occurs in *P. maidis* on maize (Napompeth 1973).

The factors affecting wing-morph differentiation in *N. lugens* have been well researched in laboratory-based studies (Kisimoto 1959, 1965). In the rice fields of Japan, the first generation of adults produced on susceptible varieties early in the season are predominantly brachypterous, the females showing a higher degree of brachyptery than the males. As the crop ages and *N. lugens* density increases, the proportion of macropters produced increases sharply for both sexes and emigrating macropters can be monitored over rice fields from 60 days after transplanting (Cook and Perfect 1985b). These authors reported a similar pattern for *S. furcifera* females, though macropter production tended to increase earlier in the season. *S. furcifera* males are usually macropterous, but brachypterous individuals occasionally occur (J. Cheng personal communication).

Thus, the expression of wing-dimorphism in planthopper populations is closely associated with habitat stability and continuity. Fecundity is higher in short-winged morphs and they characteristically predominate in the first generation of species in transient habitats such as annual crops; this leads to rapid resource utilization and eventually to the production of large numbers of macropterous forms in response to habitat cues. In more permanent habitats, selection favors brachypterous forms due to their increased fecundity and longevity, which allow for the spread of reproductive risks through time (Denno et al. 1989).

Temperate and Tropical Population Dynamics

Migration

N. lugens illustrates the ability of planthoppers to exploit patchy habitats particularly well and has been widely researched in both temperate and tropical parts of its geographical range. *N. lugens* is widely distributed in Asia, occurring throughout the year in the tropics and subtropics and expanding its range northward following the path of seasonal rice cultivation to northern India, China, Korea, and Japan. The source of annual populations of *N. lugens* in these temperate areas was strongly debated for a number of years, with some authors reporting diapause-like phenomena in the egg stage (Miyake and Fujiwara 1962; Okumura 1963); it is now clear that they originate from long-distance migrants derived from subtropical and tropical populations and may have covered distances of several hundred kilometers (Kisimoto 1979). Migration from southern China, and possibly parts of the southeast Asian archipelago, occurs in association with particular frontal systems during the early part of the growing season (Asahina and Tsuruoka 1968; Kisimoto 1976; Iwanaga et al. 1987). There is, thus, a fairly narrow "window" through which crop invasion can occur, leading to a high degree of synchrony in the resultant populations. The low levels and short duration of immigration reflect the distance of the source and the very particular nature of the transporting winds. It is probable that successful colonists represent an extremely small proportion of the migrating population (Iwanaga et al. 1987).

Similar meteorological considerations apply in China during the northward spring migration into new plantings, though the process is more gradual and the impact of long-distance movement less clear-cut than for Japan because there is no crop-free zone analogous to the South China Sea acting as a barrier to colonization. Seasonality does, however, impose a linearity on the process (Cheng et al. 1979) and there is no doubt that temperature and the position of the Intertropical Convergence Zone are the primary determinants of northward progress.

The early preoccupation with long-distance movement arising from the known capacity of the insect to move from mainland China to Japan led to an expectation that movements of this magnitude would be important in the dynamics of *N. lugens* in the tropics. Rosenberg and Magor (1987) used trajectory models to investigate the relationships between potential sources and sinks for *N. lugens* in southeast Asia and concluded that the potential for regular interchange within and between southeast and eastern Asia exists, particularly in association with the tracks of major typhoons. Assuming a maximum flight duration of 30 hours and observed patterns of wind speed and direction, these authors showed that displacements in excess of 750 km were likely to occur regularly. Prevailing winds suggest that no interchange of *N. lugens* takes place between either southeast or eastern Asia and southern Asia (India, Pakistan, and Sri Lanka). Although the trajectory analyses clearly demonstrate that the potential for *N. lugens* to undertake long-distance migration exists within the tropics, the extent to which this actually occurs is unclear and work is in progress to assess the probability of such intercountry exchanges.

Because of the year-round availability of rice within the rice-growing regions of the tropics, *N. lugens* immigration is highly variable in time, but can nevertheless be described as seasonal (Perfect and Cook 1987). It reflects local cropping patterns and is not associated with particular meteorological conditions; indeed, a striking feature of the tropical situation is the predominance of local effects such as the degree of synchrony in local plantings. Available evidence suggests that most dispersal of *N. lugens* in the tropics occurs over distances of 5–30 km; for example, Perfect and Cook (1987) showed variation in immigration rate and abundance of *N. lugens* between sites separated by 30 km that could be related to cropping pattern and inferred that most populations could arise from local sources. Loevinsohn (1984) came to similar conclusions in his study of an irrigation system in Central Luzon, Philippines and the radar studies of Riley et al. (1987) suggested a maximum normal range of dispersal of 30 km during the dry season in the Philippines. Loevinsohn (1984) calculated a normal dispersal range of 2 km for *N. lugens*.

Other indirect evidence to support the importance of short-distance migration is provided by Claridge et al. (1982a) who showed considerable variation between *N. lugens* populations collected from different rice cultivars separated by approximately 20 km in Sri Lanka. The slow spread of virulent *N. lugens* genotypes capable of exploiting previously resistant rice varieties further supports the case for short-distance migration, as does the occurrence of insecticide-resistant *N. lugens* populations in restricted areas. For example, populations at the International Rice Research Institute showed resistance to carbamates and carbofuran in the early 1980s when this was not detected elsewhere (see Heinrichs Chapter 16).

Napompeth (1973) studied the population dynamics of *P. maidis* in Hawaii and reported that adult *P. maidis* were frequently recorded in light traps several miles from the nearest cornfields. He also provided data suggesting similar effects of maize cropping systems on *P. maidis* migration patterns to those observed for rice and *N. lugens*. In areas of continuous cultivation, short-range flights between plantings predominate. Where there is no marked fallow period, immigrants may arise from survivors on nonpreferred alternate hosts; if the fallow period is extended, then wind-borne migrants are important for population establishment.

Dense, swarming populations of *Perkinsiella saccharicida* Kirk were observed 30 km off the coast of the Bundaberg district, Australia during studies on the population dynamics of this species (Bull 1972). Such flights were frequently seen on land at distances exceeding 10 km from the sugarcane crops when climatic conditions were appropriate, and the author suggests that such swarms could be transported for several hundreds of kilometers (Bull 1981).

Few data are available on the range of migratory or dispersal flights of other species of planthoppers. *Prokelisia marginata* undertakes migratory flights between different habitats, and although macropters have been recorded more than 30 km off shore, most redistribution occurs over a few kilometers (Denno and Grissell 1979; Denno 1983).

To summarize, the overall pattern that emerges for *N. lugens* is one of a mosaic of populations in the tropics closely associated with local cropping patterns and in which colonization of new rice plantings is achieved largely by short-distance migration. Long-range movement of individuals between rice-growing areas within the tropics, analogous to the long-range tropical to temperate displacements, certainly takes place but it is difficult to quantify. It is unlikely that long-distance migration contributes significantly to population initiation against the background of fairly massive local movement. It may be significant, however, in gene flow or virus transmission to distant locations.

Population Establishment

Immigration into the crop may be a passive process arising through fallout of individuals as they reach the end of their flight range, or result from responses to host plant cues. Cook and Perfect (1985a) compared water trap catches of *N. lugens* and *S. furcifera* at rice and nonrice sites which were either alternative crops, natural low-level vegetation, or recently ploughed uncultivated soil. Both planthoppers species were caught at all sites and the authors concluded that neither species showed a rice-orientated landing response. In contrast, Napompeth (1973) concluded that immigration of *P. maidis* into maize in Hawaii ceases after about 8 weeks due to lack of attractiveness of the maturing crop, thereby inferring a positive landing response to a young crop for this species.

The timing and magnitude of immigration is an important determinant of the pattern of population growth, peak density, and, thus, potential crop damage caused by rice planthoppers (Holt et al. 1989). The timing and magnitude of immigration is influenced by the spatial and temporal distribution of source populations, insect flight range, and climatic factors. The spatial distribution of source populations is influenced by cropping patterns and the temporal distribution by the degree of cropping asynchrony, within the flight range of the planthoppers. For the rice planthoppers, population establishment differs considerably in the temperate and tropical regions of their distribution.

In temperate Asia, immigration may occur at specific times during the cropping cycle associated with particular synoptic weather conditions such as the annual immigration of *N. lugens* and *S. furcifera* from mainland China into Japan and Korea (Kisimoto 1976; Kim et al. 1987). Immigrant densities of *N. lugens* in Japan are low, averaging 0.01/hill (a rice hill is the unit of plant stand in transplanted rice) and with a range of 0.001–0.14/hill (Kisimoto 1979), whereas those of *S. furcifera* are higher, averaging 0.12/hill with a range of 0.008–0.6/hill. Of these two rice planthopper species, *S. furcifera*, is, thus, considered to be the more successful long-distance colonizer (Kisimoto 1981) although it has lower pest status because subsequent population buildup within the crop is less rapid.

In temperate Asia, the size of the initial immigrant densities of rice planthoppers provides a good indicator of subsequent peak population size and this relationship has been studied to develop forecasting systems. Data from an 8-year study are summarized by Kuno (1979, 1984). For *N. lugens*, the correlation between immigrant density and that at the peak population (generation 3) was 0.76; for *S. furcifera*, a significant correlation could not be established even though the population peak occurs at the second generation.

In tropical Asia, the relationship between climatic factors and immigration patterns is less clear. Temperature is not a limiting factor either for rice cultivation or for the population development of the rice planthoppers, although, as in temperate regions, take-off is inhibited by rainfall and high wind speeds (Ohkubo and Kisimoto 1971; Chen and Cheng 1980). Double or multiple rice cropping is therefore possible and asynchrony in planting dates is common, particularly in the large irrigation systems, resulting in potential sources of immigrant rice planthoppers throughout the year. The potential for immigration year-round has a profound impact on patterns of population development of *N. lugens* and these vary both from year to year and in different rice growing habitats (Cook and Perfect 1989).

In the large irrigation systems in the Philippines, for example, where the variance of planting dates is often pronounced, early season immigration rates of *N. lugens* can be high and prolonged, resulting in rapid population growth with overlapping generations. Detailed studies were conducted by Kenmore (1980) on high-density populations on the experimental farm of the International Rice Research Institute, Philippines and on nearby farmers' fields within the major immigration system of Laguna Province. High levels of immigration of up to 80/hill were recorded and these could often be associated with nearby *N. lugens* outbreaks. Studies by Cook and Perfect (1985a) also in the Philippines show that immigration rates at sites within irrigation systems were higher than those at more isolated rice farms approximately 20 km away. Few data on immigration rates exist for other planthopper species, though Bull (1981) recorded immigration rates of 2.25 *Perkinsiella saccharicida* per stalk in an isolated area of sugarcane, 8 km from other planthopper source areas in Australia.

Cook and Perfect (1985a) investigated the relationship between immigration rates of *N. lugens* and *S. furcifera* on population development by comparing water trap catches and population growth over a range of rice habitats where immigration rates differed within a season. The study showed that, unlike work in Japan, there was no clear relationship; often high peak densities were associated with lowest immigration rates and vice versa. They concluded that in the tropics within-field factors such as natural enemy numbers were more significant determinants of peak population size during a season.

Holt et al. (1989) carried this work further using a simulation model of *N. lugens* population growth to investigate the impact of different patterns of immigration on peak densities. They investigated the effects of different patterns of immigration defined as background (prolonged, low rate) and pulsed (short duration, high rate) to simulate the type of immigration that might be experienced in younger plantings adjacent to hopper-burned fields. Results showed that manipulating the patterns of immigration could induce *N. lugens* outbreaks, though the rates required to achieve these effects were greater than rates recorded in the water trap study described above. Background immigration rates of 0.2/hill/day starting early in the season (before 35 days after transplanting) could result in damaging *N. lugens* populations later in the season. When background immigration started later, the rate had to be increased twofold to produce the same peak densities. With pulsed immigration, pulses arriving later than 30 days after transplanting (DAT) needed to be 10 times larger than those arriving earlier in the season to cause an outbreak. Simulations incorporating both immigration patterns showed that background immigration rates less than 0.01/

hill/day did not change the effect of pulses but higher background rates had an additive effect.

The reproductive status of immigrant insects is an important factor in population establishment. The majority of immigrant *N. lugens* arriving in the crop are reproductively immature, conforming to the flight–oogenesis syndrome whereby protein is either invested in flight musculature or egg production, depending on circumstance (Johnson 1969). However, a few gravid females of *N. lugens* do undertake some dispersive flights, as evidenced by their recovery from light trap samples taken at a height of 4 m above the crop (Perfect unpublished data). Napompeth (1973) suggested that colonizing females of *P. maidis* are often gravid, allowing oviposition to begin immediately.

Immigration is clearly a key factor in creating the potential for population establishment, but of itself is not sufficient to ensure that this occurs. In the seasonal northerly migrations that occur in temperate Asia, climatic conditions contribute to the success of population establishment by creating latitudinal zones of new plantings that provide suitable conditions for emigrants carried by prevailing winds from older rice further to the south. Here, immigration is generally successful as evidenced by the close relationship between immigrant densities and the ensuing population. In tropical rice, and to some degree during the southerly migrations at temperate latitudes, immigrants may arrive in plantings at any stage of crop development. Population establishment will depend on the characteristics of the crop and the enemy complex it supports, and there is, therefore, no clear relationship between level of immigration and population establishment.

Population Growth

There is considerable variation in the number of annual generations produced by planthoppers; this is determined in the main by the linked variables of habitat availability and temperature (Denno and Roderick 1990). Most species have a generation time of approximately 1 month under optimal conditions but this may be extended by poor host quality or low temperature. The variation is clearly shown in species with a wide geographic distribution; *Laodelphax striatellus* is bivoltine in Sweden, undergoing nymphal diapause during winter, but produces eight generations annually in Israel without any break in population development (Harpaz 1972). In the tropics, the normal generation time for *N. lugens* is 24 days, giving the potential for some 12 annual generations which can be achieved in South East Asia (Dyck et al. 1979), whereas in Northern Japan the species is trivoltine (Kuno 1979). In temperate populations of *N. lugens*, generation time is significantly longer at the lower, early season temperatures (Mochida and Okada 1979); this is an important parameter in determining the potential impact of pest species and for predicting likely control needs in China (Cheng and Holt 1990). Favorable temperature regimes increase the peak density at the second generation by some 30%; coupled with appropriately timed immigration, they also promote the production of a third generation within the growing period of the crop.

Population development of *N. lugens* has been studied extensively in its temperate habitat in Japan and is characterized by rapid growth during three generations, with peak densities occurring toward the end of the growing season (Kuno 1979; Kisimoto 1981). Peak densities may be more than 500 times the initial immigrant density representing an eightfold increase per generation and this high growth rate can be attributed to high fecundity (Kuno 1979). In contrast, the population size of *S. furcifera* increases only four times over the course of three generations. Therefore, despite higher initial immigrant densities, this species has lower growth rates and rarely attains sufficiently high field densities to cause economic damage. The pattern of population development is also similar from year to year though it is not possible to predict peak densities from immigrant densities with the same accuracy as *N. lugens* because of lower and more variable reproductive rates (Kuno 1979).

Kuno and Dyck (1985) compared population growth characteristics of *N. lugens* from Japan, based on detailed data reported by Kuno and Hokyo (1970), and from data collected using a similar sampling regime on irrigated rice in the Philippines (see Cariño et al. 1979). They demonstrated marked differences in the pattern of population development that were proposed as characteristic of temperate and tropical regions. The dynamics of Philippines populations differed from those in Japan by showing initial immigrant densities an order of magnitude higher (0.37 compared to 0.01 adults per rice hill), growth rates per generation approximately halved, and earlier peak densities. Peak densities in Japan were always in the third generation, whereas in the Philippines, peak density commonly occurred in the second. The overall rate of net population growth was strikingly different at 3.2 for tropical populations compared to 513 for those in Japan. Differences in growth patterns between the two populations can be partly explained by differences in predation pressure and, in particular, by the impact of spiders, mirids and veliids. Climatic and rice varietal differences between the two habitats may also play a role. Both Philippines and Japanese populations, however, showed large variation in later generations both among fields and among years, suggesting that the growth processes were density-independent, typical of outbreak-type pests.

Cook and Perfect (1989) confirmed these observations of higher immigrant densities, lower growth rates, and earlier peak densities in irrigated habitats in the Philippines but found interesting differences between irrigated and rain-fed rice habitats. In rain-fed rice, the pattern of population development was similar to that described above for Japanese *N. lugens* populations showing distinct generation peaks which increased throughout the season. The rates of population increase were lower, though very variable probably due to differences in predation pressure and the interaction between predator numbers and immigration rates of hoppers (Cook and Perfect 1985a). The dissimilarity in growth pattern among *N. lugens* populations in different rice habitats is associated with lower immigration rates in the rain-fed, more-isolated site, where planting is governed by rainfall patterns and is, therefore, more synchronous. Kuno and Dyck (1985) suggested that the differences observed between Japanese and Philippines populations were a function of their location within the distribution range of *N. lugens*, the latter representing a perennial and the former a seasonal population. A similar relationship existing between Philippines populations separated by only 12 km suggest habitat characteristics and flight range are crucial factors in determining population growth patterns.

A study of the population dynamics of *Peregrinus maidis* throughout a season on maize in Hawaii (Napompeth 1973) showed overlapping generations and no distinct generation peaks, such that all stages of *P. maidis* were present throughout the cropping period. Peak populations occurred mid-season, and fluctuations in density were not as great within a season as the year-to-year fluctuations.

Because of the differences in fecundity of brachypters and macropters, the factors affecting wing-morph determination will have an indirect but important influence on the rate of population increase. The switch between brachyptery and macroptery in dimorphic populations has been studied for a number of species (Denno and Roderick 1990). Brachyptery would be expected to predominate where habitat conditions remain favorable, leading to high populations and rapid exploitation of food resources; for most planthoppers, the early resident generations in transient habitats display this characteristic. The result is commonly deterioration of the resource, conditions which favor dispersal, and the factors most frequently demonstrated as triggering the production of macropterous forms are either directly or indirectly connected with plant quality. For field populations of *N. lugens* and *S. furcifera*, the degree of nymphal crowding during the early instars has been shown to influence wing-morph (Johno 1963; Ohkubo 1967; Watanabe 1967; Mori and Kiritani 1971; Takagi 1972); the nutritional status of the host plant also plays a significant part (Kisimoto 1965; Saxena et al. 1981).

Cook and Perfect (1982) reared isolated individuals of both species as part of an investigation into determining adult wing-morph from nymphal wing pad length and concluded that though wing-morph was determined in the early instars; the mechanism was surprisingly flexible because nymphs destined to become macropters could give rise to brachypters, presumably through the change in status involved in isolating the individual on a young plant of high nutritional status. The reverse effect did not occur under the conditions studied. Using nymphal wing pad length to determine patterns of wing-morph production in field populations, the authors were able to show that substantial proportions of brachypters were still being generated in senescing crops. Thus, though the proposed relationship between plant quality and wing-morph is logical and has been demonstrated for other species such as *J. pellucida* (Mochida 1973) and *P. marginata* (Denno et al. 1985), there are likely to be other factors involved.

Dispersion

The spatial distribution of planthoppers within rice fields has been studied extensively to develop effective sampling procedures for ecological studies, pest monitoring, and the establishment of economic thresholds (Kuno 1977). Kuno (1977) showed that the distribution of the immigrant population of *N. lugens* in Japan in young rice plantings was random and that the distribution could be described by the Poisson distribution. By comparison, the distribution of the breeding population shows departure from the Poisson toward aggregation and is best described by the negative binomial distribution. Kuno (1977) used an aggregation index C_A to express the degree of aggregation, which is the reciprocal of the parameter k of the negative binomial and is calculated from the following equation: $C_A = (\sigma^2 - m)/m^2$, where m is the mean density and σ^2 the variance. When the distribution is random, C_A equals zero; positive values greater than zero indicate aggregated distributions, whereas negative values indicate uniform distribution. Over an 8-year study in Japan, the mean value of C_A for *N. lugens* adults and nymphs combined was 1.33 (range 0.45–3.37) suggesting a very aggregated distribution.

Kuno (1968) also collected data on spatial distribution for two other rice planthoppers, *S. furcifera* and *Laodelphax striatellus* in Japan, and showed that distribution in immigrant populations was similar to the Poisson distribution, but breeding populations conformed to the negative binomial distribution patterns. The mean values of C_A for *S. furcifera* and *L. striatellus* were 0.437 and 0.297, respectively (Kuno 1984). Wu et al. (1987) showed that the degree of aggregation of field populations of *S. furcifera* in China increased progressively from adults to early-instar and late-instar nymphs.

Kushmayadi et al. (1990) monitored the spatial distribution of the different life stages for *N. lugens* populations in Indonesia using the C_A index of aggregation (Kuno 1968) and Iwao's model (Iwao 1968). Values of C_A were compared for *N. lugens* populations in nine locations in West Java during single wet and dry seasons. Their data suggested that the degree of aggregation as expressed by C_A was independent of mean density. In most locations, values of C_A were higher for nymphs than for adults, indicating more aggregated nymphal populations than adult populations; this is to be expected, given the limited dispersal of emerging nymphs. More surprisingly, values for macropterous adults were similar to or higher than those for brachypterous adults. Because brachypters normally remain on the rice hill on which they are produced and macropters are clearly able to disperse, these results are difficult to interpret in the absence of more detailed data. In the latter part of the growing season, macropters may show aggregated distribution as teneral adults emerge prior to emigration. Studies by Perfect et al. (1985) in the Philippines suggested that such adults undertake few dispersive flights within fields and if data over the whole season are analyzed, this transient but very marked effect may dominate the outcome. In contrast to the earlier work in Japan, this study showed that the degree of aggregation did not vary significantly with the initial immigrant density.

Kuno and Dyck (1985) compared within-field spatial distribution patterns of temperate *N. lugens* populations from Japan and tropical populations from the Philippines. Although the distribution patterns of both populations conformed to the negative binomial distribution, the Philippines populations were significantly less aggregated than the Japanese populations. This was thought to be associated with more prolonged random immigration throughout the season in the Philippines.

Few data are available on spatial distribution patterns for other delphacids. Takara and Nishida (1983) found that immigrant *Peregrinus maidis* in Hawaii were aggregated along the borders of maize fields, density decreasing with distance from the edge of the fields; this contrasts with the data of Napompeth (1973) also from Hawaii where an initial random distribution of this pest species across fields gave rise to a contagious distribution later in the crop. Differences observed are likely to arise from local influences. Despite the abundant evidence of random distribution of *N. lugens* immigrants, very strong edge effects can be observed where new plantings are adjacent to older, heavily infested crops (T.J. Perfect personal observation).

Population growth characteristics appear to be dominated by temporal patterns of immigration, themselves closely linked to a composite function of flight range and the size and distribution of sources of immigrants. Where habitat availability is strongly seasonal (e.g., Japan) or confined for other reasons (e.g., rain-fed rice in the Philippines), population growth tends toward exponential with very clear generation peaks and a preponderance of brachypterous adults until habitat cues induce the production of emigrant macropters. Over the growing period, distribution changes from random to strongly aggregated. Where continuous immigration is possible, founder populations are often higher and incoming macropters maintain a more even balance of wing-morphs throughout the growth period. Similar spatial patterns can be detected, but these are profoundly influenced by the level and timing of continuing immigration. Growth patterns are consequently much more difficult to predict in such populations, but the rate of increase from generation to generation is often much lower than for seasonal habitats.

Population Suppression by Natural Enemies

Once colonization of the habitat has taken place, a number of factors operate to determine the subsequent size of the population, the most significant being host plant suitability (Cook and Denno Chapter 2) and natural enemy populations (Döbel and Denno Chapter 10; Cronin and Strong Chapter 11; Benrey and Lamp Chapter 14). For most planthopper species, egg mortality is the key mortality factor involved in population suppression (Waloff 1980; Kenmore et al. 1984; Cook and Perfect 1989; Denno and Roderick 1990) and despite the abundance and diversity of specific egg parasitoids, there is evidence that generalist predators such as mirid bugs and spiders often have a much greater impact. In the concluding section of this chapter, we will confine ourselves to a brief comparison of the natural enemy fauna of temperate and tropical regions and its potential for suppressing planthopper populations.

Egg mortality plays a key role in the population suppression of many temperate planthopper species (Denno and Roderick 1990). High percentages of egg parasitism have been recorded for *Prokelisia marginata* (Stiling and Strong 1982b), but associated with low egg densities. More commonly, egg parasitism shows inverse density dependence associated with the high intrinsic rate of increase commonly shown by planthopper populations, and despite the considerable diversity of specific parasitoids, they rarely have a key impact on population development (see Denno and Roderick 1990). Egg mortality rates of 55% were recorded for *Stenocranus minutus* (May 1971) as a result of egg parasitoids, predators, and hatching failure and 70% mortality for *Javesella pellucida*, largely due to egg predators (Raatikainen 1967). Egg predators of the rice planthoppers include the mirid genera *Cyrtorhinus* and *Tytthus* and some species of Nabidae (Chiu 1979).

Nymphs are commonly parasitized by Dryinidae, Elenchidae, and Pipunculidae, often maturing in adults and thus dispersed to affect new populations, whereas the major nymphal predators in most habitats appear to be spiders, with Lycosidae featuring prominently in this role (Hassan 1939; Kuno 1973b; Drosopoulos 1977; Ôtake 1977; Chiu 1979; Waloff 1980). Döbel (1987) showed that *Prokelisia marginata* is the preferred diet for the wolf spider *Pardosa* in North American salt marsh communities. Spiders are by far the most important predators of both nymphs and adults, generally of the families Lycosidae, Linyphiidae, and Tetragnathidae; coccinellid larvae may sometimes be important. *Cyrtorhinus* also preys on nymphs and the veliid bug *Microvelia* can sometimes be important for rice planthoppers (Chiu 1979; Nakasuji and Dyck 1984).

Several authors have reported the effects of natural enemies on tropical populations of rice planthoppers; many are common to both temperate and tropical habitats. The early studies by Hinckley (1963) in Fiji and Stapely (1976) in the Solomon Islands suggested that the mirid *Cyrtorhinus lividipennis*, a predator of eggs and early-instar nymphs contributed significantly to population suppression of *N. lugens*. Kenmore (1980) and Kenmore et al. (1984) collected detailed data on natural enemies of *N. lugens* in the Philippines. They reported that egg predation was only significant where *N. lugens* densities were high and when egg density exceeded 20/plant. They also demonstrated a highly significant numerical response of spiders (a grouping that included species of the genera *Lycosa*, *Callitrichia*, *Tetragnatha*, *Araneus*, *Argiope*, and *Oxyopes*) to planthopper abundance. Juvenile and adult spiders were distinguished. The response was due to either increased fecundity or higher net immigration and survival of ballooning spiders. Kenmore et al. (1984) reported a preadult survivorship of only 4% and showed that 65–91% of the mortality occurred before the first instar.

Cook and Perfect (1989) examined *N. lugens* survivorship in field studies over a 4-year period in the Philippines. Levels of egg predation and parasitism were variable but the highest egg predation level observed was 74% compared with a maximum of 32% of eggs parasitized. Over three wet seasons and two dry seasons, the survivorship of *N. lugens* ranged between 1 and 12%, higher in some seasons than the 4% survivorship reported by Kenmore et al. (1984) though both studies clearly show that in all seasons most mortality occurred between egg and first instar life stages.

Fowler et al. (1991) report studies on *N. lugens* in Sri Lanka where up to 54% egg parasitism by *Anagrus* and *Oligosita* was observed together with egg predation by larvae of *Panstenon* (Hymenoptera: Pteromalidae) that reached 18%; otherwise, egg predation was unimportant at the sites studied, possibly associated with low populations of *Cyrtorhinus*. Previous studies in Sri Lanka by Ôtake et al. (1976) showed good suppression of *N.*

lugens populations in the absence of high populations of predators and it may be that parasitoids played a larger role in suppression under these circumstances. Although there is limited evidence for density-dependent parasitism by the parasitoids of planthoppers (Cronin and Strong Chapter 11; Stiling Chapter 13), several predators, particularly mirid bugs, show strong density-dependent numerical responses (Döbel and Denno Chapter 10).

Cook and Perfect (1985a) showed a strong correlation between flight activity of *N. lugens* and *Cyrtorhinus lividipennis* in the Philippines associated with synchronous invasion of rice plantings and were further able to demonstrate a significant relationship between the rate of population change and the level of immigration of *Cyrtorhinus* early in the crop for *N. lugens*; a similar relationship existed for *S. furcifera* at a later stage of crop growth.

High percentages of egg parasitism have been recorded for *Saccharosydne saccharivora* (Metcalfe 1971) in Jamaican sugarcane fields and Napompeth (1973) implicated the same mirid species as a major factor in controlling *P. maidis* on sugarcane in Hawaii, suggesting that populations only build up where *Cyrtorhinus* is not abundant. *Tytthus mundulus* was also present, but ineffective in regulating the population. It is likely that the characteristics of the faunal complex existing on the host plant and adjacent vegetation at the time of colonization have an important bearing on the impact of predation and parasitism on egg mortality, and thus on final population size. A diverse fauna of generalist predators and sources of parasitoids should make a major contribution to egg mortality. Some evidence to support this view comes from studies on the resurgence of *N. lugens* in response to insecticide treatment (Reissig et al. 1982a), although the picture is confounded by the stimulant effect of some insecticides on egg production. It is clear that the principal cause of the extremely rapid population growth observed is related to natural enemy mortality. The resurgence phenomenon is most pronounced when insecticide treatments are applied in response to immigrants observed in the field and it is reasonable to suppose that this is due to reduced mortality of the protected eggs arising from the impact of the insecticide on the mobile and exposed predators and parasites.

Although *Cyrtorhinus* can inflict heavy mortality on the eggs of *N. lugens*, it does not appear to exert much effect on nymphal populations (Chiu 1979). Nakasuji and Dyck (1984) studied the impact of *Microvelia douglasi atrolineata* on populations of *N. lugens* in the Philippines. Very high densities of *Microvelia* are sometimes observed, reaching a recorded peak of 450 m^{-2} (Kenmore 1980), and Nakasuji and Dyck (1984) concluded that this species was very effective in attacking insects falling on the water surface. *Mesovelia* spp. was present at lower densities; in neither case were populations correlated with planthopper abundance.

It seems clear that for rice planthoppers there is a major difference in the role of natural enemies in population suppression between temperate and tropical regions. This is highlighted in the work of Kuno and Dyck (1985), but is also evident from the difference in efficacy of insecticidal control against *N. lugens* in the temperate versus the tropical rice-growing regions of Asia. Chemical control has been (and remains) effective for many years in Japan and Korea (Heinrichs 1979). Rice production technologies associated with the "green revolution" in Southeast Asia resulted in the introduction of pesticides to counter a growing threat from *N. lugens* with disastrous results and it seems these can be directly associated with the induced resurgence due to high mortality of natural enemies (Reissig et al. 1982b; Heinrichs Chapter 16). The fact that no such resurgence occurs in temperate rice systems underlines the lesser importance of natural enemies as regulators of planthopper populations in temperate rice. It is likely that this phenomenon is associated with seasonality. Although predators are known to migrate with their host planthoppers (Kisimoto 1981), there is likely to be an establishment lag that often precludes their effective function in population suppression.

Discussion

The dynamics of planthopper populations present a pattern of changing abundance that reflects habitat distribution and quality in space and time, overlain by interspecific and intraspecific interactions that determine variability within the broad framework of resource availability. There are temporal and spatial components of population dynamics and interpretation of both is strongly dependent on scale. The temporal component may relate to seasonality in host plant quality and/or availability and climate. The spatial component is a function of the mobility of the insect species and its ability to exploit changing patterns of host plant distribution. The spatial distribution of resources may often act as a key determinant of subsequent population growth by influencing the level of initial colonization (Kareiva 1983). Natural enemies also play a crucial role in determining the population dynamics of planthoppers.

Much of the material considered in this chapter concerns rice planthoppers of economic importance because their study has been well funded in both temperate and tropical habitats. Comparable effort and resources have not been invested in understanding the biology and ecology of most other planthopper species. Where possible, data have been quoted for other species, and for the most part these do not suggest that generalizations based on *N. lugens* and *S. furcifera* would present an atypical picture.

In studies of the population dynamics of *N. lugens*, there has been a tendency to distinguish distinct types of temperate and tropical population dynamics with the former characterized by synchronous, very low levels of invasion, and discrete generation peaks increasing progressively in size through the season. Tropical populations tend to show much greater variability in the timing and level of invasion associated with the greater diversity of potential sources of immigrants, and no consistent relationship between the sizes of different generation peak is shown; the second resident generation is often present at higher density than the third, and for some data (Kenmore 1980), the generation peaks are difficult to distinguish at all. Such a situation should logically exist if the same level of mobility is inferred for tropical and temperate populations because this would create the potential for a continuous source of immigrants. Analysis of data from the Philippines suggests that the distinction is not so clear-cut.

Habitat continuity, in both space and time, is greater in the tropics than for temperate rice, but seasonality is nonetheless an important factor in population dynamics. The bulk of rice cultivation occurs in the major wet season, and where irrigation facilities are poorly developed, regional cropping synchrony is still characteristic of large areas of smallholder rice. Studies in such situations (Cook and Perfect 1989) show patterns of population development that relate more closely to the expectation for temperate rice; similar low levels of invasion are experienced; generation peaks are distinguishable and show a progressive increase through the season. The principal difference lies in the much lower rate of increase between generations.

The data from Perfect and Cook have been used in the construction of a simulation model (Holt et al. 1987) which suggests the period of early immigration is critical for subsequent population development and that there is a complex interaction between level and timing of immigration and natural enemy action. It appears that higher levels of immigration may favor the early establishment of viable natural enemy populations which hold subsequent *N. lugens* populations in check, whereas lower levels do not. Populations of the planthopper are, therefore, able to erupt more easily where short bursts of low-level immigration occur. Clearly, this is a difficult situation to analyze because many of the important predators are generalists and other components of the fauna must play an important role in sustaining natural enemies; it is also somewhat difficult to relate to outbreak dynamics because even within this scenario we are concerned with population levels that are probably not economically significant. The authors have yet to encounter a case of severe *N. lugens* infestation where associated pesticide use can be ruled out as a primary factor. It is tempting to draw the conclusion that the use of pesticides in tropical rice causes planthopper population outbreaks for the same reasons they occur in temperate regions through their disabling effect on natural enemy action.

Summary

1. Planthoppers are frequently dimorphic and such species are effective in exploiting transient habitats; dispersive macropterous adults often alternate with the more fecund brachypterous morph. Migration is particularly well developed in rice planthoppers allowing for population continuity and the exploitation of seasonally available habitats at both temperate and tropical latitudes.
2. The ability of planthoppers to migrate long distances (750 km) is highlighted by the annual invasion of Japan across the South China Sea. Such extreme movements are associated with specific meteorological conditions, but similar linear migrations also occur for the temperate populations and are related to the seasonal extension of rice cultivation. In the tropics, long-distance movement may occur but its effects are overshadowed by local dispersive flights over the range of a few kilometers.
3. Immigration occurs over a fairly short period in temperate rice and leads to distinct generations with sharp population peaks and showing a tendency to exponential increase. Cropping synchrony has a key effect on tropical population development. In isolated plantings, similar patterns to those for temperate rice are observed; large irrigated systems with asynchronous plantings are subject to continuous immigration, have indistinct generations, and populations are often at a maximum earlier in the crop. The latter have been characterized as typical "tropical" populations.
4. Temperate populations show clear succession of wing-morphs, with the first resident adult generation dominated by brachypters; macropters are present in abundance only at emigration. In the tropics, similar considerations probably apply to the resident population but population structure is markedly influenced by the level and timing of dispersive flights from local sources. The difference is chiefly associated with planting synchrony.
5. Founder populations are generally randomly distributed; this probably arises through passive fallout and appears to be the norm for most planthoppers. Low dispersal of ensuing nymphs and brachypters subse-

quently gives rise to a strongly aggregated distribution; this is less pronounced in tropical populations as a result of the higher and continuing immigration rates commonly observed.

6. Despite the abundance of specific parasitoids, generalist predators make the major contribution to mortality in both temperate and tropical situations. In temperate regions, natural mortality is ineffective in population regulation and explosive population development is the norm. In the tropics, natural regulation is normally effective but is readily disarticulated through differential mortality arising from pesticide use.

ACKNOWLEDGMENTS

The authors gratefully acknowledge the support provided by funding from the Natural Resources and Environment Department of the UK Overseas Development Administration to a long-term research program that has contributed much to our knowledge and understanding of brown planthopper population dynamics. The views expressed in this chapter have been much influenced by discussion and contact with colleagues at the Natural Resources Institute and collaborating institutions in the UK and Asia. The synthesis presented here owes much to the support and encouragement of Dr. Robert F. Denno.

9

Long-Distance Migration in Delphacid Planthoppers

Ryoiti Kisimoto and L. Jane Rosenberg

Introduction

Insect migration has been succinctly defined by Dingle (1989) as "a syndrome evolved for the displacement of the individual in space beyond its normal station-keeping or foraging movements." Migration is characterized by the interaction of specialized behavioral responses which serve a specific ecological function, namely, the choice of where the next generation will breed: the "here" or "elsewhere" options described by Southwood (1977a). Defined as such, the term migration encompasses a range of spatial and temporal scales, from the microscale (<1 km and lasting <1 hour) through the mesoscale (1–100 km with a duration up to 48 hours), to the macroscale which covers intracontinental and intercontinental displacements in excess of 100 km and lasting up to several days.

In this chapter, we concentrate on migrations where the extent of insect movement is determined by both the insect's intrinsic flight capabilities and the weather in the planetary boundary layer. Because planthoppers have relatively low airspeeds, they must use the wind to traverse any significant distances, and we are concerned only with migrations in the air above the flight boundary layer [i.e., outside the layer of air near the earth's surface within which the insect has control over its movements relative to the ground (Taylor 1974)]. Once above the flight boundary layer, insect displacements are determined by the windfields in which they are flying, the direction of movement being determined by wind direction and the extent of movement by the wind speed, and the insect's own flight duration. The effects of air temperature and humidity determine whether take-off is possible and influence the height of flight (Ohkubo 1981; Rosenberg and Magor 1987).

We discuss the factors which lead to the development of migrant individuals in planthopper populations and the adaptive behavior of such migrants which enables them to leave their flight boundary layer and sustain flight in changing windfields for varying lengths of time. Finally, we consider existing evidence supporting long-distance movements by planthoppers. Inevitably, most of the discussion will be focused on species of planthoppers from eastern Asia, where some of the most comprehensive studies of planthopper population dynamics and migration have been conducted (e.g., Cheng et al. 1979; Kisimoto 1981) and, in particular, on the rice brown planthopper, *Nilaparvata lugens*, a pest of major agricultural importance throughout Asia.

Diapause and wing-dimorphism are also key features in the life histories of planthoppers, giving them great flexibility in the exploitation of resources and resulting in some species being major agricultural pests (Denno and Roderick 1990). As these aspects of planthopper life history are reviewed in detail elsewhere in this volume (Denno, Chapter 4), we will confine our discussion to the conditions which promote the production of long-distance migrants and lead to pronounced seasonal differences in the distribution of some planthopper species.

Production of Migrants

Ecological Factors and Host Plants

Insect migrations are often seasonal in nature and variable in extent, but the ecological and behavioral mechanisms which control them are considered to have a common theme. They allow insects to exploit resources widely scattered in space and time by bringing individuals to those habitats most suitable for the development of their offspring and enabling them to escape from unfavorable conditions (Dingle 1982). Hence, the proportion of migrants in an insect population has been correlated both with the degree of spatial and temporal impermanence of the habitat and the duration of habitat suitability in relation to generation time (Southwood 1977a; Denno et al. 1991). Most planthopper species have a short generation time (approximately 1 month) and many also inhabit transient habitats, both of which favor the development of traits that maximize the ability to disperse, arrive in new habitats successfully and exploit them effectively through rapid reproduction before the habitat deteriorates (Southwood 1977a).

Most planthopper species are monophagous or oligophagous, feeding on gramineous plants, including cereal crops which form a simple and uniform vegetation cover, often artificially managed (Wilson et al. Chapter 1). There is no evidence that planthoppers obligately or regularly alternate host plants, although *Javesella pellucida* annually migrates to oats from other hosts in Finland (Raatikainen 1972). Also, the small brown planthopper, *Laodelphax striatellus*, can breed on various cereal crops and annual and perennial weed grasses, moving between plants as the host ripens or withers (Kisimoto 1969). Along the Atlantic coast of North America, the host grass, *Spartina alterniflora*, of the planthopper *Prokelisia marginata* occurs both in high marsh where the vegetation persists throughout the year and on tidal streamsides where the grass grows only in the summer. Flight by macropters in June allows the planthopper to escape the crowded high marsh stands and exploit the streamside vegetation where high-quality food and oviposition sites are available. In September, *P. marginata* escapes from this declining habitat and moves back to the high marsh where offspring can overwinter (Denno et al. 1980). Thus, the high level of macroptery (about 80%) and the seasonal migration cycle of these temperate saltmarsh planthoppers enables different generations to maximize exploitation of transient population development sites and more permanent overwintering habitats (Denno and Roderick 1990). In contrast, the sibling species *Prokelisia dolus*, which remains mostly on the high marsh along the Atlantic coast, exhibits only 30% macroptery, reflecting the more sedentary nature of this species in a more persistent habitat (Denno et al. 1989). Along the Gulf coast of North America, *S. alterniflora* provides a more stable habitat for *P. marginata* throughout the year and the planthopper populations are mostly (90%) brachypters and similar low levels of macroptery (range 0 to 26%) have been recorded for 14 other species of planthopper which inhabit relatively persistent and uniform stands of saltmarsh vegetation in North America (Denno 1985b). McCoy and Rey (1981) have suggested that the ratio of migrants to nonmigrants among salt marsh-inhabiting planthoppers may reflect strategies for dealing with differences in the stability and distribution of their respective habitats. They concluded that for *P. marginata*, macroptery predominates where habitats are temporary and extensive but is low where habitat patches are persistent and extensive throughout the year and where resource patches are small, infrequent, and isolated. For other species, such as *N. lugens* and the whitebacked planthopper *Sogatella furcifera*, which feed exclusively on rice and have no diapause, seasonal migrations by different generations are closely associated with spatial changes in the distribution of their temporary but extensively cultivated host plant.

A common pattern of population development for several planthopper species, which facilitates maximum exploitation of temporary habitats, is colonization by low densities of macropters, followed by rapid population growth with more fecund brachypters and, finally, exodus from the deteriorating host plants by macropters. This pattern is demonstrated clearly by *N. lugens* when it colonizes the newly transplanted rice crop in Japan each spring and early summer. The initial density of macropterous immigrants is low (0.01 planthoppers per rice hill) (Kuno 1979) and subsequent population development is characterized by an exponential density increase over three distinct generations in which the proportions of macropters and brachypters varies. In the first generation, in response to the low initial density of the immigrant population, most female adults (90%) are brachypters, whereas the males are entirely macropters, thus facilitating mate location (Denno et al. 1991). Brachypters again form the majority (80%) of the female adult population in the following generation when there is also a peak of male brachypters (25–33%). By the final generation, only 58% of females are brachypters and all males are again macropters (Kisimoto 1965, 1977; Kuno 1979). In tropical populations of *N. lugens*, similar changes in the proportions of macropters and brachypters occur although distinct population peaks are not always evident, particularly in areas where irrigation enables asynchronous rice planting (Hirao 1986; Cook and Perfect 1989).

For species such as *N. lugens* which are closely linked with a cultivated host, the proportion of migrants is greatest when the crop is senescing and near harvest (Jeffrey 1982) and Padgham (1983a) has suggested that macropters emerging on young rice in an early generation on the crop are often nonfliers, whereas migrants are common in later generations when the host plant reaches senescence. Dyck et al. (1979) and Jeffrey (1982) identified only two seasonal population peaks for macropterous *N. lugens* in the Philippines, coinciding with the harvest of the two major rice crops, and suggested that the size of the dispersing macropterous population is related to the cropping pattern in the surrounding area.

The proportion of macropters in a population, however, is not always a reliable indicator of the presence of individuals with the ability to sustain flight necessary for long-distance migration. Although macropters form only a small proportion of *P. marginata* populations around the Gulf of Mexico, these planthoppers have been caught on islets 1 km or more offshore (Antolin and Strong 1987) and on an oil platform 32 km offshore (Sparks et al. 1986). In eastern Asia, some species of planthoppers inhabiting perennial hosts, such as *Stenocranus matsumurai* on *Phragmites japonica* and *Hosunka hakonensis* on *Miscanthus sinensis*, produce mainly macropters in both sexes but few are ever attracted to light-traps and they appear to be relatively sedentary and undertake only trivial flights (Table 9.1, Group A). For other planthoppers, brachypters predominate and macropters appear only when the host plants are severely stressed or population density is high. In *Nilaparvata muiri*, for example, brachypterous forms are usual, but when macropters are produced, they appear to be very

Table 9.1. Flight activity of planthopper species (based on annual light-trap catches) in relation to wing-morph composition of populations in Japan.

	Wing Polymorphism[b]			
Flight Activity[a]	A M female, M male	Ca MB female, M male	Cb MB female, MB male	B B female, B male
High flight activity [High light-trap catch]	*Saccharosydne procerus*	*Sogatella furcifera*	*Nilaparvata lugens* *Laodelphax striatellus* *Toya propinqua* *Paracorbulo sirokata* *Opiconsiva sameshimai*	*Nilaparvata muiri*
↑	*Epeurysa nawaii*	*Sogatella vibex* *Sogatella kolophon*	*Garaga nagaragawana* *Paradelphacodes paludosa* *Dicranotropus tikuzenensis* *Kosswigianella exigua*	*Nilaparvata bakeri*
	Stenocranus matsumurai	*Terthron albovittatum*	*Cemus nigromaculosus*	*Cemus nigropunctatus*
Low flight activity (seldom caught in light-traps)	*Terauchiana singularis* *Tropidocephala brunipennis* *Tropidocephala nigra* *Tropidocephala festiva* *Hosunka hakonensis*	*Chloriona tateyamana*		*Hirozunka japonica* *Zuleica nipponica* *Muellerianella extrusa* *Unkanodes albifascia*

Note: A = male and female macropterous; B = male and female brachypterous, Ca = male macropterous and female dimorphic, Cb = male and female dimorphic. (Nasu 1965; Hirao 1969; Okada and Hirao 1981).

[a]Flight activity classified on the basis of relative abundance in annual light-trap catches.

[b]M = macropter, B = brachypter.

mobile because they are regularly caught in light-traps and on ships on the East China Sea (Table 9.1, Group B) (Kisimoto 1981). Other species of planthopper exhibit wing-dimorphism either in the females only (e.g., *S. furcifera*, Table 9.1, Group Ca) or in both sexes (Table 9.1, Group Cb).

Most of the species showing wing-dimorphism in both sexes or a male bias in macroptery (Groups Ca and Cb Table 9.1) appear to have the greater flight potential (Denno et al. 1991) and even these account for a very small proportion of active fliers. *N. lugens* and *S. furcifera* (Table 9.1, Groups Ca and Cb) dominate the microfauna caught in high-level aerial netting (Riley et al. 1987, 1991; Reynolds and Wilson 1989) and in net catches on ships on the South China Sea and between the Philippine islands (Mochida 1974; Saxena and Justo 1984; Rosenberg unpublished data). Seven species of planthopper are regularly caught on ships on the East China Sea, approximately 400 km east of mainland China. These are *N. lugens*, *S. furcifera*, *L. striatellus*, *S. vibex* (= *S. longifercifera*), *S. kolophon*, *Toya propinqua*, and *N. muiri* (Kisimoto 1981), and of these species, only *T. propinqua* inhabits a perennial host, the grass *Cynodon dactylon*, which occurs in more or less permanent stands.

Genetic and Environmental Determinants of Wing Form

Many planthopper species display polymorphic (genetically controlled) and polyphenic (environmentally controlled) variation in the production of migrants, and in both cases, the variation within and between species ranges from behavioral differences in flight ability among individuals or populations to the occurrence of winged and wingless forms (Dingle 1989). The development of migration strategies depends both on environmental conditions which determine selection pressures and on the underlying genetic and developmental factors which determine the response to such pressures (Dingle 1989).

Many species exhibit wing-dimorphism and field populations of planthoppers contain both macropters and brachypters, the proportions of which vary geographically and seasonally (Kisimoto 1977; Kuno 1979; Denno et al. 1980; Cook and Perfect 1989). Intrinsic variation among individuals is high and contributes significantly to the evolution of migration, and gene differences have been found which influence both wing-polymorphism and flight performance in some species. In selection experiments, Mahmud (1980) showed an increase in the proportion of macropters between the first and second generations of *L. striatellus* and concluded that there was a strong genetic basis for the wing-polymorphism, although the proximate cues determining alary polymorphism remained food quality and quantity and crowding. For *N. lugens*, Kisimoto (1969) considered that the factors

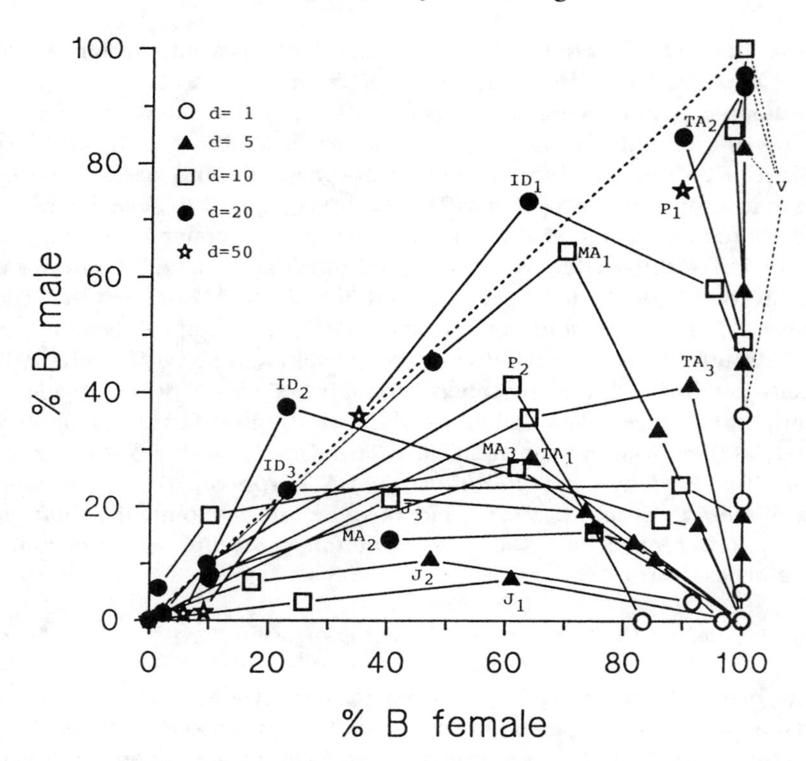

Figure 9.1. Geographic variation in *N. lugens* brachypter production at five nymphal densities (○—$d=1$; ▲—$d=5$; □—$d=10$; ●—$d=20$; ★—$d=50$) (Kisimoto 1965, 1989). Key to *N. lugens* populations and dates of collection:
Indonesia: ID_1 Deli Serdang, 1987; ID_2 Pekalongan, 1987; ID_3 Cerebon, 1987.
Japan: J_1 Chikugo, 1978; J_2 Ishugaki, 1988; J_3 Hachijyozima, 1988.
Malaysia: MA_1 and MA_2 Alor Setar, 1987; MA_3 Sungai Petani 1987.
Philippines: P_1 Laguna, 1987; P_2 Davao, 1987.
Thailand: TA_1 Cheng Mai, 1989; TA_2 Pisanuk, 1987; TA_3 Bangkok, 1989.
Vietnam: V Cantho, 1989.

controlling the proportion of wing forms in a population were inherited, but that wing-form differentiation at the individual level was affected principally by environmental factors, such as population density and host plant condition.

A comparison of wing-morph production in *N. lugens* populations in eastern Asia (Kisimoto 1965, 1989) showed that all colonies produced predominantly female brachypters and male macropters when nymphs were reared in isolation. The proportion of male and female wing forms changed with increasing nymphal density (increased macropters in both sexes) due to the combined effects of crowding and host plant deterioration. The effects of crowding and host plant nutrition varied among different colonies of planthoppers (Fig. 9.1) and it is likely that the switch between brachyptery and macroptery varies between different populations of *N. lugens*. For example, the proportion of brachypters was high in some areas where there was the cultivation of high-yielding varieties and asynchronous planting with high applications of fertilizers (e.g., Vietnam [V] and the Philippines [P_1], but low in others with similar intensive rice cultivation systems (e.g., Malaysia [MA_2]) (Fig 9.1).

Differences between Japan and the tropics in the proportions of brachypters and macropters in *N. lugens* and *S. furcifera* populations have been interpreted by some authors to indicate that migration from the tropics to Japan is unlikely (Nagata and Masuda 1980; Nagata 1982). Iwanaga et al. (1985, 1987) argue, however, that wing-morph variability among Japanese planthopper populations shows that immigrants to Japan can come from a variety of source areas, including the tropics, and that similar wing-morph variations in other parts of eastern Asia can be attributed to populations developing from planthoppers derived from different source areas.

Denno et al. (1991) concluded from a study of 35 planthopper species that not only did levels of migration increase as habitat persistence decreased but also that the switch from brachyptery to macroptery tended to occur at lower densities among populations inhabiting temporary habitats. Thus, the relationship between population density, plant quality, and wing-morph appears to be complex. Genetic factors are almost certainly involved and these vary in their influence throughout the distribution area of the species, manifesting themselves in effects on flight behavior and on the production of macropters.

Flight Capacity of Planthoppers

Age of Migrants

An important aspect of planthopper migration is the capacity to exploit newly colonized habitats by reproduction, as described by Johnson's (1969) oogenesis–flight syndrome of prereproductive migration. Reproduction is delayed in macropterous females of several planthopper species including *P. marginata*, *P. dolus* (Denno et al. 1989), and *J. pellucida* (Waloff 1973). Female *N. lugens*, *S. furcifera*, and *L. striatellus* caught on the East China Sea during the seasonal migration to Japan are sexually immature (Noda 1986a; Kisimoto unpublished data), as are *N. lugens* collected in net-traps or light-traps on land (Kisimoto 1976). Ovary development is delayed in macropterous *N. lugens* and *S. furcifera*, whereas brachypterous females are sexually mature soon after molting to adults and males respond to mating signals on the same day as the adult moult (Kisimoto 1965; Ohkubo 1967; Mochida 1970; Ichikawa 1979).

In general, macropterous *N. lugens* begin emigrating from host plants after a teneral period of 1 to 2 days (Ohkubo and Kisimoto 1971) and their flight willingness (indicated by the percentage of insects flying >0.5 hour in tethered flight experiments) reaches a maximum 80% on the third day after molting. The same level of flight willingness is maintained by males for the next 2 days.

Take-off Behavior

Light intensity appears to be the dominant factor for initiating take-off for many planthopper species, but in some species, this can be overridden by temperature if the latter falls below the threshold for flight. In both temperate and tropical regions, *N. lugens* and *S. furcifera* have a bimodal crepuscular take-off periodicity, with the former species showing a major take-off peak (75%) in the evening (Perfect and Cook 1982). In temperate areas, however, *N. lugens* take-off behavior can be modified in the late summer to late autumn by low temperatures at dawn and a single take-off peak occurs in the afternoon or at dusk (Ohkubo and Kisimoto 1971; Chen and Cheng 1980; Ye et al. 1981). Other Delphacids such as *S. furcifera*, *L. striatellus*, *S. vibex*, and *T. propinqua* also have a bimodal periodicity of take-off (Ohkubo and Kisimoto 1971; Kisimoto 1987). *J. pellucida* flight activity is enhanced when minimum and maximum temperatures are at or exceed 15 and 20°C, respectively (Waloff 1980) and this planthopper and several *Prokelisia* species in temperate areas have been recorded as migrating during the daytime (Raatikainen 1967; Denno and Roderick 1990). A single *P. marginata* has also been caught at night in a light-trap in the Gulf of Mexico (Sparks et al. 1986).

The relative advantages of taking-off at either dusk or during the day are not fully understood. It appears that for species which have evolved in tropical habitats, changes in light intensity which occur at dawn and dusk are the principal cues for synchronizing flight, whereas in temperate areas, warm daytime temperatures are necessary for flight. Padgham et al. (1987) suggested that long-distance migration in the tropics occurred primarily at night to avoid the high radiation levels experienced during daylight. There are definite advantages for insects to synchronize the initiation of flight prior to movement, including increasing the chances of finding a mate when a more suitable habitat is found.

Although planthoppers have relatively low airspeeds, they do not rely on ascending air currents to carry them aloft. Both *J. pellucida* and *N. lugens* have been observed actively leaving their host plants and flying vertically upward in calm conditions (Raatikainen 1967; Ohkubo 1981; Riley et al. 1987) and radar observations have shown that *N. lugens* in China sustain their ascent to heights of 700 m or more within 1 hour of take-off in the absence of updrafts (Riley et al. 1991). Take-off by *N. lugens* and *J. pellucida* can be inhibited by strong winds (Ohkubo and Kisimoto 1971; Chen and Cheng 1980; Waloff 1980; Ye et al. 1981) and may be confined to lulls between gusts in windy conditions (Riley et al. 1987). There is no evidence of *N. lugens* or *S. furcifera* selecting a particular wind direction for migration as has been postulated for the leafhopper *Empoasca fabae* (Taylor and Reling 1986). In the tropics, there is some indication from light-trap catches that *N. lugens* flight activity is enhanced at full moon (Jeffrey and Dyck 1983), although this periodicity in mass flights may also reflect a generation period which is similar to the lunar cycle.

Flight Performance

Three flight categories have been identified for macropters (Ohkubo: 1981); nonflight, medium, and long distance; in this section, we are concerned with the flight behavior of planthoppers in the medium- and long-distance groups because these are the individuals responsible for the seasonal changes in distribution of various species. Baker et al. (1980) demonstrated that in *N. lugens* longer fliers, produced more lift than short fliers, and in a series of tethered flight experiments, Ohkubo (1981) found that average flight duration ranged from 2 to 4 hours for *L. striatellus* males and females, respectively, 8 to 11 hours for *S. furcifera*, and 9 to 12 hours for *N. lugens*. The maximum duration of tethered flight recorded for *S. furcifera* was 32 hours (Ohkubo 1981), whereas for *N. lugens*, it varied from 11 to 23 hours (Baker et al. 1980; Ohkubo 1981; Padgham et al. 1987). These experiments also showed that laboratory-reared insects flew for shorter periods than field collected ones (Baker et al. 1980). Padgham et al. (1987), using field-collected *N. lugens*, showed that there was a diurnal variation in flight duration with evening flights lasting longer than those starting in the morning (average flight duration at dusk was 4.9 and 6.3 hours for males and females, respectively, compared to only 2 hours following a dawn take-off).

The increase in flight willingness in *N. lugens* on the third day after molting appears to be related to the deposition of flight fuels (Padgham 1983a). Flight duration is governed by carbohydrate and lipid reserves (Weis-Fogh 1952; Cockbain 1961; Chen 1983), the former being a readily available fuel source early in flight and the lipid being used to sustain flight. Lipid levels and body weight increase rapidly in *N. lugens* brachypterous females in the first 3 days after adult emergence, reflecting ovarian development, whereas the lower rate of lipid increase in macropters with no significant increase in body weight is probably associated with the deposition of lipid for flight (Padgham 1983a). In contrast, glycogen reserves were highest in the fifth instar stage and decreased at adult emergence. Body weight loss was 28% after a 8.5-hour flight for *N. lugens* (Baker et al. 1980) and 20% for a 5–6-hour flight (Ohkubo 1973), with a mean rate of lipid utilization in free flight of 0.0032 + (0.0069 × live weight) mg/hour (Padgham 1983b). Padgham (1983a, 1983b) confirmed that lipid was the most important fuel for sustaining *N. lugens* flight and suggested that the potential maximum flight times for Philippine populations ranged from 2 to 26 hours with a mode at 10–14 hours. A similar range of flight duration values was obtained from an analysis of catches on ships on the East China Sea in 1973 and 1981, in which trajectories representing the wind-borne displacement of planthoppers from the ships back to the nearest sources were used to derive flight times of 9 to 30 hrs (Rosenberg and Magor 1983a).

The ability to sustain continuous flight varies considerably among individuals reared on the same rice cultivar and among insects reared on different rice varieties (Padgham 1983a). Insects reared on IR20, a cultivar susceptible to *N. lugens* feeding, could sustain flight for up to 26 hours, whereas the maximum flight for insects reared on resistant IR36 was only 16 hours.

Flight is also influenced by atmospheric conditions, most notably temperature and humidity. In tethered flight experiments, most *N. lugens* macropters sustained wing beating when the temperature was 17°C, although a few maintained flight at 10°C (Ohkubo 1981). In the field, the maximum height at which *N. lugens* and *S. furcifera* have been caught over China and the East China Sea varies between 1.5 km in autumn, 2 km in the spring, and 2.5 km in summer (Dung 1981; Zhu et al. 1982; Hidaka 1989), reflecting seasonal variations in the vertical distribution of temperature in the planetary boundary layer. It appears from radar observations in China that migrating *N. lugens* do not fly in the warmest part of the vertical air-temperature profile (i.e., the top of the surface temperature inversion), but tend to actively climb at a rate of about 0.2 m/sec until a height is reached at which air temperatures approach those too low to sustain flight (Riley et al. 1991). Direct observations of mass flights of microinsects (principally *N. lugens*) in the autumn often revealed a distinct flight ceiling which ranged from 600 to 1300 m above ground, corresponding to an air temperature of about 16°C and reflecting the limiting effect of temperature on height of flight (Riley et al. 1991).

The effects of humidity on flight have not been so clearly demonstrated, although Ohkubo (1981) found in laboratory experiments that flights of *N. lugens* were longer under high humidity. It is reasonable to assume, however, that insects undertaking flights of several hours duration require certain humidity levels to prevent desiccation. Humidity certainly affects survival rates because macropters of *N. lugens*, *S. furcifera*, and *L. striatellus* lived only 1 day when deprived of water and food, whereas survival for 2 to 7 days was recorded when relative humidity was maintained at >90% after a 1.5-day feeding period (Hirao 1979a).

Flight capacity is clearly highly variable, but radar observations (Riley et al. 1987) and analysis of lipid reserves (Padgham 1983a) suggest that most planthoppers will land in the first few hours after take-off and that few would remain airborne for longer periods. This assumption is confirmed by Japanese studies which show a negative linear correlation between *S. furcifera* and *N. lugens* immigrant densities and distance northeastward from the west Kyushu coast (Kisimoto 1979).

Landing

The behavioral processes by which planthopper migration comes to an end are unknown. Many species of insects (e.g., locusts) make a succession of migratory flights before the migration phase ends, but it is unlikely that this is the case with planthoppers. Female *N. lugens* macropters increase body weight and lipid and glycogen levels after the end of prolonged flight (i.e., they then resemble the brachypterous form), suggesting that they are ready for reproduction and that further migration is unlikely (Padgham 1983a; Padgham et al. 1987). It is not clear to what extent small insects such as planthoppers actively control their landing other than to reduce lift and descend to their flight boundary layer (Pedgley 1982). More planthoppers appear to land in lower wind speeds because catches of *S. furcifera* and *N. lugens* on the East China Sea are negatively correlated with wind speed at approximately 500 m during periods of mass migration (Kisimoto 1991). The initial random distribution of colonizing *N. lugens* macropters in both temperate and tropical rice fields (Kisimoto 1965; Kuno 1968; Cook and Perfect 1985a) which leads to an aggregated spatial distribution in following generations suggests a random fallout at the end of migratory flight with only trivial movements occurring subsequently.

There is circumstantial evidence to suggest that local mesoscale meteorological features may lead to the concentration of planthoppers in specific areas. In Japan, putative landing sites for *N. lugens* have been identified in lee positions in valleys transverse to the prevailing wind direction and at valley heads in valleys parallel to the wind (Noda and Kiritani 1989), whereas in both China and Japan, landing of migrant *N. lugens* has been associated with rain belts in frontal systems and with descending air in cold fronts (Kisimoto 1976; Jiang et al. 1981, 1982; Hu 1983; Tan et al. 1984).

Seasonal Migrations

Seasonal changes in climate provide regular and generally predictable fluctuations over an annual time scale and are reflected in the changing suitability of habitats for reproduction and population growth over time. The most pronounced changes occur between winter and summer in the temperate zone and between wet and dry in the tropics, and seasonal movements can occur either within the temperate or tropical zones, or between the tropics and temperate areas (Dingle 1982). The presence of essentially "new" habitats in the spring which offer unexploited resources and reduced predation makes migration into the temperate zone from the tropics particularly worthwhile to those species able to do so (Dingle 1989). The distances traveled by seasonally migrating insects vary from a few meters to hundreds of kilometers, but the ecological and physiological patterns are similar, regardless of scale.

Flights of *P. marginata* in June and September along the Atlantic coast of North America are an example of intrazonal seasonal movements between optimal breeding and survival habitats. Population growth is favored in the nutritionally rich tall form of *S. alterniflora* which grows by streams in the spring and summer, whereas survival of overwintering nymphs is greater in the short-growth form of the grass which persists on high ground in the winter (Denno et al. 1980). Similar intrazonal movements, involving downwind flights of several kilometers between alternate hosts, have been recorded in the spring and early summer for *J. pellucida* in Finland and *L. striatellus* in Japan (Raatikainen 1967, Kisimoto 1969). In the tropical dry season, most *N. lugens*, *S. furcifera*, and other microinsects appear to migrate between rice crops over distances varying between 6 and 30 km after an evening take-off (Riley et al. 1987). This limited intrazonal movement is considered sufficient to ensure the insects survival between crops throughout the year in areas of asynchronous rice cultivation where rice at various stages of growth are present over relatively short distances. Although the insects undertaking these mesoscale (1 to 100 km) movements comprise the majority of the migrant population, it is likely that the small percentage of planthoppers reported in flight for longer periods are representative of migrants capable of sustaining flight for several hours and undertaking longer-distance migrations (Riley et al. 1987).

Most of the evidence for long-distance migration by planthoppers is circumstantial, consisting of the simultaneous arrival of several insects in an area and these arrivals coinciding with an airstream which had passed over at least one known source of the insects. The well-documented wind-borne migration of planthoppers to Japan and Korea in spring and early summer each year is part of the final three stages of an interzonal migration which extends the distribution areas of *N. lugens* and *S. furcifera* from the tropics to about 42 to 44°N (Fig. 9.2). Using trap catches on land and sea, and mark and recapture experiments, five northward migration waves have been identified in China between mid-April and early August (Fig. 9.2, C_1 to C_5), associated with the prevailing southerly and southwesterly winds of the summer monsoon (Cheng et al. 1979). The movements to Japan and Korea

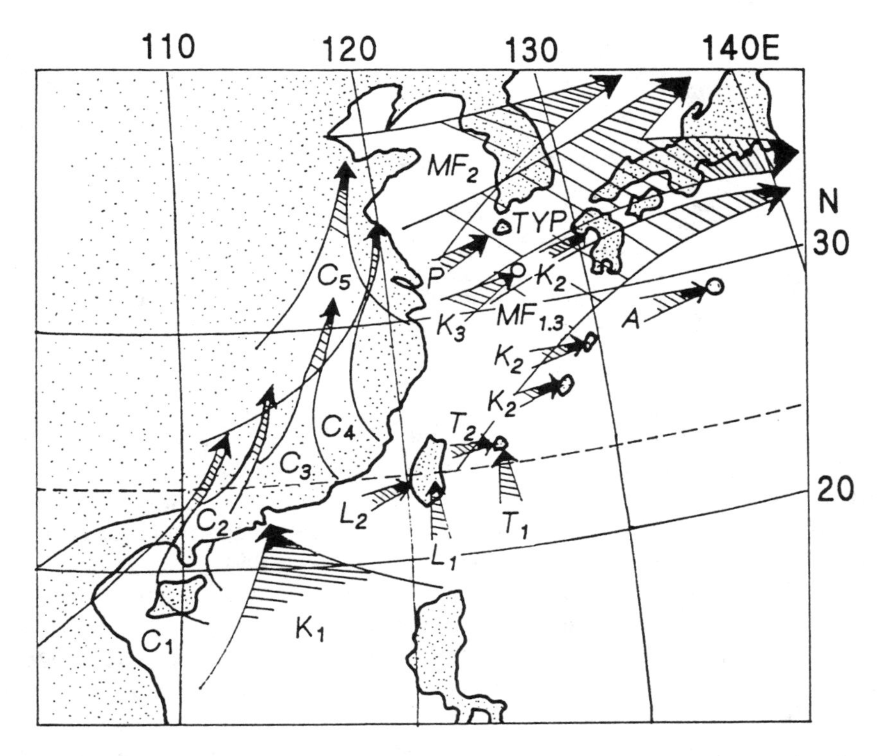

Figure 9.2. Schematic diagram of the northward migration of *N. lugens* and *S. furcifera* in eastern Asia. A = Tango (Asahina and Tsuruoka 1968); C_1–C_5 = China (Cheng et al. 1979); K_1 = Hong Kong (Kisimoto and Dyck 1976); K_2 = Okinawa, Amami and Chikugo (Kisimoto et al. 1982); K_3 = survey ship on East China Sea (Kisimoto 1991); L_1,L_2 = Taiwan (Liu 1985); $MF_{1\text{-}3}$ = tracks of frontal depressions associated with minor immigrations to Japan (Kisimoto 1976); P = Che-joo Island (Park 1973); T_1,T_2 = Ishigaki Island (Tsurumachi and Yasuda 1989); TYP = track of frontal depressions associated with mass immigrations to Japan (Kisimoto 1976).

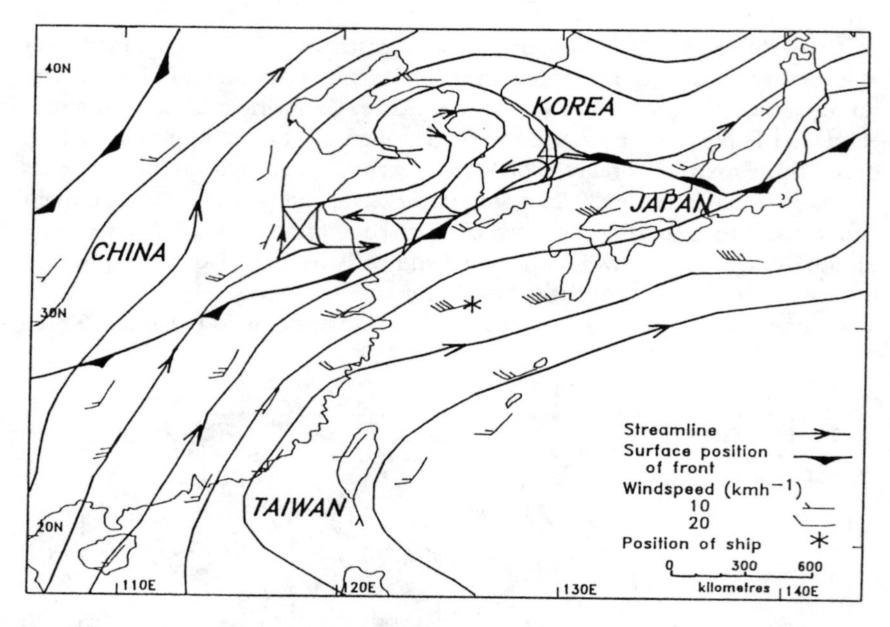

Figure 9.3. A 1.5-km windfield on 25 June 1981 during a mass immigration of *N. lugens* to Japan (Rosenberg and Magor 1984).

are mainly linked with strong southwesterly low-level jets occurring up to 1000 m above the surface in the warm sectors of eastward-moving frontal depressions (Fig. 9.2, $MF_{1,2,3}$ TYP; Fig. 9.3) (Asahina and Tsuruoka 1968; Satomi and Itakura 1970; Kisimoto 1971, 1976, 1979, 1991; Matsumoto et al. 1971; Seino et al. 1987; Rosenberg and Magor 1984).

Movements of *N. lugens* from the Philippines to Taiwan and southern China and between Taiwan and southern China have been postulated as part of the seasonal redistribution of the species (Kisimoto and Dyck 1976; Liu 1985; Rosenberg and Magor 1986)(Fig. 9.2, K_1, L_1,L_2). Evidence for long-distance migration by planthoppers in the tropics is less well documented than in subtropical and temperate areas, although the capture of planthoppers on the South China Sea and on ships near the Philippines (Saxena and Justo 1984; Rosenberg unpublished data), combined with the results of flight performance experiments on tropical populations (Baker et al. 1980; Padgham et al. 1987), suggest that *N. lugens,* at least, is equally mobile throughout its distribution range.

The occurrence of a return equatorward migration in the autumn is less well established, partly because of the difficulty in distinguishing immigrants from locally bred populations. In the late summer and autumn, the Asian summer monsoon is gradually replaced by the north and northeast winds of the winter monsoon and the single crop of more northerly rice-growing regions becomes unsuitable for feeding and reproduction as it matures and is harvested. Three southward migration waves have been identified in China between late August and late October (Cheng et al. 1979), and *N. lugens* and *S. furcifera* have been reported damaging the second rice crop in Okinawa in mid-September to early October after immigration on northerly and northwesterly winds (Tsurumachi and Yasuda 1989). Further southward movements to the tropics in October and November have been postulated from trap catches in Taiwan (Liu 1985). As with the summer immigration to Japan, emigrants in the autumn are associated with passage of frontal depressions in temperate areas (e.g., Asahina and Tsuruoka 1969, 1970; Itakura 1973). Under these meteorological conditions, Wada et al. (1987) found that some *N. lugens* continued to arrive in Japan from China in the autumn, although this movement is apparently nonadaptive because neither the migrants nor their offspring would survive the winter. In general, however, over most of Asia, the autumn and winter are marked by the prevalence of northeasterly winds which tend to carry insects in a southerly direction and this equatorward movement has been confirmed by direct radar observations of microinsects in flight night after night in the autumn in central China (Riley et al. 1991). At present, it is not clear what proportion of these migrants reach the overwintering areas of southern China and adjacent rice-growing regions.

Although long-distance migratory flights can be broadly divided into two categories, poleward in the spring and equatorward in the autumn (Dingle 1982), the resultant expansion and contraction of the insects' distributional area is the product of a series of movements in several directions. Modeling of potential wind-borne movements under different weather conditions which occur periodically in Asia has indicated that not all movements are either poleward or equatorward (Rosenberg and Magor 1983b). Wada et al. (1987) have shown that movements to the northeast occur when frontal depressions disturb the northeast monsoon in the autumn, whereas in summer, the passage of a cold front across Japan can lead to a southeastward migration of *S. furcifera* and *N. lugens* across Japan to the ocean weather station Tango in the Pacific (Asahina and Tsuruoka 1968; Kisimoto 1979) (Fig. 9.4).

A regular feature of many seasonal migrations both in Asia and North America is the presence of parasites and predators migrating with their hosts, suggesting a link between dispersal and parasite/predator–prey dy-

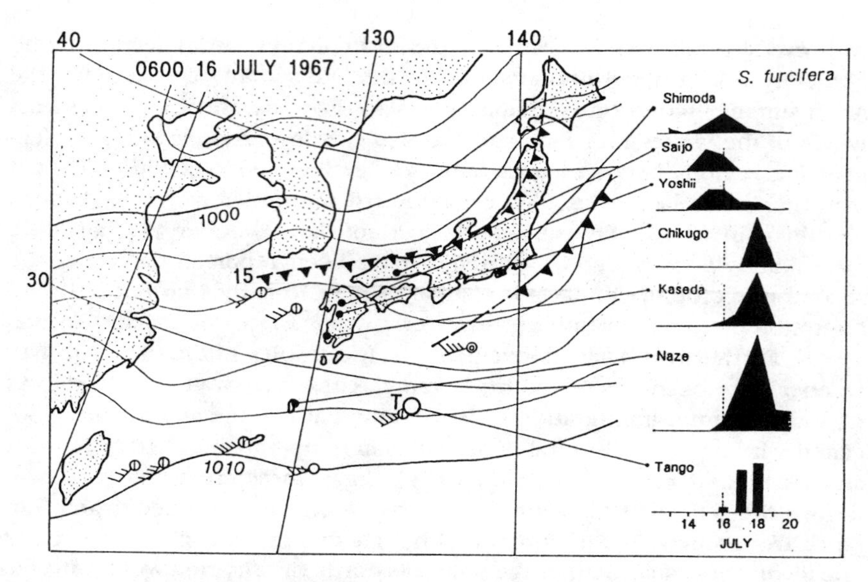

Figure 9.4. Light-trap catches of *S. furcifera* in Japan at Tango (T), associated with the passage of a southeastward-moving cold front (surface position of front shown for 15 and 16 July 1967) (Kisimoto 1979).

namics. *P. marginata* and the egg parasite *Anagrus delicatus* regularly disperse to offshore islands in the Gulf of Mexico (Antolin and Strong 1987), whereas in Asia, the predatory mirid, *Cyrtorhinus lividipennis*, is frequently captured in aerial nets (Riley et al. 1987, 1991), airplane traps (Dung 1981), and on ships at sea (Kisimoto 1979, 1981; Oya and Hirao 1982; Liu et al. 1983; Rosenberg unpublished data). These movements would tend to synchronize predator and prey population dynamics and have some stabilizing effect on planthopper populations.

Discussion

In this chapter, we have confined our consideration of migration to the factors, both biological and environmental, which lead to seasonal changes in planthopper distributions and are the product of the species own intrinsic flight capabilities and the prevailing winds at the time of dispersal. Once airborne, the displacements of microinsects such as planthoppers are largely determined by the windfields they encounter while in flight. Every airborne organism tends to fall to the ground because it is denser than air (Pedgley 1982), and even small insects must maintain flight by wing flapping or gliding to stay airborne. Although planthoppers have generally low airspeeds and tend to move with the wind, several complex behavioral mechanisms are involved in leaving and returning to the ground. The best studied species, *N. lugens*, does not conform to the notion of drifting, passively transported "aeroplankton" because it has been observed actively directing its flight upward after take-off and then moving with the wind at a height where the temperature allows sustained flight (Riley et al. 1991). Individuals then end their migration by ceasing wing flapping and returning to their flight boundary layer. There are occasions, however, when the movement takes the migrants to areas where habitat conditions are unsuitable for survival or population development (e.g., Asahina and Tsuruoka 1970; Wada et al. 1987), but, in general, over a period of time, a sufficient proportion of migrants are displaced by winds to favorable habitats.

The diversity of habitats inhabited by planthoppers is a reflection of their adaptability and ability to exploit and colonize (Denno and Roderick 1990). The emphasis on migration in this chapter should not disguise the fact that in many planthopper species, the most favored survival strategy is a complete absence of migration for most of the time and a promoted evolution of brachypterous generations (e.g., *P. dolus*, *N. muiri*). Several species of planthoppers are agricultural pests, however, because of their mobility and capacity for colonizing new and, sometimes, distant habitats and they exemplify a migration–life history strategy based on the "division of labor between different morphs in a polymorphic species" (Dingle 1980). Many planthoppers undergo regular migrations between habitats in which departure and return is by different individuals. The interhabitat migrations of several planthopper species including *J. pellucida*, *L. striatellus*, and *P. marginata* involve individuals of different generations migrating between population development sites and habitats where the population overwinters in diapause (Denno and Roderick 1990).

Much of the debate about whether the movements of "aeroplankton" or microinsects could be termed migration was centered on the existence of a return migration to sustain the genetic base for migrants in the permanent breeding areas. The strongly oriented flights of some butterflies indicated an ability to return, but this did not seem applicable to microinsects, which were seen as drifting at the mercy of the wind (Taylor 1985). Return migrations are difficult to substantiate because of the problem in distinguishing immigrants from locally produced populations. There has also been a less-concerted research effort in studying return movements because, in the short term, it is the annual expansion of the insects' distribution area which is important for agricultural protection agencies and not its contraction. Recent evidence from China and North America suggests, however, that the absence of return migrations is more apparent than real. Radar observations in China of the first return migration postulated by Cheng et al. (1979) clearly demonstrated that *N. lugens* macropters produced in the autumn in east central China were highly adapted to long-distance migration on the prevailing northeasterly winds toward the overwintering areas of the species (Riley et al. 1991). Taylor and Reling (1986) have also produced evidence suggesting that in North America the leafhopper *E. fabae* migrates toward their southern overwintering sites on northerly winds which occasionally replace the prevailing westerly winds. In both of these examples, the reported return migrations do not take the insects to their overwintering areas, but to intervening breeding sites where the progeny of the immigrants would be able to complete their development before migrating further south toward the permanent breeding areas (Taylor and Reling 1986; Riley et al. 1991). Similarly, the northward migration of *N. lugens* in the spring and summer takes place in a series of steps, the early stages displacing most migrants from the tropics to the subtropics, where subsequent generations infest the newly planted rice crop and, later in the season, produce migrants which undertake movements to temperate rice-growing areas. As Riley et al. (1991) concluded, further studies are needed to determine whether return migrations continue as far south as the overwintering areas, thereby establishing the importance of the returning migrants in maintaining the gene pool of long-distance fliers in the overwintering areas.

The success of many planthoppers both in terms of the range of host plants they inhabit and their status as agricultural pests suggests that they have evolved a successful migration strategy. Most of the evidence presented here is derived from a unique set of data concerning one species, *N. lugens*. The intensive research carried out on this single species could act as a model system for directing research efforts on other species of migratory planthopper.

Summary

1. Insect migration enables individuals to exploit spatially and temporally scattered habitats and many planthopper species have evolved traits such as wing-dimorphism and flight, which allow them to colonize and exploit transient habitats, persist in these while they are favorable, and escape when the habitat deteriorates.
2. Many planthopper species are composed of two wing forms, flightless brachypters and macropters that can fly, the proportion of which varies in time and in space according to environmental and genetic interactions.

3. Reproduction is suppressed in female macropters and dispersal usually begins after a 1–2-day teneral period. Take-off is largely controlled by light intensity and temperature and is an active behavioral response by the insect, involving movement to the top of the host, and flight from the plant.
4. Not all macropters are migrants, however, but among those that do fly, more lift is produced by long fliers than by short fliers. The ability to sustain flight also varies among individuals reared on the same host between insects reared on different varieties of the same host, and between insects reared on plants of different ages.
5. Seasonal changes in distribution reflect the changing suitability of habitats. In some tropical species of planthoppers, the distribution range expands into temperate areas in summer and contracts again during winter. There are also interhabitat migrations of some species in spring and autumn involving different generations alternating between reproduction sites and overwintering sites.
6. A common feature of many migrations is the presence of parasites and predators with their hosts which has led to suggestions of a link between migration and parasite/predator–prey population dynamics.
7. Because planthoppers possess relatively low airspeeds, the distances moved are determined by the windfields in which the insects are flying and the length of time that they can remain airborne through wing flapping and gliding on updrafts.
8. Radar observations of mass migrations by the rice brown planthopper *Nilaparvata lugens* have demonstrated that the height of flight is limited by temperature. This is consistent with trapping from aircraft which shows that insects are caught at greater heights in the summer than in the autumn. Certain humidity levels are considered necessary to prevent desiccation during flights of several hours duration.
9. The best documented long-distance migrations are those by the rice pests *N. lugens* and *S. furcifera* in southeastern Asia. Five northward migration waves in the spring and early summer have been identified between the tropics and temperate rice-growing areas. Other migrations within the tropics have been postulated and all these movements have been associated with seasonal changes in the prevailing windfields and the occurrence of specific synoptic weather patterns.
10. There is a growing body of evidence supporting the occurrence of return migrations to overwintering areas in the autumn in both southeastern Asia and North America, but further work is needed to assess the importance of return migrants in maintaining the genes for long-

distance migration in the permanent breeding areas. It is suggested that *N. lugens* could be used as a model for research on other migratory species.

ACKNOWLEDGMENT

The authors wish to thank Dr. D.R. Reynolds for his helpful comments and suggestions during the preparation of this chapter.

PART FOUR

Species Interactions and Community Structure

In addition to direct density effects, predators, parasitoids, and pathogens may also contribute to planthopper population change. Besides the theoretical considerations of predator/prey and parasitoid/host interactions, such as population "regulation" and "stability," an equally important and closely related concern is the ability of natural enemies to suppress and maintain pest populations below economically acceptable density thresholds. We have chosen to group in Part Four those chapters whose first concern is a theoretical treatment of species interactions, though points relevant to pest suppression and management are emphasized throughout. Issues in biological control and the compatibility of natural enemies and resistant varieties as management strategies are more explicitly addressed in Chapters 14–18.

Part Four begins with an examination of "general" predators and their ability to suppress planthopper populations in both natural and agricultural settings. Döbel and Denno (Chapter 10) identify the important predators of planthoppers, survey their functional and numerical responses, and discuss their implications for prey suppression and population stability. The life history traits of predators and planthopper prey are contrasted in the context of predator's ability to track spatial and temporal changes in planthopper populations. Factors which uncouple and destabilize predator/planthopper interactions and promote outbreaks are stressed. Last, management options are considered including the use of predator complexes.

The focus then switches to parasitoids in Chapter 11 where Cronin and Strong review interactions between these natural enemies and their planthopper and leafhopper hosts. Parasitoid searching behavior and spatial and temporal patterns of parasitism with respect to host density are examined. These data are then used to assess the importance of parasitoids in the regulation, stability, and persistence of parasitoid/host populations.

From planthoppers as victims of predation and parasitism, the emphasis shifts to species interactions characterized by planthoppers as vectors of plant pathogens. In Chapter 12, Nault concentrates on plant viruses and their transmission biology and vector–pathogen relationship with planthoppers. Characteristics of planthopper-vectored viruses are reviewed as are mechanisms of transmission. The origin and evolution of planthopper-vectored viruses is conjectured and their pathogenicity to planthoppers is addressed. He concludes with a discussion of the characteristics of some plant viruses that predispose them for transmission by delphacids.

Because planthoppers co-occur with a variety of other herbivorous insects, it is impossible to single out delphacids in a discussion of community structure. Nevertheless, planthoppers along with leafhoppers (Cicadellidae) are often the most abundant herbivores by far in grassland habitats. For this reason there is strong justification and precedent for considering these taxa together and the forces which structure the Auchenorrhyncha communities on grasses, a tack taken by Stiling in Chapter 13. Factors which influence the number of hopper species utilizing a particular host plant, the likelihood for interspecific competition between hopper species, and mechanisms for coexistence are considered. The number of parasitoids per host and the frequency of attack by egg and nymphal parasitoids are examined in the context of biological control.

10

Predator–Planthopper Interactions

Hartmut G. Döbel and Robert F. Denno

Introduction

Parasitoids have been selected often over predators in classical biological control programs because they are generally more host-specific and possess life history traits such as high fecundity, high larval survival, high searching efficiency, short handling time, and low mutual interference, which are thought to be more conducive to pest suppression (see Waage and Hassell 1982; Greathead 1986; Waage 1990). Invertebrate predators, due to their presumed polyphagous feeding habit (Rothschild 1966; Waage 1990), generally lower reproductive potential (Murdoch et al. 1985), variable ability to effect a numerical response (Renault and Miller 1972), tendency toward cannibalism (Heong and Rubia 1990b), and Type II functional responses (Holling 1966; Nakamura 1977), have been considered less attractive candidates as control agents (Huffaker et al. 1977; Greathead 1986). Nevertheless, there is growing awareness and evidence for the importance of invertebrate predators in the biological control and population dynamics of herbivorous insects (Murdoch 1975; Ehler 1977, 1990; Ehler and Miller 1978; Luff 1983; Hanks and Denno 1993).

In part, the greater attention to predators may be attributable to the erosion of classical biological control and population dynamics theory which favored parasitoids for reasons including aggregation, the density-dependent suppression of hosts, and stability (see Comins and Hassell 1979; Waage and Hassell 1982; Murdoch 1990; Waage 1990). The failure to find widespread evidence for positive spatial density-dependent mortality (PSDD) in parasitoids (Morrison and Strong 1980; Stiling 1987; Walde and Murdoch 1988), the observation that PSDD may not be perceived at all levels of patchiness (Heads and Lawton 1983), and the realization that inverse density-dependent parasitism may also stabilize parasitoid–host systems (Hassell 1985b; Chesson and Murdoch 1986) have all contributed to the controversy over the criteria for the selection of biological control agents (Waage 1990). Some have even argued that an aggregating biological control agent may not be the kind that maximizes pest suppression (Murdoch et al. 1985), and that aggregation as an independent criterion should not be promoted in the choice of a control agent (Waage 1990). Perhaps now the door has been opened for a more serious consideration of general predators, which historically have been overshadowed by parasitoids in pest management programs and during the process of selecting biological control agents. Nonetheless, there is extensive agreement that ecological theory, albeit evolving, has general relevance to the selection process used in biological control (Murdoch et al. 1984; Greathead 1986; Kareiva 1990; Murdoch 1990; Waage 1990).

Invertebrate predators are very abundant and often conspicuous in many of the agricultural as well as natural habitats occupied by planthoppers (Rothschild 1966; Yasumatsu and Torii 1968; Chiu 1979; Kiritani 1979; Waloff 1980; Greathead 1983; Döbel 1987; Ooi 1988; Manti 1989; Throckmorton 1989; Denno and Roderick 1990; Döbel et al. 1990). Moreover, there is almost unanimous agreement that predation is a much more important source of mortality than is parasitism in planthopper populations (Hinckley 1963; Rothschild 1966; Napompeth 1973; Otake 1977; Chiu 1979; Bull 1981; Kenmore 1980; Waloff and Thompson 1980; Denno and Roderick 1990). Our general objective in this chapter is to elucidate which invertebrate taxa are the important predators of planthoppers, to determine the extent to which predators are able to suppress planthopper populations in both agricultural and natural settings, and to identify the ecological characteristics of effective predators.

Murdoch (1990) has outlined several features of natural enemies that are likely to enhance the control and suppression of prey. These include (1) aggregation in response to local host prey density (provided the enemies can rearrange themselves in response to a changing pest distribution), (2) a short generation time of the enemy compared to that of the pest, (3) enemy dispersal greater than that of the pest in a spatially heterogeneous environment, and (4) complementary enemies (species that attack different stages of the pest, or species that attack at different times, so avoiding temporal refuges) (Murdoch 1990). Furthermore, Ehler (1990) has argued that temporal persistence, such that an enemy can maintain a population when preferred prey are rare, is an important consideration in biological control. We use these criteria as a framework for exploring the characteristics of predators and predator complexes which may lead to the suppression of planthopper populations.

The information we use to assess predator–planthopper interactions derives from a variety of sources including the ecological literature on the experimental manipulation and removal of predators (e.g., Kenmore 1980; Döbel 1987; Ooi 1988; Manti 1989; Döbel and Denno in preparation b, c), the entomological literature on planthopper resurgence following pesticide applications (e.g., Heinrichs et al. 1982a; Reissig et al. 1982b; Heinrichs and Mochida 1984), the biological control literature on planthoppers (e.g., Swezey 1936; Fullaway 1940; Matsumoto and Nishida 1966), and the pest management literature on the delphacid pests of agricultural crops (e.g., Raatikainen 1967; Kiritani et al. 1971; Napompeth 1973; Greathead 1983; Ooi 1988; Manti 1989). Specifically, we investigate the functional and numerical (both aggregative and reproductive components) responses of predators and discuss their implications for population suppression and stability. Also, we examine the dietary breadth and feeding preferences of planthopper predators, and the relative fecundity, generation time, and dispersal capabilities of planthoppers and their predators. Further, we explore factors that uncouple or destabilize predator–planthopper interactions and promote outbreaks. We also discuss how planthopper habitats might be managed to maximize refuges for enemies and synchronize predator–planthopper interactions. Last, we emphasize how complexes of predators attacking different stages or at different times might better suppress planthopper populations.

Major Invertebrate Predators of Delphacid Planthoppers

Even though a wide diversity of invertebrate predators (Araneae, Phalangida, Acarina, and nine orders of Insecta Collembola, Odonata, Orthoptera, Dermaptera, Hemiptera, Neuroptera, Coleoptera, Diptera, and Hymenoptera) are known to attack and feed on delphacid planthoppers (Swezey 1936; Rothschild 1966; Ôtake 1977; Chiu 1979; Waloff 1980), only a few predator taxa appear to play an important role in planthopper population dynamics and/or pest suppression (Napompeth 1973; Kenmore et al. 1984; Döbel 1987; Ooi 1988). Based on our survey of the literature (Table 10.1), spiders, hemipteran predators in the families Miridae and Nabidae, and occasionally coccinellid beetles and phalangids were considered to be the most important predators of delphacids in both natural (e.g., Rothschild 1966; Waloff 1980; Waloff and Thompson 1980; Döbel 1987; Döbel et al. 1990) and agricultural systems (e.g., Swezey 1936; Metcalfe 1971; Kiritani et al. 1972; Napompeth 1973; Kenmore 1980; Ooi 1982, 1988). Our criteria for determining the "importance" of particular predator species included experimental evidence for the suppression of planthopper populations,

Table 10.1. Important invertebrate predators feeding on delphacid planthoppers in agricultural and natural ecosystems. The importance of predator taxa was assessed from the literature on the basis of (1) experimental evidence for the suppression of planthopper populations and functional and numerical responses, (2) correlations between predator and planthopper density in field populations, and (3) author judgment.

Predator Taxon	Planthopper Prey	Stage Attacked[a]	Host Plant (Location)	Evidence for Predator Importance			Reference
				Exp[b]	Cor[c]	Author[d]	
ARANEAE							
Lycosidae							
Lycosa modesta	*Prokelisia dolus* and	(N,A)	*Spartina alterniflora* (USA)	C,F,FR	+[c]		Döbel (1987)
	Prokelisia marginata	(N,A)				X	Döbel et al. (1990)
				FR			Döbel and Denno (in preparation a)
				R,NR	+[c]		Döbel and Denno (in preparation c)
Lycosa pullata	*Conomelus anceps*	(N,A)	*Juncus effusus* (UK)	PR		X	Rothschild (1966)
Pagiopalpus atomarius	*Perkinsiella saccharicida*	(N,A,)	*Saccharum officinarum* (USA)			X	Swezey (1936)
Pardosa floridana	*Prokelisia marginata*[e]	(N,A,)	*Spartina alterniflora* (USA)			X	Vince et al. (1981)
Pardosa laura	*Sogatella furcifera* and	(N,A)	*Oryza sativa* (Japan)	FR			Nakamura (1977)
	Laodelphax striatellus	(N,A)					

Pardosa littoralis	*Delphacodes detecta* and *Tumidagena minuta*	(N,A) (N,A)	*Spartina patens* (USA)	PP		X	Raupp and Denno (1979); Döbel (1987); Döbel et al. (1990)
	Prokelisia dolus and *Prokelisia marginata*	(N,A) (N,A)	*Spartina alterniflora* (USA)	C,F,FR	+[c]		Döbel (1987)
						X	Döbel et al. (1990)
				FR			Döbel and Denno (in preparation a)
				R,NR	+[c]		Döbel and Denno (in preparation c)
Pardosa pseudo-annulata	*Nilaparvata lugens*	(N,A)	*Oryza sativa* (Japan)	I,S			Kiritani et al. (1972)
		(N,A)	*Oryza sativa* (Philippines)			X	Thomas et al. (1979)
				C,I	+[c]	X	Kenmore (1980)
				C,I		X	Dyck and Orlido (1977)
						X	Barrion et al. (1981)
					ns[f]	X	Kuno and Dyck (1985)
				FR			Heong and Rubia (1989, 1990b)
				FR		X	Heong et al. (1991b)
		(N,A)	*Oryza sativa* (Malaysia)		+[c]		Ooi and Shepard (in press)
					+[c]	X	Heong et al. (1990a)
				FR	ns[f]	X	Ooi (1988)
					ns[f]	X	Ooi (1982)
	Nilaparvata lugens and *Sogatella furcifera*	(N,A) (N,A)	*Oryza sativa* (Malaysia) *Oryza sativa* (China)			X	Chiu (1984)

(*continued*)

Table 10.1. (*Continued*)

Predator Taxon	Planthopper Prey	Stage Attacked[a]	Host Plant (Location)	Evidence for Predator Importance: Exp[b]	Cor[c]	Author[d]	Reference
Pardosa T-insignita	Nilaparvata lugens and	(N,A)	Oryza sativa (Korea)	PP		X	Lee and Park (1977)
	Sogatella furcifera	(N,A)					
Pirata subpiraticus	Nilaparvata lugens	(N,A)	Oryza sativa (Korea)	PP,I			Lee and Park (1977)
Pirata sp.	Nilaparvata lugens	(N,A)	Oryza sativa (Japan)	S		X	Kobayashi (1975)
			Oryza sativa (China)			X	Chiu (1984)
Linyphiidae							
Atypena formosana	Nilaparvata lugens and Sogatella furcifera	(N,A)	Oryza sativa (Philippines)		+[c]	X	Reddy and Heong (1991)
Erigonidium graminicolum	Nilaparvata lugens and Sogatella furcifera	(N,A)	Oryza sativa (China)			X	Chiu (1984)
Gnathonarium dentatum	Nilaparvata lugens	(N,A)	Oryza sativa (Korea)	PP,I			Lee and Park (1977)
Grammonota inornata	Prokelisia marginata[e]	(N,A)	Spartina alterniflora (USA)	PP		X	Vince et al. (1981)
Lepthyphantes tenuis	Javesella pellucida	(N,A)	Grasses (UK)		+[f]		Waloff (1980)
	Conomelus anceps	(N,A)	Juncus effusus (UK)	PR		X	Rothschild (1966)
Linyphia clathrata	Conomelus anceps	(N,A)	Juncus effusus (UK)	PR		X	Rothschild (1966)
Linyphia triangularis	Conomelus anceps	(N,A)	Juncus effusus (UK)	PR		X	Rothschild (1966)

Tetragnathidae							
Tetragnatha laboriosa	Prokelisia marginata	(N,A)	Spartina alterniflora (USA)	C		X	Throckmorton (1989)
Tetragnatha mandibulata	Perkinsiella saccharicida	(N,A)	Saccharum officinarum (USA)			X	Swezey (1936)
	Nilaparvata lugens	(N,A)	Oryza sativa (Phillippines)			X	Barrion et al. (1981)
Tetragnatha sp.	Nilaparvata lugens	(N,A)	Oryza sativa (SE Asia)	PP			Ôtake and Hokyo (1976)
Micryphantidae							
Oedothorax insecticeps	Nilaparvata lugens	(N,A)	Oryza sativa (Philippines)	C,I		X	Kenmore (1980)
Salticidae	Perkinsiella saccharicida	(N,A)	S. officinarum (Mauritius)			X	Williams (1957)
Araneidae	Laodelphax striatellus	(N,A)	Oryza sativa (Japan)			X	Kobayashi and Hiwada (1968)
"Spiders"	Prokelisia marginata[e]	(N,A)	Spartina alterniflora (USA)	F	–[c], ns[f]		Vince et al. (1981)
	Nilaparvata lugens	(N,A)	Oryza sativa (SE Asia)			X	Chiu (1979)
			Oryza sativa (Sri Lanka)	C		X	Fowler (1987)
			Oryza sativa (Malaysia)			X	Ng (1978)

(continued)

Table 10.1. (*Continued*)

Predator Taxon	Planthopper Prey	Stage Attacked[a]	Host Plant (Location)	Evidence for Predator Importance			Reference
				Exp[b]	Cor[c]	Author[d]	
	Sogatella furcifera	(N,A)	*Oryza sativa* (Philippines)	C,I	+[f]	X	Manti (1989)
	Prokelisia marginata	(N,A)	*Spartina alterniflora* (USA)			X	Pfeiffer and Wiegert (1981)
	Dicranotropis hamata	(N,A)	*Holcus mollis* (UK)			X	Waloff and Thompson (1980)
	Peregrinus maidis	(N,A)	*Zea mays* (Hawaii)			X	Napompeth (1973)
	Javesella pellucida	(N,A)	Cereal crops			X	Raatikainen (1967)
	Stenocranus minutus	(N,A)	*Dactylis glomerata*			X	Waloff and Thompson (1980)
PHALANGIDA							
Phalangidae							
Leiobunum blackwalli	*Conomelus anceps*	(N,A)	*Juncus effusus* (UK)	PR		X	Rothschild (1966)
Oligolophus agrestis	*Conomelus anceps*	(N,A)	*Juncus effusus* (UK)	PR		X	Rothschild (1966)
INSECTA							
HEMIPTERA							
Miridae							
Cyrtorhinus fulvus	*Tarophagus colocasiae*	(E)	*Colocasia esculenta* (Hawaii)	BC,R	+[f]		Matsumoto and Nishida (1966)

Cyrtorbinus lividipennis	*Nilaparvata lugens*	(E,N)	*Oryza sativa* (Japan)			X	Suenaga (1963)
			Oryza sativa (Philippines)	C,I	$+^{f}$	X	Dyck and Orlido (1977)
					$+^{c}$, $+^{f}$	X	Cook and Perfect (1985a)
				C,FR	$+^{f}$	X	Manti (1989)
				FR			Manti and Shepard (1990)
					$+^{c}$	X	Heong et al. (1990a)
				FR		X	Heong et al. (1990b)
			Oryza sativa (Japan/Philippines)		$+^{f}$	X	Kuno and Dyck (1985)
			Oryza sativa (Malaysia)	C,I	$+^{f}$	X	Ooi and Shepard (in press)
				C,I,FR	$+^{f}$	X	Ooi (1988)
				FR			Sivapragasam and Asma (1985)
						X	Ng (1978)
			Oryza sativa (India)			X	Bentur and Kalode (1987)
			Oryza sativa (China)	FR			Qingcai and Jervis (1988)
			Oryza sativa (SE Asia)			X	Chiu (1979)
			Oryza sativa (Solomon Is.)	PP		X	Stapley (1976)
			Oryza sativa (Fiji)	I		X	Hinckley (1963)

(*continued*)

Table 10.1. (*Continued*)

Predator Taxon	Planthopper Prey	Stage Attacked[a]	Host Plant (Location)	Evidence for Predator Importance			Reference
				Exp[b]	Cor[c]	Author[d]	
			Oryza sativa (Indonesia)		+[c]	X	Ôtake and Hokyo (1976)
	Sogatella furcifera	(E)	*Oryza sativa* (Malaysia)		ns[f]	X	Ooi (1982)
		(E)	*Oryza sativa* (Philippines)		+[c], +[f]	X	Cook and Perfect (1985a)
				FR	+[f]	X	Manti (1989)
	Sogatella kolophon	(E)	Pasture grass (Fiji)			X	Hinckley (1963)
	Peregrinus maidis	(E)	*Zea mays* (Hawaii)		+[c], +[f]	X	Napompeth (1973)
Tytthus alboornatus	*Delphacodes detecta*	(E)	*Distichilis spicata* (USA)			X	Tallamy and Denno (1979)
			Spartina patens (USA)	PP		X	Raupp and Denno (1979)
Tytthus chinensis	*Perkinsiella vitiensis*	(E)	*S. officinarum* (Fiji)			X	Hinckley (1963)
Tytthus mundulus	*Perkinsiella saccharicida*	(E)	*S. officinarum* (Hawaii)	BC		X	Swezey (1936)
				BC		X	Zimmerman (1948)
			S. officinarum (Australia)			X	Osborn (1974)
					+[c]	X	Bull (1981)
	Peregrinus maidis	(E)	*Zea mays* (Hawaii)			X	Zimmerman (1948)
					ns[f]	X	Napompeth (1973)

Tytthus parviceps	*Saccharosydne saccharivora*	(E)	*S. officinarum* (Jamaica)			X	Metcalfe (1971)
	Perkinsiella saccharicida	(E)	*S. officinarum* (Jamaica)		$+^{c}$	X	Bull (1931)
Tytthus pygmaeus	*Conomelus anceps*	(E,N)	*Juncus effusus* (UK)	PR		X	Rothschild (1966)
Tyrrhus vagus	*Prokelisia marginata*	(E)	*Spartina alterniflora* (USA)			X	Denno (1983)
Nabidae							
Dolichonabis limbatus	*Conomelus anceps*	(N,A)	*Juncus effusus* (UK)	PR		X	Rothschild (1966)
Nabis capsiformis	*Laodelphax striatellus*	(N)	*Zea mays* (Israel)			X	Harpaz (1972)
	Perkinsiella saccharicida	(E)	*S. officinarum* (Hawaii)			X	Swezey (1936)
Nabis flavo marginatus	*Conomelus anceps*	(N,A)	*Juncus effusus* (UK)	PR		X	Rothschild (1966)
Veliidae							
Mierovelia douglasi (including ssp. *atrolineata*)	*Nilaparvata lugens*	(N,A)	*Oryza sativa* (Philippines)		$+^{c}$, $+^{f}$	X	Cook and Perfect (1985a)
					$+^{c}$	X	Heong et al. (1990a)
				C,I		X	Kenmore (1980)
					ns^{f}	X	Kuno and Dyck (1985)
				FR		X	Nakasuji and Dyck (1984)
			Oryza sativa (SE Asia)			X	Chiu (1979)

(*continued*)

Table 10.1. (*Continued*)

Predator Taxon	Planthopper Prey	Stage Attacked[a]	Host Plant (Location)	Evidence for Predator Importance			Reference
				Exp[b]	Cor[c]	Author[d]	
Microvelia sp.	*Nilaparvata lugens*	(N,A)	*Oryza sativa* (Malaysia)		ns[f]	X	Ooi (1982)
Mesoveliidae							
Mesovelia vittigera	*Nilaparvata lugens*	(N,A)	*Oryza sativa* (Philippines)		+[c]	X	Heong et al. (1990a)
COLEOPTERA							
Coccinellidae							
Brumoides suturalis	*Sogatella furcifera*	(N,A)	*Oryza sativa* (India)			X	Garg and Sethi (1983)
	Peregrinus maidis	(N,A)	*Sorghum halepense* (India)			X	Fisk et al. (1981)
Casnoidea interstitialis	*Nilaparvata lugens*	(N,A)	*Oryza sativa* (Malaysia)			X	Lim et al. (1978); Ooi (1988)
Ceratomegilla maculata	*Tagosodes orizicolus*	(N,A)	*Oryza sativa* (Central America)			X	King and Saunders (1984)
Coccinella arcuata	*Nilaparvata lugens*	(N,A)	*Oryza sativa* (India)			X	Abraham et al. (1973)
						X	Abraham and Mathew (1975)
					+[c]		Israel and Prakasa Rao (1968)
Coccinella septempunctata	*Laodelphax striatellus*	(N)	*Oryza sativa* (Russia)			X	Harpaz (1972)
	Conomelus anceps	(N)	*Juncus effusus* (UK)	PR		X	Rothschild (1966)

Coelophora inaequalis	*Perkinsiella saccharicida*	(N)	*S. officinarum* (Hawaii)		X	Swezey (1936)
Illeis indica	*Peregrinus maidis*	(N,A)	*Sorghum bicolor* (India)		X	Fisk et al. (1981)
Micraspis discolor	*Nilaparvata lugens*	(N,A)	*Oryza sativa* (Thailand)	PP		Yasumatsu et al. (1975)
			Oryza sativa (Malaysia)	PP		Ôtake and Hokyo (1976)
Micraspis vincta	*Nilaparvata lugens*	(N,A)	*Oryza sativa* (Thailand)	PP		Yasumatsu et al. (1975)
Synharmonia octomaculata	*Nilaparvata lugens*		*Oryza sativa* (Malaysia)		X	Lim et al. (1978); Ooi (1988)
Staphylinidae						
Paederus fuscipes	*Nilaparvata lugens*	(N,A)	*Oryza sativa* (Malaysia)		X	Lim et al. (1978); Ooi (1988)

[a]E = egg; N = nymph; A = adult.
[b]Experimental evidence: BC = effective biological control; C = suppression of planthopper populations caged with predators; F = numerical response of predator to increasing prey density brought about by host plant fertilization; FR = functional response of predator to increasing planthopper density in laboratory; NR = numerical response of predator to manipulated planthopper density in the field; I = planthopper resurgence following predator kill from insecticide application; PP = high predator-to-prey ratio in field populations; PR = evidence for predation from precipitin test; R = predator removal followed by increase in planthopper population in the field; S = increase in predator population and decrease in planthopper population following prey supplementation in the field.
[c]Published correlation between predator and planthopper density in the field; ns = not statistically significant; + = significant positive correlation; − = significant negative correlation.
[d]Author concluded predator to be an important source of planthopper mortality: based on survey data, feeding observations, field experiments (e.g., predator exclusion), and/or high predator–prey ratios.
[e]Research conducted prior to the description of *Prokelisia dolus* (Wilson 1982a); planthopper populations at this location consist primarily of *Prokelisia dolus* and not *P. marginata* (Denno and Roderick unpublished data).
[f]Our correlation analysis and/or reanalysis of published data: ns = not statistically significant; + = significant positive correlation; − = significant negative correlation: no serial autocorrelations were found using the Durbin–Watson statistic (Draper and Smith 1981).

strong functional and numerical responses to increases in planthopper density, significant correlations between predator and planthopper density in field populations, high feeding rates on planthoppers, and the judgement, of the author (Table 10.1).

Of the 17 species of delphacids for which information was available, spiders were considered to be major predators of 13 species and were identified as the most important predator for 9 species of planthoppers (Table 10.2). Both hunting spiders (e.g., *Lycosa* and *Pardosa* in the Lycosidae) and web-builders (Linyphiidae and Tetragnathidae) were regarded as important predators (Table 10.1). Habitats in which spiders inflicted heavy mortality on planthoppers included terrestrial grasslands (Waloff 1980) as well as flooded agricultural crops (rice) and marshlands (Kenmore 1980; Ooi 1988; Döbel 1987).

There is a perception in much of the planthopper literature that due to their general feeding habits spiders may not be as effective predators as more specialized feeders (Swezey 1936; Williams 1957; Hinckley 1963; Matsumoto and Nishida 1966; Rothschild 1966). In fact, some spider species may have a more restricted diet than was once thought. For example, in the salt marshes along the Atlantic coast of North America, wolf spiders are the most abundant predators and 60% of the prey caught by *Pardosa littoralis* and 65% of the victims of *Lycosa modesta* were *Prokelisia* planthoppers (Döbel 1987). Dietary specialization in these two marsh-inhabiting wolf spiders apparently reflects a feeding preference or high capture efficiency for certain delphacid prey and is not singularly a reflection of relative prey abundance (Döbel 1987; Döbel and Denno in preparation a). However, in the rice fields of Japan, *Nilaparvata lugens* and *Nephotettix* leafhoppers formed 25% and 53% of the diet of *Pardosa* (=*Lycosa*) *pseudoannulata* and the proportion of *N. lugens* in the spider's diet varied according to its relative abundance in the field (Kiritani et al. 1972).

Hemipterans in the genera *Cyrtorhinus* and *Tytthus* (Miridae) and bugs in the Nabidae were cited as the most effective predators for seven species of planthoppers (Table 10.2). For example, *C. lividipennis* is a voracious predator on the eggs of *N. lugens* in rice fields throughout Asia and the Pacific islands (Hinckley 1963; Stapley 1976; Kuno and Dyck 1985; Ooi 1988; Manti 1989; Heong et al. 1990a), and the introduction of *T. mundulus* into Hawaii resulted in the effective biological control of *Perkinsiella saccharicida* on sugarcane (Swezey 1936; Zimmerman 1948). Similarly, populations of *Tarophagus colocasiae* declined dramatically following the introduction of *C. fulvus* into the taro fields of Hawaii (Matsumoto and Nishida 1966) (see Asche and Wilson 1989b for taxonomic revision of *Tarophagus*). Like spiders, hemipterans appear to be influential predators in both terrestrial and semiaquatic habitats (Hinckley 1963; Napompeth 1973; Waloff 1980; Ooi 1988).

Both hemipterans and spiders were mentioned as important predators for 12 species of planthoppers including *Dicranotropus hamata, Nilaparvata lugens, Peregrinus maidis, Perkinsiella saccharicida*, and *Sogatella furcifera* (Swezey 1936; Napompeth 1973; Waloff 1980; Kenmore 1980; Ooi 1982; Cook and Perfect 1985a; Table 10.2). Because mirids feed primarily on eggs and/or early instars of planthoppers and spiders prey mainly on nymphs and adults (Table 10.1), these two groups of predators have the potential to augment each other's impact on planthopper populations. However, this potential may be offset to some extent because wolf spiders are known to capture mirid egg predators (Hinckley 1963; Heong et al. 1989).

Cyrtorhinus and *Tytthus* are relatively specialized in their dietary habits, feeding primarily on the eggs and young nymphs of Delphacidae and Cicadellidae (Swezey 1936; Hinckley 1963; Matsumoto and Nishida 1966; Rothschild 1966; Napompeth 1973; Greathead 1983; Bentur and Kalode 1987; Manti 1989). In the few cases in which predation on delphacid and cicadellid eggs has been compared, delphacid eggs (*Nilaparvata lugens* and *Sogatella furcifera*) were more susceptible than cicadellid eggs (*Nephotettix virescens*), and of the two planthopper species, eggs of *N. lugens* were most preferred (Manti 1989; Heong et al. 1990b). However, there is some contrary evidence suggesting that leafhopper eggs are preferred over those of *N. lugens* (see Greathead 1983). There is abundant evidence suggesting that *C. lividipennis* switches between delphacid and cicadellid hosts if one becomes scarce (Hinckley 1963; Napompeth 1973; Stapley 1976; Bentur and Kalode 1987). Also, if prey are unavailable, these mirids will feed on their own eggs (Hinckley 1963).

Besides mirids, other hemipterans such as nabids (*Dolichonabis limbatus* and *Nabis flavomarginatus* on *Conomelus anceps*; Rothschild 1966) and veliids which feed on delphacids that have fallen on the waters' surface in rice fields (e.g., *Microvelia douglasi* and *M. douglasi atrolineata* on *N. lugens*; Kenmore 1980; Nakasuji and Dyck 1984; Cook and Perfect 1985a; Heong et al. 1990a) may be locally important predators on planthoppers. Phalangids, which accounted for 60% of the total predation on the nymphs and adults of *Conomelus anceps*, were identified as the most important predators of this marsh-inhabiting delphacid (Rothschild 1966). The larvae of two pteromalid wasps (*Panstenon oxylus* and *Mesopolobus aequus*) feed primarily on the eggs of *Javesella pellucida* and were viewed as the most important natural enemy of this pest in cereal crops (Raatikainen 1967). Even though coccinellid beetles are consistently found in association with many delphacid species (Tables 10.1 and 10.2) and are occasionally implicated as important factors in reducing planthopper populations

Table 10.2. Invertebrate predator taxa responsible for inflicting significant mortality on populations of 17 species of planthoppers. The most important predator taxa identified by the various authors are indicated (MI).

Predator Taxon	Habitat	Araneae	Miridae	Nabidae	Coccinellidae	Other	Selected References
Conomelus anceps	Marsh	+	+	+	+	Phalangida^{MI}	Rothschild (1966)
Delphacodes detecta	Salt marsh	+^{MI}	+				Denno (1979); Tallamy and Denno (1979); Döbel (1987); Döbel et al. (1990)
Dicranotropis hamata	Grassland	+^{MI}		+^{MI}			Waloff (1980); Waloff and Thompson (1980)
Javesella pellucida	Cereal crops	+				Pteromalidae^{MI}	Raatikainen (1967)
Laodelphax striatellus	Grasses and rice	+^{MI}			+		Kobayashi and Hiwada (1968); Harpaz (1972)
Nilaparvata lugens	Rice field	+^{MI}	+^{MI}		+	Veliidae	Kiritani et al. (1971); Kenmore (1980); Ooi (1982, 1988); Manti (1989)
Peregrinus maidis	Corn/sorghum	+	+^{MI}		+		Napompeth (1973); Fisk et al. (1981)
Perkinsiella saccharicida	Sugarcane	+	+^{MI}	+	+		Swezey (1936); Zimmerman (1948); Bull (1981)

Prokelisia dolus	Salt marsh	+[MI]	+			Denno (1983); Döbel (1987)
Prokelisia marginata	Salt marsh	+[MI]	+			Denno (1983); Döbel (1987)
Saccharosydne saccharivora	Sugarcane		+[MI]			Metcalfe (1971); Bull (1981)
Sogatella furcifera	Rice field	+[MI]	+[MI]			Ooi (1982); Cook and Perfect (1985a)
Sogatella kolophon	Grassy field		+[MI]			Hinkley (1963)
Stenocranus minutus	Grassland	+[MI]		+		Waloff (1980); Waloff and Thompson (1980)
Tagosodes orizicolus	Rice field				+	King and Saunders (1984)
Tarophagus colocasiae	Taro field	+	+[MI]		+	Matsumoto and Nishida (1966)
Tumidagena minuta	Salt marsh	+[MI]	+			Denno (1977); Tallamy and Denno (1979)

(Israel and Prakasa Rao 1968; Chiu 1979; Fisk et al. 1981), these predators of planthopper nymphs and adults generally inflict less mortality than do either spiders or hemipterans (Table 10.2). Some coccinellids (*Harmonia* and *Verania*) are more effective in their attack of adult prey (Dyck and Orlido 1977), whereas others (*Coccinella arcuata*) feed primarily on nymphs (Ng 1978). Other predaceous beetles in the Carabidae and Staphylinidae (Rothschild 1966; Ôtake 1977; Chiu 1979; Ooi 1982), earwigs in the Forficulidae (Terry 1905; Swezey 1936), lacewings in the Chrysopidae (Harpaz 1972), and conocephaline grasshoppers (Swezey 1936) are occasionally mentioned as predators of planthoppers. Even though the densities of these predators can on occasion be high, they rarely inflict much mortality on planthopper populations (Ôtake 1977; Ooi 1988).

Ants are often found in association with planthoppers (Fullaway 1918; Hirashima et al. 1979; King and Saunders 1984), but their role as predators is not generally clear. Some ant species (*Myrmica laevinodis*) are unequivocal predators of leafhoppers and planthoppers in certain habitats such as the grasslands of Poland where Auchenorrhyncha (mostly nymphs) form such a high percentage of their diet (Andrzejewska et al. 1971). However, other ant species (*Pheidole megacephala*) have been reported tending *Peregrinus maidis* and feeding on honeydew, and the interaction was considered to be primarily mutualistic (Napompeth 1973; Fisk et al. 1981).

In summary, various species of spiders and hemipterans are implicated repeatedly as the most important predators of planthopper species. In the sections that follow, we concentrate on these taxa and critically analyze their ability to limit the population growth of planthoppers. We especially focus on experimental studies that shed light on factors influencing the coupling, stability, and persistence of planthopper–predator interactions.

Functional Responses of Planthopper Predators

If a predator is unable to effectively capture and kill prey in a confined laboratory setting, it is unlikely that such a predator will play an important role in prey population dynamics in the field (Kareiva 1990). Kareiva (1990) goes on to point out that functional response experiments in the laboratory are the logical starting point to identify the potential of a predator to suppress prey populations.

The **functional response** defines the *per capita* ability of a predator to attack hosts at different densities (Holling 1959a, 1959b), and depending on its form, the response has different theoretical implications for predator–prey population dynamics (Holling 1965; Hassell and May 1973; Murdoch and Oaten 1975). The Type II response, characteristic of many invertebrate predators, rises continually at a decreasing rate to an upper asymptote set by handling time, satiation, or egg (fecundity) limitation (Holling 1959a; Murdoch 1990). Under most circumstances, Type II functional responses are thought to destabilize predator–prey interactions because an increasingly smaller fraction of prey are taken as prey density increases, resulting in inverse density-dependent mortality (Hassell and May 1973; Murdoch 1990).

In contrast, predators (some invertebrates) that search more actively or efficiently as prey density increases exhibit a sigmoidal Type III functional response (Murdoch and Oaten 1975; Hassell et al. 1977). Because the Type III response is density dependent up to some prey density threshold, it is thought to lend stability to predator–prey interactions (Holling 1965; Murdoch and Oaten 1975; Oaten and Murdoch 1975a; Hassell 1978; Murdoch 1990).

Under most conditions the major predators of planthoppers all exhibit Type II functional responses. For example, Type II responses have been shown for the wolf spiders *Pardosa pseudoannulata* (Ooi 1988; Heong and Rubia 1989, 1990a; Heong et al. 1991b), *P. laura* (Nakamura 1977), *P. littoralis* and *Lycosa modesta* (Döbel 1987; Döbel and Denno in preparation a) and *Pirata subpiraticus* (Wu and Wang 1987), both sexes of the mirid bug *Cyrtorhinus lividipennis* (Sivapragasam and Asma 1985; Qingcai and Jervis 1988; Manti 1989; Heong et al. 1990b), and the veliid *Microvelia douglasi* (Nakasuji and Dyck 1984) to increasing densities of the planthoppers *Nilaparvata lugens*, *Prokelisia dolus*, and *P. marginata* (Figs. 10.1A–D). For planthopper predators, however, levels of hunger vary with satiation and physiological state and can dramatically influence the amplitude of the functional response (Nakamura 1977). For example, *Pardosa laura* captures more prey and ingests more food when it is starved than when its "gut" is filled to capacity (Nakamura 1977). Also, females of the wolf spider *Pardosa laura* carrying egg cases have a high hunger threshold and feed much less than females without eggs (Miyashita 1963; Nakamura 1977).

With an increase in prey density, there was a decrease in the proportion of prey taken resulting in a pattern of inverse density-dependent predation, a pattern which has been shown for *Pardosa pseudoannulata* (Kiritani et al. 1972; Ooi 1988; Fig. 10.1A), *Cyrtorhinus lividipennis* (Sivapragasam and Asma 1985; Qingcai and Jervis 1988; Manti 1989; Fig. 10.1B), and *Microvelia douglasi* (Nakasuji and Dyck 1984; Fig. 10.1D). For *C. lividipennis*, handling time and satiation limited its functional response even though this predator exhibited a clear aggregative response and spent more time in high-density patches of *N. lugens* (Qingcai and Jervis 1988; Manti 1989).

Functional responses of predators can also be used to elucidate prey

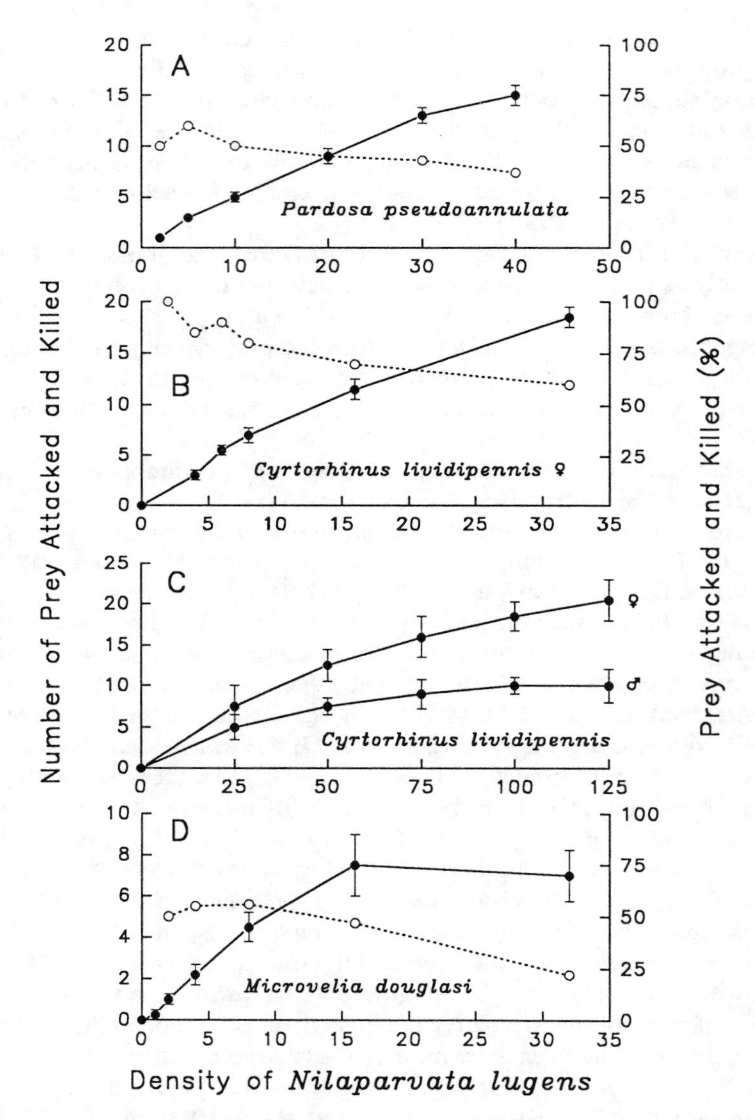

Figure 10.1. Functional responses of several invertebrate predators to increasing densities of *Nilaparvata lugens* in the laboratory. The average number of prey attacked and killed in 24 hours (± one standard error) is shown (solid lines) as is the percentage of total prey taken (dashed lines). **(A)** *Pardosa pseudoannulata* (adults starved for 3 days) offered macropterous planthopper adults (from Ooi 1988); **(B)** *Cyrtorhinus lividipennis* (adult females starved for 8 hours) offered planthopper eggs (from Sivapragasam and Asma 1985); **(C)** *Cyrtorhinus lividipennis* (adults of both sexes starved for 14 hours) offered planthopper eggs (from Manti 1989); **(D)** *Microvelia douglasi* (adults starved for 1 day) offered second instar planthoppers (from Nakasuji and Dyck 1984).

susceptibility and predator efficiency (Holling 1965). In rice systems, for example, the wolf spider *P. pseudoannulata* captured the nymphs and adults of *Nilaparvata lugens* more frequently than those of the green leafhopper *Nephotettix virescens* (Heong and Rubia 1989; Fig. 10.2A). In the field, however, the capture rate of leafhoppers (*Nephotettix cincticeps*) by *P. pseudoannulata* can be greater than that for *N. lugens* if leafhoppers are more abundant (Kiritani et al. 1972; Sasaba et al. 1973). Heong and Rubia (1989) suggested that because of its preference for *N. lugens*, *P. pseudoannulata* plays a more important role in the dynamics of the brown planthopper compared to other naturally co-occurring prey species. Similarly, the eggs of *N. lugens* were more susceptible to adult predation from *Cyrtorhinus lividipennis* than the eggs of *N. virescens* (Heong et al. 1990b; Fig. 10.2B). Females of *C. lividipennis* consumed significantly more eggs of the brown planthopper *N. lugens* than did males demonstrating a significant difference in prey capture potential between the sexes of this mirid (Manti 1989; Manti and Shepard 1990; Fig. 10.1C). Also, the adults of *C. lividipennis* handled the eggs of *N. lugens* more rapidly than did mirid nymphs, and both developmental stages displayed a Type II functional response (Sivapragasam and Asma 1985).

Pardosa pseudoannulata will feed on mirid predators as well as planthoppers. In laboratory studies, the attack rate of *P. pseudoannulata* was higher and the handling time lower when caged with the mirid predator *C. lividipennis* compared to *N. lugens*, results which suggest a "preference" for *C. lividipennis* (Heong et al. 1991b; Fig. 10.2C). These results imply that the effect of predation of *C. lividipennis* on *N. lugens* may be decreased in the presence of the naturally co-occurring wolf spider *P. pseudoannulata* (Heong et al. 1989, 1991b). However, in mixed prey experiments (10 : 50, 20 : 40, 30 : 30, 40 : 20, and 50 : 10 individuals of *N. lugens* : *C. lividipennis*), *P. pseudoannulata* fed on *N. lugens* over *C. lividipennis*, and the spider's preference for *N. lugens* varied inversely with *N. lugens* density (Heong et al. 1991b). Because the spider fed more on *N. lugens* at low relative densities, the authors suggested that this response was the antithesis of "predator switching" described by Murdoch and Oaten (1975) and Hassell (1978). The authors went on to suggest that in a mixed prey environment the spider chose the easiest prey to catch. However, in the presence of another predator (the highly mobile *C. lividipennis*), the behavior of *N. lugens* may have changed, making it relatively more susceptible to the visually searching wolf spider. If this were the case, the behavior of *N. lugens* would be modified most at low planthopper-to-mirid ratios, making the planthopper most susceptible to spider predation and perhaps giving the illusion of preference. This argument may also explain the apparent discrepancy in wolf spider preference when it was evaluated in pure

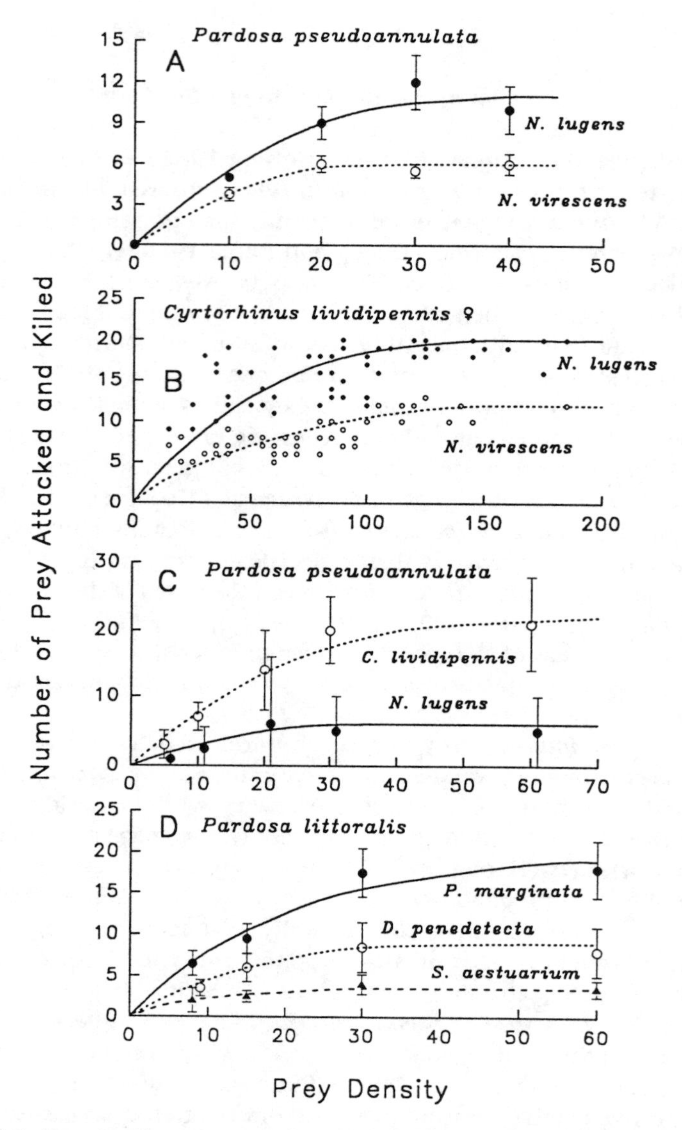

Figure 10.2. Functional responses of several invertebrate predators to increasing densities of planthoppers and other prey species (leafhoppers and *Cyrtorhinus lividipennis*) in the laboratory. The average number of prey attacked and killed in 24 hours (± one standard error) is shown. Responses show that planthoppers are generally more susceptible to invertebrate predation than leafhoppers. (**A**) *Pardosa pseudoannulata* (adult females starved for 3 days) offered adults of *Nilaparvata lugens* and *Nephotettix virescens* in separate experiments (mean ± SE; from Heong and Rubia 1989); (**B**) *Cyrtorhinus lividipennis* (newly emerged adult females) offered eggs of *N. lugens* and *N. virescens* (from Heong et al. 1990b); (**C**) *Pardosa pseudoannulata* (adult females starved for 3 days) offered adult females of *N. lugens* and *C. lividipennis* in separate experiments (mean ± 95% confidence limits; from Heong et al. 1991b); (**D**) *Pardosa littoralis* (adult females starved for 2 days) offered adults of the planthoppers *Prokelisia marginata* and *Delphacodes penedetecta* and the leafhopper *Sanctanus aestuarium* in separate experiments (means ± SD; from Döbel 1987).

(Fig. 10.2C) and mixed species conditions. Most importantly, these results point to the complexity of predator–prey relationships and the potential significance of multiple species interactions which are not necessarily elucidated by simple one-on-one experiments in the laboratory.

We investigated the functional response of the most abundant salt marsh-inhabiting wolf spider, *Pardosa littoralis*, to three naturally co-occurring herbivore species: the planthoppers *Prokelisia marginata* and *Delphacodes penedetecta*, and the leafhopper *Sanctanus aestuarium* (Döbel 1987; Döbel and Denno in preparation a). *Prokelisia marginata* was the most frequently captured prey species by far, followed by *D. penedetecta* and *S. aestuarium* (Fig. 10.2D). This pattern of prey capture was essentially the same for the larger, but less abundant, wolf spider *Lycosa modesta* (Döbel 1987). Because *P. marginata*, *D. penedetecta*, and *S. aestuarium* are all about the same size (3 mm in length), body size cannot be used to explain the differences in capture rate among these three prey species.

Behavioral differences among the prey species help to explain variation in their susceptibility to spider predators. With the approach of a spider, *S. aestuarium* readily escapes by jumping from the plant, *D. penedetecta* moves to the back of the grass stem out of site from the spider, and *P. marginata* simply moves down the stem in full view of the predator. Because wolf spiders respond to motion and have a higher capture efficiency for active prey (Nakamura 1977), *P. marginata* was the most vulnerable to attack. These functional response experiments confirm field observations and show that *Prokelisia* planthoppers are the most frequent prey of wolf spiders and are under the highest risk from predation. However, the functional response of *P. littoralis* to *P. marginata* and its sympatric congener *P. dolus* were the same, suggesting that these two planthoppers of similar size (3 mm) were equally vulnerable to wolf spider predation (Döbel 1987). Furthermore, there is a slight but significant tendency for the brachypters of *Prokelisia* planthoppers to be at higher risk from wolf spider predation than macropters (Döbel 1987).

Of the salt marsh-inhabiting hunting spiders, *L. modesta* was most effective at capturing *P. marginata* followed by *P. littoralis* and the salticid *Marpissa pikei* (Döbel 1987; Fig. 10.3A). In part, capture rate can be attributed to the body size of the spider species because the two variables are roughly positively related. For instance, at the highest experimental prey density (120/cage), *L. modesta*, the largest spider (100–200 mg live adult weight), captured 70 prey/24 hours followed by the smaller wolf spider *P. littoralis* (25–40 mg) which captured 42 planthoppers; the smallest spider *M. pikei* (10–15 mg) captured only 10 prey/24 hours. Larger individuals of *L. modesta* (85 mg) also have a much higher capture rate of *Prokelisia* planthoppers than smaller individuals (12 mg) (Fig. 10.3B). However, with

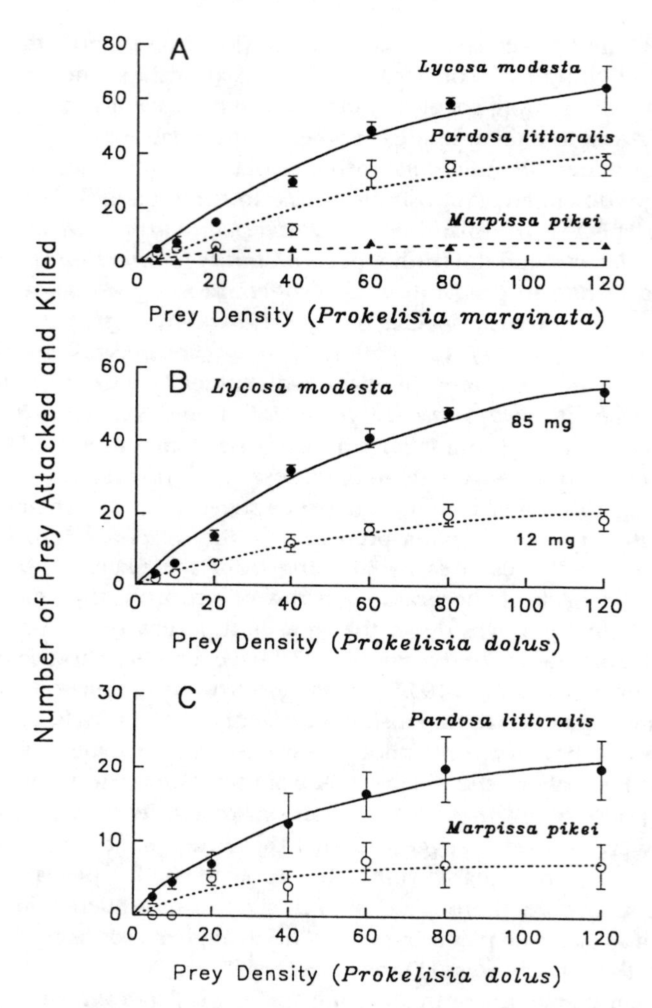

Figures 10.3. Functional responses of lycosid (*Lycosa modesta* and *Pardosa littoralis*) and salticid (*Marpissa pikei*) spiders to increasing densities of *Prokelisia* planthoppers. The average number of prey attacked and killed in 24 hours (mean ± SD) is shown (from Döbel 1987). (**A**) *L. modesta*, *P. littoralis*, and *M. pikei* (adult females starved for 2 days) offered adults of *P. marginata*; (**B**) two weight classes of *L. modesta* (85- and 12-mg spiders starved for 2 days) offered adults of *P. dolus*; (**C**) same weight class of *P. littoralis* and *M. pikei* (12-mg spiders starved for 2 days) offered adults of *P. dolus*.

spider weight controlled at 12 mg, the capture rate of *P. littoralis* was nearly three times that of *M. pikei* at the highest prey density, suggesting that the wolf spider was a far more effective predator (Fig. 10.3C).

Of all the planthopper predators we surveyed, the highest daily capture/consumption rates recorded were for adult wolf spiders (e.g., *Lycosa modesta* = 43–70 planthoppers/24 hours, *Pardosa littoralis* = 20–69, *Pardosa pseudoannulata* = 12–45) followed by mirid bugs (*Cyrtorhinus lividipennis* = 20–24, *Tytthus mundulus* = 10–12) and coccinellid beetles (*Coccinella arcuata* = 18–29) (Table 10.3). High capture rates for these voracious predators are explained in part by high instantaneous search rates (a = 0.239 and 0.491 days) and short handling times (T_h = 0.063 and 0.031 days) for *P. pseudoannulata* and *C. lividipennis*, respectively (Heong et al. 1991b; Manti 1989). Generally, capture rates were higher for females than males and adults compared to immature predators (Table 10.3). Again, the relative body size of the predator appears to explain in part this pattern.

It is essential to realize that most of the data on prey capture rates were determined in the laboratory after starving the predators. As a consequence, most of these capture rates should be interpreted as potential maxima. Under field conditions, hunger levels of planthopper predators are usually much less as are capture rates (see Rothschild 1966; Nakamura 1974). That capture rate is positively related to hunger level is confirmed by an experiment performed by Dyck and Orlido (1977). Following a period of initial starvation, capture rates decreased for *P. pseudoannulata* the longer they were caged with either nymphs or adults of *N. lugens* (Table 10.3). Nevertheless, the relative capture rates of the various planthopper predators determined in the laboratory appear to retain this ranking even though predators have not been starved. For example, when predators were provided an ample supply of *Conomelus anceps*, the highest capture rates recorded were for mirid bugs (3.7 eggs/day) and wolf spiders (3.3 nymphs/day) followed by nabids (2.5–3.1 nymphs/day), coccinellid larvae (2.5 nymphs/day), phalangids 1.8–1.9 nymphs/day), a variety of web-building spiders (0.4–2.0 nymphs/day) and carabid beetles (1.4 nymphs/day), (Rothschild 1966).

In addition to the morphology and behavior of the predator and prey species, the structure of the habitat may also influence the functional response and predator–prey interactions in general (Robinson 1981; Uetz 1991). For example, by adding thatch (dead plant litter which occurs naturally on the salt marsh) to experimental arenas, the functional response of the wolf spider *Pardosa littoralis* to increasing *Prokelisia* density was changed from a Type II to a Type III curve (Fig. 10.4). At low densities, planthoppers were taken significantly less frequently in the thatch-contain-

Table 10.3. Daily (24 hours) individual capture/consumption rate of the major predators of delphacid planthoppers under conditions of unlimited prey availability in the laboratory.

Predator Taxon	Planthopper Prey	Daily Consumption Rate (Stage Attacked)	Reference
Lycosidae			
Lycosa modesta[AD]	*Prokelisia marginata*	60–70 (adults)[FR]	Döbel (1987)
	Prokelisia dolus	55–60 (adults)[FR]	
	Delphacodes penedetecta	43 (adults)[FR]	
Lycosa modesta[Imm]	*Prokelisia dolus*	20 (adults)[FR]	
Lycosa sp.[AD]	*Conomelus anceps*	3 (adults), 2 (Nymphs)	Rothschild (1966)
Marpissa pikei[AD]	*Prokelisia marginata*	10 (adults)[FR]	Döbel (1987)
Pardosa littoralis[AD]	*Prokelisia marginata*	43 (adults)[FR]	Döbel (1987)
		69 (adults[B])[FR]	
		58 (adults[M])[FR]	
	Prokelisia dolus	44 (adults[B])[FR]	
		54 (adults[M])[FR]	
Pardosa littoralis[Imm]	*Prokelisia dolus*	20 (adults)[FR]	Döbel (1987)
Pardosa pseudoannulata[AD]	*Nilaparvata lugens*	20 (adults[M])	Ooi (1988)
		15–20 (adults)	Samal and Misra (1975)
		40 (adults)[1]	Dyck and Orlido (1977)
		17 (adults)[2]	
		30 (adults)[3]	
		41 (nymphs)[1]	
		45 (nymphs)[2]	
		25 (nymphs)[3]	
Pardosa pseudoannulata[♀]	*Nilaparvata lugens*	12 (adults)[FR]	Heong and Rubia (1989)
		6 (adults[B])	Heong et al. (1991b)
		13 (adults)	Chiu (1979)
Pardosa pseudoannulata[♂]	*Nilaparvata lugens*	11 (adults)	Chiu (1979)
Pardosa pseudoannulata[Imm]	*Nilaparvata lugens*	3–4 (adults)	Chiu (1979)
Pardosa T-insignita	*Nilaparvata lugens*	3 (nymphs), 4 (adults)	Lee and Park (1977)
Pardosa T-insignita	*Sogatella furcifera*	1 (nymph), 6 (adults)	Lee and Park (1977)

Miridae			
Cyrtorbinus lividipennis♀	*Nilaparvata lugens*	24 (eggs)	Manti (1989)
		20 (eggs)[FR]	Heong et al. (1990b)
		9 (eggs)	Manti and Shepard (1990)
		22 (eggs)[FR]	Sivapragasam and Asma (1985)
		3 (1st instar nymphs)	Manti (1989)
		4 (1st instar nymphs)	Sivapragasam and Asma (1985)
Cyrtorbinus lividipennis♂	*Nilaparvata lugens*	8 (eggs)	Manti (1989)
		8 (eggs)[FR]	Heong et al. (1990b)
		2 (eggs)	Manti and Shepard (1990)
		21 (eggs)	Sivapragasam and Asma (1985)
		1 (1st instar nymph)	Manti (1989)
Cyrtorbinus lividipennis[Imm]	*Nilaparvata lugens*	6 (eggs)	Sivapragasam and Asma (1985)
Tytthus pygmaeus[AD]	*Conomelus anceps*	4 (eggs)	Rothschild (1966)
Tytthus pygmaeus[Imm]	*Conomelus anceps*	1 (2nd or 3rd instar)	Rothschild (1966)
Tytthus mundulus[AD]	*Peregrinus maidis*	10–12 (eggs)	Usinger (1939)
Tytthus sp.♀	*Nilaparvata lugens*	14 (eggs)	Basilio and Heong (1990)
Nabidae			
Nabis flavomarginatus[AD]	*Conomelus anceps*	2 (adults), 3 (nymphs)	Rothschild (1966)
Nabis flavomarginatus[Imm]	*Conomelus anceps*	3 (nymphs)	Rothschild (1966)
Veliidae			
Microvelia douglasi[AD]	*Nilaparvata lugens*	7 (2nd instar nymphs)	Nakasuji and Dyck (1984)
Coccinellidae			
Coccinella arcuata[AD]	*Nilaparvata lugens*	20 (2nd and 3rd instars)	Abraham et al. (1973)
		29 (nymphs)	Abraham and Mathew (1975)
Coccinella arcuata[Imm]	*Nilaparvata lugens*	18–27 (nymphs)	Abraham and Mathew (1975)
Coccinella septempunctata[AD]	*Conomelus anceps*	2 (adults), 3 (nymphs)	Rothschild (1966)
Coccinella septempunctata[Imm]	*Conomelus anceps*	2 (adults)	Rothschild (1966)

Note: AD = Adult predator; Imm = Immature predator; B = Brachypterous adult; M = Macropterous adult; 1, 2, 3 = Capture rate determined 1, 2, and 3 days after caging. FR = Maximum capture/consumption rate determined from functional response at highest prey density.

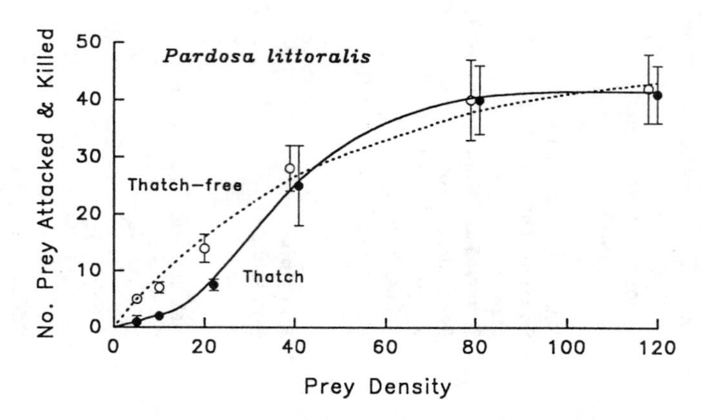

Figure 10.4. Functional response of *Pardosa littoralis* to increasing densities of *Prokelisia* planthoppers in two structurally different experimental arenas: (1) thatch-free arenas containing only living *Spartina* plants and (2) thatch-rich arenas containing living plants as well as dead plant litter. The average number of prey attacked and killed in 24 hours (mean ± SD) is shown (from Döbel 1987). Adding thatch to the habitat changed the functional response from a Type II to a Type III.

ing compared to the thatch-free environment, suggesting that planthoppers were able to escape predation apparently by finding refuge in the thatch. By providing refuges for planthoppers, the presence of thatch should tend to stabilize *Prokelisia*/wolf spider interactions (Holling 1965; Murdoch and Oaten 1975; Oaten and Murdoch 1975a; Hassell 1978).

The prevalence of Type II functional responses among the major predators of planthoppers implies that outbreak and escape should be properties of the population dynamics of planthoppers. However, a strong numerical reponse on the part of a predator may offset the destabilizing effects of a Type II functional response. Promise for planthopper management is also afforded by the results of several functional response experiments which show that planthoppers are among the most susceptible prey species compared to a variety of other naturally co-occurring prey including leafhoppers and mirid bugs.

Numerical Responses of Planthopper Predators

Some predators respond numerically to an increase in the population size of their prey by increasing reproduction and/or by aggregating in areas of high prey density (Holling 1959a, 1959b, 1961; Hassell 1978; Reichert and

Lockley 1984). In such cases, the numerical response of the predator is density-dependent and is directly influenced by changes in prey density (Hassell and May 1973; Comins and Hassell 1979; Waage and Hassell 1982). Positive correlations between predator and prey density are often used as evidence for the numerical or aggregation response of a predator (see Murdoch 1990). Although a positive correlation may indicate a numerical response to prey density, it may not. For instance, predators may aggregate in response to some environmental factor independent of host density (Murdoch 1990). Thus, positive density correlations between predator and prey species suggest, but do not rigorously demonstrate, a numerical response.

Although there is controversy over the extent to which aggregation imposes stability on predator–prey interactions (Hassell and May 1973; Murdoch and Oaten 1975; Hassell 1978; Chesson and Murdoch 1986; Murdoch and Stuart-Oaten 1989; Murdoch 1990; Waage 1990), the prevailing view in the literature is that a strong numerical response constitutes a potentially important attribute of a natural enemy for effective biological control (Beddington et al. 1978; van Lenteren 1986; Kareiva 1990; Murdoch 1990). Murdoch (1990) argues that there is a trade-off between pest suppression and stability and that, in this context, aggregation is a characteristic of natural enemies which leads to a lower pest equilibrium density, but at the expense of stability.

Most evidence for the numerical response of planthopper predators comes from field studies in which the densities of predator and prey species were correlated (Table 10.1). In some cases, density correlations were based on temporal data from the same plot (e.g., Matsumoto and Nishida 1966; Napompeth 1973), whereas in other instances, correlations were spatial and based on data taken from many plots during similar time periods (Kenmore 1980; Cook and Perfect 1985a; Manti 1989). There were only a few experimental studies which attempted to link the behavioral aggregation of predators (numerical response) with density patterns of planthoppers and predators in the field and planthopper suppression (Matsumoto and Nishida 1966; Döbel and Denno in preparation c).

Of the major predators of planthoppers, mirid egg predators showed the strongest and most consistent numerical responses to planthopper density (Napompeth 1973; Bull 1981; Manti 1989; Ooi 1988). Positive density correlations occur among *Cyrtorhinus lividipennis* and *Peregrinus maidis* (Napompeth 1973), *Nilaparvata lugens* (Kuno and Dyck 1985; Ooi 1988; Manti 1989; Heong et al. 1990a), and *Sogatella furcifera* (Manti 1989) (Figs. 10.5A, B). Similarly, the density of *Cyrtorhinus fulvus* is strongly and positively correlated with the density of the taro planthopper, *Tarophagus colocasiae* (Matsumoto and Nishida 1966; Fig. 5C), as are the densities of

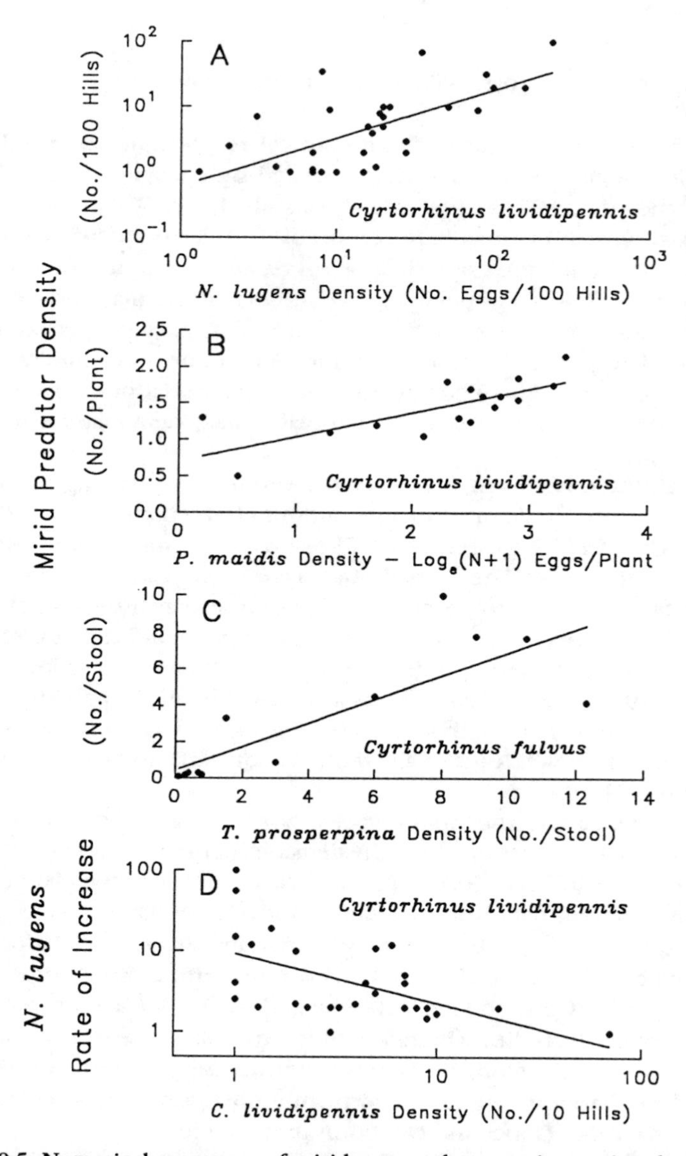

Figure 10.5. Numerical responses of mirid egg predators to increasing densities of planthopper prey in the field. (**A**) Relationship between the densities of *Cyrtorhinus lividipennis* and *Nilaparvata lugens* (No./100 hills) in rice at Tanjung Karang, Malaysia; $Y_{\mathrm{PRED}} = 2.29 + 0.34X_{\mathrm{PREY}}$ for nontransformed data, $R^2 = 0.58$, $P<0.01$ (from Ooi 1988). (**B**) Relationship between the density of *C. lividipennis* and the egg density of *Peregrinus maidis* [$\log_e (N+1)$/plant] in corn at Kualoa Ranch, Oahu, Hawaii; $Y_{\mathrm{PRED}} = 0.80 + 0.29X_{\mathrm{PREY}}$ for $\log_e$-transformed data, $R^2 = 0.55$, $P<0.01$ (from Napompeth 1973). (**C**) Relationship between the densities of *Cyrtorhinus fulvus* and *Tarophagus colocasiae* (No./stool) in taro at Kahaluu Valley, Oahu, Hawaii; $Y_{\mathrm{PRED}} = -0.44 + 0.64X_{\mathrm{PREY}}$, $R^2 = 0.64$, $P<0.01$ (from Matsumoto and Nishida 1966). (**D**) Relationship between the rate of population increase of *N. lugens* (generational growth from one rice-growing period to the next, 1 = no growth) and the density of *C. lividipennis* (No./10 hills) in rice at Tanjung Karang, Malaysia; $Y_{\mathrm{PREY\ INCREASE}} = 0.99 - 0.67X_{\mathrm{PRED}}$ for $\log_{10}$-transformed data, $R^2 = 0.37$, $P<0.01$ (from Ooi 1988).

Tytthus mundulus and *T. parviceps* with *Perkinsiella saccharicida* (Bull 1981).

The case of *Cyrtorhinus fulvus* provides convincing evidence for a strong numerical response as a requisite for prey suppression. After the artificial introduction of this mirid into the taro fields of Hawaii, populations of the taro planthopper, *T. colocasiae*, declined spectacularly, and permanent control was achieved (Fullaway 1940). Years after this introduction, subsequent investigations into this predator–prey relationship revealed that *C. fulvus* was very efficient in tracking taro planthopper populations. For example, both spatial and seasonal changes in the abundance of this mirid mirrored changes in the population size of *T. colocasiae* (Matsumoto and Nishida 1966). The numerical response of *C. fulvus* has both aggregative and reproductive components. Also, following the experimental removal of mirids from field plots of taro, populations of *T. colocasiae* increased dramatically and significantly compared to control plots which contained predators (Matsumoto and Nishida 1966).

Similarly, in sugarcane, *Tytthus mundulus* and *T. parviceps* effected a strong numerical response to *Perkinsiella saccharicida* by dispersing effectively to spring plantings where their populations rapidly proliferated (Bull 1981). At this location in Australia, predation from *Tytthus* egg predators was density-dependent and was the major mortality factor responsible for the collapse of the planthopper population.

In the rice fields of Asia and the Pacific region, *Cyrtorhinus lividipennis* is an important predator implicated in the suppression of *Nilaparvata lugens* populations (Hinckley 1963; Stapley 1976; Kuno and Dyck 1985; Ooi 1988; Manti 1989). This mirid exhibits a clear aggregative response, spending more time in high-density patches of planthoppers (Qingcai and Jervis 1988). A strong numerical response in this mirid species (Fig. 10.5A) is also associated with its ability to control planthopper populations (Ooi 1988; Manti 1989). Strong temporal correlations in density between *Cyrtorhinus lividipennis* and *Nilaparvata lugens* (Ôtake and Hokyo 1976; Ooi 1988; Manti 1989) and predator removal experiments (Ooi 1988; Manti 1989) document the ability of this mirid predator to track and suppress populations of brown planthopper. Furthermore, there is a negative relationship between the density of this mirid and the rate of increase in the populations of its planthopper prey (Cook and Perfect 1985a; Kuno and Dyck 1985; Ooi 1988; Fig. 10.5D).

Veliids such as *Microvelia douglasi atrolineata* attack rice planthoppers that have fallen or been dislodged onto the waters' surface. They too can show a numerical response to prey density (Cook and Perfect 1985a; Heong et al. 1990a), but not in all cases (Kenmore 1980; Kuno and Dyck 1985; Kenmore et al. 1984; Fig. 10.6A). Part of the difficulty in consistently demonstrating a numerical response for veliids may result from the behavior of these highly mobile predators on the surface of the water. Veliids move rapidly to areas of surface disturbance such as a struggling planthopper, and, as a consequence, their density and distribution may be altered during sampling and may not always reflect any previous small-scale numerical response to prey density (Kenmore 1980). Thus, it is not surprising that populations of *M. douglasi* are not always closely correlated with those of

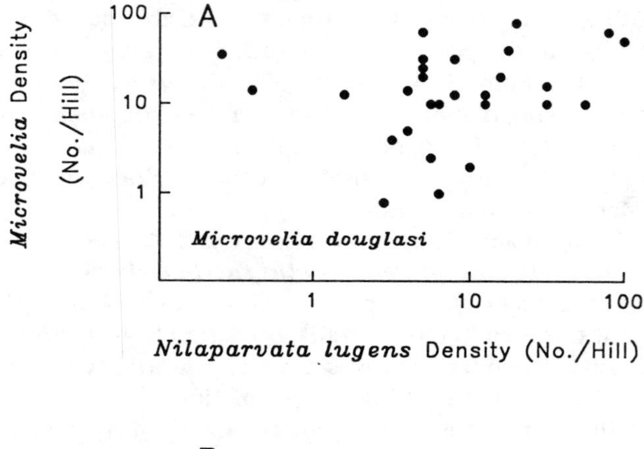

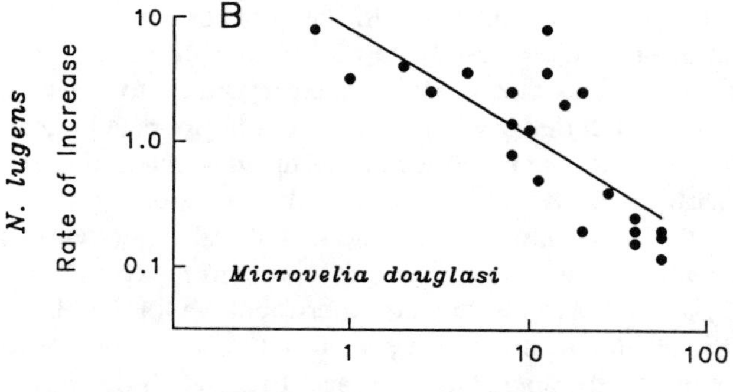

Figure 10.6. (**A**) Relationship between the densities of *Microvelia douglasi* and *Nilaparvata lugens* (No./hill) in rice at IRRI, Los Baños, Philippines; $Y_{PRED} = 15.98 + 0.20X_{PREY}$ for nontransformed data, $R^2 = 0.35$, ns. (**B**) Relationship between the rate of population increase of *N. lugens* (population growth from one generation to the next, 1 = no growth) and the density of *M. douglasi* (No./hill) in rice at IRRI, Los Baños, Philippines; $Y_{PREY\ INCREASE} = 0.90 - 0.86X_{PRED}$ for $\log_{10}$-transformed data, $R^2 = 0.79$, $P<0.01$ (from Kuno and Dyck 1985).

N. lugens (Kenmore 1980; Kenmore et al. 1984). Nonetheless, in some rice-growing areas, the density of these surface predators can be inversely related to the rate of increase in *N. lugens* populations (Fig. 10.6B) and they are considered important predators in the rice ecosystem (Nakasuji and Dyck 1984; Kuno and Dyck 1985).

One reason for the efficiency of veliid predators may involve their "group attacking" behavior. *M. douglasi* was more successful in killing individual fifth instar nymphs of *N. lugens* when up to 8 conspecific veliids were present and assisting each other in the attack (Nakasuji and Dyck 1984). Thus, conspecifics facilitate prey capture, reduce handling time, and apparently do not exhibit mutual interference in groups of eight; above this group size, the percentage of successful attacks declines dramatically.

Spiders, particularly wolf spiders, are thought to be among the most effective natural enemies of planthoppers (Dyck and Orlido 1979; Kenmore 1980; Table 10.2), a reputation based on their abundance, large body size, and high consumption rate (Rothschild 1966; Kiritani et al. 1972; Nakamura 1977; Chiu 1979; Kenmore 1980; Kuno and Dyck 1985; Döbel 1987; Ooi 1988). Nevertheless, their tendency to aggregate in response to prey increase is more variable than mirid predators such as *Cyrtorhinus lividipennis*. For example, although strong numerical responses for lycosid spiders such as *Pardosa pseudoannulata*, *P. littoralis*, and *Lycosa modesta* have been reported (Kenmore 1980; Döbel 1987; Heong et al. 1990a; Döbel and Denno in preparation c; Figs. 10.7A,B), spiders do not always appear to aggregate in areas of high prey densities (Ooi 1982, 1988; Kuno and Dyck 1985; Fig. 10.7C). The difficulty of accurately sampling these skittish and highly mobile predators may contribute in part to the inconsistency in demonstrating a numerical response for wolf spiders. This may also explain why numerical responses of wolf spiders are more frequently documented at larger spatial scales (see Kenmore 1980; Döbel 1987).

Furthermore, at high densities, conspecifics of *P. pseudoannulata* exhibit mutual interference which results in a reduction in searching efficiency and eventually leads to a decline in the predation rate on *Nilaparvata lugens* (Kiritani and Kakiya 1975; Heong and Rubia 1990b). Nevertheless, several studies show that the rate of increase in planthopper populations is inversely related to the density of wolf spiders (Kuno and Dyck 1985; Ooi 1988; Döbel and Denno in preparation c; Fig. 10.8), suggesting that spiders can play an important role in the suppression of planthopper populations.

For spiders, there is evidence that numerical responses to planthopper density can result from increases in both reproduction (Suzuki and Kiritani 1974; Kenmore et al. 1984) and immigration (Döbel 1987; Döbel and Denno in preparation c). At higher prey densities, the wolf spider *Pardosa laura* ingests and assimilates more food (Nakamura 1977), and as a conse-

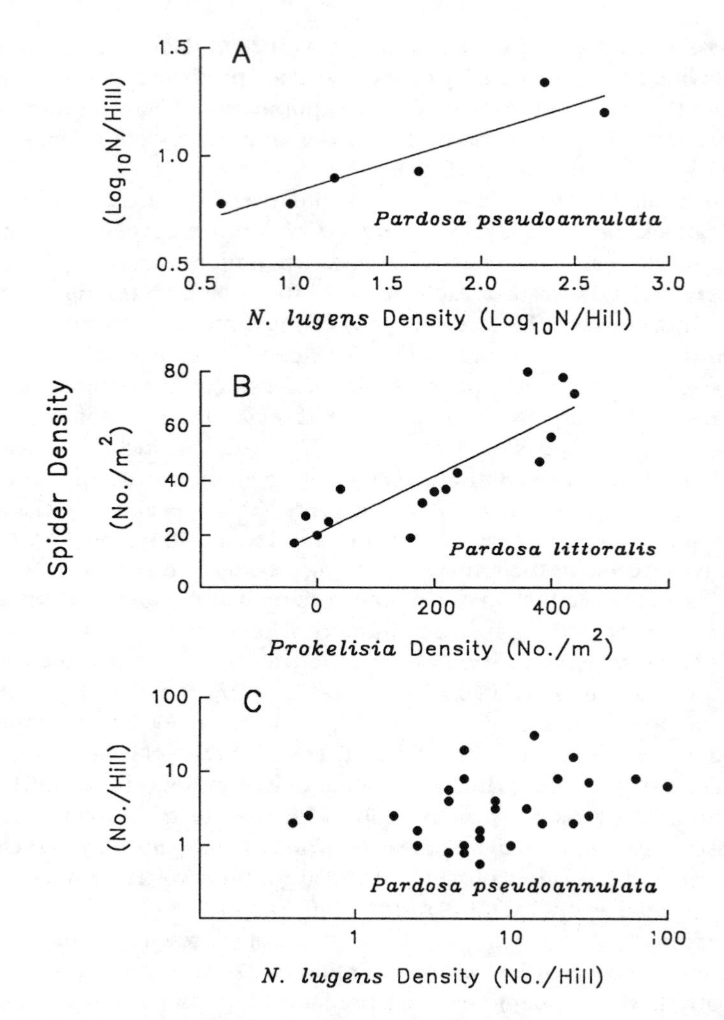

Figure 10.7. Numerical responses of lycosid predators to increasing densities of planthopper prey in the field. (**A**) Relationship between the densities of *Pardosa pseudoannulata* (saturation density) and *Nilaparvata lugens* ($\log_{10}$peak *N*/hill) in rice at IRRI, Los Baños, Philippines; $Y_{PRED} = 0.57 + 0.27X_{PREY}$ for $\log_{10}$-transformed data, $R^2 = 0.81$, $P<0.01$ (from Kenmore 1980). (**B**) Relationship between the density of *Pardosa littoralis* and experimentally manipulated densities of *Prokelisia* planthoppers (No./m^2) in small plots of *Spartina* on a salt marsh at Tuckerton, New Jersey, USA. Density relationships were assessed after 5 days, indicating that the spider's numerical response was due to immigration and not reproduction. $Y_{PRED} = 20.97 + 0.10X_{PREY}$, $R^2 = 0.66$, $P<0.01$ for nontransformed data (from Döbel and Denno in preparation c). (**C**) Relationship between the densities of *Pardosa pseudoannulata* and *Nilaparvata lugens* (No./hill) in rice at IRRI, Los Baños, Philippines; $Y_{PRED} = 0.92 + 0.30X_{PREY}$ for $\log_{10}$−transformed data, $R^2 = 0.10$, ns (from Kuno and Dyck 1985).

quence allocates more energy to reproduction as has been shown for *Pardosa psedoannulata* (Suzuki and Kiritani 1974). Similarly, under conditions of high prey availability, the web-building spider, *Linyphia triangularis*, shows enhanced reproduction (Turnbull 1962).

Döbel and Denno (in preparation c) examined the short-term numerical

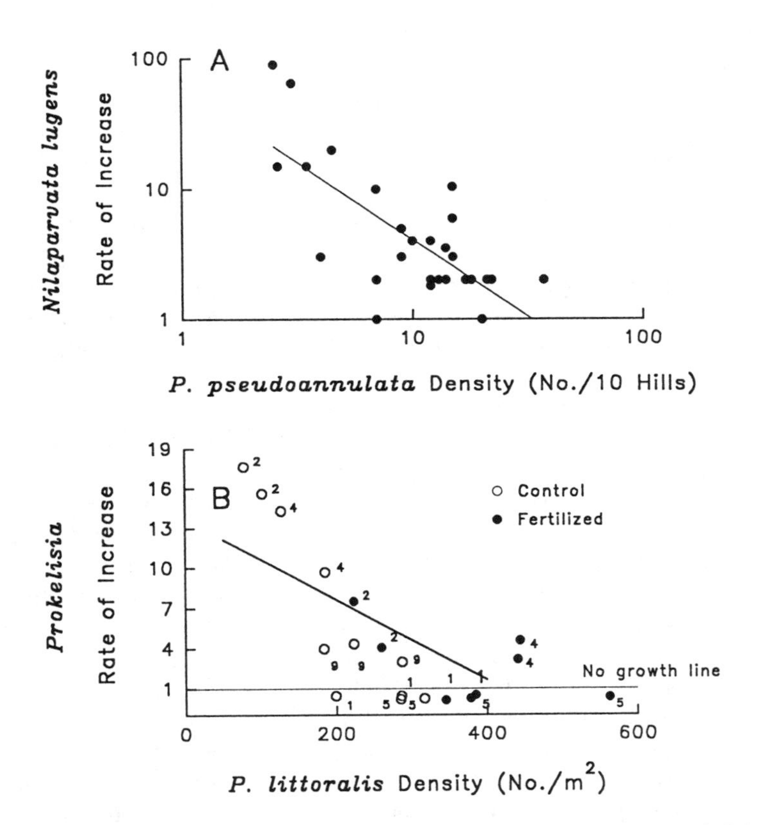

Figure 10.8. (**A**) Relationship between the rate of population increase of *N. lugens* (population growth from one generation to the next, 1 = no growth) and the density of spiders (*Pardosa pseudoannulata*) (No./10 hills) in rice, Tanjung Karang, Malaysia; $Y_{PREY\ INCREASE} = 1.82 - 1.22X_{PRED}$ for $\log_{10}$-transformed data, $R^2 = 0.55, P<0.01$ (from Ooi 1988). (**B**) Relationship between the rate of population increase of *Prokelisia* planthoppers (density in generation 3/density in generation 2, 1 = no growth) and the density of *Pardosa littoralis* (No./m^2) in fertilized and control plots of *Spartina* on a salt marsh at Tuckerton, New Jersey, USA; plot means from 1982 (2) through 1989 (9) are shown; $Y_{PREY\ INCREASE} = 13.65 - 0.03X_{PRED}$ for nontransformed data, $R^2 = 0.48, P<0.01$ (from Döbel and Denno in preparation c).

response of *Pardosa littoralis* to experimentally manipulated densities of *Prokelisia* planthoppers (0, 200, and 400 individuals/m^2) on a North American salt marsh. Their results showed that at the end of the 5-day experimental period, spider density was significantly and positively related to the density of planthoppers in plots providing direct evidence for a numerical response (Fig. 10.7B). Thus, due to the short duration of the experiment, the numerical response of the wolf spider must necessarily have resulted from immigration and not reproduction.

Also, by increasing the nutrient quality of *Spartina* via fertilization, populations of *Prokelisia* planthoppers can be elevated (Denno 1983). Over the short term, spiders (mostly *Pardosa littoralis* and *Lycosa modesta*) aggregated in fertilized plots where prey was most abundant (Döbel 1987). Similarly, Vince et al. (1981) argued that spiders prevented a long-term increase of *Prokelisia* populations by responding numerically to the initial buildup of planthopper populations in fertilized plots of *Spartina*. The resulting pattern across all plots several years after the initiation of fertilizer treatments was an inverse relationship between spider and planthopper density.

Further evidence for the ability of spiders (Lycosidae, Micryphantidae, and Clubionidae) to respond numerically to prey populations comes from prey supplementation experiments in the field. For example, Kobayashi (1975) released *Drosophila* flies as an additional prey source for spiders into dikes surrounding rice fields in Japan. Subsequently, spider density increased on the release dikes and in rice fields adjacent to them, documenting that spider density could be enhanced by supplementing their diet with alternative prey items. The numerical response of spiders was attributed to both aggregative and reproductive responses. Coincident with the increase in spider density in the rice fields was a decrease in populations of planthoppers (*Laodelphax striatellus* and *Nilaparvata lugens*) and leafhoppers (*Nephotettix cincticeps*). However, because hopper populations later rebounded, Kobayashi (1975) concluded that spiders were unlikely to control hopper populations over longer periods of time. This conclusion seems overly harsh given the slight nature of the planthopper rebound which may have resulted from a generational event such as egg hatch.

Perhaps the most convincing evidence for the role of predators in suppressing planthopper populations is the significant negative relationship between planthopper population increase and predator density (mirids, veliids and spiders; Figs. 10.5D, 10.6B, and 10.8) (Cook and Perfect 1985a; Kuno and Dyck 1985; Ooi 1988; Döbel and Denno in preparation c). The inverse nature of the relationship also provides strong support for the view that the action of some predators on planthopper populations is density-dependent.

Comparative Reproductive Capability and Dispersal Between Planthoppers and Their Invertebrate Predators

Weak numerical responses of predators can result in the underexploitation of prey in high-density patches and inverse density-dependence or density-independence (Smith and Maelzer 1986; Walde and Murdoch 1988). A weak numerical or aggregative response can result from searching time or egg/fecundity limitations (Walde and Murdoch 1988), submaximal oviposition in areas of high prey density (Cronin and Strong 1990a), and/or a discrepancy in dispersal capability between predator and prey with the predator being the relatively less mobile (Murdoch 1990). In this section, we contrast the reproductive potential and dispersal capability of planthoppers and their predators and explore how the interaction of these traits affects the numerical response of predators and the subsequent ability of predators to limit population growth in planthoppers.

A comparison of fecundities between delphacids and their predators reveals a decided reproductive advantage for planthoppers in most cases (Table 10.4). Most planthopper species (e.g., *Nilaparvata lugens*, *Sogatella furcifera*, and *Peregrinus maidis*) are two to three times as fecund (lifetime fecundities of 300–600 eggs/♀) as their mirid (100–300 eggs/♀), veliid (25–190 eggs/♀), and coccinellid predators (200 eggs/♀). Also, in temperate regions where many wolf spiders (*Lycosa* and *Pardosa*) are univoltine, their fecundities are comparatively low, ranging from 35 to 160 eggs/female (Kaston 1948; Döbel et al. 1990; Döbel unpublished data). The univoltine nature of temperate spiders along with their low fecundity have led authors to suggest that these predators are unlikely to effect a numerical response sufficient to limit the population growth of their prey (Renault and Miller 1972). In tropical and subtropical regions, lycosids such as *Pardosa pseudoannulata* produce several egg sacs (3–6) throughout adult life and are apparently much more fecund, producing 176–600 eggs/female (Gavarra and Raros 1975; Wang et al. 1982). Interestingly, the fecundities of both *P. pseudoannulata* and *Microvelia douglasi* are elevated substantially by feeding on a mixture of prey species compared to when they are fed only a single species of planthopper of leafhopper (Suzuki and Kiritani 1974; Sonoda et al. 1992; B. Barrion personal communication).

As adults in a laboratory setting, planthoppers generally live slightly longer (24–35 days) than their mirid predators (17–24 days). However, veliids, coccinellids, and lycosid spiders outlive their planthopper prey with adult longevities of 18–89 days, 35 days, and 60–257 days, respectively. In general, the generation times of mirids are a bit shorter than those of their planthopper hosts (Table 10.4), a trait which to some extent offsets

Table 10.4. Comparison of life history traits and potential for population growth between delphacid planthoppers and their invertebrate predators.

Taxa of Planthopper Prey and Predator	Fecundity (eggs/♀)[a]	Adult Longevity (days)[b]	Generation Time (Days)	Generations per Year[c] (Location)	Body Length (mm)	R_0	r_m	Reference
Planthopper Prey								
Nilaparvata lugens	300–600	24	23–29	$3–5^{JAP}$	3–4	98.9	0.16^{E}	Mochida and Okada (1979); Kuno (1979)
	—	—	29	12^{SEA}	—	15.4^{GG}	0.11^{E}	Heong (1982); Denno et al. (1989)
Sogatella furcifera	484	25	—	3^{JAP}	3–4	2.6^{GG}		Dyck et al. (1979) Kisimoto (1965); Kuno (1979)
Predators								
Cyrtorbinus lividipennis	21 ± 19	8 ± 3	25	8^{MAL}	3–4	11.0	0.10	Sivapragasam and Asma (1985); Ôtake (1977)
Microvelia douglasi	129	21	29	12^{PHIL}	—	44.1	0.13	Manti (1989)
	25 ± 2	18 ± 2	30	12^{PHIL}	2	22.8	0.01	Ôtake (1977); Nakasuji and Dyck (1984)
Pardosa pseudoannulata	95–190	38–89	—	—	—	—	—	Sonoda et al. (1992)
	176 ± 74	134 ± 35	264	—	—	—	—	Gavarra and Raros (1975)
	300 ± 55	—	—	—	—	—	—	B. Barrion (personal communication)
Coccinella arcuata	601	104–257	141	$2–3^{CHI}$	8–10	—	—	Wang et al. (1982)
	210	35	20–22	12^{IND}	7	—	—	Abraham et al. (1973)
Planthopper Prey								
Prokelisia dolus	103 ± 11	28 ± 2	40	3^{USA}	3	51.3	0.10^{E}	Denno et al. (1987,

Prokelisia marginata	67 ± 9	16	42	3[USA]	3	18.7	0.07[E]	Roderick (1987); Denno et al. (1989)
Predators								
Pardosa floridana	43 ± 9	—	—	1[USA]	5–7	—	—	Kaston (1948)
Pardosa littoralis	35 ± 7	60–90[L]	330	1[USA]	5–6	—	—	Döbel (1987); Döbel (unpublished data)
Lycosa modesta	55 ± 21	60–90[L]	330	1[USA]	8–10	—	—	Döbel (1987); Döbel (unpublished data)
Planthopper Prey								
Peregrinus maidis	605 ± 190	35 ± 21	49	7[HAW]	3	31.9	0.07	Napompeth (1973) Tsai and Wilson (1986)
Predator								
Cyrtorhinus lividipennis	306	24 ± 4	44	7[HAW]	3	15.9	0.06	Napompeth (1973)
Tytthus mundulus	107 ± 57	17 ± 5	18–31	10[HAW]	3	—	—	Williams (1931); Verma (1955)
Planthopper Prey								
Perkinsiella saccharicida	300	30	48–56	5[HAW]	3–4	—	—	Swezey (1936); Fennah (1969)
Predator								
Tytthus mundulus	107 ± 57[PM]	17 ± 5[PM]	18–31[PM]	10[HAW]	3	—	—	Williams (1931); Verma (1955)

Note: E: r_m estimated from generation time and R_0 or Generational Growth Rate (GG) [see Kuno (1979) for calculation of Generational Growth Rate]. L: Adult survival in the laboratory under conditions of sporadic feeding. PM: Determined for *Tytthus mundulus* when offered *Peregrinus maidis* as prey.

[a]Reported fecundities of planthoppers are for brachypters; fecundities of macropters are generally lower (see Denno et al., 1989).

[b]Reported longevities of planthoppers are for brachypters; longevities of macropters are generally the same (see Denno et al., 1989).

[c]Maximum number of generations for planthoppers in tropical locations (China, India, Malaysia, Philippines, and SE Asia) estimated from generation time on the basis of continuous cropping.

the fecundity advantage of planthoppers. For example, the generation times of *Cyrtorhinus lividipennis* and *Tytthus mundulus* range from 18 to 44 days, whereas those of *Nilaparvata lugens* (23–29 days), *Peregrinus maidis* (49 days), and *Perkinsiella saccharicida* (48–56 days) are longer. The generation time of *M. douglasi* is approximately 30 days, similar to that for *N. lugens*, and the generation times for spider predators are decidedly longer (141–264 days for *P. pseudoannulata*), particularly for univoltine species in temperate regions (330 days for *Lycosa modesta* and *Pardosa littoralis*).

In the few cases where enough life history data exists to calculate an intrinsic rate of increase (r_m), planthoppers are projected to have higher reproductive potentials than their mirid and veliid predators (Table 10.4). Due to very long generation times and in some cases low fecundities, the reproductive potential of lycosid predators is much less than their planthopper prey. Although to a lesser extent, this is true for coccinellid beetles as well, at least for *Coccinella arcuata* with its comparatively low fecundity.

An overall comparison of reproductive traits suggests that on the surface planthoppers have the potential to escape from many of their predators with mirids presenting the most formidable chase. For example, the successful biological control of *Perkinsiella saccharicida* on sugarcane by *Tytthus mundulus* has been attributed to the mirid's short generation time and continuous reproduction, where in Hawaii it undergoes 10 generations/year compared to only 3 or 4 for the planthopper (Williams 1931; Swezey 1936; Rothschild 1966). Similarly, *Cyrtorhinus lividipennis* was able to "control" or dramatically suppress populations of *Nilaparvata lugens* and *Peregrinus maidis* if both predator and prey simultaneously colonized the crop (Napompeth 1973; Manti 1989).

However, other predators with combinations of life history traits that do not obviously favor rapid population growth should not be discounted when considering the suppression of planthopper populations. In particular, lycosid spiders fall into this category. As adults, these predators are large (5–10 mm in length; Table 10.4), and because capture rate is roughly related to body size, they can consume many more planthoppers/day than smaller predators (Table 10.3). Thus, their voracious appetite coupled with long adult life may compensate to some extent for their extended generation time (delayed age to first reproduction) and reduced fecundity compared to planthoppers.

That high capture rate alone can be sufficient enough to balance population growth in planthoppers is demonstrated by caging studies in which planthopper populations have been suppressed when confined with *Pardosa* spiders at natural field densities (Döbel 1987). Caging studies in general show that planthopper population growth can be drastically limited in confined settings with predators (Dyck and Orlido 1977; Kenmore et al. 1984; Döbel 1987; Fowler 1987; Ooi 1988; Manti 1989). Nonetheless, it is their suite of life history traits (low fecundity and long generation time) that has undoubtedly perpetuated the notion that spiders in particular and predators in general are effective at checking population growth only at low planthopper densities (Metcalfe 1971; Ôtake 1977).

We argue that the apparent discrepancy in reproductive potential between planthoppers and their predators is not the single most important factor contributing to the outbreak of planthopper populations. Our contention is based on the observations that some predators (*Cyrtorhinus lividipennis*) have reproductive potentials very similar to those of their prey, that other predators (e.g., Lycosidae) have ravenous appetites and partially offset their reproductive disadvantage with high consumption rates, and that, under confined conditions, planthopper populations are often checked by predators. Rather, we argue that inherent differences in dispersal capability between planthoppers and their predators and/or other external forces (e.g., habitat persistence/disturbance) are more important in limiting the numerical (aggregative) response of predators, uncoupling predator–prey interactions, and allowing planthopper populations to grow relatively unchecked.

Let us consider how intrinsic differences in dispersal capability may lead to asynchrony in planthopper–predator interactions. Specifically, we consider how wing-dimorphism in planthoppers promotes the spatial and temporal escape of planthopper populations from their predators. Most species of planthoppers are wing-dimorphic with macropterous and brachypterous adults in the same population (Denno and Roderick 1990). Macropterous adults are the colonizing wing form and are known to migrate great distances (Kisimoto 1976; 1981), but they pay a reproductive penalty for their flight capability as evidenced by delayed oviposition and reduced fecundity (Denno et al. 1989). Brachypterous adults, on the other hand, reproduce at an earlier age and are generally more fecund than macropters; however, they are flightless and have traded off dispersal ability for increased reproductive potential (Denno et al. 1989).

A common "colonization syndrome" exists for several multivoltine species of planthoppers (e.g., *Nilaparvata lugens, Peregrinus maidis,* and *Sogatella furcifera*) exploiting temporary habitats, which illustrates the advantage of a polymorphic life history (Denno and Roderick 1990). In a new crop, macropters colonize at low densities, their offspring are mostly brachypterous, and rapid population growth ensues; in the following generation, nymphs molt to macropterous adults which can escape deteriorating conditions by dispersing to other habitats (Kisimoto 1965; Napompeth

1973; Kuno 1979; Fisk et al. 1981). The sequence of wing-form change is mediated largely by increasing density and deteriorating plant quality (Denno and Roderick 1990). Thus, planthoppers are able to maximize both dispersal capability as well as reproduction by partitioning these functions to some extent between two wing forms (Denno et al. 1989).

For predators, the upshot of this polymorphic life history is that it is makes planthopper populations very difficult to track. Because most planthopper predators are monomorphic, reproduction and dispersal are compromised in one morph. Consequently, polymorphic planthoppers are "preadapted" to escape their enemies in both space (efficient colonization of new habitats by macropters) and time (rapid establishment and population growth resulting from the high fecundity of brachypters; see Kisimoto 1965).

To a great extent, habitat persistence dictates the overall proportion of migratory and flightless forms in delphacid populations (Denno et al 1991). Species in temporary habitats are polymorphic but with high proportions of macropterous adults, whereas those in permanent habitats are mostly brachypterous (Denno et al. 1991). Thus, the discrepancy in dispersal capability between planthoppers and their predators may be amplified in ephemeral habitats (agricultural crops) where, due to their polymorphic life style, planthoppers will often own the advantage.

It comes as no surprise that our worst agricultural pests (e.g., *Nilaparvata lugens* and *Peregrinus maidis*) are wing-dimorphic, and outbreaks in these species have been attributed to the asynchrony with predator life histories which is promoted by their flexible, polymorphic life-style (see Kuno 1979; Kenmore 1980; Denno and Roderick 1990). An example from temperate Asia serves to underscore this point. Neither *Nilaparvata lugens* nor *Sogatella furcifera* overwinter successfully in temperate Asia and the rice fields of Japan and Korea are colonized annually by macropters emigrating from China (Kisimoto 1976). With the second-generation appearance of brachypterous females, exponential population growth ensues which often terminates during the third generation with an outbreak population composed primarily of macropterous adults (Kisimoto 1965; Kuno 1979). *Cyrtorhinus lividipennis* also fails to overwinter in Japan and migrates transoceanically from China (Kisimoto 1981), but it colonizes later and at lower densities and, as a consequence, never plays a significant role in the suppression of rice planthoppers in the region (see Kuno and Hokyo 1970, Kuno 1979; Kenmore 1980). By contrast, natural enemies are thought to play a much greater role in suppressing *N. lugens* in tropical Asia where the opportunity for planthoppers and predators to synchronously colonize new rice fields is much higher (Cook and Perfect 1985a; Ooi 1988; Manti 1989).

Both species of planthoppers which have been effectively controlled with introduced biological agents (mirids) have life history characteristics predisposing them for suppression. *Perkinsiella saccharicida* is a primarily macropterous species with a fecundity (300 eggs/♀; Swezey 1936; Fennah 1969) much less than that reported for the brachypters of *Nilaparvata lugens* and *Peregrinus maidis* (up to 600 eggs/♀; Kisimoto 1965; Napompeth 1973; Tsai and Wilson 1986; Denno et al. 1989). We would argue that although *P. saccharicida* is an excellent colonist (see Swezey 1936), its comparatively low fecundity has provided the opportunity for *Tytthus mundulus* to reproductively track and effectively suppress this sugarcane pest. This same mirid predator proved ineffective in suppressing populations of the very fecund, wing-dimorphic planthopper *Peregrinus maidis* (Verma 1955; Napompeth 1973). Thus, a predator which proves effective against one planthopper species may not be for another, and it is the interaction of a predator's combination of life history traits with those of a particular prey species which must be considered.

The taro planthopper (*Tarophagus colocasiae*) is a primarily brachypterous species and has been effectively controlled with the introduction of *Cyrtorhinus fulvus* into Hawaii (Zimmerman 1948; Matsumoto and Nishida 1966). The mirid predator is a stronger flier than the planthopper and is able to effect a rapid numerical response (Matsumoto and Nishida 1966). Thus, although *T. colocasiae* undoubtedly enjoys a reproductive advantage over *C. fulvus* due to the elevated fecundity of brachypterous planthoppers, it owns a decided dispersal disadvantage which allows *C. fulvus* to effectively track spatial changes in planthopper populations.

Over long distances, migratory planthoppers have a clear colonizing advantage over their predators (Kisimoto 1965, 1981; Kuno 1979), and the swamping of a new habitat by many colonizing macropters can lead to the escape from natural enemies and population outbreak (Kenmore 1980; Denno 1983; Ooi 1988). However, if new habitats are not far (meters) from source areas, many of the predators of planthoppers can colonize simultaneously with their planthopper prey. For example, mirids can colonize new fields of rice, corn (*Cyrtorhinus lividipennis*; Stapley 1976; Napompeth 1973; Ooi 1988; Manti 1989), taro (*Cyrtorhinus fulvus*; Matsumoto and Nishida 1966), and sugarcane (*Tytthus mundulus*; Williams 1931; Rothschild 1966) with their planthopper hosts and, in so doing, effectively suppress planthopper populations. Similarly, lycosid spiders can effect rapid numerical responses over short distances by colonizing new habitats shortly after their establishment (Kobayashi 1975; Kenmore 1980; Döbel and Denno in preparation c). Colonization may result from the immigration of mobile adults or the ballooning of spiderlings (Kenmore et al. 1984; Ooi 1988). However, ballooning is poorly documented in most systems (D.

Wise personal communication) and dispersal by walking should be considered the major means by which wolf spiders effect a numerical response until proven othewise. Nonetheless, due to their mobility, lycosids are considered important predators in the suppression of planthoppers during the early stages of a crop (Ooi 1988; Ooi and Shepard in press). However, it is wise to point out here that there are very few hard data on the relative dispersal and colonization rates of predators and planthoppers into new habitats and how habitat structure influences these rates. Such data hold the key to understanding and managing predator/planthopper interactions.

The above examples point to the importance of temporal and spatial synchrony between the life histories of planthoppers and their predators as criteria for the effective suppression of planthopper populations. Whereas there are inherent reproductive and dispersal advantages for polymorphic planthoppers that tend to desynchronize and uncouple their population dynamics from most predators, some predators (e.g., *Cyrtorhinus lividipenis* and *Pardosa pseudoannulata*) possess life history traits that allow them to rather effectively chase and subdue planthopper populations, at least under certain conditions. Although intrinsic life history characteristics of some predators (e.g., poor dispersal capability, low reproductive potential, and small body size) undoubtedly preclude them from ever becoming important sources of planthopper mortality, more often than not it is external factors (e.g., climate, habitat disturbance, and pesticide applications) that are responsible for uncoupling the life histories of otherwise effective predators.

Life Cycle Synchrony, Altered Numerical Responses, and the Suppression of Planthopper Populations

Even those predators regarded as the most important in planthopper systems (e.g., *Pardosa*, *Microvelia*, and *Cyrtorhinus*) vary in their numerical response to planthopper populations (Kuno and Dyck 1985; Heong et al. 1990a). A weak numerical response can result in life cycle asynchrony and often leads to population outbreak (see Manti 1989). Numerous factors, including the structure of the habitat, the distance of new habitats from predator refuges, the presence and identity of alternative prey (potential for switching), mutual interference among conspecific predators, and natural (harsh winters) and artificial perturbations (pesticides), can all delay or weaken the numerical response of predators and/or desynchronize predator/prey interactions (Hassell 1978; Walde and Murdoch 1988). In this section we provide evidence for the importance of life cycle synchrony for prey suppression and discuss factors which alter the numerical response and uncouple predator–planthopper population dynamics.

Delayed Numerical Response and Planthopper Escape

In general, if predators colonize habitats late or in low numbers relative to their planthopper prey or if their life histories become asynchronous, predators are unlikely to inflict sufficient mortality to suppress planthopper populations (Table 10.5). The strongest case for the importance of a rapid numerical response to prey suppression comes from studies in which the arrival time of predators to the planthopper system was experimentally manipulated. Using field cages, Manti (1989) staggered the arrival time of *Cyrtorhinus lividipennis* and then measured its subsequent ability to suppress populations of *N. lugens*. Cages were stocked with four and two pairs of *N. lugens* and *C. lividipennis*, respectively. However, the timing of predator "arrival" varied across four treatments; predators were added 0, 7, 14, and 21 days after planthopper inclusion. Simultaneous "arrival" of the mirid and "arrival" after 7 days resulted in the effective control planthopper populations. In contrast, when mirids arrived 14 and 21 days late, planthopper populations exploded and were not significantly different from control cages lacking predators.

A similar picture emerges in the field where the failure of mirid bugs to control planthopper populations can usually be attributed to a delayed numerical response and examples include *C. lividipennis* on *N. lugens* (Dyck and Orlido 1977; Bentur and Kalode 1987) and *Peregrinus maidis* (Napompeth 1973), and *Tytthus mundulus* on *Perkinsiella saccharicida* (Osborn 1974). Most notably, the consistent failure of *Tytthus mundulus* to suppress populations of *P. maidis* was credited to the mirid's late colonization of corn (Verma 1955; Napompeth 1973).

Coccinellid beetles often colonize habitats late (*Illeis indica* in sorghum, Fisk et al. 1981; *Coccinella arcuata* in rice, Ng 1978) or appear late during the season (*Coccinella septempunctata* in *Juncus*; Rothschild 1966) and, as a consequence, contribute little to planthopper mortality and the suppression of planthopper populations (Rothschild 1966; Ng 1978). Observations for mirids and coccinellids confirm the importance of a rapid and strong numerical response as an important ingredient for planthopper suppression.

Predator/Prey Ratios and Planthopper Suppression

Not only is the timing of the numerical response important for effective prey suppression, but so is the strength of the numerical response or the

Table 10.5. Factors affecting the coupling of planthopper–predator interactions (relative dispersal ability, timing of habitat colonization, and/or degree of life history synchrony between planthoppers and their predators) and the resulting effect on planthopper population dynamics.

Planthopper Taxon	Predator Taxon	Crop/Host Plant (Location)	Predator Dispersal (D), Colonization (C), and Life History (LH) Synchrony Relative to Prey	Population Response of Prey	Reference
Conomelus anceps[B]	*Oligolophus agrestis*[OP]	*Juncus* marsh (UK)	Synchronous LH	Population control	Rothschild (1966)
	Tytthus pygmaeus	*Juncus* marsh (UK)	Synchronous LH	High nymph mortality	Rothschild (1966)
		Juncus marsh (UK)	Asynchronous LH	High egg survival	Rothschild (1966)
	Coccinella septempunctata	*Juncus* marsh (UK)	Asynchronous LH	Low mortality	Rothschild (1966)
Nilaparvata lugens[M]	*Cyrtorhinus lividipennis*	Rice (Philippines)	Early C by predator[E]	Population control	Manti (1989)
			Late C by predator[E]	Outbreak	
		Rice (Philippines)	Early C by prey	Outbreak	Ooi (1988)
			Early C by predator[I]	Population control	
		Transplanted rice (Fiji)	Pred D<Prey D	Outbreak	Hinckley (1963)
		Rice (India)	Late C by predator	No suppression	Bentur and Kalode (1987)
		Rice (Philippines)	Asynchronous LH Late C by predator	Population growth	Dyck and Orlido (1977)
	Pardosa pseudoannulata	Rice (Malaysia)	Synchronous LH	Population decline	Ng (1978)
		Rice (Solomon Is.)	Early C by predator[AH]	Population control	Stapley (1976)
	Coccinella arcuata	Rice (Philippines)	Early C by predator[I]	Population control	Ooi (1988)
		Rice (Malaysia)	Asynchronous LH	Outbreak	Ng (1978)

Peregrinus maidis[M]	*Cyrtorhinus lividipennis*	Corn (Hawaii)	Synchronous C	Suppression	Napompeth (1973)
			Late C by predator	Outbreak	
Perkinsiella saccharicida[M]	*Tytthus mundulus*	Corn (Hawaii)	Asynchronous LH	No suppression	Napompeth (1973)
	Illeis indica[OC]	Sorghum (India)	Late C by predator	No suppression	Fisk et al. (1981)
	Tytthus mundulus	Sugarcane (Hawaii)	Synchronous LH	Biological control	Williams (1931); Rothschild (1966)
		Sugarcane (Australia)	Asynchronous LH	No suppression	Osborn (1974)
		Sugarcane (Australia)	Synchronous LH	Suppression	Bull (1981)
Prokelisia dolus[B,MP]	*Pardosa littoralis*	*Spartina* marsh	Late C by predator[NR]	Outbreak	Döbel and Denno (in preparation c)
			Early C by predator	Suppression	
Sogatella furcifera[M]	*Cyrtorhinus lividipennis*	Rice (Philippines)	High C rate by pred	Low pop. growth	Cook and Perfect (1985a)
Tarophagus colocasiae[B]	*Cyrtorhinus fulvus*	Taro (Hawaii)	Pred D>Prey D	Suppression	Matsumoto and Nishida (1966)

NOTE: AH Early season presence of predators in rice depended on the close proximity of weeds with alternate prey species.

B Planthopper adults mostly brachypterous (>90%).

E Determined by experimentally manipulating the arrival time of predators relative to prey in field cages.

I With insecticide kill of early colonists (predators), planthopper populations are likely to outbreak.

M Planthopper adults mostly macropterous (>75%); predator failed to respond numerically to within-field changes in planthopper distribution; reproductive response by planthoppers also implicated.

MP Planthopper populations were mixed and were comprised of both *Prokelisia dolus* (70%) and *P. marginata* (30%).

NR Populations of planthoppers increased when the numerical response of spiders was experimentally impeded in the field.

OC Other coccinellids with the same delayed colonization included *Brumoides saturalis* and *Menochilus sexmaculatus*.

OP Other Phalangida with the same response included *Leiobunum blackwalli* and *Platybunus triangularis*.

relative density of predators. For aphids, natural enemy/prey ratios must be relatively high soon after the colonization of a new habitat if outbreaks are to be prevented (Wellings and Dixon 1987). This begs the specific question of exactly how high predator/prey ratios must be to deter population growth in planthoppers. Some data, mostly from agricultural systems, are available to roughly address this question (Table 10.6). For instance, pop-

Table 10.6. Predator/prey ratio and its effect on the mortality and/or population dynamics of *Nilaparvata lugens.*

Predator Taxon	Predator/ Planthopper Prey Ratio	Population Response of Planthopper	Reference
Lycosidae			
Pardosa pseudo-annulata	1:10	100% daily nymphal mortality[C]	Dyck and Orlido (1977)
	1:50	80% daily nymphal mortality[C]	
	1:100	41% daily nymphal mortality[C]	
	1:10	100% daily adult mortality[C]	
	1:50	46% daily adult mortality[C]	
	1:100	40% daily adult mortality[C]	
Lycosa sp.	1:8	No outbreak[F]	Chiu (1984)
Pirata sp.	1:8	No outbreak[F]	Chiu (1984)
Linyphiidae			
Erigonidium graminicolum	1:4	No outbreak[F]	Chiu (1984)
Miridae			
Cyrtorhinus lividipennis	1:1	79% daily nymphal mortality[C]	Dyck and Orlido (1977)
	1:20	23% daily nymphal mortality[C]	
	1:1	73% daily mortality[C]	Chiu (1979)
	1:2	Effective control[C]	Manti (1989)
	1:20	Effective control	Stapley (1976)
Coccinellidae			
Coccinella (=*Harmonia*) *arcuata*	1:4	77% daily nymphal mortality[C]	Dyck and Orlido (1977)
	1:4	95% daily adult mortality[C]	
"Predators"[a]	1:50	Effective control[F]	Ooi (1988)

Note: F = field study; C = cage study.
[a]Predators included *Cyrtorhinus lividipennis,* veliids, spiders, and predatory beetles.

ulations of *Nilaparvata lugens* in rice are "effectively controlled" in the field at predator/prey ratios of 1 : 4–1 : 8 for lycosid spiders and 1 : 2–1 : 20 for *Cyrtorhinus lividipennis* (Stapley 1976; Chiu 1984; Manti 1989). Laboratory estimates derived from cage studies (all players were simultaneously introduced) show that many more *C. lividipennis* (1 : 1–1 : 2, predator to prey) and coccinellid beetles (1 : 4) are necessary to inflict the same high levels of mortality (>70%) on *N. lugens* as do a relatively few lycosid spiders (1 : 50–1 : 100) (Dyck and Orlido 1977). Ooi (1988) estimates that all predators combined (primarily spiders mirids and predaceous beetles) at a predator/prey ratio of 1 : 50 will provide effective suppression of planthopper populations in the field.

What little data are available for natural systems suggest a slightly different pattern. In the Atlantic coast salt marshes of North America where wolf spiders (*Pardosa littoralis* and *Lycosa modesta*) are the dominant predators on *Prokelisia* planthoppers, a spider/planthopper (adults or nymphs) ratio of 3 : 1 and higher prevents population growth (Fig. 10.9). Thus, higher predator/prey ratios may be required for the suppression of wing-dimorphic planthopper species in this relatively harsh and heterogeneous environment.

These ratios are only general estimates and should not be taken as fool-proof values for guaranteed planthopper suppression. Further, these ratios represent averages and little is known about how temporal variation in ratios will influence planthopper population dynamics. However, these ratios do identify two important points. First, if mirids and other predators with relatively high suppression ratios (e.g., Coccinellidae) are to be effective, they must colonize new habitats not only early, but in relatively high numbers; conversely, low densities of predators with low suppression ratios (e.g., some spiders) may effectively retard population growth. Second, low colonization densities by certain predators (e.g., *Cyrtorhinus lividipennis*) may be compensated for if other predators simultaneously found new habitats.

Habitat Structure

Outbreaks of aphids in patchy compared to continuous stands of goldenrod (*Solidago*) resulted from the reduced ability of coccinellid predators to aggregate and control aphid populations in patchy configurations of hosts (Kareiva 1987). Similarly, in a wolf spider-dominated salt marsh system, a discontinuous habitat disrupted the numerical response of *Pardosa littoralis*, destabilized interactions with *Prokelisia* planthoppers, and led to population outbreaks (Döbel and Denno in preparation c). For instance, spider–planthopper interactions were studied on small "islands" of *Spartina*

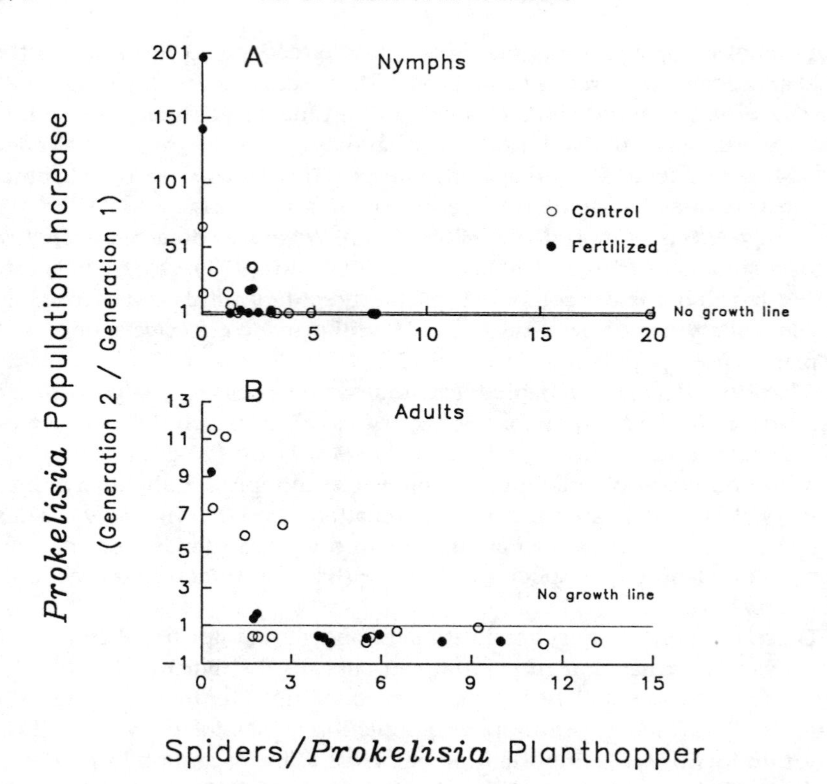

Figure 10.9. Relationship between the rate of population increase of *Prokelisia* planthoppers (density in generation 2/density in generation 1, 1 = no growth) and the ratio of *Pardosa littoralis* to planthopper density (No./m^2) in fertilized and control plots of *Spartina* on a salt marsh at Tuckerton, New Jersey, USA; ratios were calculated using both planthopper nymphs (**A**) and adults (**B**); planthopper populations do not increase at ratios of 3 or more (spiders : planthoppers); from Döbel and Denno (in preparation c).

surrounded by tidewaters. Islands were about 10 m^2 in area and were separated from each other and the "mainland" marsh by at least 10 m. Spider and planthopper densities could be manipulated easily on these islands, but more importantly, the numerical response of spiders could be limited because spider immigration from the "mainland" was restricted across the open water. All spiders and other arthropods were removed by insect vacuum from all experimental islands during late July when most of the planthopper population was in the egg stage. Because the eggs of *Prokelisia* are inserted into the blades of *Spartina*, they were not removed by the vacuum sampler. Nymphs, after hatching from eggs, served as prey for spiders.

Four days after spider removal, two spider treatments were applied to the experimental islands. On half the islands, wolf spiders were replaced at an ambient marsh density of 100 spiders/m^2. On the remaining islands, spiders were not added and their arrival on the island was dependent on natural, but very impeded immigration. Islands from which spiders were removed carried low densities of spiders throughout the season, confirming that immigration was dramatically reduced, whereas islands on which spiders were added carried the replacement density of 100 spiders/m^2 through the season, suggesting that spiders remained on these islands (Fig. 10.10A). On islands where spiders were removed and never replaced, planthopper populations increased rapidly, whereas on islands where spiders were removed and replaced, planthopper populations remained significantly smaller and increased very slowly (Fig. 10.10B).

To measure the effects of spiders on planthoppers in a situation where the numerical response of spiders was not impeded by open water, the same experiment was conducted on the mainland in large, continuous patches of *Spartina* (Döbel and Denno in preparation c). Here, instead of experimental plots being isolated islands surrounded by water, plots were surrounded by an expanse of undisturbed *Spartina*. As before, spiders and arthropods were removed from all plots, and on half of the plots, spiders were replaced and on the other half, they were not. Thus, for spiders to recolonize these experimental plots, they had only to move a few meters from surrounding vegetation. Spider density did not differ between the two treatments and even though all spiders were removed from one set of plots, they had quickly recolonized. Similarly, planthopper populations did not differ between the spider treatment plots. These results underscore the interaction between the numerical response of wolf spiders, the ability to suppress planthopper populations, and habitat structure. In the experimental island situation, the removal of wolf spiders led to planthopper outbreak because their numerical response was interrupted by the discontinuous nature of the habitat. In contrast, spider movement was not impeded in the unbroken mainland setting and spiders quickly reentered experimental plots where they checked the population growth of *Prokelisia*. Thus, the large-scale structure of the habitat can influence predator search, alter the numerical response, and affect the coupling of spider–planthopper interactions.

Small-scale habitat structure (the amount of thatch or litter associated with the host plant of the planthopper) can also influence the strength of the numerical response of *Pardosa littoralis* to its *Prokelisia* prey. For example, following defaunation, Döbel and Denno (in preparation b) ap-

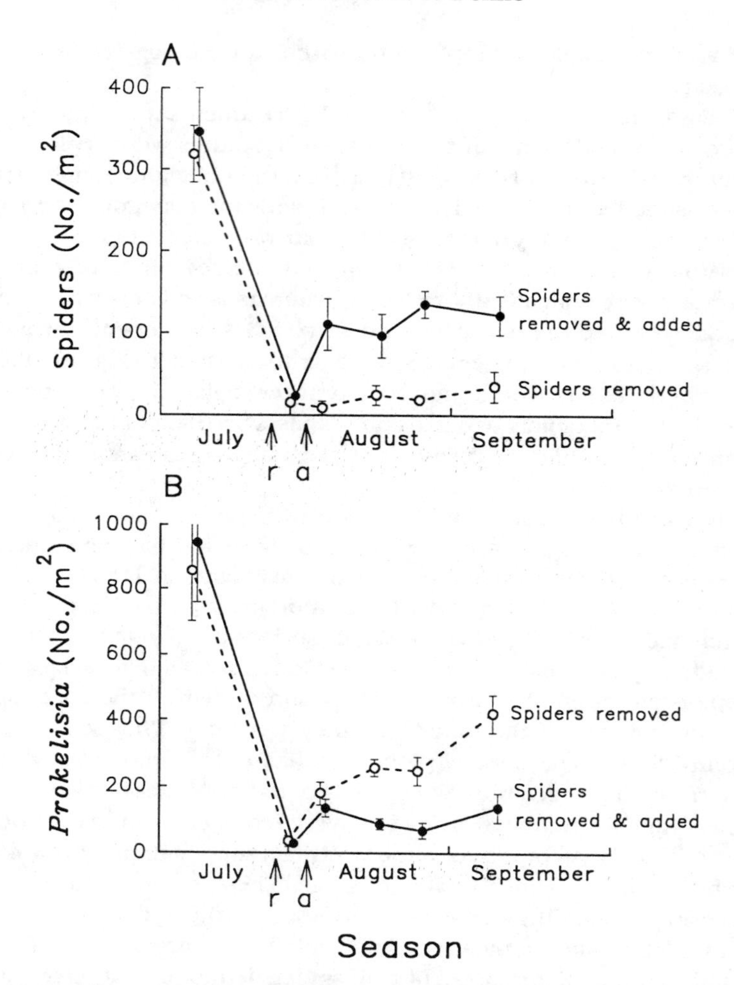

Figure 10.10. Effect of predator removal (*Pardosa littoralis*) on the population size of *Prokelisia* planthoppers on small islands of *Spartina* (surrounded by water) on a salt marsh at Tuckerton, New Jersey, USA. (**A**) Islands from which spiders were removed (r) carried low densities of spiders throughout the season confirming that immigration was dramatically reduced; islands on which spiders were removed and added (a) carried the replacement density of 100 spiders/m^2 through the season, suggesting that spiders remained on these islands (ANOVA, $F_{1,63}$ = 160.56, $P<0.01$). (**B**) Islands from which spiders were removed and never replaced carried significantly higher populations of planthoppers compared to islands on which spiders were removed and replaced (ANOVA, $F_{1,63}$ = 108.01, $P<0.01$); from Döbel and Denno (in preparation c).

plied three thatch treatments (no thatch, normal thatch, and extra thatch) and three planthopper density treatments (0, 200, and 400 hoppers/m^2) to small plots of *Spartina* grass. After 5 days, there was a significant and rapid numerical response of spiders to increasing planthopper density for all three thatch treatments (Fig. 10.11). However, there was a highly significant density × thatch interaction (P=0.046); the numerical response of spiders to planthopper density was much stronger in plots containing thatch than those without thatch and was especially strong in plots with extra thatch. Moreover, spiders accumulated more in thatch-containing plots even though planthoppers were absent. This density pattern suggests that thatch affected the numerical response of spiders directly in addition to

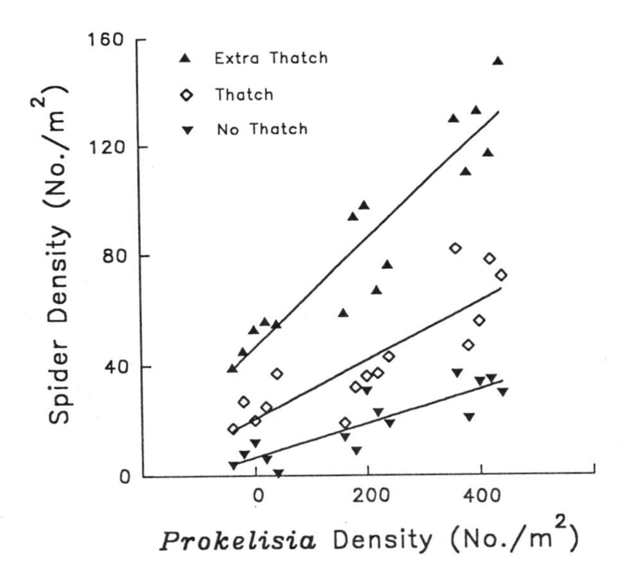

Figure 10.11. Relationship between the density of *Pardosa littoralis* and experimentally manipulated densities of *Prokelisia* planthoppers (No./m^2) in small plots of *Spartina* varying in habitat structure on a salt marsh at Tuckerton, New Jersey, USA. Plots contained either no thatch, normal thatch, or twice the normal amount of thatch. Density relationships were assessed 5 days after the application of density and thatch treatments, indicating that the spider's numerical response was due to immigration and not reproduction. The numerical response of spiders to planthopper density was much stronger in plots with thatch; $Y_{\text{PRED NO THATCH}} = 6.3 + 0.06X_{\text{PREY}}$, $R^2 = 0.74$, $P<0.01$ for nontransformed data; $Y_{\text{PRED THATCH}} = 20.97 + 0.10X_{\text{PREY}}$, $R^2 = 0.66$, $P<0.01$; $Y_{\text{PRED EXTRA THATCH}} = 46.27 + 0.20X_{\text{PREY}}$, $R^2 = 0.84$, $P<0.01$; significant Thatch × Density Interaction, ANOVA, $F_{2,44} = 14.03$, $P<0.01$ (from Döbel and Denno in preparation c).

the effects of prey density. Spiders aggregated in vegetation with more thatch, perhaps because thatch acts as a refuge for spiders allowing them to hide from their own avian predators (see Kale 1965; Pfeiffer and Wiegert 1981). Thus, the aggregative response of spiders to planthopper density is modified in a major way by variation in the amount of thatch.

That a strong numerical response of wolf spiders in thatch-rich habitats influences planthopper population dynamics was demonstrated by a long-term field experiment in which levels of thatch were experimentally manipulated in large plots of *Spartina* on the salt marsh (Döbel and Denno in preparation b). Three thatch treatments (no thatch, normal thatch, and extra thatch) were applied to field plots during early spring and the density of *Prokelisia* planthoppers and wolf spiders were monitored throughout the year. Populations of spiders were significantly higher and planthoppers significantly lower in plots with thatch compared to thatch-free plots (Figs. 10.12A,B). Furthermore, the ratio of spiders to planthoppers (spider load) was much higher in stands of *Spartina* with thatch (Fig. 10.12C).

Interactions Between Predators and Host Plants

The extent to which the architecture, surface structure, and chemistry of host plant species or varieties interact to influence predator–planthopper dynamics is poorly investigated. Generally, host plant characteristics are likely to influence predators directly by altering their ability to search and locate prey (e.g., effect a numerical response) or indirectly by modifying prey suitability (e.g., size and palatability) and/or availability (e.g., development rate and dispersion) (Barbosa 1988; Roland 1990; Hare 1992).

For instance, the searching efficiency of *Cyrtorhinus lividipennis* varies across rice cultivars, resulting in differences in the mortality of *Nilaparvata lugens* and *Sogatella furcifera* (Cheng et al. 1992). Similarly, *Pardosa pseudoannulata* inflicts more mortality on *Nilaparvata lugens* when it is feeding on resistant compared to susceptible rice varieties (Kartohardjono and Heinrichs 1984). Consequently, resistance- and predator-inflicted mortality were synergistic. In contrast, mortality from *C. lividipennis* did not vary across varieties, and planthopper mortality from plant resistance and predation was merely additive. Because *N. lugens* probes more and is generally more active on resistant varieties (Sogawa 1982), Kartohardjono and Heinrichs (1984) suggested that *N. lugens* was more susceptible to predation by visually responsive predators such as lycosid spiders that react only to moving prey.

When raised on resistent compared to susceptible varieties of rice, *Nilaparvata lugens* survives poorly, develops more slowly, molts at a smaller body size to the adult stage, and has reduced fecundity (Choi et al.

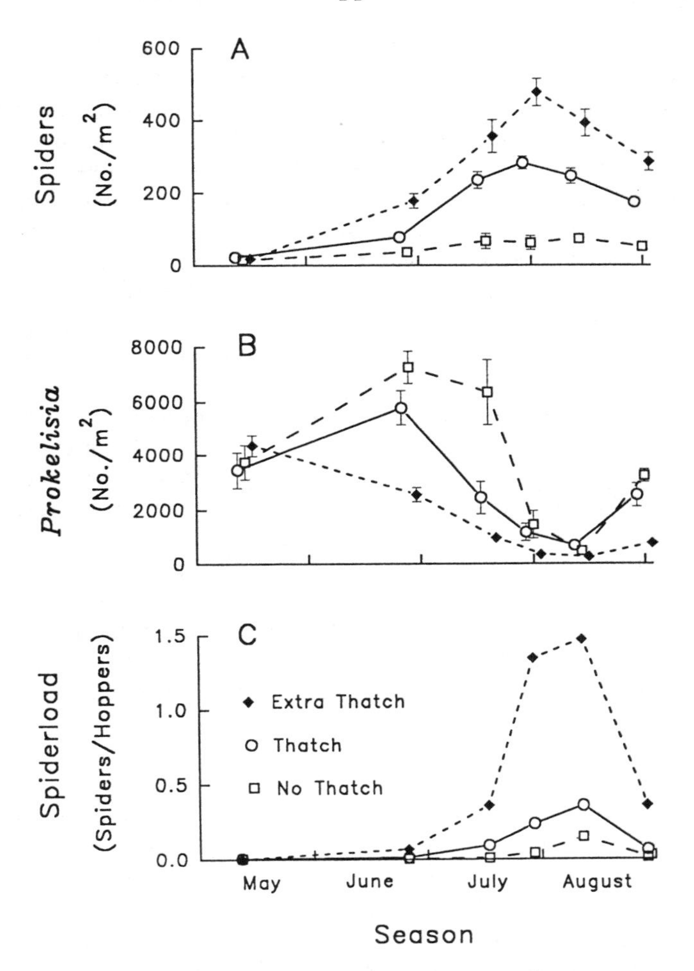

Figure 10.12. (**A**) The density of *Pardosa littoralis* (No./m^2), (**B**) *Prokelisia* planthoppers, and (**C**) the ratio of spider to planthopper density (spider load) throughout the season in structurally different plots of *Spartina* on a salt marsh at Tuckerton, New Jersey, USA. Habitat structure was manipulated in spring by raking and replacing thatch; plots contained either no thatch, normal thatch, or twice normal thatch. Spiders were significantly more abundant, spider load was higher, and planthoppers were less abundant in plots with thatch compared to thatch-free controls. There was a significant thatch effect on spider density ($F_{2,133}$ = 664.4, $P<0.01$, ANOVA), planthopper density ($F_{2,133}$ = 174.4, $P<0.01$, ANOVA), and the ratio of spiders to planthoppers ($F_{2,133}$ = 27.1, $P<0.01$, ANOVA); from Döbel and Denno in preparation b).

1979; Pongprasert and Weerapat 1979; Saxena and Pathak 1979; Claridge and Den Hollander 1982; Pathak and Heinrichs 1982; Wu et al. 1986). Consequently, populations of planthoppers remain small and predators such as spiders, mirids, and veliids do not effect a strong numerical response (Stapley et al. 1979; Kenmore et al. 1984; Manti 1989). Thus, strong varietal resistance may suppress prey populations to a level that precludes any significant response by predators. Furthermore, lower predator density on resistant rice varieties can result in an increased net survival rate of *N. lugens* (Kenmore et al. 1984). Observations like these have led authors to suggest that the use of "moderately resistant varieties" will maintain planthopper prey at a density which will encourage predators by promoting a numerical response (Stapley 1976).

Provided that densities are not too low, resistant varieties tend to offset the intrinsic reproductive advantage of planthoppers allowing predators to more effectively track and suppress planthopper populations. Likewise, because of their additive effects, most authors feel that resistant varieties and predators are often complementary management strategies (Stapley 1976; Dyck and Orlido 1979; Kenmore et al. 1984; Kartohardjono and Heinrichs 1984; Salim and Heinrichs 1986). However, the extent to which natural enemies increase or decrease rates of herbivore adaptation to resistant hosts will depend on a variety of factors, including enemy behavior, and the population dynamics of the enemy herbivore system (Gould et al. 1991).

Proximity of Predator Refuge to New Habitat

The distance a predator must move to effect a numerical response is influenced not only by its inherent dispersal capability but also by the proximity of the source (refuge) to the new habitat. If the source area is too distant, the numerical response of the predator will be nonexistent or delayed, and planthopper populations can erupt. As an example, *Tytthus mundulus*, the major predator of *Perkinsiella* planthoppers, was more likely to invade rice fields and feed on *Nilaparvata lugens* if sugarcane fields were growing in close proximity (Hinckley 1963). Also, whether or not *Cyrtorhinus lividipennis* invaded rice and effectively suppressed populations of *N. lugens* depended on the proximity of *Digitaria*, an abundant grass which apparently provided alternate hosts (Stapley 1975). In the same system, weeds (with alternate prey) growing in rice fields during intercropping periods allowed *C. lividipennis* to multiply and rapidly colonize the subsequent rice crop on which populations of *N. lugens* were checked (Stapley 1976).

Similarly, lycosid spiders (*Pardosa pseudoannulata*) remained in rice straw following grain harvest, but moved to bordering levees after plow-down in preparation for the next crop (Kenmore 1980). The early colonization of the new rice crop in the same field 3 weeks later was attributed to the proximity of the levee which acted as a refuge for spiders during the crop-free period (Kenmore 1980). Thus, in a planthopper management context, the recommendation may be to encourage vegetation on levees to conserve predators. However, the advantages of such decisions may have to be weighed against possible disadvantages. For example, grasses (*Leersia*) on the levees in China serve as overwintering hosts for other rice pests (the gall midge, *Orseola oryzae*) and weed removal is considered an effective control (Chiu 1984).

What is poorly investigated is how various natural (e.g., storms and fires) and artificial events (e.g., agricultural management and cropping practices) and the spatial scale at which they occur influence habitat structure which, in turn, alters the numerical response of predators and influences the synchrony of predator–prey interactions. For example, there has been much debate over the effects of continuous, rotated, synchronized, and staggered plantings of rice on planthopper population dynamics and outbreak (Dyck et al. 1979; Oka 1979; Kenmore 1980; Cook and Perfect 1985a; Ooi 1988). Double or continuous cropping with staggered planting times are thought by many to favor outbreaks of *Nilaparvata lugens* because this highly mobile species can disperse so efficiently among crops (see Dyck et al. 1979). Breaks in the cropping season with synchronous planting and harvesting have been suggested as possible pest control measures by some (Dyck et al. 1979; Oka 1979), but have been discounted by others (Ooi 1988). Regardless, interactions with natural enemies have been largely ignored in these arguments. Greathead (1983) reasons that continuous breeding at the same location favors the establishment of natural enemies and subsequent suppression of planthopper populations in rice. Further, he argues that closed seasons desynchronize enemy–planthopper associations which become reestablished only after an extended period of time. Nevertheless, to our knowledge, there have been no rigorously conducted field experiments which have been designed to test the specific effects of cropping pattern and its spatial scale on planthopper–predator dynamics. The rice system offers an excellent opportunity for experimentally examining the effects of planting and harvesting pattern, plot size, proximity of predator refuge (levee, bund, and alternate crop), and management practices (flooding, fertilization, and rice variety) on the relative colonization rates of planthoppers and their predators. Given the complex interactions which must exist among these factors, it is no wonder that a clear picture of the effects of cropping pattern on planthopper outbreak fails to prevail!

Alternative Prey

To further complicate matters, the presence of alternate prey in either source areas or new habitats can influence the timing and strength of a predator's numerical response. Interactions between *Cyrtorhinus lividipennis* and *Peregrinus maidis* in the corn fields of Hawaii serve to demonstrate this point (Napompeth 1973). A key factor analysis revealed that the suppressing mechanisms acting on *P. maidis* populations operated in a density-dependent manner and that egg mortality from *C. lividipennis* was the factor contributing most to overall mortality. Nevertheless, there was annual variation in the ability of this mirid to control planthopper populations. In years when *C. lividipennis* colonized corn fields late, several weeks after the arrival of *P. maidis*, planthopper populations grew rapidly. During the year when a large population of *C. lividipennis* was already present in the fields prior to the arrival of *P. maidis*, populations of the corn planthopper were effectively suppressed. The early colonization of corn on this one year by *C. lividipennis* was attributable to the presence of the eggs of a cicadellid host, *Draculacephala minerva*.

In this case, it was the very early presence of an acceptable alternate prey in the new habitat which encouraged the premature colonization of corn by *C. lividipennis* and accelerated its numerical response to *P. maidis*. Whereas neighboring habitats may act as refuges for predators due to the presence of alternative prey, the extent to which predators leave an occupied habitat and colonize a new one will depend on the predator's foraging behavior and the preference spectrum of the prey involved.

Natural Perturbations

Climatic perturbations (e.g., harsh winters, cold snaps, storms, fires) can lead to outbreaks of herbivorous insects by shifting the growth advantage to the herbivore and preventing high predator/prey ratios from being achieved (Wellings and Dixon 1987). Such may be the case for a guild of salt marsh-inhabiting planthoppers, leafhoppers, and their predators (Denno et al. 1981). For planthopper species which overwintered as exposed nymphs (e.g., *Delphacodes detecta* and *Tumidagena minuta*), there was a significant negative relationship between the severity of the previous winter and population density the following spring. No such relationship existed for species which overwintered as eggs embedded in their host plants (*Megamelus lobatus*), and there was a positive relationship between winter severity and spring density for some species with concealed eggs. Such positive relationships may result from the greater activity levels of predators (lycosid spiders and the mirid egg predator *Tytthus alboornatus*) which inflicted more mortality during mild winters (Denno et al. 1981).

Similarly, lycosid spiders were unable to limit their *Prokelisia* prey during years when the initial spring ratio of spiders to planthoppers was low, even though spiders responded numerically to *Prokelisia* populations (Döbel 1987). Only during years with high spring ratios of spiders to planthoppers did spiders suppress *Prokelisia* populations (see Fig. 10.9). Climate was suspected as the major factor influencing among-year variation in spring ratios.

Planthopper species on the salt marsh which embed their eggs in vegetation also buffer the effects of severe storms and high tidewaters (Raupp and Denno 1979; Denno et al. 1981), perturbations which probably have a much greater effect on predators which lay their eggs in exposed positions on the host plant (e.g., Coccinellidae) or actively carry their young (e.g., Lycosidae). At the very least, these predators may be drastically rearranged on the marsh following a severe storm.

The normal rising of tidewaters in low marsh habitats may also serve to uncouple predator–prey interactions. The planthopper *Prokelisia marginata* is able to withstand submergence on site, whereas the usual response of spiders is to retreat to higher ground (Throckmorton 1989). Thus, spiders may be forced repeatedly to relocate planthopper prey which exhibit a very aggregated distribution on the marsh (Denno et al. 1980).

At a larger geographic scale, seasonal climates in general may serve to uncouple planthopper–predator interactions. The harsh winters characteristic of boreal and north temperate latitudes interrupt predator–prey interactions annually and may serve to keep predator populations at bay. In contrast, opportunities are greater for continuous reproduction in the tropics and subtropics, and, as a result, prey populations may be easier for predators to track.

In support of this reasoning, predators appear to be more important in suppressing populations of *Nilaparvata lugens* in tropical Asia than they are in temperate Japan (Kuno and Hokyo 1970; Kuno 1979; Cook and Perfect 1985a). Another example comes from the *Spartina* marshes of North America. Whereas *Prokelisia* planthoppers are the dominant herbivore on both Gulf coast (30°N, Cedar Key, Florida) and Atlantic coast marshes (40°N, Tuckerton, New Jersey), they are on average 10 times as abundant along the Atlantic coast (Denno 1983). Along the Gulf coast, planthoppers are multivoltine and breed continuously, whereas on the Atlantic coast of New Jersey there are only three generations per year and planthoppers overwinter from November though March (Denno et al. 1987). Predator/prey ratios differ significantly between the two locations. On the Gulf coast where planthoppers are much less abundant, the ratio of spiders to planthopper is higher (1 : 2.5) than that which occurs on mid-Atlantic marshes (1 : 5.5) (Denno 1983). A similar pattern exists for *Tytthus vagus*, a mirid predator which is relatively more abundant along the Gulf coast 1 : 14 (mirid : planthopper) than the Atlantic 1 : 45. Thus, lower predator/planthopper ratios are associated with fluctuating, harsh-winter climates which selectively advantage planthoppers by deterring the action of predators.

Population Resurgence Following Insecticide Application

Populations of *Nilaparvata lugens*, *Saccharosydne saccharivora*, and *Sogatella furcifera* have been observed to outbreak following the application of insecticides in agricultural crops (Metcalfe 1971; Chiu 1979; Dyck and Thomas 1979; Heinrichs et al. 1982a, 1982b; Reissig et al. 1982b; Kenmore et al. 1984; Ooi 1988; Shepard et al. 1990). Insecticide-induced resurgence of planthopper populations is often attributed to the decimation of invertebrate predators, particularly lycosid spiders, mirid (*Cyrtorhinus lividipennis*), and veliid (*Microvelia douglasi atrolineata*) bugs (Itô et al. 1962; Kiritani et al. 1971; Hirano and Kiritani 1975; Dyck and Orlido 1977; Kiritani 1977; Stapley et al. 1977; Kenmore 1980, 1991; Heinrichs et al. 1982b; Chiu 1984; Heinrichs and Mochida 1984; Ooi 1988; Manti 1989; Table 10.7). For example, following repeated pesticide applications in rice, the number of spider and veliid predators was significantly reduced, whereas brown planthopper populations exhibited an 800-fold increase in density compared to the control plots (Kenmore et al. 1984; Fig. 10.13). Furthermore, independent studies have shown certain insecticides to be highly toxic to planthopper predators including *Pardosa pseudoannulata* (Chiu and Cheng 1976; Chu et al. 1976; Heinrichs et al. 1982b), *C. lividipennis* (Chiu and Cheng 1976; Dyck and Orlido 1977; Reissig et al. 1982b; Heinrichs et al. 1982b; Fabellar and Heinrichs 1984), *Tytthus mundulus* (Verma 1956), and *M. douglasi atrolineata* (Fabellar and Heinrichs 1984; Heinrichs et al. 1982b). Also, the biomagnification of certain pesticides (BHC) has been shown to occur in spiders (*Pardosa pseudoannulata*) that have fed on poisoned rice leafhoppers (Kiritani and Kawahara 1973).

Nevertheless, other mechanisms such as the stimulation of planthopper reproduction and increased plant growth are also implicated in the insecticide-induced resurgence of *Nilaparvata lugens* populations (Chelliah and Heinrichs 1980; Chelliah et al. 1980; Heinrichs et al. 1982a; Table 10.7). Also, in nonplanthopper systems, there are several instances in which insect pests increased in numbers due to the development of insecticide resistance at the same time natural enemies were decimated (Luck et al. 1977).

Table 10.7. Mechanisms for insecticide-induced resurgence in populations of *Nilaparvata lugens* in rice.

Location of Outbreak/ Population Increase	Insecticide	Mechanism	Implicated Natural Enemy	Reference
Indonesia	Modern insecticides	Predator kill	Invertebrate predators	Gallagher et al. (this volume)
Japan	Benzene hexachloride	Predator kill	Spiders	Itô et al. (1962)
Korea	Several insecticides	Predator kill	Spiders (*Pirata* and *Gnathonarium*)	Lee and Park (1977)
Malaysia	Endosulphan, BPMC, and trichlorphon	Increased P/P ratio[a]	*Cyrtorhinus* and *Lycosa*	Ooi (1988)
Philippines	Diazinon, methyl parathion	Predator kill	*Cyrtorhinus*	Dyck and Orlido (1977)
	Diazinon, decamethrin and methyl parathion	Stimulation of reproduction[b], increased plant growth		Chelliah and Heinrichs (1980)
	Diazinon and decamethrin	Predator kill	Spiders (Lycosidae, Micryphantidae) and Veliidae	Kenmore (1980); Kenmore et al. (1984)
	Decamethrin and methyl parathion	Stimulation of reproduction,[b] predator kill	Spiders (*Lycosa*)	Heinrichs et al. (1982a)
	Organophosphates, carbamates, and syn. pyrethroid	Stimulation of reproduction,[b] predator kill?	*Cyrtorhinus* and spiders	Reissig et al. (1982b)
	Carbofuran, decamethrin, diazinon, methyl parathion, and FMC	Stimulation of reproduction,[b] predator kill	*Cyrtorhinus* and *Microvelia* spiders (*Lycosa*)	Heinrichs et al. (1982b); Heinrichs and Mochida (1984)
	Monocrotophos	Predator kill	*Cyrtorhinus* and spiders	Manti (1989)
Solomon Islands	Parathion	Predator kill	*Cyrtorhinus*	Stapley (1976)
Thailand	Modern insecticides	Predator kill	Spiders	Kenmore (1991)
Taiwan	Insecticides	Predator kill	*Lycosa* and *Oedothorax*	Chiu (1979)
Vietnam	Insecticides	Predator kill	*Spiders*	Kenmore (1991)

[a]Predator/prey ratio.

[b]Application of certain insecticides stimulates reproduction in *Nilaparvata lugens* (increases adult longevity, length of oviposition period, and reproductive rate) (Chelliah et al. 1980).

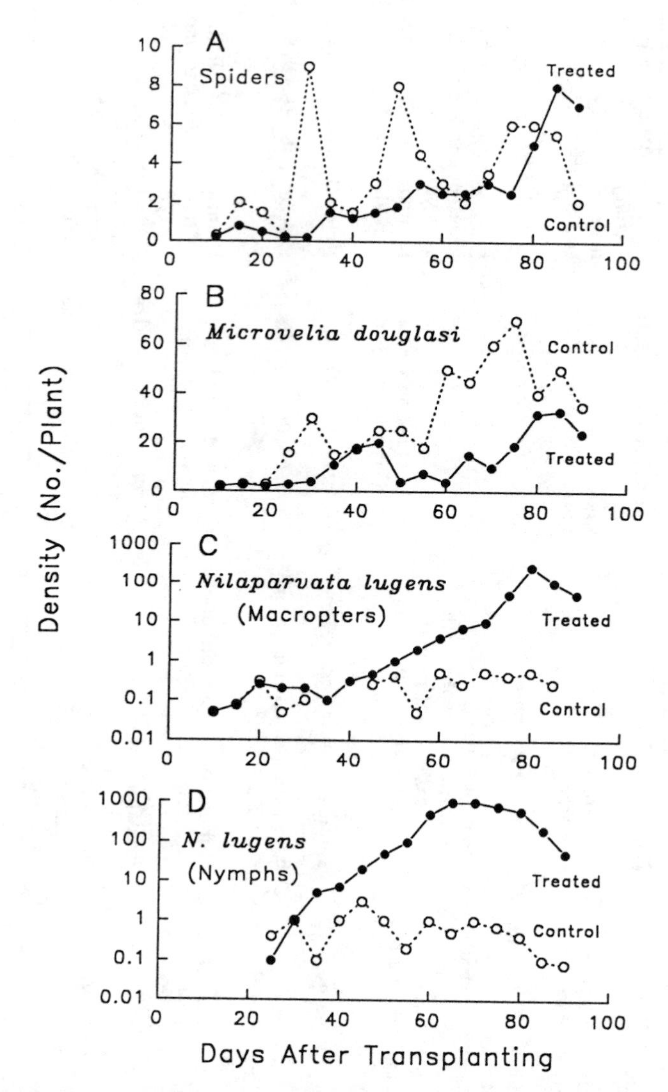

Figure 10.13. Changes in the population size of (**A**) spiders, (**B**) *Microvelia douglasi* and the (**C**) macropterous adults, and (**D**) nymphs of *Nilaparvata lugens* following insecticide applications to rice (IR20) at IRRI, Los Baños, Philippines. Thirty-four days after transplanting, rice was treated with diazinon and at 47, 58, and 69 days after transplanting (DAT) rice was treated repeatedly with deltamethrin. Predator populations decreased dramatically while planthopper populations resurged almost three orders of magnitude (from Kenmore et al. 1984).

Despite the complicated nature of the mechanisms involved in insecticide-induced resurgence, there is strong evidence that insecticides uncouple or disrupt predator–planthopper interactions and promote outbreaks (Kenmore 1980; Kenmore et al. 1984; Ooi 1988; Heinrichs Chapter 16; Gallagher et al. Chapter 17). In agricultural systems experiencing poorly timed insecticide applications, planthoppers are strongly advantaged over their predators due to their superior dispersal ability and high fecundity. Any insecticide-induced increase in reproductive rate could further advantage the planthopper. Also, because planthoppers embed their eggs in plant tissues, they can remain hidden from some insecticides. By contrast, no life stages of predators such as lycosid spiders are able to avoid a direct pesticide application. All of these life history characteristics tend to favor the planthopper in an environment where pesticides desynchronize interactions with predators.

Stability and Persistence of Predator–Planthopper Interactions

Dispersal is thought to be essential for the persistence of unstable predator–prey interactions because, without immigration, local populations would go extinct (den Boer 1968; Reddingius and den Boer 1970; Crowley 1981; Reeve 1988). Whereas some planthopper species (e.g., *Javesella pellucida*, *Nilaparvata lugens*, and *Prokelisia marginata*) have high dispersal capability (>70% macroptery in populations) (Denno et al. 1991), commonly disperse among habitats (Raatikainen 1967; Denno and Grissell 1979; Waloff 1973, 1980; Cook and Perfect 1985a), and clearly have the migratory ability to outdistance their predators (Kisimoto 1976, 1981), it is very difficult to show that local extinction would result were it not for interhabitat dispersal (but see Antolin and Strong 1987; Cronin and Strong Chapter 11).

Other planthopper species (*Conomelus anceps*, *Pissonotus quadripustulatus*, and Pacific coast populations of *Prokelisia dolus*) disperse very little by flight (>99% brachyptery) and appear to be associated with the same very isolated habitat for years (Rothschild 1966; Denno et al. 1991). Yet, both predators and parasitoids are abundant in the habitats of these flightless species (Rothschild 1966; Denno unpublished data; P. Stiling personal communication). Although dispersal by flight cannot be ruled out altogether for these very immobile species, it is clearly negligible and other mechanisms (e.g., temporal and spatial refuges) probably allow for their persistence with natural enemies. It is important to note that in the equitable climate of Southern and Baja California, populations of *P. dolus* and

Gulf coast populations of *P. quadripustulatus* reproduce continuously throughout the year and interactions with natural enemies are not uncoupled by a harsh winters.

Planthoppers tend to exhibit aggregated distributions (Rothschild 1966; Metcalfe 1969; Sogawa 1982; Denno and Roderick 1992) and several of their major predators (particularly mirids) show strong numerical responses to increased planthopper density. Because many predators aggregate in areas of high host density, some planthoppers in low-density habitats will escape. Thus, a partial refuge results, extinction is prevented, and interactions are to some extent stabilized (Murdoch and Oaten 1975; Hassell 1978; Hassell and May 1973; Murdoch et al. 1989).

Spatial and temporal refuges for prey may also confer stability to predator–prey interactions (Murdoch and Oaten 1975; Hassell 1978; Reeve and Murdoch 1986). Factors such as prey size, stage, and/or age can act as refuges from natural enemies (Murdoch 1990; Thorarinsson 1990; Cronin and Strong Chapter 11). Stage-specific predators (mirid bugs for the most part) result in refuges for planthoppers because older nymphs and adults are largely invulnerable to attack. For example, compared to the eggs of *Nilaparvata lugens*, first instar nymphs are much less susceptible to predation by *Cyrtorhinus lividipennis* (Sivapragasam and Asma 1985). Thus, once eggs hatch, the risk of mirid predation is substantially reduced. Also, for *Microvelia douglasi,* the percentage of successful attacks on *N. lugens* decreased as the developmental stage (nymph to adult) of the planthopper increased (Nakasuji and Dyck 1984).

Predator size also interacts with prey size to influence the rate of predation and the extent prey stage or age provides refuge. For instance, the capture rate of *Prokelisia dolus* is much greater for large individuals of *Lycosa modesta* (85 mg) than that for smaller spiderlings (Döbel 1987). In this case, the refuge from predation for prey will be greatest when prey are large (adults) and predators are small, or when prey are very small and predators are large. Thus, the degree of life cycle synchrony between spiders and planthoppers will dictate the extent to which prey size functions as a refuge from predation.

The structure of the habitat including the host plant can provide refuges for planthoppers in two ways. First, habitat structure can alter the foraging behavior of predators, making certain habitats less risky for planthoppers. For instance, the salt marsh-dwelling spider, *Pardosa littoralis*, avoids thatch-free habitats independent of prey density (Döbel 1987). Many tetragnathid spiders selectively construct their webs in the upper canopy of rice, and a refuge for *Nilaparvata lugens* exists at the basal portion of plants (Ôtake and Hokyo 1976). At a larger spatial scale, *P. littoralis* and *Lycosa modesta* avoid foraging in low marsh habitats which are regularly flooded by tidewaters (Denno 1983; Döbel et al. 1990). *Prokelisia marginata*, by migrating annually to low marsh habitats, finds refuge from wolf spider predation (Denno and Grissell 1979; Denno 1983).

Second, particular microhabitats may be inaccessible to certain predators, thereby providing planthopper prey with a refuge (Table 10.8). For example, eggs of *Cyrtorhinus lividipennis* and *Peregrinus maidis* in the base, crown, and roots of their host plants experience low levels of mirid predation compared to eggs embedded in leaves in the upper strata of the plant (Verma 1955; Manti 1989). Similarly, eggs of *Javesella pellucida* deposited in thick-walled stems of oats and wheat escape predation from *Panstenon oxylus* (Raatikainen 1967). For two reasons, eggs placed in thin-walled stems are at risk. First, wasps are able to reach the hollow interior of thin stems with their short ovipositors where they place their eggs. Second, if stem walls are too thin, the eggs of *J. pellucida* protrude into the hollow of the stem where they are easily attacked by the free- living larvae of *P. oxylus*.

Changes in the foraging behavior of predators may also create partial temporal/spatial refuges for prey. As an example, *Pardosa pseudoannulata* forages among the basal stems of the rice plant during the day, but moves at night to the middle and upper portions of the rice canopy in search of prey. Because *Nephotettix* occurs on the upper leaves of the plant and *Nilaparvata lugens* inhabits the crown, the change in foraging behavior may explain why the diet of *P. pseudoannulata* consists primarily of planthoppers during the day and leafhoppers at night (Sasaba et al. 1973).

Temporal refuges for planthoppers can result if the occurrence of a particular predator is very seasonal. For instance, in the upland salt marshes of temperate North America, populations of the egg predator *Tytthus vagus* peak in late May and decline rapidly through summer (Raupp and Denno 1979). Thus, the multivoltine planthoppers, *Delphacodes detecta* and *Tumidagena minuta* obtain partial refuge from mirid predation during the summer and especially autumn months. Many of the invertebrate predators of *Conomelus anceps* also appear during specific seasons of the year (Rothschild 1966).

Planthoppers are iteroparous, depositing a few eggs each day throughout their lifetime (Denno et al. 1989). By reproducing continuously, the risk from predation can be spread though time, temporal refuges for prey can result, and interactions with predators stabilized (see den Boer 1968). Brachypterous adults, in particular, are able to spread reproductive risks in time because they reproduce earlier in life and live longer than macropters (Denno et al. 1989). Thus, temporal refuges from predation may occur for brachypterous species and populations (e.g., *Pissonotus quadripustulatus*, some populations of *Prokelisia dolus*) which are not realized for mac-

Table 10.8. Spatial refuges for planthoppers from their invertebrate predators.

Planthopper Taxon	Predator Taxon	Planthopper Stage/Refuge	Reference
Javesella pellucida	*Panstenon oxylus*	Eggs/Thick stems of oats and wheat	Raatikainen (1967)
Nilaparvata lugens	*Cyrtorhinus lividipennis*	Eggs/Basal portion of rice stem	Manti (1989)
	Microvelia douglasi	Nymphs and adults/Rice plant	Nakasuji and Dyck (1984)
Peregrinus maidis	*Tytthus mundulus*	Eggs/Roots of young corn	Verma (1955)
Prokelisia dolus	*Pardosa littoralis*	Nymphs and adults/Habitats with thatch	Döbel (1987); Döbel and Denno (in preparation b)
Prokelisia marginata	*Pardosa littoralis*	Adults/Migration to low marsh habitats	Denno (1983); Döbel et al. (1990)

ropterous species that deposit most of their eggs within a shorter time period.

Whereas dispersal and refuges may allow prey to partially escape from predators, other factors such as switching to alternative prey may permit polyphagous predators to persist during periods of low prey availability, thereby stabilizing the interaction (Murdoch 1969; Oaten and Murdoch 1975b; Hassell 1978; Table 10.9). When populations of *Nilaparvata lugens* decline, *Pardosa pseudoannulata* will switch over and feed on *Nephotettix* leafhoppers (Kiritani et al. 1972). Similarly, *Cyrtorhinus lividipennis* switches between leafhopper and planthopper eggs when one becomes less available (Hinckley 1963; Napompeth 1973; Bentur and Kalode 1987). Often, switching to alternative prey involves predator dispersal to neighboring habitats (Hinckley 1963; Kenmore 1980; Bentur and Kalode 1987; Ooi 1988).

Some predators such as the spiders *P. pseudoannulata* and *Oedothorax insecticeps* and the mirids *C. lividipennis* and *C. fulvus* resort to cannibalism when planthopper prey are rare (Table 10.9). Other predators may partially buffer periods of low planthopper abundance by feeding on the leaves (*C. lividipennis* and *Tytthus pygmaeus*) or pollen of plants (*Coccinella septempunctata* and *Micraspis sp.*) (Rothschild 1966; Sivapragasam and Asma 1986; Ooi 1988), and spiders are noted for their ability to withstand long periods of starvation (Nakamura 1977; Döbel unpublished data).

A variety of stabilizing factors appear to operate in predator–planthopper systems. The high dispersal capability of many planthopper species and the presence of spatial and temporal refuges tend to prevent the extinction of local populations. Furthermore, prey switching, cannibalism, the ability of some predators to withstand long periods of low prey availability all lend to a predator's local persistence. Mutual interference among conspecifics such as that observed for *Pardosa pseudoannulata* should also have a weak stabilizing role by preventing the overexploitation of planthopper populations (Heong and Rubia 1990b).

Stability and Planthopper Pest Suppression

Oaten and Murdoch (1975a) and Murdoch (1990) argue that spatial refuges, an invulnerable stage class, and Type III functional responses, although they confer stability, interfere with biological control because a higher equilibrium density of the pest inevitably results. An example of a physical refuge interfering with control comes from California Red Scale (*Aonidiella aurantii*) on citrus, where scales on the interior of trees are

Table 10.9. Factors enhancing the persistence of planthopper predators under conditions of low prey availability.

Predator Taxon	Planthopper Prey	Persistence Mechanism	Reference
Pardosa pseudoannulata	*Nilaparvata lugens*	Cannibalism	Kiritani et al. (1972)
		Withstands starvation	Nakamura (1977)
		Movement to levee following rice harvest	Kenmore (1980)
		Movement to bunds following rice harvest	Ooi (1988)
		Switching to alternate prey	Kiritani et al. (1972); Ooi (1988)
Pardosa littoralis	*Prokelisia dolus* and *Prokelisia marginata*	Movement to thatch habitats reduced avian predation	Kale (1965); Döbel (1987); Döbel and Denno (in preparation b)
Oedothorax insecticeps	*Nilaparvata lugens*	Cannibalism	Kiritani et al. (1972)
Argiope sp.	*Nilaparvata lugens*	Cannibalism	Barrion et al. (1981)
Cyrtorhinus fulvus	*Tarophagus prosperpina*	Cannibalistic on eggs	Matsumoto and Nishida (1966)
Cyrtorhinus lividipennis	*Nilaparvata lugens*	Switching to eggs of *Nisia atrovenosa*	Bentur and Kalode (1987)
		Cannibalistic on eggs	Hinckley (1963)
		Switching to eggs of Cicadellidae	Hinckley (1963)
		Switching to alternate prey	Stapley (1976)
		Switching to phytophagy	Sivapragasam and Asma (1985)
		Switching to eggs of Lepidoptera	Ooi and Shepard (in press)
	Peregrinus maidis	Switching to eggs of Cicadellidae	Napompeth (1973)
Tytthus pygmaeus	*Conomelus anceps*	Switching to phytophagy	Rothschild (1966)
Coccinella septempunctata	*Conomelus anceps*	Switching to phytophagy (pollen-feeding)	Rothschild (1966)
Micraspis sp.	*Nilaparvata lugens*	Switching to phytophagy (pollen-feeding)	Ooi (1988)

relatively free from parasitism (Reeve and Murdoch 1986). When the refuge was eliminated by removing scales from the interior of the tree, the overall equilibrium density on the periphery of trees was reduced (Murdoch 1990).

However, the extent to which a refuge elevates the equilibrium density of an herbivore will depend on the predator's response in the system. An example from the natural planthopper–wolf spider system demonstrates this point. Levels of thatch (litter) vary tremendously within and among stands of *Spartina alterniflora* on the mid-Atlantic salt marshes of North America (Denno 1983; Döbel 1987). At low densities, thatch provides a spatial refuge for *Prokelisia* planthoppers as evidenced by the change in the shape of the functional response of the wolf spider, *Pardosa littoralis*, from a Type II to a Type III curve when thatch is added to experimental arenas (Döbel 1987; Fig. 10.4). However, thatch-rich habitats also act as refuges for spiders, perhaps from their avian predators (see Kale 1965). Spiders accumulate in thatch-rich habitats independent of prey density and their numerical response to planthopper density is significantly higher in thatch-rich compared to thatch-free habitats (Döbel and Denno in preparation b; Figs. 10.11 and 10.12A).

Thus, thatch-rich habitats provide refuges for both planthoppers and wolf spiders. Stability is imposed because some fraction of planthoppers escape spider predation by hiding in the thatch; spiders apparently reduce their risk of attack from avian predators, yet spiders show their strongest numerical response to planthoppers in these habitats (Hassell 1978; Murdoch 1990). Nevertheless, the equilibrium density of *Prokelisia* planthoppers is significantly lower and the density of spiders higher in thatch-rich field plots of *Spartina* compared to thatch-poor plots (Döbel and Denno in preparation b; Figs. 10.12A,B). Thus, there may be some cases (instances where the same habitat offers refuge to both prey and predator) in which habitat structure bestows both stability and low prey equilibrium density on predator–prey interactions.

Predator Complexes and Enemy Interactions

In managed agricultural systems, control should be enhanced by the use of complementary natural enemies that attack different stages of the pest or attack at different times, thus avoiding invulnerable stage classes and temporal refuges (Murdoch 1990). This view is supported by the assertion that it is the combination of mirid egg predators (*Cyrtorhinus lividipennis*) and spiders (*Pardosa pseudoannulata*) feeding on planthopper nymphs and adults which effectively suppresses populations of *Nilaparvata lugens* and

Sogatella furcifera (Ooi 1982, 1988; Manti 1989; Ooi and Shepard in press). Other predator combinations such as spiders and veliids have also been implicated in the successful control of rice delphacids (Kenmore et al. 1984).

Similarly, in natural systems (*Juncus* marsh), it is a succession of predators attacking the various stages of *Conomelus anceps* which "exercise control" over populations (Rothschild 1966). Mirid egg predators appeared first in spring followed by nabids and phalangids, and spiders which were most abundant in summer and autumn.

Because egg mortality is the key factor driving the population dynamics of several species of planthoppers (Napompeth 1973; Waloff 1980; Waloff and Thompson 1980; Cook and Perfect 1989; Denno and Roderick 1990), egg predators should be targeted as essential members of predator ensembles for control purposes. Another consideration is to encourage a group of predators whose members exhibit different capacities for an aggregative response. The combination of a very density-responsive predator (e.g., *Cyrtorhinus lividipennis*) with others which are less so (e.g., *Pardosa pseudoannulata*) may simultaneously deter local outbreaks of prey and, at the same time, interfere with spatial refuges that might occur were there only a single aggregative predator. The encouragement of parasitoids (e.g., *Anagrus*), many of which exhibit density-independent parasitism or inverse density-dependent parasitism (Stiling and Strong 1982a; Cronin and Strong Chapter 11), also may be important in deterring population growth in planthoppers, particularly at low prey densities.

Most data suggest that several predators acting in concert with parasitoids can more effectively suppress planthopper populations than a single predator species alone (Rothschild 1966; Kenmore et al. 1984; Ooi 1988). However, their effects may not always be additive [see Polis et al. (1989) for a general discussion of intraguild predation]. For instance, *Pardosa pseudoannulata* as well as other spider species will prey on *Cyrtorhinus lividipennis*, potentially reducing the impact of both of these predator species on *Nilaparvata lugens* (Hinckley 1963; Heong et al. 1989; Heong et al. 1991b). However, that *P. pseudoannulata* prefers to feed on *N. lugens* should partially offset the potential negative interaction between these two predators (Heong et al. 1991b). Also, *Tytthus mundulus* feeds indiscriminately on both the parasitized (Hymenoptera) and unparasitized eggs of *Perkinsiella saccharicida* (Swezey 1936).

Just how important antagonistic interactions among predators are in the context of planthopper suppression and population dynamics remains to be seen. However, one observation from the taro fields of Hawaii suggests that such interactions may have a relatively weak effect. Coccinellids and spiders, for example, have been observed feeding on *Cyrtorhinus fulvus*, yet this predator has been primarily responsible for the successful biological control of *Tarophagus colocasiae* (Matsumoto and Nishida 1966). Currently, data are very superficial but they do suggest that the combined suppressing effects of a complex of predators preying on planthoppers outweigh any consequences of antagonistic interactions among predators. Nevertheless, details of the interaction are poorly known and deserve future attention.

Conclusions

The most important predators of planthoppers are mirid and nabid bugs, spiders, and, infrequently, phalangids and coccinellid beetles (Table 10.1). Predator (mirids and spiders) removal experiments in the field (Matsumoto and Nishida 1966; Döbel 1987), caging studies in which predators (mirids, veliids, and spiders) were either included or excluded (Dyck and Orlido 1977; Kenmore 1980; Kenmore et al. 1984; Döbel 1987; Fowler 1987; Ooi 1988; Manti 1989; Throckmorton 1989), the resurgence of planthopper populations following the selective kill of predators with insecticides (Table 10.7), and the effective biological control of pest planthoppers with introduced predators (Swezey 1936; Zimmerman 1948; Matsumoto and Nishida 1966) all provide evidence for the potential suppressing role of predators in planthopper population dynamics.

Furthermore, the strong numerical response of some predators (mirids) is associated with their ability to suppress planthopper populations (Matsumoto and Nishida 1966; Napompeth 1973; Ooi 1988). Other predators such as some lycosid spiders show less consistency in their tendency to aggregate in areas of high prey density (Kuno and Dyck 1985; Ooi 1988), yet they too are able to substantially subdue planthopper population growth under certain conditions (Döbel 1987; Ooi 1988; Döbel and Denno in preparation c).

Although there is a suggestion in the management literature that predators play a more important role in tropical compared to temperate agroecosystems (Kuno 1968; Kuno and Hokyo 1970; Cook and Perfect 1985a; Cheng et al. 1990), our assessment of studies in both natural and managed systems suggests that predators can be very important suppressors of planthoppers at both tropical and temperate latitudes (Rothschild 1966; Kobayashi 1975; Miyai et al. 1978; Döbel 1987; Döbel and Denno in preparation a,b,c).

Failure of predators to realize their potential in suppressing planthopper populations usually results when predator–planthopper interactions are uncoupled, either by natural (harsh winters or structural features of the habitat) or unnatural phenomena (insecticides or harvesting practices). The result is either spatial or temporal asynchrony in the life cycles of predators and planthoppers which often leads to outbreak due to the inherent reproductive and dispersal advantages that many wing-dimorphic planthoppers manifest over their predators (Tables 10.4 and 10.5).

Thus, a focus on factors which are known to uncouple predator–planthopper interactions provides insight into population dynamics and allows us to offer several considerations for the more successful management of pest planthoppers. First, to successfully track planthopper populations, the best predators should possess relatively high dispersal and reproductive capability (e.g., mirid bugs) or have high consumption rates (e.g., lycosid spiders).

Second, the use of more than one predator may avoid the occurrence of an invulnerable prey stage (refuge) and lead to better suppression. Thus, the combined action of several predators (also parasitoids) could compensate for the inadequacies of a single predator species unless there are strong antagonistic interactions among natural enemies. It may be especially important to encourage egg predators in the complex because egg mortality is often the key mortality factor in planthopper population dynamics (Denno and Roderick 1990). Also, to avoid refuges resulting from highly density-responsive predators, a mix of predators comprised of species exhibiting both strong and weak numerical responses should be encouraged.

Third, cropping pattern can also result in the uncoupling of predator–planthopper interactions and promote outbreaks. Following harvest, predators must disperse to alternative habitats with suitable prey to survive crop- free periods. If refuges (e.g., levees or bunds with vegetation or other crops in rice-growing areas) occur at too great a distance from the crop, predators will not be able to effect a numerical response sufficient to suppress planthoppers in the new crop. Furthermore, predator populations will shrink in neighboring refuges if either the quantity or quality of alternative prey species is low. The upland rice-growing areas of Indonesia provide an ideal landscape for testing the effects of paddy size, distance of refuge, quality of refuge (vegetation structure), suitability and abundance of alternate prey, and prey augmentation on the numerical response of the important predators of planthoppers.

A commonly held view in rice systems is that continuous cropping with staggered planting times promote outbreaks of *Nilaparvata lugens* because this highly mobile species can disperse so efficiently among crops (see Dyck et al. 1979). As a consequence, synchronized planting has been recommended as a control tactic for *N. lugens* (Oka 1979), but this strategy has not proved very effective (Ooi 1988). Part of the explanation for why this approach has not been successful may rest with natural enemies. The synchronized planting and harvest of large rice fields could seriously disadvantage predators due to the large-scale elimination of spatial refuges (see Greathead 1983). Nevertheless, the issue of cropping pattern is far from settled and more research is needed to understand the effects of the rice landscape (field size, polyculture, planting strategy, and refuges), habitat structure (varietal architecture, plant spacing, stubble removal, and bund vegetation) and alternative prey (in rice and neighboring refuges) on the detailed movements and foraging behavior of planthopper predators and planthopper dispersal and population growth. It is also very worth noting that this sort of information is badly wanting in natural planthopper systems as well and a focus on factors which uncouple predator–planthopper interactions may prove essential to the understanding planthopper populations dynamics.

Last and perhaps most important is the judicial use of pesticides to conserve natural enemies. Of all the factors that can uncouple predator–planthopper interactions and lead to outbreak, insecticides are perhaps the most severe.

The general call for more detailed information on predators by modelers who attempt to predict planthopper outbreaks (Heong 1982; Holt et al. 1987) underscores the need for more meticulous field studies on the invertebrate predators of planthoppers. Despite the recent attention that predators have received in the management literature, we know very little about the actual predation rates of most predators in the field, their life history traits and interactions among factors which influence their dispersal and aggregation behavior.

Summary

1. In both natural and managed systems, invertebrate predators inflict much more mortality on planthopper populations than do parasitoids. The important predators responsible for this mortality include spiders (e.g., Lycosidae), hemipterans in the families Miridae, Nabidae, and Veliidae, and occasionally phalangids and coccinellid beetles.
2. Under most conditions, the major predators of planthoppers such as lycosid spiders (*Lycosa* and *Pardosa*), mirids (*Cyrtorhinus*), and veliids (*Microvelia*) exhibit Type II functional responses. In one case, however, the addition of thatch (plant litter) to experimental arenas containing living plants created a refuge for *Prokelisia* planthoppers and changed the functional response of its major spider predator (*Pardosa littoralis*) from a Type II to a Type III curve. Functional response

experiments with mirids and spiders suggest that delphacid planthoppers are generally more susceptible to predation than leafhoppers (Cicadellidae).

3. Mirid egg predators (*Cyrtorhinus*) show the strongest and most consistent numerical response to planthopper density, which, in many cases, is associated with prey suppression. Although less consistent, lycosid spiders can also aggregate in areas of high prey density. In the case of *Pardosa littoralis*, habitat structure influences its short-term numerical response to *Prokelisia* density; spiders aggregate more in complex-structured habitats and more effectively suppress planthopper populations there.
4. A variety of factors tend to stabilize predator–planthopper interactions. For planthoppers, high dispersal capability and the presence of temporal and spatial refuges promote the persistence of populations. For planthopper predators, prey switching, cannibalism, and the ability to withstand starvation tend to prevent local extinction.
5. There are several examples from a variety of natural and agricultural systems demonstrating a negative relationship between predator density (*Cyrtorhinus, Microvelia,* and *Pardosa*) and the rate of planthopper population increase.
6. Wing-dimorphism in planthoppers promotes escape from predators and population outbreaks. Weak numerical responses of predators often result from an inherently low reproductive potential and/or poor dispersal capability compared to planthoppers. A comparison of fecundities between most planthopper species (brachypterous morph) and their major predators reveals a decided advantage for planthoppers. Similarly, the high dispersal capability afforded by the macropterous wing form further contributes to the ability of planthoppers to escape their predators. The combination of life history traits possessed by *Cyrtorhinus* (short generation time and moderately high fecundity and dispersal capability) places it high in the ranks as a predator potentially capable of effectively tracking and suppressing planthopper populations. The life history disadvantages of wolf spiders (relatively low fecundity and long generation time) are partially offset by the high consumption rates of these important predators.
7. Life cycle synchrony between predators and planthoppers is essential for prey suppression. External factors such as habitat structure which interferes with the foraging behavior of predators, the absence of nearby refuges and alternate prey, harsh winters, and pesticide applications uncouple predator–planthopper interactions and promote planthopper outbreaks.

8. Complexes of predators and parasitoids which attack different prey stages and at different times are often more effective at suppressing planthopper populations than a single predator because invulnerable stage classes and temporal refuges can be avoided. Also, to encourage both density-responsive predators (e.g., mirids) and those which are less so (e.g., some spiders) may provide more complete control.
9. The effective biological control of pest planthoppers (*Perkinsiella saccharicida* and *Tarophagus colocasiae*) with mirid predators, predator removal experiments in the field followed by planthopper outbreak, the effective suppression of planthopper populations in confined cages with predators, and the resurgence of planthopper populations following the selective kill of predators with insecticides together strongly suggest that invertebrate predators play an important suppressing role in the population dynamics of planthoppers.
10. Future research in both natural and managed systems should concentrate on the relative dispersal rates of planthoppers and their predators and factors (e.g., habitat structure and host plant architecture/chemistry) which either influence functional and numerical responses or uncouple predator–planthopper interactions. As a last consideration, predator–predator interactions and their effect on planthopper population dynamics deserve much more attention.

ACKNOWLEDGMENTS

We thank B. Benrey and D. Wise for their comments on earlier drafts of this report. Also, we gratefully acknowledge B. Barrion for providing us with unpublished data on the life history traits of *Pardosa pseudoannulata* and M. Shepard for allowing us access to manuscripts in press. This research was supported by National Science Foundation Grants BSR-8206603 and BSR-8614561 to RFD. This is Contribution Number 8578 of the Maryland Agricultural Experiment Station, Department of Entomology.

11

Parasitoid Interactions and Their Contribution to the Stabilization of Auchenorrhyncha Populations

James T. Cronin and Donald R. Strong

Introduction

The Fulgoroidea are common herbivores in both agricultural and natural systems, often causing severe damage to their host plants (Chapters 14–20). As a result of their economic importance, much research has been conducted on the ecology of planthoppers and the role that natural enemies play in limiting their damage. Researchers have, through extensive collections and many hours of field work, identified the majority of natural enemies associated with planthoppers.

The Fulgoroidea are attacked by two major groups (guilds) of parasitoids which are distinguished broadly by the host stage they exploit. These are the egg parasitoids and the nymphal/adult parasitoids which span a number of families in primarily three insect orders: the Hymenoptera, Strepsiptera, and Diptera (Waloff and Jervis 1987; Stiling Chapter 13). Most egg parasitoids are Hymenoptera, although wasps in the family Dryinidae attack nymphs and adults as do the Strepsiptera and Diptera (Pipunculidae) (Table 11.1). On average, each planthopper species is host to approximately two species of egg parasitoids and two nymphal/adult parasitoids (Stiling Chapter 13). Parasitism by these natural enemies often accounts for substantial mortality to their planthopper hosts; rates average over 36% for egg parasitoids and 17% for nymphal parasitoids. Although researchers have painstakingly documented rates of parasitism for host species, often from samples taken over a wide spatial range and for a number of years, relatively few studies have attempted to assess the role parasitoids play in planthopper population dynamics. Even fewer studies have taken an experimental ap-

Table 11.1. Parasitoid families known to attack the Fulgoroidea (from Waloff and Jervis, 1987).

Taxonomic Group	Stage Attacked
Diptera	
Pipunculidae	Nymph/adult
Strepsiptera	
Elenchidae	Nymph/adult
Halictophagidae	Nymph/adult
Hymenoptera	
Aphelinidae	Egg
Dryinidae	Nymph/adult
Encyrtidae	Egg, nymph/adult
Eulophidae	Egg
Mymaridae	Egg
Scelionidae	Egg
Trichogrammatidae	Egg

proach in the field in attempting to elucidate interactions between planthoppers and their parasitoids.

Critical data on the interaction between planthoppers and their parasitoids (aside from seasonal rates of parasitism) are rare. The paucity of ecological data on the Fulgoroidea led us to extend our review to include information on leafhoppers (Cicadellidae) as well. Because of the close taxonomic and ecological similarity between planthoppers and leafhoppers, we feel that combining these two groups is warranted. We review the interactions between the Auchenorrhyncha and their parasitoids; in particular, we examine parasitoid searching behavior and spatial and temporal patterns of parasitism with respect to host density. The resulting information will be used to assess the importance of parasitoids to the regulation, stability, and persistence of planthopper and leafhopper populations. A comprehensive review by Waloff and Jervis (1987) on the systematics, taxonomy, biology, and ecology of the Auchenorrhyncha provided a valuable initial source of information for the present review.

Parasitoid Behavioral Patterns and Their Influence on Host and Parasitoid Stability

In the broadest ecological sense, stability is avoidance of the two extremes of population eruption and extinction. Between extremes, stability may be density vague (Strong 1988, 1992) and without forces that impel densities toward some specific "equilibrium," this situation could lead to stochastic boundedness (Chesson 1978). One distinctly different situation is simple equilibrium stability, with a single equilibrium point. Between these two extremes lies a vast spectrum of possibilities (Connell and Sousa 1983). One interesting possibility is that simple equilibrium stability is so weak that the forces toward the specific equilibrium value are extremely difficult to detect given the inevitable stochasticity of the environment (Hassell 1985a, 1986a, 1987). In this context, an important consideration is whether stability is strong enough to prevent extinctions in the face of environmental stochasticity (Morrison and Barbosa 1987; Reeve 1988). Another possibility focuses on multiple point equilibria (Peterson 1984). Additionally, a very topical possibility is chaos, in which deterministic forces are so extremely reactive as to create bewilderingly complicated dynamics. Against this background of possibliites, we are able to weigh evidence largely in the realm of deterministic point equilibrium and density vagueness.

Typically, studies of the interactions between planthoppers and their parasitoids come from field collections of parasitized hosts, and most deal with economically important plant species. There are a few notable exceptions, however, involving planthoppers of grasslands (Waloff 1975; Waloff and Thompson 1980) and those inhabiting salt marshes (Stiling and Strong 1982a; Strong 1989; Cronin and Strong 1990a). Much of the experimental work is laboratory bound, but several field experiments with egg parasitoids have been conducted (Ôtake 1967; Becker 1975; Cronin and Strong 1990a). In the following section, we review the searching behavior of these parasitoids and how these behaviors contribute to host–parasitoid stability.

Parasitoid Functional Response

The number of hosts attacked by each parasitoid female per unit time, the parasitoid functional response, is the per–capita parasitoid effect as a function of host density (Holling 1959a). A Type I response is simply density independent. A Type II response, previously thought to be typical of arthropod predators and parasitoids (Hassell 1978), does not stabilize the interaction because parasitoids are less efficient at higher host densities; it produces inverse density-dependence (Hassell and May 1973). A Type III functional response, however, leads to density dependence up to some host density threshold and can contribute to population stability (Murdoch and Oaten 1975; Oaten and Murdoch 1975a; Murdoch 1979). However, time delays, such as those described by the Nicholson–Bailey difference equations, tend to negate the stabilizing effect of a Type III response. The Type

III response is typical of vertebrate predators, but some arthropods exhibit it as well (Murdoch and Oaten 1975; Hassell 1978).

It should be noted at the outset that discriminating among the categories of functional responses has not involved much statistical rigor (Trexler et al. 1988). It can be tenuous, in practice, to distinguish especially between Type II and Type III responses because any trends in real data are usually awash in great amounts of variability. Discerning nonlinearity, which is the essence of the difference between Type II and Type III responses, is particularly problematical in a typical background of high variance. The difficulty is amplified by the fact that the nonlinearity involves position; where does a Type III response begin and end in the range of host densities? It is possible that a Type III curve occurs over a range of host densities lower than the one spanned by a particular data. If so, is this range below typical natural densities? A functional response evident in the lab might have little or no effect on population dynamics in nature, where fluctuations and variability from other "external" forces are so great that any regulatory forces from the response are overwhelmed.

Because of the need for data on the number of searching parasitoids as well as the parasitism rate per host, only four studies have examined the functional response of planthopper and leafhopper parasitoids, of which three deal with *Anagrus* spp. attacking the egg stage (Moratorio 1977; Chantarasa-ard et al. 1984a; Cronin and Strong present study). The remaining study addresses the response by a dryinid which parasitizes nymphs and adults (Chua et al. 1984). Moratorio (1977) examined the behavior of the egg parasitoids *A. mutans* and *A. silwoodensis* on their shared host, the cicadellid *Cicadella viridis* (L.). Both species exhibited a Type II response (Fig. 11.1A), but *A. mutans* had a greater handling time (T_h), resulting in fewer hosts parasitized at higher host density. A Type II response was also found by Chantarasa-ard et al. (1984a) for *A. incarnatus* Haliday attacking *Nilaparvata lugens* (Stål) (Fig. 11.1B) and by Cronin and Strong (present study) for *A. delicatus* Dozier attacking *Prokelisia marginata* Van Duzee (Fig. 11.1C). Only the dryinid, *Pseudogonatopus flavifemur* Esaki attacking *N. lugens* showed a functional response with sufficient accelerating mortality to be reasonably described by a sigmoid curve (Fig. 11.1D) (Chua et al. 1984). It is important to note that the acceleration occurs at very low densities, so any regulatory effect would not act over most of the range of host densities used in this experiment.

Parasitoid Density and Interference

As parasitoid density increases, the probability that searching females will encounter each other becomes high, especially if parasitoids aggregate at

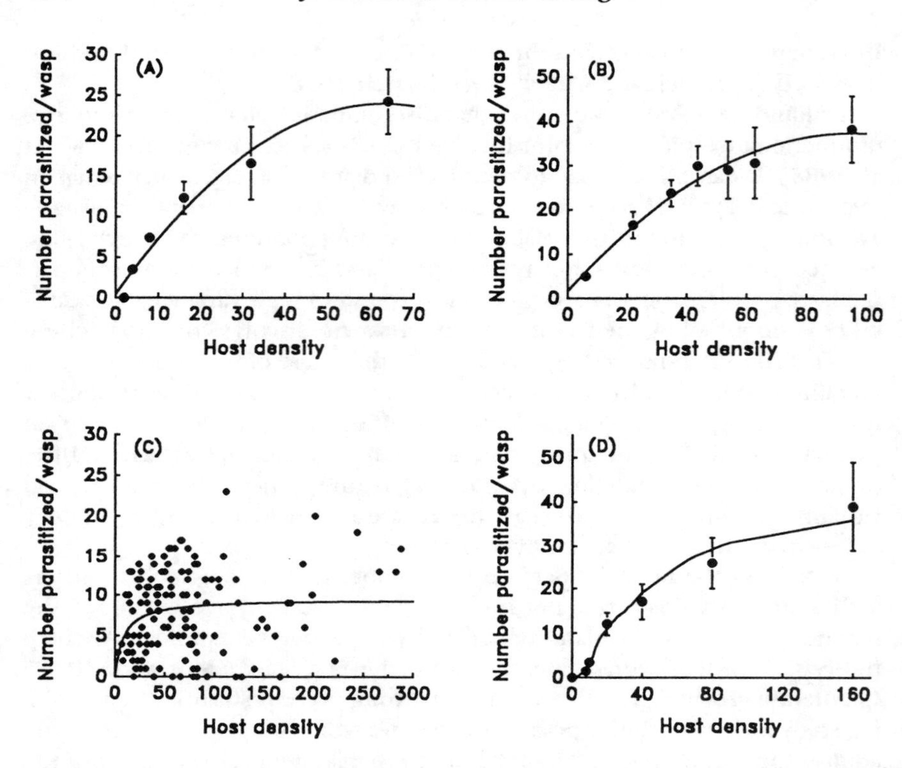

Figure 11.1. Functional responses for parasitoids attacking the Auchenorrhyncha. **(A)** The mymarid *Anagrus silwoodensis* (Mymaridae) attacking *Cicadella viridis*, Type II (Moratorio 1977). **(B)** *A. incarnatus* attacking *Nilaparvata lugens*, (Chantarasa-ard et al. 1984a). **(C)** *A. delicatus* attacking *Prokelisia marginata*, Type II. **(D)** *Pseudogonatopus flavifemur* (Dryinidae) attacking *N. lugens*, Type III (Chua et al. 1984).

high host density as mainstream models would suggest (Hassell and May 1973). The effects of such encounters can reduce the per capita number of eggs parasitized by causing a decrease in searching efficiency or an increase in handling time as parasitoid density increases (Hassell 1978). In addition, high parasitoid density can lead to a shift in offspring sex ratio toward a male bias (Wylie 1965; Viktorov 1968). Interactions among parasitoids at high parasitoid density is termed mutual interference. Mutual interference can strongly stabilize the otherwise unstable Nicholson–Bailey model, provided the interference constant (m) falls between 0 and 1 (Hassell and May 1973).

Five studies (the same four cited in the previous section on parasitoid functional response, plus Pitcairn et al. 1990) have examined the effects of parasitoid density on parasitoid behavior. Among the egg parasitoids, Moratorio (1977) found no effect of the density of *A. mutans* or *A. silwoodensis* on their searching efficiency in the laboratory (Fig. 11.2A). These data agreed with previous observations that searching females do not interact with each other. On the other hand, Chantarasa-ard et al. (1984a) and Chua et al. (1984) found significant effects of egg parasitoid and nymphal/adult parasitoid density, respectively, on their searching efficiency in the laboratory (Figs. 11.2B,C).

The remaining two studies (Pitcairn et al. 1990; Cronin and Strong present study) looked for interference among parasitoids in the field. On the basis of a detailed knowledge of host egg and parasitoid larval development rates (degree-day development), Pitcairn et al. (1990) were able to predict the number of adult *A. epos* emerging from blackberry leafhopper (*Dikrella californica*) eggs at time $t + 1$ from leaf samples collected in the field at time t. Their clever method provided an indirect estimate of wasp density at each sample date. During the course of 12 months of biweekly sampling, a marginally significant negative effect of *A. epos* density on performance was found (Fig. 11.2D).

In our study with *A. delicatus*, parasitoid density was directly estimated by trapping of wasps on host plants in a field experiment spanning 20 replicate weeks. Experimentally established patches of host eggs were exposed concurrently with trap plants, providing estimates of both wasp density and parasitism. Further details on this experiment can be found in Cronin (1991) and Cronin and Strong (unpublished data). Among our temporal replicates, we found a significant decline in parasitoid performance as density increased (Fig. 11.2E). A great deal of variability surrounds the line, however.

We (Cronin and Strong unpublished data) performed a series of laboratory experiments to elucidate the mechanism underlying interference in *A. delicatus*. One cause of the decline in wasp searching efficiency is that *A. delicatus* indiscriminantly attacks hosts; wasps do not avoid ovipositing in hosts that have already been parasitized (superparasitism). Interference is indirect ("pseudointerference" sensu Free et al. 1977) and is a consequence of the increasing likelihood of wasting eggs on previously parasitized hosts as the ratio of parasitized to unparasitized hosts increases. *Anagrus delicatus* also demonstrates significant direct ("mutual") interference that involves the nonharmful disruption of conspecific searching on a patch of hosts.

Although data are limited, it appears that interference may be common in the parasitoids of planthoppers and leafhoppers. Four out of six species examined showed strong intraspecific interactions at high parasitoid density. In all three cases of interference, m fell within the range where stability is possible (m = 0.32–0.59; Fig. 11.2). Evidence from *A. epos* and *A.*

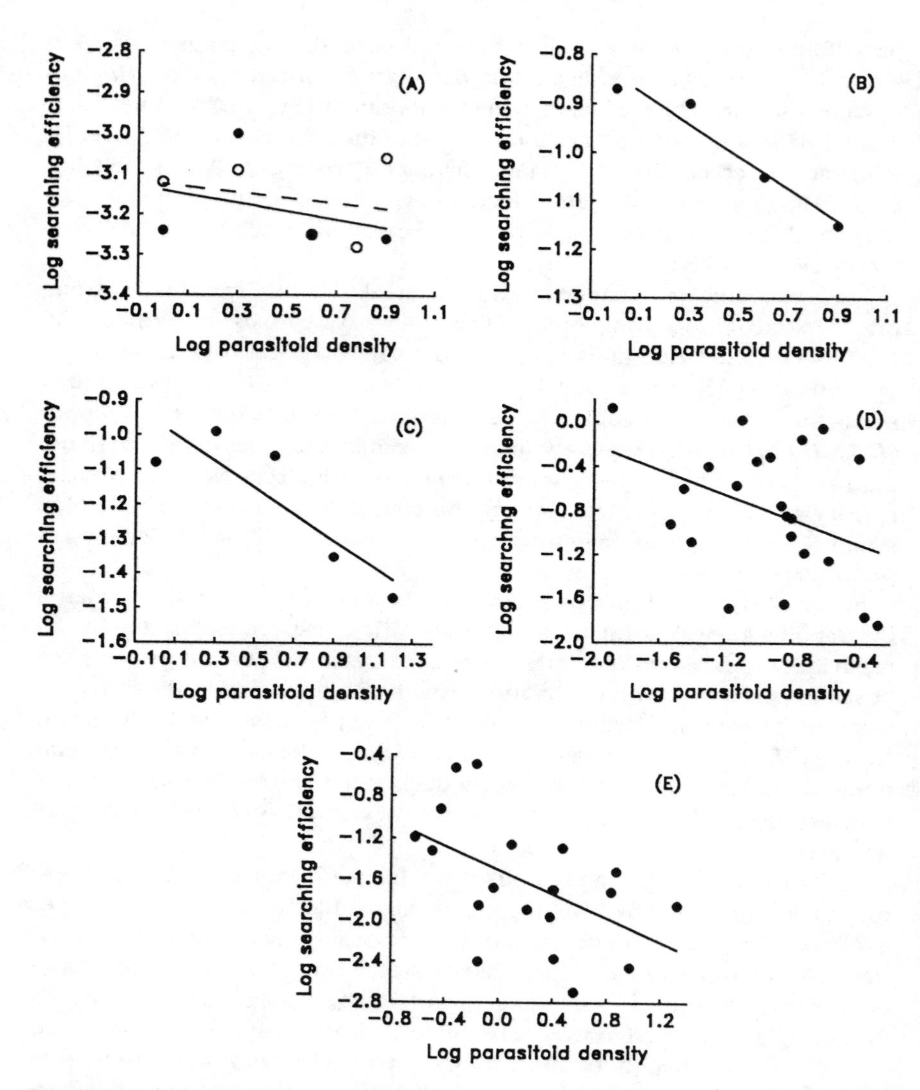

Figure 11.2. Interference among searching parasitoids of the Auchenorrhyncha. **(A)** The mymarid *Anagrus mutans* (solid circles and line, interference constant; m = 0.12) and *A. silwoodensis* (open circles, dashed line; m = 0.09) attacking *Cicadella viridis* (no relationship, $P > 0.05$) (Moratorio 1977). **(B)** *A. incarnatus* (m = 0.32) attacking *Nilaparvata lugens* (R^2 = 0.97, $P < 0.05$) (Chantarasa-ard et al. 1984a). **(C)** The dryinid *Pseudogonatopus flavifemur* (m = 0.36) attacking *N. lugens* (R^2 = 0.58, $P < 0.001$) (Chua et al. 1984). **(D)** *A. epos* (m = 0.56) attacking *Dikrella californica* (R^2 = 0.16, $P < 0.07$) (Pitcairn et al. 1990). **(E)** *A. delicatus* (m = 0.59) attacking *Prokelisia marginata* (R^2 = 0.27, P = 0.02).

delicatus in the field tends to argue against the conclusion of Griffiths and Holling (1969) that interference is mainly a laboratory phenomenon. Clearly, more detailed laboratory and field studies are needed before we can generalize to all parasitoids of leafhoppers and planthoppers.

Spatial and Temporal Patterns of Parasitism

It has long been recognized that spatial or temporal density-dependent parasitism can stabilize host–parasitoid interactions and maintain populations at low densities (Howard and Fisk 1911; Huffaker 1969; DeBach 1974), a goal that is often sought in biological control programs (Huffaker et al. 1971; Batra 1982). The pattern of density dependence is thought to result from parasitoid aggregation at patches of high host density (Hassell and May 1973). Unfortunately, the evidence supporting the prevalence of spatial (Stiling 1987; Walde and Murdoch 1988) and temporal density dependence in insect populations (Dempster 1983; Stiling 1987, 1988) is equivocal. Independent reviews of field studies during the past several decades show that cases of density dependence and inverse density dependence occur with equal frequency (25%), whereas density independence occurs in about half the studies (Stiling 1987; Walde and Murdoch 1988). On the basis of these reviews, population stability based on density dependence does not appear to be common (but see Hassell 1985a, 1986a, 1987; Hassell et al. 1989).

Recently, however, it has been noted that inverse density dependence can bring about population stability (Hassell 1985b; Chesson and Murdoch 1986). This counterintuitive hypothesis can be easily understood when one examines the distribution of parasitism. In the case of inverse density dependence, hosts are relatively free of parasitism (a partial refuge occurs) at high host density. Some proportion of the host population survives parasitism, imparting stability on the system. As long as the distribution of parasitism is clumped (heterogeneous) in space, such that some proportion of the population is relatively free of attack, stability can occur (see Heterogeneous Parasitism Rates and Stability section below for further discussion). Equilibrium population densities when parasitism is inversely density-dependent are quite high because many hosts (in high-density patches) escape parasitism (Murdoch 1992). As has been noted (Chesson and Murdoch 1986), stability from this sort of interaction can operate counter to the objective of biological control, which is to prevent high host densities.

We examined the literature for the Auchenorrhyncha (combining data from both the planthopper and the leafhopper literature) for spatial and temporal patterns of parasitism. We found a total of 9 published papers (with 10 host–parasitoid comparisons) that explicitly examined spatial density dependence (Table 11.2) and 11 (with 21 comparisons) that examined temporal density dependence (Table 11.3). In each case, we ac-

Table 11.2. Case studies of spatial relationships between host density and parasitism of the Auchenorrhyncha.

Host Species	Parasitoid	Analysis	Study Type	Response Type	Reference
EGG PARASITOIDS					
Mymaridae					
Eupteryx urticae	*Anagrus* sp. nr. *atomus*	Regr.	C	DD	Stiling (1980b)
Laodelphax striatellus	*Anagrus nr. flaveolus*	Regr.*	E	DI	Ôtake (1967)
Laodelphax striatellus	*Anagrus optabilis*	Regr.*	C	DI	Miura et al. (1981)
Macrosteles sexnotatus	*Anagrus holci*	Regr.	E	DD	Becker (1975)
Nephotettix cincticeps	*Gonatocerus* sp.	Regr.*	C	DI	Miura et al. (1981)
Nilaparvata lugens	*Anagrus optabilis*	Regr.*	C	DI	Miura et al. (1981)
Nilaparvata lugens	*Anagrus* sp.	Regr.*	E	DI	Ôtake (1967)
N. lugens/S. furcifera	*Anagrus incarnatus*	Regr.*	C	DI/INV	Chantarasa-ard et al. (1984b)
Prokelisia marginata	*Anagrus delicatus*	Regr.	C	DI	Strong (1989)
Prokelisia marginata	*Anagrus delicatus*	*k*-value	C	DI-INV	Roderick (1987)
Prokelisia marginata	*Anagrus delicatus*	*k*-value	C	INV	Stiling and Strong (1982a)
Prokelisia marginata	*Anagrus delicatus*	Regr.	E	DI	Cronin and Strong (1990a)
Sogatella furcifera	*Anagrus* sp.	Regr.*	E	DD	Ôtake (1967)
Sogatella furcifera	*Anagrus optabilis*	Regr.*	C	DI	Miura et al. (1981)
Typhlocyba pomaria	*Anagrus epos*	Regr.	C	DI/DD	Seyedodeslami and Croft (1980)
Trichogrammatidae					
Nephotettix cincticeps	*Paracentrobia andoi*	Regr.	E	DD	Sasaba and Kiritani (1972)
Nephotettix cincticeps	*Paracentrobia andoi*	Regr.*	C	DI	Miura et al. (1981)
Nilaparvata lugens	*Paracentrobia andoi*	Regr.*	C	DI	Miura et al. (1981)

NYMPHAL/ADULT PARASITOIDS					
Dryinidae					
Nephotettix spp.	Dryinidae	Regr.*	C	DI	Peña and Shepard (1986)
Nilaparvata lugens	Dryinidae	Regr.*	C	DI	Peña and Shepard (1986)
Sogatella furcifera	Dryinidae	Regr.*	C	DI	Peña and Shepard (1986)
Prokelisia marginata	Dryinidae	Regr.	C	DI	Stiling et al. (1991b)
Pipunculidae					
Nephotettix nigropictus	Pipunculidae	Regr.*	C	DI	Peña and Shepard (1986)
Nephotettix virescens	Pipunculidae	Regr.*	C	DI	Peña and Shepard (1986)
Strepsiptera					
Nephotettix spp.	Strepsiptera	Regr.*	C	DI	Peña and Shepard (1986)
Nilaparvata lugens	Strepsiptera	Regr.*	C	DI	Peña and Shepard (1986)
Prokelisia marginata	*Elenchus koebelei*	Regr.	C	DI	Stiling et al. (1991b)
Sogatella furcifera	Strepsiptera	Regr.*	C	DI	Peña and Shepard (1986)

Note: Study type: C = data from field collections, E = data from field experiments. Response type: DD = density dependence, DI = density independence, INV = inverse density dependence. * indicates our own analysis of the author's data.

Table 11.3. Case studies of temporal relationship between host density and parasitism of the Auchenorrhyncha. Duration of study is reported in months.

Host Species	Parasitoid	Duration	Analysis	Study Type	Response Type	Reference
EGG PARASITOIDS						
Eulophidae						
Saccharosydne saccharivora	*Tetrasticus sp.*	24	Life table	C	INV	Metcalfe (1972)
Mymaridae						
Dikrella californica	*Anagrus epos*	36	Life table	C	DDD†	Williams (1984)
Javesella pellucida	*Anagrus atomus*	48	Regr.*	C	DI	Raatikainen (1967)
Laodelphax striatellus	*Anagrus* nr. *flavelous*	3	Regr.*	E	DI	Ôtake (1967)
Prokelisia marginata	*Anagrus delicatus*	12	Regr.	C	INV†	Strong (1989)
Prokelisia marginata	*Anagrus delicatus*	12	*k*-value	C	INV	Stiling and Strong (1982a)
P. marginata and *P. dolus*	*Anagrus delicatus*	3	*k*-value	C	DI	Benrey and Denno (unpublished data)
Prokelisia marginata	*Anagrus delicatus*	10	Regr.	E	INV	Cronin and Strong (1990a)
Saccharosydne saccharivora	*Anagrus flaveolus*	24	Life table	C	DI	Metcalfe (1972)
Typhlocyba froggatti	*Anagrus armatus*	5	Regr.*	C	INV†	Teulon and Penman (1986)
Trichogrammatidae						
Nephotettix cincticeps	*Paracentrobia andoi*	36	Life table	C	DD	Sasaba and Kiritani (1972)
NYMPHAL/ADULT PARASITOIDS						
Dryinidae						
Cicadellid complex	Dryinidae	48	Regr.*	C	DI	Waloff (1975)
Delphacid complex	*Dicondylus bicolor*	36	Regr.*	C	DI	Waloff (1975)
Dichoptera hyolinata	*Dryinus* sp. A	12	Regr.*	C	DI	Swaminanthan and Ananthakrishnan (1984)

Dicranotropis hamata	*Dicondylus bicolor*	72	*k*-factor	C	DDD	Waloff and Thompson (1980)
Javesella pellucida	*Dicondylus bicolor*	36	Regr.*	C	DI	Waloff (1975)
Javesella pellucida	*Dicondylus lindbergi*	72	Regr.*	C	DI	Raatikainen (1967)
Muirodelphax exiguus	*Dicondylus bicolor*	36	Regr.*	C	DD	Waloff (1975)
Nephotettix spp.	Dryinidae	10	Regr.*	C	DI	Peña and Shepard (1986)
Nilaparvata lugens	Dryinidae	10	Regr.*	C	DI	Peña and Shepard (1986)
Psammotettix confinus	Dryinidae	48	Regr.*	C	DI	Waloff (1975)
Sogatella furcifera	Dryinidae	10	Regr.*	C	DI	Peña and Shepard (1986)
S. furcifera/N. lugens	*Haplogon. orientalis*	3	Regr.*	C	DI	Ôtake et al. (1976)
Prokelisia marginata	Dryinidae	12	Regr.	C	DI	Stiling et al. (1991b)
Nematoda						
S. furcifera/N. lugens	Nematoda	3	Regr.*	C	DI	Ôtake et al. (1976)
Pipunculidae						
Cicadellid complex	Pipunculidae	48	Regr.*	C	DI	Waloff (1975)
Delphacid complex	*Cephal. semifumosus*	36	Regr.*	C	DD	Waloff (1975)
Errastunus ocellaris	Pipunculidae	72	*k*-factor	C	DI	Waloff and Thompson (1980)
Javesella pellucida	*Cephal. semifumosus*	36	Regr.*	C	DI	Waloff (1975)
Laodelphax elegantulus	*Cephal. semifumosus*	36	Regr.*	C	DI	Waloff (1975)
Muirodelphax exiguus	*Cephal. semifumosus*	36	Regr.*	C	DI	Waloff (1975)
Neophilaenus lineatus	*Verrallia aucta*	108	*k*-factor	C	DD	Whittaker (1971)
Nephotettix cincticeps	Pipunculidae	24	Regr.*	C	DDD	Kiritani et al. (1970)
Nephotettix nigropictus	Pipunculidae	10	Regr.*	C	DI	Peña and Shepard (1986)
Nephotettix virescens	Pipunculidae	10	Regr.*	C	DI	Peña and Shepard (1986)
Philaenus spumarius	*Verrallia aucta*	108	*k*-factor	C	DD	Whittaker (1973)
Psammotettix confinus	Pipunculidae	48	Regr.*	C	DI	Waloff (1975)
S. furcifera/N. lugens	Pipunculidae	3	Regr.*	C	DI	Ôtake et al. (1976)

(*continued*)

Table 11.3. (*Continued*)

Host Species	Parasitoid	Duration	Analysis	Study Type	Response Type	Reference
Strepsiptera						
Delphacid complex	*Elenchus tenuicornis*	36	Regr.*	C	DI	Waloff (1975)
Javesella pellucida	*Elenchus tenuicornis*	36	Regr.*	C	DI	Waloff (1975)
Javesella pellucida	*Elenchus tenuicornis*	72	Regr.*	C	DI/DD	Raatikainen (1967)
Laodelphax elegantulus	*Elenchus tenuicornis*	36	Regr.*	C	DD†	Waloff (1975)
Muirodelphax exiguus	*Elenchus tenuicornis*	36	Regr.*	C	DI	Waloff (1975)
Nephotettix spp.	Strepsiptera	10	Regr.*	C	DI	Peña and Shepard (1986)
Nilaparvata lugens	Strepsiptera	10	Regr.*	C	DI	Peña and Shepard (1986)
Nilaparvata lugens	*Elenchus* sp.	12	Regr.*	C	DI	Kathirithamby (1985)
Paraliburnia dalei	*Elenchus tenuicornis*	36	Regr.*	C	DD	Waloff (1975)
Poophilus costalis	*Halictoph. pontifex*	48	Regr.	C	DD	Greathead (1970)
Poophilus latiusculus	*Halictoph. pontifex*	48	Regr.	C	DI	Greathead (1970)
Poophilus consperus	*Halictoph. pontifex*	48	Regr.	C	DI	Greathead (1970)
Cordia peragrans	*Halictoph. pontifex*	48	Regr.	C	DI	Greathead (1970)
Clovia centralis	*Halictoph. pontifex*	48	Regr.	C	DI	Greathead (1970)
Clovia quadrispinosa	*Halictoph. pontifex*	48	Regr.	C	DI	Greathead (1970)
Prokelisia marginata	*Elenchus koebelei*	12	Regr.	C	DI	Stiling et al. (1991b)
Ribautodelphax angulosus	*Elenchus tenuicornis*	36	Regr.*	C	DI	Waloff (1975)
Saccharosydne saccharivora	*Stenocranophilus sp.*	24	Life table	C	DD	Metcalfe (1972)
Sogatella furcifera	Strepsiptera	10	Regr.*	C	DI	Peña and Shepard (1986)
S. furcifera/N. lugens	*Elenchus* sp.	3	Regr.*	C	DI	Ôtake et al. (1976)

Note: Study type: C = data from field collections, E = data from field experiments. Response type: DD = density dependence, DDD = delayed density dependence, DI = density independence, INV = inverse density dependence. * indicates our own analysis of the author's data. † indicates significant serial correlation based on Durbin–Watson statistic (Montgomery and Peck 1982).

cepted the author's interpretation of the data in terms of the response of parasitism to host density. The agricultural literature that we examined contained many additional studies that did not examine the relationship between host density and parasitism, even though information on both host density and levels of parasitism were given. We took the liberty of analyzing these data as well. In all, we found an additional 4 papers (with 18 comparisons) that contained enough data to support examination for spatial patterns and 10 (with 37 comparisons) that we could examine for temporal patterns of parasitism. We attempted to be inclusive in our literature search but may have overlooked some studies; for that we apologize. Regardless of possible oversights, we feel that there are sufficient data in these papers (28 spatial and 58 temporal comparisons total) to reveal the general patterns in the Auchenorrhyncha.

We attempted to be conservative in our reanalysis of others' data. Spatial and temporal patterns of parasitism were examined by means of using simple linear regressions of host density on parasitism. All temporal data were examined for any delayed density dependence by lagging of data by one or two sample dates. Only papers with four or more simultaneous or sequential estimates of host density and parasitism were included in our analysis.

A number of statistical queries have been raised concerning the validity of the use of regression and correlation in detecting density dependence from field data (Reddingius 1971; Pielou 1974; Royama 1977, 1981). Of most significance are the assumptions that measurements of density and parasitism rate are made without error and that no serial correlation (autocorrelation) exists between sample dates (that density values are independent). Parameter estimates can be problematic, especially for measurement of the parasitism rate of hosts with overlapping generations (Van Driesche 1983; Van Driesche et al. 1991). We made no attempt to control for biases in authors' estimates of host density and parasitism. We assume them to be accurate or at the very least to be biased consistently throughout each study such that the bias is independent of host density. Serial correlations in temporal data, on the other hand, can introduce bias that leads to spuriously significant density relationships (a Type I error) (Varley and Gradwell 1968). To identify possible errors of this type, we use the Durbin–Watson statistic (Montgomery and Peck 1982) to test for serial correlations. This statistic is applied to all cases of significant temporal density relationships (density dependence and inverse density dependence) found in Table 11.3. For studies with a significant response to density and a significant serial correlation of densities through time, we do not reverse response type (to density independence). Instead, we simply urge caution in their interpretation.

SPATIAL PATTERNS

Approximately three-quarters of the cases examined showed no spatial density dependence (Table 11.4). The remaining 25% is split fairly evenly between cases of density dependence and inverse density dependence. It is interesting to note that all cases of inverse density dependence come from one system: the egg parasitoid *A. delicatus* and its host *P. marginata*. Only the egg parasitoids demonstrated any density dependence (35%); cases involving nymphal/adult parasitoids showed only density independence. Significant relationships (density dependence and inverse density dependence) occurred more frequently among the egg parasitoids than among nymphal/adult parasitoids (Fisher's exact test, $P = 0.026$). In total, fewer significant relationships occurred among the Auchenorrhyncha (26%) than for all host–parasitoid systems (50%) reviewed by Stiling (1987) (Fisher's exact test, $P = 0.011$).

For *A. incarnatus* parasitizing *N. lugens* (Chantarasa-ard et al. 1984b) and *A. delicatus* parasitizing *P. marginata* (Stiling and Strong 1982a; Strong 1989; Cronin and Strong 1990a; Roderick 1987; Benrey and Denno unpublished data), the spatial response to host density supports the laboratory findings of a Type II functional response for each species. A parasitoid that exhibits a Type II response should have a density-independent attack rate at

Table 11.4 Summary of studies examining spatial and temporal density dependence (see Tables 11.2 and 11.3) for egg and nymphal/adult parasitoids, separately and combined. Numbers in parentheses are proportions of studies showing response to density.

Pattern	Parasitoid Guild	Response to Host Density DD	DI	INV
Spatial	Egg	5 (0.24)	13 (0.62)	3 (0.14)
	Nymphal/adult	0 (0.0)	10 (1.0)	0 (0.0)
	Total	5 (0.16)	23 (0.74)	3 (0.10)
Temporal	Egg	2 (0.18)	4 (0.36)	5 (0.45)
	Nymphal/adult	11 (0.23)	37 (0.77)	0 (0.0)
	Total	13 (0.22)	41 (0.69)	5 (0.09)

Note: DD = density dependence, DI = density independence, INV = inverse density dependence.

lower host densities, but as the response reaches a plateau, parasitism should tend toward inverse density dependence. For both parasitoids, parasitism was either inversely density dependent or density independent.

The fact that egg parasitoids are more responsive to host density is puzzling. We could advance many ad hoc reasons why the sedentary egg stage may incur a higher frequency of significant responses, but no single reason appears compelling enough to justify invoking it over other alternatives. In the absence of detailed studies of parasitoid foraging behavior, it would be fruitless, at this point, to speculate about the cause of this pattern among the Auchenorrhyncha.

We can offer no concrete reasons for why fewer cases of density dependence or inverse density dependence should occur among planthoppers and leafhoppers. The lack of statistical power that results from the typically small sample sizes may have inflated the number of nonsignificant patterns. Results may also be biased toward density independence because appropriate spatial scales were not examined (Heads and Lawton 1983, but see Strong 1989). Only 5 of the 27 comparisons examined more than 1 spatial scale, 4 of which dealt with the same host–parasitoid system (Stiling and Strong 1982a; Chantarasa-ard et al. 1984b; Strong 1989; Cronin and Strong 1990a; Roderick 1987). Three of the 5 studies showed a significant density response (all inverse) at some spatial scale (Stiling and Strong 1982a; Chantarasa-ard et al. 1984b; Roderick 1987). For example, Stiling and Strong (1982a) examined the effects of parasitism by the egg parasitoid *A. delicatus* in a north Florida salt marsh at a range of spatial scales from within leaves to among plants and finally among small islands. The pattern of parasitism, however, was inversely density dependent at all spatial scales. Ecologists today appear to be more aware of the potential problem of studying only one spatial scale, and it is likely that more studies in the future will address this issue. Until then, we can only suggest that doing so may reduce the likelihood of finding a significant response. According to Hassell (1985a, 1986a, 1986b, 1987), it may be very difficult to extract true patterns of parasitism from field data as a result of the great environmental variability so typical of natural populations. Underlying density dependence may be hidden by stochastic processes or time delays, resulting in an overestimate of the prevalence of density-independent processes in field collections. However, if density dependence in parasitism rate is so feeble, then population dynamics are not likely to be affected by it.

We have attempted to circumvent this problem by performing well-controlled experiments that are replicated in time and over spatial scales that are relevant to the searching behavior of the parasitoid involved (Cronin and Strong 1990a). We controlled host age, patch size, host plant quality, and exposure time by experimentally forcing *Prokelisia* spp. to lay eggs on cordgrass leaves in well-defined patches over a 2-day period. Plants were exposed to searching *A. delicatus* in the field for a 3-day period and then returned to the lab. This procedure permitted accurate estimates of parasitism and was repeated on 20 separate occasions to account for any temporal variation in parasitism. Despite these precautions, we still found no evidence for density dependence, a result that corresponds closely with observational data collected by Stiling and Strong (1982a) and Strong (1989). We suggest that the observational data reviewed in this chapter can provide valid inferences about density relationships and that the results of the reviews should not be lightly dismissed.

TEMPORAL PATTERNS

Temporal density independence was the pattern most frequently observed among the 58 available comparisons (Table 11.4). Sixty-nine percent of these studies showed no density dependence, whereas 22% showed direct density dependence and 9% revealed inverse density dependence. Of those 18 significant responses, only 4 showed significant serial correlations among sample densities (Table 11.3). Three come from egg parasitoids and one from a nymphal-adult parasitoid. Because omission of these studies from subsequent analysis would only further magnify the preponderance of nonsignificant responses, we elected to include them. The proportion of significant temporal responses was significantly lower than that found in Stiling's (1987) review for all insect parasitoids (Fisher's exact test, $P = 0.008$). Among the Auchenorrhyncha, egg parasitoids showed more significant responses (63%) than did nymphal/adult parasitoids (23%) (Fisher's exact test, $P = 0.011$). Again, the difference between egg and nymphal/adult parasitoids is difficult to explain with the data at hand. Bias, however, may have been introduced into this analysis because 40% of the egg parasitoid data are from one host–parasitoid interaction (*P. marginata* and *A. delicatus*), which represented three of the five inverse responses. Other forms of bias may be introduced if host and parasitoid species are represented by multiple comparisons. In any event, this result corresponds with the evidence from the spatial data that egg parasitoids more often exhibit significant responses to host density.

Hassell et al. (1989; see also Solow and Steele 1990) have recently argued that the preponderance of density independence in temporal studies may simply be the result of the short duration typical of most analyses. They argue that as the experimental period increased, so did the probability that a significant density response would occur. We categorized the temporal data from Table 11.3 by duration (<1, 2, 3, 4, and >5 years) and plotted the proportion of reports in each category of a significant density response. We found that, as the duration of the study increased, significant density patterns became more likely (Fig. 11.3), but much of this pattern is driven by the low proportion of density responses for studies of less than 1 year. Beyond 1 year, the proportion of density responses is not strongly affected by time.

The data from our review of the literature do not provide strong support for stability due to simple point equilibrium of host–parasitoid interactions. However, environmental variability, lack of statistical power, and the short duration of these studies may be obscuring density-dependent patterns of parasitism (Hassell 1985a, 1986a, 1987; Hassell et al. 1989; but see Mountford 1988). Before we can draw any sound conclusions about the universality of density-dependent processes in the Auchenorrhyncha, more detailed studies will have to be performed. In particular, long-term studies involving a hierarchy of spatial scales can be most informative. In addition, experimental studies manipulating hosts such as those by Ôtake (1967), Sasaba and Kiritani (1972), Becker (1975), and Cronin and Strong (1990a) would help ferret out any underlying density dependence in natural populations. Until studies like these can be performed, the controversy surrounding the role of density dependence in population regulation will remain unresolved.

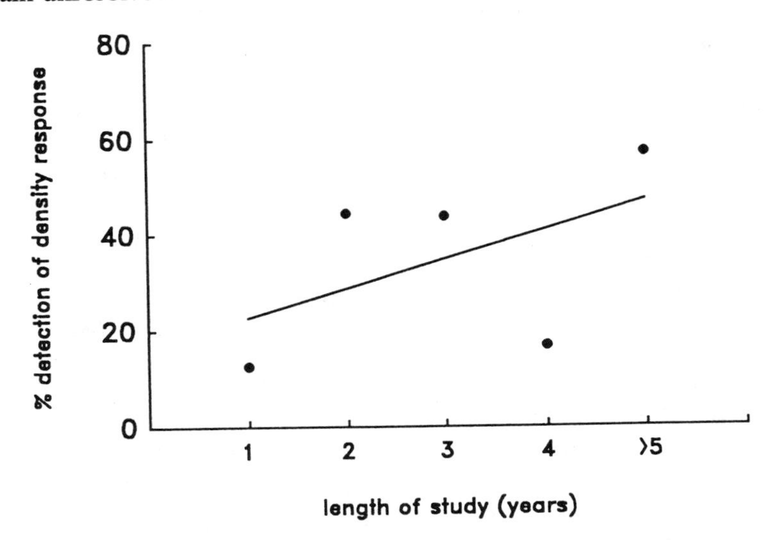

Figure 11.3. The relationship between the length of study and the probability of detecting a significant response of parasitism to host density (density dependence and inverse density dependence). Data were angular transformed and weighted by the inverse of the variance ($1/4n$, where n = number of studies in that time period). Line fit by least squares regression ($R^2 = 0.42, P < 0.001$).

Heterogeneous Parasitism Rates and Stability

The proponderance of cases of spatial density independence (74%) found in our review of the Auchenorrhyncha is not *prima fascie* evidence for the instability of planthopper– and leafhopper–parasitoid interactions. Many of these density-independent relationships are characterized by high variance in parasitism rate per host patch, which is considered to be a powerfully stabilizing force (Hassell 1985b; Chesson and Murdoch 1986). Recently, Pacala et al. (1990) and Hassell et al. (1991) have provided a general criterion, the $CV^2 > 1$ rule, for determining whether sufficient heterogeneity in parasitism exists to theoretically stabilize the host–parasitoid interaction. In their review, Pacala and Hassell (1991) found that 9 of 34 case studies exhibited sufficient heterogeneity to satisfy the $CV^2 > 1$ rule. For eight of these nine cases, the principal source of heterogeneity was density-independent.

We do not evaluate the role of heterogeneous parasitism rates in stabilizing all 31 case studies in Table 11.2 (see below). Instead, we exemplify the CV^2 approach with a parasitoid whose biology and ecology is most familiar to us: the fairyfly *A. delicatus*. J. D. Reeve, J. T. Cronin, and D. R. Strong (unpublished data) estimated the CV^2 from data on *A. delicatus* parasitism obtained in Strong (1989) and Cronin and Strong (1990a). On 24 of 28 sampling dates, the CV^2 exceeded 1 (14 were significantly greater than 1), and over all dates averaged 3.34 ± 0.78 ($\pm$ 1SE). Because parasitism was independent of host density in these data (Table 11.2) the source of heterogeneity was also density-independent. However, stability, in a theoretical sense, is not proved in this example. A critical assumption of the stability criterion is violated: The life cycle of *P. marginata* and *A. delicatus* are not completely synchronous. Asynchronous development tends to destabilize host–parasitoid interactions and generally requires heterogeneity (CV^2) greater than 1 to stabilize the interaction (Godfray and Hassell 1989; J. D. Reeve, J. T. Cronin, and D. R. Strong unpublished data).

Serious consideration of the heterogeneity in parasitoid foraging is an important advancement in the study of host–parasitoid interactions; it may well prove to be an important cause of stability. However, the $CV^2 > 1$ rule for evaluating the contribution of heterogeneity to stability must be used judiciously. The assumptions of random parasitoid search and oviposition within a patch, discrete nonoverlapping host and parasitoid generations, and synchronous host and parasitoid development must be met before the $CV^2 > 1$ rule for stability can be considered valid. The robustness of this stability criterion is largely unexplored with respect to deviations from most of these assumptions. Unless we are certain the assumptions are met, or until we more fully understand the consequence of violating these assumptions, the CV^2 rule may be of little practical use. We, therefore, advise a fundamental understanding of host–parasitoid biology and parasitoid foraging behavior **before** a test of the stabilizing effect of heterogeneity is considered. In light of the paucity of these kinds of data for host–parasitoid systems listed in Table 11.2, we decline to evaluate the stability criterion for these systems.

Refugia from Parasitism

We have already mentioned that aggregation of parasitism produces a partial refuge in which some hosts may escape parasitism and that this process can confer stability on the host–parasitoid interaction or prevent extinctions (Murdoch et al. 1989). Many other types of refugia can exist in natural populations that confer some protection from parasitism. In general, any biological or physical factor that makes the host species less vulnerable can contribute to the stability of the system (Bailey et al. 1962; Murdoch and Oaten 1975; Hassell 1978; but see McNair 1986). Such factors as host size, age, and distribution can act as refuges from parasitism, as can physical refugia.

Some planthopper hosts are relatively invulnerable to parasitism beyond a certain developmental stage. May (1971) found that *A. stenocrani* completely avoids parasitizing *Stenocranus minutus* (Fabricius) eggs later in development. In addition, Cronin and Strong (1990b) found that successful parasitism of host eggs declined with egg age; well-developed eggs produced almost no offspring. Ôtake (1968) found more subtle differences with hosts age; parasitoid development was simply protracted in older hosts. Size-dependent parasitism (which is often correlated with age) has also been documented. Greathead (1970) found that parasitism by the strepsipteran *Halictophagus pontifex* Fox increased with the size of its cercopid host, allowing smaller hosts to be relatively free of parasitism. Not only would host age or size provide a refuge from parasitism, but it may also reduce a parasitoid's efficiency and/or increase handling time, by causing parasitoids to waste eggs or time on an invulnerable developmental stage or size class (Walde et al. 1989). Both could contribute to the stability (or persistence) of the system.

Environmental factors have also contributed to the protection of planthoppers from their parasitoids. These refugia either exclude natural enemies from the host habitat or interfere with or prevent the detection of the host. Stiling and Strong (1982a), for example, found that host eggs of *P. marginata* were concentrated in the basal portions of their cordgrass host, whereas parasitism by *A. delicatus* was concentrated in the apical leaves. They argued that, as a result of the tides, basal leaves were more protected from parasitoids because they were submerged for longer periods of time. Tides, then, could effectively reduce the time available for parasitizing the abundant hosts at the basal portions of the plant, which could lead to the inverse pattern of parasitism found by Stiling and Strong (1982a). Recently, however, Roderick (1987) concluded that the tides had little effect on parasitism rates in a California salt marsh. More information will be needed to determine the importance of tidal inundation as a refuge from parasitism. Tay (1972) also found distributional differences of the cicadellid *Cicadella viridis* within its host plant *Juncus* that were similar to those found in the salt marsh. Leafhoppers were concentrated low on the Juncus stems, and egg parasitism in these areas was much lower than at the stem tips. A similar situation was found by Arzone (1974). Either parasitoids do not search as often near the ground or they are less effective at high host density (inverse density dependence); either way, a partial refuge is created. Differential susceptibility of hosts to parasitism within different microhabitats of the host plant appears to be a relatively common phenomenon among parasitoid species (Lawton 1986).

There is also evidence that the host plant can confer some protection from parasitoids. Parasitoids may locate hosts by first cueing in on the host plant (Price et al. 1980; Vinson 1981). Hosts on plants that are outside of the sensory range of a parasitoid would, therefore, be overlooked. Hosts on different plant species may also come to differ markedly in quality through ingestion of secondary metabolites (Vinson 1976, 1981; Price et al. 1980; Gibson and Mani 1984; Barbosa 1988), which can influence the host selection process. Abdul-Nour (1971) found that the strepsipteran *Halictophagus languedoci* in grasslands attacked a complex of leafhoppers that fed on the grass *Brachypodium phaenicoides* but did not attack those feeding on either *B. ramosum* or *Bromus erectus*, even though host plants occurred within the same localized habitat. In common garden experiments, A. M. Rossi, J. D. Reeve and J. T. Cronin, (unpublished data) found that parasitism of *Carneocephala floridana* eggs by *Gonatocerus* sp. was significantly higher on grasses (*S. alterniflora* and *Distichlis spicata*) than on herbaceous perennials (*Salicornia virginica* and *Borrichia frutescens*). Whether variations in host susceptibility result from differences in plant nutrition that affect leafhopper quality, and ultimately parasitoid quality, or from differences in parasitoid foraging behavior among host plants is not known. Regardless of the cause, the less-susceptible host plants act as a partial refuge from parasitism.

The plant may also function as a refuge by limiting access of searching parasitoids to their hosts. For example, host eggs that are buried deep within stems, leaves, or buds may escape attack from egg parasitoids that have ovipositors too short to reach them. Planthoppers that can insert eggs deeper into the plant tissue may be relatively immune from attack by a larger portion of the parasitoid population. Host-species differences in the placement of eggs apparently result in different parasitoid complexes attacking each species (Claridge and Reynolds 1972; Ali 1979). Settle and Wilson (1990) demonstrated that parasitism of the grape leafhopper is higher than that of the variegated leafhopper because the former's eggs are more readily detected by foraging *Anagrus epos*. The greater depth at which variegated leafhopper eggs are buried is presumably the cause of their reduced detection. Intraspecific differences in susceptibility to parasitoid attack have also been found. Raatikainen (1967) found that *Javesella pellucida* eggs in cereal stems were invulnerable to attack by *A. atomus*, whereas eggs in leaf blades incurred high levels of parasitism. In oats, thin-walled stems had a higher incidence of parasitism than thick-walled stems, suggesting an inability of the wasp to detect or gain access to more deeply concealed hosts. We, on the other hand (Cronin 1991; Cronin and Strong unpublished), find that the host plant provides no physical barrier to *A. delicatus* searching for *Prokelisia* spp. eggs concealed within the leaf. Even the wasps with the shortest ovipositors had free access to the most deeply embedded host eggs. We are aware of no other studies that have considered differential susceptibility of individual planthoppers or leafhoppers within a population to their natural enemies. However, the phenomenon has been found among other host–parasitoid systems involving such diverse groups as gall-making and other endophytic flies (Weis 1983; Price and Clancy 1986; Weis and Abrahamson 1985; Price 1988; Romstöck-Völkl 1990), moths (Graham and Baumhofer 1927; Cronin and Gill 1989), and bark beetles (Ball and Dahlsten 1973).

The potential for refugia among the Auchenorrhyncha is certainly much greater than the evidence at hand would suggest. Few researchers have explicitly considered looking for refugia in their field studies, and the necessary experiments needed to identify them have rarely been performed. Because a refuge can exist in so many different forms and can have a large effect on the stability and/or persistence of host–parasitoid populations, more emphasis should be placed on experimental testing for mechanisms of escape from parasitism.

Role of Alternate Hosts in Host–Parasitoid Stability

A broad host range can contribute significantly to the persistence and stability of host–parasitoid systems. When densities of preferred hosts fall to low levels, polyphagous parasitoids can switch (not necessarily exclusively) to more abundant alternate hosts. This frequency-dependent switching from one host to another should result in a Type III, sigmoid functional response (Murdoch 1969; Hassell 1978). Because a host and its parasitoid are less coupled and because of the parasitoid's broader diet, time delays should be absent from the functional response (Murdoch and Oaten 1975). In the absence of time delays, a Type III response can be strongly stabilizing because mortality can be directly density dependent (Hassell 1986b). In essence, a broader host range can buffer the parasitoid population against fluctuations in the abundance of hosts. When one host becomes locally extinct because of local catastrophes or simply seasonal changes in abundance of susceptible host stages, alternate hosts may be the only means by which a parasitoid can persist. That is to say, not only can alternative hosts contribute to population stability, but more fundamentally, they can allow parasitoid populations to persist.

Host records for the parasitoids of planthoppers and leafhoppers are largely incomplete and, therefore, probably reflect an underestimate of the degree of polyphagy among these parasitoids. Existing records suggest that the egg parasitoids of the Auchenorrhyncha are relatively monophagous or oligophagous (Waloff and Jervis 1987). Data from the Mymaridae suggest that the oviposition site of the host is an important determinant of whether an egg will be parasitized (Claridge and Reynolds 1972; Ali 1979). Claridge and Reynolds (1972) suggested that *Polynema euchariforme* and *P. bakkendorfi* were restricted in host range by the oviposition sites of their *Oncopsis* hosts. *Polynema euchariforme* was restricted to eggs of host species that were laid deep within birch buds, whereas *P. bakkendorfi* was restricted to hosts that laid eggs closer to the surface. Similar results were found by Ali (1979) with *Anagrus* spp. In support of host site specificity, Ali (1979) found that some *Anagrus* spp. rejected oviposition sites of unnatural hosts but accepted these unnatural hosts when they were placed among eggs of their natural hosts. Successful development was possible from these unnatural host species, although survival of offspring was not always as high (Ali 1979; Walker 1979). Monophagy may be the rule among egg parasitoids, but until more complete records can be obtained and the taxonomy of the Mymaridae improved (Huber 1986), we must view this statement with some skepticism.

Polyphagy appears to be the rule among the Pipunculidae, Strepsiptera, and Dryinidae (Waloff and Jervis 1987). For example, in a grassland community, Waloff (1975) and Waloff and Thompson (1980) found that the majority of parasitoids were generalists. In woodland forests of South Wales, Jervis (1980) found that all five pipunculids and three of the four dryinids that attack the typhlocybine leafhoppers were polyphagous. The polyphagy appears to extend to one host family, which appears to be a general pattern among nymphal/adult parasitoids; few exceptions exist (Waloff and Jervis

1987). In addition, few of these parasitoids are restricted to attacking hosts on a single plant species (Waloff and Jervis 1987).

A number of studies involving the Auchenorrhyncha suggest the importance of alternative hosts for the persistence and stability of host–parasitoid interactions. Evidence for frequency-dependent host selection comes from the studies by Whittaker (1971, 1973) in grasslands of southern England. When the cercopid *Neophilaenus lineatus* was more abundant than the alternate host, *Philaenus spumarius*, it incurred higher rates of parasitism by the pipunculid *Verrallia aucta* (Fig. 11.4). In years when *P. spumarius* was more abundant, it incurred higher rates of parasitism. This switching between hosts involves switching habitats as well because the two hosts feed on different plant species. Adult mortality of both host species was density-dependent when the effects of the pipunculid were included, and the degree of density dependence was greatest when the two species were treated together (Whittaker 1973). This result led Whittaker (1973) to conclude that populations of *P. spumarius* and *N. lineatus* were more stable in the presence of each other than when occurring alone.

A classic example of the importance of alternative hosts in host–parasitoid interactions comes from the agricultural literature on the pests of

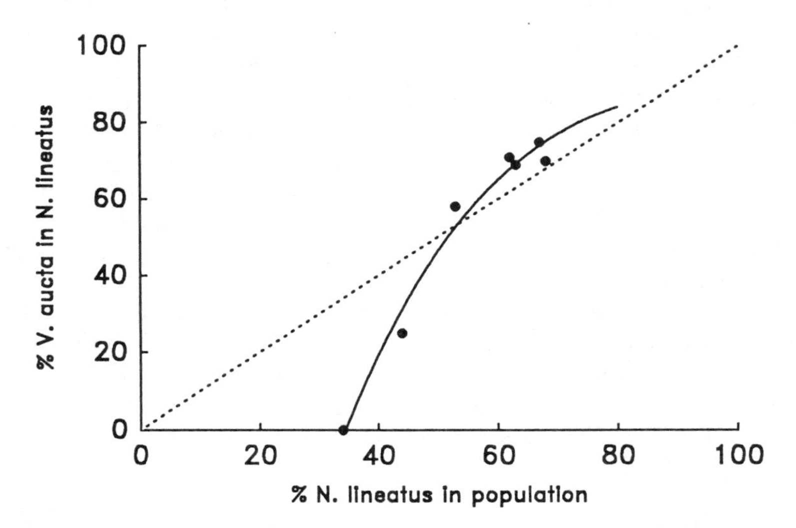

Figure 11.4. Host switching behavior by the pipunculid *Verrallia aucta* between its two cercopid hosts, *Neophilaenus lineatus* and *Philaenus spumarius*. Percentage of *V. aucta* in *N. lineatus* is greater than expected on the basis of random foraging (straight line) in years when *N. lineatus* is most abundant. In years when *N. lineatus* is rare (< 50%), the parasitoid is underrepresented in this host. Data are from Whittaker (1973), and switching curve is fit by eye.

grapes in California, USA. The grape leafhopper, *Erythroneura elegans*, became a serious pest of the Old World grape, *Vitis vinifera*, soon after the grape's introduction into the New World in 1781. The leafhopper's eggs are attacked by the mymarid *Anagrus epos*, which has significantly contributed to the control of this pest. The success of this parasitoid in controlling the grape leafhopper is directly tied to the availability of its alternative host, *Dikrella cruentata*, which feeds on wild *Rubus* spp. (Doutt and Nakata 1965, 1973; Williams 1984). As temperatures warm in the spring and grape leafhoppers resume egg laying, *A. epos* numbers respond rapidly. By late summer, parasitism rates can reach levels greater than 90%. During late fall and winter, adult leafhoppers cease oviposition and undergo reproductive diapause. The wasps do not undergo diapause, and grape leafhopper eggs are absent during the winter months (Doutt and Nakata 1973). In winter, *A. epos* parasitizes only *D. cruentata*, whose eggs can be found year round. Parasitism of this alternative host in northern California is temporally density-dependent (Williams 1984), as might be expected when frequency-dependent host selection occurs (Hassell 1978, 1986b).

It is clear from the above data that both host species are necessary for *A. epos* to utilize the large populations of the grape leafhopper. A number of other studies implicate the lack of synchronization (the presence of a suitable host developmental stage) as a cause of host switching (Raatikainen 1967; Claridge and Reynolds 1972; Waloff 1975; Hulden 1984). Jervis (1980), on the other hand, found little evidence that polyphagy among dryinids and pipunculids of typhlocybine leafhoppers was the result of poor synchronization with one or several hosts. Suitable stages (nymphs and adults) of all host species were available during periods when foraging parasitoids were active. Instead, he suggested that host range was delimited by discriminative oviposition or defensive behavior of nymphs, among other things.

There is little doubt of the importance of these alternative hosts in buffering parasitoid populations against potential extinction. However, support for the role of alternative hosts in contributing to interaction stability is mostly anecdotal. More detailed analyses of the regulatory power of parasitoids on their entire host ranges are needed.

Parasitoid Dispersal

There is little question that parasitoid dispersal contributes to the persistence of the interactions between the Auchenorrhyncha and their parasitoids. Evidence from the importance of alternate hosts to polyphagous parasitoids have already demonstrated this point. *Anagrus epos* must disperse from vineyards (Doutt and Nakata 1965, 1973), and *A. atomus* must disperse from oat fields in search of suitable overwintering hosts (Raatikainen 1967). Some parasitoids travel great distances in search of suitable host habitats. An example is the egg parasitoid *A. delicatus*, which regularly disperses up to several kilometers in search of new hosts (Antolin and Strong 1987; Cronin unpublished data). Several other examples exist that document the dispersiveness of parasitoids of leafhoppers and planthoppers (Ôtake 1970, 1976; Raatikainen 1972).

The role of dispersal has been addressed in the context of formal stability models. In the simplest abstract models without environmental variability, migration among spatially subdivided units of a population has either no effect or even a detrimental one on stability (Allen 1975). Introducing some forms of environmental variability and the notion of "nonequilibrium coexistence," Caswell (1978) showed how the disturbance caused by a predator could greatly prolong the coexistence of unstable combinations of competitors. For predator–prey and parasitoid–host models, perhaps the major key to persistence of deterministically unstable interactions is desynchronization of subpopulations; local populations that go extinct are refounded by immigrants (Crowley 1981; Nachmann 1987). In the metapopulation model, local extinctions must be complemented by other, extant subpopulations that can supply immigrants. Reeve (1988) has elaborated this principle in hybrid, equilibrium with nonequilibrium, modeling, showing how greatly the relative levels of dispersal and environmental variability can affect persistence of subdivided populations. Persistence of unstable ensembles of subpopulations can be greatly enhanced when variability caused by environmental stochasticity is complemented by low to moderate rates of migration among subpopulations. This is to say that groups of local populations, which would all become extinct if isolated, can persist if they are sufficiently desynchronized.

There is a long tradition in ecology of verbal–graphic models that rely on dispersal for the long-term persistence of parasitoid–host populations (Andrewartha and Birch 1954; den Boer 1968; Reddingius and den Boer 1970; Roff 1974a, 1974b). Population dynamics are much affected by random external forces that generate extinction and asynchrony among subpopulations. Predation, parasitism, and the weather act in a nonregulatory fashion to depress population densities and can cause local extinctions, but because subpopulations are no longer synchronous, extinction of the metapopulation (the collection of subpopulations) is rare. Vacant patches are eventually recolonized by individuals dispersing from nearby local subpopulations. The fitness of organisms in this type of environment could be increased by dispersal among a number of subpopulations, that is, by "spreading the risk" of extinction among a number of host patches (den Boer 1968).

Such a situation appears to exist among populations of the salt marsh-inhabiting delphacid *P. marginata* and its mymarid egg parasitoid *A. delicatus* (Strong 1988; Cronin and Strong unpublished data). Hosts are dispersed among many discrete patches of salt-marsh cordgrass, isolated as a result of man-made development, erosion, and the physical contour of the marsh. In addition, oyster bars and islands along the Gulf coast of north Florida contain populations separate from those on the mainland. Among these heterogeneous islands, planthopper population densities and net reproductive rates differ from island to island (Strong et al. 1990). Although extinctions can be difficult to document absolutely, evidence does suggest that island populations occasionally approach extinction of hosts and parasitoids (Antolin and Strong 1987; Strong unpublished data). Recolonization of vacant island habitats occurs by the regular dispersal of both macropterous adult planthoppers and female parasitoids (Strong and Stiling 1983; Antolin and Strong 1987).

Evidence for risk spreading by *A. delicatus* in such a variable environment comes from their pattern of oviposition in host patches (Cronin and Strong unpublished data). In nature, parasitoids lay only a small fraction of their eggs in patches containing abundant hosts. An average of two eggs were laid per parasitoid per leaf containing experimentally placed host eggs, and then wasps dispersed. This low number of eggs laid per host patch is not the result of egg depletion or sperm limitation, nor is it related to the density of hosts per leaf. In the laboratory, time on a patch prior to dispersal is quite variable but also independent of host density. Dispersing wasps are very adept at locating new host patches, even at long distances within the marsh. At least a few wasps locate experimentally isolated host patches (> 300 m from other hosts) within just a few days and can cause high per capita rates of parasitism (Antolin and Strong 1987; Cronin and Strong unpublished data). Submaximal oviposition rates coupled with frequent bouts of dispersal, despite the presence of abundant hosts, appear to be a common feature of this system and fit well within the framework of risk-spreading theory. Presently, we have no data on the risks involved in remaining and ovipositing within a single or few host patches, but the data would suggest that there has been strong selection against such a strategy. A risk-spreading strategy, in concert with weakly stabilizing factors such as mutual interference (Fig. 11.2D), and heterogeneous parasitism can contribute greatly to long-term persistence of the *P. marginata–A. delicatus* interaction.

Discussion and Conclusions

Data on the ecology of the Auchenorryncha are presently incomplete. Only limited evidence exists that addresses the regulatory power of the parasitoids of planthoppers and leafhoppers, and the data are equivocal. Most evidence comes from an analysis of the spatial and temporal relationship between host density and parasitism from field collections of hosts. More than 70% of the studies reviewed found density independence, and an additional 10% found inverse density dependence, leaving only about 20% of the studies that showed some regulatory power. This is probably a slight underestimate of the importance of density dependence because of low sample sizes, short durations of study, and environmental noise, all of which could bias results toward density independence. However, if density dependence exists in these populations, it is probably a weak stabilizing force in the face of other environmental factors.

A myriad of other factors can contribute to the overall stability of planthopper–parasitoid interactions, despite any or all being weak forces. Parasitoid interference, heterogeneous parasitism, the presence of refugia, alternate hosts, and alternate strategies of risk spreading can all contribute toward the long-term persistence of natural populations. For example, in *A. delicatus*, parasitoid interference or heterogeneous parasitism rates may weakly contribute to the regulation and stability of the system, but neither alone is likely sufficient to cause host–parasitoid persistence. The presence of an invulnerable developmental stage (Cronin and Strong 1990b), the availability of alternate hosts (*Prokelisia dolus*), and an apparent risk-spreading strategy (Strong 1988; Cronin and Strong unpublished data) can act together with these weak regulatory mechanisms to provide strong population stability. This situation probably exists in other planthopper systems where no strong regulatory role is played by their parasitoids.

Until more critical experiments are performed on the ecology of planthopper–parasitoid interactions, we can only speculate as to the role parasitoids play in stabilizing host populations. We need to progress beyond basic sampling programs that provide only indirect data on parasitoid behavior, through estimates of rates of parasitism. Only long-term manipulative field experiments, spanning many generations and spatial scales, will aid in resolving the controversy of the regulatory power of parasitoids. Experimental studies of the dynamical behavior of parasitoids within host patches will be most useful in providing information on the broad-scale effects of parasitism on population persistence. Patch-level studies can reveal information on the host location and selection process, presence of refugia, patterns of parasitoid foraging behavior, and so on. Such an approach has been fruitfully employed by Kareiva and Odell (1987) in their study of the role of coccinellids in controlling their aphid prey. Without such studies, we will not be able to further our understanding of the role of parasitoids in the population dynamics of their planthopper and leafhopper hosts.

Summary

1. Parasitoids are common natural enemies of planthoppers and leafhoppers and often contribute substantial mortality to host populations. High mortality in itself does not, however, provide evidence for population regulation or control.
2. We review the interactions between the Auchyenorrhyncha and their parasitoids in terms of their effect on stability or population persistence. Our review concentrates on parasitoid searching behavior and the spatial and temporal patterns of parasitism with respect to host density.
3. Direct studies of paratisoid functional response and interference are few and provide equivocal evidence for the regulatory power of parasitoids. On the other hand, spatial and temporal relationships between host density and parasitism are predominantly density independent (> 70% of the studies reviewed), indicating no or weak regulation by parasitoids.
4. Other potentially stabilizing factors, such as heterogeneous parasitism rates, the presence of refugia for hosts, host switching by parasitoids, and less traditional nonequilibrial strategies by parasitoids, have received very little consideration.
5. The critical data needed to assess the regulatory power of planthopper and leafhopper parasitoids are largely unavailable. It is imperative that long-term manipulative field experiments over many generations and spatial scales, coupled with detailed examination of parasitoid searching behavior, be performed if we hope to resolve this issue.

12

Transmission Biology, Vector Specificity and Evolution of Planthopper-Transmitted Plant Viruses

Lowell R. Nault

Introduction

Special attention has been paid to a number of planthopper species because of the impact they have on crop production. Few would argue that one of the best studied planthoppers, *Nilaparvata lugens* (Stål), is also one of the world's most destructive insects. Each year, *N. lugens* causes more than $1.23 billion in losses to rice in Southeast Asia (Herdt 1987). These losses are caused by damage from feeding injury and by plant viruses transmitted by the planthopper. *N. lugens* is one of more than two dozen planthopper species known to transmit plant pathogens.

No book on planthoppers would be complete without a discussion of their role as vectors. This coverage of the subject mainly is limited to transmission of the plant viruses because currently little is know about the vector–pathogen relationships of planthopper-transmitted, mycoplasma-like organisms. The chapter is written with two groups of scientists in mind. The first is insect biologists, including those familiar with planthoppers, who do not have a background in plant virology. The second is virologists and plant pathologists unfamiliar with the biology of planthoppers and their homopteran relatives. Accordingly, in this chapter the characteristics of plant virus groups that have planthopper vectors are reviewed and the mechanisms by which planthoppers transmit them are discussed. My discussion on vector specificity is founded on this understanding of transmission mechanisms. The origin and evolution of planthopper-transmitted viruses are also speculated. Finally, a discussion of why planthoppers are ill-suited to transmit some plant viruses and better-suited to transmit others

Table 12.1. Planthopper transmitted plant viruses, their vector species and distribution.[a]

Plant Virus	Planthopper Vector	Geographic Distribution
Tenuivirus		
Rice stripe*	*Laodelphax striatellus* (Fallen)	Japan, Korea, Taiwan, USSR
	Unkanodes sapporonus (Matsumura)	
	U. albifascia (Matsumura)	
	Tertbron albovittatus (Matsumura)	
Maize stripe*	*Peregrinus maidis* (Ashmead)	Tropics, worldwide
Rice grassy stunt	*Nilaparvata lugens* (Stål)	South, Southeast, and East Asia
	N. bakeri (Muir)	
	N. muiri China	
Rice hoja blanca*	*Tagosodes orizicolus* (Muir)	Neotropics
	T. cubana (Crawford)	
European wheat striate mosaic*	*Javesella pellucida* (F.)	Europe
	J. dubia (Kirschbaum)	
	J. obscurella (Boheman)	
Rhabdovirus		
Barley yellow striate mosaic*	*Laodelphax striatellus* (Fallen)	Europe
Colocasia babone disease	*Tarophagus proserpina* (Kirkaldy)	Pacific Islands
Cynodon chloritic streak	*Toya propinqua* (Fieber)	Mediterranean
Digitaria striate	*Sogatella kolophon (Kirkaldy)*	Australia
Finger millet mosaic	*Sogatella longifurcifera* (Esaki & Ishihara)	India
	Peregrinus maidis (Ashmead)	
Iranian maize mosaic	*Ribautodelphax notabilis* (Logvinenko)	Iran, USSR
	Peregrinus maidis (Ashmead)	
Maize mosaic	*Peregrinus maidis* (Ashmead)	Tropics and subtropics, worldwide
Maize sterile stunt	*Sogatella kolophon* (Kirkaldy)	Australia
	S. longifurifera (Esaki & Ishihara)	
	Peregrinus maidis (Ashmead)	
Northern cereal mosaic	*Laodelphax striatellus* (Fallen)	Japan
	Ribautodelphax albifascia (Matsumara)	
	Unkanodes sapporanus Matsumura	
	Meullerianella fairmairei (Perris)	

Wheat chlorotic streak*	*Laodelphax striatellus* (Fallen)	France
Wheat rosette stunt	*Laodelphax striatellus* (Fallen)	China
Reoviridae (Fijivirus)		
Fiji disease*	*Perkinsiella saccharicida* Kirkaldy	Madagascar, Australia, South Pacific Is.
	P. vastatrix Breddin	
	P. vitiensis Kirkaldy	
Maize rough dwarf* (includes cereal tillering disease and Mal de Rio Cuarto	*Laodelphax striatellus* (Fallen)	Central Europe, Scandinavia, Mediterranean, South America, Asia
	Toya propinqua (Fieber)	
	Javesella pellucida (F.)	
	Sogatella vibix (Haupt)	
	Ribautodelphax notabilis (Logvinenko)	
	Dicranotropis namata (Boheman)	
Oat sterile dwarf* (includes Arrhenatheram blue dwarf and cereal tillering disease)	*Javesella pellucida* (F.)	Central and northern Europe
	J. dubia (Kirschbaum)	
	J. obscurella (Boheman)	
	J. discolor (Boheman)	
	Dicranotropis hamata (Boheman)	
Pangola stunt	*Sogatella furcifera* (Horvath)	South America, Fiji, Taiwan
Rice black streaked dwarf	*Laodelphax striatellus* Fallen	Southeast Asia, USSR
	Unkanodes albifascia Matsumura	
	U. sapporona Matsumura	
Reoviridae (rice ragged stunt)		
Rice ragged stunt	*Nilaparvata lugens* (Stål)	Southeast Asia, Taiwan
Unclassified virus		
Coconut foliar decay disease virus	*Myndus taffini* Bonfils	Vanuatu

Note: An asterisk following the virus name indicates that transovarial transmission has been reported in some vectors.

[a]Sources include Brunt et al. (1990), Conti (1984), Gingery (1988), Jackson et al. (1987), Julia (1982) and Nault and Ammar (1989).

is presented. Whenever possible, I draw on examples of transmission from the planthopper literature. However, from time to time, the better-studied leafhoppers are referred to for key points.

Planthopper Vector Taxa

Of the 20 planthopper families, only the Delphacidae and Cixiidae have species demonstrated to be virus vectors. Relatively few of the 1100 described Delphacidae species are known vectors. Experimentally, 27 species from 14 delphacid genera have been documented to experimentally transmit plant viruses (Table 12.1). The majority of Delphacidae have monocot hosts, principally the Gramineae (Kisimoto 1973; O'Brien and Wilson 1985). Thus, nearly all known viruses transmitted by planthoppers have graminaceous hosts. Most cause diseases of maize, rice, wheat, and other cereal crops and forage grasses. This is reflected in the common names of these viruses (Table 12.1). These viruses and their delphacid vector species are distributed throughout all temperate and tropical regions and many oceanic islands. One cixiid species transmits a virus infecting coconut.

Plant Virus Taxonomy

There is no universally accepted set of characters that define a virus species. A virus can be defined as a collection of virus isolates whose properties are similar and can be given the same name. The properties for characterizing plant viruses have been summarized recently (Brunt et al. 1990). Among the most useful characters for describing a virus species are virion morphology, physical properties, nucleic acid chemistry, cytopathology, replication strategy, and serology. Another stable virus character is vector taxa. For example, among the homopteran-transmitted viruses, vectors of a given virus always are restricted to a single family. Using these and other characters, the plant virus subcommittee of the International Committee on Taxonomy of Viruses (ICTV) have assembled the plant viruses into 36 well-defined groups (Brown 1989; Matthews 1991). Two of the groups, the Rhabdoviridae and Reoviridae (some of which have planthopper vectors), have been given family status. The taxonomic status (hierarchy) of other plant virus groups is uncertain.

In this chapter, vernacular names are used. These have been modified, as suggested by Fenner (1976), to indicate when it is known that a virus is a definitive member of a group. For example, the planthopper-transmitted maize stripe virus is modified to the maize stripe tenuivirus and the maize mosaic virus to the maize mosaic rhabdovirus. For additional information on viruses discussed in this chapter, refer to *Viruses of Tropical Plants* (Brunt et al. 1990) or to the CMI/AAB "Descriptions of Plant Viruses," published by the Commonwealth Mycological Institute, Ferry Lane, Kew, Surrey, England.

Viruses Transmitted by the Delphacidae

Tenuivirus Group

Unlike the Reoviridae and Rhabdoviridae that have broad host ranges, the members of the tenuivirus group are known to infect only the Gramineae and their delphacid vectors. The tenuivirus genomes appear to be composed of five RNAs; however, whether the RNA genome is single-stranded, double-stranded, or both, presently is unknown (Falk et al. 1989). The fine, filamentous nucleoprotein strand associated with the infectious particles is the main distinguishing feature of the tenuiviruses (Gingery 1988). In extracts from infected plants, the particles are 3 nm in diameter and assume several configurations, including branched filamentous structures that may be supercoiled circular filaments. These structures are of undetermined length and are unstable and pleomorphic in vitro. Rice hoya blanca is the only tenuivirus whose virions have been observed by electron microscopy in infected delphacid vectors or plants (Shikata and Galvez 1969). Immunogold labeling has been used to detect the maize stripe tenuivirus (MStV) and its noncapsid protein in the cytoplasm of infected plant cells (Ammar et al. 1985). The technique has not been applied to locate this or other tenuiviruses in delphacids (Ammar, personal communication), but enzyme-linked immunosorbent assay (ELISA) has been used to detect MStV in individual delphacid organs and eggs (Nault and Gordon 1988). Tenuiviruses are also characterized by the presence of large amounts of a low molecular weight noncapsid protein in infected plant tissue (Gingery 1988). The noncapsid protein has not been detected in MStV inoculative *Peregrinus maidis* (Falk et al. 1987).

Rhabdoviridae

The Rhabdoviridae have hosts that include a diverse array of organisms (Calisher et al. 1989). Included are mammals, birds, reptiles, fish, insects, plants and protozoans. Two vertebrate-infecting rhabdovirus genera are recognized, *Vesiculovirus*, which includes the viruses causing vesicular stomatitis and antigenically related viruses, and *Lyssavirus*, which includes the rabies virus and rabies-related viruses. Many of the viruses of these two genera also have hematophagous arthropod vectors. Interestingly, one of these viruses, vesicular stomatitis, has been shown experimentally to multiply in the delphacid, *P. maidis* (Lastra and Esparza 1976).

Plant rhabdoviruses have complex bullet-shaped or bacilliform virions 45–100 nm in diameter and 100–430 nm in length (Francki et al. 1985; Jackson et al. 1987). The virions are surrounded by a host-derived membrane and contain four to six structural proteins and a single molecule of negative sense single-stranded (ss) RNA. Eighteen rhabdoviruses have Auchenorrhyncha vectors with 11 transmitted by delphacids (Table 12.1) and 7 by cicadellids. Other plant rhabdoviruses have aphid, piesmid, or mite vectors (Jackson et al. 1987). The serological relationships used to establish genera for the vertebrate-infecting rhabdoviruses (Calisher et al. 1989) have not been useful for plant rhabdovirus taxonomy. Virologists have proposed that plant rhabdoviruses be divided into two subgroups based on properties of their virion proteins, the kinetics of their transcriptase activities, and the site of virion maturation in plant cells (Francki et al. 1985). This division of the rhabdoviruses does not correspond with vector groups or families.

Reoviridae

The viruses classified as members of the Reoviridae include members that infect a broad range of hosts, including mammals, birds, plants, and arthropods (Francki et al. 1985). Included among the arthropod-infecting reoviruses are the cytoplasmic polyhedrosis viruses. Reoviruses are among the largest viruses with icosahedral particles and have a diameter of 65–70 nm. Reoviridae genera or groups are separated by such traits as the number of double-stranded (ds) RNA segments comprising the genome, the morphology of the protein shells, and serological relationships. The plant reoviruses include five that are delphacid-transmitted and placed in the Fijivirus group (Table 12.1) and three that are cicadellid- (leafhopper) transmitted and placed in the phytoreovirus group. The two groups further differ in that fijiviruses are characterized by virions containing 10 ds-RNA genome segments and the presence of spikes on the virion outer and inner protein shells, whereas phytoreovirus virions have 12 ds-RNA genome segments and no spikes on the protein shells. Rice ragged stunt virus is an unclassified reovirus with 10 ds-RNA genome segments and a delphacid vector.

Coconut Foliar Decay Disease Virus

The coconut foliar decay disease virus (FDDV) has icosahedral virions 20 nm in diameter and a single-stranded circular DNA (Randles and Hanold

1989). This virus apparently represents a novel group of plant viruses. The vector is the cixiid planthopper *Myndus taffini* (Julia 1982). Nothing is known about the transmission relationship between this virus and its vector.

Vector–Virus Relationships

Homopteran vectors exhibit one of three basic types of relationships with plant viruses. In one type, virus acquired from plants attaches to the stylets, probably the lumen of the food canal formed by the maxillary stylets and to the lining of the foregut of leafhopper, aphid, and perhaps other vectors such as whiteflies (Harrison and Murant 1984; Berger and Pirone 1986; Nault and Ammar 1989). Many of these viruses are known or suspected to require the presence of a noncapsid, viral-coded protein which apparently assists in attachment of virus to the vector (Harrison and Murant 1984; Thornbury et al. 1985; Hunt et al. 1988). It is not known how these viruses detach from vectors or how they are returned back to plants. It is believed that vectors extravasate (McLean and Kinsey 1984), that is, flush virus from the foregut back through the food canal and into the plant as they sample plant cells for ingestion sites. Vector saliva may play a role in transmission but no evidence has been forwarded to support this hypothesis. Nault and Ammar (1989) introduced the term *foregut-borne* to describe this mode of transmission for the leafhopper-borne maize chlorotic dwarf machlovirus and related rice tungro machlovirus. Here I expand on use of the term to include all viruses transmitted by homopterans that are lost after a vector molts, for example, when the stylets and lining of the foregut are lost. Matthews (1991) has adopted this terminology for the most recent edition of his text, *Plant Virology*. Some of the foregut-borne viruses, particularly those transmitted by the Sternorrhyncha Homoptera (e.g., aphids), are acquired primarily from and transmitted to the plant epidermis (see later discussion). Transmission is nonpersistent, that is, the half-life or retention time of virus in the vector is measured in minutes (Sylvester 1958). Other foregut-borne viruses are acquired from and transmitted to the phloem. Transmission of these viruses is semipersistent, that is, the half-life is measured in hours.

The remaining viruses transmitted by the Homoptera have very different relationships with their vectors. Virus is acquired from plant phloem and/or mesophyll, passes through the gut wall and into the vector's haemolymph, and then passes to the salivary glands to be discharged with the salivary secretions. For the geminiviruses, luteoviruses and pea enation mosaic virus, evidence indicates that they do not multiply in their vectors. The vector relationship of these viruses is classified as *circulative* (Sylvester 1980). For the tenuiviruses, rhabdoviruses and reoviruses, including those vectored by the delphacids, proof of multiplication has been documented for at least some members of each group (Table 12.2) (Nault and Ammar 1989). The vector relationship of these viruses is classified as *propagative* (Sylvester 1980). Transmission of circulative and propagative viruses is persistent, that is, the half-life is measured in days, weeks, or months. All plant viruses in a family or group that have homopteran vectors are transmitted by the same transmission mechanism or mode. Of the three basic types of transmission, delphacids are known to transmit only propagative viruses.

Evidence for Virus Multiplication in Vectors

Evidence of multiplication of plant viruses in delphacid vectors has come from three different experimental approaches. In the first, virus is transferred experimentally between insects, either by needle injection or naturally, by transovarial transmission (see later discussion), until the dilution attained in the final inoculative insects exceeds the dilution end point of virus in the initial inoculum (Whitcomb 1972). For instance, data have been presented for rice stripe tenuivirus (RSV) in its delphacid vector, *Laodelphax striatellus*, based on serial passage by injection of vectors resulting in a final dilution of 1.25×10^{-6} that exceeds the 10^{-2} to 10^{-3} dilution end point (Okuyama et al. 1968).

A second experimental approach giving evidence of multiplication is electron microscopic observation of aggregates of virus particles and sites of viral assembly and/or accumulation in vector cells. Shikata (1979) summarizes studies with several delphacid-transmitted fijiviruses. A detailed study of the maize mosaic rhabdovirus (MMV) in *P. maidis* gives unequivocal evidence of multiplication in its vector (Ammar and Nault 1985). MMV particles assemble (bud) through three types of membranes in vector cells, including the inner and sometimes outer nuclear membranes of cells in most tissues examined, and in the endoplasmic reticulum and plasma membranes of salivary glands.

A third approach that recently provided a rapid, easy-to-use, sensitive, and quantitative method for demonstrating plant virus multiplication in vectors is by following increase in virus titre using ELISA. This approach has been used to demonstrate that MMV (Falk and Tsai 1985) and MStV (Falk et al. 1987; Nault and Gordon, 1988) multiply in *P. maidis*.

These approaches have not been applied to all viruses transmitted by delphacids but other experimental criteria, especially a long latent period

Table 12.2. Mean latent periods for propagative and circulative viruses in their Homopteran vectors.[a]

Plant Virus	Virus group	Vector Species and Family[c]	Mean Latent Period (hour)	Reference
Propagative viruses[b]				
Digitaria striate	Rhabdoviridae	*Sogatella kolophon* (D)	144	Greber (1979)
Maize mosaic*	Rhabdoviridae	*Peregrinus maidis* (D)	440	Carter (1941)
Barley yellow striate m.	Rhabdoviridae	*Laodelphax striatellus* (D)	378	Conti (1980)
Maize stripe*	Tenuivirus	*Peregrinus maidis* (D)	374	Gingery et al. (1979)
Europ. wheat striate m.	Tenuivirus	*Javesella pellucida* (D)	396	Slykhuis and Watson (1958)
Rice grassy stunt	Tenuivirus	*Nilaparvata lugens* (D)	204	Rivera et al. (1966)
Rice hoya blanca*	Tenuivirus	*Tagosodes orizicolus* (D)	744	Galvez (1968)
Wound tumor*	Reoviridae	*Agallia constricta* (C)	473	Maramorosch (1950)
Rice gall dwarf	Reoviridae	*Nephotettix cincticeps* (C)	355	Inoue and Omura (1982)
Rice transitory yellowing	Reoviridae	*Nephotettix cincticeps* (C)	338	Chen and Chiu (1980)
Maize rayado fino*	Marafivirus	*Dalbulus maidis* (C)	384	Nault et al. (1980)
Oat blue dwarf*	Marafivirus	*Macrosteles fascifrons* (C)	408	Banttari and Zeyen (1970)
Stawberry crinkle*	Rhabdoviridae	*Chaetosiphon jacobi* (Ap)	149	Sylvester et al. (1974)
Circulative viruses				
Beet curly top	Geminivirus	*Circulifer tennellus* (C)	17	Magyarosy and Sylvester (1979)
Maize streak	Geminivirus	*Cicadulina mbila* (C)	23	Storey (1928)
Beet pseudo curly top	Geminivirus	*Micrutalis mellifera* (M)	24	Simons (1962)
Cowpea yellow mosaic	Geminivirus	*Bemesia tabaci* (Al)	12	Anno-Nyako et al. (1983)
Tobacco leaf curl	Geminivirus	*Bemesia tabaci* (Al)	4	Varma (1963)
Filaree red-leaf	Luteovirus	*Acyrthosiphon pisum* (Ap)	48	Sylvester and Osler (1977)
Zinnia yellow net	Luteovirus	*Bemesia tabaci* (Al)	19	Srivastava et al. (1977)
Potato leaf roll	Luteovirus	*Myzus persicae* (Ap)	17	Tanaka and Shiota (1970)
Barley yellow dwarf	Luteovirus	*Sitobion avenae* (Ap)	45	van der Broek and Gill (1980)
Pea enation mosaic	PEMV	*Acyrthosiphon pisum* (Ap)	25	Sylvester (1965)

[a]For some reports, a median rather than mean latent period is reported.

[b]For viruses shown with an asterisk (*), proof of multiplication has been demonstrated by filution end point, electron microscopy, or ELISA.

[c]Letters in parentheses are D = Delphacidae, C = Cicadellidae, Ap = Aphididae, Al = Aleyrodidae, M = Membracidae.

and the occurrence of transovarial transmission (Sinha 1981), are considered as strong indicators that these viruses are propagative. The latent period is the time it takes for the virus to complete the cycle in the vector, that is, from its uptake (acquisition) from the plant to its discharge from vector saliva. Experimentally, this is the period from first exposure of a vector to an infected plant until the vector can inoculate a healthy plant. The mean latent periods of known or suspected propagative viruses (including those transmitted by delphacids) are compared to those of several nonpropagative, circulative viruses transmitted by various Homoptera in Table 12.2. The mean ± SE latent period for 13 propagative viruses from 4 virus groups is 368 ± 41 hours. The comparable figure for 10 circulative viruses from 3 virus groups is 23 ± 4.1 hours. The reason(s) for the consistent differences in lengths of latent periods between circulative and propagative viruses is unknown. However, it is probable that the dynamics of the receptor-mediated endocytosis for transport of the luteoviruses and perhaps other circulative viruses through the gut wall and salivary gland membranes (Gildow 1987) differ substantially from the underlying processes responsible for virus recognition, multiplication, and transport of the propagative viruses.

Transmission Biology

Transovarial transmission has been reported with vectors of nine delphacid-transmitted viruses (Table 12.1) and from leafhoppers of several propagative viruses but never from circulative or foregut-borne viruses transmitted by the Auchenorrhyncha (Nault and Ammar 1989). Transovarial transmission occurs when infected females pass virus to their progeny through the egg (Sinha 1981). Sexual transmission from infected males to noninfected females has not been reported. Failure of MStV to multiply or accumulate in the testes may be a reason why MStV is not sexually transmitted in its delphacid vector (Nault and Gordon 1988).

High rates (33–100%) of transovarial transmission have been reported for tenuiviruses in their delphacid vectors (Gingery 1988) compared to low rates for Fijiviruses (0.2–17.0%) and delphacid-transmitted rhabdoviruses (Sinha 1981; Conti, 1984; Nault and Ammar 1989). In experiments that lasted over 6 years, rice stripe tenuivirus was shown to pass to 40 successive generations in vectors kept on virus-immune hosts (Shinkai 1962). Ninety-five percent of progeny insects of the fortieth generation were infective. Insects that received virus transovarially remained infective throughout their nymphal stages and for 2 to 3 weeks of early adult life, but some adults lost inoculative capability later. When virus is acquired transovarially by nymphs from infected females, the latent period is brief in newly hatched nymphs (Sinha 1981) compared to longer latent periods when virus is acquired by immatures from plants.

The mechanism involved in transovarial transmission of delphacid-transmitted viruses has not been studied, but the process has been investigated in the cicadellid-transmitted rice dwarf phytoreovirus (RDV) (Nasu 1965). RDV virions in the infected leafhoppers' ovarioles were found attached to the surface of bacterial L-symbionts. Virus apparently is carried with these symbionts when the ovariole mycetocytes penetrate neighboring oocytes to form the egg mycetome. The mycetocytes are specialized fat bodies where symbionts reside; the mycetome is a collection or body of mycetocytes (Brooks 1985). The manner whereby virus enters the egg in delphacid-transmitted viruses may be similar to that in leafhoppers. In delphacid adults there is no mycetome and the symbionts are yeast-like eukaryotes, rather than bacteria, that are scattered throughout the abdominal fat body (Brooks 1985). However, in the egg stage, these yeast-like symbionts of delphacids are contained in a well-defined mycetome (Ammar 1985).

When delphacids acquire virus from plants, minimum acquisition thresholds are usually several minutes which is the time for planthoppers to reach the phloem with their stylets where virus uptake occurs (Nault and Ammar 1989). Inoculation threshold periods are similar. Usually virus is more efficiently transmitted by delphacids when virus is acquired from plants when insects are nymphs rather than later as adults (Table 12.3). As an extreme example of the phenomenon, sugarcane Fiji disease fijivirus is transmitted by *Perkinsiella saccharicida* only if the virus is acquired by first-instar nymphs, but it is not transmitted when acquired by older instars or adults (Hutchinson and Francki 1973). Sinha (1967) suggests that susceptibility and/or permeability of the gut to the leafhopper-transmitted

Table 12.3. Transmission efficiency by adult delphacids when virus is acquired from plants either while as nymphs or as adults.

Virus	Vector Species	Percent Transmission		Reference
		Nymph	Adult	
Maize rough dwarf	*Laodelphax striatellus*	43.0	20.0	Conti (1966)
Wheat rosette stunt	*Laodelphax striatellus*	21.5	11.5	Zhang et al. (1982)
Digitaria striate	*Sogatella kolophon*	20.5	10.0	Greber (1979)
Maize stripe	*Peregrinus maidis*	44.7	16.1	Tsai and Zitter (1982)
European wheat striate	*Javesella pellucida*	47.9	13.9	Sinha (1960)

wound tumor phytoreovirus decreases as the vector ages. The same may be true for delphacid vectors. That the gut is involved in transmission success is suggested from studies with the delphacid-transmitted Iranian maize mosaic virus. The field vector is *Ribautodelphax notabilis* (Izadpanah et al. 1983) but an experimental vector, *P. maidis*, transmits inefficiently when virus is acquired from plants, but transmits efficiently when virus when is injected into the delphacid's haemocoel (Nault and Ammar 1989). This evidence suggests that vector specificity is associated with gut barriers. Another study with maize rough dwarf fijivirus (MRDV) also implicates the gut as a barrier to transmission. The delphacids *Toya propinqua* and *Sogatella vibix* cannot transmit MRDV when fed on infected plants but if their gut is needle-punctured following virus acquisition, virus is transmitted (Harpaz and Klein 1969). Similarly, if infective sap from maize is injected into the haemocoel, these species are "transformed" into vectors.

Vector Specificity

Factors other than the gut also are involved in vector specificity for delphacid-transmitted viruses. Failure of virus to replicate in an insect is an obvious transmission barrier. Studies of this phenomenon have been conducted with cicadellid but not delphacid-transmitted viruses. A classic example for rhabdoviruses is provided by two strains of potato yellow dwarf rhabdovirus (PYDV). One strain is transmitted by *Aceratagallia sanguinolenta* and by other *Aceratagallia* species, but not by *Agallia constricta*, whereas the second strain is transmitted by *A. constricta* but not by *A. sanguinolenta* (Black 1970). Adam and Hsu (1984) reported that differences in the G (glycosylated) protein might be related to selective transmission of these strains. The G protein, which protrudes from the virion envelope, functions in attachment of animal rhabdoviruses to host recognition sites on the plasma membrane during the early stages of infection (Jackson et al. 1987). Recent findings have indicated that the same is probably true for PYDV in insect cell cultures (Gaedigk-Nitschko et al. 1988). It is surprising then that in vector cell monolayers (VCM), Hsu et al. (1977) found no differences in susceptibility of cell lines from vector and nonvector agalline species to PYDV strains. However, in another PYDV nonvector, *Dalbulus elimatus*, the *A. constricta*-transmitted strain did not multiply, whereas the *A. sanguinolenta* strain did, but much less efficiently than in the cell line of vector species.

The ability to infect VCM also has been used to explain specificity in transmission of phytoreoviruses. Wound tumor phytoreovirus (WTV) readily infected cell lines from two vector species, *A. constricta* and *A.*

novella, but only with difficulty infected the cell lines of one nonvector, *A. sanguinolenta*, and failed to multiply in the cell line of a second nonvector, *D. elimatus* (Hsu et al. 1977; Black 1979).

Leafhopper-transmitted WTV isolates can be converted to nontransmissible ones by long-term maintenance in host plants by vegetative propagation (Whitcomb and Black 1969). Electrophoretic analyses of the genomes from several WTV isolates showed that transition to the nontransmissible state is always accompanied by deletion mutations in one or more of four ds-RNA segments (Reddy and Black 1974). Nuss (1984) suggested that the gene products of segments S2 and S5 are needed for leafhopper transmission of WTV. The products of these two segments comprise the outer capsid (shell) of the virus, which indicates that capsid proteins may be involved in the recognition of host cells, in virus penetration into cells, or both (Adam 1984).

Evidence from an electron microscopic examination of *P. maidis* or maize infected with MMV suggests that the sites in cells where virus budding occurs are important for transmission. In plant cells and in most vector tissues examined, MMV particles bud on nuclear and cytoplasmic membranes and accumulate in perinuclear spaces and in dilated cisternae connected with endoplasmic reticulum (Ammar and Nault 1985; McDaniel et al. 1985). In secretory cells of the principal salivary glands, however, MMV particles bud mainly on the plasma membranes and accumulate in intercellular and extracellular spaces; these spaces apparently are connected with the extracellular vacuoles and canaliculi that lead to the salivary ductules and ducts (Ammar 1986; Ammar and Nault 1985). A similar difference occurs in the budding sites of the rabies virus in fox brain and salivary gland cells; plasma membrane budding in salivary gland cells is considered essential for bite transmission of this virus (Murphy and Harrison 1980). Ammar (1987) suggested an intercellular and extracellular route for MMV transport in the principal salivary glands of *P. maidis* that allows efficient discharge of virus with the saliva during feeding.

Are Viruses Pathogenic to Delphacid Vectors?

The best studied examples of effects of propagative plant pathogens on their vectors involve leafhopper-transmitted mollicutes (Purcell and Nault 1991). In the genus *Dalbulus*, it has been concluded that in apparently well-adapted, co-evolved, vector species, mollicutes not only are nonpathogenic, but they may benefit their vectors (Madden and Nault 1983; Madden et al. 1984; Nault 1985, 1990). In contrast, poorly adapted *Dalbulus* species frequently are inefficient vectors and their life span and fecundity are greatly reduced compared to noninfected leafhoppers.

Reports of effects of plant viruses on their delphacid vectors are contradictory. Nasu (1963) reported that RSV reduces the life span and fecundity of *Laodelphax striatellus*, but later Kisimoto (1973) concluded that the virus has no effect on the vector. Similarly, Watson and Sinha (1959) first reported that European wheat striate mosaic tenuivirus (EWSMV) was pathogenic to *Javesella pellucida*, but later such effects were attributed to inbreeding of the vector in the laboratory (Kisimoto and Watson 1965). Ammar (1975a) found that when *J. pellucida* acquired EWSMV from their mothers, they did not live as long as uninfected controls, but when this vector acquired virus from plants, there was no effect on nymphal survival, adult longevity, fecundity of females, or survival of eggs (Ammar 1975b). In other studies, (Ammar and Nault, unpublished), no effects were found on adult longevity of *P. maidis* infected with MMV or MStV. Numbers of MMV virions found in *P. maidis* (Ammar and Nault 1985) were smaller than the massive, often crystalline aggregates found in plant cells (McDaniel et al. 1985). It was concluded that compared to maize, *P. maidis* better regulates (lowers) the rate of virus replication in infected cells, which could explain pathogenicity in maize plants and not in vectors.

Sinha (1981) and Nault and Ammar (1989) have concluded the delphacid-transmitted plant viruses have not been shown unequivocally to harm their vectors and that more work is needed. There is reason to believe, as has been shown in leafhopper–mollicute associations, that in well-adapted delphacid virus vectors, genotypes naturally have been selected that minimize or eliminate pathogenicity of plant virus to vectors.

Origins of Delphacid-Transmitted Plant Viruses

Matthews (1991) remarks that the rhabdoviruses and reoviruses are of special evolutionary interest because members of each group infect vertebrates, invertebrates, or higher plants. Regardless of the hosts infected, members of a virus group share many characteristics and are likely to have a common ancestor. A feature of both rhabdoviruses and reoviruses is their replication in insects. Matthews (1991) considers that plant rhabdoviruses and reoviruses most likely have originated in insects and become secondarily adapted to plants. I suggest that tenuiviruses may have originated similarly. Many viruses from these groups can survive in their vectors by transovarial passage. However, Purcell (1982) argues that unless propagative phytopathogens are transmitted transovarially to 100% of vector offspring, they cannot be maintained indefinitely in their vectors without horizontal (insect to plant to insect) transmission. No maternally transmitted reovirus, rhabdovirus, or tenuivirus is transmitted from an infected female to all her progeny. Similarly, none of these viruses is seed-borne. Thus, these viruses use phylogenetically unrelated hosts, plants and insects, as necessary parts of the virus life cycle. For viruses in these groups that are not transovarially transmitted by their vectors (i.e., dead-end insect hosts), it is assumed that selection for this characteristic has been replaced entirely by dependence on alternate plant hosts.

Several nonplant infecting insect viruses of delphacids, leafhoppers, or aphids provide further weight to arguments that propagative plant viruses originated in their vectors. A reovirus, named the peregrinus-maidis virus (PgMV) after its delphacid vector, is one of the few such viruses to have been studied in detail (Falk et al. 1988). The virus was inadvertently discovered when Falk and his associates were looking for MMV and MStpV in *P. maidis* vectors. The PgMV is unlike any other reovirus. It has 12 ds-RNAs that might suggest it is related to the phytoreoviruses, but unlike a phytoreovirus, PgMV does not infect plants or leafhoppers. It has two more ds-RNA segments than a fijivirus, but the virion morphology and cytopathology in insects resemble those of a delphacid-transmitted fijivirus.

Another nonplant infecting reovirus isolated from an auchenorrhynchan is the leafhopper A virus (LAV) (Hatta and Francki 1982). This virus is structurally similar to the fijiviruses but the number of genome segments is unknown. The virus is vertically transmitted from infected females to ~ 20% of their progeny. Additionally, LAV can pass horizontally from infected to noninfected leafhoppers that feed on the same maize plant (Ofori and Francki 1985). A similar example of an insect virus for which a plant serves as a transitory reservoir is the aphid-infecting *Rhopalosiphum padi* virus (RPV) (D'Arcy et al. 1981). RPV is vertically transmitted to 28% of offspring by parthenogenetic females but maintained in nearly 90% of the population by supplemental horizontal transmission in plants (Gildow and D'Arcy 1988).

Horizontal transmission, in which plants serve as "vectors" or reservoirs but not as hosts of homopteran viruses, may be commonplace. Circumstantial evidence for this is suggested by the appearance of isometric and rod-shaped virus-like particles (VLP) in plants also infected with tenuiviruses (Gingery 1988). These isometric and rod-shaped viruses first were thought to be the causal agents of plant disease until the filamentous tenuivirus particles associated with pathogenicity were discovered. Instead, these VLP may be insect viruses inserted into plant vascular tissue by feeding delphacids. For instance, the reovirus-like particles found in plants fed on by *P. maidis* (Trujillo et al. 1974; Lastra and Carballo 1983) are morphologically similar to PgMV and may be the same virus. It is from these reservoirs of delphacid viruses in plant vascular tissues that the first plant-infecting, delphacid-transmitted viruses may have evolved.

The antiquity of delphacid-transmitted viruses may never be known because, unlike insect vectors, plant viruses have no fossil record. Nault and Ammar (1989) speculate that plant-infecting reoviruses date back to the divergence of major taxonomic groups of the Homoptera ~ 180 million years ago (Hennig 1981). In the ancestral state, the Reoviridae probably had 10 ds-RNA genome segments as is common for most reoviruses (Francki et al. 1985); the state with 12 segments appears to be apomorphic, having evolved in phytoreovirus and PgMV. Similarly, the absence of spikes on capsids, the most common condition, is probably ancestral, and the spikes in Fijiviruses and the insect cytoplasmic polyhedrosis virus are probably derived characters. The different traits in phytoreoviruses and fijiviruses may have evolved after the divergence of the Cicadellidae and Delphacidae and, therefore, remained isolated and distinct. Assuming that these two reovirus genera are of insect origin, could PgMV or LAV be ancestral viruses? It is not clear that either virus is ancestral to phytoreoviruses or fijiviruses because, as was noted earlier, the two insect viruses have characteristics of both plant virus groups.

Whereas a strong case for co-evolution can be argued for the origin and divergence of reoviruses in their cicadellid and delphacid vectors, the same cannot be said for the Rhabdoviridae. As was noted earlier, no consistent association can be made for rhabdovirus traits and vector families. As new rhabdovirus traits evolved, so too must have the adoption of new vector groups. Rather than co-evolution, the colonization hypothesis of Mitter and Brooks (1983), which explains how related parasites occupy different hosts, provides a better explanation for the distribution of plant rhabdovirus vectors among Homopteran families (Nault 1987). According to this hypothesis, the plant host serves as a conduit whereby a rhabdovirus can transfer from a vector of one family to that of another via a common host. Such events must be rare because, at present, no plant virus is known to be transmitted by vector species from more than one Homopteran family. Consistent with this hypothesis, it also must be assumed that once a virus is established in its new vector family, mutations and resultant genomic changes isolate it rapidly from the former vector group.

Do Planthoppers Transmit Only Propagative Viruses?

The delphacids, unlike the leafhoppers, aphids, and whiteflies, are known only as vectors of propagative, persistently-transmitted viruses (Table 12.4). Is this the extent of the range of viruses transmitted by planthoppers? Are there planthoppers that transmit circulative or foregut-borne viruses?

Table 12.4. Numbers of plant viruses transmitted by the Delphacidae and five other Homopteran families among virus groups and families recognized by the plant virus subcommittee of the International Committee on Taxonomy of Viruses (ICTV). Virus groups are arranged by transmission mode and persistence of virus in vectors

Virus group	Transmission		Auchenorrhyncha			Sternorrhyncha		
	Mode	Persistence[a]	Delphacidae	Cicadellidae	Membracidae	Aphididae	Aleyrodidae	Pseudococcidae
Tenuivirus	Propagative	P	5					
Rhabdoviridae	Propagative	P	11	7		9		
Reoviridae	Propagative	P	6	3				
Marafivirus	Propagative	P		3				
Geminivirus	Circulative	P		8	1	(1)	13	
Luteovirus	Circulative	P				6(5)	(3)	
Pea enation mosaic	Circulative	P				1		
Machlovirus	Foregut-borne	S		1(1)				
Badnavirus	Foregut-borne	S		1		(1)		5(1)
Closterovirus	Foregut-borne	S				8(4)	(5)	(1)
Parsnip yellow fleck	Foregut-borne	S				2(1)		
Caulimovirus	Foregut-borne	N/S				6(3)		
Carlavirus	Foregut-borne	N				29(6)	(3)	
Potyvirus	Foregut-borne	N				68(87)	(1)	
Potexvirus	Foregut-borne	N				(2)		
Alfalfa mosaic	Foregut-borne	N				1		
Cucumovirus	Foregut-borne	N				3(2)		
Fabavirus	Foregut-borne	N				4		
TOTAL			22	23(1)	1	145(112)	13(12)	5(2)

Note: Data taken from Brown (1989), Brunt et al. (1990), Casper (1988), Duffus (1987), Harrison (1985), Jackson et al. (1987), Milne (1988), Murant et al. (1988), Nault and Ammar (1989), Matthews (1991), and R. Hull (personal communication). Numbers in parentheses are viruses considered provisional members of groups indicated.

[a] P = persistent; S = semipersistent; N = nonpersistent.

My answer to these questions is that new planthopper-borne viruses probably will be discovered and some very likely will be circulative and persistently transmitted. The best candidates are the geminiviruses that have members transmitted by aphids, whiteflies, treehoppers, and leafhoppers (Table 12.4). If systematic relationships are relevant, planthopper-transmitted geminiviruses will be from subgroup A (Brown 1989), which include viruses transmitted by leafhoppers and treehoppers that contain a monopartite genome. Perhaps planthoppers also will be found that transmit luteoviruses (Table 12.4) or uncharacterized circulative viruses not now recognized by the ICTV.

Although it is likely that planthoppers transmit circulative viruses, the chances that planthoppers transmit foregut-borne viruses is more remote. Of 48 viruses transmitted by the Auchenorrhyncha, only 2 that are transmitted by leafhoppers have this relationship and both are semipersistently transmitted (Table 12.4). The great majority of foregut-borne viruses are nonpersistently transmitted by aphids and only a few by whiteflies (Table 12.4). In the case of aphids, nonpersistent transmission, including virus acquisition and inoculation, occurs within seconds when vectors probe the plant epidermis or underlying parenchyma (Sylvester and Richardson 1963; Zettler 1967; Nault and Bradley 1969; Markham et al. 1987). The ability of aphids and whiteflies but not delphacids or cicadellids to transmit nonpersistently transmitted viruses (Table 12.4) may be attributed to the profound differences in feeding behavior between these homopteran taxa. When aphids and whiteflies explore plants with their stylets to determine host status, they do so by making brief, shallow, intercellular probes in the epidermis and sometimes the underlying parenchyma (Pollard 1973; Walker 1988). This probing is accompanied by a rupturing of the plasma membrane and penetration of the protoplast by the stylet tips where sap is sampled (Hashira 1969; Spiller et al. 1985; Kimmins 1986; Lopez-Abella et al. 1988). Little damage is done to cells which is especially critical if virus is to be inoculated and become established. It is during exploratory probing that the nonpersistently transmitted, foregut-borne viruses are transmitted. After test probing and acceptance of a host, aphids make deep intercellular probes. The anticlinal walls are followed by the stylets until the principal feeding site, the phloem, is reached (Pollard 1973). It is during these deep feeding probes that the foregut-borne, circulative, and propagative viruses are acquired from and inoculated to plant cells by aphids.

In contrast to the aphids and whiteflies, delphacids and other auchenorrhynchans rapidly pierce plant tissues with their stylets. A high proportion of first probes are prolonged and many result in feeding from vascular tissue (Naito 1977; Sogawa 1982; Backus 1985). The path of the stylets of leafhoppers and delphacids is intracellular through the epidermis and underlying tissues (Chang 1978; Backus 1988). Unlike aphids that use mandibular stylets as a guide for the intercellular penetration of the maxillary stylets, leafhoppers and planthoppers insert the mandibles a short distance, anchor them in the parenchyma, and then use the maxillaries to penetrate a straight or curved intracellular path to target tissues (Day et al 1951; Naito 1977; Backus 1988). This mode of feeding by hoppers damages the epidermis and parenchyma, causing rupture (Smith 1925; Chang 1978; Sogawa 1982) or complete destruction (Withcombe 1926) of penetrated cells. This feeding behavior is inimical to virus transmission as inoculated cells must survive to sustain virus replication and subsequent translocation. Although it cannot be said unequivocally that the reason why planthoppers may not transmit foregut-borne viruses is because of their feeding behavior, this appears to be the best explanation.

Concluding Remarks

Progress toward understanding the delphacid-transmitted viruses has lagged behind that of the cicadellid-transmitted viruses, although it is just a few well-studied leafhopper-transmitted viruses that make this difference. This is especially true for WTV that historically has been a model system for study of ds-RNA viruses. Surprisingly, WTV has never been found in plants in nature and was isolated only once from field-collected leafhoppers. Another well-studied leafhopper-borne virus that causes important disease is the rice dwarf phytoreovirus. Among the leafhopper-transmitted rhabdoviruses, PYDV has been investigated in depth. Knowledge of these leafhopper-transmitted viruses has benefited extensively from use of vector cell monolayers. Significant progress in understanding the delphacid-transmitted viruses will not be forthcoming until vector cell cultures of delphacids can be established and used to study the infection process.

Clues concerning the putative insect origin of delphacid-transmitted viruses most likely will be revealed when more attention is paid to nonplant pathogenic viruses of the Homoptera. Unfortunately, these viruses are ignored by most plant virologists because they do not cause plant disease and they are neglected by insect pathologists because of their apparent lack of potential for biological control.

Almost certainly, many planthopper-transmitted plant viruses remain to be discovered. Because of the economic importance of the Gramineae, it is no surprise that it is the delphacid-transmitted viruses that have been the first to be isolated and studied. Thus, the recent discovery of the novel FDDV is noteworthy because it is the first planthopper-transmitted virus isolated from a tree species (coconut) and transmitted by a cixiid. Most planthoppers, especially those from nondelphacid families, have tropical distributions and feed on noncultivated dicotyledonous plants. Perhaps with the increasing attention being given to disappearing tropical habitats, efforts will be made to catalog the tropical insect fauna (including planthoppers), as well as associated microorganisms and viruses.

Summary

1. The delphacids are vectors of 22 viruses from three taxonomically distinct groups of RNA plant viruses; a cixiid is the vector of a unique DNA virus.
2. The delphacid-transmitted reoviruses, rhabdoviruses, and tenuiviruses have, in common, persistent transmission and a propagative relationship with their vectors. Also, many are transmitted transovarially.
3. The passage of plant-acquired virus through the gut wall of planthoppers and the multiplication of virus in vector cells explains some cases of vector specificity for the delphacid-transmitted viruses.
4. There is no good evidence that planthopper-transmitted plant viruses are pathogenic to delphacid vectors.
5. Planthopper-transmitted plant viruses most likely evolved from viruses that originated in their insect vectors.
6. The feeding behavior of planthoppers, which is destructive to plant epidermal and parenchyma cells, likely precludes these insects as vectors of nonpersistently transmitted, foregut-borne viruses. Feeding behavior should not preclude delphacids as vectors of persistently transmitted circulative viruses, although to date no such plant viruses have been discovered.
7. Progress toward a better understanding of planthopper-transmitted viruses will benefit the study of virus infections in cultured planthopper cells, additional studies of planthopper (insect) viruses, and a search for plant viruses and planthopper vectors from nondelphacid, fulgoroid families in the tropics.

ACKNOWLEDGMENTS

I especially appreciate the assistance of two of my graduate students, Kirk Larsen and Astri Wayadande, who helped with the search for literature and with whom I have had productive discussions on the biology and transmission of plant viruses by the Homoptera. I also thank E. D. Ammar, B. W. Falk, R. E. F. Matthews, and M. Thresh for helpful reviews of early drafts of the manuscript, and M. R. Wilson for assistance with nomenclature. Salaries and research support for the author were provided by state and federal funds appropriated to The Ohio State University/Ohio Agricultural Research and Development Center. This is journal article no. 75-92.

13

Interspecific Interactions and Community Structure in Planthoppers and Leafhoppers

Peter Stiling

Introduction

In habitats as diverse as salt marshes, temperate woodlands, grassland meadows, and rice fields, some of the most abundant insect herbivores are leafhoppers and planthoppers. Often, many species are found utilizing the same host plants. Thus, Denno (1976, 1977, 1978) found 10 sap-feeding Auchenorrhyncha on *Spartina patens* (salt meadow hay), 9 on *Distichlis spicata* (salt grass), 6 on *Spartina alterniflora* (salt marsh cord grass), and 2 on *Juncus roemerianus* (needle rush) in Atlantic coastal marshes. Claridge and Wilson (1976, 1981) provide a comprehensive list of the leafhoppers of British trees, which vary in their leafhopper load from 1 species up to 10 (*Quercus petraea*) and even 11 (*Alnus glutinosa*). On grasslands dominated by *Holcus mollis*, Waloff (1979) has commonly recorded five species of Auchenorrhyncha, and at least 34 species of leafhoppers and planthoppers have been recorded from rice fields in southeast Asia (Wilson and Claridge 1985). What factors influence how many species utilize a particular host plant? What structures the guild? Does competition between members of a leafhopper or planthopper community occur frequently? How do species avoid competing, or do they? This chapter will address some community-wide phenomena of planthopper and leafhopper assemblages, including the effects of competition, the frequency of parasitoid attack, and the influence of these sources of mortality on communities of Auchenorrhyncha.

How Many Species Per Plant?

In recent years, there has been much discussion of why different types of plants accrue different loads of herbivores. Explanations have ranged from the taxonomic age of the plant (Southwood 1960, 1961) to the area of distribution of the plant (Strong 1979). Of British typhlocybine leafhoppers, 62 species are known to feed on the leaf mesophyll of trees and shrubs. Claridge and Wilson (1981) listed the host records for 55 of them feeding on a total of 36 species of native and introduced trees and shrubs. Most trees had low similarities in their leafhopper faunas, and taxonomic relationships between trees appeared relatively unimportant in determining the similarity of leafhopper faunas. There was a significant regression between area of host range and number of associated leafhoppers, but only 16% of the variation in numbers was explained. Thus, on a gross level, numbers of leafhoppers on a species of plant are reasonably predicted by host range, though other factors are obviously of major importance in determining community structure, and Claridge and Wilson (1981) suggested host plant chemistry was paramount among these, though the data are lacking. Another important feature emerging from these studies was that recently introduced non-native trees did not have a depauperate fauna as expected, but had accrued large numbers of leafhopper species (see also Connor et al. 1980). *Acer pseudoplatanus* (sycamore), introduced into Britain 1800 years ago, had a larger leafhopper fauna than *A. campestre* and harbored species not found on the latter. Southern beech, *Nothofagus*, has attracted a leafhopper fauna almost as large as that of *Acer pseudoplatanus*. The *Nothofagus* fauna in Britain is drawn exclusively from the northern beech, *Fagus*, and *Quercus* species.

On a smaller spatial scale, Denno et al. (1981) studied the relationship of patch sizes of a salt marsh grass, *Spartina patens*, and the abundance of its sap feeders in New Jersey salt marshes over a range of patch sizes from just a few square meters to over 20 ha. Of the resident species of Auchenorrhyncha, two became less abundant as patch size decreased, but four others showed no differences in density.

Within a particular stand of plants or group of hosts, the actual abundance of planthoppers can be influenced by the plant itself and even by the community of plants in the area. For example, the leafhopper *Empoasca fabae* was reported to decline in abundance as host plant density increased (Mayse 1978). It is not known whether it did so because the nitrogen content of leaves declines in denser plantings. Densities of the same hopper, *E. fabae*, are higher on pure stands of alfalfa than on plants mixed with broad-leaved weeds or grasses (Lamp et al. 1984; Oloumi-Sadeghi et al. 1987). Structural diversity of a host plant can also affect the size of the planthopper and leafhopper community feeding on it. Denno (1980) showed that the salt marsh grasses *Spartina patens* and *D. spicata*, each of which is structurally complex and forms a persistent thatch, support eight and six species of planthoppers and leafhopper, respectively. *Spartina alterniflora*, a structurally simple marsh plant with no thatch that lives in the same areas as the other two species, supports only five (Denno and Roderick 1991). Tallamy and Denno (1979) and Denno (1977) elegantly showed that, on removal of such thatch, the architectural complexity was lessened and the diversity of sap suckers was reduced. Similar experiments in England, in which complexity was reduced by cutting and burning in terrestrial grasslands, have reduced population diversity of Auchenorrhyncha (Morris 1973; Morris and Lakhani 1979). Direct positive correlations appear to exist between planthopper diversity and architectural complexity in many communities (Morris 1971, 1974; Murdoch et al. 1972; Denno 1977; Tallamy and Denno 1979).

Do Leafhoppers Compete?

Given that more than one leafhopper species often occur on a host plant, does competition between them commonly occur, and if so how does it affect community structure? Alternatively, are coexisting leafhoppers present at such low densities as to be noncompetitors?

One of the most frequently cited unexplained examples of coexistence without apparent competitive exclusion is from studies by Ross (1957, 1958) on coexisting *Erythroneura* leafhoppers on sycamore (*Platanus*) trees in Illinois. Ross reported that six species occupied the same ecological niche without competing and that weather, a density-independent factor, was the most important regulatory influence on their populations. In a reanalysis, McClure and Price (1975) showed that the species did, in fact, compete severely. Even at relatively low population densities of two individuals of each of two competing species per leaf, the progeny produced per female was reduced. At 6 females per leaf, reproductive success was reduced up to 70%, and at 14 females per leaf, no progeny matured. Mean peak field densities at two sites in Illinois were seven individuals per leaf and less than one. McClure and Price (1975) felt not only that competition for food on leaves was likely to be limiting but also that oviposition sites along veins and eggs themselves were likely to be destroyed in the face of heavy feeding by leafhoppers. McClure and Price (1976) later proposed that each *Erythroneura* species had its peak abundance in a different latitudinal zone. Thus, geographical overlap is reduced and interspecific competition lessened. Because these peak abundances were the same in the Ross

(1957) study as in McClure and Price's (1975) work, they were reasoned to be meaningful and stable. The strong dispersal capability of *Erythroneura* and other leafhoppers (Johnson 1969) was argued to explain the continued existence of different species of leafhopper away from their centers of distribution, though no data from dispersal studies or sticky traps was ever provided.

In a parallel study, Stiling (1980a, 1980b) investigated competition and coexistence among *Eupteryx* leafhoppers living on British stinging nettles. In earlier work, Le Quesne (1972) reported no obvious temporal or spatial differences between the three species *Eupteryx aurata*, *E. cyclops*, and *E. urticae* on nettles. Stiling (1980a) showed that *E. aurata* moved onto other host plants in the summer, reducing interspecific contact with the other two species. *Eupteryx urticae* and *E. cyclops* were specific to nettles year-round, and experimental studies demonstrated severe reduction in the number of eggs laid per female of each species under mixed-species conditions. Coexistence may have been promoted to some degree by a differential distribution within the nettle canopy of second-generation leafhoppers. Individuals of *E. urticae* occurred higher up in the canopy than those of *E. cyclops*. Differential distributions of leafhoppers and planthoppers within the canopy of vegetation appears to be the major factor that separates coexisting species and possibly allows coexistence (Table 13.1). Vertical zonation of leafhoppers has been found to occur in seven out of the eight investigations on this subject so far.

Whether or not planthoppers and leafhoppers regularly compete with other herbivorous insects occupying the same plants is open to debate. There have been few studies on the subject to date. In perhaps the first study of this nature, McClure (1974) showed how *Erythroneura lawsoni* (Cicadellidae) feeding on leaves of sycamore (*Platanus occidentalis*) in the United States avoided competition with a tingid bug and a mirid by feeding at different times of the year and on different parts of the leaves. Evans (1989) experimentally increased the densities of grasshoppers, *Phoetaliotes nebrascensis* (Acrididae), at Konza Prairie, Kansas, USA, but could find no effect on the phytophagous Homoptera, which included at least 110 species of Cicadellidae. Grasshoppers are certainly one of the most common and destructive herbivores of many grassland systems. In north Florida salt marshes, we have discovered that grasshoppers are present on mainland sites, where they cause much damage, chewing over 90% of *Spartina* leaves. In this system, we believe grasshoppers can and do have a substantial effect on *Prokelisia* planthoppers, depressing population levels (Stiling et al. 1991a). The reason may be that grasshoppers chew the tops off the leaves, depriving young *Prokelisia* nymphs of a refuge where they may normally hide from spiders, other predators, or abiotic extremes.

Table 13.1. Mechanisms that enable potentially competing species of Auchenorrhyncha to coexist.

Species	Host Plant	Prime "Mechanism of Coexistence"	Reference
Erythroneura (8)	*Platanus occidentalis*	Geographical zonation and subsequent adult dispersal	McClure and Price (1976)
Eupteryx (3)	*Urtica dioica*	(i) Use of different host plants (ii) Vertical zonation	Stiling (1980a)
Various (8)	*Spartina patens*	Vertical zonation	Denno (1980)
Cercopids (4)	Grassland communities	(i) Use of different host plants (ii) Vertical zonation	Halkka et al. (1977)
Various (many)	Grassland communities	Vertical zonation	Andrzejewska (1965)
Various (6)	*Holcus mollis*	(i) Temporal zonation (ii) Vertical zonation	Waloff (1979)
Philaenus and *Lepyronia* (2)	Various	Vertical zonation	McEvoy (1986)
Cicadas (9)	Various	Vertical zonation	MacNally and Doolan (1986)

Important Mortality Sources in Leafhopper and Planthopper Communities

Although both McClure and Price (1975) and Stiling (1980b) documented substantial competitive effects between different species of leafhoppers occupying the same host plant, it is hard to know whether such effects are the strongest in communities of Auchenorrhyncha or whether competitive effects are unimportant in the face of mortality caused by parasitoids, predators, or other sources. The comparative studies are lacking, not only for planthoppers and leafhoppers but for insects in general. Stiling et al. (1984) showed that competition only accounted for an additional 11% mortality in *Hydrellia valida* leaf miners on *Spartina* grass, as opposed to around 90% for that caused by parasitoids (Stiling et al. 1982). Karban (1989) has shown that for each of three species of herbivores on *Erigeron glaucus*, the seaside daisy, spittlebugs, plume moth caterpillars, and thrips, the major mortality factor is different. Competitive effects were important in spittlebugs, predation affected caterpillars the most, and host plant variation was most important for thrips. Such information can only be obtained after a comparative study of all mortality factors and, as previously stated, such studies have not often been done for leafhoppers. What is needed is a key-factor type study. Stiling (1988) has summarized key-factor studies for 58 species of insects. Again, different sources of mortality were important for different systems.

For the Auchenorrhyncha of British grasslands, variation in adult fecundity was thought to be responsible for the large variation in population density (May 1971; 1978; Tay 1972; Waloff and Thompson 1980). Adult dispersal has been deemed most important for planthoppers on sugarcane (Metcalfe 1972). Factors acting in a density-dependent manner that would return population densities to equilibrium (e.g., competition) have been more difficult to determine. Stiling (1988) showed that, for 9 out of 13 species of planthoppers and leafhoppers, no density-dependent mechanisms were found. Density-dependent parasitism was found for one species, *Saccharosydne saccharivora* (Metcalfe 1972), and density-dependent alterations in fecundity were found in two others (Kuno and Hokyo 1970).

Planthopper Communities and Biological Control

Despite the absence of knowledge on the full effects of parasitoids and predators on planthopper and leafhopper communities, there has been a general interest in releasing natural enemies against pest species in biological and integrated control campaigns. Against the pest planthoppers and leafhoppers attacking rice, for example, various workers have looked for effective natural enemies (Dutt and Giri 1978; Miura et al. 1979).

Many authors have implied that natural enemies in general, and parasitoids in particular, can have substantial impacts on host populations and communities. Thus, Meyerdirk and Hussein (1985) recorded egg parasitization levels of over 90% for eggs of *Circulifer tenellus* on sugar beets in southern California and reasoned that parasitoids had a "major impact on leafhopper population density." Some dryinid parasitoids may also feed on a host or several hosts before parasitization, increasing the mortality levels. Chiu et al. (1984) studied the searching behavior and functional response of the dryinid *Pseudogonatopus flavifemur*. They reported that the dryinid showed a sigmoidal functional response and, thus, had potential as a biocontrol agent of brown planthopper even though the rate of parasitism in the field was rather low (20% or less).

Are such generalizations valid? Do many parasitoids of leafhoppers act in density-dependent fashion? I have gathered together published studies that looked explicitly at host densities and concurrent parasitism levels in the field (Table 13.2). Although the sample size is small in six out of nine cases,

Table 13.2. Evidence for the density-dependent attack of Auchenorrhyncha eggs and nymphs by parasitoids.

Species	Stage	Parasitoid	Density Dependence?	Reference
Eupteryx urticae	Egg	*Anagrus* sp. nr. *atomus*	Yes	Stiling (1980b)
Macrosteles sexnotatus	Egg	*Anagrus holci*	Yes	Becker (1975)
Nephotettix cincticeps	Egg	Complex of species	Yes	Sasaba and Kiritani (1972)
Prokelisia marginata	Egg	*Anagrus delicatus*	No	Stiling and Strong (1982a) Strong (1989); Benrey and Denno (unpublished) Cronin and Strong (1990a); Roderick (1987)
	Nymph	*Elenchus koebelei*	No	Stiling et al. (1991b)
Saccharosydne saccharivora	Egg	*Anagrus flaveolus* and *Tetrastichus* sp.	No	Metcalfe (1972)
	Nymph	*Stenocranophilus* sp.	Yes	
Typhlocyba pomaria	Egg	*Anagrus epos*	Yes, but only in 1 of 3 cases	Seyedodeslami and Croft (1980)
Poophilus costalis	Nymph	*Halictophagus pontifex*	Yes	Greathead (1970)
TOTAL (7 species)			Yes in 6 of 9 cases	

leafhopper and planthopper eggs and nymphs are attacked in a density-dependent manner. This result appears to conflict with the general pattern of insect–host parasitism in which a little less than 25% of the nearly 200 cases report density dependence (Stiling 1987; Walde and Murdoch 1988). At first it appears that even if one requires density dependence as a necessary prerequisite for biological control, parasitoids of the Auchenorrhyncha fit the bill. However, Cronin and Strong have conducted a more exhaustive review of parasitism in leafhoppers and planthoppers (Chapter 11), reanalyzing many data from original publications. Their conclusions, based on 30 spatial relationships and 54 temporal ones, conformed to Stiling's (1987) original findings of about 25% density-dependent responses in insect host-parasitoid systems.

How have releases of parasitoids performed against leafhopper and planthopper pests? The answer is difficult to quantify, but according to the relevant information in a world review of biological control campaigns by Clausen (1978; summarized in Table 13.3), parasitoids released against Auchenorrhyncha have not done very well. Against 6 different species of Auchenorrhyncha, 26 species of parasitoids had been released by the time Clausen's review on biocontrol was completed. In only 10 cases (38.5%) was establishment successful, and in only 2 instances (7.7%) was some degree of control achieved. Against *Circulifer tenellus*, at least 13 different species of egg parasitoids were released in the United States, but none became established. Could such a large number of parasitoids increase the degree of interspecific interactions and reduce the likelihood of control as suggested by Ehler and Hall (1982)? What is the normal load of parasitoids on a species of planthopper or leafhopper? Which type of parasitoid inflicts a higher percentage of mortality on the host population? Which type of parasitoid should be released? In the next section, I will try to answer these questions and point the way for future studies of Auchenorrhyncha communities and biological control.

Mortality Caused by Egg and Nymphal Parasitoids

There is no doubt that parasitism (%) caused by egg or nymphal parasitoids in the field can often be misleading if equated with the magnitude of mortality. For instance, egg parasitism can be recorded as artificially high if unparasitized eggs have already hatched, leaving a large proportion of parasitized eggs behind. Van Driesche (1983) has elaborated on this point. Nevertheless, with careful and frequent sampling over the course of a generation, it is possible to estimate accurately the "real" mortality due to egg and nymphal parasitoids. Stiling (1979) was able to measure average

Table 13.3 Evidence for the successful establishment of parasitoids released against pest leafhoppers and planthoppers and for the control of these pests. Data from Clausen (1978).

Host Species and Geographic Location of Problem	Parasitoid Released	Stage Attacked	Establishment Successful? (Y/N)	Control Successful? (Y/N)
Clastoptera undulata spittlebug (Cecropidae) Bermuda	*Carabunia myersi*	Nymph	Y	?
Circulifer tenellus beet leafhopper (Cicadellidae) U.S.A.	*Ootetrastichus beatus*	Egg	N	—
	Aphelinoidea anatolica	Egg	N	—
	Aphelinoidea "A"	Egg	N	—
	Aphelinoidea "M"	Egg	N	—
	Aphelinoidea "O"	Egg	N	—
	Aphelinoidea "S"	Egg	N	—
	Gonatocerus "A"	Egg	N	—
	Gonatocerus "B"	Egg	N	—
	Gonatocerus "E"	Egg	N	—
	Gonatocerus "X"	Egg	N	—
	Gonatocerus "Y"	Egg	N	—
	Polynema "A"	Egg	N	—
Edwardsiana froggatti apple leafhopper (Cicadellidae) Australia and New Zealand	*Anagrus armatus*	Egg	Y	Y
	Aphelopus typhlocybae	Nymph	N	—
Peregrinus maidis corn planthopper (Delphacidae) Hawaii	*Anagrus osborni*	Egg	Y	N
Perkinsiella saccharicida sugarcane planthopper (Delphacidae) Hawaii	*Anagrus frequens*	Egg	Y	?
	Anagrus optabilis	Egg	Y	Y
	Ootetrastichus beatus	Egg	Y	?
	Ootetrast. formosanus	Egg	Y	?
	Haplogonatopus vitiensis	Nymph	Y	?
	Perinsiella sinensis	Nymph	Y	?
Tarophagus proserpina taro leafhopper (Delphacidae) Hawaii	*Anagrus perforator*	Egg	?	—
	Ootetrastichus megameli	Egg	Y	—
	Gonatopus sp.	Nymh	?	—
6	26		10	2

Table 13.4. Average levels of egg parasitism for different host–parasitoid systems.

Host Species		Percent Parasitism	Reference
Cicadella viridis	1969	9.1	Tay (1972)
	1970	6.1	
	1971	4.7	
Conomelus dehneli	1st gen.	30	Arzone (1977)
	2nd gen.	80	
Conomelus anceps	1954–55	0.1	Whalley (1956)
	1959–60	0.3	Rothschild (1967)
	1960–61	0.1	
Eupteryx urticae	1st gen.	50.9	Stiling (1980a, 1980b)
	2nd gen.	20.3	
Macrosteles sexnotatus		44.6	Becker (1975)
Nephotettix cincticeps		38.9	Orita (1969)
		46.5	Sahad and Hirashima (1984)
		69.5	Orita (1972)
		55.0	Miura (1976a, 1976b)
Sogatella furcifera		31.7	Miura et al. (1981)
Stenocranus minutus	1969	15.9	May (1971)
	1970	23.0	
Typhlocyba froggatti	Winter	66.0	Dumbleton (1934, 1937)
		50.0	Teulon and Penman (1986)
Typhlocyba froggatti	Summer	85.0	Dumbleton (1934, 1937)
		74.2	Teulon and Penman (1986)
Prokelisia marginata		23.70	Stiling et al. (unpublished data)
		Mean = 35.90 ± 27.19	

nymphal parasitism accurately in *Eupteryx* leafhoppers over the course of each 2-month generation by taking extensive weekly samples. In Tables 13.4 and 13.5, I have gathered together similar estimates of average parasitism levels in the egg and nymphal stages of Auchenorrhyncha from a variety of other carefully investigated systems.

Several results emerge from a comparison of these studies of parasitism rates. First, average egg parasitism rate was 36%, more than twice the level of nymphal parasitism (t_{75} = 3.04, $P < 0.01$). This result suggests that release of egg parasitoids in biological control programs may be more effective than the release of nymphal parasitoids. Also, egg parasitoids kill pests before the harmful nymphal or adult stages are reached. Second, univoltine host species seem to suffer lower rates of nymphal parasitism (12.3 ± 8.2%, n = 11) than either generation of bivoltine species (17.1 ± 13.6%, n = 19, generation 1; 21.2 ± 21.0%, n = 19, generation 2) (Table 13.5),

Table 13.5. Average levels of nymphal parasitism (%) for different host–parasitoid systems (parasitism by pipunculids, dryinids, and strepsipterans combined). Calculated from Tables 13.6, 13.7, and 13.8.

Host Species	Generation	Percent Parasitism
Arthraldeus pasceullus	I	25.0
	II	12.1
Euscelis plebejus	I	1.5
	II	17.8
Jassargus distinguendus	I	21.9
	II	21.0
Psammotettix confinis	I	28.7
	II	26.3
Streptanus sordidus	I	6.2
	II	4.9
Javesella pellucida	I	6.0
	II	4.6
Laodelphax elegantulus	I	15.5
	II	10.1
Paraliburnia dalei	I	1.6
	II	0.2
Muirodelphax exiguus	I	7.7
	II	8.4
Ribautodelphax angulosus	I	9.8
	II	0
Dicranotropis hamata	I	32.4
	II	25.8
Doratura stylata		13.5
Balclutha punctata		4.4
Mocydiopsis paricauda		3.7
Diplocolenus abdominalis		11.9
Elymana sulphurella		15.8
Recilia coronifera		7.7
Adarrus ocellaris	I	11.3
	II	27.7
Alebra sp.		10.7
Fagocyba cruenta	I	42.1
	II	22.2
Edwardsiana sp.	I	31.6
	II	29.0
Lindbergina aurovittata		8.7
Ribautiana ulmi		29.7
Alnetoidia alneti		23.9

(*continued*)

Table 13.5. (*Continued*)

Host Species	Generation	Percent Parasitism
Eupteryx aurata	Site I, gen. I	9.0
	Site I, gen. II	76.6
	Site II, gen. I	0
	Site II, gen. II	3.0
Eupteryx cyclops	Site II, Gen. I	21.0
	Site II, gen. II	26.2
Eupteryx urticae	Site I, gen. I	44.3
	Site I, gen. II	71.4
	Site II, gen. I	10.0
	Site II, gen. II	16.3
Nilaparvata lugens	Wet season	11.0
	Dry season	11.5
Nephotettix sp.	Wet season	24.2
	Dry season	20.9
Sogatella furcifera	Wet season	14.4
	Dry season	19.0
Prokelisia marginata		7.2
		Mean = 17.6 ± 15.3

though the difference is not significant for either generation (generation 1, $t_{28} = 1.21, P > 0.1$; generation 2, $t_{28} = 1.65, P > 0.1$). More than one host generation may allow for a quicker buildup of parasitoid numbers from one generation to the next than is possible on a univoltine species, in which overwintering mortality may act to knock back populations of parasitoids each year. The implications for biological control are clear; multivoltine Auchenorrhyncha should be adversely affected more by the release of parasitoids than univoltine ones.

Among the egg parasitoids, mymarids and trichogrammatids seem to be predominant, but there is insufficient information to compare parasitism rates by the two groups accurately. Besides, their mode of action is similar; both are internal, solitary egg parasitoids. Of the nymphal parasitoids, however, three major groups are evident, dryinid wasps, pipunculid flies, and strepsipterans.

Dryinid wasps (Dryinidae) are nymphal parasitoids that have some interesting features. First, the first pair of legs are chelate, a modification for capturing nymphs. Second, the larva develops semiexternally with its head and mouthparts buried in the host's abdomen and its own abdomen enclosed in a characteristic dryinid sac (Stiling and Strong 1982b). Pipunculid flies (Pipunculidae) develop exclusively as parasitoids of the Auchenorrhyncha, feeding completely within the host's abdomen and the adults are superb fliers. Strepsipteran parasitoids have a number of interesting features as well, including neotenic females and "triungulinid" larvae (Waloff 1981). In Tables 13.6, 13.7, and 13.8, I have compared parasitism rates (%) of Auchenorrhyncha hosts among these three groups of parasitoids. The result for pipunculids averages out to 13.5% for 30 species of hosts, nearly twice that of the dryinid wasps, 7.5% (t_{101} = 2.28, $P < 0.05$). Levels of parasitism by Strepsiptera are also fairly high but have probably been overestimated because of the inclusion of data from three separate years when rates of parasitism were exceptionally high (60%) on *Ulopa reticulata* (Waloff 1981). If only one of these years is included, as is the case for the other host–parasitoid systems, then average parasitism by Strepsiptera is only 11.59 ± 19.04%. As before, the second generation of a bivoltine host suffers greater nymphal parasitism from both dryinids and pipunculids. A notable exception to this trend is provided by recent studies of nymphal and adult parasitism of *Prokelisia* planthoppers in north Florida salt marshes. In these studies, thousands of individuals (n = 88,354) were collected from eight different sites every day for a year (Stiling et al. 1991b, 1991c). The average level of parasitism by strepsipterans was P. 7.06% by strepsipterans but only 0.18% by dryinids. It is possible that levels of parasitism by pipunculids are higher than those by dryinids because their hosts are more easily captured, are more abundant, or are in some other way more prone to parasitism. I have, therefore, compared the ranks of parasitism levels by these parasitoids on shared hosts, hosts that support both groups of parasitoids (Table 13.9). I have then ranked parasitoids according to their rates of parasitism. The rankings confirm that, on average for common hosts, pipunculids show the higher parasitism levels.

Numbers of Parasitoids Per Host Species

Many species of the planthoppers and leafhoppers are attacked not by just one species of dryinid or pipunculid but by several. *Nilaparvata lugens* is attacked by at least six egg parasitoids and six nymphal parasitoids, and *Sogatella furcifera* by four egg parasitoids and five nymphal parasitoids. Perhaps the record goes to *Nephotettix cincticeps*, from which over 24 parasitoids have been recorded worldwide, 12 from the eggs and 12 from the nymphs (see Appendix 1). Is nymphal parasitism by pipunculids greater than that by dryinids because more species of pipunculids attack a given host? Perhaps more importantly, is egg parasitism greater than nymphal parasitism because of greater numbers of egg parasitoids per host species? Do numbers of parasitoids per host species differ with geographical location of host or with host taxa (Delphacidae versus Cicadellidae)? With these questions in mind, I have tabulated the numbers of species of egg and nymphal parasitoids per host species for planthoppers and leafhoppers from around the world (Table 13.10).

Table 13.6. Average levels of nymphal parasitism (%) by pipunculids.

Host Species	Generation	Percent Parasitism	Reference
Arthraldeus pasceullus	I	19.5	Waloff (1975)
	II	9.9	
Euscelis plebejus	I	0.9	
	II	12.2	
Jassargus distinguendus	I	18.1	
	II	19.8	
Psammotettix confinis	I	15.3	
	II	22.8	
Streptanus sordidus	I	1.2	
	II	2.2	
Javesella pellucida	I	0.2	
	II	0.4	
Laodelphax elegantulus	I	2.3	
	II	4.8	
Muirodelphax exiguus	I	0.7	
	II	4.1	
Paraliburnia dalei	I	1.3	
	II	0	
Ribautodelphax angulosus	I	0	
	II	0	
Balclutha punctata		0	
Dicranotropis hamata		0	
Doratura stylata		6.6	
Mocydiopsis paricauda		3.5	
Diplocolenus abdominalis		11.8	Waloff and Thompson (1980)
Elymana sulphurella		14.5	
Recilia coronifera		5.2	
Adarrus ocellaris	I	9.4	Waloff and Jervis (1987)
	II	26.3	
Alebra sp.		5.5	Jervis (1980)
Fagocyba cruenta	I	25.5	
	II	14.7	
Edwardsiana sp.	I	14.5	
	II	7.7	
Lindbergina aurovittata		2.2	
Ribautiana ulmi		1.9	
Alnetoidia alneti		21.3	

(*continued*)

Table 13.6. (*Continued*)

Host Species	Generation	Percent Parasitism	Reference
Eupteryx aurata	Site I, gen. I	9.0	Stiling (1979)
	Site I, gen. II	43.0	
	Site II, gen. I	0	
	Site II, gen. II	2.0	
Eupteryx cyclops	Site II, gen. I	21.0	
	Site II, gen. II	18.0	
Eupteryx urticae	Site I, gen. I	43.9	
	Site I, gen. II	66.1	
	Site II, gen. I	10.0	
	Site II, gen. II	10.0	
Nephotettix sp.	Wet season	22.7	Peña and Shepard (1986)
	Dry season	18.7	
Philaenus spumarius		26.8	Whittaker (1969b)
Neophilaneus lineatus		27.2	

Mean = 13.56 ± 13.87
Average for species with one generation = 10.5 ± 11.49 (n = 13).
Average for species with two generations, generation 1 = 10.71 ± 8.5 (n = 18).
Average for species with two generations, generation 2 = 14.67 ± 17.05 (n = 18)

Do delphacids and cicadellids support more egg parasitoids than nymphal parasitoids? One point becomes immediately clear; more studies have examined nymphal parasitoids than have examined egg parasitoids. I compared the numbers of egg parasitoids to the number of nymphal parasitoids for cases in which at least one type of either parasitoid had been recorded. If I ignored cases in which no egg parasitoids were found, the number of species of egg parasitoids per host was 2.23 ± 2.09 (S.D.) (n = 120). The average number of species per host was 2.02 ± 1.88 (n = 375) for nymphal parasitoids. I used a contingency table grouping hosts with 1, 2, 3, 4, or 5 or more species of parasitoids to statistically compare the numbers of parasitoid species per host for both egg and nymphal parasitoids. The number of egg parasitoids per host were not significantly higher than those of nymphal parasitoids ($\chi_4^2 = 9.22, P < 0.10$) (Fig. 13.1), and the difference between means of 2.23 and 2.02 species of parasitoids per host seems biologically marginal at best.

What is the average parasitoid load (number of species) for an individual species of Auchenorrhyncha? If all records are taken into account, it is 2.30 (Table 13.10). However, many studies concentrate either on egg parasitoids or, more commonly, on nymphal parasitoids and ignore the parasi-

Table 13.7. Average levels of nymphal parasitism (%) by dryinids.

Host Species	Generation	Percent Parasitism	Reference
Arthraldeus pasceullus	I	6.9	Waloff (1975)
	II	2.4	
Euscelis plebejus	I	0.6	
	II	6.4	
Jassargus distinguendus	I	4.6	
	II	1.5	
Psammotettix confinis	I	15.8	
	II	4.6	
Streptanus sordidus	I	5.1	
	II	2.8	
Javesella pellucida	I	2.0	
	II	2.6	
Laodelphax elegantulus	I	0	
	II	0	
Paraliburnia dalei	I	0	
	II	0	
Muirodelphax exiguus	I	1.2	
	II	0.6	
Ribautodelphax angulosus	I	0	
	II	0	
Balclutha punctata		4.4	
Doratura stylata		7.3	
Mocydiopsis paricauda		0.2	
Diplocolenus abdominalis		0.1	Waloff and Thompson (1980)
Elymana sulphurella		1.5	
Recilia coronifera		2.6	
Adarrus ocellaris	I	2.1	
	II	1.9	
Dicranotropis hamata	I	32.4	Waloff and Jervis (1987)
	II	25.8	
Alebra sp.		5.5	Jervis (1980)
Fagocyba cruenta	I	22.3	
	II	8.8	
Edwardsiana sp.	I	20.0	
	II	23.1	
Lindbergina aurovittata		6.7	
Ribautiana ulmi		28.3	
Alnetoidia alneti		3.4	
Eupteryx aurata	Site I, gen. I	0	Stiling (1979)

(*continued*)

Table 13.7. (*Continued*)

Host Species	Generation	Percent Parasitism	Reference
	Site I, gen. II	59.0	
	Site II, gen. I	0	
	Site II, gen. II	1.0	
Eupteryx cyclops	Site II, gen. I	0	
	Site II, gen. II	10.0	
Eupteryx urticae	Site I, gen. I	0.7	
	Site I, gen. II	23.7	
	Site II, gen. I	0	
	Site II, gen. II	7.0	
Dichoptera hyalinata		36.6	Swaminathan and Ananthakrishnan 1984
Eurybrachys tomentosa		20.0	
Nilaparvata lugens	Wet season	9.7	Peña and Shepard (1986)
	Dry season	6.4	
	Site I	7.5	Greathead (1983)
	Site II	2.1	
	Site III	7.8	
Sogatella furcifera	Wet season	7.6	Peña and Shepard (1986)
	Dry season	11.2	
	Site I	4.9	Greathead (1983)
	Site II	11.4	
Nephotettix sp.	Wet season	0.5	Peña and Shepard (1986)
	Dry season	0	
	Site I	0.2	Greathead (1983)
	Site II	0.2	
Macrosteles sexnotatus		0.4	Becker (1975)
Macrosteles viridigriseus	I	2.0	Khafagi (1986)
	II	1.2	
Oncopsis flavicollis		30.0	Ponomarenko (1986b) [as cited by Waloff and Jervis (1987)]
Erythroneura sp.	Site I	2.0	McClure and Price (1975)
	Site II	9.8	
Prokelisia marginata		0.18	Stiling et al. (unpublished data)

Mean = 7.52 ± 10.91
Average for species with one generation = 10.5 ± 12.60 (n = 14).
Average for species with two generations, generation 1 = 6.23 ± 9.20 (n = 20).
Average for species with two generations, generation 2 = 9.55 ± 14.20 (n = 20).

Table 13.8. Average levels of nymphal parasitism (%) by Strepsiptera.

Host Species	Generation	Percent Parasitism	Reference
Javesella pellucida	I	4.1	Waloff (1975)
	II	1.7	
Laodelphax elegantulus	I	13.5	
	II	5.6	
Paraliburnia dalei	I	0.3	
	II	0.2	
Muirodelphax exiguus	I	5.9	
	II	3.9	
Ribautodelphax angulosus	I	9.8	
	II	0	
Dicranotropis hamata		4.2	
Javesella pellucida		27.0	Pekkarinen and Raatikainen (1973)
Nilaparvata lugens	Wet season	1.4	Peña and Shepard (1986)
	Dry season	5.5	
Nephotettix sp.	Wet season	1.5	
	Dry season	2.7	
Sogatella furcifera	Wet season	7.4	
	Dry season	9.8	
Prokelisia marginata		7.06	Stiling et al. (unpublished data)
Goldeus harpago		68.8	Abdul-Nour (1971) [as cited by Waloff and Jervis (1987)]
Ulopa reticulata	1969	63.0	Waloff (1981)
	1970	68.0	
	1979	65.0	
		Mean = 16.36 ± 24.08	

toids of the other life history stages completely. This practice would tend to artificially depress the numbers of species of parasitoid per host. I have calculated the average number of parasitoids per host for cases in which both egg and nymphal parasitoids were recorded (Table 13.10). When this procedure is followed, the average number of parasitoids per host increases to 4.68 ± 3.97. This result underscores the importance of using only those host records in which the parasitoids of all life history stages were addressed when assessing the number of parasitoid species per host.

Do leafhoppers have fewer parasitoids than planthoppers? The average number of parasitoid species per host (all host records) is 2.38 ± 2.52 for cicadellids ($n = 335$) and 2.71 ± 2.45 ($n = 68$) for delphacids. A chi-

Table 13.9. Rank parasitism rates by pipunculids, dryinids, and strepsipterans on shared hosts, from 1, highest rate of parsitism, to 3, lowest rate of parasitism. Data are from Tables 13.6, 13.7, and 13.8.

Host Species	Generation	Rank of Parasitism Rate		
		Dryinid	Pipunculid	Strepsipteran
Arthraldeus pasceullus	I	2	1	
	II	2	1	
Euscelis plebejus	I	2	1	
	II	2	1	
Jassargus distinguendus	I	2	1	
	II	2	1	
Psammotettix confinis	I	1	2	
	II	2	1	
Streptanus sordidus	I	1	2	
	II	1	2	
Javesella pellucida	I	2	3	1
	II	1	3	2
Laodelphax elegantulus	I		2	1
	II		2	1
Muirodelphax exiguus	I	2	3	1
	II	3	1	2
Paraliburnia dalei	I		1	2
Doratura stylata		1	2	
Mocydiopsis paricauda		2	1	
Diplocolenus abdominalis		2	1	
Elymana sulphurella		2	1	
Recilia coronifera		2	1	
Adarrus ocellaris	I	2	1	
	II	2	1	
Fagocyba cruenta	I	2	1	
	II	2	1	
Edwardsiana sp.	I	1	2	
	II	1	2	
Lindbergina aurovittata		1	2	
Ribautiana ulmi		1	2	
Alnetoidia alneti		2	1	
Eupteryx aurata	Site I, II	1	2	
	Site II, II	2	1	
Eupteryx cyclops	Site II, II	2	1	
Eupteryx urticae	Site I, I	2	1	
	Site I, II	2	1	
	Site II, I	2	1	
	Site II, II	2	1	
Nephotettix sp.	Wet season	3	1	2
Nilaparvata lugens	West season		1	2
	Dry season		1	2
Sogatella furcifera	Wet season		1	2
	Dry season		1	2
Prokelisia marginata		2		1
MEAN		1.78	1.39	1.75

Table 13.10. Numbers of Auchenorrhyncha host species attacked by various numbers of parasitoids (see Appendix 1 for sources).

Host Stage Attacked = Egg																
	0	1	2	3	4	5	6	7	8	9	10	11	12	13	14	15	. . . 24
Delphacidae A																	
Europe	39	13	2	0	1	1	1										
Rest of world	0	5	3	1	3	1											
World total	39	16	4	2	4	2	1										
Delphacidae B																	
Europe		10	1		1	1											
Rest of world		2	1		3		1										
World total		12	2		4	1	1										
Cicadellidae A																	
North America	114	18	4	3	1	3			1								
Europe	109	17	4	2	2					1			1				
Rest of world	38	11	3	1	2	2			1				1				
World total	258	46	9	6	6	5			2	1			2				
Cicadellidae B																	
North America		7	2	2		1			1								
Europe		12	3	2	2												
Rest of world		2	1		2	1							1				
World total	22	5	4	5	2				1				1				
Other Auchenorrhyncha A	25	8	2	2	2												
Other Auchenorrhyncha B	1	1															
TOTAL A	322	70	15	10	12	7	1		2	1			2				
TOTAL B		35	8	4	9	3	1		1				1				

Host Stage Attacked = Nymph	0	1	2	3	4	5	6	7	8	9	10	11	12	13	14	15	. . . 24
Delphacidae A																	
Europe	4	24	18	3	5		2			1							
Rest of world	6	4	1				2										
World total	14	27	14	3	4	1	4			1							
Delphacidae B																	
Europe		3	4	1	2		2				1						
Rest of world		4	1				2										
World total		7	5	1	2		4				1						
Cicadellidae A																	
North America	17	78	26	15	5	2	1										
Europe	8	61	27	17	8	5	3	1	1		1	2		1		1	
Rest of world	13	28	9	4	2		2						1				
World total	36	167	59	36	16	7	6	1	0		2	2	1	1		1	
Cicadellidae B																	
North America		8	3	1	1												
Europe		8	3	3	2	1	1					1					
Rest of world		3		1		2							1				
World total		19	7	5	3	1	3					1	1				
Other Auchenorrhyncha A	17	21	1														

(*continued*)

Table 13.10. (*Continued*)

	0	1	2	3	4	5	6	7	8	9	10	11	12	13	14	15	... 24
Other Auchenorrhyncha B		2															
TOTAL A	67	215	74	39	20	8	10	1	0	1	2	2	1	1	0	1	
TOTAL B		28	12	6	5	1	7			1		1	1				

Host Stage Attacked = Egg or Nymph

	0	1	2	3	4	5	6	7	8	9	10	11	12	13	14	15	... 24
Delphacidae A																	
Europe		24	18	4	4	2	1	2	1	1							
Rest of world		3	3	3		2				1		1					
World total		26	21	7	3	3	1	2	1	2	1		1				
Delphacidae B																	
Europe			3	2	1	2	1	2	1		1						
Rest of world			1	2		2					1		1				
World total			4	4	1	4	1	2	1		2		1				
Cicadellidae A																	
North America		80	31	17	9	4	2						1				
Europe		58	30	18	8	8	3	2	2	1	1	1	1	1	1	1	
Rest of world		33	13	4	2	3			1		1	1					1
World total		170	69	40	20	14	5	3	2	1	3	2	2	1	1	1	1

Taxon	0	1	2	3	4	5	6	7	8	9	10	11	12	13	14	15
Cicadellidae B																
North America		5	2	4		1						1				
Europe		5	4	2	4	1	1	1						1		
Rest of world		2			2					1	1					1
World total		12	7	6	5	2	2	1		1	1	1		1		1
Other Auchenorrhyncha A	31	3	3	2												
Other Auchenorrhyncha B		1	1													
TOTAL A	227	93	50	25	17	6	5	3	3	4	2	3	1	1	1	1
TOTAL B		17	12	7	9	3	4	2	0	3	1	2	0	1		1

Host Stage Attacked = Egg or Nymph

Taxon	Mean	S.D.	Total Number of Species	Taxon	Mean	S.D.	Total Number of Species
Delphacidae A				**Delphacidae B**			
Europe	2.36	1.91	57	Europe			13
Rest of world	4.15	3.74	13	Rest of world			7
World total	2.71	2.45	68	World total			20
Cicadellidae A				**Cicadellidae B**			
North America	1.90	1.45	144	North America			13
Europe	2.76	2.72	136	Europe			19
Rest of world	2.49	3.53	59	Rest of world			7
World total	2.38	2.52	335	World total			40
Other Auchenorrhyncha A			39	**Other Auchenorrhynca B**			2
TOTAL A	2.30	2.37	442	**TOTAL B**	4.68	3.97	62

Note: Figures in the first column are numbers of species attacked by no species of parasitoids, those in the second column by 1 species, and so on. Column totals are not always the same because some species are found in more than one geographic region. A = cases where egg or nymphal parasitoids have been recorded; B = cases where egg and nymphal parasitoids have been recorded.

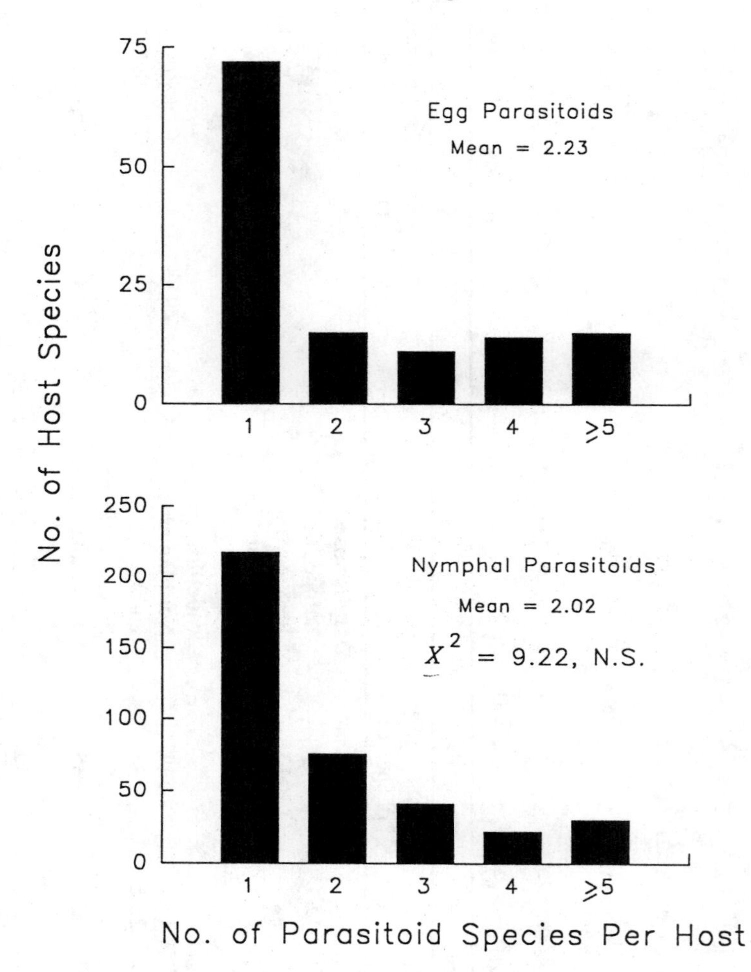

Figure 13.1. Number of species of egg and nymphal parasitoids supported per species of Auchenorrhyncha host.

squared contingency table reveals no significant difference between the two ($\chi_4^2 = 5.49, P < 0.3$) (Fig. 13.2). Thus, cicadellids and delphacids tend to support about the same number of parasitoids per host species.

Do hosts in some parts of the world support more parasitoids than hosts in other parts? The best records for testing this question are those for the Cicadellidae. In a three-way contingency table for hosts from Europe, North America, and the rest of the world combined that supported 1, 2, 3, 4, or 5 or more parasitoid species, there was no significant difference in parasitoid load ($\chi_{13}^2 = 13.48, P < 0.5$) (Fig. 13.3).

Any difference between delphacids and cicadellids in the mortality caused by parasitoids or that caused by geographical location is likely attributable to the characteristics of specific parasitoid species (e.g., host locating ability and attack success) rather than to the number of species of parasitoids a given host can support. In Figure 13.4, I have plotted overall nymphal parasitism rates (%) against the number of recorded parasitoid species per host. The data are based mainly on European records (see Table 13.11). There is no significant relationship between the two; $r = 0.152$, P

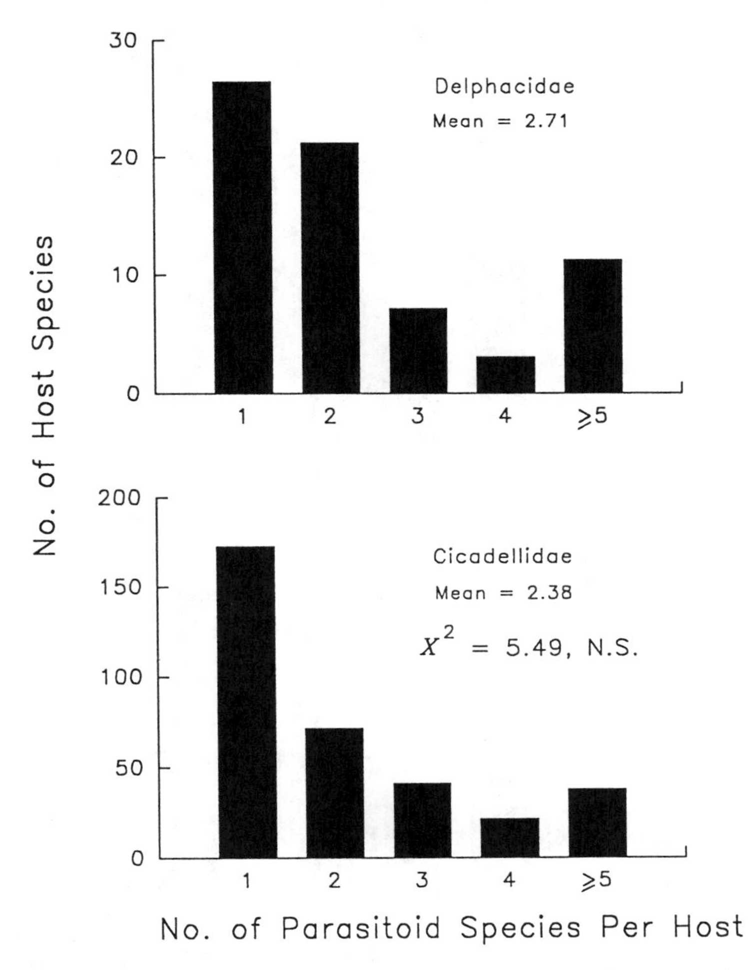

Figure 13.2. Number of parasitoid species (egg and nymphal) supported per delphacid and cicadellid host species.

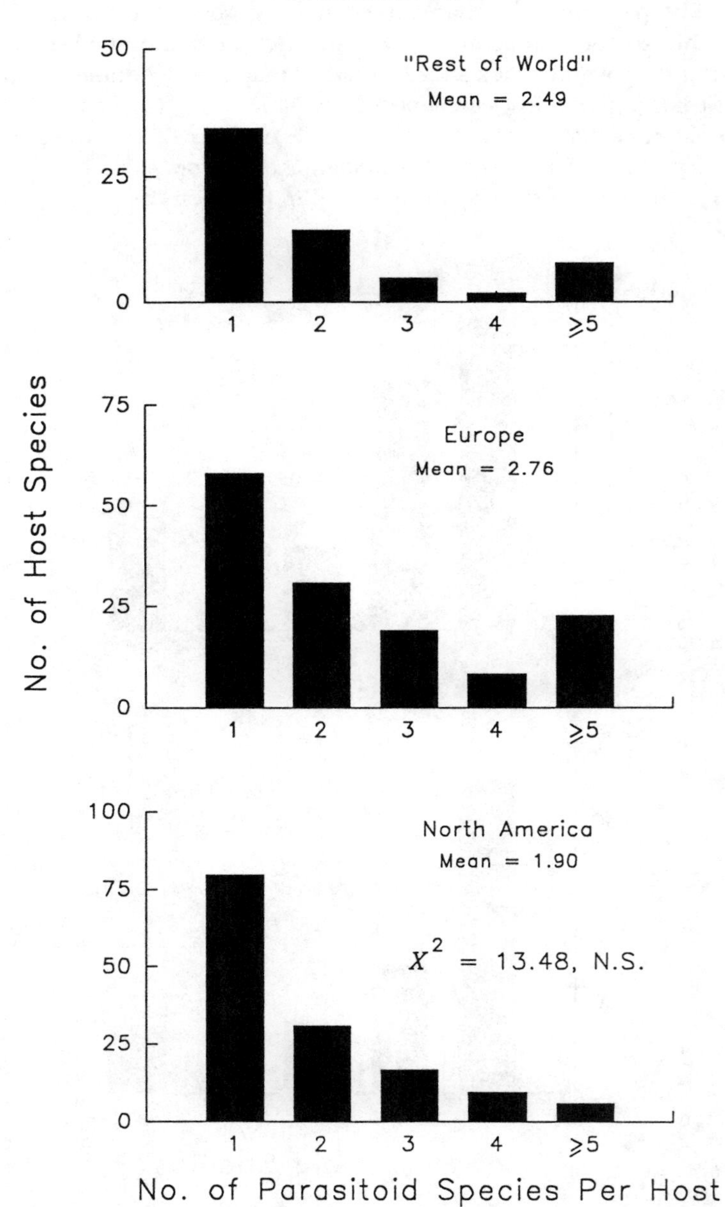

Figure 13.3. Geographic variation in the number of parasitoid species supported per cicadellid host species from around the world. Hosts were divided into three regional categories (North America, Europe, and "Rest of World") for analysis.

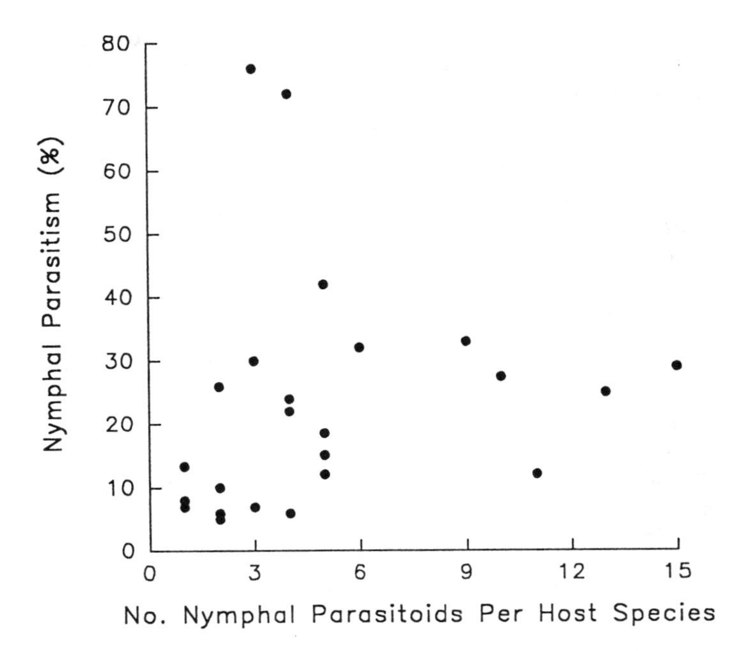

Figure 13.4. The relationship between total nymphal parasitism (%) and the number of attacking species of nymphal parasitoids per host. Data are largely European and are from Table 13.11.

> 0.2, $n = 24$. The same is true for egg parasitism and the number of egg parasitoids (Table 13.12, Fig. 13.5) ($r = -0.305, P > 0.2, n = 9$). Success rates of biological control campaigns against planthoppers and leafhoppers may, therefore, depend not so much on the community ecology of host complexes and the number of parasitoid species released, but rather on the vigor and natural history of the natural enemies released.

Summary

1. Commonly, more than one species of planthopper or leafhopper is found on any given species of host plant. Therefore, interspecific interactions may commonly occur. This chapter addresses the question of their coexistence and other community-wide phenomena.
2. In seven of eight cases examining coexistence among species of Auchenorrhyncha feeding on the same host plant, differing vertical zonation within the canopy is argued to reduce competition and promote coexistence.
3. In several instances, planthopper and leafhopper eggs and nymphs are attacked in a density-dependent fashion by parasitoids. Nevertheless,

Table 13.11. Percent nymphal parasitism and numbers of recorded parasitoid species for bivoltine species of hosts. Percent parasitism was taken as the highest parasitism of either generation.

Species	Percent Nymphal Parasitism	Number of Species of Nymphal Parasitoids
Arthraldeus pasceullus	25.0	13
Jassargus distinguendus	21.9	4
Psammotettix confinis	28.7	15
Streptanus sordidus	6.2	3
Javesella pellucida	6.0	4
Ribautodelphax angulosus	9.8	2
Dicranotropis hamata	32.4	9
Doratura stylata	13.5	1
Balclutha punctata	4.4	2
Diplocolenus abdominalis	11.9	11
Elymana sulphurella	15.8	5
Recilia coronifera	7.7	1
Adarrus ocellaris	27.7	10
Fagocyba cruenta	42.1	5
Edwardsiana sp.	31.6	6
Lindbergina aurovittata	8.7	1
Ribautiana ulmi	29.7	3
Alnetoidia alneti	23.9	4
Eupteryx aurata	76.6	3
Eupteryx cyclops	26.2	2
Eupteryx urticae	71.4	4
Nilaparvata lugens	11.5	5
Sogatella furcifera	19.0	5
Prokelisia marginata	5.0	2

Table 13.12. Percent egg parasitism and numbers of egg parasitoid species per host, calculated as in Table 13.11.

Species	Percent Egg Parasitism	Number of Species of Egg Parasitoids
Cicadella viridis	9.1	12
Eupteryx urticae	50.9	1
Macrosteles sexnotatus	44.6	4
Sogatella furcifera	31.7	4
Conomelus anceps	0.3	5
Conomelus dehneli	80.0	1
Stenocranus minutus	23.0	2
Nephotettix cincticeps	69.5	12
Typhlocyba froggatti	85.0	1

releases of parasitoids for the control of pest species have more often resulted in failure than success.

4. The potential of egg parasitoids as biological control agents seems to be much greater than that of nymphal parasitoids because they cause about twice the degree of mortality to host populations.
5. Among nymphal parasitoids, parasitism by pipunculid flies is usually greater than that by dryinid wasps.
6. Planthoppers and leafhoppers tend to support the same number of parasitoid species per host. Host species tend to support roughly the same number of egg parasitoid species as nymphal parasitoid species. There is no tendency for host species from one part of the world to support more species of parasitoids than hosts from another region.
7. There is no significant relationship between parasitism rate (%) and the number of recorded parasitoid species (egg and nymphal) per host.
8. Success in the biological control of Auchenorrhyncha is more likely to result from the vigor of released enemies rather than community-related factors such as the number of parasitoid species released.

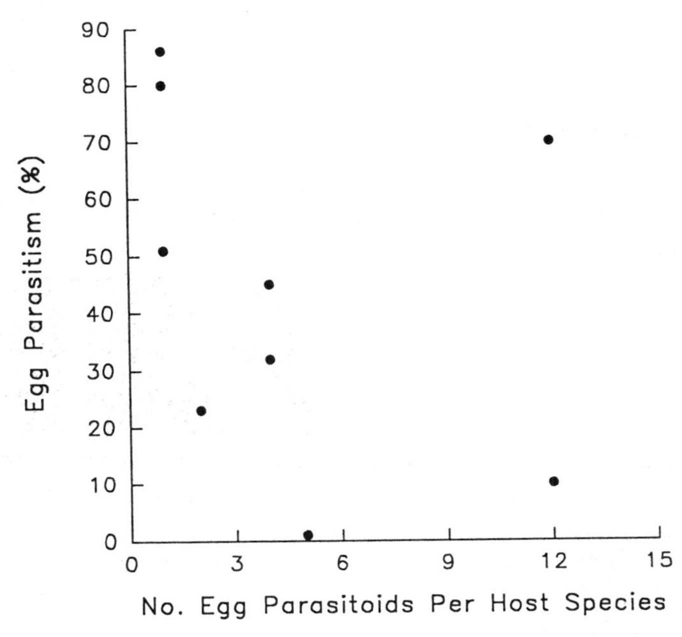

Figure 13.5. The relationship between total egg parasitism (%) and the number of attacking species of egg parasitoids per host. Data are largely European and are from Table 13.11.

APPENDIX 1. Individual host–parasitoid records for Auchenorrhyncha used in Table 13.10.

Host Species	Family[a]	Egg Parasitoid	Nymphal Parasitoid	Geographic Location[b]	Reference[c]
Acanthodelphax denticauda	D		*Elenchus tenuicornis*	E	2
Anakelisia fasciata	D	*Anagrus incarnatus*		E	2
Chloriona glaucescens	D		*Cephalops chlorionae*	E	2
			Elenchus tenuicornis	E	2
Chloriona propinqua	D		*Elenchus tenuicornis*	E	2
Chloriona smaragdula	D		*Elenchus tenuicornis*	E	2
Chloriona unicolor	D		*Elenchus tenuicornis*	E	2
Conomelus anceps	D	*Anagrus ensifer*		E	2
		Anagrus incarnatus		E	2
		Anagrus sp. "c"		E	2
		Anagrus sp. "c3"		E	2
		Anagrus sp. "d"		E	2
			Cephalops semifumosus	E	2
			Cephalops ultimus	E	2
			Elenchus tenuicornis	E	2
Conomelus dehneli	D	*Anagrus incarnatus*		E	2
Criomorphus albomarginatus	D		*Dicondylus bicolor*	E	2
			Elenchus tenuicornis	E	2
Criomorphus bicarinatus	D		*Dicondylus bicolor*	E	2
Criomorphus williamsi	D		*Cephalops semifumosus*	E	2
			Cephalops ultimus	E	2
			Elenchus tenuicornis	E	2
Delphacinus mesomelas	D		*Dicondylus bicolor*	E	2
			Elenchus tenuicornis	E	2
Delphacodes venosus	D		*Cephalops semifumosus*	E	2
			Cephalops ultimus	E	2

Dicranotropis carpathica	D		*Elenchus tenuicornis*	E	2
Dicranotropis divergens	D		*Pseudogonatopus rosellae*	E	2
Dicranotropis hamata	D	*Anagrus* sp.		E	14
			Dicondylus bicolor	E	2
			Pseudogonatopus distinctus	E	2
			Pseudogonatopus rosellae	E	2
			Cephalops carinatus	E	2
			Cephalops perspicuus	E	2
			Cephalops semifumosus	E	2
			Cephalops subultimus	E	2
			Cephalops ultimus	E	2
			Elenchus tenuicornis	E	2
Ditropis pteridis	D	*Anagrus* sp.	*Dicondylus bicolor*	E	2
			Pseudogonatopus distinctus	E	2
			Cephalops oberon	E	2
			Cephalops semifumosus	E	2
			Cephalops subultimus	E	2
			Elenchus tenuicornis	E	2
Eurybregma nigrolineata	D		*Elenchus tenuicornis*	E	2
Eurysa lineata	D		*Cephalops ultimus*	E	2
			Halictophagus sp.	E	2
Eurysa lurida	D		*Halictophagus* sp.	E	2
Florodelphax leptosoma	D		*Elenchus tenuicornis*	E	2
Gravesteiniella boldi	N		*Dicondylus bicolor*	E	2
			Pseudogonatopus distinctus	E	2
Hyledelphax elegantulus	D		*Dicondylus bicolor*	E	2
			Pseudogonatopus distinctus	E	2
			Cephalops semifumosus	E	2
			Elenchus tenuicornis	E	2

(*continued*)

APPENDIX 1. (*Continued*)

Host Species	Family[a]	Egg Parasitoid	Nymphal Parasitoid	Geographic Location[b]	Reference[c]
Javesella discolor	D		*Pseudogonatopus distinctus*	E	2
			Cephalops semifumosus	E	2
			Elenchus tenuicornis	E	2
Javesella dubia	D		*Elenchus tenuicornis*	E	2
Javesella forcipata	D		*Cephalops* sp.	E	2
			Elenchus tenuicornis	E	2
Javesella obscurella	D	*Anagrus* sp.		E	14
			Dicondylus bicolor	E	2
			Elenchus tenuicornis	E	2
Javesella pellucida	D	*Anagrus incarnatus*		E	2
			Dicondylus bicolor	E	2
			Pseudogonatopus distinctus	E	2
			Cephalops semifumosus	E	2
			Elenchus tenuicornis	E	2
Kostwigianella exigua	D		*Pseudogonatopus distinctus*	E	2
			Elenchus tenuicornis	E	2
Laodelphax striatellus	D	*Anagrus* sp.		E	14
			Agonatopoides solidus	E	2
			Echthrodelphax hortuensis	E	2
			Haplogonatopus oratorius	E	2
			Stenocranophilus anomalocerus	E	2
Megadelphax quadrimaculatus	D		*Elenchus tenuicornis*	E	2
Megadelphax sordidulus	D	*Anagrus* sp.		E	14
			Dicondylus bicolor	E	2
			Dicondylus dichromus	E	2

			Echthrodelphax hortuensis	E	2
			Haplogonatopus oratorius	E	2
			Pseudogonatopus rosellae	E	2
			Elenchus tenuicornis	E	2
			Agonatopoides solidus	E	2
Megadelphax sp.	D				
			Pseudogonatopus augustae	E	2
			Pseudogonatopus septemdentatus	E	2
Metadelphax propinqua	D				
				E	2
Muellerianella brevipennis	D	*Anagrus* sp.			
			Elenchus tenuicornis	E	2
				E	2
Muellerianella fairmairei	D	*Anagrus ensifer*			
				E	2
		Anagrus incarnatus			
				E	2
		Anagrus mutans			
				E	2
		Anagrus sp.			
			Donisthorpina pallida	E	2
			Elenchus tenuicornis	E	2
				E	2
Muellerianella sp.	D	*Anagrus* sp. "b"			
				E	2
		Anagrus sp. "g"			
			Agonatopoides solidus	E	2
Muirodelphax aubei	D				
			Elenchus tenuicornis	E	2
Muirodelphax denticauda	D				
			Elenchus tenuicornis	E	2
Paraliburnia adela	D				
			Elenchus tenuicornis	E	2
Ribautodelphax albostriatus	D				
			Pseudogonatopus distinctus	E	2
Ribautodelphax angulosus	D				
			Elenchus tenuicornis	E	2
			Dicondylus bicolor	E	2
Ribautodelphax collinus	D				
			Elenchus tenuicornis	E	2
			Echthrodelphax hortuensis	E	2
Ribautodelphax imitans	D				
			Pseudogonatopus albosignatus	E	2
			Pseudogonatopus ligusticus	E	2
			Elenchus imitans	E	2

(*continued*)

APPENDIX 1. (*Continued*)

Host Species	Family[a]	Egg Parasitoid	Nymphal Parasitoid	Geographic Location[b]	Reference[c]
Ribautodelphax pungens	D		*Dicondylus bicolor*	E	2
Ribautodelphax sp.	D		*Dicondylus bicolor*	E	2
Stegelytra putoni	D		*Elenchus tenuicornis*	E	2
Stenocranus major	D	*Anagrus incarnatus*		E	2
Stenocranus minutus	D	*Anagrus incarnatus*		E	2
		Anagrus stenocrani		E	2
			Cephalops curtifrons	E	2
			Elenchus tenuicornis	E	2
Stiroma affinis	D		*Elenchus tenuicornis*	E	2
Stiroma bicarinata	D	*Anagrus* sp.		E	14
			Dicondylus bicolor	E	2
			Elenchus tenuicornis	E	2
Stiroma sp.	D		*Cephalops carinatus*	E	2
			Cephalops obtusinervis	E	2
			Cephalops sp.	E	2
			Dicondylus bicolor	E	2
Struebingianella dalei	D		*Pseudogonatopus distinctus*	E	2
			Elenchus tenuicornis	E	2
Struebingianella lugubrina	D	*Anagrus incarnatus*		E	2
			Elenchus tenuicornis	E	2
Unkanodes excisa	D		*Dicondylus bicolor*	E	2
			Dicondylus dichromus	E	2
Xanthodelphax flaveolus	D	*Anagrus* sp.		E	14
			Elenchus tenuicornis	E	2
Xanthodelphax stramineus	D		*Donisthorpina pallida*	E	2
			Elenchus tenuicornis	E	2

Harmalia albicollis	D	*Anagrus incarnatus*		R	5
Laodelphax striatellus	D	*Anagrus incarnatus*		R	5
		Anagrus optabilis		R	3
		Oligosita sp.		R	14
		Mymar sp.		R	4
			Elenchus japonicus	R	6
Nilaparvata bakeri	D	*Anagrus incarnatus*		R	5
		Anagrus flaveolus		R	3
Nilaparvata lugens	D	*Paracentrobia andoi*		R	4
		Anagrus incarnatus		R	5
		Anagrus flaveolus		R	3
		Anagrus optabilis		R	3
		Gonatocerus sp.		R	14
		Mymar sp.		R	14
			Elenchus japonicus	R	6
			Pseudogonatopus sp.	R	8
			Haplogonatopus sp.	R	8
			Echthrodelphax fairchildii	R	8
			Elenchus yusamatsui	R	8
Nilaparvata muiri	D	*Anagrus incarnatus*		R	5
		Anagrus optabilis		R	3
			Elenchus japonicus	R	6
Prokelisia marginata	D	*Anagrus delicatus*		R	13
			Elenchus koebelei	R	13
			Dryinidae	R	13
Saccharosydne procerus	D	*Anagrus optabilis*		R	3
		Anagrus nigriventris		R	3
Sogatella furcifera	D	*Anagrus incarnatus*		R	5
		Anagrus optabilis		R	3

(*continued*)

APPENDIX 1. (*Continued*)

Host Species	Family[a]	Egg Parasitoid	Nymphal Parasitoid	Geographic Location[b]	Reference[c]
		Anagrus flaveolus		R	3
		Gonatocerus sp.		R	14
			Elenchus japonicus	R	6
			Pseudogonatopus sp.	R	8
			Haplogonatopus sp.	R	8
			Echthrodelphax fairchildii	R	8
			Elenchus yusamatsui	R	8
Sogatella longifurcifera	D	*Anagrus incarnatus*		R	5
			Elenchus japonicus	R	6
Sogatella panicola	D	*Anagrus incarnatus*		R	5
		Anagrus frequens		R	3
		Anagrus flaveolus		R	3
		Anagrus panicicolae		R	3
			Elenchus japonicus	R	6
Stenocranus minutus	D	*Anagrus flaveolus*		R	3
Terthron albovittatum	D	*Anagrus incarnatus*		R	5
Zuleica nipponica	D	*Anagrus incarnatus*		R	5
		Anagrus optabilis		R	3
		Anagrus nigriventris		R	3
Adarrus geniculatus	C		*Gonatopus sepsoides*	E	2
			Halictophagus languedoci	E	2
Adarrus ocellaris	C		*Lonchodryinus ruficornis*	E	2
			Gonatopus sepsoides	E	2
			Tomosvaryella sylvatica	E	2
			Pipunculus campestris	E	2

		Eudorylas fascipes	E	2
		Eudorylas jenkinsoni	E	2
		Eudorylas obliquus	E	2
		Eudorylas obscurus	E	2
		Eudorylas subfascipes	E	2
		Eudorylas subterminalis	E	2
Adarrus taurus	C	*Gonatopus distinguendus*	E	2
		Gonatopus lunatus	E	2
		Gonatopus sepsoides	E	2
		Gonatopus spectrum	E	2
		Gonatopus pulicarius	E	2
		Tetrodontochelys pulicarius	E	2
		Halictophagus languedoci	E	2
Adarrus sp.	C	*Gonatopus lunatus*	E	2
Agallia consorbrina	C	*Halictophagus agalliae*	E	2
Agallia laevis	C	*Halictophagus agalliae*	E	2
Aguriatiara germani	C	*Aphelopus melaleucus*	E	2
Alebra albostriella	C	*Aphelopus atratus*	E	2
		Aphelopus serratus	E	2
		Chalarus parmenteri	E	2
		Chalarus sp. B nr. *spurius*	E	2
Alebra wahlbergi	C	*Aphelopus atratus*	E	2
Alnetoidea alneti	C	*Aphelopus melaleucus*	E	2
		Aphelopus serratus	E	2
		Chalarus exiguus	E	2
		Chalarus fimbriatus	E	2
Aphrodes bicinctus	C	*Gonatopus bilineatus*	E	2
		Gonatopus striatus	E	2

(*continued*)

APPENDIX 1. (*Continued*)

Host Species	Family[a]	Egg Parasitoid	Nymphal Parasitoid	Geographic Location[b]	Reference[c]
Aphrodes sp.	C	*Gonatocerus litoralis*		E	2
		Polynema sp.		E	14
			Gonatopus striatus	E	2
Araldus propinquus	C		*Gonatopus distinguendus*	E	2
			Gonatopus sepsoides	E	2
			Gonatopus spectrum	E	2
			Halictophagus languedoci	E	2
Arocephalus longuidus	C		*Gonatopus sepsoides*	E	2
Arocephalus punctum	C		*ınteon pubicorne*	E	2
			Gonatopus sepsoides	E	2
Arocephalus sagittarius	C		*Gonatopus sepsoides*	E	2
			Halictophagus languedoci	E	2
Arthraldus pasceullus	C		*Lonchodryinus ruficornis*	E	2
			Anteon pubicorne	E	2
			Gonatopus lunatus	E	2
			Gonatopus sepsoides	E	2
			Tomosvaryella kuthyi	E	2
			Dorylomorpha xanthopus	E	2
			Pipunculus campestris	E	2
			Eudorylas fascipes	E	2
			Eudorylas jenkinsoni	E	2
			Eudorylas obliquus	E	2
			Eudorylas obscurus	E	2
			Eudorylas subfascipes	E	2
			Eudorylas subterminalis	E	2

Athysanus argentarius	C		*Pipunculus* sp.	E	2
Balclutha punctata	C		*Eudorylas fuscullus*	E	2
			Eudorylas opacus	E	2
Batrachomorphus sp.	C		*Neocladia tibialis*	E	10
Chloriata sp.	C		*Aphelopus camus*	E	2
Cicadella viridis	C	*Anagrus atomus*		E	2
		Anagrus sp. nr. *atomus*		E	2
		Anagrus incarnatus		E	2
		Anagrus mutans		E	2
		Anagrus silwoodensis		E	2
		Anagrus sp. "a"		E	2
		Anagrus sp. "g"		E	2
		Gonatocerus ater		E	2
		Gonatocerus cicadellae		E	2
		Polynema woodi		E	2
		Polynema sp.		E	2
		Ooctonus sp.		E	14
Cicadula albigensis	C		*Dorylomorpha hungarica*	E	2
Cicadula flori	C		*Dorylomorpha hungarica*	E	2
Cicadula frontalis	C	*Anagrus* sp.		E	14
			Dorylomorpha hungarica	E	2
Cicadula quadrinotata	C		*Dorylomorpha hungarica*	E	2
			Pipunculus campestris	E	2
			Eudorylas fascipes	E	2
			Eudorylas montium	E	2
			Eudorylas obliquus	E	2
Cicadula saturata	C		*Dorylomorpha hungarica*	E	2
Cicadula sp.	C		*Gonatopus formicarius*	E	2

(*continued*)

APPENDIX 1. (*Continued*)

Host Species	Family[a]	Egg Parasitoid	Nymphal Parasitoid	Geographic Location[b]	Reference[c]
Conosanus obsoletus	C		*Dorylomorpha xanthopus*	E	2
			Pipunculus campestris	E	2
			Lonchodryinus ruficornis	E	2
Deltocephalus assimilis	C		*Gonatopus sepsoides*	E	2
Deltocephalus sp.	C		*Gonatopus distinguendus*	E	2
			Mystrophorus formicaeformis	E	2
Diplocolenus abdominalis	C		*Gonatopus formicarius*	E	2
			Gonatopus sepsoides	E	2
			Mystrophorus formicaeformis	E	2
			Tomosvaryella palliditarsis	E	2
			Tomosvaryella sylvatica	E	2
			Dorylomorpha xanthopus	E	2
			Pipunculus campestris	E	2
			Pipunculus sp.	E	2
			Eudorylas obscurus	E	2
			Eudorylas subterminalis	E	2
			Eudorylas sp.	E	2
Diplocolenus nigrifrons	C		*Gonatopus formicarius*	E	2
Doliotettix lunulatus	C		*Eudorylas* sp.	E	2
Doratura stylata	C		*Pipunculus fonsecae*	E	2
Doratura sp.	C		*Gonatopus graecus*	E	2
Edwardsiana avellanae	C		*Aphelopus melaleucus*	E	2
Edwardsiana bergmani	C		*Aphelopus melaleucus*	E	2
			Chalarus sp. A nr. *spurius*	E	2
Edwardsiana crataegi	C		*Aphelopus atratus*	E	2

			Aphelopus melaleucus	E	2
			Aphelopus serratus	E	2
Edwardsiana flavescens	C		*Aphelopus melaleucus*	E	2
Edwardsiana geometrica	C		*Aphelopus melaleucus*	E	2
			Aphelopus serratus	E	2
			Chalarus pughi	E	2
			Chalarus sp. A nr. *spurius*	E	2
Edwardsiana hippocastani	C		*Aphelopus atratus*	E	2
			Aphelopus melaleucus	E	2
			Chalarus sp. A nr. *spurius*	E	2
Edwardsiana lethierryi	C		*Aphelopus atratus*	E	2
			Aphelopus melaleucus	E	2
			Aphelopus serratus	E	2
Edwardsiana menzbieri	C		*Aphelopus melaleucus*	E	2
Edwardsiana plebeja	C		*Aphelopus melaleucus*	E	2
Edwardsiana prunicola	C	*Anagrus nigriceps*		US	1
		Anagrus epos		US	1
Edwardsiana rosae	C	*Anagrus atomus*		E	2
		Anagrus bartheli		E	2
			Aphelopus atratus	E	2
			Aphelopus melaleucus	E	2
			Chalorus sp. A nr. *spurius*	E	2
Edwardsiana sp.	C		*Aphelopus atratus*	E	2
			Aphelopus melaleucus	E	2
			Aphelopus serratus	E	2
			Chalarus fimbriatus	E	2
			Chalarus pughi	E	2
			Chalarus sp. A nr. *spurius*	E	2

(*continued*)

APPENDIX 1. (*Continued*)

Host Species	Family[a]	Egg Parasitoid	Nymphal Parasitoid	Geographic Location[b]	Reference[c]
Elymana sulphurella	C	*Anagrus* sp.		E	14
			Lonchodryinus ruficornis	E	2
			Gonatopus sepsoides	E	2
			Pipunculus campestris	E	2
			Pipunculus thomsoni	E	2
			Eudorylas fuscipes	E	2
Empoasca vitis	C	*Anagrus* sp. nr. *atomus*		E	2
		Anagrus sp.		E	2
		Stethynium triclavatum		E	2
			Aphelopus atratus	E	2
			Aphelopus camus	E	2
			Aphelopus melaleucus	E	2
			Aphelopus nigriceps	E	2
			Aphelopus serratus	E	2
			Chalarus sp. A nr. *parmenteri*	E	2
			Chalarus sp. nr. *pughi*	E	2
			Chalarus spurius	E	2
			Chalarus sp. A nr. *spurius*	E	2
			Chalarus sp. C nr. *spurius*	E	2
			Chalarus sp. D nr. *spurius*	E	2
Empoasca sp.	C		*Aphelopus melaleucus*	E	2
Enantrocephalus cornutus	C		*Gonatopus lunatus*	E	2
Erythyria aureola	C		*Chalarus* sp. A nr. *spurius*	E	2
Eupelix cuspidata	C		*Halictophagus* sp.	E	2
Eupterycyba jucunda	C		*Aphelopus serratus*	E	2

Eupteryx aurata	C	*Anagrus* sp. nr. *atomus*		E	2
			Aphelopus atratus	E	2
			Chalarus pughi	E	2
			Chalarus spurius	E	2
Eupteryx cyclops	C	*Anagrus* sp. nr. *atomus*		E	2
			Aphelopus atratus	E	2
			Chalarus spurius	E	2
Eupteryx mellissae	C		*Aphelopus atratus*	E	2
			Chalarus sp. A nr. *spurius*	E	2
Eupteryx stachydearum	C		*Aphelopus atratus*	E	2
Eupteryx urticae	C	*Anagrus* sp. nr. *atomus*		E	2
			Aphelopus atratus	E	2
			Chalarus fimbriatus	E	2
			Chalarus pughi	E	2
			Chalarus spurius	E	2
Eurhadina concinna	C		*Aphelopus nigriceps*	E	2
			Chalarus sp. B nr. *spurius*	E	2
Eurhadina pulchella	C		*Chalarus* sp. B nr. *spurius*	E	2
Euscelis incisus	C		*Lonchodryinus ruficornis*	E	2
			Anteon pubicorne	E	2
			Gonatopus lunatus	E	2
			Gonatopus sepsoides	E	2
			Dorylomorpha xanthopus	E	2
			Pipunculus campestris	E	2
			Eudorylas ruralis	E	2
			Eudorylas subterminalis	E	2
			Eudorylas sp.	E	2
Eucselis sp.	C		*Gonatopus sepsoides*	E	2
Fagocyba carri	C		*Aphelopus melaleucus*	E	2
			Aphelopus serratus	E	2

(*continued*)

APPENDIX 1. (*Continued*)

Host Species	Family[a]	Egg Parasitoid	Nymphal Parasitoid	Geographic Location[b]	Reference[c]
Fagocyba cruenta	C		*Aphelopus melaleucus*	E	2
			Aphelopus atratus	E	2
			Aphelopus serratus	E	2
			Chalarus parmenteri	E	2
			Chalarus sp. A nr. *spurius*	E	2
Fagocyba sp.	C		*Aphelopus atratus*	E	2
			Chalarus sp. A nr. *spurius*	E	2
			Chalarus sp. B nr. *spurius*	E	2
Goldeus harpago	C		*Gonatopus sepsoides*	E	2
Graphocraerus ventralis	C		*Pipunculus zugmayeriae*	E	2
Hauptidia moroccana	C	*Anagrus atomus*		E	2
Iassus lanio	C		*Anteon infectum*	E	2
			Anteon scapulare	E	2
Idiocerus confusus	C	*Polynema* sp. A		E	2
		Gonatocerus sp.		E	14
			Anteon flavicorne	E	2
Idiocerus distinguendus	C	*Polynema* sp. D		E	2
Idiocerus lituratus	C	*Polynema* sp. B		E	2
Idiocerus populi	C	*Gonatocerus* sp. nr. *maga*		E	2
		Gonatocerus tremulae		E	2
		Polynema atratum		E	2
			Verrallia aucta	E	2
			Eudorylas sp.	E	2
Idiocerus stigmaticalis	C	*Polynema* sp. C		E	2
			Anteon flavicorne	E	2

Idiocerus sp.	C	*Gonatocerus tremulae*		E	2
		Gonatocerus sp.		E	2
		Polynema euchariforme		E	2
		Polynema vitripennis		E	2
			Anteon flavicorne	E	2
Jassargus bisubulatus	C		*Lonchodryinus ruficornis*	E	2
Jassargus distinguendus	C		*Lonchodryinus ruficornis*	E	2
			Gonatopus sepsoides	E	2
			Tomosvaryella sylvatica	E	2
			Eudorylas subfascipes	E	2
			Eudorylas subterminalis	E	2
Jassargus flori	C		*Lonchodryinus ruficornis*	E	2
			Gonatopus sepsoides	E	2
Jassargus obtusivalis	C		*Lonchodryinus ruficornis*	E	2
			Gonatopus sepsoides	E	2
			Halictophagus languedoci	E	2
Jassargus repletus	C		*Gonatopus sepsoides*	E	2
Kybos butleri	C		*Chalarus latifrons*	E	2
Kybos smaragdula	C		*Aphelopus serratus*	E	2
			Chalarus latifrons	E	2
Lindbergina aurovittata	C		*Aphelopus serratus*	E	2
Linnavouriana decempunctata	C		*Aphelopus melaleucus*	E	2
Macropsis sp.	C		*Anteon ephippiger*	E	2
			Anteon gaullei	E	2
			Anteon pubicorne	E	2
Macrosteles laevis	C		*Lonchodryinus ruficornis*	E	2
			Tetrodontochelys pedestris	E	2
			Pipunculus campestris	E	2
			Eudorylas fuscipes	E	2
Macrosteles sexnotatus	C	*Anagrus atomus*		E	2

(*continued*)

APPENDIX 1. (*Continued*)

Host Species	Family[a]	Egg Parasitoid	Nymphal Parasitoid	Geographic Location[b]	Reference[c]
		Anagrus bolci		E	2
		Gonatocerus litoralis		E	2
		Gonatocerus sp. nr. *paludis*		E	2
			Anteon ephippiger	E	2
			Anteon pubicorne	E	2
			Pipunculus campestris	E	2
			Eudorylas fuscipes	E	2
Macrosteles variatus	C		*Eudorylas fuscipes*	E	2
Macrosteles viridigriseus	C		*Anteon pubicorne*	E	2
Macrosteles sp.	C		*Anteon pubicorne*	E	2
			Gonatopus lunatus	E	2
			Eudorylas fascipes	E	2
Megamelus notula	D	*Anagrus incarnatus*		E	2
Mocuellus collinus	C		*Gonatopus formicarius*	E	2
			Eudorylas subterminalis	E	2
Mocydia crocea	C		*Lonchodryinus ruficornis*	E	2
			Anteon ephippiger	E	2
			Anteon fulviventre	E	2
			Anteon pubicorne	E	2
			Eudorylas longifrons	E	2
			Eudorylas obliquus	E	2
Oncopsis alni	C	*Polynema euchariforme*		E	2
			Verrallia beatricis	E	2
Oncopsis flavicollis	C	*Polynema euchariforme*		E	2

			Anteon brachycerum	E	2
			Anteon juvineanum	E	2
			Verrallia setosa	E	2
Oncopsis subangulata	C	*Polynema euchariforme*		E	2
			Verrallia pilosa	E	2
			Verallia setosa	E	2
Oncopsis tristis	C	*Polynema bakkendorfi*		E	2
Oncopsis sp.	C		*Anteon brachycerum*	E	2
			Verrallia pilosa	E	2
Opsius stactogalus	C	*Polynema* sp.		E	14
			Tomosvaryella frontata	E	2
Ossiannilissonola callosa	C		*Aphelopus melaleucus*	E	2
Paluda elongata	C		*Gonatopus distinguendus*	E	2
			Gonatopus sepsoides	E	2
			Gonatopus spectrum	E	2
Populicerus albicans	C		*Anteon flavicorne*	E	2
Populicerus confusus	C		*Anteon flavicorne*	E	2
Populicerus laminatus	C		*Anteon flavicorne*	E	2
Populicerus populi	C		*Anteon flavicorne*	E	2
Psammotettix alienaus	C		*Gonatopus formicarius*	E	2
			Gonatopus lunatus	E	2
			Gonatopus sepsoides	E	2
Psammotettix cephalotes	C		*Lonchodryinus ruficornis*	E	2
Psammotettix confinis	C		*Lonchodryinus ruficornis*	E	2
			Anteon flavicorne	E	2
			Anteon pubicorne	E	2
			Gonatopus formicarius	E	2
			Gonatopus graecus	E	2
			Gonatopus sepsoides	E	2
			Tomosvaryella kuthyi	E	2

(*continued*)

APPENDIX 1. (*Continued*)

Host Species	Family[a]	Egg Parasitoid	Nymphal Parasitoid	Geographic Location[b]	Reference[c]
			Tomosvaryella sylvatica	E	2
			Dorylomorpha xanthopus	E	2
			Pipunculus campestris	E	2
			Eudorylas fascipes	E	2
			Eudorylas obliquus	E	2
			Eudorylas obscurus	E	2
			Eudorylas subterminalis	E	2
			Eudorylas subfascipes	E	2
Psammotettix kolosvarensis	C		*Eudorylas subterminalis*	E	2
Psammotettix nodosus	C		*Lonchodryinus ruficornis*	E	2
			Anteon pubicorne	E	2
			Gonatopus sepsoides	E	2
			Eudorylas subfascipes	E	2
			Eudorylas subterminalis	E	2
Psammotettix notatus	C		*Gonatopus sepsoides*	E	2
			Halictophagus languedoci	E	2
Psammotettix putoni	C		*Gonatopus sepsoides*	E	2
			Halictophagus languedoci	E	2
Psammotettix striatus	C		*Anteon ephippiger*	E	2
			Gonatopus campestris	E	2
			Gonatopus distinguendus	E	2
			Gonatopus horvathi	E	2
			Gonatopus lunatus	E	2
			Gonatopus rhaensis	E	2
			Gonatopus sepsoides	E	2
			Halictophagus languedoci	E	2

Recilia coronifera	C	*Gonatopus sepsoides*	E	2
Rhytidodus decimaquartus	C	*Anteon arcruetum*	E	2
		Anteon flavicorne	E	2
Ribautiana tenerrima	C	*Aphelopus atratus*	E	2
		Aphelopus serratus	E	2
		Chalarus sp. A nr. *spurius*	E	2
Ribautiana ulmi	C	*Aphelopus atratus*	E	2
		Aphelopus melaleucus	E	2
		Chalarus sp. A nr. *spurius*	E	2
Sorboanus assimilis	C	*Gonatopus sepsoides*	E	2
		Dorylomorpha platystylis	E	2
Sorboanus xanthoneurus	C	*Gonatopus sepsoides*	E	2
Speudotettix subfusculus	C	*Dorylomorpha maculata*	E	2
		Cephalops obtusinervis	E	2
		Cephalops sp.	E	2
		Eudorylas obliquus	E	2
Streptanus marginatus	C	*Dorylomorpha maculata*	E	2
Streptanus sordidus	C	*Lonchodryinus ruficornis*	E	2
		Anteon pubicorne	E	2
		Gonatopus sepsoides	E	2
Thamnotettix confinis	C	*Gonatopus sepsoides*	E	2
		Dorylomorpha maculata	E	2
		Eudorylas sp.	E	2
Thamnotettix dilutior	C	*Gonatopus bilineatus*	E	2
Thamnotettix sp.	C	*Gonatopus popovi*	E	2
Tremulicerus distinguendus	C	*Anteon flavicorne*	E	2
Turrutus socialis	C	*Gonatopus sepsoides*	E	2
		Eudorylas subterminalis	E	2
Typhlocyba bifasciata	C	*Aphelopus atratus*	E	2

(continued)

APPENDIX 1. (*Continued*)

Host Species	Family[a]	Egg Parasitoid	Nymphal Parasitoid	Geographic Location[b]	Reference[c]
Typhlocyba quercus	C	*Anagrus* sp.		E	4
			Aphelopus atratus	E	2
			Aphelopus melaleucus	E	2
			Aphelopus serratus	E	2
			Chalarus argenteus	E	2
			Chalarus parmenteri	E	2
			Chalarus sp. A nr. *spurius*	E	2
Typhlocyba sp.	C		*Aphelopus atratus*	E	2
			Aphelopus melaleucus	E	2
Ulopa reticulata	C		*Halictophagus silwoodensis*	E	2
Zygina flammigera	C		*Aphelopus atratus*	E	2
			Aphelopus melaleucus	E	2
Zygina suavis	C	*Anagrus atomus*		E	2
			Chalarus sp. B nr. *parmenteri*	E	2
Zygina sp.	C		*Aphelopus atratus*	E	2
			Aphelopus serratus	E	2
Aceratagallia neosignata	C		*Halictophagus* sp.	R	1
Athysanella sp.	C		*Pipunculus koebelei*	R	1
Austroagalloides sp.	C		*Halictophagus schwarzii*	R	1
Balclutha sp.	C		*Pipunculus anthracias*	R	1
Batrachomorphus sp.	C		*Pipunculus erinys*	R	1
			Pipunculus hylaeus	R	1
Bothrogonia ferruginea	C		*Halictophagus piperi*	R	1
			Halictophagus chinensis	R	1
Campbellinella fatigondus	C		*Pipunculus lamellifer*	R	1
Cicadella parthaon	C	*Ooctonus australensis*		R	1
			Halictophagus australensis	R	1

Cicadella spectra	C	*Ooctonus australensis*		R	1
			Halictophagus australensis	R	1
Cicadella sp.	C		*Halictophagus* sp.	R	1
Cicadulina mbila	C	*Anagrus cicadulinae*		R	1
Cicadulina pastusae	C		*Eudorylas absonditus*	R	1
Curtara pagina	C		*Membracixenos placula*	R	1
Dalbulus maidis	C		*Halictophagus* sp.	R	1
			Dryinidae	R	1
Delphacodes samestrimai	D	*Anagrus flaveolus*		R	3
Deltocephalus sp.	C		*Halictophagus* sp.	R	1
			Pipunculus beneficiens	R	1
			Gonatopus optabilis	R	1
			Gonatopus vitiensis	R	1
			Anteon pallidicornis	R	1
Divitracus hospes	?		*Tomosvaryella synadelphus*	R	1
Draeculacephala minerva	C	*Ootetrastichus beatus*		R	1
		Gonatocerus mexicanus		R	1
		Oligosita caerulocephala		R	1
		Paracentrobia lutea		R	1
			Halictophagus acutus	R	1
Empoasca fabialis	C	*Anagrus empoascae*		R	1
		Anaphes sp.		R	4
Empoasca facialis	C	*Anagrus scassellatii*		R	1
Empoasca minuenda	C	*Camptoptera minutissima*		R	1
Eurinoscopus sp.	C		*Anteon leiosomus*	R	1
			Anteon nitidus	R	1
			Anteon dimidiatus	R	1
Erythroneura mori	C	*Alaptus* sp.		R	1

(*continued*)

APPENDIX 1. (*Continued*)

Host Species	Family[a]	Egg Parasitoid	Nymphal Parasitoid	Geographic Location[b]	Reference[c]
Euscelis sp.	C		*Gonatopus koebeli*	R	1
			Gonatopus melanias	R	1
Eutettix sp.	C		*Halictophagus* sp.	R	1
			Anteon nigricornis	R	1
Graminella comata	C		*Halictophagus* sp.	R	1
Hecalus immaculatus	C		*Halictophagus phaeodes*	R	1
			Pipunculus cruciator	R	1
			Tomosvaryella pseudophanes	R	1
Hirozuunka japonica	C	*Anagrus flaveolus*		R	3
		Anagrus perforator		R	3
Idioscopus clypealis	C		*Halictophagus indicus*	R	1
			Pipunculus annulifemer	R	1
			Dryinidae	R	1
			Epipyrops fuliginosa	R	1
Idioscopus niveosparsus	C	*Centroda idioceria*		R	1
		Gonatocerus sp.		R	14
			Pipunculus annulifemer	R	1
			Dryinidae	R	1
			Epipyrops fuliginosa	R	1
Inazuma dorsalis	C		*Gonatopus andoi*	R	1
Ladoffa ignota	C		*Halictophagus schwarzii*	R	1
Ledra auditura	C		*Anteon esakii*	R	1
Macropsis sp.	C		*Anteon cognatus*	R	1
Macrosteles orientalis	C	*Anagrus incarnatus*		R	5
Macrosteles sp.	C		*Anteon coriaceus*	R	1
			Anteon parvulus	R	1
			Gonatopus vitiensis	R	1
			Gonatopus melanias	R	1

Nagra nagaragawana	C	*Anagrus flaveolus*		R	3
		Anagrus optabilis		R	3
		Anagrus incarnatus		R	3
Neodartus bellus	C	*Centroda penthimiae*		R	1
Nephotettix bipunctatus	C		*Gonatopus sakaii*	R	1
Nephotettix cincticeps	C	*Ooctonus* sp.		R	1
		Anagrus sp.		R	1
		Gonatocerus cincticeptis		R	1
		Gonatocerus miurai		R	1
		Anagrus flaveolus		R	1
		Anaphes sp.		R	1
		Gonatocerus sp.		R	1
		Mymar indica		R	1
		Paracentrobia andoi		R	1
		Oligosita shibuyae		R	1
		Oligosita nephotettixicum		R	1
		Oligosita sp.		R	1
			Halictophagus orientalis	R	1
			Eudorylas distocruciator	R	1
			Eudorylas mutillatus	R	1
			Tomosvaryella oryzaetora	R	1
			Eudorylas javaensis	R	1
			Eudorylas orientalis	R	1
			Eudorylas ruralis	R	1
			Tomosvaryella sylvatica	R	1
			Tomosvaryella subvirescens	R	1
			Dorylomorpha lini	R	1

(*continued*)

APPENDIX 1. (*Continued*)

Host Species	Family[a]	Egg Parasitoid	Nymphal Parasitoid	Geographic Location[b]	Reference[c]
			Eudorylas cruciator	R	1
			Tomosvaryella epichalea	R	1
Nephotettix nigropictus	C	*Gonatocerus* sp.		R	1
		Paracentrobia sp.		R	1
		Paracentrobia andoi		R	1
		Oligosita sp.		R	1
		Anagrus sp.		R	14
			Halictophagus munroei	R	8
			Tomosvaryella sp.	R	8
			Pipunculus sp.	R	8
			Pipunculus mutillatus	R	1
			Tomosvaryella oryzaetora	R	1
			Tomosvaryella subvirescens	R	1
Nephotettix plebius	C		*Gonatopus pulcherrimus*	R	1
			Gonatopus dubiosus	R	1
Nephotettix virescens	C	*Gonatocerus* sp.		R	1
		Paracentrobia sp.		R	1
		Paracentrobia andoi		R	1
		Anagrus sp.		R	14
			Halictophagus munroei	R	8
			Tomosvaryella sp.	R	8
			Pipunculus sp.	R	8
			Pipunculus mutillatus	R	1
			Tomosvaryella oryzaetora	R	1
			Tomosvaryella subvirescens	R	1

Nephotettix sp.	C	*Anagrus optabilis*		R	1
		Mymar taprobanicum		R	1
		Polynema sp.		R	1
		Gonatocerus sp.		R	1
		Anagrus sp.		R	1
		Paracentrobia yusamatsui		R	1
		Paracentrobia garuda		R	1
		Oligosita sp.		R	1
Oncometopia sp.	C		*Halictophagus obtusae*	R	1
Paradorydium menalus	C		*Halictophagus stenodes*	R	1
Phlepsius sp.	C		*Halictophagus* sp.	R	1
Phrynophyes sp.	C		*Pipunculus beneficiens*	R	1
			Dryinidae	R	1
Planicephalus flavicostus	C		*Halictophagus* sp.	R	1
			Pipunculidae	R	1
			Gonatopus ashmeadi	R	1
Psammotettix striatus	C		*Tomosvaryella itoi*	R	1
			Tomosvaryella sylvatica	R	1
Rhotidus sp.	C		*Pipunculus eucalypti*	R	1
Tartessus fulvus	C	*Pterygogramma acuminata*		R	1
Tartessus syrtidis	C		*Gonatopus gigas*	R	1
Tartessus sp.	C		*Pipunculus cruciator*	R	1
			Pipunculus picrodes	R	1
Thaumatoscopus sp.	C		*Pipunculus koebelei*	R	1
Thomsoniella areunta	C		*Halictophagus paradeniya*	R	1
Tituria chinensis	C		*Halictophagus chinensis*	R	1
Typhlocyba froggatti	C	*Anagrus armatus*		R	12
Xerophloea viridis	C		*Halictophagus insularum*	R	1
Aceratagallia accola	C		*Halictophagus americanus*	US	1

(continued)

APPENDIX 1. (*Continued*)

Host Species	Family[a]	Egg Parasitoid	Nymphal Parasitoid	Geographic Location[b]	Reference[c]
Aceratagallia californica	C		*Halictophagus americanus*	US	1
Aceratagallia curta	C		*Halictophagus americanus*	US	1
Aceratagallia fuscoscripta	C		*Halictophagus americanus*	US	1
Aceratagallia helveola	C		*Halictophagus americanus*	US	1
Aceratagallia sanguinolenta	C		*Halictophagus americanus*	US	1
			Pipunculidae	US	1
Aceratagallia uhleri	C		*Halictophagus americanus*	US	1
			Pipunculidae	US	1
Aceratagallia vulgaris	C		*Halictophagus* sp.	US	1
Aceratagallia sp.	C		*Halictophagus omani*	US	1
			Gonatopus sp.	US	1
			Pipunculidae	US	1
Agallia constricta	C		*Halictophagus serratus*	US	1
			Pipunculidae	US	1
Agallia quadripunctata	C		*Halictophagus americanus*	US	1
Agallia sp.	C		*Chalarus latifrons*	US	1
Amblysellus curtisi	C		*Gonatopus ashmeadi*	US	1
			Pipunculidae	US	1
Amritodes atkinsoni	C		*Halictophagus indicus*	R	1
			Pipunculus annulifemer	R	1
			Dryinidae	R	1
			Epipyrops fuliginosa	R	1
Aphrodes sp.	C		*Halictophagus* sp.	US	1
Arundanus sp.	C		*Pipunculus cressoni*	US	1
			Pipunculus arundani	US	1

			Dryinidae	US	1
Athysanella acuticauda	C		*Halictophagus mackayi*	US	1
Athysanella argenteola	C		Pipunculidae	US	1
			Dryinidae	US	1
Athysanella attenuata	C		Pipunculidae	US	1
			Dryinidae	US	1
Athysanella emarginata	C		*Halictophagus* sp.	US	1
			Pipunculidae	US	1
			Dryinidae	US	1
Athysanella reducta	C		Pipunculidae	US	1
			Dryinidae	US	1
Athysanella texana	C		*Halictophagus bidentatus*	US	1
			Tetrodontochelys sp.	US	1
Athysanella utahna	C		*Halictophagus bidentatus*	US	1
Athysanella sp.	C		*Halictophagus mackayi*	US	1
			Pipunculidae	US	1
			Gonatopus sp.	US	1
Balclutha impicta	C	*Paracentrobia subflava*	*Acrodontochelys americanus*	US	1
Balclutha neglecta	C		*Halictophagus* sp.	US	1
			Dryinidae	US	1
Ballana sp.	C		Pipunculidae	US	1
Ceratagallia bigeloviae	C		*Halictophagus* sp.	US	1
Ceratagallia vastitatis	C		*Halictophagus americanus*	US	1
Ceratagallia sp.	C		*Acrodontochelys mimus*	US	1
Chlorotettix glabanatus	C		Dryinidae	US	1
Chlorotettix spatulatus	C		Pipunculidae	US	1
Chlorotettix unicolor	C		*Anteon osborni*	US	1
Chlorotettix viridis	C		Dryinidae	US	1
Circulifer tenellus	C	*Polynema eutettixi*		US	1
		Anagrus giraulti		US	1

(*continued*)

APPENDIX 1. (*Continued*)

Host Species	Family[a]	Egg Parasitoid	Nymphal Parasitoid	Geographic Location[b]	Reference[c]
		Parallelaptera rex		US	1
		Gonatocerus capitaias		US	1
		Paracentrobia subflava		US	1
		Aphelinoidea plutella		US	1
		Aphelinoidea anatolica		US	1
		Erythmelus sp.		US	14
			Halictophagus americanus	US	1
			Tomosvaryella vagabunda	US	1
			Pipunculus subopacus	US	1
			Gonatopus sp.	US	1
Colladonus geminatus	C		*Tomosvaryella lepidipes*	US	1
Cuerna costalis	C	*Gonatocerus ashmeadi*		US	1
		Ufens niger		US	1
		Paracentrobia acuminata		US	1
			Halictophagus sp.	US	1
Deltocephalus fuscinervosus	C		*Neogonatopus pallidiceps*	US	1
Deltocephalus minutus	C	*Paracentrobia subflava*		US	1
Destria fumida	C		Dryinidae	US	1
Dikraneura mali	C		*Aphelopus* sp.	US	1
Dikraneura sp.	C		Pipunculidae	US	1
			Aphelopus arizonicus	US	1
Diplocolenus configurates	C		*Halictophagus mackayi*	US	1
Dorycephalus platyrhynchus	C	*Polynema longipes*		US	1
		Oligosita americana		US	1
			Halictophagus insularum	US	1

Dorycephalus sp.	C		Dryinidae	US	1
Draeculacephala antica	C		*Halictophagus* sp.	US	1
Draeculacephala delongi	C		*Halictophagus* sp.	US	1
Draeculacephala minor	C	*Paracentrobia acuminata*		US	1
Draeculacephala mollipes	C	*Gonatocerus mexicanus*		US	1
		Gonatocerus koebelei		US	1
		Ufens niger		US	1
		Paracentrobia acuminata		US	1
		Paracentrobia prima		US	1
			Halictophagus acutus	US	1
Draeculacephala protola	C	*Paracentrobia acuminata*		US	1
			Halictophagus sp.	US	1
Draeculacephala producta	C	*Gonatocerus koebelei*		US	1
Draeculacephala sp.	C		*Halictophagus acutus*	US	1
Driotura robusta	C		*Halictophagus* sp.	US	1
			Dryinidae	US	1
Edwardsiana commisuralis	C		*Aphelopus* sp.	US	1
Edwardsiana rosae	C	*Anagrus nigriceps*		US	1
		Anagrus armatus		US	1
Empoa albicans	C		Dryinidae	US	1
Empoasca fabae	C	*Anagrus armatus*		US	1
		Anagrus nigriventris		US	1
		Anagrus epos		US	1
		Polynema sp.		US	14
Empoasca solana	C	*Anagrus* sp.		US	1
Empoasca sp.	C	*Paracentrobia subflava*		US	1
			Chalarus latifrons	US	1
			Chalarus spurius	US	1
			Aphelopus sp.	US	1

(*continued*)

APPENDIX 1. (*Continued*)

Host Species	Family[a]	Egg Parasitoid	Nymphal Parasitoid	Geographic Location[b]	Reference[c]
Endria inimica	C		*Halictophagus mackayi*	US	1
			Pipunculidae	US	1
			Gonatopus ashmeadi	US	1
Erythroneura comes	C	*Anagrus epos*		US	1
			Aphelopus comesi	US	1
Erythroneura obliqua	C		*Aphelopus* sp.	US	1
Erythroneura plena	C	*Anagrus epos*		US	1
Erythroneura tricincta	C		*Aphelopus* sp.	US	1
Erythroneura vitis	C	*Polynema striaticorne*		US	1
		Paranagrus sp.		US	1
		Anagrus sp.		US	1
Erythroneura vulnerata	C		*Aphelopus comesi*	US	1
Erythroneura sp.	C		*Aphelopus microleucus*	US	1
			Aphelopus pulcherrimus	US	1
Exitianus exitrosus	C		*Halictophagus* sp.	US	1
			Pipunculidae	US	1
			Dryinidae	US	1
Flexamia abbreviata	C		*Halictophagus* sp.	US	1
			Dryinidae	US	1
Flexamia arizonensis	C		Dryinidae	US	1
Flexamia atlantica	C		*Halictophagus* sp.	US	1
			Pipunculidae	US	1
			Dryinidae	US	1
Flexamia curvata	C		*Halictophagus* sp.	US	1
			Dryinidae	US	1

Flexamia frigidus	C	*Halictophagus bidentatus*	US	1
Flexamia graminea	C	*Halictophagus* sp.	US	1
Flexamia inflata	C	Dryinidae	US	1
Flexamia pectinata	C	*Halictophagus mackayi*	US	1
		Dryinidae	US	1
Flexamia picta	C	*Halictophagus bidentatus*	US	1
		Pipunculidae	US	1
		Dryinidae	US	1
Flexamia prairiana	C	*Halictophagus mackayi*	US	1
		Dryinidae	US	1
Flexamia reflexa	C	*Halictophagus mackayi*	US	1
		Dryinidae	US	1
Forcipata loca	C	Pipunculidae	US	1
		Aphelopus arizonicus	US	1
Forcipata sp.	C	*Pipunculus ater*	US	1
Frigartus frigidus	C	*Halictophagus bidentatus*	US	1
Gilletiella atropuncta	C	Pipunculidae	US	1
		Dryinidae	US	1
Gilletiella sp.	C	Dryinidae	US	1
Graminella mohri	C	*Halictophagus bidentatus*	US	1
		Halictophagus mackayi	US	1
		Pipunculidae	US	1
		Dryinidae	US	1
Graminella nigrifrons	C	*Gonatopus ashmeadi*	US	9
		Neogonatopus agropyrus	US	9
		Tomosvaryella sylvatica	US	9
		Pipunculus ater	US	9
		Pipunculus aeguus	US	9

(*continued*)

APPENDIX 1. (*Continued*)

Host Species	Family[a]	Egg Parasitoid	Nymphal Parasitoid	Geographic Location[b]	Reference[c]
			Halictophagus sp.	US	9
Graminella sonora	C	*Paracentrobia subflava*		US	1
			Halictophagus sp.	US	1
			Gonatopus sonora	US	1
Graminella villica	C		*Halictophagus* sp.	US	1
Graphocephala confluens	C	*Ufens niger*		US	1
Graphocephala versuta	C	*Gonatocerus novifasciatus*		US	1
			Tomosvaryella subvirescens	US	1
Gyponana sp.	C	*Gonatocerus marilandicus*		US	1
Hardya dentata	C		Dryinidae	US	1
Hebecephalus rostratus	C		Dryinidae	US	1
Hecalus flavidus	C		Dryinidae	US	1
Hecalus viridis	C		Dryinidae	US	1
Hecalus sp.	C		Dryinidae	US	1
			Tomosvaryella brevijuncta	US	1
			Pipunculidae	US	1
			Gonatopus sp.	US	1
Helochara communis	C	*Trichogramma minutum*		US	1
Homalodisca coagulata	C	*Cosmocomoidea morrilli*		US	1
		Gonatocerus ashmeadi		US	1
		Gonatocerus fasciatus		US	1
		Anagrus giraulti		US	1
		Ufens spiritus		US	1
Homalodisca insolita	C	*Paracentrobia acuminata*		US	1

		Oligosita americana		US	1
		Ufens niger		US	1
		Acmopolynema sp.		US	14
		Gonatocerus sp.		US	14
Hymetta trifasciata	C		*Aphelopus bicolor*	US	1
Idiocerus alternatus	C		Dryinidae	US	1
Idiocerus apache	C		*Halictophagus* sp.	US	1
Idiocerus rotundens	C		*Halictophagus* sp.	US	1
			Dryinidae	US	1
Idiocerus snowi	C		Dryinidae	US	1
Idiocerus sp.	C		*Halictophagus callosus*	US	1
			Chalarus sp.	US	1
			Dryinidae	US	1
			Anteon cognatus	US	1
Idiodonus heidemanni	C		Dryinidae	US	1
Laevicephalus acus	C		Dryinidae	US	1
Laevicephalus parvulus	C		*Halictophagus* sp.	US	1
			Pipunculidae	US	1
			Dryinidae	US	1
Laevicephalus unicoloratus	C		*Halictophagus* sp.	US	1
Laevicephalus sp.	C		*Halictophagus mackayi*	US	1
			Pipunculidae	US	1
			Dryinidae	US	1
Latalus sayi	C		Pipunculidae	US	1
			Neogonatopus ombrodes	US	1
			Neogonatopus contortulus	US	1
Limnotettix sp.	C		Dryinidae	US	1
Macrosteles lepidus	C		Dryinidae	US	1

(*continued*)

APPENDIX 1. (*Continued*)

Host Species	Family[a]	Egg Parasitoid	Nymphal Parasitoid	Geographic Location[b]	Reference[c]
Macrosteles quadrilineatus	C		*Halictophagus mackayi*	US	1
			Tomosvaryella sylvatica	US	1
			Gonatopus ashmeadi	US	1
			Tetrodontochelys plesius	US	1
Mesamia coloradensis	C		*Halictophagus mackayi*	US	1
			Dryinidae	US	1
Mocuellus caprillus	C		*Halictophagus* sp.	US	1
Mocuellus collinus	C		Pipunculidae	US	1
			Dryinidae	US	1
Nanopsis distinctus	C		Dryinidae	US	1
Nesophrosyne maritima	C	*Paracentrobia perditrix*		US	1
Nesophrosyne sp.	C		*Pipunculus juvator*	US	1
			Pipunculus obscuratus	US	1
			Pipunculus timberlakei	US	1
			Pipunculus ulube	US	1
			Pipunculus macrothrix	US	1
Norvellina seminuda	C		Dryinidae	US	1
Oncometopia orbona	C	*Gonatocerus ashmeadi*		US	1
		Ufens spiritus		US	1
			Halictophagus oncometopiae	US	1
Oncopsis sobrius	C	*Polynema striaticorne*		US	1
Opsius stactogalus	C	*Barypolynema saga*		US	1
Orocastus perpusillus	C		Dryinidae	US	1
Paraphlepsius irroratus	C		*Halictophagus* sp.	US	1
			Pipunculus ater	US	1

			Dryinidae	US	1
Paraphlepsius sp.	C		*Gonatopus pseudochromus*	US	1
Polyamia satur	C		Dryinidae	US	1
Polyamia weedi	C		*Gonatopus ashmeadi*	US	1
Polyamia sp.	C		*Halictophagus* sp.	US	1
			Dryinidae	US	1
Ponana pectoralis	C		Dryinidae	US	1
Psammotettix lividellus	C		*Halictophagus bidentatus*	US	1
			Halictophagus mackayi	US	1
			Pipunculidae	US	1
			Gonatopus ashmeadi	US	1
			Neogonatopus agropyrus	US	1
Rytidodus decimaquartus	C	*Gonatocerus ovicenatus*		US	1
Scaphoideus paludosus	C		*Neogonatopus niger*	US	1
Scaphytopius acutus	C		*Pipunculus subopacus*	US	1
Scaphytopius nitridus	C		*Tomosvaryella appendipes*	US	1
			Pipunculidae	US	1
Sorhoanus orientalis	C		Dryinidae	US	1
Sorhoanus uhleri	C		*Halictophagus mackayi*	US	1
Stirellus bicolor	O		*Halictophagus mackayi*	US	1
			Halictophagus sp.	US	1
			Dryinidae	US	1
Thamnotettix sp.	C		*Dorylomorpha tridentata*	US	1
Typhlocyba pomaria	C	*Anagrus epos*		US	1
		Anagrus nigriventris		US	1
		Anagrus armatus		US	1
			Aphelopus typhlocybae	US	1
Typhlocyba quercus	C	*Anagrus epos*		US	1
		Anagrus nigriceps		US	1

(continued)

APPENDIX 1. (*Continued*)

Host Species	Family[a]	Egg Parasitoid	Nymphal Parasitoid	Geographic Location[b]	Reference[c]
Xerophloea vanduzeei	C		*Halictophagus insularum*	US	1
Xerophloea sp.	C		Dryinidae	US	1
Cixius nervosus	O		*Dryinus collaris*	E	2
Cixius contaminatus	O		*Dryinus collaris*	E	2
Caliscelis sp.	O		*Bocchus europaeus*	E	2
Dictyophora europaea	O		*Dryinus terraconensis*	E	2
Fulgora europaea	O		*Halictophagus kuehnelti*	E	2
Hysteropterum flavescens	O		*Richardsidryinus corsicus*	E	2
Hysteropterum latifrons	O		*Dryinus sanderi*	E	2
Issus coleoptratus	O		*Dryinus collaris*	E	2
Issus lavri	O		*Halictophagus* sp.	E	2
Neophilaenus lineatus	O		*Verrallia aucta*	E	2
Philaenus spumarius	O	*Ooctonus* sp.		E	14
			Verrallia aucta	E	2
Strictocephala bisonia	O	*Polynema striaticorne*		E	2
Tachycixius pilosus	O		*Dryinus collaris*	E	2
			Cephalops furcatus	E	2
Tettigometra concolor	O		*Halictophagus tettigometrae*	E	2
Tettigometra impressiformis	O		*Halictophagus tettigometrae*	E	2
Tettigometra impressopunctata	O		*Halictophagus tettigometrae*	E	2
Tettigometra obliqua	O		*Halictophagus tettigometrae*	E	2
Tettigometra pincta	O		*Halictophagus tettigometrae*	E	2
Aetalion reticulatum	O	*Abelloides marquesi*		R	1
		Lathromerella sp.		R	1
Carynota mera	O		Dryinidae	US	1

Ceresa malina	O		*Membracixenos desantisi*	R	1
Ceresa nigripectus	O		*Membracixenos desantisi*	R	1
Dichoptera hyalinata	O		*Dryinus* sp. (A)	R	7
Enchenopa binotata	O	*Polynema consabrinus*		R	1
		Polynema enchenopae		R	1
		Gonatocerus sp.		R	14
Eurybrachys tomentosa	O	*Proleurocerus fulgoridis*		R	7
		Tetrastichus sp.		R	7
			Dryinus sp. B	R	7
Glossonatus univittatus	O		Dryinidae	US	1
Ipo sp.	O		*Anteon myrmecophilus*	R	1
Membracis tectigera	O	*Uscanopsis carlylei*		R	1
Otinotus sp.	O	*Centroda azizi*		R	1
Oxyrachis sp.	O	*Centrodora mumtazi*		R	1
Platycotis tuberculata	O	*Paracentrobia platycotis*		R	1
Pyrilla perpusilla	O	*Ooencyrtus papilionis*		R	11
Strictocephala bifasciata	O		*Membracixenos desantisi*	R	1
Strictocephala borealis	O	*Paracentrobia ceresarus*		US	1
Strictocephala bisonia	O	*Polynema howardi*		US	1
		Polynema striaticorne		US	1
		Gonatocerus dolichocerus		US	1
		Paracentrobia ceresarus		US	1
Strictocephala festina	O	*Polynema imitatrix*		US	1
		Gonatocerus ornatus		US	1
		Paracentrobia perditrix		US	1
Strictocephala inermis	O	*Polynema striaticorne*		US	1
		Gonatocerus sp.		US	1
		Paracentrobia ceresarum		US	1
		Tetrastichus sp.		US	1
Strictocephala taurina	O	*Polynema striaticorne*		US	1

(*continued*)

APPENDIX 1. (*Continued*)

Host Species	Family[a]	Egg Parasitoid	Nymphal Parasitoid	Geographic Location[b]	Reference[c]
		Paracentrobia platycotis		US	1
Telamona sp.	O		Dryinidae	US	1
Thelia bimaculata	O		*Aphelopus theliae*	US	1

[a]Family: C = Cicadellidae; D = Delphacidae; O = other (e.g., Aetalionidae, Membracidae, Dictyopharidae, Eurybrachidae, Eurymelidae, Fulgoridae).
[b]Geographic location: E = Europe; US = North America, usually the United States, more rarely Canada; R = rest of world (e.g., Australia, South America, Asia, West Indies).
[c]References:
1 = Freytag (1985)
2 = Waloff and Jervis (1987)
3 = Sahad and Hirashima (1984)
4 = Miura et al. (1981)
5 = Chantarasa-ard and Hirashima (1984)
6 = Kifune (1986)
7 = Swaminathan and Ananthakrishnan (1984)
8 = Peña and Shepard (1986)
9 = Freytag (1986)
10 = Quartau (1981)
11 = Yadav and Chaudhary (1984)
12 = Teulon and Penman (1986)
13 = Stiling and Strong (1982b)
14 = Huber (1986)

Note that Huber (1986) lists records only of parasitoid genera, not of parasitoid species. For example, *Nephotettix cincticeps* is attacked by at least two species of *Gonatocerus*; Huber lists only the genus *Gonatocerus* (once). *Cicadella viridis* is attacked by seven species of *Anagrus* (Waloff and Jervis 1987); Huber lists the genus only once: attacked by *Anagrus*. Huber's data, therefore, cannot be used to generate host–parasitoid ratios. However, for host species already listed, additional parasitoid genera recorded by Huber can be used as additional host records.

PART FIVE

Ecological Approaches to Planthopper Management

Earlier sections of this volume have reviewed the host plant relationships of planthoppers, their life history characteristics, population ecology, and interactions with natural enemies. Against this basic backdrop we now move on to consider those aspects of delphacid ecology, genetics, and behavior that interact with management practices to influence their status as agricultural pests. In particular, chapters in this section emphasize how research findings can point the way to ecologically sound management approaches. Such approaches primarily concern the use of natural enemies, resistant varieties, the judicial application of insecticides, and the interaction of these management tactics. The compatibility of these management practices, their probability for success given regional management and social constraints, and their shortcomings are emphasized. Given their threat to Asian agriculture, it is not surprising that rice planthoppers provide a recurrent basis for many discussions.

Benrey and Lamp (Chapter 14) begin with a review of classical biological control attempts against pest planthoppers using introduced predators and parasitoids. Interwoven is a discussion of the most important endemic natural enemies implicated in the suppression of pest planthoppers. The action of natural enemies, both predators and parasitoids, is examined in the context of ecological theory, and the conditions which maximize enemy potential for planthopper pest management are highlighted.

Following a direct treatment of natural enemies, the emphasis changes to the use of resistant varieties and insecticides as management tactics, but the interaction of these strategies with natural controls remains as a chronic theme. Roderick (Chapter 15) introduces the subject of host plant adaptation in planthoppers and, by default, arguments rely heavily on the database of virulence evolution in rice planthoppers to novel varieties. Evidence is sought for genetic-based variation in planthopper traits that influence host plant use. He continues with a discussion of the ecological and genetic factors that promote host plant adaptation and specialization.

Heinrichs (Chapter 16) follows with a review of the chronology of insecticidal approaches to rice planthopper control and the reasons for their failure in containing populations of *Nilaparvata lugens* in tropical Asia. In particular, he analyzes the development of insecticide resistance and its contribution to the widespread phenomenon of resurgence. The interaction between insecticide and varietal resistance and its implications for resistance management strategies are stressed.

Varietal breakdown and insecticide-induced resurgence continue as the major theme in Chapter 17. Here, Gallagher, Kenmore, and Sogawa detail the genetics of rice varietal resistance to *N. lugens* and stress the importance of considering genetic variation in field populations of planthoppers when assessing the potential for virulence evolution. They move on to discuss how frequent and excessive applications of pesticides kill natural enemies and accelerate the breakdown of novel varieties. A case study in Indonesia demonstrates how management techniques can be altered to extend the life of a resistant variety and deter planthopper outbreaks in rice.

Hare (Chapter 18) concludes this section with a specific focus on the form and magnitude of potential interactions among several major management tactics used to control planthoppers in Asian rice. Strategies for the deployment of resistant varieties, the compatibility of varietal resistance and biological control, and the interaction of cultural manipulations with biological control are major subjects of debate.

14

Biological Control in the Management of Planthopper Populations

Betty Benrey and William O. Lamp

Introduction: Biological Control and Population Dynamics Theory

Biological control is "the utilization of natural enemies to reduce the damage caused by noxious organisms" (DeBach and Rosen 1991). Although the ultimate goal is damage reduction, biological control of insects relies on the suppression of pest populations through the action of natural enemies, that is, predators, parasites, and pathogens (Smith 1919). Ecological theory has provided a context for understanding the dynamics of pest–enemy interactions and examples of biological control have been used to test ecological theory (Hassell and May 1974; Ehler 1976; Beddington et al. 1978; Waage 1990). Population dynamics theory concerns just how a wide variety of factors including natural enemies interact to influence population size (Price 1984). The practice of biological control involves manipulation of natural enemies within agroecosystems to achieve a level of suppression defined on socioeconomic criteria. Despite this difference, population dynamics theory has proved useful in the development of practical biological control. For example, the prediction of pest population dynamics has been improved (e.g., *Nilaparvata lugens* Stål by Hirao 1979b) and testable hypotheses concerning pest–enemy dynamics have been provided (e.g., Chesson and Murdoch 1986).

Biological control has proved successful for many insect pests (for reviews, see Huffaker and Messenger 1976; DeBach and Rosen 1991). Overall, three types of biological control are employed in pest management: (1) importation of exotic species of natural enemies and their establishment in a new habitat (also called classical biological control); (2) augmentation of established species of natural enemies through direct manipulation of their densities, such as through mass production followed by periodic releases; and (3) conservation of existing natural enemies through habitat modification or altered agronomic practices. Although importation of exotic species has received the greatest attention in the past, there is a growing focus on natural enemy augmentation and conservation for insect pest management (Dent 1991).

Recent analyses concerning rates of success and establishment of natural enemies for biological control have identified broad patterns that may be useful in developing control strategies against planthoppers (e.g., Hall and Ehler 1979; Hokkanen and Pimentel 1984; Stiling 1990; Gross 1991). For example, in an analysis of Clausen's (1978) summary of biological control attempts around the world, Hall et al. (1980) compared the rate of success, and Hall and Ehler (1979) compared the rate of establishment of natural enemies in several categories, including host taxa, predators versus parasites, islands versus continents, and native versus exotic pests. These studies suggested that the rate of enemy establishment in new habitats for pests in the taxa Homoptera and other Exopterygota was higher than that for other orders of Endopterygota. Within the Homoptera, attempts in the Aleyrodoidea and Coccoidea predominate. The results of these assessments provide encouragement for a future focus on the natural enemies of planthoppers and their ability to suppress populations.

Our understanding of the conditions that result in successful biological control and of its interaction with other management strategies has grown in recent years. Despite the relative paucity of biological control efforts against planthopper pests compared to other pest taxa, several examples of successful biological control do exist (Swezey 1936; Fullaway 1940). In this chapter, we first review efforts for planthopper biological control by briefly describing planthopper pest–natural enemy–crop systems and presenting selected case histories. Second, we discuss the potential for biological control in the context of ecological theory. Finally, we summarize the status and potential of biological control for planthopper pest management.

Biological Control Efforts for Planthopper Pests

Several species of delphacid planthoppers are of global economic importance. Because of the direct feeding damage they cause and their ability to transmit plant viruses, these insects can cause severe crop loss (Nault and Rodriguez 1985). In addition, their relatively high reproductive potential and dispersal capability allows planthoppers to track changes in favorable resources and predisposes them to become agricultural pests (Denno and

Roderick 1990). As a result of their pest status, planthoppers have been the focus of several important biological control efforts.

Our review of planthopper–natural enemy interactions has concentrated on studies of five planthopper pest species: the brown planthopper, *Nilaparvata lugens* (Stål); the taro planthopper, *Tarophagus colocasiae* (Kirkaldy); the sugarcane planthopper, *Perkinsiella saccharicida* Kirkaldy; the corn planthopper, *Peregrinus maidis* (Ashmead); and the cereal planthopper, *Javesella pellucida* (Fabricius). Other aspects of planthopper–natural enemy interactions are discussed elsewhere in this volume. Döbel and Denno (Chapter 10) and Cronin and Strong (Chapter 11) provide more information on planthopper predators and parasitoids and their role in population dynamics. The studies reviewed here are summarized in Table 14.1 and outlined by pest species below.

Although the range of planthopper natural enemies is extensive, most control attempts have involved enemies that attack eggs (Table 14.1). In fact, egg mortality is thought to be the key factor in the population dynamics of planthoppers (Waloff and Thompson 1980; Denno and Roderick 1990). Planthopper eggs are attacked by a variety of parasitoids and predator species. Hymenopteran parasitoids in the families Mymaridae, Eulophidae, and Trichogrammatidae have been commonly used in biological control. Important egg predators include Hemiptera in the families Miridae and Nabidae, and Hymenoptera in the family Pteromalidae.

Nymphs and adults are also attacked by a wide range of parasitoids and predators (Table 14.1). Although nymphal mortality may be high, studies indicate that it is not a key mortality factor for several species of planthoppers (Waloff and Thompson 1980; Denno and Roderick 1990). The nymph and adult stages are attacked by several species of endoparasitoids in the Pipunculidae (Diptera), Dryinidae (Hymenoptera), and Strepsiptera. These parasitoids usually kill their hosts and may sterilize the adults either by feeding on internal reproductive organs or by deforming genitalia (Kathirithamby 1978). Spiders are the most frequent predators of nymphs and adults in both natural and agricultural systems (Kenmore et al. 1984). Beetles in the families Coccinellidae and Carabidae and true bugs in the families Miridae and Veliidae may also be significant sources of nymphal and adult mortality. Ants may prey on nymphs and adults, but may also feed on the honeydew secreted by the planthoppers and protect them from other natural enemies (Napompeth 1973).

Compared with the use of predators and parasitoids, the use of pathogens for biological control has received little attention to the present (Soper 1985 and references therein). Because of their piercing-sucking feeding habits, planthoppers are not generally susceptible to viral, bacterial, or protozoan pathogens. However, entomopathogenic fungi do represent a

Table 14.1. Natural enemies implicated in the suppression of planthopper pests.

Natural Enemy	Planthopper Host	Host Plant	Location	Stage Attacked	Reference
INSECT PREDATORS					
Hemiptera: Miridae					
Cyrtorhinus fulvus	*Tarophagus colocasiae*	Taro	Hawaii, USA	Egg	Fullaway (1940)
Cyrtorhinus lividipennis	*Peregrinus maidis*	Corn	Hawaii, USA	Egg	Fullaway and Krauss (1945)
	Nilaparvata lugens	Rice	Philippines, Malaysia	Egg	Chiu (1979); Ooi (1988)
Tytthus mundulus	*Perkinsiella saccharicida*	Sugarcane	Hawaii, USA	Egg	Swezey (1936)
Tytthus mundulus	*Peregrinus maidis*	Corn	Hawaii, USA	Egg	Fullaway and Krauss (1945)
Hemiptera: Veliidae					
Microvelia douglasi	*Nilaparvata lugens*	Rice	Philippines	Nymph	Kenmore et al. (1984)
Hymenoptera: Pteromalidae					
Panstenon oxylus	*Javesella pellucida*	Cereals	Finland	Egg	Raatikainen (1967)
Mesopolobus aequus	*Javesella pellucida*	Cereals	Finland	Egg	Raatikainen (1967)
SPIDER PREDATORS					
Lycosidae					
Pardosa pseudoannulata	*Nilaparvata lugens*	Rice	Philippines, Malaysia	Nymph, adult	Kenmore et al. (1984); Ooi (1988)
MITE PREDATORS					
Achloroloplus gracilipes	*Javesella pellucida*	Cereals	Finland	Nymph, adult	Raatikainen (1967)

INSECT PARASITOIDS					
Hymenoptera: Eulophidae					
Ootetrastichus beatus	*Peregrinus maidis*	Maize	Hawaii, USA	Egg	Napompeth (1973)
Ootetrastichus spp.	*Peregrinus maidis*	Maize	Hawaii, USA	Egg	Napompeth (1973)
Ootetrasitichus beatus	*Perkinsiella saccharicida*	Sugarcane	Hawaii, USA	Egg	Swezey (1936)
Ootetrastichus formosanus	*Perkinsiella saccharicida*	Sugarcane	Hawaii, USA	Egg	Swezey (1936)
Hymenoptera: Mymaridae					
Anagrus atomus	*Javesella pellucida*	Cereals	Finland	Egg	Raatikainen (1967)
Anagrus sp.	*Nilaparvata lugens*	Rice	Taiwan	Egg	Lin (1974); Chiu (1979)
			Malaysia	Egg	Ooi (1988)
Anagrus optabilis	*Nilaparvata lugens*	Rice	Thailand	Egg	Yasumatsu et al. (1975)
Anagrus osborni	*Peregrinus maidis*	Maize	Hawaii, USA	Egg	Napompeth (1973)
Anagrus frequens	Peregrinus maidis	Maize	Hawaii, USA	Egg	Napompeth (1973)
	Perkinsiella saccharicida	Sugarcane	Hawaii, USA	Egg	Swezey (1936)
Mymar taprobanicum	*Nilaparvata lugens*	Rice	Thailand	Egg	Yasumatsu et al. (1975)
Polynema sp.	*Nilaparvata lugens*	Rice	Thailand	Egg	Yasumatsu et al. (1975)
Gonatocerus sp.	*Nilaparvata lugens*	Rice	Thailand	Egg	Yasumatsu et al. (1975)
Oligosita sp.	*Nilaparvata lugens*	Rice	Malaysia	Egg	Ooi (1988)
Paranagrus optabilis	*Perkinsiella saccharicida*	Sugarcane	Hawaii, USA	Egg	Swezey (1936)
Paranagrus perforator	*Perkinsiella saccharicida*	Sugarcane	Hawaii, USA	Egg	Swezey (1936)
Hymenoptera: Dryinidae					
Dycondilus lindbergi	*Javesella pellucida*	Cereals	Finland	Nymph, adult	Raatikainen (1967)
Haplogonatopus vitiensis	*Peregrinus maidis*	Maize	Hawaii, USA	Nymph, adult	Napompeth (1973)
Haplogonatopus vitiensis	*Perkinsiella saccharicida*	Sugarcane	Hawaii, USA	Nymph, adult	Swezey (1936)
Pseudogonatopus hospes	*Perkinsiella saccharicida*	Sugarcane	Hawaii, USA	Nymph, adult	Swezey (1936)
Strepsiptera: Elenchiidae					
Elenchus japonicus	*Nilaparvata lugens*	Rice	Japan	Nymph, adult	Esaki and Hashimoto (1937)

potential source of biological control agents. For example, *Erynia radicans*, a member of the order Entomophthorales, is known to attack *Nilaparvata lugens* (Shimazu 1979). This species may be significant in inoculative or inundative releases in the future because it can be cultured artificially and because of its ability to kill insect pests (McGuire 1989). Other fungal pathogens include *Beauveria bassiana* and *Metarhizium* spp. Field applications of these fungi have been shown to suppress *N. lugens* populations on rice in Korea (Aguda et al. 1987). A complete inventory of known planthopper pathogens is provided by Soper (1985).

Nematode parasites of planthopper pests have largely been overlooked. One species, *Agamermis unka* (Nemata: Mermithidae), has been reported from field populations of *N. lugens*, infecting up to 50% of adults collected (Choo et al. 1989). These authors further reported that infection is influenced by planthopper feeding location on the plant, and that incidence is higher in brachypterous forms compared to macropterous forms. However, the role of nematodes in suppressing planthopper populations has not been investigated.

Several planthopper species are pests in a variety of agricultural crops including rice, maize, oats, sugarcane, and taro. These crops occur in temperate and tropical regions on continental and island location, and planting cycles range from multiple crops per year (e.g., rice) to one crop in 15 months (taro). Rice alone may be grown on lowland and upland habitats, under rain-fed only to irrigated to deep-water conditions, and within tropical to temperate regions. Furthermore, crop production ranges between labor-intensive, subsistence, and large-scale mechanized farms, with each type having a different set of production and crop protection practices. As a result, the type of crop system may have a significant impact on biological control efforts as well as the outcome for pest management (Greathead and Waage 1983).

Biological control efforts for planthopper pest management have yielded highly successful results in several cases, but, in general, the outcome has been variable to date. The following sections focus on individual planthopper species and summarize published information on biological control and the role of natural enemies in the suppression of specific delphacid pests.

Nilaparvata lugens, Brown Planthopper

N. lugens is a widely-distributed, major pest of tropical rice in South, Southeast, and East Asia, the South Pacific Islands, and Australia (Dyck and Thomas 1979; Kenmore et al. 1984; Ooi 1988). In addition, adult planthoppers migrate into temperate regions such as Japan and Korea each summer (Dyck and Thomas 1979; Kuno 1979; Kisimoto 1981). Its feeding activities on vascular tissues cause a cascade of plant physiological responses known symptomatically as hopperburn, leading, under severe conditions, to plant death (Dyck and Thomas 1979; Kenmore 1980). Furthermore, *N. lugens* is a known vector of several viruses (Conti 1985). *N. lugens* has been called a pest of the green revolution because of the adoption of modern agricultural practices (e.g., flooding fields, increased use of nitrogenous fertilizer, and the use of high yielding varieties). These practices have not only contributed to the actual pest status of *N. lugens* but have also made its control more difficult (Dyck et al. 1979; Kenmore et al. 1984).

Because it is an endemic pest of Asian rice in the tropics, biological control of *N. lugens* has focused on the conservation of natural enemies, particularly as a consequence of changing agricultural practices (Kenmore et al. 1984; Ooi 1988). More than 200 natural enemies have been recorded for *N. lugens* (Ooi 1988). Although many have some role in population suppression, their impact varies (Chiu 1979). Studies on biological control of *N. lugens* have generally been confined to surveys of natural enemy species, the ecology and biology of specific natural enemies, and the estimation of predation and parasitism rates (e.g., Chiu 1979; Kenmore 1980; Ooi 1988; Chua and Mikil 1989).

Egg parasitism is often very high, whereas nymph and adult parasitism are generally low throughout most Asian countries including Thailand (Miura et al. 1979), Sri Lanka (Ôtake et al. 1976), the Philippines (Peña and Sheppard 1986), and Malaysia (Ooi 1988). Various parasitoids predominate at different times during the crop-growing season and attack *N. lugens* at different developmental stages (Chiu 1979). Among the egg parasitoids are several species in the families Mymaridae, Trichogrammatidae, and Eulophidae. In Taiwan, Lin (1974) found that *Anagrus* sp. constituted 93% of the egg parasites in the paddy field, and Chiu et al. (1984) reported up to 38% parasitism at the end of the season. Yasumatsu et al. (1975) reported that four mymarid parasitoids, *Anagrus optabilis*, *Mymar taprobanicum*, *Polynema* sp., and *Gonatocerus* sp., have reduced the densities of planthoppers and leafhoppers in rice in Thailand. In Malaysia, Ooi (1988) found that percent parasitism by two species of *Anagrus* and two species of *Oligosita* varied widely across sites from 0 to 75.5%. He concluded that egg parasitoids could be important mortality factors.

The nymphs and adults of *N. lugens* are attacked by several species of dryinids, strepsipterans, and pipunculids (Table 14.1). To date, six species of dryinids and four species of strepsipterans have been reported on *N. lugens* from Asia (Hinckley 1963; Ôtake et al. 1976; Chandra 1980; Kathirithamby 1982; Greathead 1983; Ooi 1988). The incidence of parasitism by dryinids and strepsipterans is quite variable (Waloff 1980). For example,

Ooi (1988) found that in Malaysia the incidence of parasitism by dryinids and strepsipterans was, in general, low and never exceeded 15%. However, in northern Thailand, Yasumatsu et al. (1975) reported up to 90% parasitism and concluded that nymph and adult parasitoids could be significant factors in suppressing *N. lugens* populations. In general, however, parasitoids have been considered of minor importance in controlling *N. lugens* (Chiu et al. 1974; Ôtake and Hokyo 1976; Ooi 1988).

Predators of eggs, nymphs, and adults are more important than parasitoids in controlling *N. lugens* populations (Hinckley 1963; Kenmore et al. 1984; Ooi 1988). The most commonly reported predator is a mirid bug, *Cyrtorhinus lividipennis*. *C. lividipennis* is widely distributed in many of the countries where *N. lugens* occurs, and it feeds on all developmental stages of *N. lugens*, but primarily on eggs (Hinckley 1963; Lin 1974; Chiu 1979; Ooi 1988). Stapley (1975) indicated that a critical factor in the maintenance of *C. lividipennis* was the nearby presence of the grass, *Digitaria*, which allowed for the multiplication of the mirid. In Malaysia, Ooi (1988) found that *C. lividipennis* was present early in the season before the arrival of *N. lugens*. When the planthopper population began to increase, *C. lividipennis* showed a positive numerical response.

In addition to *C. lividipennis*, about 16 species of spiders have been identified as preying on *N. lugens*, among which *Lycosa* (=*Pardosa*) often predominate (Kobayashi and Shibata 1973; Sasaba et al. 1973; Mochida and Dyck 1977; Kiritani et al. 1972; Kenmore et al. 1984; Ooi 1988). At the International Rice Research Institute, *Lycosa* (=*Pardosa*) species have been considered to be the most important predators of *N. lugens* (Chiu 1979; Kenmore 1980). Ooi (1988) speculated that the presence of spiders in rice fields might determine the establishment and subsequent increase of *N. lugens* populations. However, Ooi (1988) also indicated that large immigrations of macropters into new rice fields can swamp spiders and ultimately lead to outbreak. Other predators include several species of coccinellids and veliids that have been reported to play an important role in suppressing populations of *N. lugens* (Ooi 1988; Kenmore 1980). Among the coccinellids, *Coccinella arcuata* is the most important and common predator of *N. lugens* and *Sogatella furcifera* in India, Fiji, Australia, and Papua, New Guinea. Studies by Kenmore et al. (1984) and Nakasuji and Dyck (1984) demonstrated that veliids, particularly *Microvelia douglasi*, were very important mortality factors of *N. lugens*.

Two studies suggest that predators are important in suppressing *N. lugens* populations (Kenmore et al. 1984; Ooi 1988). Exclusion experiments in both studies demonstrated that *N. lugens* densities increased in the absence of predators (Fig. 14.1). In one of these studies, Ooi (1988) concluded that predators have the potential to control *N. lugens* based on the rates of

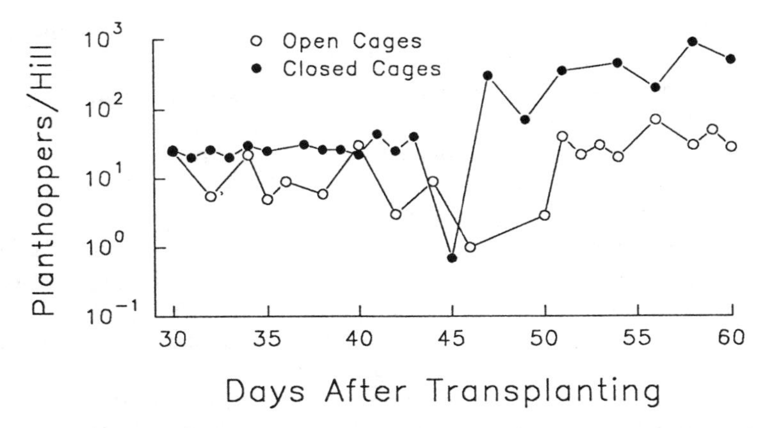

Figure 14.1. Comparison of the nymphal densities of *Nilaparvata lugens* between cages in which predators were excluded (closed cages) and open cages to which predators had access. The major predators implicated in the suppression of planthopper populations were spiders and veliid bugs. All arthropods were initially removed from the caged rice, and after 25 days, cages were stocked with young planthopper nymphs. Initial planthopper densities equaled 25 per hill and each cage contained 4 hills of rice. The experiment was conducted during the dry season at IRRI, Los Baños, Philippines (from Kenmore 1980).

change in *N. lugens* and predator densities. In spite of their demonstrated potential, biological control with predators is not always effective. The question arises as to whether *N. lugens* outbreaks are caused by reduced effectiveness of natural enemies.

Mochida and Dyck (1977) reported six likely causes of *N. lugens* outbreaks: introduction of high yielding varieties, cultural practices (irrigation, double, or continuous cropping and staggered planting), increased use of nitrogenous fertilizer, increased use of insecticides, climatic factors, and weeds. However, outbreaks in Japan occurred long before the introduction of high yielding varieties and outbreaks in Malaysia only occurred sporadically despite the presence of these high-yielding varieties (Ooi 1988). Some of these management practices, and especially that of insecticide use, may be traced to their adverse effects on natural enemies.

One common cause of insect pest outbreaks is the resurgence of insect pest populations in response to insecticide applications (Ripper 1956), and *N. lugens* is one of the most intensively studied cases (Chiu 1979; Heinrichs et al. 1984b; Heinrichs Chapter 16). One of the primary factors responsible for pest resurgence has long been believed to be the elimination of natural enemies as a result of insecticide applications (DeBach 1974; Croft and Brown 1975). Several studies have shown that insecticides disrupt natural biological control, allowing *N. lugens* populations to increase (Itô et al. 1962; Kenmore et al. 1984; Ooi 1988). However, alternative mechanisms have been proposed to explain pest resurgence, especially with *N. lugens*. For example, changes in host plant quality, such as increased plant growth, have been implicated in the resurgence of *N. lugens* populations (Chelliah and Heinrichs 1980; Heinrichs and Mochida 1984). Another implicated resurgence mechanism is the enhancement of insect fecundity by chemical stimulation. Researchers have demonstrated that certain chemicals, such as decamethrin, methyl parathion, and diazinon, stimulated reproduction of planthopper females (Chelliah and Heinrichs 1980; Heinrichs et al. 1982b; Reissig et al. 1982a). Thus, the disruption of biological control may promote outbreaks, and chemical stimulation of *N. lugens* fecundity exacerbates the situation. Identification of key mechanisms promoting *N. lugens* resurgence will allow pest managers to implement improved strategies for its control.

Because the loss of natural enemies may result in pest resurgence, conservation of natural enemies should be a key component of the management of resurgent pests. For example, Reissig et al. (1982a) evaluated insecticides for their relative toxicity to *N. lugens* and its natural enemies and concluded that certain insecticides were less toxic to some natural enemies than to others. A number of studies have shown that natural enemies are more susceptible to insecticides than their herbivorous hosts (Abdelrahman 1973; Braun et al. 1987; Plapp and Bull 1978). Furthermore, insecticides may directly affect natural enemies by altering their fecundity, walking speed (Hoy et al. 1984), and food intake (Heinrichs and Mochida 1984; O'Brian et al. 1985). Indirectly, insecticides may also affect natural enemies by altering the quality of the prey, directly or through the host plant (e.g., Hussey and Huffaker 1976), by eliminating alternate hosts or prey that might be essential for the completion of the life cycle (Caron 1981; Kiritani 1979), or by the elimination of other food sources, such as honeydew produced by homopterans (Coppel and Mertins 1977).

N. lugens outbreaks are also frequently associated with certain cultural practices, such as irrigation, mainly by increasing host plant availability through the practice of double and triple cropping (Kenmore 1980). Natural enemies may be able to track population increases associated with some of these practices, but the extensive use of insecticides along with these other practices precludes any conclusion concerning their effect on the pest–natural enemy interaction its subsequent influence on outbreak. Oka (1979) has proposed the use of synchronous culture of rice and a defined rice-free period within regions to reduce *N. lugens* densities. This practice may also reduce the effect of specialist natural enemies, such as some *Anagrus* spp., without consideration of alternate hosts during the rice-free period. In addition, the use of rice-free periods may interrupt the numerical response of generalist predators, such as spiders, in response to planthopper densities (Kenmore 1980). Current knowledge is insufficient to predict the outcome of interactions between cultural practices and biological control for *N. lugens*.

In most of the countries where *N. lugens* is an important pest, rice production often depends on pesticides for crop protection. Only recently have control strategies shifted to include an integrated approach in which the role of natural enemies in suppressing *N. lugens* densities is recognized (see Heinrichs Chapter 16; Gallagher et al. Chapter 17). Based on our review, *N. lugens* biological control has relied primarily on manipulating naturally occurring species of natural enemies. Although specific guidelines for the incorporation of natural enemy densities into management decisions are generally lacking, In Malaysia, Ooi (1988) suggested that a predator/prey ratio of 1 : 50 (determined by scouting) resulted in the retention of planthopper populations below the economic threshold. Just how such predator/prey ratios should be modified to maintain effective planthopper control under local management conditions is yet to be determined.

Perkinsiella saccharicida, Sugarcane Planthopper

P. saccharicida is widely distributed in tropical areas, recorded from Africa, Asia, Australasia and Pacific Islands, North America, and South America where it feeds on sugarcane (CAB 1987). Damage results from feeding by heavy infestations and *P. saccharicida* also vectors Fiji disease (DeBach 1974).

The case of *P. saccharicida* in Hawaii is considered one of the classic examples of successful biological control (DeBach and Rosen 1991), and the information that follows is based primarily on accounts by Swezey (1936) and Zimmerman (1948). The planthopper was first found in Hawaii on Oahu in 1900 by R.C.L. Perkins, and most likely was transported by sugarcane cuttings from Australia. By 1904, *P. saccharicida* had become a serious pest in Hawaii. Damage was so extensive that the economic foundation of Hawaii, rooted in sugarcane, was under threat from planthopper attacks. The Hawaiian Sugar Planters' Association began a campaign for the exploration and introduction of natural enemies from several countries.

In his remarks following an fact-finding visit to Australia, Perkins mentioned the following (Swezey 1936): "At Bundaberg, is an extensive cane district with our leaf-hopper everywhere present, but never in numbers such as we are accustomed to in these islands. In fact we never saw the hoppers nearly as numerous as they are on our least affected plantations." He concluded that the planthopper was kept under control by a number of predators and parasites. From 1904 through 1916, natural enemies were cultured and shipped to Hawaii from Queensland, Australia and later on from Fiji, Formosa, China, the Philippines, and Japan. By 1907, moderately effective planthopper control had been achieved (Zimmerman 1948).

The most important of the egg parasitoids introduced was a mymarid wasp, *Paranagrus optabilis*, discovered in Queensland in 1904 (Zimmerman 1948). Although the first introduction consisted of the transfer of small cuttings of cane leaf midribs with parasitized eggs, very few parasitoids survived. Later, successful shipments were made in cages containing cane plants with breeding cultures of planthoppers and parasitoids. During 1905, *P. optabilis* was released throughout the cane district, and by 1906, it had become established on most of the plantations. Several life history characteristics of *P. optabilis* contributed to its success, including its short life cycle, which was less than half as long as the life cycle of the planthopper, parthenogenesis which aided in population increase in the field and culturing in the lab, and its dispersal ability which allowed the wasp to colonize new fields simultaneously with the planthopper.

Other hymenopterous parasitoids that were introduced included two additional mymarid species, *Paranagrus perforator* and *Anagrus frequens* (Zimmerman 1948). The later was the first to become established and it became very abundant and widely distributed until *P. optabilis* became prevalent. Subsequently, *A. frequens* almost entirely disappeared from the cane fields. A eulophid wasp, *Ootetrastichus beatus*, was introduced from Fiji in 1905, became established, and generally spread throughout the plantations. However, it interfered with the effectiveness of *P. optabilis* because *O. beatus* larvae eat all remaining planthopper eggs, parasitized or not, when they emerge from the egg clusters. In 1916, another eulophid wasp, *Ootetrastichus formosanus*, was introduced from Formosa, and by 1919, egg parasitism rates were higher than those for other parasitoids in several locations.

A colony of the dryinid wasp, *Haplogonatopus vitiensis*, was initiated from a single female in an attempt to introduce this parasitoid from Fiji in 1906. This parasitoid was established in several plantations, and for a time it was a factor in reducing pest densities. However, by 1918 it became scarce due to the diminished density of planthoppers and by the action of hyperparasites. Another dryinid wasp, *Pseudogonatopus hospes*, was introduced from China during 1906 and 1907. Establishment at effective densities failed; however, it was recovered at several locations far from introduction sites in 1927. Hyperparasites suppressed populations and prevented *P. hospes* from becoming a useful biological control agent.

Twelve years after the introduction of these parasitoids, some plantations still experienced damage as a consequence of local outbreaks (Zimmerman

1948). Consequently, a search was conducted for additional natural enemies in 1919, leading to the introduction of the mirid egg predator, *Tytthus mundulus*. *T. mundulus* quickly established and became widely distributed throughout the cane district by 1923. The feeding habits of *T. mundulus* may have contributed to its success in suppressing planthopper populations. This predator punctures many eggs, but only feeds completely on and kills a few. A transovarially transmitted fungus commonly occurs in eggs. This fungus is not harmful to unattacked eggs, but rather infects the egg whenever the egg chorion is punctured. The fungal infection resulting from feeding punctures completes the destruction of those eggs not eaten by *T. mundulus*.

A number of natural enemies already existing in Hawaii aided in the reduction and control of *P. saccharicida* (Zimmerman 1948). Most were predators and became very abundant during planthopper outbreaks. Some were native species inhabiting the forests in the vicinity of cane fields, and others had been introduced previously to control other pests. For example, the coccinellid, *Coelophora inaequalis*, was introduced from Australia to control aphids, and at the time of maximum planthopper infestations was nearly always present in planthopper-infested cane fields. Also, several species of spiders increased greatly in cane fields where unlimited supplies of food prevailed during planthopper infestations.

At present, *P. saccharicida* is effectively controlled in Hawaii and Mauritius mainly by *T. mundulus* (DeBach and Rosen 1991). Whenever there has been a local increase of planthopper densities, the subsequent increase of this egg predator reduced its numbers (Zimmerman 1948). In Hawaii, planthopper control by this predator was so successful that had this predator been found early in the search for natural enemies, it alone could have resulted in complete control, and no other introductions would have been necessary (Zimmerman 1948). However, information is lacking on the role of other species of natural enemies on *P. saccharicida*. Even so, the control of *P. saccharicida* by *T. mundulus* stands as one of the most illustrious cases of classical biological control (DeBach and Rosen 1991).

Tarophagus colocasiae, Taro Planthopper

Three species of *Tarophagus* attack taro, *Colocasia esculenta* (L.) Schott, an important root crop in the Pacific. Suspected vectors of alomae and bobone virus diseases (Jackson 1980), these planthoppers also cause injury by feeding. In Fiji, feeding during severe local outbreaks causes stunting and wilting of the plant, particularly during dry periods [Swaine (1971) cited by Waterhouse and Norris (1987)].

Early studies on the pest status of *Tarophagus* species were all published under the name of *Tarophagus proserpina*. However, in a recent review of the genus, Asche and Wilson (1989b) recognized three species: *Tarophagus proserpina*, *Tarophagus persephone*, and *Tarophagus colocasiae*, each with a different distribution in Asia and the Pacific. The three species are sympatric only on Papua New Guinea. *T. colocasiae* is the only species occurring in the North Pacific. In Southeast Asia, the distributions of *T. colocasiae* and *T. persephone* overlap. Alone, *T. proserpina* inhabits the South Pacific, whereas *T. persephone* is found in Australia. In several countries in the Pacific, attempts were made to control *T. colocasiae* by introducing mirid predators and hymenopterous parasitoids. Here we summarize one of the successful attempts in the Hawaiian Islands.

T. colocasiae became established in the Hawaiian Islands in 1930. Populations increased rapidly and spread to all other islands where taro was grown, resulting in near destruction of the crop (Fullaway 1940). In 1930, a mirid egg predator, *Cyrtorhinus fulvus*, was introduced from the Philippines where it had been a significant source of mortality. Shortly after its introduction, *C. fulvus* increased in abundance, became widely distributed, and successfully controlled *T. colocasiae* (Waterhouse and Norris 1987).

Matsumoto and Nishida (1966) also reported that *C. fulvus* was effective in suppressing *T. colocasiae* populations in Hawaii. Populations of *C. fulvus* were highly correlated with those of the planthopper in time. Similarly, the spatial distributions of the prey and predator were highly correlated, indicating that the predator was effectively tracking its prey. In an analysis of the age structures of predator and prey populations, they found that although first instar nymphs represented a high proportion of the total predator population, first instar nymphs comprised only a small fraction of the total planthopper population. From this observation, the authors concluded that egg predation by *C. fulvus* was an important contributing factor to the rarity of first instar planthoppers. Finally, that planthopper densities showed a marked increase in experimental plots from which predators were removed provides even stronger evidence for the control potential of *C. fulvus* (Fig. 14.2).

After its success in Hawaii, *C. fulvus* was later introduced into and became established on Guam, the Marianas Islands, Ponape and the Caroline Islands (Clausen 1978), Fiji, Samoa, French Polynesia, and the Solomon Islands (Waterhouse and Norris 1987). Because different species of *Tarophagus* co-occur in some of these areas, the taxonomic identity of the planthopper species in these earlier studies is not clear. Based on a summary of eight attempts to control "*T. proserpina*" by *C. fulvus* in the Pacific, Asche and Wilson (1989b) established that five introductions led to success (Table 14.2). They reported that the pattern of success was not related to the *Tarophagus* species involved. Furthermore, success was not associated

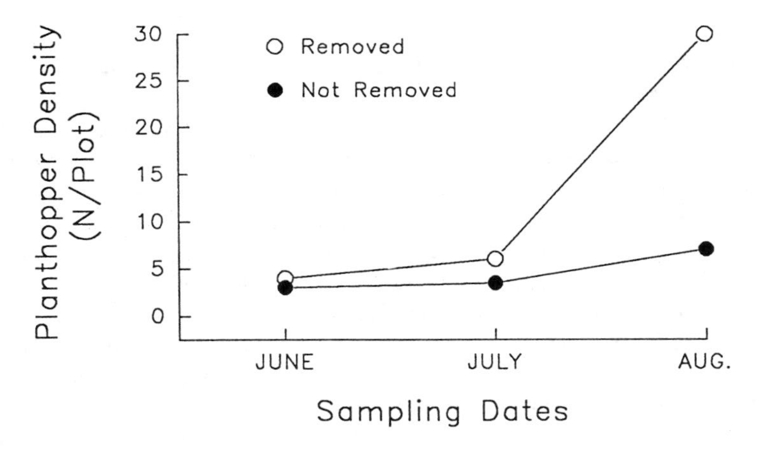

Figure 14.2. Comparison of the density (number per plot) of *Tarophagus colocasiae* between plots of taro from which the mirid predator *Cyrtorhinus fulvus* was removed (by aspirator) and not removed at Kahaluu Valley, Oahu, Hawaii (from Matsumoto and Nishida 1966).

Table 14.2. Introductions of *Cyrtorhinus fulvus* for the biological control of *Tarophagus* species [from Asche and Wilson (1989b)].

Introduction Site	*Tarophagus* sp.	Collected from	*Tarophagus* sp.	Result of Introduction
Tahiti	*T. proserpina*	Hawaii	*T. colocasiae*	–
Guam	*T. colocasiae*	Hawaii	*T. colocasiae*	+
Hawaii	*T. colocasiae*	Philippines	*T. colocasiae* or *T. persephone*	+
Ponape (Caroline Is.)	*T. colocasiae*	Guam	*T. colocasiae*	+
Western Samoa	*T. proserpina*	Hawaii	*T. colocasiae*	+
Solomon Is.	*T. persephone or T. colocasiae*	Sikaiana	*T. persephone* or *T. colocasiae*	+
Solomon Is.	*T. persephone* or *T. colocasiae*	Samoa	*T. colocasiae*	–
Tuvalu	*T. proserpina*	Fiji	*T. proserpina*	–

with the relationship between the target species of *Tarophagus* and the source species of *Tarophagus* in the country from which *C. fulvus* had been exported. They concluded that possible taxonomic confusion among *C.*

fulvus populations as well as that for the *Tarophagus* species have partially clouded the precise identity of the players involved in some of these biological control successes.

Nevertheless, *T. colocasiae* in Hawaii represents a second case of successful biological control among delphacids. Furthermore, as in the case of *Perkinsiella saccharicida*, the most important agent was a mirid egg predator. Matsumoto and Nishida (1966) listed four attributes of *C. fulvus* that may have contributed to its effectiveness in suppressing *T. colocasiae*. First, this mirid is associated only with the taro plant where it is primarily monophagous on planthopper eggs. Second, by resorting to cannibalism on its own eggs, it is capable of surviving under conditions of low planthopper egg density. Third, females lay eggs in the vicinity of the planthopper eggs, and shortly after hatching, nymphs begin feeding on the planthopper eggs. Fourth, the adults have high dispersal capability and can fly efficiently among plants. In addition, *C. fulvus* consumes only a portion of an egg before it moves on to attack another, a behavior that facilitates the transmission of a fungus which contributes to egg death [Williams (1932) as cited by Matsumoto and Nishida (1966)]. A clearer understanding of the taxonomic relationships among *C. fulvus* populations and their relationship with specific *Tarophagus* species on other islands should increase the chances for the future biological control of *Tarophagus* planthoppers with introduced mirid predators.

Peregrinus maidis, Corn Planthopper

P. maidis is a pantropical and subtemperate pest of corn and sorghum, but is most abundant at tropical latitudes (Napompeth 1973). It is not only the vector of maize mosaic virus, which is found in Hawaii, the Caribbean, South America, and possibly the southern United States, but also maize stripe and maize line viruses in East Africa, an uncharacterized corn virus new to Florida, and sorghum chlorosis virus from India (Clausen 1978). The plant damage caused by the virus is often greater than that caused by planthopper oviposition and feeding.

Several species of hymenopterous parasitoids attack *P. maidis*, including two mymarid egg parasitoids, *Anagrus* (*Paranagrus*) *osborni* from Australia and *Anagrus frequens* from the Philippines, two eulophid egg parasitoids, *Ootetrastichus beatus* and another undescribed *Ootestrastichus* sp., from Fiji and Formosa, respectively, and a dryinid nymphal and adult parasitoid, *Haplognatopus vitiensis* (Clausen 1978). Many predators, such as spiders, ladybugs, reduviids, and earwigs, also attack *P. maidis* (Fullaway 1918). The mirid egg predator, *Tytthus mundulus*, which was introduced to control the sugarcane leafhopper, also attacks eggs of *P. maidis*. However, it has not been effective in reducing *P. maidis* densities (Napompeth 1973). The only predator purposely introduced to control *P. maidis* was the mirid, *Cyrtorhinus lividipennis*, introduced to Hawaii in 1939 from Guam (Fullaway and Krauss 1945).

Little information exists regarding the impact of these natural enemies on *P. maidis*. The highly effective control of the sugarcane leafhopper by *T. mundulus*, and the observation that this predator frequently attacked *P. maidis* eggs in the field, suggested that it might be equally effective against *P. maidis* (Napompeth 1973). However, control was not achieved for several reasons (Verma 1955). First, the stages of corn suitable for oviposition are different for the predator and the planthopper. Second, in the early stages of corn development, the planthopper oviposits in the roots, making the eggs unavailable to the mirid. Finally, by the time the predator population builds up, the corn crop dies.

In a study conducted by Napompeth (1973) in Hawaii, key-factor analyses indicated that both density-independent and density-dependent factors were regulating the population density of *P. maidis*. Density-independent factors, such as weather and cultural practices, indirectly affected the condition of the corn plants, which, in turn, acted in a density-dependent manner on the planthopper. Natural enemies acted in a density-dependent manner as a whole component, and among them, *C. lividipennis* and spiders were the most important. The analysis revealed that *C. lividipennis* was the most important predator, whereas *T. mundulus* was of little importance. However, the performance of *C. lividipennis* as well as that for *T. mundulus* was not always satisfactory due to the lack of synchronization with the planthopper population. Napompeth (1973) suggested that mass-rearing of the mirid and its release at a time early in crop development could be useful as part of an integrated control program.

P. maidis has been reported sporadically as a pest of sorghum in southern India (Fisk 1980; Fisk et al. 1981). In a study conducted by Fisk et al. (1981), populations of coccinellid beetles were observed to build up in cornfields in response to large infestations of aphids. The coccinellids were observed feeding on planthopper adults and nymphs and were considered a potentially important factor in reducing densities during some sampling periods. In this study, ants, *Camponotus compressus*, commonly inhabited cornfields where they were observed "attending" the planthoppers and feeding on the honeydew. The authors suggested that ants may had been a factor in reducing predation and emigration of the planthoppers. Similarly, Khan and Rao (1956) reported positive correlations between the presence of ants and *P. maidis* densities.

Efforts to control *P. maidis* through introductions of natural enemies have not been successful. Evidence reviewed above suggests that existing natural enemies are not compatible with the pest biology and crop culture. The addition of basic information regarding specific interactions between the planthopper and its natural enemies may lead to specific crop practices to enhance biological control. Alternatively, the search for other natural enemy species may lead to a more effective complex of natural enemies.

Javesella pellucida, Cereal Planthopper

J. pellucida is a serious pest of cereals, particularly oats, in Europe (Kanervo et al. 1957). As a result of the damage in Finland, oat production ceased in several regions during 1956 (Kanervo et al. 1957). The distribution of *J. pellucida* is limited to the temperate areas of the Northern Hemisphere (Harpaz 1972).

J. pellucida reproduces on small-grain cereals and related grass species (Raatikainen 1967). While feeding, the insect may transmit any of several plant viruses, each of which may reduce yield. The use of spring cereals as cover crops for grass fields seems to offer especially favorable conditions for propagation of the planthopper. During early summer, newly molted macropterous adults move in swarms from timothy fields to cereal fields where they mate and oviposit. Eggs are usually laid into stems and sometimes in leaves (Raatikainen 1967). Adults of *J. pellucida* are usually slow flyers, but when temperatures are high, flights can cover many kilometers with the aid of the wind, resulting in the spread of the population to new localities (Kanervo et al. 1957).

The most common natural enemies are the pteromalid egg predators *Panstenon oxylus* and *Mesopolobus aequus*. A mymarid wasp, *Anagrus atomus*, also parasitizes the eggs. Nymphs and adults are parasitized by the dryinid, *Dicondylus lindbergi*, and the strepsipteran endoparasite, *Elenchus tenuicornis*, and preyed upon by a polyphagous mite, *Achlorolophus* (=*Erythraeus*) *gracilipes* (Kanervo et al. 1957; Raatikainen 1967).

In an extensive review on *J. pellucida*, Raatikainen (1967) discussed the life history of each of its natural enemies and their role in the population dynamics of this planthopper in Finland. The planthopper was sampled from oats, spring wheat, and barley. *P. oxylus* was the main source of mortality for eggs deposited in stems for seven of the eight years of the study, and *M. aequus* was the principal mortality factor in the remaining year. *Anagrus atomus* was very effective at parasitizing eggs deposited in the leaves but not stems; at least 90% of the eggs found on leaves at the end of the oviposition period were parasitized, whereas most of the eggs in the stems remained undiscovered by *A. atomus*. The proportion of eggs destroyed by the pteromalids *P. oxylus* and *M. aequus* was greater in spring wheat than in oats (70 and 50%, respectively). The number of internodes inhabited by pteromalid larvae was positively correlated with the number of internodes containing delphacid eggs. Whether parasitism was density-dependent or density-independent was not clear. Parasitism of the nymphs by dryinid and strepsipteran parasitoids averaged 18% in oats and 22% in wheat at all population densities and in the different localities. *P. oxylus*, *M. aequus*, and *A. atomus* were active dispersers and followed *J. pellucida* from one habitat to the next. The endoparasitoids *E. tenuicornis*, *D. lindbergi*, and *A. gracilipes* were easily carried to new sites by the parasitized macropters of *J. pellucida* and other delphacids. These egg and nymphal parasitoids together attacked all developmental stages of *J. pellucida* at different times and were able to destroy them in most habitats; however, eggs located within thick stem walls (but not thin ones) escaped attack because oviposition was impeded. When the density of one of these natural enemies decreased, the density of the other increased, as exemplified by the pteromalid egg predators. Raatikainen (1967) suggested that these enemies were important in the suppression of the planthopper because their action accounted for 50 to 60% of the total mortality.

Although the Raatikainen (1967) study did not address the role of these or other natural enemies as potential biological control agents, he did consider the effect of altering crop management practices on natural enemy populations. For example, he found that 96% of *P. oxylus* occurred in the stems at the time of cereal harvest and that the majority of these wasps occurred above 10–20 cm in the canopy. Because machines used for harvest varied in cutting height, he calculated that when a "mowing machine" was used, 60–70% of the larvae were killed, whereas only 40% were killed when a "combine harvester" was used. Thus, *J. pellucida*, like *Nilaparvata lugens*, represents a case of an endemic pest species in which suppression by natural enemies is potentially high. However, particular crop management practices may promote outbreaks due to their negative effects on natural enemies.

Ecological Theory and Biological Control

Ecological theory is relevant to many of the principles of biological control and can influence decisions taken at different stages of a biological control program. The underlying assumption of classical biological control is that in its native habitat, a pest population is maintained below a certain level by natural enemies (Greathead 1986). Outbreaks of pest species are the result of a lack of suppression either because they have been introduced to new regions without their natural enemies or because the environment has changed the relationship between the pest and its enemies. Thus, "stable regulation of pest numbers" at low densities has been proposed as a prerequisite for effective control (see Hassell and May 1974; Murdoch and Oaten 1975). However, the notion of stability as a requisite for successful biological control is not universally accepted (Murdoch 1990), and other criteria have been proposed (Murdoch et al. 1984; Chesson and Murdoch 1986).

Attempts to improve the practice of biological control based on theoretical principles have been common during the last decade, though not without controversy. In this section, we examine some of the ecological principles and debates relevant to the biological control of planthoppers. The goal of this examination is to address the following questions: Where and why has biological control been successful for planthoppers? Which features of natural enemies have contributed to successful biological control? What life history characteristics of the pest have contributed to the success or lack of biological control? We do not aim to comprehensively review the theoretical basis for biological control. A considerable number of reviews that discuss these theoretical aspects can be found in the literature (e.g., Greathead 1986; Ehler 1990; Miller and Ehler 1990; Murdoch 1990; Waage 1990).

Attributes of Natural Enemies

The criteria for the identification and selection of the "best" natural enemies occupy a considerable body of literature (Greathead 1971; Zwölfer 1971; Ehler 1976, 1982; Huffaker et al. 1977; Pschorn-Walcher 1977; Hall and Ehler 1979; among others). Many of these studies have focused on particular attributes of natural enemies that are believed to maximize the depression and stability of a pest population. Such attributes are usually included in analytical population models as discrete, measurable life history parameters (e.g., generation time, fecundity, aggregation, interference, searching efficiency, handling time, and sex ratio) (Waage 1990). Laboratory studies have proven useful in identifying potential biocontrol agents for planthopper pests. For example, Manti (1989) examined the role of *Cyrtorhinus lividipennis* as a predator of *Nilaparvata lugens*. His studies indicated that *N. lugens* populations may be controlled effectively through the combined action of *C. lividipennis* and several species of spiders. His conclusions are supported by results from experiments that examine individual attributes of this mirid predator, such as its high reproductive capability and potential to effect a strong numerical response.

Functional response models are a standard method used to assess density-related changes in the attack rate of natural enemies (Waage 1990). Although these responses tend to be simple laboratory evaluations (Waage

1990), they are useful in identifying those natural enemies most capable of responding to changes in hosts densities and, therefore, most likely to suppress pest populations (Kareiva 1990). Most studies on functional responses of predators and parasitoids of planthoppers have shown a Type II response (Moratorio 1977; Chantarasa-ard et al. 1984b; Sivapragasam and Asma 1985; Ooi 1988; Manti 1989). For example, Chantarasa-ard et al. (1984b) analyzed the functional response of a mymarid egg parasitoid, *Anagrus incarnatus*, which parasitizes the eggs of *N. lugens*. Parasitism rates increased with increasing egg densities resulting in a Type II functional response. However, at high host densities, the proportion of hosts parasitized decreased, resulting in a pattern of inverse density-dependent parasitism.

Even though these studies as well as others (Nakasuji and Dyck 1984; Chua and Mikil 1989; Heong et al. 1990b) have provided basic information on the potential of particular natural enemies to suppress planthopper populations, their application in biological control programs requires further scrutiny. For instance, the potential killing power of an enemy must be considered in the context of other factors in the field. For example, the highly effective control of *Perkinsiella saccharicida* by *Tytthus mundulus*, due to several biological attributes discussed earlier, suggested that it might be equally effective against *Peregrinus maidis* (Napompeth 1973). However, effective control was not achieved due to the lack of synchrony between the predator and the pest populations. Similarly, studies on the functional response of *Lycosa* spiders demonstrated that they have the potential to suppress *N. lugens* populations (Ooi 1988); however they sometimes occur in relatively low numbers. Thus, the lack of a sufficient numerical response may prevent this predator from being effective in controlling a planthopper outbreak in spite of its high planthopper consumption potential.

Single Versus Multiple Introductions

The controversy over single versus multiple introductions has been an important and long-standing issue in biological control and has its basis in the competitive exclusion principle. Some have argued that the best strategy is to introduce many enemy species at once and hope that the best species or the best combination of species will be sorted out in the field (Ehler 1990). The alternative view states that competition among co-introduced natural enemies may eliminate a potentially good biological control agent if that species is a poor competitor. Supporting this view are studies which document that the competitive ability of a parasitoid may be inversely related to traits important in biological control, such as high fecundity and good colonizing ability (Force 1972; Price 1975). However, the available evidence indicates that multiple introductions often result in greater mortality (Beirne 1975; Ehler and Hall 1982).

Some studies on predators of planthoppers have demonstrated that the presence of other natural enemies may not always result in higher mortality rates. In laboratory studies on *Nilaparvata lugens*, Heong et al. (1991b) showed that the attack rate of the spider, *Pardosa pseudoannulata*, on this planthopper was lower in the presence of the mirid, *Cyrtorhinus lividipennis*. Although these examples support the view that negative interactions may result between co-existing natural enemies, other studies indicate that when a more complex community of natural enemies has been used, better control results (Nakasuji and Dyck 1984; Ooi 1988). For example, the successful control of *Perkinsiella saccharicida* in Hawaii was attributed to the multiple introductions of several species of predators and parasitoids (Swezey 1936). Similarly, Ooi (1988) suggested that low populations of *N. lugens* in some localities in Malaysia were due to the action of different types of natural enemies, some of which were present early in the season and prevented the planthopper from realizing its outbreak potential. Thus, the combined action of several natural enemies that operate on the different life history stages of the pest should be encouraged for a successful management of planthopper populations.

New Versus Old Associations

The hypothesis of enemy associations predicts that effective agents are more likely to be found on host–prey species closely related to the target pest (a new association) where there has not been the opportunity for co-evolution (Pimentel 1963) rather than on old associations which are highly co-evolved. Hokkanen and Pimentel (1984) analyzed 286 successful introductions of insect and pathogen agents used against insect pests and weeds and concluded that new associations were 75% more successful than old ones. However, these conclusions have been challenged, and different criteria for the definition of new and old associations have been discussed (Goeden and Kok 1986; Greathead 1986).

A common practice for planthopper pests has been to search for natural enemies in the area of origin of the pest (e.g., *Perkinsiella saccharicida*). For other planthopper species, control practices have concerned the conservation of existing natural enemies (e.g., *Nilaparvata lugens*). Thus, although no particular criterion has been adopted, control practices for planthoppers have involved "old associations" in which the natural enemies used have had previous contact with the pest in question. One possible exception is the case of the three *Tarophagus* species on the various Pacific islands. Although the Asche and Wilson (1989b) study was not conclusive, in part because of the potential confusion regarding the taxonomic identity of the mirid predator, *Tarophagus* may represent a series of both new and old associations. Further study of this system may be able to sort out the evolutionary history of the host–enemy interaction and its influence on the success of the enemy introduction. In any case, future searches for new natural enemies should not be dismissed on the basis of existing information.

Islands Versus Continents

Because some of the most remarkable successes of biological control have occurred in islands, such as Hawaii and Fiji, islands may provide a particularly favorable environment for classical biological control (Imms 1931; Taylor 1955). One suggestion for this pattern stems from the lower species diversity on islands (Greathead 1986). Organisms introduced onto islands may be able to establish themselves and become pests or successful natural enemies more easily in an environment with fewer predators and/or competitors. It has also been suggested that islands provide climatic conditions that are favorable for biological control because natural enemies are not subjected to climatic extremes (Waterhouse and Norris 1987). In their review, Hall and Ehler (1979) found that the rate of establishment of natural enemies was higher on islands. However, the rate of biological control success on islands was not generally greater compared to the mainland (Huffaker et al. 1977; Hall et al. 1980).

Because all classical biological control attempts with planthoppers have taken place on islands, it is not possible to compare their success with continental endeavors. Although both of the classic biological control successes were achieved for the most part on Hawaii (*Perkinsiella saccharicida* and *Tarophagus colocasiae*), attempts on other islands to control *Tarophagus* using the same mirid predators have achieved various levels of success. Furthermore, although *Tytthus mundulus* successfully controlled *P. saccharicida* on Hawaii, it failed to effectively suppress *Peregrinus maidis* on the same island. Thus, a focus on the specific attributes of the predator, planthopper, cropping system, and their interaction may prove more useful in attempting to elucidate those factors affecting success.

Tropical Versus Temperate Regions

Greathead (1986) raised the issue of whether or not biological control was more suited to tropical than to temperate regions. He found no evidence supporting the view that more successful biological control programs have been achieved in North America and Australasia than in the tropics. For planthopper systems, the classical examples of biological control occurred on tropical islands, but there are no data for a temperate comparison. However, several authors have suggested that endemic natural enemies play a more important role in the suppression of *N. lugens* populations in tropical regions (Kuno and Hokyo 1970; Kenmore et al. 1984; Cook and Perfect 1985a; Ooi 1988; Manti 1989).

Regional differences in the role of natural enemies in suppressing *N. lugens* populations stem from the dissimilarity in planthopper population dynamics between temperate and tropical Asia. In temperate areas (Japan and Korea) where it is unable to survive winter, planthopper populations are replaced every year by immigrants from southern regions, whereas in tropical areas, populations remain in paddy fields throughout the year (SE Asia) (Kisimoto 1977; Dyck et al. 1979; Kuno 1979; Cook and Perfect 1985a). Enemies are apparently less able to track planthopper populations in temperate Asia. For example, *Cyrtorhinus lividipennis*, although it migrates to Japan from China simultaneously with *N. lugens* (Kisimoto 1981), never attains sufficient densities to effectively suppress planthopper populations (Kuno 1979; Kenmore 1980). In tropical Asia, natural enemies show a consistent presence in pesticide-free rice and the probability for planthopper outbreak is less (Cook and Perfect 1985a).

Density Dependence

The issue of density dependence has been a major concern in the theory of biological control, especially in the development of host–parasitoid models (May and Hassell 1988; Murdoch 1990). However, recent studies that have examined host–parasitoid systems indicate that density-dependent relationships are not that frequent (Stiling 1987; Walde and Murdoch 1988). Whether or not successful biological control can be achieved without density dependence is controversial. Population models have suggested that host regulation by natural enemies is achieved by the density-dependent response of natural enemies to temporal and spatial changes in host density (Cook and Hubbard 1977; Beddington et al. 1978; Hassell 1978). Beddington et al. (1978) concluded that the differential exploitation of host patches by parasitoids provided the key to most successful cases of biological control. However, other investigators (Murdoch et al. 1984; Murdoch 1990) have concluded that stability provided by parasitoid aggregation in high host density patches is not a necessary condition for successful biological control and that other criteria, such as low mean pest density and low temporal variability in pest density, may be more important.

Cronin and Strong (Chapter 11) analyzed 27 spatial and 53 temporal relationships to determine the general patterns of density-dependent mortality in the Auchenorrhyncha. Overall, 75% of the cases showed no evidence for density dependence. Additionally, many studies that have examined the relationship between host density and parasitism in planthoppers have found this relationship to be inverse density-dependent (Stiling and Strong 1982a; Chantarasa-ard et al. 1984b; Roderick 1987; Antolin and Strong 1987; Benrey and Denno unpublished data). Egg depletion, a low numerical response of the parasitoids to the planthoppers population increase, and parasitism of alternate hosts are some of the reasons that have been implicated in these inverse density-dependent patterns of parasitism. However, several studies on the response of predators to planthopper densities show positive correlations between predator and prey density, suggesting a strong numerical response (Matsumoto and Nishida 1966; Napompeth 1973; Ooi 1988; Manti 1989; Döbel and Denno Chapter 10). In particular, the strong numerical response of *Cyrtorhinus lividipennis* (Ooi 1988; Manti 1989) and *Cyrtorhinus fulvus* (Matsumoto and Nishida 1966) has been considered an important feature for suppressing populations of *Nilaparvata lugens* and *Tarophagus colocasiae*, respectively.

Alternate Hosts

Alternative prey species may be essential for a natural enemy to survive in a region from year to year. Such is the case for the mymarid wasp, *Anagrus epos*, an egg parasitoid of the grape leafhopper *Erythroneura elegantula*. The wasp, which does not overwinter in grapes, depends on an alternate host, the blackberry leafhopper (*Dikrella cruentata*) for survival during the winter (Doutt and Nakata 1965, 1973; Williams 1984). Dispersal of *A. epos* from blackberry bushes to grapes in early spring results in the effective control of leafhoppers during summer (Doutt and Nakata 1973). Thus, the presence of alternate prey may determine whether a natural enemy will successfully control a planthopper pest.

One example of the role of alternate prey in the biological control of *Peregrinus maidis* involves the egg predator *Cyrtorhinus lividipennis* and the leafhopper *Draeculacephala minerva* in Hawaiian cornfields (Napompeth 1973). Only during 1971 did *D. minerva* colonize corn and deposit its eggs, thereby providing an early season food source for *C. lividipennis* (Fig. 14.3). During this year, *C. lividipennis* colonized fields early, fed on the eggs of the alternate host, switched later to the eggs of *P. maidis*, and ultimately suppressed populations of this planthopper. In the two years when *D. minerva* failed to colonize corn (1970 and 1972), *C. lividipennis* moved into fields late, by which time populations of *P. maidis* had escaped (Fig. 14.3). Thus, the presence of *D. minerva* facilitated the early colonization of *C. lividipennis* into cornfields by serving as an alternate source of

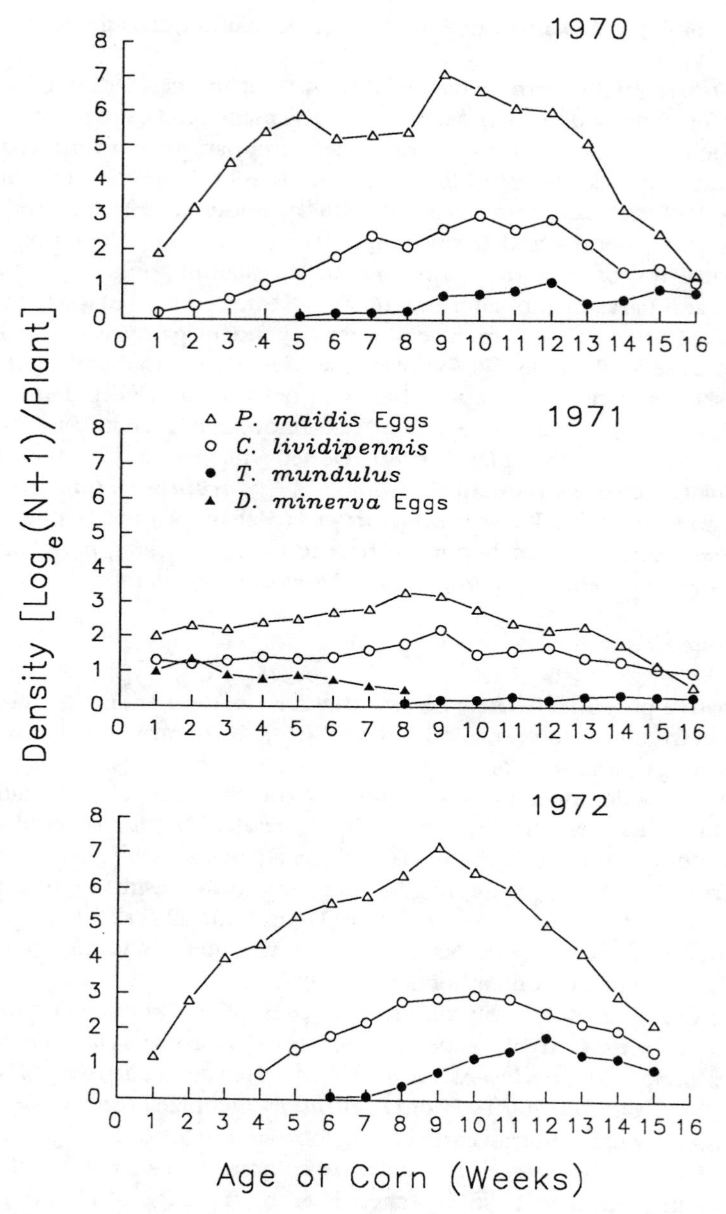

Figure 14.3. Seasonal variation in the density [log$_e$(N+1)/corn plant] of the eggs of *Peregrinus maidis*, the nymphal and adult densities of the mirid predators *Cyrtorhinus lividipennis* and *Tytthus mundulus*, and the egg density of the leafhopper (alternate prey) *Draeculacephala minerva* at Kualoa Ranch, Hawaii during 1970, 1971 and 1972. The early season occurrence of *D. minerva* eggs in 1971 encouraged the early colonization of corn by *C. lividipennis*, which successfully suppressed populations of *P. maidis* only that year (from Napompeth 1973).

food. Alternate hosts may be potentially important factors in the biological control of other planthopper pests as well, especially in those cases where the pest and natural enemy are not synchronized due to cropping practices. Prey augmentation, as a means of retaining enemies in habitats when the target pest is rare, has not been rigorously pursued as a control tactic in planthopper systems.

Parasitoids Versus Predators

The conventional belief that predators cannot be effective control agents has been challenged during the past decade (Hall and Ehler 1979; Murdoch et al. 1984; Murdoch et al. 1985, Greathead 1986). In an analysis of Clausen's (1978) summary of biological control attempts around the world, Hall and Ehler (1979, 1980) found no significant difference in the rate of establishment between predators and parasitoids. However, they did find the rate of success was greater for predators than for parasitoids, at least in part because a large proportion of successes was attributed to the coccinellid beetle *Rodolia cardinalis*. When the introductions of *R. cardinalis* were removed from the analysis, the rate of success for parasitoids was greater than that for predators. From these studies, Hall and Ehler (1979, 1980) concluded: "... the belief that parasites are somehow 'better' than predators in classical biological control is not consistent with the available evidence and should be regarded as a questionable assumption." Also, Murdoch et al. (1984) reviewed the mathematical theory pertaining to insect pests and natural enemies within the context of successful biological control programs. They concluded that successful biological control could be achieved through the action of polyphagous natural enemies such as predators. Moreover, the two classical biological control successes for planthoppers involved mirid egg predators, although they are relatively specialized feeders. As reviewed earlier, specialized feeding is considered by some to be an important enemy feature for effective pest suppression.

Another characteristic of predators, that they often kill more prey than is required for food, may be either advantageous or disadvantageous for biological control, depending on prey–host density. For example, the habit of partial egg feeding by *T. mundulus* has been implicated in the effective control of the *P. saccharicida*. On the other hand, by requiring less food to complete development, some parasitoids are able to build up large populations on relatively small host populations, and consequently can persist when the host population is small. This parasitoid characteristic has long been assumed to be desirable for successful biological control (Beddington et al. 1978).

For planthoppers, evidence indicates that predators are more effective than parasitoids in population suppression. Even though a great number of parasitoid species have been identified which attack all planthopper stages, more often predator species have been implicated in controlling their populations (Dyck and Orlido 1977; Kiritani 1979; Kuno 1979; Kenmore 1980; Ooi 1988; Denno and Roderick 1990 and references therein). For example, mirid predators have been held responsible for the successful control of *T. colocasiae* and *P. saccharicida* in Hawaii, whereas various predators (e.g., mirids, veliids, and spiders) have been considered important in suppressing populations of *N. lugens* (see the case history). The notion that predators are more important than parasitoids in suppressing planthopper populations is further supported by results from cage experiments, in which the exclusion of predators (Fig. 14.1) resulted in an increase of planthopper density, and by studies on the resurgence of planthopper populations following the application of insecticides that selectively kill predators (Heinrichs and Mochida 1984).

Conclusions: Evidence and Potential for Biological Control

Efforts to control planthopper pests with biological agents have yielded mixed results. For example, releases of natural enemies against *Perkinsiella saccharicida* and *Tarophagus colocasiae* in Hawaii have been considered successful examples of classical biological control, whereas other releases against these species have not been fruitful (e.g., Asche and Wilson 1989b). In particular, natural enemies have been less than reliable sources of control for endemic planthopper pests in an atmosphere of insecticides and disruptive cropping practices. To conclude this chapter, we review some of the salient features of planthopper pest–natural enemy systems and address some of the factors that influence the utilization of natural enemies in integrated pest management programs.

Several factors may account for the difficulty in controlling pest populations of planthoppers with natural enemies. In general, most pest planthoppers are considered r-type organisms, characterized by a high fecundity, a short life span, and high mobility, which enable them to increase rapidly in numbers beyond their carrying capacity (Southwood 1977b; Waloff 1980). For example, *Nilaparvata lugens* has been described as an outbreak or r-type pest in both temperate and tropical regions (Kuno and Hokyo 1970; Kuno and Dyck 1985). Under favorable conditions, low numbers of macropterous adults can initiate an exponential increase, resulting in an outbreak (Kisimoto 1981). Invasion of temporary habitats by the pest usually precedes its natural enemies; thus, suppression of pest population growth is delayed. Evidence from *Peregrinus maidis* suggests that prey augmentation could be used to synchronize enemy–planthopper interactions and promote control; the presence of leafhoppers encouraged the early colonization of corn by mirid egg predators, resulting in planthopper suppression (Napompeth 1973). In rice systems, the encouragement of nonpest alternatives (e.g., aquatic Diptera) may retain natural enemies during periods of low pest abundance or promote their early colonization of paddies (K. Gallagher and W. Settle personal communication). Thus, improved synchronization between highly mobile, r-type planthoppers and their natural enemies is very likely to improve biological control.

From the case studies presented, predators played a more important role than parasitoids in controlling planthopper populations. Parasitism was often low and of limited value in suppressing planthopper populations. Yet, hard evidence for this conclusion was lacking in many studies. In the case of *N. lugens*, low parasitism by one species was sometimes overcome by the combined effects of several parasitoid species. Similarly, even though the successful control of *P. saccharicida* and *T. colocasiae* in Hawaii was attributed to the action of a mirid predator, several parasitoid species were introduced in the initial stages of the control program. These parasitoids may have played a combined role along with predators in suppressing planthopper populations. Thus, a combination of natural enemies rather than only one should be considered for both classical biological control and enemy conservation programs.

Biological control is one of several tactics employed for pest management, and any of these tactics can interact with and influence the effectiveness of biological control. Early developments of integrated pest management focused on the combined use of chemical and biological control measures for pest management (Stern et al. 1959). More recently, researchers have also considered the interaction of biological control with other control tactics, in particular the use of resistant crop varieties (Boethel and Eikenberry 1986). Such interactions include both the compatibility and potential problems of combining tactics. With importation and conservation of natural enemies, the primary concern for integration with other control tactics is the conservation of natural enemies through preventive pest management.

The use of insecticides for planthopper management has long been recognized as promoting pest outbreaks (Heinrichs et al. 1984b). An integrated approach that couples the types of insecticides and the timing of their application with an enhancement in the effectiveness of natural enemies would decrease the probability of pest resurgence. This approach has proven effective for the control of *N. lugens* (Ooi 1988) and may be applicable to other planthopper pests. However, the use of insecticides may have complex and often unpredictable effects on ecosystem dynamics; thus, the use of nonchemical, preventive control tactics may be necessary to ensure natural enemy effectiveness.

The interaction of varietal resistance and biological control is not very well understood, and recent studies have indicated a need for information on the cause-and-effect relationships concerning the compatibility of these two strategies of pest control (Herzog and Funderburk 1985). Although host plant resistance is designed to reduce pest impact, resistance may interact with natural enemies in either positive or negative ways. Resistant cultivars may negatively affect the action of natural enemies by altering the nutritional quality of the prey to natural enemies as a result of feeding on "suboptimal" diets (El-Shazly 1972), by exposing the natural enemies to toxic resistant factors through the host (Campbell and Duffey 1979; Barbosa et al. 1982), or by reducing the searching efficiency of natural enemies through morphological features (e.g., glandular trichomes) (Rabb and Bradley 1968). For example, resistant varieties have been found to affect the searching behavior of *Cyrtorhinus lividipennis*, resulting in different predation rates (Cheng et al. 1992). However, most of the information available supports the hypothesis that the use of resistant varieties is compatible with biological control (van Emden and Wearing 1965; Bergman and Tingey 1979).

Habitat management for pest control may also influence natural enemy–pest interactions. Many cultural practices modify key components of natural enemy ecology, including cultivation, irrigation, and harvest practices. Farmers can manipulate many habitat features that may affect planthopper pests, ranging in scale from the morphology and chemistry of the crop plants, to the spatial and temporal pattern of crop growth, to the landscape-level interactions between habitats within a region. Each of these manipulations may have a profound effect on the ecology of natural enemies and biological control, but few controlled studies are available.

Speight (1983) reviewed four methods of habitat management for pest control: (1) altering the availability of the crop for pest exploitation in space and time (e.g., changing planting date or plant density), (2) altering the availability of the crop in relation to other types of vegetation (e.g., intercropping and adjacent habitats), (3) influencing pest survival in and around the site between consecutive crops (e.g., cultivation and cover crops), and (4) increasing natural enemy abundance and efficiency. Although all of these methods can have direct detrimental effects on the pest, the first three have the potential to adversely affect the conservation of natural enemies. Because some prey are required to maintain natural enemies in the habitat, any practice leading to local extinction of the pest (and therefore the natural enemy) may exacerbate pest problems in the future.

For example, Oka (1979) has proposed the use of synchronous culture of rice and a defined rice-free period within regions to reduce *N. lugens* densities. However, this practice may also reduce the effect of specialist natural enemies without consideration of alternate hosts during the rice-free period. Furthermore, the use of rice-free periods may interrupt the numerical response of generalist predators such as spiders in response to planthopper densities (Kenmore 1980).

Thus, as new crop and pest management practices are developed, researchers need to consider the impact of these methods on the natural enemy–pest interaction. From past experience, practices that positively affect natural enemy survival and ensure their synchronization with pest phenology are especially important for maximizing the potential of natural enemies to suppress planthopper populations.

Summary

1. A review of natural enemy interactions for five species of planthopper pests (the brown planthopper, *Nilaparvata lugens*; the taro planthopper, *Tarophagus colocasiae*; the sugarcane planthopper, *Perkinsiella saccharicida*; the corn planthopper, *Peregrinus maidis*; and the cereal planthopper, *Javesella pellucida*) shows that biological control efforts have been highly variable, but with several very successful endeavors.
2. Successful attempts of classical biological control against planthopper pests have involved enemies that attack eggs. Most notably, the effective biological control of *Perkinsiella saccharicida* and *Tarophagus colocasiae* in Hawaii was achieved with the introduction of egg predators (Miridae), although parasitoids (Mymaridae and Eulophidae) are implicated in the control of *Perkinsiella saccharicida* as well.
3. Studies of systems in which natural enemies are endemic (e.g., rice) also suggest that predators provide better biological control than parasitoids. Even though a great number of parasitoid species have been identified attacking all planthopper stages, predators have been implicated far more often in population suppression.
4. Nymphs and adults of planthoppers are attacked by a wide range of parasitoids (Dryinidae, Pipunculidae, and Strepsiptera) and predators (spiders, Nabidae, and Coccinellidae). Although nymphal mortality may be high, independent studies indicate that egg mortality, rather than nymphal mortality, is the key factor in population dynamics of planthoppers. Nevertheless, these predators and parasitoids should be conserved as part of the enemy complex.

5. Agronomic practices (e.g., pesticide applications and cropping patterns) that adversely affect natural enemy survival or desynchronize enemy and pest life histories are incriminated in planthopper outbreaks.
6. The combined action of several natural enemies that operate on the different life stages should be encouraged for the successful management of planthopper populations.
7. Future research should concentrate on modifications of crop and pest management practices that conserve natural enemies. For example, maintenance of refuges, prey augmentation, and the encouragement of nonpest prey should be promoted.

ACKNOWLEDGMENTS

We thank P. Gross, J.D. Hare, D.B. Hogg, D.W. Williams, and M.R. Wilson for their comments on earlier drafts of this chapter. This is Contribution Number 8589 of the Maryland Agricultural Experiment Station, Department of Entomology.

15

Genetics of Host Plant Adaptation in Delphacid Planthoppers

George K. Roderick

Introduction

Delphacid planthoppers, by any measure, are very successful plant-feeding insects. Indications of their success include a worldwide distribution (O'Brien and Wilson 1985), an extensive range of habitats occupied (Denno and Roderick 1990), and status as major agricultural pests (Wilson and O'Brien 1987). Characteristic of most planthoppers is a monophagous feeding habit and a close association with their host plants for feeding, mating, oviposition, and substrate-borne acoustic communication (Wilson et al. Chapter 1). Thus, understanding the factors which promote or constrain adaptation to a particular host plant species or variety is central to explaining patterns of host plant use (e.g., specialization) and deterring their continuing status as major agricultural pests.

Adaptation to novel plant species and/or crop varieties by planthoppers is accompanied by increases in survival, development time, body weight, honeydew production, oviposition rate, and/or reproduction (Sogawa 1977, 1982; Claridge and Den Hollander 1982; Pathak and Heinrichs 1982; Denno 1983; Denno and McCloud 1985; Gallagher 1988; Denno et al. 1989; Denno and Roderick 1992). Such traits, either singly or in concert, determine patterns of host plant use (Via 1990). For planthoppers and other pests on agricultural crops, the ability to feed and inflict damage is termed virulence, which is commonly used as a general measure of adaptation to novel cultivars (Claridge and Den Hollander 1982; Heinrichs Chapter 16; Gallagher et al. Chapter 17; Hare Chapter 18).

Over time, planthoppers may evolve different patterns of host plant use and/or virulence. The evolution of traits associated with host plant use, as for any traits, will depend not only on selection on those traits but also on the nature of the genetic variation underlying those traits (Futuyma and Peterson 1985; Falconer 1989; Via 1990; Hare Chapter 18). In both natural and agricultural settings, there is a greater understanding of the ecological factors influencing host plant selection by planthoppers than the underlying genetic mechanisms (Denno and Roderick 1990). From an ecological perspective, planthopper–host plant interactions are well known, as evidenced by studies on life history variation (Denno 1978, 1983; Roderick 1987; Denno et al. 1989, 1991), population biology (Drosopoulos 1977; Booij 1982a; Denno 1983; Strong and Stiling 1983; Strong 1989; Denno and Roderick 1990, 1992; Denno et al. 1991; Silvanima and Strong 1991), community structure (Raupp and Denno 1979; Tallamy and Denno 1979; Denno 1980; Stiling and Strong 1982a; Döbel 1987), and pest management (Claridge and Den Hollander 1982; Cook and Perfect 1985; Heinrichs et al. 1986a; Gallagher 1988). By contrast, our knowledge of the genetic basis of traits associated with host plant use in planthoppers is rudimentary and nonrepresentative; almost all information derives from studies of rice varietal resistance and breakdown in the brown planthopper, *Nilaparvata lugens* Stål (Gallagher et al. Chapter 17; Hare Chapter 18). Yet, if we are to interpret current patterns of host plant use and predict how genetic variation either constrains or contributes to its evolution, then a more thorough understanding of the genetic basis of host plant use will be required.

The general threat of planthoppers to rice has resulted in a constant search for genes conferring resistance and an accelerated effort to develop resistant crop varieties (Khush 1979; Smith 1990; Bonman et al. 1992; Hare Chapter 18). The difficulty of breeding for resistance is greatly exacerbated by the ability of planthoppers to "break down" rapidly resistant varieties and become virulent on novel plant genotypes. If breeding programs are to be successful, the durability of resistance traits in plants and the potential for planthoppers to overcome them must be understood. Even simple estimates of heritability, or the proportion of the phenotypic variation that is genetic, can be useful in predicting the rate at which virulence is likely to develop (Hare Chapter 18).

Genetic variation for host plant use is essential for planthoppers to incorporate new host plant species or varieties into their diets. However, their ability to feed on new plant genotypes will depend on a hierarchy of ecological and population genetic factors that act to isolate host-associated populations sufficiently to allow local genetic differentiation (Gould 1978, 1983; Futuyma and Peterson 1985; Via 1990). Such factors include (1) pronounced differences in the ability of insect genotypes to survive on different host types, (2) independent regulation of population size on host types, (3) low mobility of planthoppers among hosts, and (4) assortative mating and habitat selection (Mitter and Futuyma 1983). Furthermore, the genetic correlation of two or more traits associated with host plant use may either promote or constrain adaptation to novel hosts (den Bieman 1987a; Thompson 1988a; Jaenike 1990; Via 1984a, 1984b, 1990; Thompson et al. 1990).

In this chapter, two key issues concerning the genetics of host plant adaptation by planthoppers are addressed: first, whether significant amounts of genetic variation exist for traits that determine host plant use, and second, the extent to which the ecological and population genetic factors outlined above interact with genetic variation to promote host plant adaptation. Data from the literature is used to compute estimates of heritability for host plant use and virulence, thereby providing evidence for significant amounts of genetic variation for host plant use in planthoppers. Because a discussion of these topics demands some understanding of genetic variation and how it is measured, relevant genetic theory is included where necessary. Finally, observed genetic variation is related to the evolution of host plant adaptation. The discussion clarifies not only why planthoppers show the patterns of host plant use that they currently do (for example, monophagy with few examples of polyphagy) but also how the process of host plant adaptation occurs in natural and managed systems.

Evidence for Genetic Variation in Host Plant Use

Despite their importance as major agricultural pests, there are surprisingly few studies of planthoppers that consider the genetic foundation responsible for host plant adaptation. The first issue explored concerning the genetics of adaptation is whether traits for host plant use are controlled by one gene (monogenic) or by several genes, each with smaller effects (polygenic). This distinction is important because it will determine not only how host plant adaptation will proceed, but also how the amount of genetic variation is best summarized (Gould 1983; Via 1990; Simms and Rausher 1992). A second consideration is whether the amount of genetic variation is large relative to the total phenotypic variation; for polygenic traits, this assessment is equivalent to determining heritability (Falconer 1989). From the heritability of a trait and an understanding of selective forces in the field, the rate of evolution can be predicted (Falconer 1989; Hare Chapter 18). I recently used data from the literature to estimate heritabilities for traits associated with host plant use and virulence and I review those results here.

The number of genes that affect a trait will influence both how genetic variation is best measured and the evolutionary consequences of selection (Falconer 1989). In nearly all studies to date on planthoppers, the genetic basis of survival and performance on a particular host plant is consistent with polygenic control (den Bieman 1987a; Gallagher 1988; Hare Chapter 18). It should be noted that a polygenic basis is not necessarily expected to underlie all plant-associated traits either in planthoppers (den Bieman 1987a) or in other insects (Guldemond 1990; Thompson et al. 1990; Via 1990). For planthoppers, the most thorough experimental work that assesses the number of genes involved in host plant use comes from studies of *Nilaparvata lugens*, the brown planthopper. In crosses between demes of *N. lugens* associated with different varieties of rice (*Oryza sativa*), Den Hollander and Pathak (1981) found that all the F_1 and F_2 phenotypes for traits that determine host plant use were intermediate between the parental types. Further, their data gave no indication of dominance or recessiveness nor evidence of sex-linkage. Similarly, in crosses between *N. lugens* populations inhabiting rice and populations (currently considered to be a distinct species) inhabiting the closely related grass *Leersia hexandra*, both F_1 and F_2 individuals showed traits that were intermediate between parental types (Claridge et al. 1985b). Although in practice it becomes difficult to determine the exact number of genes involved (Wright 1968), even for as few as three genes, one can conclude that host plant use and virulence in *N. lugens* is not inherited as a single factor, but instead is under polygenic control.

Virulence in *N. lugens* evolves quickly when populations are exposed to selection by "novel" rice varieties (Kaneda and Kisimoto 1979; Den Hollander and Pathak 1981; Sogawa 1981a; Claridge and Den Hollander 1982; Pathak and Heinrichs 1982). Because several genes are implicated in planthopper virulence, genetic variation can be summarized by heritability, h^2, which is simply the ratio of genetic variation to total phenotypic variation, or V_G/V_P. Alternatively, heritability can be expressed as a ratio of the response to selection (R, difference in mean value of the trait between the population in the generation following selection and the parent population) and the selection differential (S, difference in mean value between the selected population and the parent population), or $h^2 = R/S$. This relationship can be used to estimate a realized heritability over several generations, where R is the response to selection in each generation and S is the cumulative selection differential or the amount of selection summed over the generations. Because no direct estimates of heritability exist for host plant use by planthoppers (Hare Chapter 18), realized heritabilities are calculated from three published studies (Den Hollander and Pathak 1981; Claridge and Den Hollander 1982; Pathak and Heinrichs 1982). Before reviewing the results of these analyses, the components of selection studies are first outlined by reviewing the methods and results of Claridge and Den Hollander (1982), the longest selection study of planthoppers to date.

Claridge and Den Hollander (1982) selected a series of lines of *N. lugens* for increased virulence on different rice varieties. For each of three planthopper "biotypes" from the Philippines, they initiated lines on each of two additional rice varieties on which planthoppers were selected for increased survival and performance at each generation. For each "biotype," the established line was considered as the control line, whereas the "novel" varieties contained different genes for resistance. Another population from Australia was selected on the rice variety, TN1, to which it was not adapted but which contained no additional genes for resistance. In each line, the selected individuals consisted of the first 10 individuals of each sex to reach adulthood out of the 100 nymphs used to start each generation. In such a design, selection can potentially act on both survivorship and development time, both of which influence host plant use. Claridge and Den Hollander (1982) continued the selection for at least 10 generations. They found that survivorship increased in the experimental lines over the course of the experiment as did body weight and honeydew production. This experiment demonstrates that adaptation to a novel variety can proceed very rapidly.

I examined Claridge and Den Hollander's (1982) data for changes in both development time and survival. Development time in the three control lines actually increased throughout the duration of the experiment, perhaps reflecting either a change in environmental conditions or the effects of inbreeding. However, development time in four of the six selected lines decreased relative to controls. The Australian line also showed a significant decrease in development time. For development time, heritability could not be estimated easily from the published data because only the median and not the variance was presented. Survival in all three control lines decreased slightly during the course of the experiment, probably for the same reasons suggested above for the increased development time in these same lines. In contrast, survival increased in four of the six selected lines relative to controls, and the Australian line also showed a significant positive response to selection on survival.

To calculate the realized heritability of survival, it was assumed that survival was a threshold trait, governed by an underlying normally distributed liability (Falconer 1989). Liability values above the threshold lead to one phenotype, whereas values below the threshold lead to the other phenotype. Thus, although the resulting phenotypes of the two classes will never be normally distributed, the underlying distribution of liability, which is determined by many genetic factors, will be normally distributed and therefore within the domain of quantitative genetics. It was further assumed that the variation in liability was the same before and after selection. For each line, realized heritability was calculated from the regression of the response to selection each generation, R, on the cumulative selection differential, S, or $h^2 = R/S$ (Fig. 15.1). For threshold characters, the selection differential and the response to selection depend only on the incidence of the trait in the population (Falconer 1989). Responses to selection were adjusted to reflect changes in the control lines when present. Using this method, heritabilities for survival were calculated from data generated by Den Hollander and Pathak (1981), Claridge and Den Hollander (1982) and Pathak and Heinrichs (1982). In all three studies, "biotype 1" was selected on two rice varieties, MUDGO (resistance gene *Bph1*) and ASD7 (resistance gene *bph2*), providing three replicated measures of genetic variation for survival (virulence) with respect to two different resistance genes (Table 15.1). Here, results are summarized only from the selection experiments ("biotype" on rice variety) that were replicated across the three studies.

Heritabilities were sufficiently high to suggest that significant amounts of genetic variation exist in *N. lugens* for survival on rice varieties containing different genes for resistance (Table 15.1). Thus, under similar levels of selection in the field, populations should quickly adapt to novel varieties, a prediction that is supported by field observations (Gallagher et al. Chapter 17; Hare Chapter 18). The results also demonstrate that significant genetic variation exists for virulence against varieties containing resistance genes that show either dominant or recessive expression. Two important features of heritability warrant discussion here. First, the estimation of heritability is dependent on the environment in which it is measured (Falconer 1989; Simms and Rausher 1992). Thus, in stressful environments, estimates of heritability may be quite different from those made in more favorable environments (Ott and Roderick unpublished data), and estimates made under laboratory conditions may differ from field estimates. Second, an estimate of heritability represents the ratio of genetic variation to phenotypic variation present only at the time of the experiment. High levels of heritability may have been present in the past even in the same environment, but may no longer be evident. It is possible through selection to exhaust the available additive genetic variation, which will result in very low heritabilities (Fisher 1930). Although this premise is often used to explain why traits that are tightly associated with fitness often show little additive genetic variation, low heritability in life history traits may also reflect a large environmental component (Price and Schluter 1991).

Claridge and Den Hollander (1982) went on to score body weight and honeydew production after their experiment with *N. lugens*; these traits have been considered to be indicators of planthopper performance and fitness (Sogawa 1977; Denno and McCloud 1985; Gallagher 1988). Overall, body weight and honeydew production increased significantly over the course of the experiment, despite no direct selection on these traits. This result suggests that these traits are genetically correlated with survivorship

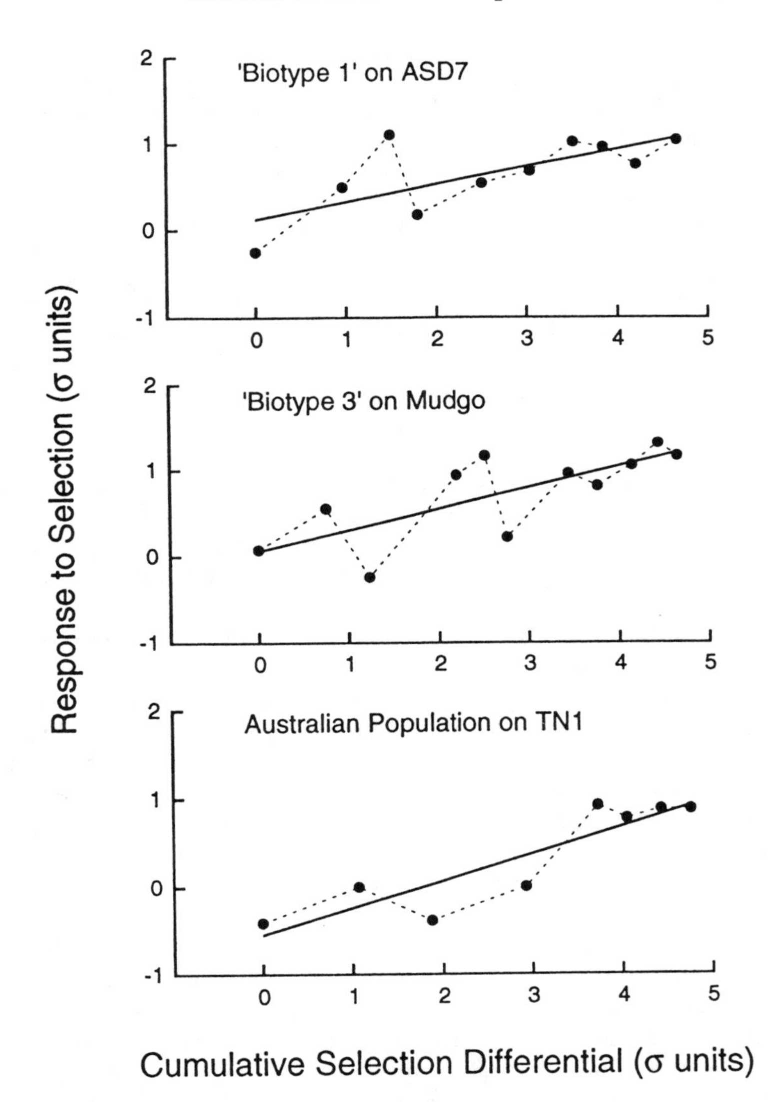

Figure 15.1. Heritability for survival of three "biotypes"/populations of *N. lugens* on three different rice varieties; calculated from data presented in a series of selection experiments by Claridge and Den Hollander (1982). Realized heritability, h^2, equals the slope from a regression of the response to selection, R, on the cumulative selection differential, S (see text). Values of h^2 for "biotype 1" on ASD7 (*bph2* gene for resistance), for "biotype 3" on MUDGO (*Bph1* gene for resistance), and for an Australian population on TN1 (no gene for resistance) were 0.21, 0.25, and 0.31, respectively (from Roderick unpublished data).

Table 15.1. Realized heritabilities for survival of *N. lugens* on different rice varieties calculated from data presented in selection experiments by Den Hollander and Pathak (1981), Claridge and Den Hollander (1982), and Pathak and Heinrichs (1982). In each study, so-called planthopper "biotypes" originally raised on variety TN1 (no gene for resistance) were selected on rice varieties MUDGO and ASD7 for 2 (Den Hollander and Pathak 1981), 10 (Claridge and Den Hollander 1982), and 6 (Pathak and Heinrichs 1982) generations (see text for details of the experiment and calculations). From these three published studies, mean heritability (h^2) and its standard error (SE) were calculated.

Rice Variety	Rice Gene for Resistance	h^2	SE
MUDGO	*Bph1*	0.279	0.086
ASD7	*bph2*	0.320	0.109

and/or development time. Provided that heritability can be measured for each trait and that laboratory settings reflect field conditions, selection experiments may provide the opportunity to measure the genetic correlations among traits (Service and Rose 1985; Falconer 1989; Simms and Rausher 1992).

For other insect taxa, significant amounts of genetic variation have been measured for traits that affect performance on host plants (Via 1990). Although the presence of significant amounts of genetic variation in planthoppers for host plant use or virulence is perhaps not surprising, the magnitude of the genetic variation relative to phenotypic variation is large; some estimates for planthoppers from unreplicated studies exceed 0.50 (Roderick unpublished data). Further, genetic variation exists for the exploitation of rice varieties with two different genes for resistance. Thus, in the field, response to selection for host plant use or virulence could proceed rapidly.

Additional estimates of heritability for quantitative traits that affect host plant-associated fitness will be especially useful for comparative purposes and because of their value as predictors of potential population performance. Because many species of planthoppers can be reared with ease in the laboratory or in glasshouse settings, half-sib breeding studies could be undertaken to estimate genetic variation for host use and dissect this variation into components attributable to maternal, dominance, and additive effects. Although studies with *N. lugens* collectively suggest that adaptation to novel varieties occurs with ease, it is not currently known how much genetic variation exists in populations of planthoppers for feeding across several host plant species. In this context, a comparison between monophagous and polyphagous species would be particularly illuminating.

Factors Promoting Host Plant Adaptation

Coupled with the presence of genetic variation for traits governing host plant use and selection for those traits in the field, a variety of ecological and population genetic factors may interact to promote host plant adaptation through the partial or complete isolation of host-associated populations. In the sections that follow, I review how four factors might promote host plant adaptation in delphacids. They are differential fitness of planthopper genotypes on across host types, independent regulation of population size on each host type, low mobility of planthoppers among host types, and assortative mating and habitat selection.

Fitness of Planthopper Genotypes on Different Hosts/Varieties

Planthopper genotypes that differ in fitness or performance across hosts or varieties have the potential to become increasingly better adapted to those particular host types (Mitter and Futuyma 1983). This process may eventually lead to genetic divergence of host-adapted populations. Here, evidence for the differential performance of planthopper genotypes across host plant varieties and its genetic basis is reviewed; assessments are made for both field populations of planthoppers as well as for laboratory-maintained "biotypes".

ADAPTATION TO DIFFERENT VARIETIES AND HOST PLANT POPULATIONS

Many studies have documented host plant-related variation in planthopper fitness such as that shown in tests of cultivar resistance (reviews by Sogawa 1982; Gallagher 1988). Only a handful of investigations, however, have recorded information on planthopper genotype in addition to data on host plant use, and most of these involve *N. lugens* on rice. Several studies have shown variation among *N. lugens* populations in performance on different host varieties, even among populations that are in close geographic proximity. For instance, Claridge et al. (1982a) examined *N. lugens* from five different cultivated rice varieties and a wild species (*Oryza rufipogon*) in Sri Lanka. They tested the performance of each population in the laboratory on each source variety and on the standard varieties TN1, MUDGO, and ASD7. Local populations differed in virulence across the various rice varieties, but generally were most vigorous on their original host variety. For example, when ranked, each population performed better on its original variety than might be expected by a random ranking. Claridge and coworkers measured both percent weight change and honeydew production over 24 hours and these performance traits were highly correlated. All populations performed well on the wild rice species. By contrast, the newest cultivar, BG379-2, showed the most resistance to all planthopper populations except to the one living on it. Finally, TN1, which contains no additional genes for resistance, was not found to be universally susceptible to all populations of *N. lugens*.

In a similar study, Gallagher (1988) reared *N. lugens* from five Philippine populations each obtained from a different rice variety. He found a surprising amount of variation among populations in both longevity and total fecundity when individuals were reared on the different varieties. Longevity and total fecundity were highly correlated. Particularly important was the spatial scale of adaptation; populations that were as little as 100 m apart (Ozimis 54 versus Pilit) showed significant differences in their abilities to use different rice varieties. Gallagher argued that because individuals of *N. lugens* probably originated that season from the same large gene pool, adaptation to the local rice varieties occurred within that growing season. All populations exhibited their lowest fitness on rice varieties containing novel genes for resistance and highest fecundity on the susceptible TN1 variety. Even though rice varieties with the *Bph1* gene for resistance had not been planted in the area for many years, the planthoppers in that area retained their ability to survive and reproduce on them. Furthermore, some populations maintained a high level of tolerance to rice varieties containing either *Bph1* or *bph2* genes for resistance, indicating that planthoppers had adapted to more than one gene for resistance. These results suggest that there is little cost to maintaining the ability to survive on varieties containing either the *Bph1* gene or those containing the *bph2* gene (see Hare Chapter 18). Gallagher (1988) also measured honeydew production on the different varieties. On all the newer varieties, results for honeydew production experiments mimicked the previous results for both longevity and egg production, whereas on the susceptible variety, TN1, honeydew production could not be used as an index of overall performance.

For species other than *N. lugens,* few studies have examined variation among planthopper populations in their performance on different host plant varieties or populations. In a study of two salt marsh planthoppers, Roderick and Denno (unpublished data) found that individuals of each species performed better on plants from their native locality than on plants of the same host plant species from another locality. In this experiment, individuals of *Prokelisia dolus* and *P. marginata* from populations in New Jersey, USA were reared in incubators on host plants of the same salt marsh grass species (*Spartina alterniflora*) transplanted from two distant locations in the United States, New Jersey (mid-Atlantic coast) and Florida (Gulf coast). All plants were maintained under constant management conditions in a greenhouse for over a year to minimize environmental effects stemming from differences between source areas. Nymphs were placed on plants at a density of 10 per plant. Individuals of both *P. dolus* and *P. marginata* developed faster on plants originating from their native New Jersey site (Fig. 15.2A). The results for survival differed between species: *P. dolus* survived equally well on plants from both sources, whereas *P. marginata* survived much better on its original host plants (Fig. 15.2B). Where significant differences in body size existed between host sources, planthoppers of both species were larger on their native hosts. These results suggest that *Prokelisia* planthoppers show some degree of regional adaptation to host plant populations.

Thus, populations of both *Nilaparvata* and *Prokelisia* show evidence for adaptation to their local plant variety or host population. However, in both systems, the degree of local adaptation was not so strong as to preclude their eventual successful performance on novel plants, at least at the varietal level. However, these comparisons among populations are necessarily circumstantial and do not provide much information on the genetic basis of host plant adaptation.

THE BIOTYPE CONCEPT AND HOST PLANT ADAPTATION

"Biotypes" have been used to describe either host-associated insect populations or, more specifically, discrete genetic entities adapted to particular host plant types (Claridge and Den Hollander 1983). Populations of *N. lugens*, which are virulent to varieties of rice carrying particular single genes for resistance have been labeled "biotypes" and have been given numbers corresponding to the rice genes for resistance (nomenclature reviewed by Gallagher et al. Chapter 17). Although the stability of these "biotypes" and the usefulness of the term has been questioned (Claridge and Den Hollander 1980, 1983; Sogawa 1981b; Claridge et al. 1984a; Gallagher 1988; Padgham et al. 1989; Heinrichs Chapter 16; Gallagher et al. Chapter 17; Hare Chapter 18), studies of "biotypes" reveal information concerning the genetic basis of varietal adaptation.

Although minor morphometric differences among "biotypes" have been identified (Saxena and Barrion 1985), there is much overlap in morphological characters, and after one generation of exposure to a susceptible variety, differences among "biotypes" became smaller yet (Claridge et al. 1984a). Thus, morphological differences among "biotypes" may be a function of the selection regime imposed by laboratory rearing. Claridge and Den Hollander (1982) found ample variation for virulence within "biotypes", and after 10 generations of selection on different rice varieties, differences between "biotypes" were greatly reduced. As noted previously,

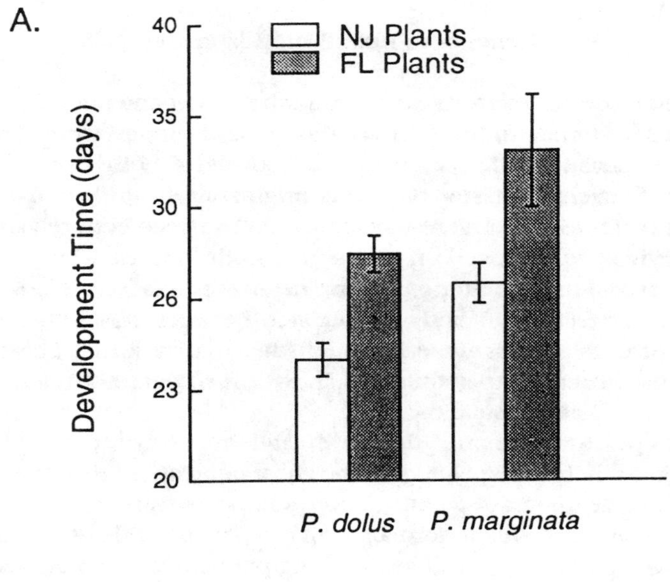

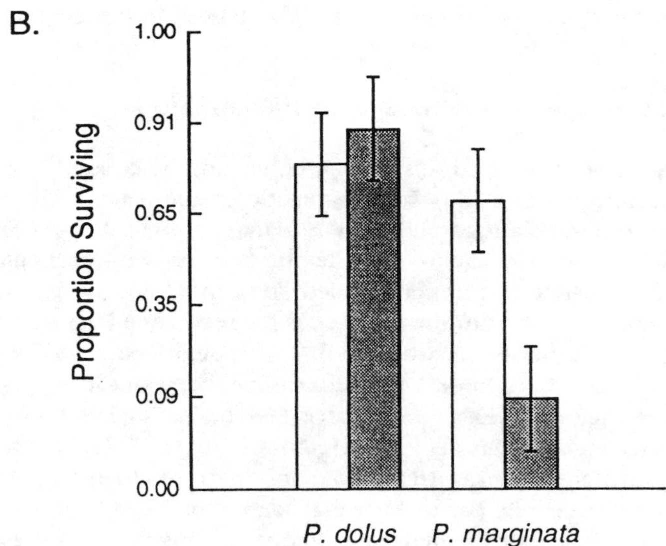

Figure 15.2. Common garden experiment showing host plant population effect on (**A**) development time and (**B**) survival of both *Prokelisia dolus* and *P. marginata*. For both species of planthoppers, individuals from New Jersey, USA were reared on plants of *Spartina alterniflora* from New Jersey (Atlantic coast) and Florida (Gulf coast Florida) (Roderick and Denno unpublished data). For development time (**A**), significant host plant and species effects existed such that planthoppers developed more quickly on native plants from New Jersey. Days to adulthood were log-transformed prior to analysis; axis for development time shows back-transformed values. For survival (**B**), a significant plant by species interaction was present; *P. dolus* showed no effect of host plant population, whereas *P. marginata* survived better on native plants from New Jersey. Data were arcsine–square root transformed prior to analysis; axis for survival shows back-transformed values.

these results indicate that sufficient genetic variation exists for rapid adaptation and concomitant change in virulence to different rice varieties. In addition, biochemical evidence for *N. lugens* showed that the so-called "biotypes" differed very little genetically and were no more different than the variants which exist in field populations (Den Hollander 1989; Demayo et al. 1990; Saxena et al. 1991).

Thus, planthopper "biotypes" cannot be considered discrete genetic entities. Whereas certain "biotypes" exhibit adaptation to particular host varieties compared to others, short periods of selection on novel varieties can alter their survival and performance. Results of Claridge and Den Hollander (1982), Gallagher (1988), and others [reviewed in Claridge and Den Hollander (1983)] illustrate that there is not a one-to-one correspondence between a single gene in a *N. lugens* "biotype" and any single gene for resistance in the rice host.

From information on both field populations and "biotypes" of *N. lugens*, two points can be made concerning the genetic constraints on host plant adaptation. First, although there is no direct evidence, traits that govern performance on rice do not appear to be integrated in ways that preclude the use of several varieties at once; if traits were negatively correlated across varieties, one would expect to see high performance on one variety associated with low performance on another. Generally, this pattern does not obtain. Second, there appears to be little cost to adaptation. For example, with the adaptation of a planthopper population to a new variety, high performance on its original variety is not usually lost. Thus, genetic factors influencing planthopper performance do not appear to constrain adaptation to novel varieties; in fact, they may promote it. Whether this pattern also results for planthoppers at the plant species level remains to be seen, but the prevalence of monophagy (feeding on only one plant species or a few closely related taxa) in planthoppers suggests that the genetic basis for host plant adaptation may operate differently at the plant species level, perhaps because many more genes for "resistance" are involved.

Regulation of Population Size on Different Hosts/Varieties

Host-associated differences in population dynamics may also lead to asynchrony, isolation among populations, and, hence, promote local adaptation to a particular host plant (Mitter and Futuyma 1983). In addition to regulation of planthopper populations as a direct effect of plant resistance (discussed above), the complex of planthopper predators and the effects of those predators on planthopper numbers can differ across host plant types. Thus, natural enemies may selectively influence planthopper densities and thereby affect host plant adaptation and patterns of host plant use. For example, Heong et al. (1991a) found geographic variation in the structure of the arthropod community on rice and suggested that rice variety was one of the contributing variables. Similarly, early maturing varieties of rice had lower predator/prey ratios (Heinrichs et al. 1986a). Also, predators were able to inflict more mortality and suppress populations of *N. lugens* more on moderately resistant compared to susceptible varieties of rice (Kartohardjono and Heinrichs 1984). These observations may explain why host plant adaptation can proceed more rapidly for *N. lugens* on novel varieties in the absence of natural enemies brought about by frequent applications of insecticides (Gallagher et al. Chapter 17).

Additional studies document differences in planthopper population dynamics across host plant species and/or habitats, although the reasons for these differences are less clear. For example, *Delphacodes campestris* produces two large generations on intermediate wheatgrass and Russian wildrye, whereas on tall fescue only the second generation was substantial (Whitmore et al. 1981). In another study of four species of planthoppers and leafhoppers, significant differences in density of each species existed across replicated weed-free, broad-leaf, grassy, and weedy plots (Lamp et al. 1984). Finally, differences in habitat structure influence the spider community and risk of predation for planthoppers associated with different host plant species and habitats on North American salt marshes (Döbel 1987). Although each of these studies demonstrates the possibility for promoting adaptation to specific host plant species or varieties, it is not obvious whether host-related differences in population "regulation" alone (e.g., suppression by natural enemies) are sufficient to foster asynchrony, isolation, and ultimately monophagy. Although regulatory factors, including natural enemies, may influence the rate of host plant adaptation, the specific effects of predators on host plant adaptation will depend on the behavior and physiology of the herbivore and predator, the dynamics of the herbivore/natural enemy system, and the degree of plant resistance (Gould et al. 1991).

Mobility of Planthoppers Among Hosts

Extensive movement of individuals among subpopulations and subsequent reproduction will tend to prevent or retard local differentiation (Wright 1968; Slatkin 1987; Roderick 1992). In some planthopper species there is considerable seasonal movement between plant species or habitats. For example, individuals of *Javesella pellucida* move between host plant species that are associated with different habitats (Prestidge 1982a; Prestidge and McNeill 1983). The salt marsh planthopper *Prokelisia marginata* moves seasonally between different growth forms of its host, *Spartina alterniflora* (Denno 1983). In such migrations, the two host-associated populations are not separate, but rather form one population that moves as an interbreeding unit between habitats.

Although most movement by planthoppers occurs over fairly short distances (Cook and Perfect 1985a, 1989; Denno et al. 1991), at least some species are capable of long-distance flight. For example, migrants of *N. lugens* colonize temperate Japan and Korea from tropical/subtropical Asia yearly (Kisimoto 1976; Kuno 1979). Immigrant populations are very small and colonists are likely to have originated from several different rice-growing areas (varieties) on mainland Asia. Therefore, the composition of the colonizing population with respect to the virulence genes they bring may determine the localized response of the colonizing population to the particular rice varieties grown in temperate Asia (see Claridge and Den Hollander 1982). Studies by Iwanaga et al. (1987) of wing-form responses to density support the contention that genetic variability among founding colonists is very high.

By contrast, in South East Asia where rice is grown continuously and there is constant, short-distance dispersal among fields (Cook and Perfect 1985a, 1989), adaptation to rice varieties may proceed differently (Gallagher 1988). For instance, in tropical Asia, dispersal of *N. lugens* may be limited to as little as a few kilometers (Dyck et al. 1979; Cook and Perfect 1985a, 1989). Demayo et al. (1990) examined the extent of population subdivision of *N. lugens* in the Philippines by a survey of variation in protein coding genes through allozyme electrophoresis. Although significant differences in allele frequencies existed among populations in the Philippines for all four loci examined, the magnitude of the differences was relatively small. In their study, only 9% of variation in allozyme frequencies was attributable to variation among populations (i.e., $F_{ST} = 0.089$). Here, genetic similarity could result from moderate to high levels of gene flow among populations or from a recent shared history of all populations. Given the genetic similarity among populations, it may seem surprising that *N. lugens* shows significant differences among populations in the ability to use rice varieties (reviewed above). Local selective forces that may contribute to differences among populations include different histories of host plant varieties, insecticide use, and climate (Demayo et al. 1990).

Whether adaptation to a local variety occurs will depend on the scale and magnitude of selection relative to planthopper movement. Strong selection for virulence may cause differentiation among populations regardless of the levels of gene flow. On the other hand, if many plant varieties are interplanted, the scale may be too fine to allow local host plant adaptation, even when dispersal is limited. At a larger regional scale, overall rates of adaptation may be accelerated by moderate levels of gene flow which could spread rare genes for virulence among the component populations (Caprio and Tabashnik 1992). A better understanding of the relative effects of gene flow and selection for host plant adaptation of planthoppers will require more detailed studies of local planthopper movement, the spatial configuration (e.g., mixed varieties versus monoculture) and genetic makeup of potential host plants, and the interaction of the two.

For instance, the host plant itself may also influence movement and gene flow. Both survival and dispersal were lower for *N. lugens* reared on a resistant rice variety compared to a susceptible one (Padgham 1983a). In *P. marginata*, host plant nutrition has an effect on wing-form development and, hence, movement (Denno et al. 1985), and individuals of both sexes are more likely to emigrate from low-quality than high-quality plants (Roderick 1987, unpublished data). Additionally, the phenology of the host plants may serve to limit gene flow among populations of Homoptera (Wood and Guttman 1982; Butlin 1990). The effects of host phenology on isolation of planthopper populations has not been studied, although there is evidence that some species synchronize their life cycles with periods of high plant nutrition (reviewed by Denno and Roderick 1990).

Finally, the wing-form composition of planthopper populations may influence patterns of gene flow and thereby affect rates of host plant adaptation. Due to reduced dispersal, brachypterous species or populations may be more genetically subdivided than macropterous species or populations. Further, adaptation to a particular host plant can proceed rapidly if that host is abundant (Hsiao 1978). It follows that the combination of brachyptery on a common host may lead to very rapid host adaptation with the establishment of local host "races" and further genetic subdivision among isolated planthopper populations (see Booij 1982a). However, in many primarily brachypterous species, a few macropters are produced seasonally or under specific conditions (Denno et al. 1991), and macropterous planthoppers may choose not to fly (Roderick 1987; Denno et al. 1991; Roderick and Caldwell 1992). Studies of whether brachypterous species or populations are indeed more genetically subdivided than their macropterous counterparts and whether this effect leads to greater rates of host plant adaptation are needed.

Assortative Mating, Habitat Selection, and Host Preference

If individuals choose mates associated with habitats or host plants on which their offspring are more likely to survive, then populations on different hosts could diverge genetically (Mitter and Futuyma 1983). For planthoppers, evidence exists for both assortative mating, habitat preference, and, in several studies, for a link between these factors and offspring performance. Use of substrate-borne acoustic signals is widespread in planthoppers and provides a possible mechanism for assortative mating (Claridge 1985a). In many planthopper lineages, acoustic differences that contribute to complete or partial isolation have been found both among species (Strübing and Hasse 1975; de Vrijer 1981; Booij 1982b; Claridge et al. 1985b; den Bieman 1987a; Claridge 1990a; Heady and Denno 1991; O'Connell 1991) and among geographically isolated populations within species (Booij 1982b; Claridge et al. 1984b, 1985a, 1988; Claridge 1990a).

Evidence also exists for the ability of planthoppers to select particular host types. For example, variety-associated populations of *N. lugens* preferred the variety on which they had been established (Domingo et al. 1983; Heinrichs and Medrano 1984; Claridge et al. 1985b), even though host preference might change over time (Kaneda and Kisimoto 1979). However, data are mixed concerning the tendency for adult females of *N. lugens* to selectively oviposit on rice varieties where offspring performance is best (Choi et al. 1979), suggesting that preference and performance are not necessarily positively associated. It is noteworthy, however, that females from populations ("species") of *N. lugens* inhabiting *Oryza* and *Leersia* both preferred to oviposit on their natal host on which offspring performance was best (Claridge et al. 1985b).

Several species of *Muellerianella* feed on different monocot species and each performed best on its natural host (Booij 1982a). For a number of species of *Ribautodelphax*, individuals preferred host grasses that enhanced offspring survival, suggesting that feeding and oviposition preferences are positively associated (den Bieman 1987a). Because such mixed-species populations prefer different hosts, contact among species (or host-associated populations) may be reduced, fostering reproductive isolation.

No single evolutionary sequence appears to account for patterns of mating, habitat preference, and host plant use for all planthopper species. Claridge et al. (1985b) suggested that host-associated species of *N. lugens* evolved differences in courtship in allopatry and that differences in host plant use were secondary. Booij (1982a) showed that for species of *Muellerianella* acoustic differences did not prevent mating in the laboratory, but that differences in hosts and habitats were of primary importance for their isolation under natural conditions. For species of *Ribautodelphax*, den Bieman (1987a) demonstrated that host choice was more restricted than subsequent survival and argued that behavioral aspects of host plant selection eclipsed any plant physiological effects. In these studies, both Booij (1982a) and den Bieman (1987a) provided evidence that host plant choice may be under the control of a single gene or only a few genes, a finding which contrasts to the apparent multiple gene control of physiological adaptation shown for other planthopper species (Claridge and Den Hollander 1982; Gallagher 1988).

In sum, the potential exists for both assortative mating and habitat preference to contribute to the isolation of populations and the promotion of host plant specialization. How many genetically different traits actually determine host plant choice is currently under debate (Jaenike 1986; Thompson 1988a; Via 1990; Thompson et al. 1990). For "*N. lugens*" populations on *Oryza* and *Leersia* and species of *Ribautodelphax* on different grass species, two aspects of divergence on new hosts, oviposition preference by females and the performance of her offspring, are positively correlated (Claridge et al. 1985b; den Bieman 1987a). It remains to be studied whether these correlations are genetically based. A positive genetic correlation between oviposition preference and offspring performance could narrow the range of potential hosts (Jaenike 1990) for planthoppers.

Conclusions

Considerable genetic variation is evident both within and among planthopper populations for the ability to use new host varieties (Claridge and Den Hollander 1982; Roderick unpublished data). Given that the potential exists for rapid exploitation of new host plant varieties, it is surprising to find that so many planthopper species are restricted to a single host plant species (Wilson et al. Chapter 1). Superficially, planthopper genotypes do not appear to show negative trade-offs in performance across host plant varieties, but such trade-offs and their potential genetic basis remain largely unstudied at the plant variety and species levels. If such negative correlations in performance across hosts have a genetic basis, this result would help to explain the monophagous feeding habit of most delphacids. Similarly, a positive genetic correlation between adult preference for and offspring survival on a particular host would promote host plant specialization (Jaenike 1990). Although such associations between adult preference and offspring performance are evident for planthoppers (Claridge et al. 1985a; den Bieman 1987a), a genetic basis for these associations has not been studied.

Evidence suggests that factors other than limited genetic variation for host use may be operating to isolate planthopper populations either partially or completely, thereby allowing for host plant specialization at the level of host plant species. These include independent population regulation on different hosts (Döbel 1987) and assortative mating coupled with habitat selection (Booij 1982a; Claridge et al. 1985a; den Bieman 1987a). In several planthopper species (e.g., *N. lugens*), extreme mobility may prevent adaptation to local rice varieties and the genetic subdivision of populations. It is interesting to note that the two most polyphagous species (*Javesella pellucida* and *Laodelphax striatellus*) are highly mobile (Denno and Roderick 1990). However, high mobility alone does not necessarily promote polyphagy at the level of host plant species because several highly mobile species are very host-specific (e.g., *N. lugens* and *Prokelisia marginata*; Sogawa 1982; Denno 1983). The occurrence of brachyptery in many species and populations may lead to a reduction in gene flow among populations and promote adaptation to a locally abundant host.

Currently, most of our information concerning the genetics and evolution of host plant adaptation in planthoppers comes from studies of *N. lugens* and varietal resistance in rice. What is clearly needed to broaden our understanding of host plant adaptation is an investigation of the genetics of host plant use by oligophagous or truly polyphagous planthopper species (e.g., *Javesella pellucida* or *Laodelphax striatellus*). For example, what is the extent of genetic variation within populations for use of different host species, what are the genetic correlations among such traits, and how much gene flow occurs between these populations? Likewise, do factors such as independent population regulation, assortative mating (perhaps through acoustical differences), and/or habitat selection promote genetic subdivision and drive the evolution of host use? Answers to such questions should be relatively straightforward to obtain and would greatly expand our limited knowledge of the genetic mechanisms which drive host plant adaptation in delphacid planthoppers.

Summary

1. The limited studies available for planthoppers indicate that significant amounts of genetic variation exist for traits that determine host plant use and/or virulence and that heritabilities for such traits can be significant.
2. Evidence both from field populations and from laboratory studies of so-called "biotypes" of *Nilaparvata lugens* suggests that despite adaptation to local host plant varieties, planthopper genotypes are able to adapt rapidly to novel varieties. Furthermore, there does not appear to be a significant cost associated with adaptation; once adapted to a novel variety, individuals do not lose there ability to perform well on their original variety.
3. Studies of host plant adaptation in planthoppers demonstrate a relative ease of incorporating new varieties into their repertoire of hosts. This

observation contrasts with patterns of host plant use at the plant species level where most delphacid species exhibit extreme monophagy. At the plant species level, there is weak evidence for negative correlations in performance across host species and positive associations between adult preference and offspring performance. A genetic basis for such correlations could contribute to the high incidence of host plant specialization in the Delphacidae, but the matter remains unstudied.

4. Other factors that may promote host plant adaptation and specialization in planthoppers include host-related differences in population dynamics (e.g., suppression by natural enemies), limited levels of gene flow among host species or habitats, and/or host-associated assortative mating perhaps coupled with oviposition preference.
5. Further studies at the level of host plant species that measure (1) genetic variation for and correlations among traits influencing host plant use and/or virulence and (2) selective forces associated with host plant use are critical to further our understanding of the factors promoting or constraining host plant adaptation in delphacid planthoppers.

ACKNOWLEDGMENTS

I gratefully acknowledge support from the USDA (90-1-34103-5105 and 58-1275-1-127), D. Bottrell and the Division of Entomology of the International Rice Research Institute, the Rockefeller Foundation, the Maryland Agricultural Experiment Station Competitive Grants program, NIH Biomedical Research Funds, the Graduate Research Board of the University of Maryland, the Department of Entomology of the University of Maryland, and the Hawaiian Evolutionary Biology Program of the University of Hawaii. Support for research on *Prokelisia* species was provided by an NSF grant (BSR-8614561) to R.F. Denno. I thank R. Denno, R. Gillespie, J. Ott, J. Perfect, and two anonymous reviewers for very helpful comments and suggestions.

16

Impact of Insecticides on the Resistance and Resurgence of Rice Planthoppers

Elvis A. Heinrichs

Introduction and Historical Perspective

Several species of planthoppers are major rice pests. *Nilaparvata lugens* (Stål) (brown planthopper) and *Sogatella furcifera* Horvath (whitebacked planthopper) occur in South and Southeast Asia throughout the year and migrate to Japan and Korea, presumably from mainland China in June and July (Nagata 1982). *Laodelphax striatellus* Fallen (small brown planthopper) is a serious pest in Japan and *Tagosodes orizicolus* Muir (the "sogata") is an important pest in the Caribbean and Central and South America (Dale in press). These planthoppers damage rice plants directly by removing phloem sap. *Nilaparvata lugens, L. striatellus,* and *T. orizicolus* are also vectors of rice viruses. *Nilaparvata lugens* and *S. furcifera* have only within the last few decades emerged as important economic pests in tropical South and Southeast Asia and *T. orizicolus* in the Americas (Heinrichs 1988). However, *N. lugens* and *S. furcifera* figure prominently in the history of rice cultivation in the temperate regions of Japan and Korea and probably China.

Records of rice planthopper outbreaks date back to 18 A.D. and have occurred in Korea and Japan for the last 1800 years (Paik 1977; Dyck and Thomas 1979; Nagata 1982). Since the Endo era (1600 to 1867), numerous famines have been attributed to the destruction of rice by these planthoppers (Nagata 1982). This includes the notorious Kyoto famine of 1732 when the planthoppers decreased rice production to 10% of normal. During the Meiji era (1868 to 1912), an outbreak in 1897 caused a severe shortage of rice, and another outbreak in 1940 forced the Japanese government to start a forecasting system for rice insects and diseases.

Rice production intensification programs have contributed to planthopper outbreaks in rice. Some of the inputs necessary for high rice yields also favor planthopper outbreaks. Nitrogen fertilizer improves host plant quality and contributes to increased populations of *N. lugens* (Heinrichs and Medrano 1985). Resistant rice varieties have been released for commercial production and have been employed as a tactic in the management of *N. lugens* and *T. orizicolus* populations, but biotypes have developed that are able to develop on such varieties (Heinrichs 1988). Insecticides have been extensively used to control rice pests and have caused severe outbreaks of *N. lugens* in tropical regions due to the resurgence of the pest after insecticide application (Heinrichs and Mochida 1984). As a result of the problems encountered with *N. lugens* control in modern rice production, control strategies in some Asian countries have been modified toward a more sustainable IPM system (Whalon et al. 1990).

The history of chemical control of rice insects in Japan has been reported by Suenaga and Nakatsuka (1958). Whale oil was the first chemical recorded as effective against *N. lugens* in 1670. By 1840, the use of whale oil in rice was common throughout Japan. After the Meiji Era, whale oil was replaced with mineral oil and fortified with pyrethrum extracts or nicotine (Nagata 1982). However, these chemicals were not sufficiently effective to prevent severe outbreaks.

Benzene hexachloride (BHC) dust was first used for rice planthopper control in Japan in 1949 and was also found to be effective against the striped rice stem borer *Chilo suppressalis*. BHC use thus spread throughout Japan and this compound was used extensively as a rice insecticide for two decades. BHC was banned for use in Japan in 1971 because it was detected in the milk of cows that consumed contaminated rice straw as a forage (Nagata 1982; Nagata and Moriya 1974).

Organophosphates replaced BHC with the first usage in 1952 of parathion which was soon replaced by diazinon. In the 1960s, carbamates were utilized to replace the organophosphates to which green leafhoppers (*Nephotettix*) had developed resistance. Carbamates such as BPMC, MTMC, and propoxur were superior to organophosphates for both *Nephotettix* and *N. lugens* control. Later, carbamates were replaced by mixtures of carbamates and organophosphates to control the carbamate-resistant *Nephotettix* leafhoppers. Buprofezin, a chitin inhibitor that is safe to humans and most natural enemies, has now been released to control *N. lugens* populations (Heinrichs et al. 1984a). Although insecticide-resistant green leafhoppers did not dictate the choice of insecticide for planthopper control in the Asian tropics, the evolution of insecticide use was similar due to the influence of the Japanese chemical industry.

Excessive use of insecticide led to the development of insecticide resistance in rice planthoppers, both in temperate and tropical Asia (Table 16.1). *Laodelphax striatellus* has been reported to be resistant to three classes of insecticides (Table 16.1). In 1964, it was found to be resistant to malathion in Hiroshima and Okayama prefectures in Japan and in 1965 resistance to other organophosphates was observed in Kagawa prefecture. Resistance of *L. striatellus* to BHC occurred next. By 1964, carbamate insecticides introduced to control organophosphate-resistant *L. striatellus* were found to be ineffective in Koyama prefecture. In 1973, resistance of *L. striatellus* to malathion was reported in Korea.

Development of resistance to BHC in *N. lugens* was first detected in Hiroshima prefecture, Japan in 1967 although BHC had been used for 15 years (Miyata 1989). Although there was some increase in the level of *N. lugens* resistance to BHC during the summer due to frequent applications, susceptibility was restored annually with the influx of susceptible immigrants. Between 1967 and 1979, immigrants of *N. lugens* developed resistance to several insecticides, including carbaryl, MTMC, MIPC, malathion, diazinon, and fenitrothion.

Compared to temperate regions, there have been fewer reports of rice

Table 16.1. Occurrence of insecticide resistance in rice planthoppers.

Planthopper Species	Insecticide Class[a]	Country	Reference
Laodelphax striatellus	CH, Carb	Japan	Miyata (1989)
	OP	Japan, Korea	Yamashina (1974); Miyata 1989
Nilaparvata lugens	CH	Fiji, Japan, Sri Lanka, Taiwan, Vietnam	Miyata (1989)
	OP	Korea	Choi et al. (1975)
	OP	Sri Lanka	Wickremasinghe and Elikawela (1982)
	Carb	Malaysia	Miyata (1989)
	OP, Carb	Japan, Philippines	Miyata (1989); Heinrichs and Tetangco (1978)
	OP, Carb, Pyr	Korea, Taiwan	Chung et al. (1982); Miyata (1989)
Sogatella furcifera	Carb	Sri Lanka	Wickremasinghe and Elikawela (1982)

[a]Carb. = Carbamate; CH = chlorinated hydrocarbon, OP = organophosphate, Pyr = pyrethroid.

planthopper resistance to insecticides in the tropics. In the Philippines, the effectiveness of diazinon for *N. lugens* control declined in 1969 after being used on 10 successive crops at the International Rice Research Institute (IRRI). Loss of effectiveness was attributed to insecticide resistance and microbial degradation. Later, resistance to carbofuran was reported at IRRI (Heinrichs 1979). During the 1970s, *N. lugens* resistance to organophosphate and carbamate insecticides was reported in Taiwan (Miyata 1989). Subsequently, Dai and Sun (1984) reported resistance to the pyrethroids in Taiwan.

Insecticide resistance and insecticide-induced resurgence (Heinrichs and Mochida 1984) have become serious problems in the management of rice planthoppers. The effectiveness of many rice planthopper insecticides has decreased and few new insecticides to control resistant hoppers have been developed and released for commercial use. Because of the increasingly stringent regulations regarding mammalian toxicity and environmental pollution, it is likely that few new insecticides will be developed to replace those to which rice planthoppers have developed resistance (Hama 1980).

Because of the constraints involved in using insecticides as a major tactic in a management system for rice planthoppers, extensive emphasis has been devoted to the development of planthopper-resistant rice varieties. However, virulence development has been a constraint in the deployment of resistant varieties.

The dependence for control on insecticides and resistant rice varieties to which *N. lugens* has been able to adapt caused this planthopper to become a pest of major economic significance in Asia in the 1970s and 1980s (Heinrichs 1988). Although their damage has not been as dramatic and widespread, *S. furcifera*, *L. striatellus*, and *T. orizicolus* also continue to be key pests of economic importance in Asia and South America.

Mechanisms of Insecticide Resistance

There are several mechanisms that regulate the selective action of insecticides; these are broadly behavioral, biochemical, or physiological (Corbett 1974). An insect may avoid contact with an insecticide through behavioral responses to a repellant effect on contact, feeding or to its vapor. Resistance can manifest through delayed or impeded penetration of the cuticle, through its metabolic detoxification, or through target site insensitivity. Whereas reduced cuticular penetration can usually confer only relatively low levels of resistance, enzymatic degradation, resulting from the action of mixed function oxidases and esterases, for example, can confer high levels of tolerance. Target site, or nerve insensitivity, also sometimes known as kdr (knock-down resistance), can also result in high tolerance levels. All mechanisms may operate to a greater or lesser extent simultaneously in a population of resistant insects.

The mechanisms by which planthoppers avoid intoxication are well understood (Hama 1980); as in many other resistant pest species, the most important mechanisms are target site insensitivity and metabolic degradation. The evolution of these resistance mechanisms results from the selection pressure exerted by the continued application of insecticides over a period of time.

The exact nature of the biochemical or physiological mechanism involved will depend on the mode of action of the particular insecticide chemical. Rotation between insecticides having different chemical activities will therefore delay the onset of resistance to any particular one. Cross-resistance between insecticides in any one chemical class (e.g., carbamates, organophosphates, pyrethroids, etc.) is, however, common, though less so between groups, so that rotation should be between classes to be effective. The increased survival of resistant planthopper "populations" or "strains" resulting from the selection process effectively reduces the commercial lifetime of a particular insecticide or insecticide group (Voss 1983).

Target-Site or Nerve Insensitivity

Organophosphate (OP) and carbamate insecticides kill planthoppers primarily through the inhibition of acetylcholinesterase (AChE), a vital enzyme in the cholinergic synapse of the insect's nervous system. AChE has been given most attention by biochemists studying the action of insecticide resistance (O'Brien 1976).

A reduced sensitivity of AChE to inhibition by OP and carbamate insecticides in rice insects was first reported in the green rice leafhopper, *Nephotettix cincticeps* Uhler, as a specific OP and carbamate resistance factor by Hama (1983). This resistance factor has since been reported in *N. lugens*. Hama and Hosoda (1983) studied the mechanisms of resistance in several strains of *N. lugens* in Japan and found that insensitivity to AChE was correlated with carbamate resistance, whereas resistance to OPs was conferred by increased aliesterase (carboxylesterases that are inhibited by organophosphates) activity. Laboratory selection with propoxur resulted in an increased frequency of AChE of low sensitivity and a corresponding increased level of carbamate resistance in *N. lugens* progeny.

Enzymatic Degradation

Studies have shown that the detoxification of insecticides in planthoppers is mediated by hydrolytic enzymes called esterases. Both esterase-catalyzed hydrolysis and insensitive AChE (target site insensitivity) were reported to be of equal importance in the resistance of *N. lugens* to propoxur (Sun et al. 1984).

Organophosphate-resistant *N. lugens* and *L. striatellus* were characterized by an abnormally high activity of aliesterases (Ozaki 1969). Similarly, Hama and Hosoda (1983) found that aliesterase activity was proportional to the level of resistance to malathion and fenitrothion and that IPB, a potent inhibitor of aliesterase, was highly synergistic with two OPs but not with the carbamate propoxur.

Studies by Chung and Sun (1983) and Sun et al. (1984) on *N. lugens* showed that detoxification by esterase hydrolysis was closely linked to malathion resistance and to a lesser extent with resistance to several other OPs. Hydrolytic degradation also contributed to resistance to isoprocarb and several other carbamates.

Chang and Whalon (1987) identified eight esterases with respectively high and moderate activities against the pyrethroids permethrin and fenvalerate in Indonesian populations of *N. lugens*. The activity of pyrethroids against resistant and susceptible *N. lugens* and their subsequent recovery were affected by the synergists piperonyl butoxide (PBO) and TBPT in a similar manner. This confirms the involvement of both hydrolytic and oxidative degradation in metabolic resistance to pyrethroids. Pyrethroids with primary esters were believed to be hydrolyzed by esterases present in planthoppers resistant to OP and carbamate insecticides (Dai and Sun 1984). This suggests that cross-resistance between different classes of insecticides may be important in *N. lugens*.

Endo et al. (1988a, 1988b) reported that *N. lugens* resistance to a commonly used rice insecticide, MTMC, resulted from low sensitivity of AChE. However, a selection study conducted over four insect generations indicated that susceptibility of the planthopper to MTMC decreased at a slow rate and that enzymatic degradation was in operation. Selection on *N. lugens* using both malathion and MTMC resulted in high aliesterase activity as well as enhanced levels of malathion degradation.

The lack of glutathione S-transferase activity and of PBO synergism with malathion, fenitrothion, or propoxur indicates that GSH S-transferases and mixed function oxidases are not important as mechanisms of insecticide resistance in the *N. lugens* strains tested by Hama and Hosoda (1983).

Rate of Insecticide Resistance Development

Numerous factors influence the rate at which insecticide resistance develops in planthoppers. These include planthopper biology and behavior, host plant quality, insecticide chemistry, and the insecticide management tactics employed.

Biology and Behavior of Planthoppers

The species of planthopper, its life cycle, sex, migratory behavior, and wing form are factors determining the rate of development of resistance to insecticides. Nagata (1982) compared the insecticide susceptibility of three rice planthopper species. The planthoppers used in his studies (*N. lugens*, *L. striatellus*, and *S. furcifera*) were collected from an area where little insecticide had been applied and where there were no known cases of insecticide resistance. The three species were most susceptible to carbamates and showed almost equal levels of sensitivity. However, *N. lugens* was the least susceptible to organophosphates of the three planthopper species.

Length of life cycle and the number of generations produced per year also influence the rate of insecticide resistance development. Short generation times and continuous reproduction provide the opportunity for the rapid development of insecticide resistance. For example, in the tropics where *N. lugens* breeds throughout the year, insecticide resistance can occur within a short period of time (Nagata 1982). Sexual differences in the degree of susceptibility to insecticide also occur. In Nagata's study the LD_{50} of female planthoppers was twice as high as that of males.

There is a distinct contrast in the rapidity of the development of insecticide resistance between migratory rice planthoppers (*N. lugens* and *S. furcifera*) and the less migratory *L. striatellus* (Nagata et al. 1979). Development of resistance in *N. lugens* and *S. furcifera* has historically been rather slow in Japan when compared to *L. striatellus* (Kilin et al. 1981). The resistance of migratory rice planthoppers to insecticides in Japan is related to the selection pressure to which insect populations are exposed in their country of origin.

BHC was one of the first of the modern insecticides used for *N. lugens* control in Japan. It was applied extensively throughout Japan because it controlled rice stemborers, *S. furcifera* and *L. striatellus* in addition to *N. lugens* (Nagata 1983). Although other rice insects developed resistance, *N. lugens* showed no significant resistance after 20 years of continuous use. One reason for the lack of resistance to insecticide was suggested to result from its propensity for long-distance migration (Nagata and Moriya 1974).

The long-distance migration of *N. lugens* to Japan was confirmed in 1967 (Kisimoto 1971). Nagata and Moriya (1969) observed that decreases in resistance to BHC in field populations of planthoppers were associated with an annual influx of insecticide-susceptible long-distance migrants during spring. Subsequently, BHC-resistant populations developed following frequent insecticide applications each summer and fall. It is thought that migrants originated from areas where pressure of selection for BHC resistance was minimal (Fig. 16.1). With increases in the use of insecticide in China, the presumed source of immigrants to Japan, there was a coincident increase in the level of insecticide resistance in the immigrant population in Japan (Fig. 16.2). Planthoppers collected from three locations in China

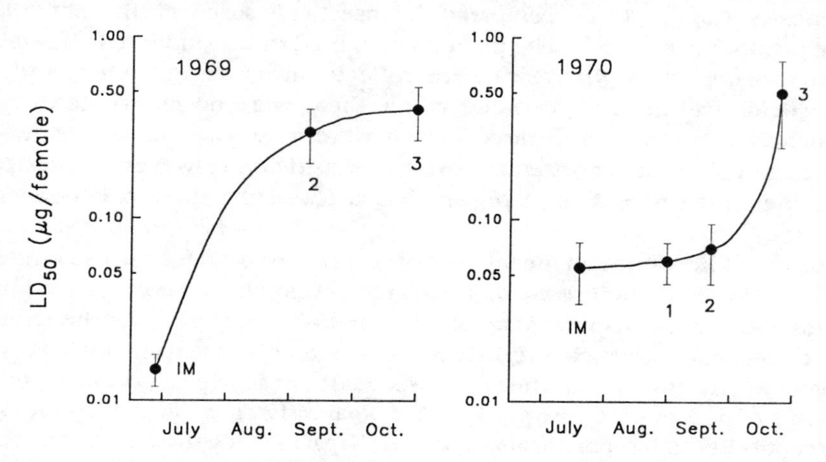

Figure 16.1. Seasonal fluctuation of BHC susceptibility of generations 1 to 3 of *Nilaparvata lugens* in BHC-treated rice paddies. LD_{50} (μg/female) as indicated by topical LD_{50} ± 95% fiducial limits. IM = immigrants (after Nagata 1982).

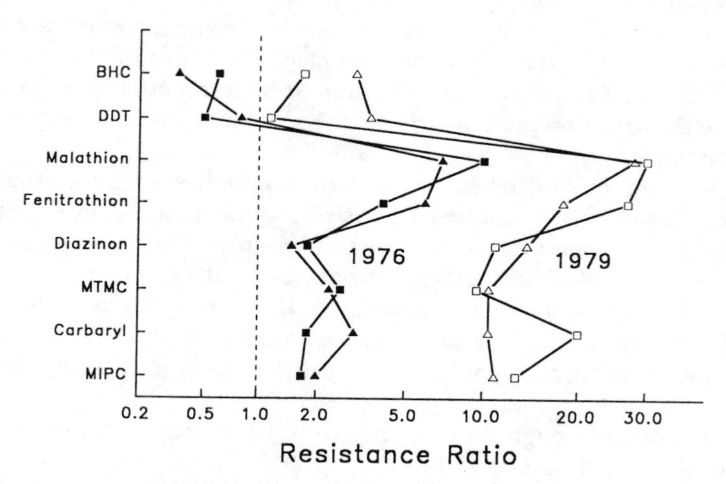

Figure 16.2. Development of insecticide resistance in immigrants of *Nilaparvata lugens* expressed as resistance ratios based on 1967 data. ▲ = 1976 Kagoshima prefecture; ■ = 1976 Nagasaki prefecture; △ = 1979 Kagoshima prefecture; □ = 1979 Nagasaki prefecture (after Nagata 1983).

showed patterns of insecticide resistance identical to the population of immigrants in Japan (Nagata 1983).

Wing-polymorphism is common in delphacid planthoppers (Denno and Roderick 1990). Brachypterous adults have reduced wings and cannot fly, whereas macropterous adults have normal wings and some can fly long distances. Brachypterous females are generally more fecund than macropterous females and begin reproduction earlier in adult life.

Nagata and Masuda (1980) studied the relationship between insecticide susceptibility and wing form in *N. lugens* and *S. furcifera* from the Philippines, Thailand, and Japan. Macropterous forms of *N. lugens* generally had an LD_{50} value 2–10 times higher than that for brachypterous forms, and the largest difference was observed with organophosphates. Results of studies conducted on *S. furcifera* were similar. Figure 16.3 illustrates the patterns of insecticide susceptibility of the wing forms of *N. lugens* from Japan, Thailand, Philippines and Taiwan (Nagata 1982). Although brachypters appear to be slightly more susceptible to insecticides than macropters, more research is needed before a clear relationship is established.

Host Plants and Interactions Between Varietal and Insecticide Resistance

The host range of planthoppers and the resistance quality of their host plants are factors that influence the rate of insecticide resistance development. For example, *N. lugens* has a narrow host range and is restricted primarily to rice (*Oryza*) (Nagata 1982). Thus, it is subjected to heavy selection from insecticides applied to rice. In contrast, *L. striatellus* is more polyphagous, feeds on rice as well as other grass hosts, and should contact rice insecticides less frequently.

Studies conducted by Heinrichs and colleagues have demonstrated the effect of host plant resistance on the level of susceptibility to insecticides. Studies on *Nephotettix* leafhoppers have shown that field populations have higher insecticide-induced mortality when feeding on leafhopper-resistant varieties than when feeding on susceptible varieties (Heinrichs et al. 1986b). As a result, the percentage of rice tungro virus-infected plants decreased with an increase in the number of diazinon applications on IR 36, a variety with moderate resistance to leafhoppers. On a susceptible variety, diazinon applications to leafhoppers were not effective, regardless of the number of applications.

Susceptibility of *N. lugens* and *S. furcifera* to insecticides is also distinctly influenced by the level of insect resistance in the host plant (Heinrichs et al. 1984b). Planthoppers reared on moderately resistant varieties were more susceptible to insecticide than those reared on susceptible varieties.

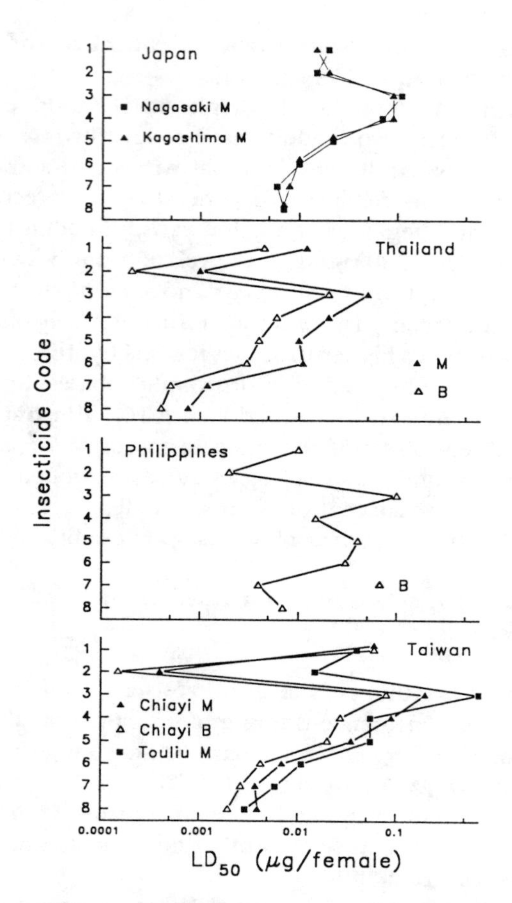

Figure 16.3. Patterns of insecticide susceptibility of *Nilaparvata lugens* collected from countries of Southeast Asia. M = macropterous form; B = brachypterous form. Insecticide code; 1 = BHC, 2 = DDT, 3 = malathion, 4 = fenitrothion, 5 = diazinon, 6 = MTMC, 7 = carbaryl, 8 = MIPC (after Nagata 1982).

Planthopper mortality in this study resulted from the interaction between insecticides and the host plant and was not attributable to any direct effect of the host plant (e.g., antibiosis). The results of this study indicated that moderately resistant rice varieties are compatible with chemical control in an integrated pest management program. Because moderate varietal resistance reduces planthopper populations, fewer insecticide applications are necessary compared to the number required to manage planthoppers on susceptible varieties. Also, when planthopper populations reach the economic threshold on moderately resistant varieties and insecticides are applied, insect mortality will be higher than on insect susceptible varieties. Thus, both (1) fewer insecticide applications and (2) higher insect mortality on moderately resistant varieties cause insecticide resistance to develop at a slower rate than on susceptible varieties.

Insecticide Chemistry and Management Tactics

The chemical nature of an insecticide, its formulation, and application frequency are all factors that influence the rate of insecticide resistance development in rice planthoppers. Nagata (1982) compared the rate of insecticide resistance development in *L. striatellus* and *S. furcifera* to organochlorine, organophosphate, and carbamate insecticides in Japan. BHC was extensively applied for rice insect control and 280,000 tons of technical material was utilized between 1961 and 1971. In spite of the extensive use of BHC, there was no detected increase in the level of resistance in studies conducted between 1967 and 1976. In fact, susceptibility increased (Fig. 16.4) as indicated by a resistance ratio (1976/1967) of 0.2. However, a high level of resistance to malathion (resistance ratio of 23.0) was detected. This indicated an inherent difference in the ability of *L. striatellus* to develop resistance against BHC and malathion.

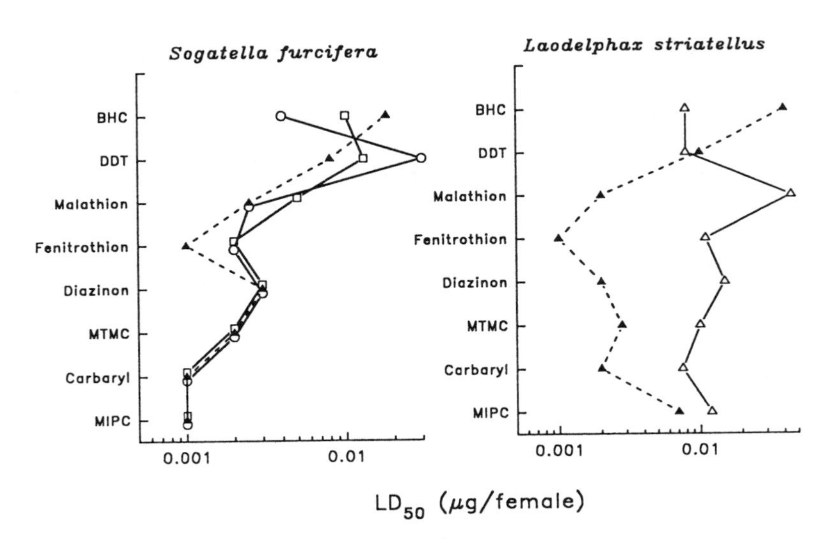

Figure 16.4. Development of insecticide resistance in *Sogatella furcifera* and *Laodelphax striatellus*, from 1967 to 1976. ▲ = Fukuoka 1967, □ = Nagasaki 1976, ○ = Kagoshima 1976, △ = Fukuoka 1976 (after Nagata 1982).

Rice planthoppers appear to develop resistance to synthetic pyrethroids rapidly. According to studies of Dai and Sun (1984), resistant strains of *N. lugens* developed within a very short period in Taiwan.

Insecticide management practices influence the rate at which resistance to insecticide develops. For example, there is a strong relationship between the number of rice insecticide applications per crop and the development of resistance to the leafhopper *Nephotettix cincticeps* in Japan. Based on an index developed by Miyata (1989), insecticide resistance occurred in 10 out of 12 prefectures when carbamates were applied more than 10 times per year. It was concluded that there was a high risk of resistance development when carbamates were applied more than eight times per year.

Iwata (1981) reported that combining insecticides in mixtures delayed the rate of resistance development. Successive selection through several generations with mixtures of propaphos and certain carbamate insecticides indicated a gradual increase in the resistance of the leafhopper population to the mixture, but the rate of increase was slow when compared to populations of leafhoppers selected against only propaphos.

Ozaki et al. (1973) found that applying a mixture of different insecticides prevented *L. striatellus* from developing resistance in Japan. Ozaki (1983) later reported that efficient suppression of resistance development in *Nephotettix* can be achieved by using mixtures that have synergistic insecticidal activity against leafhopper strains resistant to individual insecticides in the mixture.

In studies conducted in the laboratory, Wang et al. (1988) determined the rate of selection for resistance in *N. lugens* to four insecticides and their combinations. Malathion- and carbofuran-selected strains developed 7.2- and 8.0-fold resistance to malathion and carbofuran, respectively, within eight generations. The frequency of individual planthoppers with high ali-esterase activity increased gradually in each generation over eight generations with carbaryl + malathion, carbofuran + malathion, carbaryl + monocrotophos, and carbofuran + monocrotophos. The LC_{50}s were the same as those for the original susceptible strains which were not exposed to insecticide.

Cross-Resistance to Insecticides

In some cases, planthoppers that have developed resistance to one insecticide can be controlled by using another insecticide to which they are still susceptible (Ozaki 1983). However, cross-resistance (where resistance to one insecticide imparts resistance to a second insecticide even though the insect population has not been exposed to the second insecticide) has been a constraint in the management of *N. lugens*. Chung et al. (1981) determined the degree of resistance in field strains of *N. lugens* to phosphates, carbamates, and synthetic pyrethroids. Although field populations had not been previously exposed to pyrethroids, a malathion-resistant strain had a resistance ratio of 171 compared to a susceptible strain when treated with permethrin. They concluded that permethrin resistance must represent cross-resistance from planthopper exposure to other classes of insecticides (i.e., organochlorine, phosphate, and carbamate) previously used in Taiwan rice fields. It is of interest to note that all of the planthopper strains tested were susceptible to the pyrethroid fenvalerate, suggesting that permethrin has a metabolic fate different from that of fenvalerate. A survey conducted 2 years later indicated a 1000-fold resistance to permethrin, whereas resistance levels to three other pyrethroids were only 10- to 50-fold (Sun and Dai 1984). The permethrin-resistant strains were susceptible to DDT, providing circumstantial evidence that in Taiwan *N. lugens* resistance to permethrin was not related to DDT resistance (i.e., due to an altered target site) as has been reported for some other insects.

The mechanism involved in the cross-resistance to permethrin in Taiwan was not determined. Hama (1983) studied the cross-resistance of *N. lugens* to propoxur in Japan. He concluded that at least a partial cause for the cross-resistance between propoxur and other aryl N-methyl carbamate insecticides was an altered acetylcholinesterase which increased the resistance to propoxur.

Negative cross-resistance has also been reported to occur in rice planthoppers. In studies conducted by Ozaki and Kassai (1984) in Japan, various populations of *N. lugens* ranged from 19- to 710-fold resistance to malathion. LD_{50} values for fenvalerate decreased as the LD_{50} values of planthoppers treated with malathion increased (Fig. 16.5). When the LD_{50} value of malathion-treated planthoppers increased from 10 to 1000 μg/g, the LD_{50} value of fenvalerate-treated planthoppers decreased from 15 to 4 μg/g. Miyata (1989) has illustrated how insecticides having negative cross-resistance can be alternately used to manage insecticide resistant populations (Fig. 16.6). When the continued use of insecticide A leads to planthopper resistance to A, there is a subsequent increase in the susceptibility of the planthopper strain to insecticide B. Switching to insecticide B then increases the level of resistance to B and reestablishes the susceptibility to A.

Insecticide-Induced Resurgence in Planthopper Populations

National rice production programs in many developing countries have successfully achieved high yields and self-sufficiency through the planting of

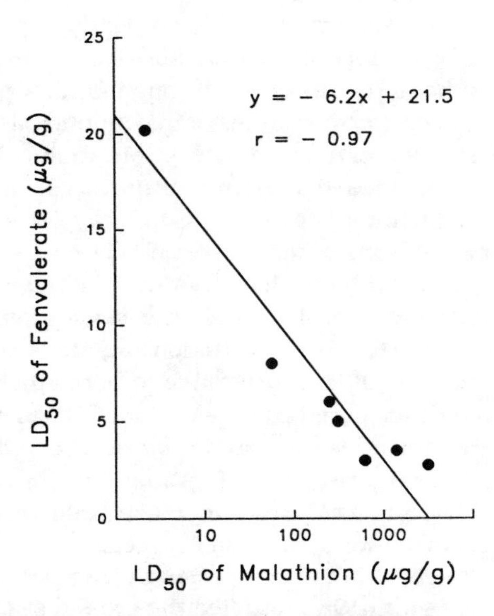

Figure 16.5. Relationship between the insecticidal activity of malathion and fenvalerate against *Nilaparvata lugens* (after Ozaki and Kassai 1984).

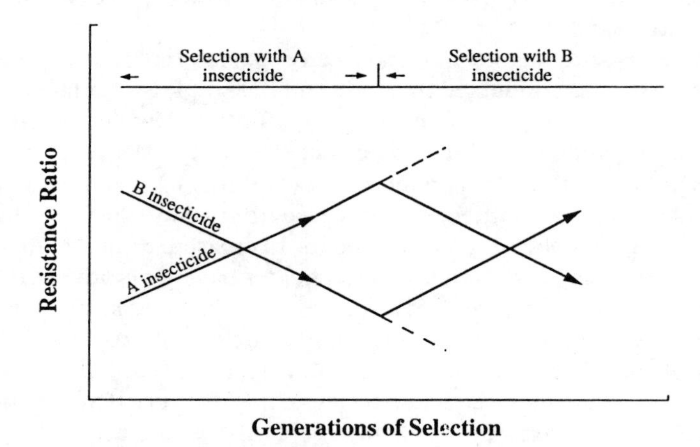

Figure 16.6. Change in susceptibility to a combination of insecticides that show negatively correlated cross resistance (after Miyata 1989).

modern varieties and the application of high levels of fertilizers and pesticides. However, the almost complete reliance on insecticides as a tactic to control rice insects has often resulted in ecological imbalances and consequent severe effects on rice production. High levels of insecticide use have resulted in the development of insecticide-resistant strains and insect resurgence.

Insecticide-induced resurgence of rice planthoppers has had a significant effect on rice production in Asia. Resurgence, a significant increase in planthopper populations after insecticide application (Heinrichs and Mochida 1984), was especially common in the 1970s and 1980s with reports from every rice-producing country in tropical Asia (Heinrichs et al. 1982a; Shepard et al. 1990).

A survey of rice entomologists in Asia indicated that one rice crop may be treated with insecticide up to 10 times (Shepard et al. 1990). Both *N. lugens* and *S. furcifera* have shown resurgence following such insecticide applications (Table 16.2). Population increases in other insects have also been reported in response to insecticide applications in rice (Shepard et al. 1990).

Indonesia is a prime example of a country where the planthopper resurgence problem led to the establishment of a more effective and ecologically sound pest management program in rice. At one time, insecticides were subsidized and 7 to 10 applications per crop were not unusual (Shepard et al. 1990). In spite of extremely heavy insecticide use from 1975 to 1979, the area of rice crop damaged exceeded 350,000 ha (Whalon et al. 1990). Even after the release of planthopper-resistant varieties in the 1980s, 50,000 ha of rice were attacked by *N. lugens* in 1986. President Soeharto, recognizing that insecticide misuse was leading to the resurgence of *N. lugens* populations, issued a Presidential Instruction (No. 3/1986) which

Table 16.2. Rice insecticide use in Asia and planthoppers reported as showing resurgence.[a]

Country	Farmers Using Insecticides (%)	Applications Per Crop (No.)	Planthopper Showing Resurgence
Bangladesh	28	1–3	*Nilaparvata lugens*
China (mainland)	90	2–3	*Nilaparvata lugens* *Sogatella furcifera*
China (Taiwan)	90	1–3	*Nilaparvata lugens*
India	50	1–6	*Nilaparvata lugens* *Sogatella furcifera*
Indonesia	85	1–5	*Nilaparvata lugens*
Korea	75	2–10	*Nilaparvata lugens*
Philippines	50	1–9	*Nilaparvata lugens*
Thailand	50	2–3	*Nilaparvata lugens*

[a]Information is based on a survey of 50 rice entomologists in Asia. Source: Modified from Shepard et al. (1990).

banned the use of 57 brands of rice insecticides. Only the highly selective insecticides such as Applaud (buprofezin), a molting inhibitor (Heinrichs et al. 1984a), and formulations containing MIPC and BPMC were maintained for planthopper control. Insecticide formulations containing carbofuran were also maintained for green leafhopper and stemborer control. Since the implementation of the Presidential Instruction in 1986, hopper-burned area has been limited to a few thousand hectares where insecticides continued to be inappropriately used.

Due to the significant economic impact of *N. lugens* in Asian rice, numerous studies have been conducted to determine the causal factors of resurgence. Destruction of natural enemies, changes in the nutritional quality of the rice plant, and the direct stimulatory effects of pesticides on the feeding and reproduction of planthoppers appear to be the major reasons that insecticides cause populations to resurge. There are numerous factors influencing the degree of resurgence such, as the chemical structure of the insecticide, its rate and formulation, its toxicity to both the planthopper and its natural enemies, and the level of rice varietal resistance to the planthopper. In the sections that follow, evidence is provided for the involvement of these factors in planthopper resurgence.

Implicated Insecticides: Their Timing and Application Rate

The impact of an insecticide on planthopper populations depends on the effect of the insecticide on the rice plant, its toxicity to the planthoppers natural enemies, and its toxicity to the target pest (Figure 16.7). Numerous insecticides belonging to the classes of organophosphates, carbamates, and synthetic pyrethroids have been shown to induce resurgence in *N. lugens* populations (Reissig et al. 1982b). Among the insecticides tested, the synthetic pyrethroids, especially decamethrin, are the most active in causing resurgence. The degree of resurgence is related to the method of insecticide application. Carbofuran injected into the root zone applied as a granular broadcast on the paddy water and when applied as a foliar spray all caused resurgence in tests conducted in the Philippines (Heinrichs et al. 1982a). However, of the various application methods, the foliar spray caused the highest degree of resurgence.

Degree of resurgence is also affected by the timing of insecticide application relative to days after transplanting (DAT) and the number of insecticide applications per crop. Timing of insecticide application relative to the life cycle of the planthopper and application frequency both apparently influence the extent of resurgence (Heinrichs et al. 1982a). Two foliar sprays of decamethrin later in population development (50 and 65 DAT)

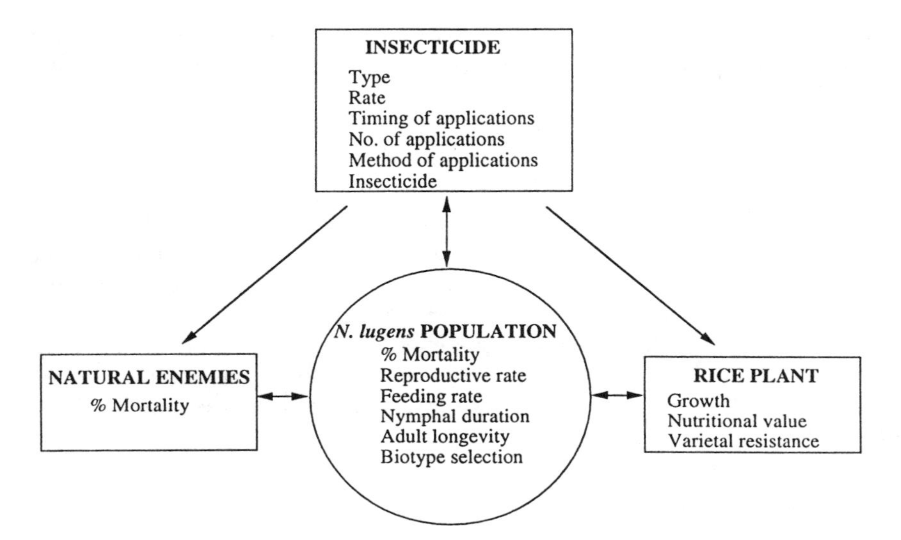

Figure 16.7. Diagram illustrating how an insecticide influences the size of *Nilaparvata lugens* populations by its direct effect on the planthopper and by its indirect effects on the planthopper via natural enemies and altered host plant characteristics (after Heinrichs and Mochida 1984).

resulted in significantly higher populations of *N. lugens* than two applications early (at 20 and 35 DAT or 35 and 50 DAT) or than one application at either 20, 35, 50, or 65 DAT.

For *N. lugens*, the application rate of an insecticide influences populations directly by its toxic effect and indirectly through effects mediated by host plant quality and natural enemies (Figure 16.7). Insecticides can act directly on the planthopper by influencing survival, reproductive rate, feeding rate, nymphal development, and adult longevity. High rates of application result in high mortality and rapid population decrease. However, sublethal rates of insecticides result in low mortality, stimulate feeding and reproduction in the survivors, increase nymphal development rate and adult longevity, and selectively kill natural enemies. The result is a high degree of resurgence in planthopper populations.

When two rates of decamethrin, diazinon, and carbofuran applied as foliar sprays were tested for their effects on resurgence, the high rate of decamethrin caused the highest degree of resurgence in *N. lugens* (Heinrichs et al. 1982b). The positive effect of sublethal rates of methyl parathion, decamethrin, diazinon, and ethylan on the feeding rate, reproductive rate, nymphal duration, and adult longevity of *N. lugens* was determined by Chelliah and Heinrichs (1980). Also, studies have shown that certain resurgence-inducing insecticides stimulate rice plant growth, which predisposes plants to increased colonization and oviposition by planthoppers (Chelliah and Heinrichs 1980).

Planthopper Feeding Rate

The feeding activity of *N. lugens* increases on plants treated with sublethal rates of resurgence-inducing insecticides. Feeding rates on plants sprayed with decamethrin, methyl parathion, and diazinon were higher than those on unsprayed plants by 61, 43, and 33%, respectively (Chelliah and Heinrichs 1980). Subsequent studies by Raman (1981) showed that the resurgence-inducing insecticides quinalphos, fenthion, cypermethrin, permethrin, and fenvalerate applied as foliar sprays on rice increased the feeding rate.

Planthopper Reproductive Rate

Studies by Chelliah and Heinrichs (1980) and Chelliah et al. (1980) demonstrated an increase in the reproductive rate of *N. lugens* when feeding on plants sprayed with sublethal rates of decamethrin, methyl parathion, and diazinon. Planthopper reproductive rates on treated plants were double those on untreated plants. Similar results were reported by Raman and Uthamasamy (1983) using other insecticides. Chelliah and Heinrichs (1980) surmised that the increase in the reproductive rate was due to the direct stimulatory effect of the insecticides or their metabolites on the feeding/reproduction of the insects and/or to a nutritional increase in the host plants treated with insecticides.

Field studies demonstrated a significant increase in the oviposition rate of *N. lugens* females on plants sprayed with decamethrin (Heinrichs et al. 1982a). The ratio of eggs per adult in the treated plants was 34 : 1, whereas the ratio on untreated plants was only 5 : 1. Similarly, Kenmore et al. (1984) reported that per capita egg production in females of *N. lugens* in an insecticide-treated field was double that expected for the observed density of females and they suggested that the oviposition response resulted from the stimulatory effects of the insecticide on planthopper reproduction.

Nymphal Duration and Adult Longevity

Resurgence-inducing insecticides alter the length of the nymphal and adult stages of *N. lugens*. The length of the nymphal period decreased by 1 day when planthoppers were fed on decamethrin-sprayed plants (Chelliah and Heinrichs 1980). Also, adult longevity was increased from 10.2 days on untreated plants to 12.7 days on treated plants. Both the decrease in nymphal duration and the increase in adult longevity (longevity and fecundity are positively correlated in planthoppers, Denno Chapter 4) contribute to the observed increase in reproductive rate observed on insecticide-treated plants.

Toxicity of Insecticides to Natural Enemies

Insecticides having a relatively higher toxicity to the natural enemies of *N. lugens* than to the target insect itself are prone to induce a high degree of planthopper resurgence. Fabellar and Heinrichs (1984) evaluated 12 insecticides for their toxicity to *N. lugens* and three of its natural enemies, the spider *Lycosa pseudoannulata* Boes et Str., the mirid bug *Cyrtorhinus lividipennis* Reuter, and the veliid bug *Microvelia atrolineata* (Bergroth). Decamethrin, the most active resurgence-inducing insecticide, caused low planthopper mortality but caused extremely high mortality of the three predators tested. Acephate, a non-resurgence-inducing insecticide was highly toxic to planthoppers but was not toxic to *L. pseudoannulata and M. atrolineata* when applied as a foliar spray.

Under field conditions where two rates of decamethrin were compared, the higher application rate caused the greatest reduction of spider, *M. atrolineata* and *C. lividipennis* populations and the greatest subsequent increase in planthopper populations (Heinrichs et al. 1982b). Treatments with high rates of ethylan, a non-resurgence-inducing insecticide, resulted in planthopper populations significantly less than controls, whereas spider populations were similar among plots treated at high and low rates of insecticide as well as the check plots.

Plant Growth and Planthopper Preference

Some resurgence-inducing insecticides stimulate rice plant growth and the change in plant physiology/morphology may be partially responsible for attracting planthoppers to plants for feeding and oviposition. Applications of methyl parathion increased tiller number and the number of leaves per plant, whereas decamethrin, a resurgence-inducing insecticide, did not (Chelliah and Heinrichs 1980). However, both methyl parathion- and decamethrin-sprayed plants attracted more *N. lugens* than the unsprayed plants. Decamethrin-sprayed plants were also reported to be more attractive to *S. furcifera* than unsprayed plants (Salim and Heinrichs 1987).

Host Plant Resistance and Biotype Selection

Host plant resistance has been utilized as a major tactic in the management of *N. lugens* in Asia. Resurgence of *N. lugens* has been shown to occur on both planthopper-resistant and planthopper-susceptible rice varieties. However, the degree of resurgence is dependent on the level of planthopper resistance expressed by a particular variety. In a field experiment, Reissig et al. (1982a) tested three rice varieties differing in their level of resistance to *N. lugens*. Planthopper populations in decamethrin-treated plants compared to the untreated check plants were 74, 50, and 5 times as large on the susceptible (IR29), moderately resistant (IR40), and highly resistant (IR42) varieties, respectively (Fig. 16.8). Thus, it is evident that varietal resistance can mask the various resurgence-inducing factors resulting from insecticide

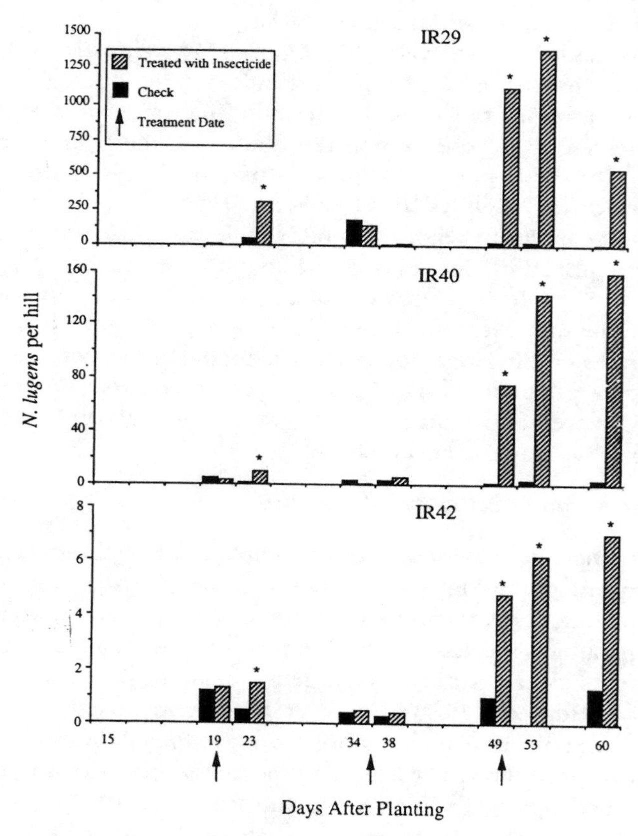

Figure 16.8. *Nilaparvata lugens* populations in decamethrin-treated and untreated field plots of a susceptible (IR29), moderately resistant (IR40), and a resistant (IR42) rice cultivar. Asterisks indicate planthopper populations significantly greater than the untreated check at the 5% level (after Reissig et al. 1982a).

application (e.g., natural enemy destruction) and dampen resurgence. However, when planthopper populations develop "resistance-breaking" biotypes which have the ability to feed and multiply on formerly resistant varieties, the application of resurgence-inducing insecticides can promote rapid growth of planthopper populations. Although biotype development allows populations to resurge, resurgence-inducing insecticides may accelerate biotype development. For example, low levels of resurgence on resistant varieties accelerates the selection of resistance-breaking genotypes in planthoppers by increasing the population size and, thus, the probability for future selection. As a result, there is a decrease in the duration that farmers can grow a particular planthopper-resistant variety before virulence evolves and it must be replaced by another resistant variety.

Biotype selection has been a severe constraint in the management of *N. lugens* populations (Claridge et al. 1982a; Pathak and Heinrichs 1982; Saxena and Barrion 1985; Heinrichs 1988). Variability in levels of virulence in planthopper populations has significantly affected the breeding and commercial utilization of planthopper-resistant cultivars in countries throughout South and Southeast Asia.

Evidence in *N. lugens* for variability in virulence was shown in studies conducted at IRRI. Distinct patterns of virulence resulted after 7 to 10 generations of selection on varieties with either the *Bph1* or *bph2* gene for resistance to *N. lugens* (Pathak and Heinrichs 1982). Claridge et al. (1982a) confirmed the variability of virulence patterns in their study of six Sri Lankan populations.

Studies on the genetics of virulence in the biotypes of *N. lugens* at IRRI indicate that virulence is inherited as a polygenic or quantitative trait and has not developed as a gene-for-gene relationship with resistance genes in rice (Den Hollander and Pathak 1981; Sogawa 1981b). The survival, feeding activity, nymphal development, and fecundity of planthoppers were the criteria used in these studies to assess virulence. Current evidence suggests that many genotypes are represented in populations that have heretofore been designated as specific biotypes. It is thus assumed that populations of the various biotypes develop through accumulation and recombination of various effective minor genes through the elimination of off-types, inbreeding among individuals selected, and/or reproductive competition among different genotypes under the selection pressure of exposure to resistant cultivars (Sogawa 1981b).

The "breakdown" of planthopper-resistant rice varieties due to the evolution of virulence allows for the expression of insecticide resistance which may have developed concurrently in *N. lugens* populations if pesticides were applied to control other target species. The "breakdown" of resistant varieties, which formerly masked the detection of insecticide resistance, is often accompanied by planthopper resurgence and ineffective chemical control.

An additional interaction involves the apparent difference in susceptibility of biotypes to insecticides. Studies on the contact toxicity of insecticides to three *N. lugens* biotypes conducted at IRRI indicated that biotype 1 (the original nonselected field strain) was least susceptible to the commonly used insecticides carbufuran, bufencarb, diazinon, and methyl parathion (Heinrichs and Valencia 1978). However, instead of inherent genetic differences in the insects, the biotypic difference in susceptibility to the insecticides may have been due to a host plant factor because each biotype was assessed on a different rice variety. In any regard, the interactions between host plant resistance and insecticide-resistant biotypes are complex. If one adds to this system the effect of insecticides on host plant nutrition and morphology and on natural enemies (Fig. 16.7), the management system for planthoppers becomes very complex. Such complexities call for an ecologically oriented and integrated approach when deploying resistant cultivars and applying insecticides.

Deployment of Rice Cultivars and Insecticides

Because rice planthopper populations have the ability to adapt to rice cultivars and insecticides, it is necessary to develop a sustainable planthopper control strategy. Such a strategy must be ecologically based and integrate biological and cultural controls with the deployment of insect-resistant rice cultivars and insecticide management techniques. A component of this strategy is resistance management which seeks to minimize the selection for virulence genes to rice varieties and resistance genes to insecticides.

Resistance management is a tool that can prolong the commercial life of a resistant host plant cultivar or an insecticide by preventing, delaying, or reverting development of insecticide resistance or varietal "breakdown" (Whalon et al. 1990). The goal of resistance management is to implement a sustainable management system that provides long-term control of rice planthoppers.

Planthopper-Resistant Rice Cultivars

Host plant resistance has and will continue to play a major role in the integrated management of rice planthopper populations. However, biotype selection will continue to be a threat to the sustainability of commercial rice cultivar releases. Various cultivar deployment strategies (e.g., gene pyramiding, horizontal resistance, tolerance, and gene rotation) have been suggested and have been employed in the management of rice planthopper populations and, in particular, *N. lugens*.

Monitoring the virulence patterns of planthopper populations throughout a given region is an essential component of a host plant resistance management program. Decisions regarding the deployment of cultivars can then be based on the reactions of cultivars to planthopper populations in the field. Several tactics related to the planting of resistant cultivars may mitigate the rate at which virulence evolves. Rotation of rice with other nonhost crops may be effective against *N. lugens* which is monophagous on rice. However, this management practice would be difficult to implement in most rice-growing regions where farmers prefer to grow rice continuously. More practical is the rotation of cultivars with different genes for resistance. This is comparable to changing the rotation of classes of insecticides between successive applications to moderate the development of insecticide resistance. A modification of the varietal rotation technique is to multiline in the same area a mosaic of cultivars each with different genes for resistance. Multilining is currently being tested in Vietnam as a resistance management technique to control rice planthoppers, but results on its effectiveness are not yet available (Heinrichs personal observations).

Sequential release of rice cultivars with major genes for resistance has been the major cultivar deployment strategy for *N. lugens* control in Asia. This technique involves releasing to farmers cultivars with one major gene incorporated into an improved plant type. This approach has not resulted in sustainable control and has led to the shifting from one cultivar to another within as few as 3 years. It also requires a highly active breeding program where donors with new genes for resistance are continuously being identified.

Stacking or the pyramiding of two or more major genes in an improved cultivar is another strategy to cope with the biotype problem. Crosses to combine two major genes have been made at IRRI, and cultivars resulting from such crosses are expected to have a longer commercial life than cultivars with a single major gene. However, this prediction has not yet been verified.

The management strategies which have been described utilize major genes for resistance. However, selection for virulence is most severe on cultivars having major genes that provide an antibiosis type of resistance. Although more difficult to breed, horizontal resistance to all biotypes should be pursued because it is of a polygenic nature having several minor genes and is more durable than major gene resistance. In contrast to antibiosis, tolerance is of a polygenic nature and is a durable type of resistance because it does not exert selection pressure on the insect population. Utri

Rajapan, a local Indonesian cultivar, is tolerant to all known biotypes of *N. lugens* in the Philippines (Panda and Heinrichs 1983). Crosses with Utri Rajapan have been made and could provide progeny with a durable type of horizontal resistance. However, developing improved germplasm with horizontal resistance is a slow process because of low heritability (Saxena and Barrion 1985). IR36 which has a major gene for resistance (*bph2*) also appears to have minor genes which provide a horizontal and stable form of *N. lugens* resistance. Although IR36 has been widely planted throughout Asia for more than a decade, biotype development on this cultivar has never been established.

Genetic engineering techniques may contribute to all of the strategies mentioned above. Numerous accessions of wild rices (*Oryza* spp.) have been identified as having resistance genes to rice planthoppers (Romena and Heinrichs 1989). Some of these may provide genes that confer a novel form of resistance but many species cannot be utilized in a conventional rice-breeding program because they are of a different genome than *O. sativa*. Recombinant DNA techniques may be useful for transferring genes from wild rices to improved high-yielding *O. sativa* cultivars.

Insecticide Resistance Management

Factors influencing selection for resistance to insecticides have been placed into three categories: genetic, biological, and operational (Georghiou and Taylor 1977). Several operational factors have been utilized in the management of insecticide resistance. Resistance management is used to prevent, delay, or reverse the development of resistance to insecticides (National Academy of Sciences 1986). An important component in an insecticide resistance management program is the ability to detect resistance in a rice planthopper population at an early stage so as to utilize techniques to reduce selection pressure (Whalon et al.1990). Monitoring is used to measure changes in the degree of resistance in time and space and is essential for the evaluation of control strategies and the implementation of IPM programs.

Rapid and simple methods have been developed to monitor resistance to insecticides in rice planthoppers (Miyata 1989). The "filter paper" and "agar plate" methods based on a measurement of total esterase activity have been used in Japan to monitor resistance in *N. lugens* and *L. striatellus*. Whalon et al. (1990) developed a microtitre plate assay and an ELISA reader to monitor resistance of *N. lugens* to insecticides in Indonesia.

Various insecticide deployment techniques have been proposed to mitigate insecticide resistance development in rice planthoppers (Ozaki 1983; Nagata and Mochida 1984; Miyata 1989; Whalon et al.1990). Minimizing the number of applications, and thus decreasing selection pressure by the planting of resistant cultivars, monitoring populations, and following economic thresholds have been major components for rice IPM programs in Asia. Local rather than area-wide application also minimizes the development of resistance and has been recommended to manage *N. lugens* in Indonesia (Whalon et al. 1990).

The use of selective insecticides that conserve natural enemies and prevent resurgence is a means for managing rice planthopper populations that is expected to delay resistance to insecticides. Buprofezin is an extremely effective insecticide for controlling *N. lugens* and is safe to natural enemies (Heinrichs et al. 1984a). It is one of only a few insecticides currently recommended for rice insect control in the Indonesian IPM program.

The sequential release of new insecticide classes to replace those to which planthoppers have developed resistance has been a commonly used resistance management technique. However, the commercialization of new insecticides is becoming increasingly uncommon due to the current cost of development.

The application of two or more insecticides, as either mixtures, mosaics, or rotations, to manage resistance, both as a preventative, but more usually as a curative measure, is now current practice and has been widely reviewed (Sawicki and Denholm 1987; Roush 1989; Tabashnik 1989; Roush and Tabashnik 1990). However, few strategies have been based on theoretical considerations and experimental work on the development of tactics to reduce resistance in the field has been sparse.

The significance of exposing pest populations to a wide range of toxic chemicals having differing modes of action may be compared with strategies using resistant crop cultivars. Both increase the challenge to the insect's capacity for adaptation, which can confer a fitness cost capable of being exploited by application of a suitably integrated strategy. For example, there is a positive interaction between susceptibility to insecticides and varietal resistance (Heinrichs et al. 1984b).

Insecticide mixtures expose individuals simultaneously to toxicants having different modes of action; this increases the likelihood of field mortality if some tolerance to either pesticide exists in the population. However, mixtures can only effectively combat resistance if it is at very low or incipient levels or to prevent its development in susceptible populations. The conditions that resistance to both insecticides should be monogenic, recessive, and rare, that there is no cross-resistance, and that each insecticide has equal persistence are seldom met under field conditions. Insecticide mixtures, applied after resistance has developed, are, therefore, generally ineffective. Detection of very low levels of resistance is, moreover, improbable using conventional techniques.

The use of mosaics, in which different insecticides are applied to adjacent areas treated simultaneously, have a similar rationale and assumptions and depend largely on the movement of insects between the adjacent treated areas or the immigration of susceptibles to delay the onset of resistance. Although there is some evidence that mosaics, including six different insecticides, may be effective, on a practical level this is unlikely to be feasible.

In rotations, two or more insecticides are alternated over time. Efficacy depends on the frequency of individuals resistant to one insecticide declining as a result of applying the alternate insecticide. This depends on there being either negative cross-resistance, a fitness cost associated with resistance, or immigration of susceptible individuals. Although a moderate fitness deficit may be present in some resistant insects, the impact of interbreeding with susceptible immigrants is perhaps more important and most field strategies employing rotations tend to rely on this element.

In general, the effectiveness of mixtures, mosaics, and rotations depends on particular conditions which are seldom encountered in the field. However, simple models, experiments, and field experience have shown that under typical conditions rotations limit selection more effectively than either mosaics or mixtures.

Specifically for planthoppers, Ozaki et al. (1973) have shown that rotating the application of conventional insecticides and applying mixtures of two chemicals prevents *L. striatellus* from developing resistance in Japan. Using two chemicals with negatively correlated cross-resistance can be utilized as a tool when rotating insecticides (Figure 16.6). Alternatively, chemicals with negatively correlated cross-resistance can be used in a mixture, as this approach will also delay the development of resistance. Miyata (1989) has reported that mixtures of synergistic insecticides delay the development of resistance. In studies conducted by Ozaki (1983) with *L. striatellus*, a mixture of malathion and carbaryl delayed the selection for resistance much longer than a mixture of malathion and MTMC. In Ozaki's studies, certain combinations of mixtures had a synergistic activity where the LD_{50} of the mixture was lower than that of either of the two chemicals in the mixture.

Conclusions

Resistance to insecticides in rice planthoppers is a more severe problem in temperate Asia than it is in the tropics. The problem has become especially acute in Japan, Korea, and Taiwan. This is due to the extensive use of insecticides for rice insect control, especially in Japan. The problem in

Japan and Korea is exacerbated because *N. lugens* and *S. furcifera* which migrate into Japan and Korea every year originate from areas of high insecticide use in mainland China.

In tropical South and Southeast Asia, insecticide-induced resurgence of *N. lugens* has become a severe problem. The degree of resurgence is, in part, related to the toxicity of an insecticide to planthoppers and the subsequent level of planthopper mortality when insecticide applications are made. Thus, the development of resistance to an insecticide can change the effect of an insecticide from one that provides control of *N. lugens* to one that causes extreme increases in planthopper populations. This phenomenon is primarily due to the poor control of the pest and the destruction of natural enemies.

Because of the insecticide-induced resurgence of *N. lugens*, some Asian countries have accelerated the implementation of IPM programs. The programs have attempted to mitigate the insecticide resistance problem by integrating host plant resistance with natural control agents and insecticides that are applied only when monitoring and surveillance systems indicate that applications are justified. The decrease in the number of insecticide applications under this system plus the application of selective insecticides which preserve natural control organisms have helped to decrease selection for insecticide resistance. Thus, the success of rice IPM programs will depend more and more on the management of resistant varieties, natural enemies, and the careful use of existing insecticides given that few new insecticides are being developed to replace those that are being lost to planthopper resistance.

Summary

1. This chapter discusses the status, mechanisms, and results of rice planthopper resistance to insecticides and the interactions between host plant resistance and insecticide resistance.
2. *Nilaparvata lugens*, *Sogatella furcifera*, and *Laodelphax striatellus* have all developed resistance to insecticides.
3. Rice planthopper resistance to insecticides is more common in temperate/subtropical regions than in the tropics.
4. The mechanisms of insecticide resistance in planthoppers include target site insensitivity and enzymatic degradation.
5. Rate of insecticide resistance development is influenced by planthopper biology and behavior, host plant quality including varietal

resistance, insecticide chemistry, and the planthopper management tactics employed.

6. Cross-resistance to insecticides has been a constraint in the management of rice planthopper populations.
7. Insecticide-induced resurgence in populations of *N. lugens* is promoted by high levels of insecticide resistance and results from a variety of factors including the selective destruction of natural enemies, changes in the nutritional quality of the rice plant, and the direct stimulatory effects of pesticides on the feeding behavior and reproduction of planthoppers.
8. "Breakdown" in varietal resistance can promote insecticide-induced resurgence in insecticide-resistant populations of *N. lugens*.
9. Levels of susceptibility of *N. lugens* to insecticides vary among different host plant "resistance-breaking" biotypes.
10. Various cultivar deployment strategies (e.g., gene pyramiding, horizontal resistance, tolerance, and gene rotation) have been suggested and have been employed in the management of rice planthopper populations.
11. Insecticide resistance management tactics include effective resistance monitoring programs, minimizing insecticide selection pressure, use of selective insecticides, sequential release of new insecticides, insecticide rotations, and insecticide mixtures.

17

Judicial Use of Insecticides Deter Planthopper Outbreaks and Extend the Life of Resistant Varieties in Southeast Asian Rice

Kevin D. Gallagher, Peter E. Kenmore, and Kazushige Sogawa

Historical Perspective

Prior to 1970, the rice brown planthopper (BPH), *Nilaparvata lugens* (Stål), was not regarded as a pest of tropical rice and received only a brief mention in an illustrated field guide to rice pests in the Philippines (Ceiba-Geigy 1968). Since the 1970s, however, this monophagous planthopper has achieved major pest status, causing massive losses in rice production throughout Asia (Dyck and Thomas 1979). Countries presently under threat include China, Thailand, and Vietnam with losses for 1990/91 alone estimated at $30 million (U.S.) in Thailand, and Vietnam. As both countries are rice exporters, these represent losses in foreign currency.

Outbreaks of *N. lugens* prompted the International Rice Research Institute (IRRI) in Los Baños, Philippines to initiate a breeding program for the development of resistant varieties of rice. In 1973, IR26 was released containing the dominant *Bph1* gene for resistance. Immediate and dramatic reductions of planthopper populations occurred in all countries where IR26 was planted. Unfortunately, field populations of *N. lugens* began to increase within 2 years; the variety had "broken down" and *N. lugens* was again giving rise to massive outbreaks on IR26 and other varieties with the *Bph1* gene. IR36 and other varieties with the *bph2* recessive resistance gene were released in 1976. The widespread planting of the new resistant varieties in Asia again led to a decline in *N. lugens* populations; within a few years, the varieties became ineffective and the brown planthopper again returned as the scourge of rice in several countries. Indonesia's "breakdown" occurred in 1985 and 1986 and Thailand's current outbreaks are occurring on Thailand-bred varieties with the *bph2* gene (Supon 60 and RD23). The dominant *Bph3* gene was released in 1982 in IR56 and other varieties; resistance in these varieties is currently "breaking down" in many countries. Recent surveys in Indonesia show that populations of *N. lugens* have adapted and are currently able to develop on these varieties.

This cycle of high rice production interrupted by planthopper outbreaks threatened the national food security of Indonesia and funds were allocated to control programs for *N. lugens* that depended on the use of resistant varieties and pesticides. Indonesia alone invested over $100 million U.S. per year between 1980 and 1987 for pesticide subsidies in rice (Fig. 17.1). Massive seed production programs were introduced to keep farmers supplied with the most recently developed resistant varieties. This control strategy did not prevent planthopper outbreaks, indeed, the measures deployed were costly and only exacerbated the problems of varietal breakdown and insecticide resistance and resurgence in *N. lugens* (Kenmore 1991).

In this chapter, we briefly review the historical development of *N. lugens* as a pest on rice in tropical Asia and discuss interactions between varietal

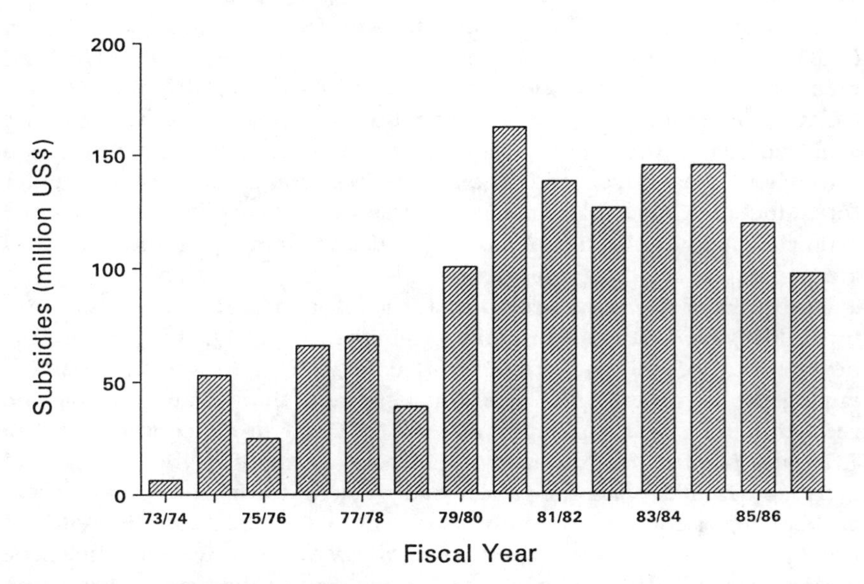

Figure 17.1. Pesticide subsidies (million U.S. $) for rice in Indonesia from fiscal years 1973–1987 (from the Ministry of Finance, Indonesia).

resistance and natural enemy kill by pesticides as they contribute to planthopper outbreaks. Furthermore, we review how governmental policy concerning pesticide subsidies can influence pest management programs to foster natural enemies and diminish the likelihood of planthopper outbreaks. We provide a case study from Central Java illustrating how government policy coupled with effective pest management and farmer training have worked in concert to extend the life of a rice variety, reduce pesticide applications, and maintain populations of BPH at acceptable levels. In the context of planthopper adaptation to resistant varieties, we emphasize the importance of considering genetic variation among individuals in field populations of *N. lugens* and question the usefulness of the "biotype concept."

Brown Planthopper Resurgence

In 1979, very little attention had been paid to the natural enemies of *N. lugens* as evidenced by the limited and inconclusive discussion of their role in a major review of brown planthopper ecology in the tropics (Dyck and Thomas 1979). Although natural enemies were thought to be important, there was little evidence at the time that *N. lugens* populations were suppressed by predators and/or parasitoids. However, pesticides had been used before 1978 to induce resurgence of *N. lugens* for research purposes (Heinrichs et al. 1978). Subsequently, mortality (lifetable) studies and enemy exclusion experiments performed at IRRI (Kenmore 1980; Kenmore et al. 1984) and in Malaysia (Ooi 1980) suggested the important role of predators and egg parasites (e.g., *Anagrus* and *Oligosita*) in suppressing *N. lugens*. Of the predators, *Lycosa* (= *Pardosa*) *pseudoannulata*, *Cyrtorhinus lividipennis*, and *Microvelia douglasi* were identified as the most important.

Research on the relationship between control by natural enemies and the toxic effects of broad spectrum pesticides on parasites and predators elucidated the mechanism of planthopper resurgence (Chelliah and Heinrichs 1980). In addition, some organophosphate and pyrethroid pesticides cause increased fecundity of BPH (Heinrichs and Mochida 1984), further increasing the probability for outbreak. Thus, when broad spectrum pesticides are applied, natural enemies can be selectively destroyed, allowing populations of *N. lugens* to increase 1000 times compared to densities when pesticides are not used; further immigration and the increased survival and fecundity of pesticide-resistant residents all contribute to population increase (Kenmore 1980; Ooi 1988).

Current research in Thailand by FAO and the Department of Agriculture of Thailand (Kenmore 1991) further documents the insecticide-induced resurgence of planthoppers (*N. lugens* and *S. furcifera* combined) due to natural enemy kill (Fig. 17.2). Rice plots treated with insecticide supported high planthopper populations (eggs and nymphs); planthopper populations in plots not receiving an insecticide treatment did not increase. Spider populations remained low in insecticide-treated plots and were significantly higher in pesticide-free plots. The rice variety used in these trials was

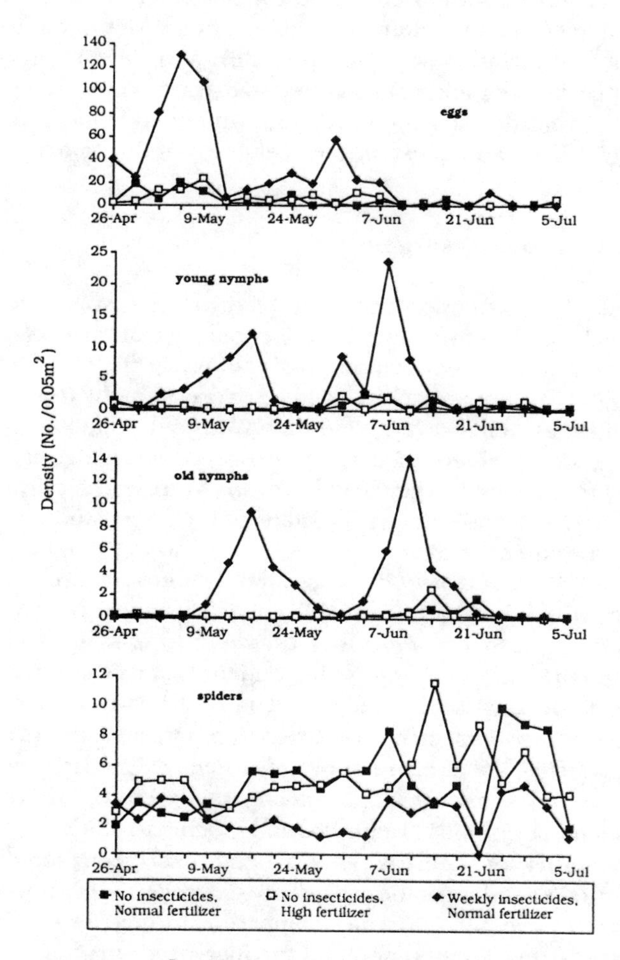

Figure 17.2. Density/0.05 m^2 of the eggs and nymphs of planthoppers (*Nilaparvata lugens* and *Sogatella furcifera* combined) and spiders on RD23 (with *bph2* gene) with and without pesticide applications and under normal and high fertilizer conditions in Chainat, Thailand. Spider density is adversely affected by insecticide treatments and planthopper populations undergo resurgence [from Kenmore (1991)].

RD23, a variety carrying the *bph2* gene which was originally released specifically to control the huge outbreaks of *N. lugens* in 1989–1990. Other studies in Vietnam, by FAO and the Department of Agriculture, produced similar results (Kenmore 1991).

What is most remarkable is that the very pesticides which induce planthopper outbreaks were commonly subsidized, often through foreign aid grants and loans, as was the case in Indonesia. In Indonesia, this cycle was broken in 1986 by a Presidential Instruction (No. 3/1986) which banned 57 pesticides from use on rice; these included primarily organophosphates and pyrethroids, which are known to result in planthopper resurgence. Scientists from national universities and research institutes were effective in promoting this change in national policy and the government began to remove subsidies for pesticides on rice. This action led to the immediate and drastic reduction in pesticide usage on rice in Indonesia. In the years following, yield remained constant and *N. lugens* has not been a major concern in Indonesia except in areas where banned pesticides are still used.

Other rice pests have also remained at low levels in Indonesia, except in one area of West Java in 1989–1991, where less than 0.1% of the national production area experienced significant yield loss due to the white rice stemborer (*Scirpophaga innotata*). This insect occurred in an area that has received perhaps the highest input of carbofuran of any rice region in the world. Current research at the Bogor Institute of Agriculture indicates that in carbofuran-treated fields survivorship of stemborers is twice as high as in untreated fields, resulting in higher damage to the crop (Triwidodo 1992). Because farmers often panic and apply banned pesticides when high stemborer damage is seen, outbreaks of *N. lugens* can result, recreating the cycle of resurgence and threatening the stability of resistant varieties.

Population Variation in Brown Planthopper and Varietal Resistance

Resistant varieties do not retain their effectiveness in the field indefinitely. Plant pathologists have called this pattern the "boom and bust cycle" (Robinson 1976) and there is an extensive literature dedicated to the study of pathogen variation and changes in pathogen virulence in the field (van der Plank 1978).

In the case of *N. lugens*, the search for population variation began with the identification of "biotypes" (e.g., Sogawa 1978a, 1978b; Saxena and Rueda 1982). The existence of individuals able to survive on varieties with the *Bph1* gene was first recognized in 1971 (IRRI 1972). Greenhouse cultures of *N. lugens* derived from field populations, and repeatedly reared on varieties with the *Bph1* gene (Mudgo) and the *bph2* gene (ASD7), were given "biotype" designations (Pathak and Saxena 1980; IRRI 1976). "Biotype 1" was not able to infest plants with any gene for host plant resistance; "biotype 2" was able to infest only plants with the *Bph1* gene; and "biotype 3" was able to infest only plants with the *bph2* gene. All biotypes could develop on plants without genes for BPH resistance.

For screening purposes, a homogeneous reaction of the variety to the biotype cultures is required. In practice, biotype cultures are used to identify genes for resistance to BPH and for evaluation of breeding lines. In such studies, plants are subjected to the three biotype cultures and the combined reaction of the cultures identifies the genes for resistance they contain (Heinrichs et al. 1985).

The term biotype must be used with caution, as this term describes laboratory populations, and not field populations. Biotypes of *N. lugens* are laboratory cultures of individuals from field populations derived by strict selection and inbreeding (Pathak and Saxena 1980). The selection or purification of the biotype populations (Medrano and Heinrichs 1984) was achieved and is maintained by continuously rearing each of the colonies on the same varieties; biotype 1 on TN1 having no genes for resistance, biotype 2 on Mudgo having the *Bph1* gene, and biotype 3 on ASD7 having the *bph2* gene. Very few individual planthoppers were used to initiate each culture, resulting in a drastic reduction in the genetic diversity within each biotype (Dobzhansky and Pavlovsky 1957; Heuttel 1976).

Biotypes have been examined for differences in morphology, genetic dominance relationships, mating behavior, and enzyme and cytological polymorphisms. Some morphometric studies show no differentiation among BPH biotypes (Sogawa 1978b), whereas others, based on a larger sample of characters, show differences (Saxena and Rueda 1982; Barrion 1985). Differences might be expected because the initial selection resulted in few founding parents and subsequent inbreeding in each colony (Dobzhansky and Pavlovsky 1957; Claridge et al. 1982a, 1982b). Several inheritance studies have been conducted without conclusive results, but most research groups have concluded that biotype characteristics are polygenically determined (Cheng and Chang 1979; Sogawa 1981b; Den Hollander and Pathak 1981). However, Saxena and Barrion (1985) suggested that biotype traits were controlled by a single major gene. Enzyme and cytological variation was found to be highest in biotype 1 colonies, but no patterns have emerged to distinguish among biotypes (Sogawa 1978a; Saxena and Barrion 1985). Furthermore, no mating barriers among biotype colonies have been found (Liquido 1978; Claridge and Den Hollander 1980).

Several selection studies have shown that each biotype can be derived through selection from the other biotypes. For example, after 12 generations of rearing biotype 1 (not adapted to any gene for resistance) on Mudgo variety with the *Bph1* gene, and on ASD7 variety with the *bph2* gene, the colonies adapted completely to the varieties (Pathak and Heinrichs 1982). Similarly, populations from biotype 1 and biotype 2 (adapted only to *Bph1* gene) cultures adapted to IR36 (*bph2* gene) and IR44 (*bph2* gene) within 8 and 11 generations, respectively (Azwar 1980). Claridge and Den Hollander (1982) demonstrated quick selection among the three biotypes: biotype 1 to biotype 2 and biotype 3 (adapted to the *bph2* gene); biotype 2 to biotype 3; and biotype 3 to biotype 2. When selecting from a biotype 2 population for a biotype 3 population, the resulting population had characteristics of both "biotypes," suggesting that adaptation was additive. Also, there is great similarity in honeydew excretion rates and survivorship among the biotypes (Claridge and Den Hollander 1980; Den Hollander and Pathak 1981). These studies question the validity of the biotype concept and its usefulness in breeding programs for BPH resistance in the field. Nonetheless, the rapid response of an *N. lugens* field population to selection provides evidence that genetic variation in host plant adaptation is common (Via 1990).

The Biotype Controversy

Proponents of the biotype approach suggest that cross-fertilization among biotypes occurs, but gene exchange is impeded, and provided with ample time and an appropriate environment, sympatric speciation may result (Saxena and Barrion 1985). Others conclude that biotypes are "simple genetic variants," are the result of rapid selection, and are interfertile (Claridge and Den Hollander 1980).

Few studies on field populations of *N. lugens* have been conducted to establish their characteristics on resistant rice varieties. Those which have been conducted indicate that field populations show little resemblance to laboratory biotype populations in their virulence characteristics (Claridge and Den Hollander 1980; Sogawa et al. 1984a, 1984b; Claridge et al. 1982). Local populations of *N. lugens* exhibit variable reactions to varieties grown in the same region and are best adapted to the plants from which they were collected (Claridge et al. 1982a). In Taiwan and in the Philippines (Mindanao), field populations of BPH can feed on rice plants containing either the *Bph1* or *bph2* gene (Cheng and Chang 1979; Medrano and Heinrichs 1985).

It has been suggested that field populations of *N. lugens* consist of sympatric populations of the various biotypes (Saxena and Barrion 1985). This conclusion was based on the repeated collection and rearing of field populations on the varieties Mudgo (*Bph1* gene) and ASD7 (*bph2* gene) and the eventual development of planthopper cultures able to survive on varieties containing each of the two genes for resistance (Pathak and Saxena 1980). Considering the lack of mating barriers and extensive planting of varieties with only one particular resistance gene, it seems unlikely that genetically distinguishable populations could exist sympatrically in the field (Claridge and Den Hollander 1983). It is more likely that plant genes for resistance act as a selection pressure on variable field populations of planthoppers, resulting in a shift in the genetic structure of populations (see Gallun et al. 1975).

The genetic changes which occur in *N. lugens* populations following the release of new resistant varieties constitute microevolution rather than macroevolution in which new species may be formed (Wright 1932). Variation among individuals, fitness differences, and inheritance of traits relevant to fitness differences are prerequisite for evolution of populations by natural selection (Endler 1986). Such differences have been demonstrated for biotypes of *N. lugens* in the laboratory (Pathak and Heinrichs 1982; Claridge and Den Hollander 1980; Den Hollander and Pathak 1981), but not so far for field populations.

Variation in Field Populations of Brown Planthopper

The persistence of a population of *N. lugens* when a rice variety with a gene for BPH resistance is newly introduced into a large area (one million hectares) will depend on the ability of some individuals to survive on the crop. The "successful" introduction of *Bph1* and *bph2* genes which dramatically reduced populations of *N. lugens* suggests that most individuals were unable to survive on varieties with these genes. However, the eventual breakdown of these varieties further suggests that genes for virulence were present and that the genetic structure of populations changed.

Gallagher (1988) investigated variation in adult longevity, fecundity, and honeydew production among and within five geographically separated populations of *N. lugens* in the Philippines. For all populations, variation was assessed on one variety with no gene for resistance (TN1 variety) and four varieties carrying resistance genes: IR26 (*Bph1* gene), IR54 (*bph2* gene), IR60 (*Bph3* gene), and Babawee (*bph4* gene). At the time of the study, varieties with the *Bph3* gene were just becoming adopted by farmers. No varieties with *bph4* gene had yet been released, and outbreaks had occurred previously in the study area only on varieties with *Bph1* and *bph2* genes.

An assessment of adult longevity and total fecundity among populations showed that some individuals from all populations were able to survive and

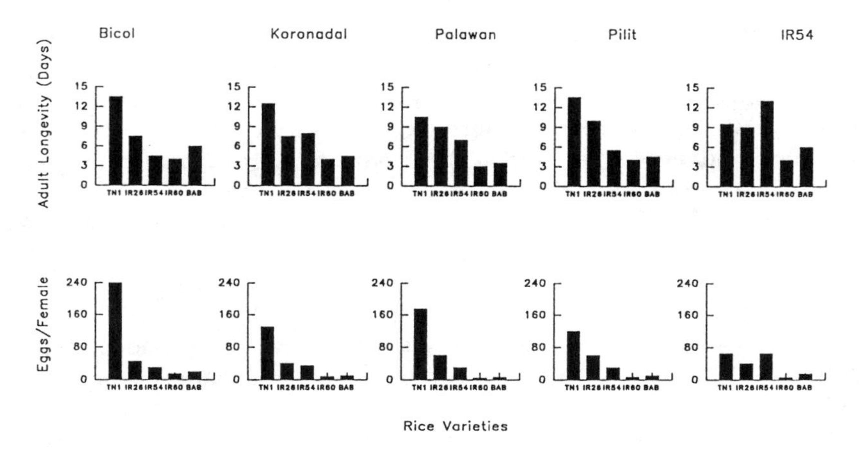

Figure 17.3. Mean adult longevity (days) and fecundity (total eggs/female) of five Philippine populations of *Nilaparvata lugens* (Bicol, Koronadal, Palawan, Pilit, and IR54) on five rice varieties (TN1, IR26, IR54, IR60 and Babawee) with different genes for BPH resistance: TN1 with no gene for BPH resistance, IR26 with the ***Bph1*** gene for resistance, IR54 with the ***bph2*** gene, IR60 with the ***Bph3*** gene, and Babawee with ***bph4*** [from Gallagher (1988)].

lay eggs on plants of every variety carrying the gene for BPH resistance (Fig. 17.3). For all populations, longevity was greatest on TN1 (no resistance genes) except for the "IR54 population" which was undergoing an outbreak in the field (Gallagher 1988). What is more interesting, however, is the individual variation contained within each population. Using the Bicol population as an example (Fig. 17.4), individuals with the greatest longevity contributed most of the eggs to the following generation. This pattern occurred on all five rice varieties, but was most evident on the newer IR60 and Babawee varieties. In the event that new varieties are placed in the field, it is clear that most BPH individuals would neither live long nor produce many eggs, and the few individuals that live the longest will contribute most to the next generation. Assuming that offspring resemble parents in their ability to feed (Sogawa 1982), mean longevity and fecundity for a population will increase with time, eventually resulting in a population able to increase rapidly on "resistant varieties."

Variation in honeydew production follows a similar pattern as longevity and egg production (Fig. 17.5). Honeydew production has been used as a measure of feeding ability in *N. lugens* (Paguia et al. 1980; Pathak et al. 1982; Sogawa 1982; Padgham and Woodhead 1986). All populations have some individuals able to feed on all varieties (resistance genes) tested, even those varieties not yet released in the field.

In the preceding studies, fecundity, longevity, and honeydew production were compared among individuals of the same population on a particular variety. The open question concerns whether single individuals in populations are able to perform on more than one variety. This was tested by measuring honeydew production of 52 individuals from the *Bicol population* when the same individuals were placed first on TN1 (no gene for resistance), second on IR26 (*Bph1*), third on IR54 (*bph2*), and last on the IR60 (*Bph3*) variety. Results showed that the single individuals tested fed as well on IR26 as on IR54, but feeding was very low on IR60.

In summary, populations of *N. lugens* have a high degree of individual variation; individuals are present which can survive and produce eggs on new varieties with new genes for resistance, and populations contain individuals able to feed on more than one type of BPH resistance gene. Other sources of plant genes for resistance to *N. lugens* have been discovered through screening methods and genetic analysis (Martinez and Khush 1974; Khush 1979; Kabir 1984; Krishna and Seshu 1984; Ikeda and Kaneda 1986). An especially fruitful source for resistance has been the Assam rice collections from northeastern India, which is thought to be one center of evolutionary diversification for *Oryza* (Khush et al. 1986). Extensive hybridization of *Oryza sativa* with other *Oryza* sp. (e.g., *O. officinalis*) through

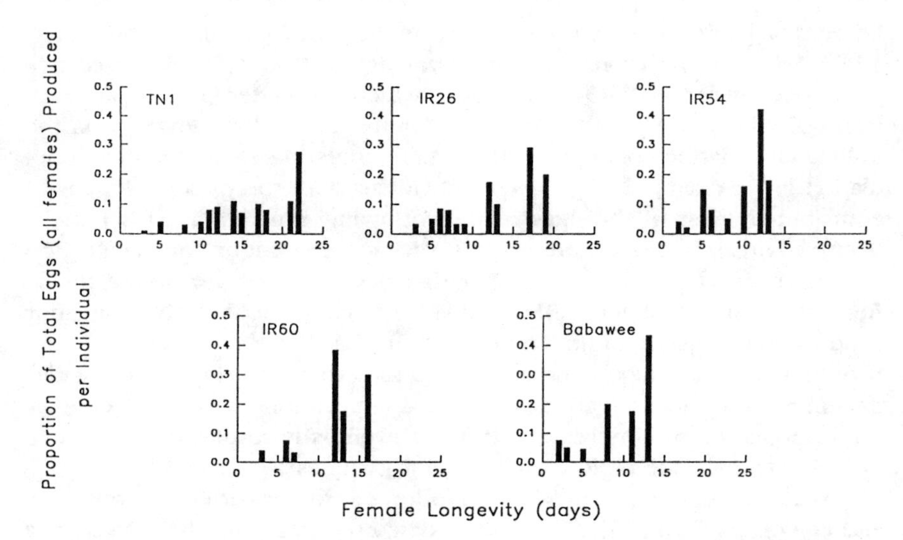

Figure 17.4. Relationship between the proportion of total eggs (all females) produced per individual and female longevity (days) for *Nilaparvata lugens* from one Philippine population (Bicol) on five rice varieties (TN1, IR26, IR54, IR60, and Babawee) [from Gallagher (1988)].

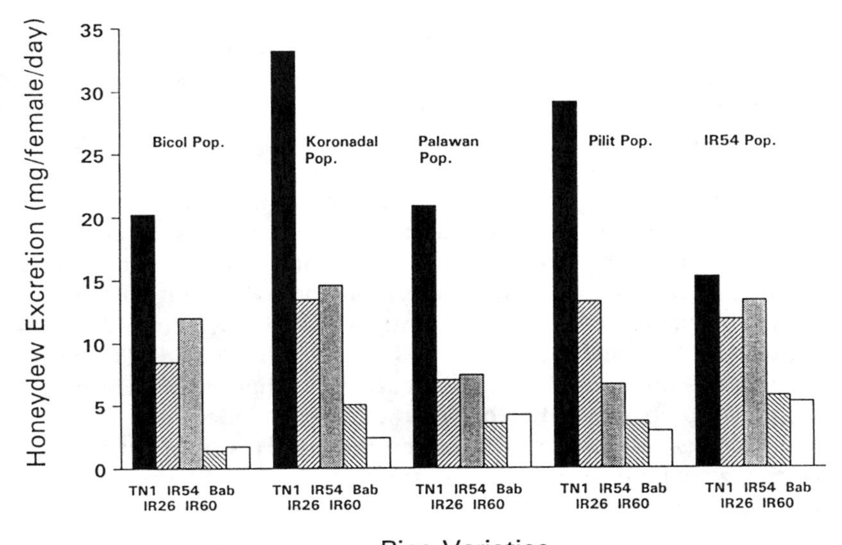

Figure 17.5. Mean honeydew excretion (mg/female/day) of *Nilaparvata lugens* from five Philippine populations (Bicol, Koronadal, Palawan, Pilit, and IR54) on five rice varieties (TN1, IR26, IR54, IR60, and Babawee) with different genes for BPH resistance [from Gallagher (1988)].

newly developed techniques of embryo rescue and protoplast fusion is being carried out at IRRI (IRRI 1990). Although these efforts will undoubtedly result in new varieties that will be initially resistant to a large number of individuals in field populations (Luong and Saxena 1989), it is unlikely that these novel varieties will remain resistant for long.

Studies on host rice resistance to other planthopper species are also being conducted. Screening for the whitebacked planthopper (WBPH), *Sogatella furcifera*, has also begun but, as is the case for *N. lugens*, these studies have largely ignored the variation among individuals in their performance on a particular variety (Ramaraju et al. 1989; Karim and Razzaque 1989).

Interaction Between BPH-Resistant Varieties and Natural Enemies

When farmers plant BPH-resistant varieties and also apply pesticides, there is never an outbreak of BPH as long as the variety remains resistant to the local planthopper population. The slow growth rate of *N. lugens* on resistant varieties masks any suppressing effect of natural enemies even though enemies are being destroyed by pesticides. Outbreaks only occur when varieties break down and pesticides kill natural enemies. Thus, farmers who grow BPH-resistant varieties in the absence of an outbreak situation and who apply pesticides will not conclude that pesticides are responsible when outbreaks eventually occur. This conclusion results because the best guide for farmers is past experience, and their past experience does not associate pesticide application with planthopper outbreak. Farmers do not have consistent feedback concerning the effect of sprays on the ecosystem. Consequently, resistant varieties mask the essential natural enemy component for *N. lugens* control, keep farmers ignorant of natural enemies, and reinforce their confidence in the success of pesticide applications. Allowing farmers to draw such a conclusion is the essential flaw in BPH management programs involving resistant varieties. Any long-term management of *N. lugens* will depend on the education of farmers concerning the role of natural enemies and the detrimental resurgent effects of broad spectrum pesticides, which are realized when varietal resistance is lost.

We suggest that pesticides actually accelerate the rate of *N. lugens* adaptation to novel varieties. Thus, investments in plant breeding are quickly lost, as is the genetic material in varieties which is discarded following breakdown. Survivorship of *N. lugens* from the egg to adult stage is very low in the presence of natural enemies (Kenmore 1980). Any differences in fitness between well-adapted individuals and poorly adapted individuals can only be expressed if the well-adapted individuals become egg-laying adults and their off-spring survive. In the presence of natural enemies, differences in fitness cannot be fully expressed because both well-adapted and poorly adapted individuals can be killed before egg laying. Thus, well-adapted individuals may not get the chance to realize their potential in the presence of natural enemies. Populations of planthoppers, therefore, will adapt to novel varieties more slowly in the presence of enemies. When pesticides are used, however, natural enemies are destroyed, resulting in a much greater survivorship of BPH immigrants and spray survivors. The fitness (fecundity) of adapted individuals will be fully realized and these individuals will produce the greatest number of offspring, thus leading to a rapid and major change in the resistance structure of the *N. lugens* population. Thus, we expect breakdown of the variety to occur more rapidly where insecticides are used given their destructive effect on natural enemies.

Central Java: A Case Study of BPH Resistance and Pesticides

The rice variety Cisadane was first released in Indonesia in the 1980–1981 wet season, and within three seasons was planted in about 40% of Central

Java's reported rice-growing area. By 1985, production of Cisadane in West and Central Java alone equaled over 1.9% of the total world rice production. The more than $123 million U.S. per year in insecticide subsidy meant that Cisadane, in these two provinces alone, received more than 11% of the world's insecticide applications on rice (Fig. 17.6). Cisadane contains the *bph2* gene, as does another variety, Krueng Aceh, the second most planted variety in Central Java. Both varieties have higher average yields than IR36 due to their longer maturation period. In addition to their increased yields, both varieties are of higher grain quality and, as a consequence, attract up to a 30% higher price.

By 1985, *N. lugens* was well-adapted to Cisadane in Central Java. We believe that the heavy application of pesticide accelerated the evolution of virulence in the planthopper population. *N. lugens* populations from Yogyakarta, Central Java, and West Java were tested for their relative ability to feed on several varieties. Feeding rates on the two most common varieties, Cisadane and Krueng Aceh, were the same as on Pelita, which contains no gene for resistance to BPH (Kamandulu and Bahagiawati 1990). Thus, Cisadane was no longer resistant to *N. lugens* in Central Java. Farmers growing Cisadane in Central Java had reduced their yields from 5.10 to 4.85

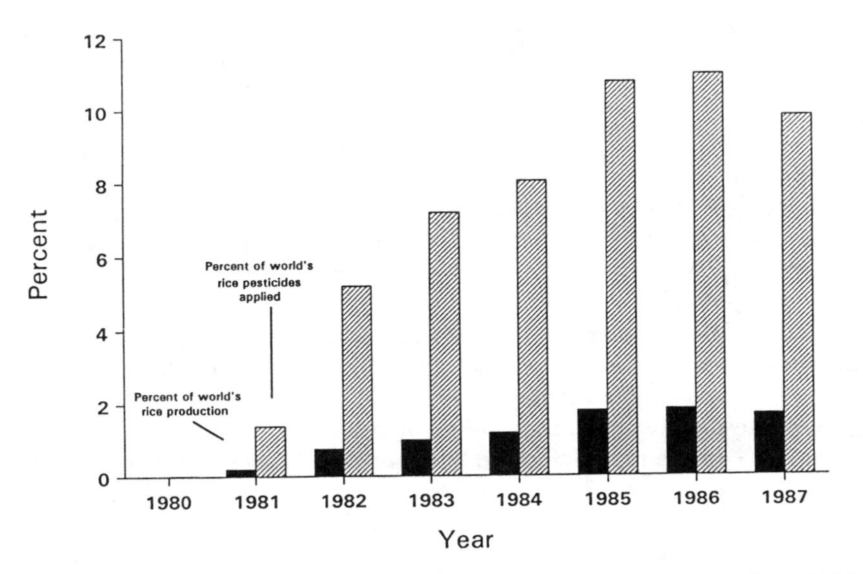

Figure 17.6. Production of Cisadane in West and Central Java, Indonesia from 1980 to 1987 expressed as a percentage of the world rice production (dark bars). Concurrent pesticide applications to Cisadane in Indonesia (percent of world total applied to rice) shown by hatched bars.

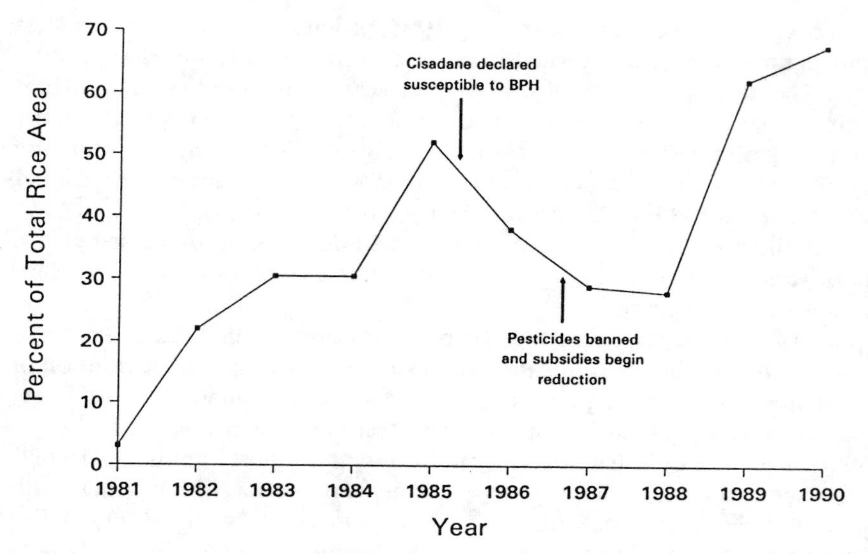

Figure 17.7. Percentage of total rice area in Central Java, Indonesia planted with Cisadane from 1981 to 1990. Government-imposed reductions in pesticide subsidies in 1986 and farmer-training programs ultimately reduced outbreaks of *N. lugens* by fostering natural enemies. The life of Cisadane was renewed and its acreage expanded. Data from Food Crop Plant Protection Center for Central Java (1991).

tons/ha (10%) and from 5.10 to 3.95 tons/ha (20%) with applications of diazinon and fenitrothion, respectively (Gadja Mada University 1987, unpublished data). The percentage of the rice-growing area in Central Java planted with Cisadane fell quickly from 48 to 27% between 1985 and 1987 as BPH adaptation and insecticide subsidies had eroded that variety's performance (Fig. 17.7).

In 1986, the banning of 57 pesticides known to cause BPH resurgence, the simultaneous withdrawal of pesticide subsidies, and the large-scale IPM training of farmers concerning the importance of conserving natural enemies led to rapid reductions of pesticide use and production. With a more than 75% drop in insecticide use by IPM-trained farmers, and a 50% drop provincewide, farmers in Central Java were able to replant Cisadane. By 1991, over 60% of the rice area was again planted in Cisadane, a remarkable testimony to the farmers' preference and to the impact of both ecologically sound policy and effective IPM training. Furthermore, the research and breeding investments in Cisadane are still providing returns.

Cisadane remains susceptible to *N. lugens* in Central Java. For instance, when farmers apply banned compounds there are localized outbreaks, but these outbreak situations can be used as "instructional examples" to educate farmers concerning the interaction among varietal breakdown, natural enemy conservation, and sound pesticide application practices.

Unfortunately, Thailand and Vietnam are currently struggling to overcome yield reductions from *N. lugens*. Both countries have exchanged visits with their Indonesian counterparts to share information concerning the management of BPH. Field research on the impact of pesticides and natural enemies on *N. lugens* populations is being repeated in both countries, but pesticide inputs remain high and adaptation of BPH to dominant varieties proceeds rapidly (Luong 1990). It remains unclear whether there is the political atmosphere conducive to the development of a national rice IPM policy that includes banning of resurgence-inducing pesticides, removal of insecticide supports, and field ecology training for farmers and extension staff. If such changes are not made, the varietal roller coaster will travel its cycle once again to the detriment of farmers, the national economy and the environment. Only the evolution and population growth of *N. lugens* will benefit from the farmer's dependence on pesticides and the removal of natural enemies.

Summary

1. Outbreaks of *Nilaparvata lugens* in Southeast Asia stimulated research to develop resistant rice varieties. However, rapid breakdown in resistance promotes the development and release of new varieties. Factors affecting this progression are discussed as are means slowing the breakdown of resistance.
2. The use of insecticides, primarily organophosphates and pyrethroids, promotes resurgence in populations of *N. lugens* due to the selective kill of natural enemies (general predators are most implicated).
3. "Biotypes" (selected laboratory populations) of *N. lugens* are used in the screening and identification of resistance genes in rice. Few differences exist in morphology and enzyme polymorphism among biotypes, and biotypes are interfertile. Furthermore, each biotype can be derived through selection from other biotypes. Field populations do not resemble laboratory-derived biotypes, and variation in *N. lugens* populations in the field is best explained by local adaptation to particular varieties. These findings question the validity of the "biotype concept."
4. Variation in longevity, fecundity, and honeydew production exists among geographically isolated field populations of *N. lugens* when raised on the same rice variety. However, for five allopatric Philippine

populations there were some individuals capable of surviving on both susceptible and resistant. Females with the greatest longevity contributed most of the eggs to the subsequent generation. Thus, in the presence of a variety with novel genes for resistance, all populations had the potential to evolve virulence rapidly resulting in varietal breakdown.

5. The slow growth rate of *N. lugens* on resistant varieties masks any suppressing effect of natural enemies even though enemies are being destroyed by pesticides. Outbreaks only occur when varieties break down and pesticides kill natural enemies. Thus, farmers growing BPH-resistant varieties in the absence of an outbreak situation and applying pesticides will not conclude that pesticides are responsible when outbreaks occur. Farmers remain ignorant of natural enemies and reinforce their confidence in the success of pesticide applications.
6. Pesticides accelerate the rate of *N. lugens* adaptation to novel varieties; survivorship and fecundity are enhanced under conditions of reduced natural enemy pressure. Also, investments in plant breeding are quickly lost as is the genetic material in varieties which are discarded following breakdown.
7. A case study from Indonesia demonstrates how pesticide reductions (reinforced by sensible government policy and implemented through farmer education) can reduce planthopper outbreaks by fostering natural enemies and extend the life of a rice variety in the field.

ACKNOWLEDGMENTS

The authors are grateful for the support and guidance provided by Dr. Andrew Gutierrez during our studies on BPH. Also, we extend our appreciation to Ir. Suyitno, head of the Food Crop Plant Protection Center for Central Java for assistance in data collection for the case study.

18

Status and Prospects for an Integrated Approach to the Control of Rice Planthoppers

J. Daniel Hare

Introduction

Some 22 species of planthoppers (Delphacidae) occur on rice in South and Southeast Asia (Wilson and Claridge 1991). Only two cause significant, widespread problems. The brown planthopper, *Nilaparvata lugens* (Stål), and the whitebacked planthopper, *Sogatella furcifera* (Horváth), damage plants by direct feeding, causing a symptom known as "hopperburn." The brown planthopper also transmits the virus diseases, grassy stunt and ragged stunt. These planthoppers became a serous pest of rice production in Asia about 30 years ago (Dyck and Thomas 1979).

Previous chapters in this volume have reviewed several tactics of planthopper pest management in rice, including host plant resistance, biological control, and cultural manipulations, among others. Presently, each of these tactics has various advantages and disadvantages. Host plant resistance, for example, has provided dramatic but short-term reductions in densities of the planthopper pests of rice with the planthoppers adapting to newly released cultivars in only a few years (Pathak and Khush 1979; Pathak and Saxena 1980; Khush 1984; Heinrichs 1986; Saxena and Khan 1989; Heinrichs, Chapter 16; Gallagher et al., Chapter 17).

In the tropics, resident natural enemies alone may be sufficient to control the planthopper pests of rice if not disrupted by insecticide applications (Kenmore et al. 1984; Heong et al. 1992); therefore, pesticides must be chosen that minimize reductions of natural enemies. Also, because the dynamics of effective natural enemy populations are coupled to the dynamics of prey populations (Cook and Perfect 1985a, 1989), it is necessary to understand how populations of natural enemies will be affected when new resistant plant cultivars are deployed (Kenmore et al. 1984).

Similarly, cultural manipulations designed to allow the crop to escape the pest in time, such as synchronous crop production and imposing areawide rice-free periods (Oka 1979), may have the additional effect of depriving natural enemy populations of their prey, thereby reducing or even eliminating resident natural enemy populations when alternate prey are unavailable. Rather than attempting to establish a pest-free agroecosystem, a more relevant approach may be to establish agroecosystems more conducive to the persistence of pest and natural enemy populations at innocuous densities. Programs to link effectively host plant resistance, biological control, and cultural manipulations for truly integrated pest management must also be practical and within the economic constraints of contemporary crop production.

An important concept to keep in mind is that the means by which particular insect species became and remain pests is through continuous adaptation to changes in the insect's ecological association with their domesticated plant hosts. When presented with a uniform and persistent environment that minimizes the variation in intensity of natural selection, then their rate of adaptation will be all the more rapid. For example, initial mortality may be high following the introduction of a new insecticide or resistant plant cultivar, but if the new materials or cultivars are used continuously and regularly, then growers will be imposing intense and methodical artificial selection favoring pest genotypes adapted to those new materials or cultivars. Given the relatively long development times for both pesticides and resistant crop cultivars, it must be disheartening for researchers to see their work undone in relatively few seasons by intense selection on adaptable pests. Therefore, to minimize the rate of adaptation to a resistant cultivar, it is necessary to manipulate the agroecosystem structure in both space and time to minimize the imposition of selection regimes favoring adaptation of pests to that cultivar.

The goal of this chapter is to evaluate the current knowledge relating to several pest management tactics for planthopper pests of rice to predict the form, if not the magnitude, of potential interactions among those tactics and to determine whether those interactions will benefit or disrupt pest management. Where appropriate, I will also attempt to specify particular research goals necessary to improve our understanding of those interactions, so that available tactics can be better integrated for planthopper pest suppression.

Genetic Considerations in the Deployment of Planthopper-Resistant Rice Cultivars

Plant breeders have had remarkable success in developing rice cultivars resistant to the brown planthopper, and planthoppers have had equally remarkable success in consistently overcoming each new form of resistance within a few seasons each and every time (Pathak and Khush 1979; Pathak and Saxena 1980; Khush 1984; Heinrichs 1986; Saxena and Khan 1989; Heinrichs, Chapter 16; Gallagher et al., Chapter 17). Briefly, rice breeders have identified at least seven major genes conferring resistance to various populations of the brown planthopper (Heinrichs 1986; Kabir and Khush 1988; Saxena and Khan 1989). Resistance conferred by three genes, *Bph1*, *Bph3*, and *Bph6*, is inherited with dominance, whereas resistance conferred by the others, *bph2*, *bph4*, *bph5*, and *bph7*, is inherited recessively (Saxena and Khan 1989). *Bph1* and *bph2* are tightly linked, as are *Bph3* and *bph4*; therefore no rice cultivars can be developed with both *Bph1* and *bph2*, or *Bph3* and *bph4*. The other three genes apparently segregate independently of these two pairs (Kabir and Khush 1988).

Rice cultivars carrying the *Bph1* gene (e.g. IR26) were released starting in 1973, but field populations of brown planthoppers adapted to these cultivars within three years. Cultivars containing the *bph2* gene (e.g. IR36 and IR42) were released starting in 1976 and gave effective brown planthopper control for another six years. These were replaced with rice cultivars with the *Bph3* gene (e.g. IR56 and IR60) in 1982, but isolated cases of adaptation of planthoppers to the *Bph3* gene have been found recently (Gallagher et al. Chapter 17.) Thus, field populations of the brown planthopper have become successively virulent to resistance conferred by each of the first three major genes incorporated into agronomically acceptable cultivars.

If host plant resistance continues to be a useful component of brown planthopper pest management, plant breeders must continue to discover new genes for resistance and incorporate them into economically useful rice cultivars. No one should assume that the supply of such genes is endless, however. As the need for a greater number of resistance genes within each plant increases, the probability decreases that they will be inherited additively and act independently of other resistance genes.

Strategies for Deployment of Resistance Genes

Breeders and growers may be able to maximize the useful lifetimes of resistance genes by varying the direction and intensity of selection for adapted planthopper genotypes by manipulating the crop's genetic structure in space and time. Several approaches have been proposed over the years, such as rotating different cultivars, planting multiple resistant lines within the same field, and combining different resistance genes within the same cultivar (Pathak and Khush 1979; Khush 1984; Heinrichs 1986; Saxena and Khan 1989). Each of these tactics may be useful, but the relative utility of each will be dependent on the genetic structure and mode of action of the virulence genes (i.e., the genes conferring adaptation to plant resistance genes) within the pest population (Gallun 1977; Kiyosawa 1982; Gould 1986).

For example, if only one resistance gene is available, the tactic of including susceptible plants with resistant plants may conserve resistance genes by minimizing the rate of pest adaptation (Atsatt and O'Dowd 1976; Gould 1986; Wilhoit 1992). Susceptible plants will not select against individuals with avirulent alleles; therefore, the avirulent allele will persist within the pest population, and selection for virulence will not go to completion. This scenario makes several assumptions about pest host-finding and mating behavior, however.

First, for accurate predictions, the resistant and susceptible cultivars should be indistinguishable to the pest (i.e., no antixenosis; Gould 1986). If so, then they will be selected in proportion to the frequency of their occurrence within the field. If the susceptible plants were more preferred, then the "reservoir" of survivors able to establish a pest population subsequently would be larger than predicted from the frequency of susceptible plants within the plant population. Alternatively, a preference for susceptible plants may allow the proportion of susceptible plants included in the mixture to be reduced.

Second, there must be no strong assortative mating (i.e., mating at random, Gould 1986) among virulent and avirulent pest individuals. Nonrandom, assortative matings could occur for a number of reasons, including behavioral recognition of similar genotypes, adult mating in the vicinity of their hosts prior to dispersal, or something as simple as large differences in development times of insects reared on different host plant genotypes. If strong assortative mating occurs, then the pest population may diverge sympatrically into virulent and avirulent entities (e.g., Bush 1974).

The assumption of nonpreference probably is not met for the brown planthopper because resistance of at least some rice cultivars is manifested through nonpreference (Cheng and Chang 1979; Cook et al. 1987; Woodhead and Padgham 1988; Kimmins 1989). Although immigrants alight equally on resistant and susceptible cultivars, immigrants eventually leave resistant cultivars (Sogawa and Pathak 1970). This simply makes the prediction of the rate of change in gene frequencies more difficult because plants are not simply chosen in the same proportion as they occur. The assumption of random mating probably is met by the brown planthopper because individuals do not discriminate against individuals reared on different cultivars in mate-choice experiments (Claridge and Den Hollander 1980), and the differences in development times of individuals reared on resistant versus susceptible cultivars (Sogawa 1981b) probably are too small to provide effective temporal isolation among those individuals.

Another complicating factor against the use of mixed stands of resistant and susceptible plants is the fact that yields are likely to be reduced from susceptible plants. The economic loss will depend on the density of the pest population, the amount of crop damage it would impose, the proportion of the total planting that is made up of susceptible plants, and the abilities of surrounding plants to compensate for the yield loss of susceptible individuals. Relatively little of this information is known at the time of planting, therefore, growers may not be able to calculate the proportion of susceptible plants that they can afford to include in their field population with any certainty.

Rather than introduce spatial genetic heterogeneity into the plant population, another approach may be to alternate resistant and susceptible plant genotypes over time. In this approach, uniform selection for virulence would be occurring every other season, and although the outcome would take longer to reach, it would be qualitatively identical with planting a uniform stand; the eventual elimination of all avirulent alleles from the pest population would result (Gould 1986).

When two or more resistance genes are available, then there are two general strategies of deployment (Gallun and Khush 1980). In the "sequential" release strategy, each gene is incorporated into a different cultivar of a crop, and the first cultivar is used until the pest becomes fully adapted to it, then the second cultivar is released. When the pest also becomes adapted to the second resistance gene, it is possible in some cases to employ "gene rotation" (i.e., re-release the first gene) and obtain effective pest suppression once more. This approach was used successfully for suppression of the Hessian fly, *Mayetiola destructor* (Say) in the United States (Foster et al. 1991).

In the "pyramided" release strategy, both resistance genes are incorporated into the same crop cultivar (Gallun and Khush 1980), and it is assumed, often without foundation (e.g., Gould 1991b), that the durability of both genes will be greater when combined than when released alone. Neither the sequential nor the pyramided strategy is inherently better in prolonging the durability of plant resistance because the effectiveness of each strategy depends on the genetic mechanism(s) by which the pests develop virulence to each resistance gene (Gallun and Khush 1980; Gould 1986).

Implicit in most predictions of the durability of sequential and pyramided release strategies is the assumption that each resistance gene is responsible for a different source of mortality, and that adaptation by the pest to the first gene has no influence on the pest's ability to adapt to the second (Gallun 1977; Gallun and Khush 1980; Gould 1986). Under the assumption of independent and additive effects, the main difference between adaptation to sequentially and simultaneously released plant resistance genes is that the sequential strategy confers selection for virulence to each resistance mechanism separately, whereas the simultaneous or pyramided strategy selects for virulence to both mechanisms at once. Although the level of pest population reduction initially may be greater for the pyramided strategy because of the combined effect of two sources of mortality, it stands to reason that the durability of two independent resistance factors will be less when the pest is being selected for virulence to both simultaneously than when being selected for virulence to each sequentially (Gould 1986).

Implicit in the gene rotation strategy is the additional assumption that adaptation (i.e., the development of virulence) occurs at some level of physiological cost. Therefore, individuals carrying avirulent genes accrue a selective advantage in the absence of resistant plant genotypes but are at a selective disadvantage in the presence of resistant plant genotypes (e.g., Atsatt and O'Dowd 1976; Kiyosawa 1982). Also assumed is that selection has not gone to completion when resistance genes are rotated, that is, there remain a few individuals in the population carrying avirulent genes when resistant cultivar #1 is withdrawn and replaced with resistant cultivar #2. Under these assumptions, the two resistance genes can be rotated as shown in Fig. 18.1. When resistant cultivar #1 is first planted, it imposes selection on the pest population for virulence. Eventually, cultivar #1 is ineffective

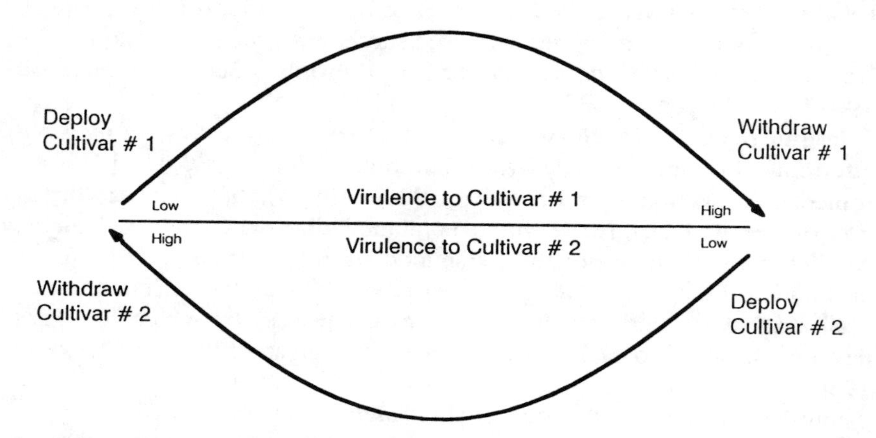

Figure 18.1. Schematic of the gene rotation strategy for maintaining two resistance genes indefinitely when the pest's ability to express virulence to each are perfectly, but negatively, correlated.

and is replaced with cultivar #2. At that time, selection for virulence to cultivar #1 ceases, and virulence to cultivar #1 will be lost simultaneously, but independently with the development of virulence to cultivar #2. Thus, at some later time, the pest population will have developed virulence to cultivar #2 and reacquired susceptibility to cultivar #1. At this point, cultivar #1 can be re-released. An example of this process is provided by the Hessian fly–wheat system in North America. The *H6* gene was first released in 1964 and provided effective suppression of Hessian fly biotypes A and B. After eight seasons, the *H6* gene was withdrawn and replaced with gene *H5* when Hessian fly biotypes C and D became dominant. Ten years later, wheat cultivars with gene *H6* were reintroduced for another eight seasons's use (Foster et al. 1991).

In the absence of any cost of adaptation, the frequency of the gene conferring virulence to cultivar #1 after that cultivar is withdrawn will change only through genetic drift, and the pest will likely maintain its virulence to cultivar #1 while developing virulence to cultivar #2. In this case, the gene rotation strategy would be unsuccessful.

The existence of such costs of adaptation often have been assumed but have been surprisingly difficult to demonstrate (see Via 1990; Jaenike 1990; and Pilson 1992 for recent reviews). Further attempts to implement the gene rotation strategy in any host plant resistance program therefore probably should not be made until firm evidence shows that there is a cost of adaptation by the pest, and that the cost is high enough to cause the rapid loss of virulence. Such data also would be critical to convince growers, regardless of their level of education, to do something as counterintuitive as to return to growing a "broken down" resistant cultivar.

All of the above scenarios assume that the genes allowing the pest to adapt to multiple resistance genes do not interact and are inherited independently. Positive correlations in the inheritance of virulence genes implies a redundancy among resistance mechanisms, and it may be unlikely that multiple plant resistance mechanisms would be recognized unless plant breeders had independent evidence of their existence. Strong negative correlations in inheritance of virulence genes, however, would promote the durability of multiple resistance mechanisms in general because the pest could never become adapted to all simultaneously (e.g., Kiyosawa 1982). Intermediate correlations, as well as the existence of other factors such as dominance, greatly complicate the prediction of durability of resistance under any strategy of gene release (Gould 1986). Therefore, the choice of a strategy for the optimal use of resistant plant cultivars cannot be made without a detailed understanding of the genetics of both plant resistance and insect virulence. The following section reviews what is known of the genetic basis of resistance and virulence in the two most important rice–

planthopper associations and evaluates the possible deployment strategies of the available resistance genes in rice.

Genetics of Planthopper Virulence to Resistant Rice Cultivars

Brown Planthopper

Coincident with the development of rice cultivars resistant to brown planthoppers was the artificial selection of three planthopper populations in the laboratory for virulence to different plant resistance genes. These populations were deemed to be "biotypes." One population, "Biotype 1," is susceptible to all resistance genes, whereas another, "Biotype 2," is virulent to *Bph1* but susceptible to the others. A third population, "Biotype 3," is virulent only to *bph2*. A South Asian biotype, "biotype 4," is also known and is virulent to both *Bph1* and *bph2*, but not *Bph3* or *bph4* (Khush 1984).

The use of the term "biotype" has probably caused more harm than good in designing and carrying out experiments to better understand the genetic basis of brown planthopper adaptation to resistant plant cultivars because it carried with it the typological concept that the pattern of inheritance of virulence of the brown planthopper must follow simple Mendelian ratios (e.g., Sogawa 1981b). The early literature also demonstrates the expectation of the "gene-for-gene" relationship, that is, for every gene in rice conferring resistance, there must be only one corresponding gene in the brown planthopper conferring virulence to that plant gene (Saxena and Barrion 1985). The available evidence (reviewed below) clearly does not support that expectation. Moreover, the general failure of investigators to find a Mendelian basis for the existence of distinct and unique "biotypes" has led to considerable acrimony among researchers, as well as a time-consuming and, in my opinion, tangential debate on the general applicability of the "biotype" concept (Saxena and Rueda 1982; Claridge et al. 1982b; Saxena and Barrion 1985; Claridge and Morgan 1987; Claridge 1990b and references therein). Once the evidence started to accumulate that even the laboratory brown planthopper populations still possessed substantial genetic variation, and that virulence might be inherited in a polygenic manner, then our understanding of planthopper genetics and the manipulation of plant genes for planthopper resistance would have been far greater if the research focus had switched from the search for Mendelian ratios to quantitative population genetics (Falconer 1981).

Initial attempts at hybridizing the laboratory populations did not show any clear-cut patterns of inheritance for abilities to utilize differentially resistant rice cultivars (Sogawa 1981b; Den Hollander and Pathak 1981; Saxena and Barrion 1985). Unfortunately, it appears that most of these studies were carried out using methods that masked individual variation within populations. For example, a group of males were taken from one population and introduced into a group of females from a second, then the performance of progeny from these group matings were examined without precisely considering their parentage. Although such procedures may be adequate to determine the *average* resemblance of offspring to their parents, these procedures do not allow the partitioning of the variation around this mean into components associated with variation among verses within families. Moreover, these procedures also do not permit the calculation of the additive component of genetic variation, which is the only component that can respond to selection (Falconer 1981). If one could calculate the ratio of the additive component of genetic variation to the total genetic variation (i.e. the heritability), then one could easily calculate the expected rate of change in the population in response to selection.

Evidence for additive genetic variation within each of the laboratory planthopper populations is provided, however, by the successful selection experiments carried out by Pathak and Heinrichs (1982) and Claridge and Den Hollander (1982). In the first series of experiments, survival of progeny of 10 pairs of individuals from a colony maintained on a rice cultivar with no known resistance genes for at least 12 years (140 generations) was assessed on a rice cultivar with the ***Bph1*** gene, and a second cultivar with the ***bph2*** gene. Mean family survival rates on plants with the ***Bph1*** gene ranged from 0% to 56%, whereas that on plants with the ***bph2*** gene ranged from 8% to 60%. Mean family survival on the totally susceptible cultivar ranged from 58% to 92% (Pathak and Heinrichs 1982). Using mass selection techniques, Pathak and Heinrichs (1982) were able to select the parental population for adaptation to either resistance gene within seven generations. Similarly, Claridge and Den Hollander (1982) were able to select three laboratory populations, each initially adapted to a rice cultivar with a different resistance gene (none, ***Bph1***, or ***bph2***), which adapted to the other cultivars within 11 generations, regardless of the previous adaptation.

These results have several important implications. First, there appears substantial variation for the ability to adapt to differentially resistant rice cultivars even after confinement on any one cultivar for more than 100 generations (Pathak and Heinrichs 1982). Second, the fact that the mass selection experiments were successful provides clear evidence that at least part of this variation has an additive genetic basis.

Two lines of evidence suggest that the cost to maintain this variation probably is low. The first is that such variation was not removed by the 140 generations of laboratory rearing on the same cultivar (Pathak and Heinrichs

1982). The second is the more direct evidence of Claridge and Den Hollander (1982), who tested their selected planthopper lines on their original host plant cultivars. After adaptation to a new plant cultivar, the planthopper populations retained most, if not all, of their ability to utilize their original cultivar. The adaptation to the *Bph1* gene was not accompanied by any loss of the adaptation to the *bph2* gene, and vice versa. It is interesting to note that the two laboratory populations virulent to both *Bph1* and *bph2* resulting from these selection experiments would be equivalent to the Southeast Asian "Biotype 4" in their rice cultivar utilization pattern.

The designs of these experiments do not lend themselves to a thorough quantitative genetics analysis (Via 1990; Simms and Rausher 1992). These studies do, however, provide enough information to make some relatively grim predictions as to the future of host plant resistance for brown planthopper management when used alone:

1. Assuming field populations of the brown planthopper are as genetically diverse as the laboratory populations, and virulence to other resistance genes also has as low of a physiological cost, then field populations are expected to continue to become adapted to any newly released rice cultivars in relatively few generations, as they have in the past.
2. Assuming that virulence to new resistance genes has as negligible a cost as does virulence to *Bph1* and *bph2*, then field populations will continue to maintain their abilities to utilize old cultivars as they adapt to new cultivars. The gene rotation strategy, therefore, is not expected to be effective, at least if only *Bph1* and *bph2* are rotated.
3. Experimental crosses were made many years ago to produce four "pyramided" rice cultivars with different combinations of *Bph1* or *bph2* with *Bph3* or *bph4* (i.e., *Bph1* with *Bph3*, *Bph1* with *bph4*, *bph2* with *Bph3*, and *bph2* with *bph4*; Khush 1979). These cultivars were to be used to determine if their resistance was more durable than single-gene resistance, but the results apparently are not yet available (Khush 1984; Heinrichs 1986; Saxena and Khan 1989; Heinrichs, Chapter 16). Given the independence of virulence alleles and their apparent lack of physiological cost, the value of pyramiding resistance genes for long-term brown planthopper management is unclear. Pyramided rice cultivars should not be released without a more detailed understanding of the genetic basis of virulence in the brown planthopper.
4. The strategy of including susceptible plant genotypes in mixtures with a single resistant genotype may be worth pursuing, but it also requires further investigation due to the behavioral preference of planthoppers for susceptible cultivars.

Whitebacked Planthopper

The whitebacked planthopper causes similar damage to that of the brown planthopper, and the two species generally co-occur (Shepard et al. 1986; Romena et al. 1986). At least five genes have been identified that confer resistance to the whitebacked planthopper (Saxena and Khan 1989) and at least four moderately resistant cultivars have been released (Heinrichs 1986). It is interesting that none of the genes conferring resistance to any laboratory population of the brown planthopper also confer resistance to the whitebacked planthopper (Romena et al. 1986). This suggests that different, largely independent factors govern resistance to each planthopper species. This obviously presents a problem to plant breeders desiring to develop cultivars resistant to both planthopper species simultaneously. Nevertheless, the four released cultivars moderately resistant to the whitebacked planthopper (IR48, 52, 60, and 62) also are resistant to at least two of the three Philippine brown planthopper laboratory populations (Heinrichs 1986). Little is known at present, about the potential abilities of whitebacked planthoppers to adapt to any of these resistant cultivars, and the same cautions and research recommendations for the brown planthopper are also applicable to the whitebacked planthopper.

Compatibility of Host Plant Resistance and Biological Control

Historically, pest managers assumed that host plant resistance and biological control were compatible and largely independent pest management strategies (Adkisson and Dyck 1980; Kogan 1982). One early justification for this view was derived from deterministic mathematical models of host–enemy interactions, which show that effective control by natural enemies is enhanced when the rate of increase (r_m) of the host population is reduced (van Emden 1966; Beddington et al. 1978; Hassell 1978; Lawton and McNeil 1979; Hassell and Anderson 1984). Such models assume that the host plant affects only the growth rate of the prey population and not the attractiveness or quality of prey individuals for discovery and utilization by the natural enemies. A substantial number of cases are now known where the assumption of uniformity of the prey population on different host plants is unwarranted. Host plant characteristics may alter the effectiveness of the natural enemies of herbivorous insects in a number of ways, but they can be grouped broadly into categories relating either to changes in the risk of discovery and/or their suitability for utilization after discovery (for reviews see Bergman and Tingey 1979; Duffey and Bloem 1986; Vinson and Barbosa 1987; Barbosa 1988; Hare 1992).

The relatively recent appreciation that host plants can indeed affect herbivore–natural enemy interactions in unanticipated ways suggests that pest suppression via host plant resistance may be incompatible with biological control in some circumstances. There is some confusion in the literature on this topic, however, because it is often difficult to translate results of short-term studies on the effects of host plants on particular life history parameters of herbivores or natural enemies into changes in herbivore–natural enemy population dynamics. A negative impact on survivorship, growth, or fecundity of a particular natural enemy does not always lead to a disruption of biological control of the herbivore population. For example, although plant resistance led to a reduction in the size and number of the parasitoid, *Lysiphlebus testaceipes* (Cresson), from greenbugs [*Schizaphis graminum* (Rodani)], overall plant damage was least and greenbug populations were smallest on resistant cultivars in the presence of parasitoids (Starks et al. 1972).

Similarly, whereas Kauffman and Flanders (1985) reported an increase in development time, reduced survival, and reduced reproduction of *Pediobius foveolatus* Crawford (Hymenoptera: Eulophidae) when reared on Mexican bean beetles, *Epilachna varivestis* Mulsant (Coleoptera: Coccinellidae), on resistant soybeans, the population growth potential of the parasitoid was reduced less than was that of the host. The important conclusion from this study was that, despite the negative impact on the parasitoid's life history parameters, the intrinsic rate of increase of the parasitoid was greatest *relative to that of its host* on the resistant cultivar. Even though host plant resistance and biological control might be judged incompatible at the individual level, host plant resistance and biological control are expected to be compatible at the population level in this system. Using a modeling approach based on energy flow, Gutierrez (1986) also concluded that the net effect of host plant resistance would benefit natural enemies if host plant resistance suppressed herbivore populations more than natural enemy populations.

Previous chapters (Döbel and Denno, Chapter 10; Benry and Lamp, Chapter 14) summarize the literature on the effects of natural enemies on brown planthopper populations. These authors conclude that predators are far more important than parasitoids in biological control of rice planthoppers. In tropical rice systems, the level of biological control provided by predators may preclude the need for any other pest management tactic (e.g., Kenmore et al. 1984; see also Cook and Perfect 1985a, 1989; Heong et al. 1992). Given the interest and (transient?) success of host plant resistance as a pest management tool for the brown planthopper, the question of compatibility between host plant resistance and biological control is a logical one to ask. The few studies addressing this question, however, have been short term in nature and did not explicitly address the issues of population growth rates of natural enemies relative to their prey, as discussed above.

For example, Kartohardjono and Heinrichs (1984) quantified mortality of brown planthopper populations from four major predators: the spiders, *Pardosa* (=*Lycosa*) *pseudoannulata* (Boesenberg et Strand) and *Callitrichia formosa* Oi, a mirid bug, *Cyrtorhinus lividipennis* Reuter, and a veliid bug, *Microvelia douglasi atrolineata* (Bergoth). A fixed number of planthoppers were released in cages on several rice cultivars differing in susceptibility to the brown planthopper. A fixed number of predators were then introduced into the cages, and planthopper mortality was assessed from 2–10 days later. By examining planthopper mortality patterns on different rice cultivars both in the presence and in the absence of predators, the authors were able to apportion the total mortality into components attributable to host plant resistance and to each predator independently. Because there was always greater planthopper mortality in the presence of predators than in their absence, the authors concluded that host plant resistance would not cause any major disruption of biological control.

The portion of the total mortality contributed by *C. formosa* and *C. lividipennis* did not vary among rice cultivars; thus, the activities of these two predators were independent of rice cultivar, and the total mortality expected from either of these predators and any resistant rice cultivar can be calculated simply as the additive combination of the expected mortality of each alone. Brown planthopper mortality from *P. pseudoannulata*, however, tended to increase with increasing host plant resistance level, thus, the two mortality sources interacted synergistically. The authors suggested that because planthoppers tend to walk around more on resistant plants as they try to locate suitable feeding sites and because *P. pseudoannulata* only attacks moving prey, the increased movement of prey on resistant plants facilitated their discovery by this hunting predator. Results for the veliid bug were not as clear because of unexpected variation in the expression of host plant resistance when this predator was examined (Kartohardjono and Heinrichs 1984).

Many of the same predators also attack the whitebacked planthopper, and the compatibility of these and other predators with host plant resistance to the whitebacked planthopper were investigated in greenhouse studies using similar experimental designs by Salim and Heinrichs (1986). As was the case with the brown planthopper, there was additive mortality of the whitebacked planthopper from host plant resistance and *C. lividipennis*. Additive whitebacked planthopper mortality occurred due to host plant resistance and *P. pseudoannulata*, in contrast to the synergistic mortality observed for the brown planthopper. No explanation for the lack of synergism between *P. pseudoannulata* and whitebacked planthopper resistance was given. However, because resistance to the two planthopper species are apparently determined by different mechanisms (Romena et al. 1986), it is unclear if a similarity in the interaction between *P. pseudoannulata* and resistance to each planthopper species should be expected. Two other predators were examined, the coccinellid, *Harmonia octomaculata* (F.), and the rove beetle, *Paederus fuscipes* Curt., and they also contributed an additive (i.e., independent) portion of mortality to that contributed by host plant resistance.

None of these studies was carried through more than a partial generation of any of the predators, therefore, no information is available to determine how host plant resistance might have affected the quality of the planthoppers for utilization by those predators. Thus, although these data strongly suggest that host plant resistance did not adversely affect the attractiveness or vulnerability of planthoppers to attack by predators, the effects, if any, of the host plants on the quality of prey for population growth of the predators remains to be determined. Further work at the population level is needed, and experiments should be designed so that population growth rates of the planthoppers and their predators can be calculated and compared over several rice cultivars differing in levels of resistance to planthoppers.

An additional cautionary note on the use of biological control with host plant resistance is provided by Kenmore et al. (1984), who found that the density of both predators (spiders and veliids) and brown planthoppers were lower in plots of moderately resistant rice than in plots of susceptible rice (i.e., predators responded numerically to reductions in prey caused by resistant host plants). This evidently led to a higher *net* rate of survival of brown planthopper nymphs on the resistant rice cultivar. Thus, the components of mortality due to host plant resistance and natural enemies were antagonistic (*sensu*, Hare 1992) and compatible for pest suppression only to the extent that they were partially compensatory. The authors suggested that host plant resistance diminished the density of brown planthoppers below that necessary for the predator populations to respond in a density-dependent manner to subsequent planthopper population growth (Kenmore et al. 1984). Possibly, host plant resistance mechanisms reduced not only the number, but also the quality of planthoppers for utilization by the predators. Clearly, these possibilities require further investigation before generalizations can be made about the compatibility of host plant resistance and biological control in the rice–planthopper association.

In addition to the potential influence of host plant cultivars and natural enemies on prey population dynamics, the activity of natural enemies may also influence the rate of adaptation of prey to resistant cultivars. Because the net effect depends strongly on such factors as the form of resistance, the life stage of prey taken, and the host finding behavior of the natural enemy, natural enemies can either significantly increase or decrease the rate of adaptation or have no effect (Gould et al. 1991).

For example, because susceptible planthoppers tend to be more mobile on resistant plants, thus more susceptible to discovery by hunting spiders (Kartohardjono and Heinrichs 1984), such spiders may be expected to capture susceptible brown planthoppers more frequently than resistant individuals. As a result, susceptible planthopper genotypes on resistant plants might be eliminated more rapidly in the presence of hunting spiders than in their absence, that is, hunting spiders would accelerate the rate of planthopper adaptation to a resistant rice cultivar. In contrast, egg predators, such as *C. lividipennis*, may not discriminate among eggs of susceptible or resistant planthopper genotypes, so egg predators probably would have no effect on planthopper adaptation. More detailed studies of predator host-finding behavior, consumption rates, and general life history (see also Döbel and Denno, Chapter 10) are required before the combined effects of predators and resistant plants on planthopper population growth and rates of adaptation can be understood.

Interaction of Cultural Manipulations with Biological Control Agents

The fact that the brown planthopper was a relatively insignificant pest of rice before the green revolution naturally suggested that perhaps the changes in rice growing practices as a result of the green revolution caused the brown planthopper to become a significant pest (Dyck and Thomas 1979). Several cultural factors were implicated in the brown planthopper's new pest status. Among them were: increasing the number of crops per year from one to continuous culture, the adoption of high-yielding cultivars with dense leaf canopies, increasing plant density, and increasing host plant quality for utilization by planthoppers through heavier fertilization and irrigation (Dyck et al. 1979). While many of these cultural factors will indeed increase planthopper density, in experimental trials, none, either alone or in combination, appeared to have the potential to cause increases as large as were commonly seen in commercial fields (Dyck et al. 1979; Kenmore et al. 1984; Heinrichs and Mochida 1984).

In tropical systems, the single most important factor associated with planthopper outbreaks has been the use of broad spectrum insecticides (Kenmore et al. 1984; Shepard et al. 1990; Ooi and Shepard in press; Heinrichs, Chapter 16; Gallagher, et al. Chapter 17). Insecticides cause planthopper populations to increase both through the removal of natural enemies (Kenmore et al. 1984) and also through direct stimulation of planthopper reproduction (Chelliah and Heinrichs 1980; Chelliah et al. 1980; Raman and Uthamasamy 1983; Heinrichs et al. 1982a; Heinrichs, Chapter 16). The success of the Indonesian decree limiting insecticide use on rice, along with other field experiments showing the absence of planthopper outbreaks in the absence of insecticide applications, suggests there may be only an occasional need for insecticides in contemporary tropical rice culture (Kenmore et al. 1984; Whalon et al. 1990; Shepard et al. 1990; Heinrichs, Chapter 16; Gallagher et al., Chapter 17).

Many cultural manipulations have been considered for planthopper suppression (e.g. Oka 1979; Shepard et al. 1990), but few have been subjected to quantitative, replicated analysis. Several of the proposed tactics, such as crop rotation and synchronizing crop production, have as their goal the escape of the crop from its pest (Oka 1979). Growers who employ these tactics, however, are also implicitly assuming that resident natural enemies will not be capable of controlling a newly established pest population.

The manipulations that eliminate pests from a particular site may also eliminate potential prey for resident natural enemy populations and predispose those sites for rapid pest population growth following colonization. A comparison between tropical and temperate rice-growing regions illustrates this point. In temperate systems, brown planthoppers do not overwinter (Kuno 1979), and the number of early-season immigrants can be used to predict accurately peak levels of infestation and timing of insecticide applications (Hirao 1979b; Cheng et al., Chapter 19). No such tight relationship exists between immigration rates and peak infestation levels in tropical systems (Cook and Perfect 1985a) presumably because the more continuous and asynchronous cropping system in the lowland tropics promotes the persistence of high predator populations within fields (see also Heong et al. 1992). The overall value of cultural manipulations to allow the crop to escape its planthopper pests may not be predictable in general, but may be determined locally by such factors as the timing of colonization of the crop and the number of colonists relative to densities of natural enemies (see Döbel and Denno, Chapter 10).

Cultural manipulations can be both expensive and difficult to implement on a geographic scale large enough to be effective, and it is probably not surprising that many of the suggestions made several years ago (e.g., Oka 1979) have not yet been tested in large-scale replicated experiments. The role of cultural manipulations in a more comprehensive rice pest management program remains difficult to forsee.

Summary

1. Research and development of pest management tactics for the planthoppers attacking rice has followed the traditional, largely independent approaches of developing new insecticides, developing plants resistant to pests, encouraging resident natural enemies, and considering some cultural manipulations to be included into the redesign of contemporary rice agroecosystems. Little attention has been given toward developing effective combinations of these approaches.
2. Host plant resistance has been an effective, though short-lived tool in planthopper pest management. The durability of all remaining resistance genes could be increased substantially with a greater effort made to understand the genetic basis of virulence in planthopper populations.
3. Rigorous, quantitative genetics studies should be performed on planthopper populations to estimate the additive component of genetic variation for adaptation to all remaining resistance genes so that the rate of adaptation to those genes can be predicted. The existence of any negative correlations in the abilities of planthoppers to utilize cultivars with different resistance genes must be ascertained.
4. Rice cultivars with multiple resistance genes (i.e., pyramided cultivars) should not be released until the genetic analyses on planthopper populations are complete and suggest that pyramided cultivars will in fact be more durable. New pyramided cultivars should be designed to incorporate only those pairs of genes whose adaptations in planthoppers are negatively correlated.
5. Short-term, intragenerational studies suggest that host plant resistance does not adversely affect the location of prey by the most important predator species. This is not sufficient, however, to conclude that host plant resistance is compatible with biological control in rice. Intergenerational population studies are required to determine the consequences of resistant rice plants on growth rates of each of the predator populations relative to that of their planthopper prey.
6. Depending on their behaviors, natural enemies may either accelerate, or retard, the process of pest adaptation to resistant plants. Detailed life table studies are required for planthoppers on different rice cultivars to identify and quantify the mortality due to host plant resistance and each of the planthopper's major predators to determine the net effect of natural enemies on planthopper adaptation rates to resistant rice cultivars. These effects could then be included into models to predict the durability of resistant rice cultivars.

7. Anecdotal evidence suggests that cultural manipulations designed to allow the rice crop to escape its planthopper pests in space and time may also allow the pest to escape its resident natural enemies. Such manipulations, therefore, may predispose a crop to pest outbreaks unless resident predators have sufficient alternate food sources. More detailed studies are required on how such manipulations as synchronous planting, crop-free periods, and stubble removal may disrupt the persistence of effective predator populations before such manipulations should be included in planthopper pest management programs.
8. Substantial opportunities still remain for contemporary ecological research on the association between rice, its planthopper pests, and the pests' natural enemies. In view of the substantial effort to develop resistant rice cultivars, it is surprising that a greater effort has not been made to understand the genetic basis of planthopper adaptation to those new cultivars. The quantity of documented genetic variation in both rice and planthoppers provides a unique opportunity to explore the genetics of insect adaptations to variable hosts.
9. Similarly, although substantial evidence implicates resident natural enemies in maintaining planthopper populations below damaging levels in the tropics, little quantitative data are available to help us understand how this actually occurs. A great deal of basic population biology on the natural enemies of planthoppers remains to be done.

ACKNOWLEDGMENTS

I thank R. F. Denno, F. Gould, B. M. Shepard, P. A. C. Ooi, and especially D. G. Bottrell for their thorough reviews of previous versions of this manuscript.

PART SIX

Definition and Implementation of Management Options

The final section of this volume addresses the definition, application, and implementation of the various management options for rice planthoppers. Cheng, Holt, and Norton (Chapter 19) review how a systems approach can be used to integrate and interpret research findings and to address management questions faced by decision makers. The computer-based simulations draw heavily on information obtained from the studies in temperate and tropical rice that are quoted in earlier sections of this volume. Simulation models are used to identify key factors in planthopper population dynamics, target conditions that promote outbreaks, and facilitate recommendations for pesticide applications. Expert systems are identified as important training and decision support tools for extension personnel, and as guides to researchers for highlighting information gaps.

The final contribution by Matteson, Gallagher, and Kenmore (Chapter 20) deals appropriately with the practicalities of implementing planthopper Integrated Pest Management in Asian rice. Extension issues relevant to planthopper management are discussed followed by a detailed description of the evolution of the rice IPM training program in Asia. Particular emphasis is placed on the quantity and quality of farmer education at the village level, where empowerment and implementation are critical. The authors provide fascinating and encouraging insight into the prospects for farmer uptake of ecologically sound management technology such as pesticide reduction and natural enemy conservation.

19

A Systems Approach to Planthopper Population Dynamics and Its Contribution to the Definition of Pest Management Options

Jiaan Cheng, Johnson Holt, and Geoff A. Norton

Introduction

Changes in pest management technology and in its perception and risk have resulted in the adoption of different control approaches. As illustrated by developments in the United States, Perkins (1982) showed that, initially, pesticides provided a cheap and effective means of control and were seen as the major, if not sole, method of control. However, with the disillusion among producers and consumers arising from problems associated with frequent pesticide use—notably, pesticide resistance, pesticide residues in food, and risks to human operators and the environment—a more ecologically-based, integrated pest management (IPM) approach has developed (Huffaker 1980; Kenmore et al. 1984). IPM requires a more sophisticated form of pest management, including a better understanding of pest population dynamics and how factors other than pesticides can be used to maintain pest populations at low levels. Because this approach demands a more holistic view of pest management, IPM has frequently embraced a systems approach from conceptual ideas (Ruesink 1976) to the use of complex quantitative computer-based tools for a number of crop pest problems, including rice pest management (Norton and Holling 1979; Conway 1984; Norton et al. 1991). This chapter concentrates on the use of two such computer-based tools, simulation models and expert systems, for brown planthopper (BPH, *Nilaparvata lugens* Stål) management in rice.

Simulation modeling provides a means by which the interactions between the components of crop–pest systems can be explored. The system is described mathematically by representing the components of the system as variables, and the relationships between them by equations. By comparing the output from the model with sets of independent data, the validity of the model as a reasonable representation of reality can be assessed. During this process, insights into the behavior of the real system can be obtained. Where the model provides a reasonable representation of reality, it can be used to evaluate management options and serve to generate testable hypotheses to facilitate problem-targeted research.

The contribution of modeling to problem solving in the field has been limited in the past, one of the problems being a lack of specialized skills to understand and apply appropriately simulation modeling to practical problems. To help bridge this gap, part of the more general void between research output and extension advice (Norton and Mumford 1982), expert systems have a potentially important role. McKinion and Lemmon (1985) define expert systems as computer software tools designed to carry out reasoning and analysis functions within narrowly defined subject areas at proficiency levels approaching that of a human expert. In many expert system applications, knowledge is represented in the form of logical relationships or rules of the form: If {certain conditions}, then {certain conclusions}. Conclusions from one set of rules can be conditions in further rules and a relatively small expert system may comprise several hundred linked rules.

Three stages can be distinguished in the application of simulation modeling and expert systems to planthopper management:

1. Problem definition and acquisition of knowledge and information
2. Analysis of management options
3. Design of recommendations for satisfactory control

Problem Definition and Acquisition of Knowledge and Information

Explicit problem definition prior to the development of simulation models, expert systems, or any other computer-based technique is an essential prerequisite for a problem-orientated approach. It allows a detailed system specification to be drawn up, helping to ensure appropriate model development with a minimum allocation of scarce resources of time and money to meet specific objectives. An alternative approach is to develop models as descriptions of systems with no particular problem in mind. This may result in models of wider applicability to problems not conceived during initial development. In IPM, such models tend to be complex descriptions of crop systems which require considerable resources to build, parameterize, and validate. In the brown planthopper modeling work described here, the former approach, involving focused specification, has been taken.

Descriptive Techniques

A number of descriptive techniques can be used as problem definition tools; they are particularly valuable when used in problem definition workshops, involving mixed groups of policy makers, researchers, and extension officers. The idea of such workshops is to determine systematically the key components and processes affecting pest problems and their control and then to determine research, extension, training, and policy priorities for improving the situation.

Recently, the International Rice Research Institute (IRRI) in the Philippines established an IPM Research Network based on this approach, involving six countries in Southeast Asia. A major objective was to facilitate analysis and exchange of information between the different countries, disciplines, and research and extension workers. In the first year, the network supported workshops in each of the six countries, bringing together national research and extension scientists to focus on the particular rice pest management problems they face (IRRI 1991). A series of techniques was employed to undertake a systematic analysis of the problem, including an historical profile of how the problem has developed, its political, social, economic, biological, and technical dimensions, and scenarios of future developments, opportunities, and constraints. Other techniques employed at this descriptive stage included flowcharts, seasonal profiles, life cycle diagrams for "brainstorming" on options for control, and pay-off matrices (Norton et al. 1991). Two brief illustrations from one of the IRRI–IPM network workshops are given in Table 19.1 and Figure 19.1; they constitute a small part of the output of a 5-day workshop held in a major rice scheme in Malaysia (IRRI 1991). Table 19.1 comes from an analysis of the rice pest surveillance system in Malaysia and shows the potential users of surveillance information and the purposes for which they might use this information. Figure 19.1 is a decision chart or "paper expert system" for BPH management, designed by a small group of participants in half a day.

In brown planthopper modeling work, another descriptive technique, the interaction matrix, has proved particularly useful (see Holt et al. 1987). To construct the interaction matrix (Fig. 19.2) the components of the system were first listed with this list of components forming both dimensions of a

Table 19.1. The possible value of rice pest surveillance in an irrigated rice scheme (Muda Agricultural Development Authority) in Malaysia.

Potential Users	Possible Value
Farmers	To prepare early for chemical control of pests
	To review the effect of cultural and biological control
Extension workers	To advise farmers on present pest population size
	To warn farmers to get ready for control when the population is approaching the economic threshold density
	To plan for the next season
Agricultural Development Authority	To mobilize pooled and supportive resources, for example, machinery and chemicals in the case of an outbreak
	To prepare a financial budget for pest control
	To ensure that control has been carried out
	To change or make new recommendations
Research	To produce predictive models and long-term correlations
Chemical industry	To give priority to the type of chemicals to be sold
	To identify short-term research needs
Treasury	To prepare funds for pest control, for example, the purchase of chemicals

symmetrical matrix. The cells of the matrix were then marked to indicate the occurrence of interactions between components of the system (Fig. 19.2). As a systematic and explicit statement of the linkages in a complex system, this matrix constitutes a valuable means of communication between modelers and biologists, particularly when judgments were being made on what components and interactions of the system were to be included in the model.

Simulation Techniques

Simulation models have been developed for both temperate and tropical rice systems. In both cases, the initial objective was to achieve a quantitative synthesis of the components and interactions within the system (Cheng and Holt 1990; Holt et al. 1987).

BPH IN ZHEJIANG PROVINCE, CHINA

In temperate rice systems, brown planthopper populations generally increase through a series of relatively discrete generations until the host becomes unsuitable due to crop ripening (Kuno 1979). The BPH model described in Cheng and Holt (1990) examines the sensitivity of simulated peak population size to changes in components of the system. The variables having most impact are the rate and pattern of BPH immigration, summer and autumn temperatures, and transplanting time. Peak density is sensitive also to BPH mortality, but a constant mortality rate was sufficient to explain differences in BPH population density between fields and years. Thus, mortality, principally attributable to natural enemies, does not appear to be a key factor in explaining BPH population fluctuation in Zhejiang Province.

For the purposes of analysis, immigration can be described by three input

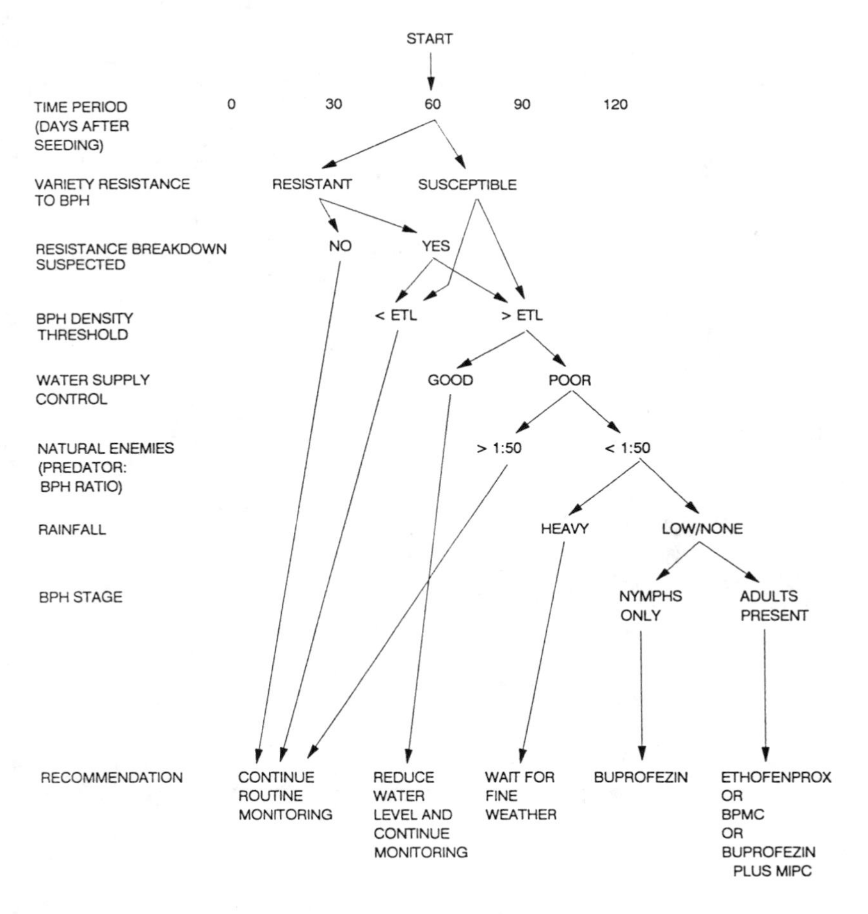

Figure 19.1. Brown planthopper (BPH) expert system for rice farmers in Malaysia. ETL (the economic threshold density) of BPH above which control is worth taking.

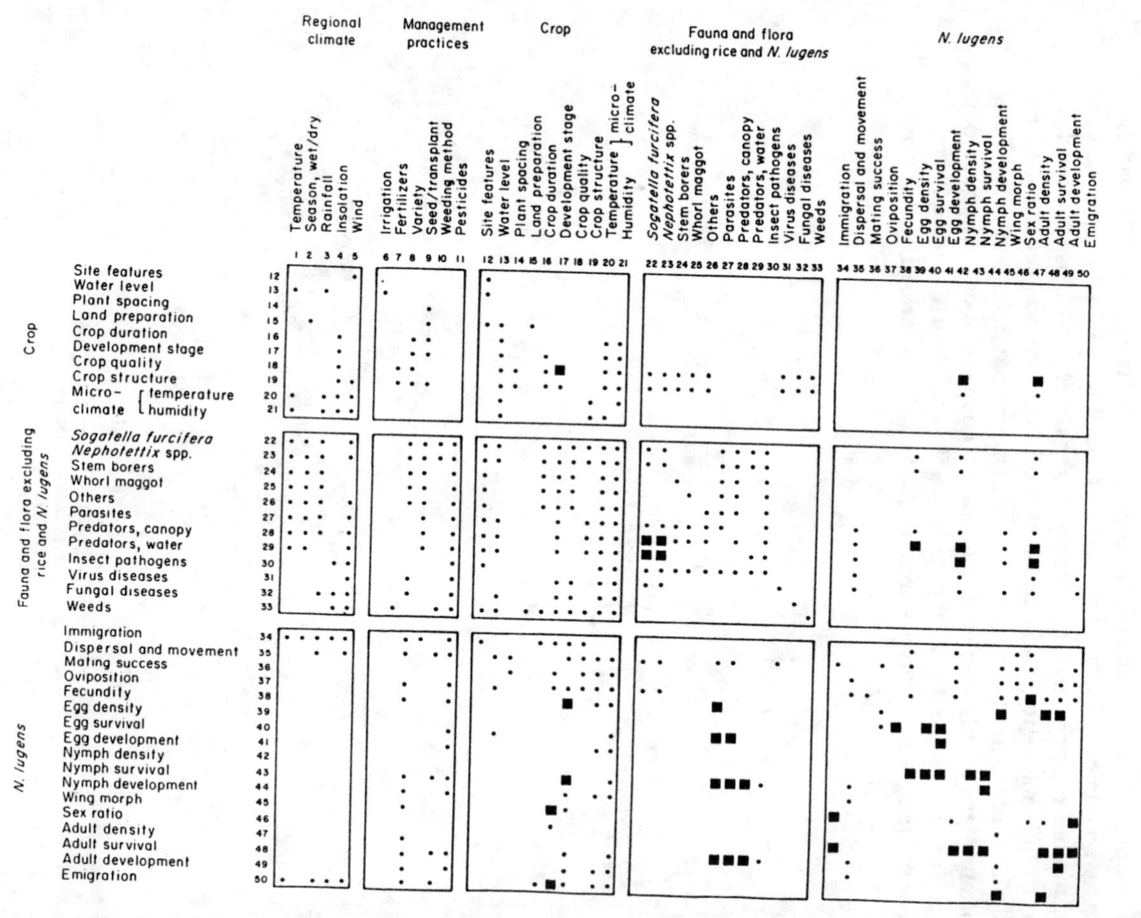

Figure 19.2. An interaction matrix for the *N. lugens*–rice crop system. A symbol (dot or square) in a particular cell indicates a direct effect of the column component on the row component. A square in a particular cell indicates an interaction explicitly included in the simulation model (from Holt et al. 1987).

variables: total number, start (colonization) time, and pattern (distribution immigrants over time). A high total number of immigrants coupled with an early start time and a concentrated pattern (most immigrants arriving over a few days shortly after transplanting; Fig. 19.3A) resulted in a high peak density in the first generation (Figs. 19.3B, C). As might be expected, total immigrant number had the most impact on first-generation density, but with start time and pattern modifying the relationship considerably.

The start of BPH immigration into the crop is usually coincident with transplanting time. Changes to transplanting time and, therefore, immigration start time change the average temperatures experienced by the BPH population developing on rice. An early transplanting time increases the growth rate of the population from the immigration to the first generation and from the first to the second generation by exposing the pest to more favorable average temperatures in autumn. The size of the population peak in the second generation is particularly sensitive to autumn temperature conditions. The combined impact of changes in transplanting time and September temperatures is shown in Fig. 19.4. In summary, the greatest risk of a BPH outbreak occurs when a relatively cool summer is followed by a warm autumn, when the crop is transplanted early, and there is a short,

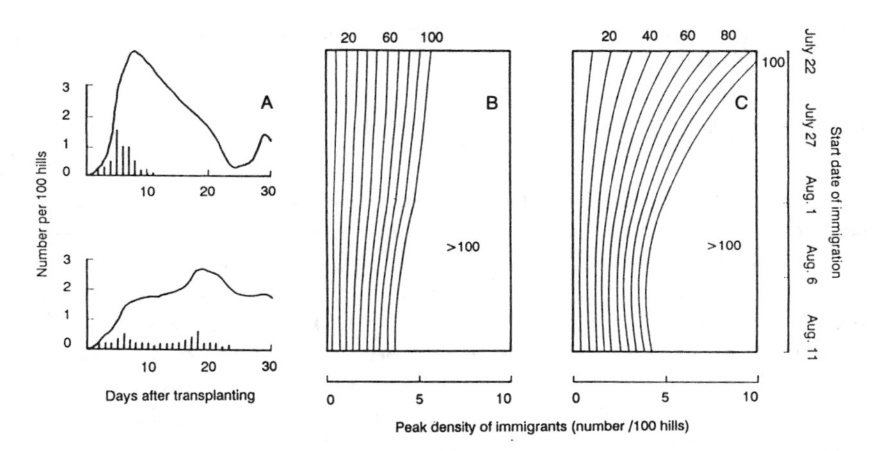

Figure 19.3. The relationship of peak density in the first generation to three components of immigration: total immigrant number, start time of immigration, and pattern of immigration over time. Two typical immigration patterns are shown (**A**), concentrated (top) and protracted (bottom); bars indicate the number of immigrants/day and lines the total immigrant density. Simulated contours of peak density of BPH in the first generation (number/100 hills) resulting from combinations of immigration start time and immigrant density, for a concentrated pattern (**B**) and for a protracted pattern of immigration (**C**).

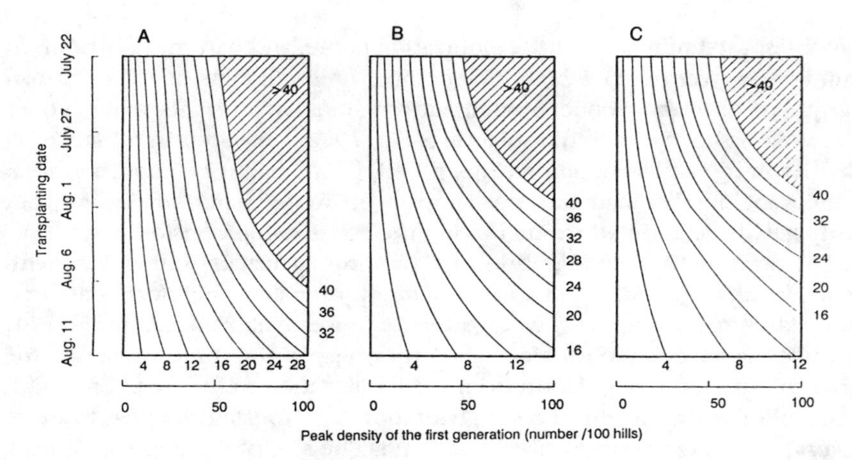

Figure 19.4. Simulated contours of peak density of BPH in the second generation resulting from combinations of transplanting time and peak density in the first generation, with temperature (**A**) favorable (autumn temperatures 1°C > average), (**B**) average (from last 10 years), and (**C**) unfavorable (autumn temperatures 1°C < average).

concentrated period of BPH immigration at the start of the season (Cheng and Holt 1990).

BPH IN LUZON, PHILIPPINES

In the Philippines, BPH populations on susceptible rice crops normally remain stable at a low level. Natural enemy action is widely regarded as the reason for population suppression (Kenmore et al. 1984), but outbreaks may be triggered by immigration, particularly when mass movements occur within asynchronously planted rice systems (Loevinsohn 1984).

Initially, a complex simulation model, involving density-dependent mortality of BPH in a multipredator–parasite system was used to account for observed changes in field populations (Holt et al. 1987). Model inputs were derived from the field studies of Cook and Perfect (1985a, 1989) and included densities of predators, other planthopper species, rates of parasitism, and rate of BPH immigration. Both predation and immigration appeared to be significant factors in observed population change, but insufficient information is available about predation processes in the field to allow detailed modeling of the interactions involved.

A second simplified version of the model was constructed to allow the model to respond dynamically when used to test management options. The major change which left BPH immigration as the single driving variable was to incorporate density-dependent mortality in a more empirical way, but one which encapsulates the essential dynamics of the BPH–natural enemy interaction (Holt et al. 1989). Density-dependent nymph survival is assumed to be a U-shaped function, with relatively high survival at low and high densities (<0.2 and >5 nymphs/hill, respectively) and low survival at medium densities (0.2–5/hill). This is supportable on theoretical grounds (Southwood and Comins 1976) and is consistent with the general observation that an otherwise stable BPH population often undergoes a dramatic increase in size in response to a perturbation. When compared with independent data, the model correctly distinguishes between low-density and high-density populations in most cases (Holt et al. 1990).

Low survival caused by natural enemies at medium nymph densities acts to maintain the BPH population in a stable state at an acceptably low-density for a range of immigration scenarios. Thus, below a certain immigration rate (0.1/hill/day), an increase in immigration has a relatively minor impact on the subsequent peak density of the BPH population in the crop. At rates greater than 0.1/hill/day, the BPH population is able to escape from natural regulatory mechanisms and reach outbreak densities (Fig. 19.5). Field observations by Cook and Perfect (1985a) suggest that immigration

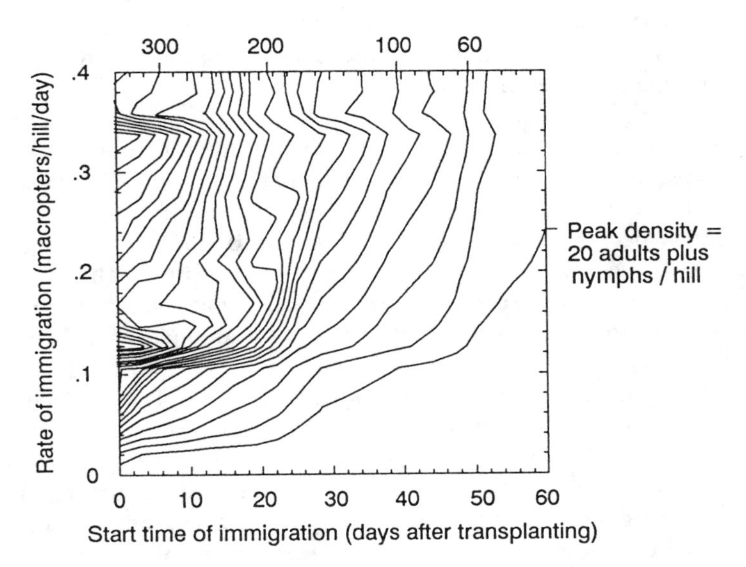

Figure 19.5. Contours of BPH peak density resulting from combinations of immigration rate and immigration start time with simulated BPH population dynamics under tropical conditions.

rates are generally below the 0.1/hill/day threshold density during the critical first 4 weeks of the season. Therefore, it would seem that within-field factors, notably mortality caused by natural enemies, are generally the key determinants of population dynamics. There are exceptions however. For example, once an outbreak has been induced by insecticides, very large numbers of emigrant macropters may be generated. Thus, the model lends support to the view that immigration rates could be sufficiently high to cause outbreaks in asynchronously planted contiguous rice systems (Loevinsohn 1984).

A Synoptic Model of BPH

Differences among the key factors which determine the dynamics of BPH populations in subtropical Zhejiang Province and the tropical Philippines are likely to be associated with seasonality. Winter temperatures are too low for BPH to overwinter in Zhejiang Province; the population is initiated by a very low density of spring immigrants from further south (Cheng et al. 1979). Natural enemy populations in and around the rice fields also decline to very low densities during the winter. Although wet and dry seasons impart a degree of seasonality in the Philippines, both BPH and its natural enemies are present throughout the year, and relative to Zhejiang Province, the densities of BPH and natural enemies are high during the first weeks of the crop (Cheng and Holt 1990).

A synoptic model representing a general hypothesis for differences in the dynamics of BPH populations between subtropical and tropical systems is shown schematically in Fig. 19.6. Conceptually, a population can be thought of as a ball on a sloping surface: The ball tends to roll to the lowest point unless perturbed in some way. The rear of the surface (Fig. 19.6) corresponds to a temperate/subtropical rice system, in which natural enemies retard but do not prevent a BPH population increase over the course of the season. In terms of the ball analogy, the population rolls steadily down hill from left to right.

The front part of the surface corresponds to a tropical rice system in which a population, maintained in a stable state by natural regulatory mechanisms, can loose its stability for two reasons. First, application of insecticide can reduce the density of natural enemies (as well as that of BPH), in effect switching the system to temperate-type dynamics. Second, a pulse of immigration can overwhelm natural regulatory mechanisms and allow the BPH population to "escape." The location of the BPH population on this surface has important implications for BPH management which are considered in the next section.

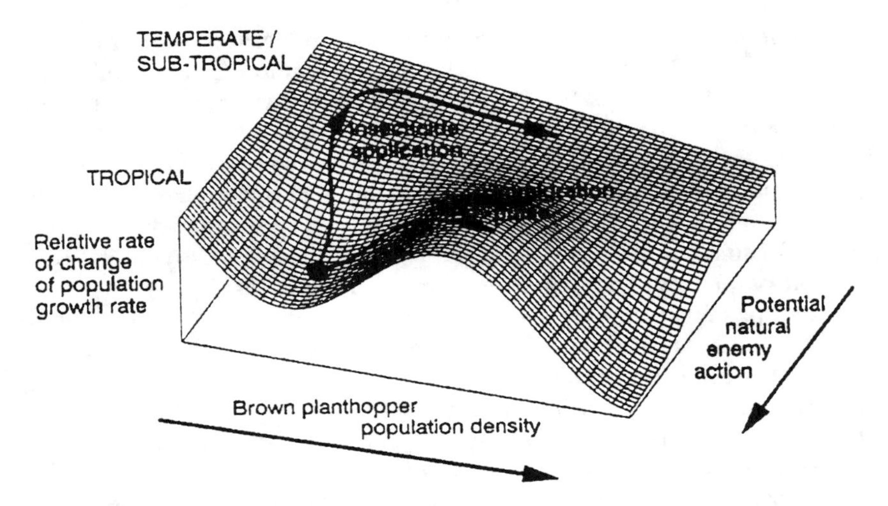

Figure 19.6. A conceptual model of brown planthopper population dynamics depicting the stability of temperate/subtropical and tropical populations and how that stability might be lost due to insecticide use or exceptional levels of immigration.

Analysis of BPH Management Options

BPH Management in Zhejiang Province, China

Insecticides remain the most important tool for BPH management in susceptible temperate rice varieties. In this case, IPM consists of making the most effective use of insecticides through careful timing of applications and the use of insecticide application thresholds. The simulation model has proved a useful tool in designing appropriate options for insecticide use. For instance, the simulation model has been used to determine the best time to apply a single insecticide treatment to suppress BPH in the second rice crop. A series of simulations were performed in which simulated populations were treated with single sprays over a range of times and with insecticides having different levels of persistence.

The results indicate that an application 30 days after transplanting (DAT) was most effective in reducing pest numbers, irrespective of the persistence of the insecticide. This time coincides approximately with the peak of the first generation and the trough of the egg population because immigration begins immediately after transplanting. Sensitivity to changes in application time is reduced as the persistence of the insecticide increases. A further series of simulations suggest that the "30 DAT is best" rule is very robust to changes in environmental and other factors, for example, temperature, transplanting time, immigration pattern, and immigration rate, provided that only two generations occur following immigration, the norm in this region. The "30 DAT is best" rule is supported by evidence from field experiments (Cheng et al. 1990).

Experience in Zhejiang Province has also shown that more than one application is sometimes needed to maintain a BPH population at an acceptable level. Thus, to investigate the best way of applying two treatments, a series of simulations was carried out as for the single-treatment case. An example of the results compares two transplanting times (Fig. 19.7). For the later transplanting time (Fig. 19.7A), an initial treatment at 30 DAT followed by a second treatment 10 days later at 40 DAT gives the lowest peak density. For the earlier transplanting time (Fig. 19.7B), an initial treatment at 30 DAT followed by a second treatment 20 days later at 50 DAT was most effective. With this earlier transplanting time, a partial third generation develops because other conditions favor BPH population increase. Thus, early transplanting is responsible for the shift in the best time for the second spray. However, in general, treating initially at 30 DAT proves to be a robust strategy for both the two-treatment and single-treatment cases.

Traditionally, economic thresholds for insecticide use are based on BPH density only, but the age structure of the BPH population, the crop stage, and temperature are also important determinants of yield loss (Kisimoto 1977; Zhang et al. 1983). In China (Zhejiang Province), the policy objective of the extension service is to limit BPH losses to less than 5%, but, increasingly, control costs are taken into account. Chemical costs vary, with more persistent insecticides being about twice the price of those with low residual action.

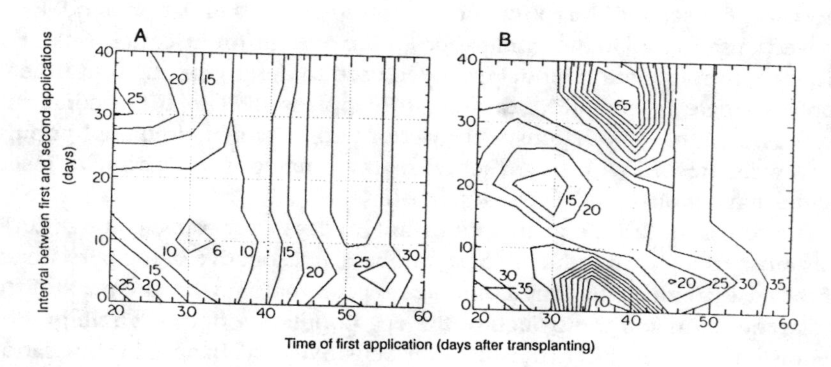

Figure 19.7. Contours of BPH peak density resulting from combinations of application time of two treatments of MTMC with rice transplanted on (A) August 1 and (B) July 22 (from Cheng et al. 1990).

To determine the economic threshold values applicable for different temperature scenarios, transplanting times, and insecticide types, a series of simulations were carried out for 15 combinations of transplanting time and temperature and for 2 types of insecticide, MTMC and Carbosulfan. The range of thresholds is shown in Fig. 19.8. For the second BPH generation (in which damage first takes place), the threshold is defined as that density which would give rise to a 5% loss in yield in the absence of control. The threshold for the first BPH generation is slightly more complicated in that it also takes into account the opportunity to apply an insecticide during the second generation. It is defined as the density which would give rise to a 5% loss in yield even with a single insecticide application during the second generation. The resulting multidimensional thresholds vary considerably with different conditions (Fig. 19.8). With late transplanting, unfavorable temperature conditions for BPH development, and when using a more persistent chemical, the thresholds are significantly higher than the current threshold values of 0.5–1/hill and 8/hill for the first and second generations, respectively. Improvements in the insecticide application decision-making process are possible, which can result in the reduction of insecticide use in some cases. Potential exists for further refinements through the inclusion of other factors such as variety and the density of natural enemies.

BPH Management in Luzon, Philippines

Given the importance of natural regulatory mechanisms in the dynamics of BPH populations in the tropics, the conservation of natural enemies must be a central concern in the implementation of any management activity. Many instances of resurgence have been cited in which insecticide use results in a higher BPH density than in the comparative but untreated case (e.g., Kenmore et al. 1984). Because insecticides are likely to remain a major component in the control of most rice pests (Lim and Heong 1984; Norton and Way 1990), the problem of how to avoid BPH resurgence is a key question for BPH management in the tropics.

To tackle this question, the "tropical" simulation model was used to assess the impact of insecticides on BPH population dynamics including resurgence. The effects of insecticide were incorporated in the model by changing BPH survival following a spray to account for both the direct increase in mortality due to the spray and an increase in survival due to mortality of natural enemies. Following an initial shift in survival at the time of the spray, both effects were modeled using sigmoid functions to describe the return of the values of the mortality components to prespray levels (Holt et al. 1992).

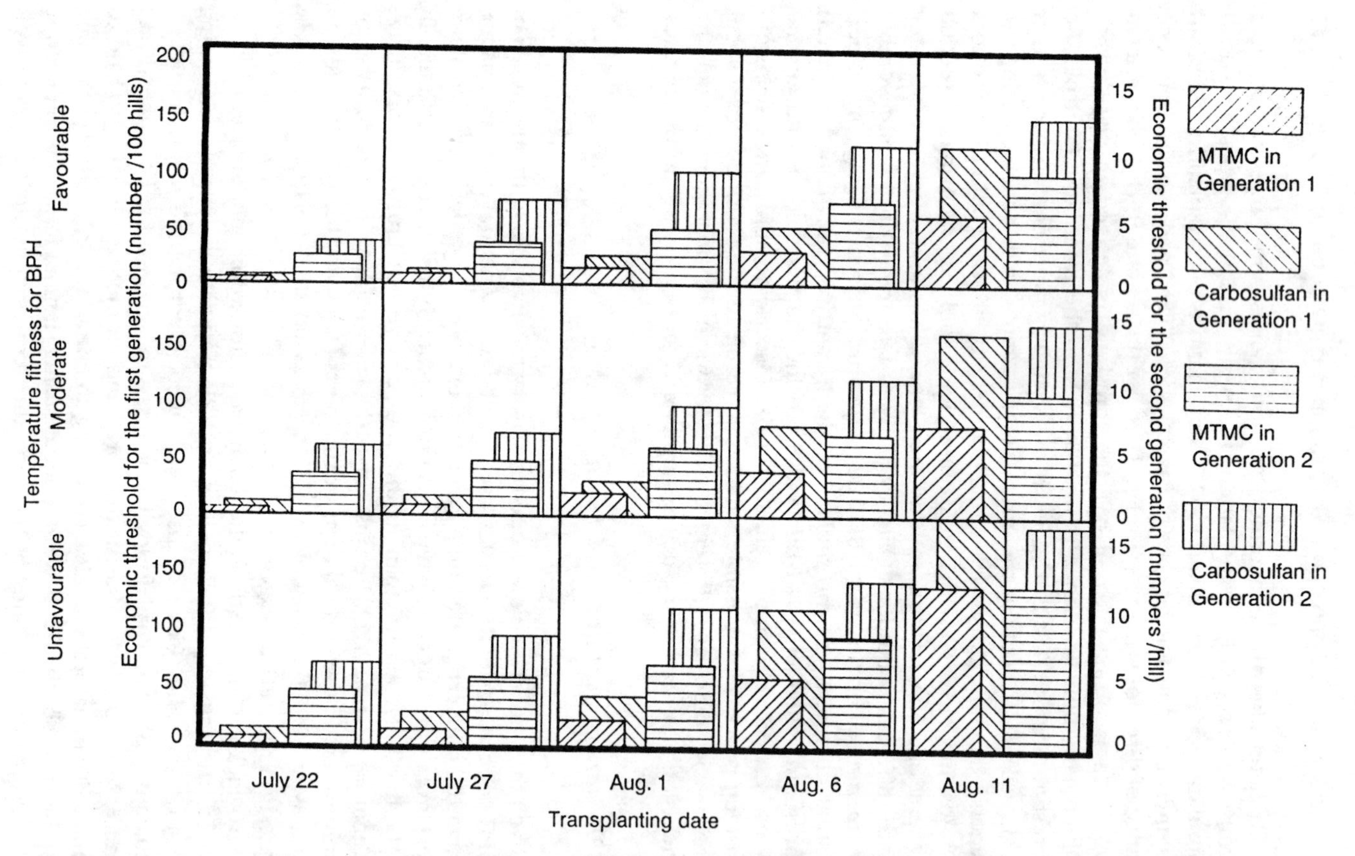

Figure 19.8. Action thresholds for the application of insecticides to control brown planthopper when treating with: MTMC in the first generation, carbosulfan in the first generation, MTMC in the second generation, and carbosulfan in the second generation.

To obtain a general picture of the effect of insecticide on BPH, two extreme situations in which insecticides cause mortality to either BPH or natural enemies only were compared with the more realistic situation in which mortality occurs to both BPH and natural enemy populations. The analysis was carried out over a range of immigration cases because of the importance of immigration in field population dynamics. Sixty immigration rate and start time combinations were examined for immigration rates ranging from 2 to 20 planthoppers per 100 hills/day and start times ranging from 5 to 30 days after transplanting. The proportion of those cases in which outbreaks occur gives an indication of the potential for BPH resurgence or for effective control following a particular insecticide application (Fig. 19.9).

For the purpose of the analysis, an outbreak is defined as a BPH population in excess of 100/hill (the precise threshold chosen has little impact on the results of the analysis). For example, we take the case of a spray applied at 30 DAT which causes mortality to both BPH and natural enemies. The proportion of cases resulting in outbreaks rose from 0.45 with no insecticide to 0.73 with insecticide, the difference (0.28) being the proportion of cases in which resurgence to outbreak levels occurred. The actual resur-

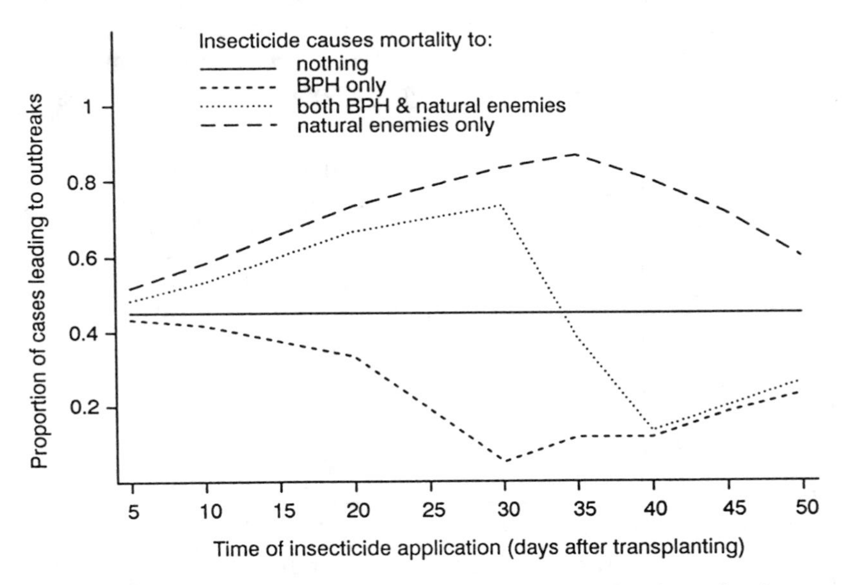

Figure 19.9. Proportion of immigration cases leading to outbreaks in the absence of insecticide and when an insecticide application causes mortality to: brown planthopper only, natural enemies only and both brown planthopper and its natural enemies.

gence risk depends on the immigration conditions at a particular place and time, but, nevertheless, Figure 19.9 gives an indication of the relative risks associated with different application times and type of mortality caused.

The main conclusions from this work suggest that very early sprays (5 DAT) appear to have very little impact in most cases. For compounds with high BPH specificity, applications between 30 and 40 DAT are most likely to be effective, but as natural enemy mortality increases, sprays at 30 DAT cause resurgence in a large number of cases. With very broad spectrum compounds which inflict high natural enemy mortality, sprays timed between 30 and 40 DAT resulted in resurgence in most cases. Sprays timed at 40–45 DAT appear to represent a reasonable compromise between low resurgence risk and potential control success provided that natural enemy mortality caused by the spray is not too great.

The analysis just described considers an insecticide with the following properties: (1) initial kill of BPH of 70%, (2) initial reduction in mortality by natural enemies from 95% to 40% (when BPH population is at densities between 0.2 and 5/hill), (3) residual mortality of BPH of 4.3 days half-life, and (4) recovery period of natural enemies of 15 days half-life.

This residual mortality corresponds to a chemical such as decamethrin or endosulfan, but the value of this and the other parameters are dependent on application efficiency and environmental variables. The impact of changes in parameter values on the general conclusions were examined by sensitivity analysis. As an example, Figure 19.10 shows the effect of changes in

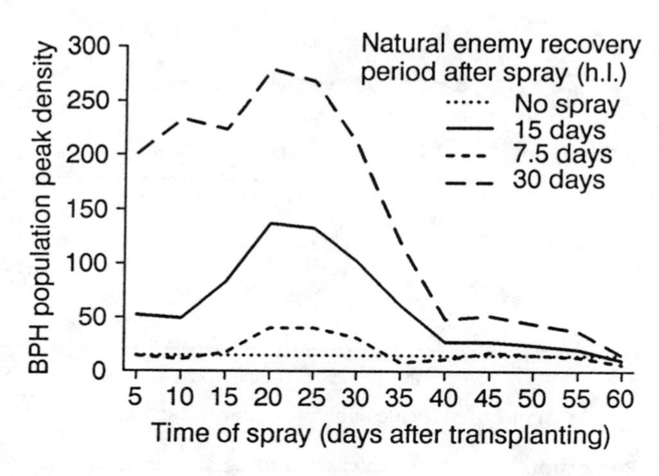

Figure 19.10. Sensitivity of the peak density of a simulated brown planthopper population to changes in the period of recovery of natural enemy populations following insecticide application.

natural enemy recovery period on the degree of resurgence resulting from spray applications at different times. With the particular immigration state selected in this example (immigration rate of 4/100 hills/day commencing on day 10), a spray application may result in anything from a small reduction in the BPH population to a 20-fold increase, depending on the rate of recovery of natural enemy populations. The extent of resurgence was also sensitive to the residual effects of the insecticide on BPH and the degree of initial disruption of natural enemies. However, despite this considerable sensitivity in the degree of resurgence, the period of greatest risk remains reasonably constant. Other things being equal, resurgence is most likely to occur when insecticide is applied at 20–30 DAT. Control efficacy is likely to be highest when insecticide is applied at 40–45 DAT.

Differences between tropical and subtropical systems in the optimal time to apply insecticides relate to the key role of natural regulatory mechanisms in tropical rice. Although, in terms of BPH population dynamics alone, an application at 30 DAT is most effective, but the consequences of destroying natural enemies at 30 DAT are very severe allowing for rapid recovery of BPH populations following the spray. Details of the interactions in BPH and natural enemy population dynamics suggest that sprays at 40 DAT have a less damaging impact (Holt et al. 1992).

Design of Recommendations

An expert systems approach can be used to incorporate these principles in a wider context for control decision making. This allows the inclusion of important practical constraints outside the scope of the simulation study, particularly that of incomplete information.

The applications of most expert systems to pest management fall into one of two categories: diagnosis (identification of the pest or other aspects of the problem) and prescription (pest control recommendations). Prescriptive systems can be designed to address problems at policy, strategy, or tactical levels, though the majority concern tactics for the implementation of pest control measures (Mumford and Norton 1985, 1989; Jones et al. 1990).

An expert system has been developed for BPH management in the Yangtze Delta area of China (Holt et al. 1990). The system prescribes tactics for BPH management in late paddy (second crop) rice. It uses knowledge in the form of the opinions of experienced rice pest control practitioners, results of simulation modeling and information available for the region (e.g., weather and pest forecasts) and the field (e.g., rice variety and transplanting time). The main aim has been to develop a system that can be used initially for training by the county- and village-level extension staff to improve the BPH control recommendations they provide.

To construct the expert system, the process of making a recommendation was broken down into its constituent parts; in this case, six steps were identified:

1. Assessment of the level of attack at the particular time in the crop season when advice is sought
2. Prediction of the growth rate of the population until peak density is reached
3. Determination of appropriate treatment thresholds for the prevailing conditions
4. Estimation of whether the peak density is likely to exceed the single- or double-treatment threshold
5. Consideration of robustness of the predictions to possible changes in relevant factors over the prediction period
6. Recommendation to the grower to do nothing, monitor the BPH population at certain times, or apply particular chemical treatments at certain times

Next, a search tree was drawn up which specifies the logic followed by the system to arrive at a recommendation. Some of the search tree branches associated with the decision period (25–40 DAT) are shown in Figure 19.11. Progress through the search tree prompts questions of the user; if the user responds with "unknown," then a recommendation is still possible, though it is necessarily less precise. The ability to deal with different levels of information, an important feature of the expert system, is illustrated by the lower branches of Figure 19.11.

Finally, the search tree was translated into rules employed by EXSYS (Exsys Inc. 1985), the expert system-building software used. For instance, the branch of the search tree shown in thick lines in Figure 19.11 is represented by the rule:

If: time period is 25–40 DAT
and variety is partially BPH resistant
and transplanting time is early
and autumn weather forecast is average temperatures
and effect of natural enemies is assessed
and mortality caused by natural enemies is average
and peak density of the first generation is known
and density > 60 per 100 hills,

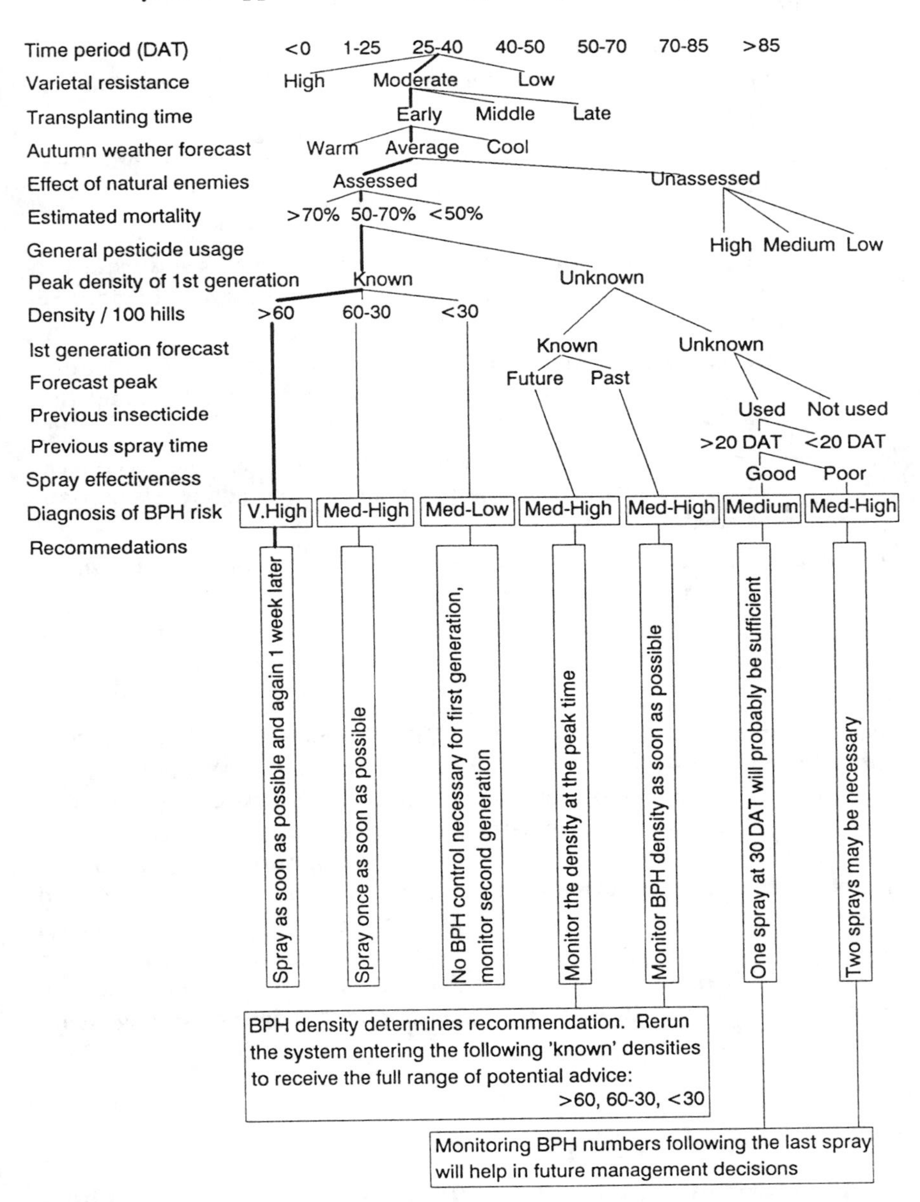

Figure 19.11. Portion of the search tree which specifies the logic followed by the expert system to arrive at a prognosis of outbreak risk and a recommendation for management of brown planthopper.

Then: BPH risk is very high
and spray once with a persistent insecticide as soon as possible and again one week later.

The development and use of this and other expert systems offer four particular benefits:

1. A forum of communication between research scientists and extension officers which helps to clarify the decision process and identify research and surveillance needs.
2. A bridge between modeling and extension providing a structure in which the multidimensional thresholds derived from the simulation study can be applied.
3. An expert system structure which encourages a rigorous, consistent approach to management over a range of situations and levels of information availability.
4. An easy-to-use reference that can serve to train inexperienced advisers.

Conclusion

In this chapter, we have attempted to show how certain systems analysis and decision analysis tools can be valuable in tackling questions relating to planthopper biology and management. The use of descriptive techniques, simulation models, and expert systems have revealed key information gaps, improved perceptions of planthopper problems, provided a means of searching for more effective control tactics, and promoted training and implementation programs. As a vital link between basic and field research on the one hand, and the practical management problems faced by farmers and their advisers on the other, we believe these tools have a critical role to play in future applied research on planthopper biology and management.

Summary

1. The brown planthopper (BPH, *Nilaparvata lugens*) is the most damaging planthopper pest throughout temperate, subtropical, and tropical rice systems. A systems approach has been used to integrate, analyze, and interpret research findings and to address key questions facing decision makers at the policy, research, extension, and farm levels.

2. Simulation models have helped to understand the key factors affecting the population dynamics of BPH and, particularly, the conditions in which outbreaks can occur.
3. Used to assess the performance of insecticide control strategies, simulation models have also provided a means of identifying the best time to apply sprays and of determining multidimensional thresholds for insecticide application.
4. Expert systems have been used to integrate these findings with knowledge obtained from experienced field research and extension specialists. They provide training and decision support tools for extension agents and highlight important information and research gaps for research managers and policy makers.

20

Extension of Integrated Pest Management for Planthoppers in Asian Irrigated Rice: Empowering the User

Patricia C. Matteson, Kevin D. Gallagher, and Peter E. Kenmore

Introduction

The key factor in sustained implementation of integrated pest management (IPM) in developing countries has been, and still is, **motivation**. Smith's (1978) description of the typical "phases" of IPM development, based on experience with cotton since the 1950s, is proving to be an enduringly accurate description of human nature as a factor to contend with in crop protection. The cycle begins with overdependence on, and misuse of, a single pest control tactic—usually insecticides—such that pests evolve the ability to overcome it (Heinrichs, Chapter 16; Gallagher et al., Chapter 17). There ensues a "Crisis" or "Disaster Phase" of crop unprofitability or failure. The spectre of frightening economic and political losses spurs government to espouse and invest meaningfully in IPM and gets farmers' attention. During the "IPM Phase" an array of tactics based on ecological crop protection principles—the use of economic threshold levels, regionwide fallow periods for the crop, and so on—are promulgated and applied. Then, after enjoyment of a period without pest control emergencies, a "Deterioration Phase" sets in. Policymakers and farmers are no longer motivated to stick to ecological principles, backsliding to routine pesticide applications that require less attention, effort, and management skill. IPM research and extension programs falter.

Thus far, experience in Asian irrigated rice corroborates those observations through the Disaster and IPM Phases. Widespread crop failure caused by the brown planthopper (*Nilaparvata lugens* Stål) was endangering economic and political stability in several countries. This threat elicited the political will, supportive policy environment, and financial backing that are essential for large-scale IPM development and extension. National governments and international development organizations working in the region devoted significant resources to rice IPM research and have collaborated in an IPM extension effort that is the largest and arguably the most innovative in the world. The history of rice IPM extension is one of experiment, evaluation, and evolution, challenged by the question raised by Smith (1978): short of crisis, is it possible to motivate developing-country rice farmers and extension agents for effective large-scale IPM extension, and then preserve the momentum of the extension effort on the long term?

Experience to date has shown that creative and ambitious measures must be taken to shatter the deeply ingrained uncritical and dependent attitude toward pesticides which prevails at all levels in developing countries, from ministries to the smallest farms (Kenmore et al. 1987). That attitude exists because the negative motivators, ignorance, fear, and passivity, have been allowed to govern pest control. Chemical company advertising perpetuates and exploits policymakers' and farmers' fear by failing to add to ecological understanding while portraying insecticides and fungicides as valuable crop insurance. After four decades of demonstrations and commercial promotion, farmers are overconfident of their mastery of pesticide application and its value (Waibel 1986).

Misunderstanding of costs and benefits makes simple prescriptions for prophylactic pesticide applications appear attractive and undemanding compared to IPM. Many pesticide applications are wasted through incorrect practices, including unnecessary preventive application. Field studies in the Philippines and Sri Lanka suggest that rice farming without insecticides is economically competitive (Marciano et al. 1981; Litsinger 1984; Matteson 1986; Kenmore 1987; Smith et al. 1989) but most farmers are too well-indoctrinated to contemplate that possibility. Instead they excuse insecticides' frequent lack of visible effect on the assumption that losses to pests would have been worse without treatment (Kenmore 1987). Not only are the benefits of rice insecticides largely imaginary, but their real cost is obscured—often by government subsidies and always through virtually unremarked acceptance of untold millions of dollars worth of "hidden" costs to human health, nontarget organisms, and the environment.

Rice IPM extension specialists in Asia hope to end this pesticide dependency with new, positive motivators for developing-country farmers: understanding, excitement at learning, and empowerment. Focus has centered increasingly on the need to teach IPM knowledge and skills in a consciousness- and confidence-raising framework that changes people's view of the natural world and of their place in it. The brown planthopper strongly influenced the technical content of IPM extension in Asian irrigated rice, and this chapter opens with a discussion of extension issues particularly relevant to planthopper management. The balance of the chapter, which is relevant to other rice pests as well as to planthoppers, describes the three-phase evolution of rice IPM training in Asia from 1979 to the present. It focuses on the village level, where agricultural scientists and administrators took farmers' and extension agents' motivation for granted too often in the past (Goodell et al. 1982). The chapter concludes with a discussion of topics needing further investigation.

Planthopper IPM and Extension

The brown planthopper had a high profile as the key insect pest when IPM development was initiated for irrigated rice in tropical Asia. In addition to *N. lugens*, the whitebacked planthopper, *Sogatella furcifera* (Horvath), is also an occasional problem. Planthoppers are actually secondary pests in tropical rice production, a problem associated with misuse of insecticides. Insecticides trigger brown planthopper outbreaks by stimulating oviposition and suppressing the natural enemies that usually keep planthopper populations in check (Kenmore et al. 1984; Ooi 1984). Because of the importance of preventing further planthopper outbreaks and because of its wider relevance, pesticide management to enhance natural pest control by ricefield predators and parasites anchored rice IPM technology, particularly at first. However, extension messages about planthopper management also cover cultural control practices, the planting of resistant cultivars, and nonchemical control measures. A list of planthopper control recommendations (Table 20.1) shows that almost every decision about growing rice influences the crop's susceptibility to pests. Therefore, an IPM training course is roughly equivalent to a course on rice production. The emphasis on planthopper-related subjects in IPM extension training has generated information about the most effective way to teach planthopper control practices, farmers' problems mastering planthopper control, and adjusting research priorities to respond better to extension requirements.

The Central Role of Natural Enemies of Rice Pests

Learning more about the natural enemies of rice pests is at once the most enjoyable and the most powerful part of the planthopper control curriculum for farmers. They have seen birds, frogs, and snakes eat pests, but generally know little about beneficial arthropods (Bentley 1989). Many

Table 20.1. Brown planthopper control recommendations (adapted from Reissig et al. (1985) and Smith et al. (1989).

1.	Grow no more than two rice crops per year.
2.	Choose early-maturing, planthopper-resistant rice cultivars.
3.	Plant neighboring fields within 3 weeks of each other.
4.	Control weeds and do not exceed recommended rice plant density.
5.	Apply nitrogen fertilizer judiciously, with split applications three times during crop growth.
6.	Visit the fields weekly from seedbed to dough grain stage, counting pests and their natural enemies and taking control action when the brown planthopper population reaches the economic threshold.
7.	Economic threshold for brown plathopper, Nueva Ecija, Philippines, 1986: one mature nymph/hill, subtracting five planthoppers for each predator encountered.
8.	To reduce planthopper populations, drain the paddy for 3 or 4 days.
9.	If draining the paddy is not feasible, apply an effective insecticide (one that does not cause planthopper resurgence) at the base of rice plants only in the infested portions of fields of susceptible cultivars.
10.	Plough down volunteer ratoon after harvest.

remark after training that they used to tear down spider webs or spray when dragonflies hovered over the paddy, fearing that any small creature was a sign of trouble.

Finding out about the life history and habits of predatory and parasitic arthropods fascinates training class participants. Trainers concentrate on major natural enemy groups: spiders, water bugs, predatory beetles and bugs, long-horned grasshoppers, dragonflies and damselflies, and wasps. Difficult scientific names are replaced by descriptive, often humorous terms that help trainees remember beneficial insects: in the Philippines, predacious and parasitic Hymenoptera with their nipped-in waists are "sexy wasps"; in Sri Lanka, *Casnoidea* spp., carabid predators with a broad, rectangular abdomen and a long, narrow thorax, are "bottle beetles" and the mirid *Cyrtorhinus lividipennis* Reuter, a major predator of planthopper eggs and nymphs, is the "planthopper destroyer." Putting a wolf spider in a jar with planthoppers and watching what happens is riveting.

This struggle for existence and, on a different plane, opposition of benign and harmful forces is a dynamic that farmers understand and respond to. With this orientation, they comprehend immediately why unnecessary insecticide use, so harmful to their "friends," must be avoided. Thus, natural enemy studies are one of the strongest motivators for farmers to change their practices.

Recognition of natural enemies is an important skill for planthopper control decision making. Economic thresholds based only on pest counts have been replaced by observation of the balance between pest and natural enemy numbers in the rice paddy (Tables 20.1 and 20.2) (Ooi 1984; Ooi and Heong 1988; Smith et al. 1989).

Table 20.2. Four IPM implementation principles of the Indonesian National IPM Programme.

1.	**Grow a healthy crop,** using varieties resistant to the local disease complex, proper fertilization, irrigation, and weeding fields as needed. A healthy plant can recover from damage better and yields more.
2.	**Conserve natural enemies,** avoiding unnecessary sprays and keeping spider habitat available on the bunds.
3.	**Observe fields weekly,** analyzing the agronomic, disease, rat, and insect situations. Thresholds are not used, but careful observation is used to detect damaging insect populations and natural enemies.
4.	**Farmers are IPM experts,** allowing field workers to provide inputs as needed.

Pest Control Decision Making and the Research/Extension Gap

The actual monitoring and decision-making process is the most difficult part of extension training for all concerned and highlights the gap between researchers' priorities and farmers' needs. As a rule, IPM researchers and agricultural administrators consider correct economic thresholds very important. Sometimes years of research effort are spent fine-tuning them and on the creation of elaborate sampling schemes. A lack of domestically developed thresholds has on occasion delayed the implementation of national IPM programs. Yet field experience indicates that these priorities are misplaced when small-scale farmers will be responsible for crop monitoring and decision making. Much of what researchers produce is rejected in practice as too complicated or inappropriate.

Typical instructions for planthopper control decision making resemble the following (IRRI 1985): Pick 20 rice hills at random in a diagonal across the paddy; count the mature planthopper nymphs on each hill; determine the average number of tillers per hill; calculate the average number of nymphs per tiller; and spray if the average number of nymphs is higher than one. More recently, a presence/absence sequential sampling method has been developed for planthoppers and their predators (Shepard et al. 1989). Sequential sampling saves much time and effort but still requires the scout to zigzag through the field, count predators and planthoppers, write down numbers, add and subtract them and consult a printed table.

Though they would no doubt be excellent for professional IPM scouts in environments where there is not much variation in crop and economic conditions, even "simplified" routines like these are too tiresome and are not appropriate for most Asian farmers. It is a small victory to persuade farmers to step off of the bund and into the paddy mud to inspect their crop. Few will take paper and pencil to the paddy for tallying numbers of insects observed. Economic thresholds and the mathematical formulas and calculations usually used to apply them are the hardest part of the IPM curriculum for both extension agents and farmers to master. Even in extension programs with good quality instruction, evaluators of threshold training sessions too often find field trainers floundering or making mistakes in front of a mystified group of trainees (E. van de Fliert unpublished). Experienced trainers have long suggested that qualitative judgments should replace quantitative ones as far as possible—such as "eyeballing" a paddy to gauge pest problems instead of using laborious calculations (Goodell et al. 1982).

Following exact instructions about walking trajectory, the number and location of rice hills to be inspected and rigorous threshold decision making pales in importance beside the need for basic changes in behavior. If, instead of spraying their whole field for insurance or as soon as they see a few insects, farmers learn to check the pest "hot spots" in their paddy, recognize insect pests and their natural enemies, and act only when and where the pests are in the vast majority, the practical battle is won. It is not appropriate to use the adoption of official scouting routines or thresholds as a criterion for judging the impact of an IPM extension program.

In fact, followup invariably shows that IPM training causes more farmers to apply the economic threshold concept, but that most do not use the recommended scouting practices or thresholds (Waibel 1986; Matteson and Senerath unpublished manuscript). This is consistent with the general adoption pattern for all technology, wherein farmers adopt research recommendations stepwise and often change them to suit local circumstances and personal experience (Byerlee and de Polanco 1986). Where thresholds are concerned, misunderstanding is obviously a factor. However, a more powerful motive is that farmers who grasp the threshold concept understand immediately that numbers handed down from above are not good enough. It is clear that only they have the necessary farm- and locality-specific information to decide wisely about applying various alternative pest control measures in their fields: access to labor, cash, and sprayers, the condition and priority of their crop, more productive alternative expenditures, current rice and pesticide prices, and so on. Ultimately, thresholds function as straw men, being rejected for farmers' more refined action criteria.

In view of the foregoing, scientists working for a small-scale farmer clientele in developing countries should reconsider the role of scouting methods and thresholds and the kind and amount of research effort they deserve. At the moment the main value of thresholds is that they supply a point of departure for farmers' personal calculations by naming a pest density considered by researchers to cause unacceptable losses. It might be more appropriate simply to include a practical range of crop loss assessment information in IPM training, so that farmers know the threat posed by different pest population levels. They habitually overestimate potential damage because they do not understand well enough the relationship between insect pest numbers or pathogens and yield loss as it varies with cultivar, stage of the crop, weather, and so on. (Waibel 1987). A presentation of crop loss assessment information in an IPM training class might follow a discussion of the degree of crop loss that farmers consider to be serious. The two types of information could then be combined via guided class discussion to create local or personal provisional action thresholds. Such thresholds would be more useful than any official recommendations, and the process would help farmers understand the threshold idea better.

Resistant Cultivars, Spot Spraying, and Confidence

Brown planthopper-resistant cultivars are one of the major products of rice pest control research and can go far toward eliminating farmers' problems. Yet farmers overwhelmingly negate the benefits conferred by their seeds and treat all rice alike with insecticides. Undoubtedly, part of the problem is farmers' experience with fraudulent "planthopper-resistant" rice, which is more widespread than is usually recognized (Goodell 1984). However, it is clear that extension programs need to deliver better training about resistant cultivars and lessen the fear which motivates farmers to hedge their bets with insecticide.

The patchy distribution of planthopper infestations makes spot spraying a particularly useful measure for reducing insecticide use in rice. However, it is another recommendation that farmers may be too fearful to adopt. Untrained farmers assume that the presence of a few insects or of a pest-infested patch of paddy is a sign that the entire crop ought to be sprayed (Kenmore 1987; van de Fliert and Matteson 1990).

Both of these examples underline the need for depth and quality in rice IPM training. Just passing along instructions, which is all that many extension systems aspire to, or even teaching farmers new skills, is not enough. Farmers' fearful attitude toward crop pests must be replaced with confident mastery. One fears what one does not understand and one is unsure in new situations. Farmers need to understand the dynamics of the paddy ecosystem and have extensive practice managing it with the guidance and reassurance of teachers.

Eyesight Problems and Planthopper Size

The small size of planthoppers (3 mm) and some of their natural enemies means that not everyone is capable of monitoring them in the rice paddy. Extension programs must take this into account by making sure that trained, sharp-sighted people are available to farm households. In Sri Lanka, rural people rarely wear eyeglasses and the average rice farmer is 50 years old (van de Fliert and Matteson 1990). Extension agents are now instructed to ask farmers who cannot see well enough at the beginning of rice IPM training to bring along another family member with good eyesight. A project in the Philippines is coping with the problem by training farm boys 10 to 13 years old as IPM scouts for hire (Adalla 1990).

Evolution of the IPM Training Approach

Evaluation of the field impact of rice IPM training has been given high priority throughout the 14 years since Asia's first experimental IPM training classes were organized. Evaluation feedback has been the basis for constant improvement of national programs. Within the region different extension systems, strategies and methodologies have been tested, and new ideas continue to be adopted. Techniques developed in other fields of endeavor, notably commercial advertising, community organizing, and participatory nonformal education, have been applied to IPM extension for the first time.

This dynamism has resulted in the constructive evolution of the extension training approach, summarized in Table 20.3. The first phase of rice IPM training was research from 1978 to 1980 on basic training principles, followed by their successful application on a pilot scale in the Philippines by master trainers using conventional teaching methods. During this period the United Nations Food and Agriculture Organization (FAO) began to coordinate and support rice IPM extension in the Asian region. In a second phase, from 1985 through 1989, an attempt was made in several countries to deliver similar training on a national scale through Training and Visit (T & V) extension systems supplemented by multimedia strategic extension campaigns. Training intensity and quality deteriorated so much that Indonesia's high-priority national rice IPM program was spurred to create an innovative dedicated training system designed to empower and motivate

Table 20.3. Evolution of rice IPM farmer training in Asia, 1979 to present.

Period	Training Approach
1978–1984	Pilot scale, Philippines. Elaboration of IPM training principles. Group field training by master trainers using conventional teaching methodology. Weekly training classes over an entire cropping season, followup for 1–2 seasons.
1985–1989	(1) Small-scale implementation of the above by NGOs, Philippines. (2) Large-scale implementation through several national T & V extension systems, sometimes supported with multimedia strategic extension campaigns.
1990–1992	Medium scale, Indonesia. Dedicated training system in which master trainers use participatory nonformal education methodology.

both extension agents and farmer-trainees. That system represents the third and current training phase, with extension workers using participatory nonformal education methods in a "learning by doing" process based on ecosystem analysis. Three multilocational training cycles have been completed since 1990 and evaluations indicate that farmer graduates of the current Indonesian system are more effective rice IPM practitioners than any trained previously. Their new ecological orientation and initiative as decision makers is meant to generate enduring "bottom-up" demand for IPM information and technology, sustaining research and extension efforts on the long term.

Initial Research and Pilot Scale Training

The requirements for effective rice IPM training were first investigated during five consecutive 1978–1980 cropping seasons in 12 villages in Nueva Ecija Province, Philippines (Goodell et al. 1981, 1982; Goodell 1984; Litsinger et al. 1984). The interdisciplinary research team included Philippine rural sociology and community organizing expertise, government extension officers, and an anthropologist and entomologists from the International Rice Research Institute (IRRI). The IPM technology extended had been developed by IRRI researchers since 1972, when the problem of brown planthopper resurgence was addressed (Hansen 1987).

The team's basic findings have been validated repeatedly by subsequent experience in several Asian counties (Whitten et al. 1990; Matteson and Senerath unpublished manuscript). Rice IPM was learned and then actually practiced when:

1. Farmers were trained in a group, with frequent discussions. Farmers learn best by talking and listening to peers. Responding to questions as part of a group is less threatening, and there was group reinforcement of the idea of relying on individual judgment. In a society where farmers are normally timid, obedient, and dependent, associating science with authority, they are initially uncomfortable when IPM trainers ask them to think on their own.
2. Farmer groups attended weekly 2-hour classes in the field during an entire 3-month-plus growing season (20–40 contact hours). Class subjects were timed to field events and addressed local problems. The subject matter was pared down to essential points, which were constantly repeated and reviewed. During the main part of the season crop monitoring and pest control decision making was a weekly exercise, both for adequate surveillance and for ensuring enough practice of these important skills with a range of pests.
3. Fancy training materials such as handouts and illustrated manuals, which turned out to be both distracting and ineffective, were deemphasized in favor of hands-on field practice with real plants and pests and maximizing personal contact between farmer and trainer.
4. Lectures, theory, and technical terms were avoided and the technology was greatly simplified.
5. Demonstrations conducted in farmer training villages engaged farmers' curiousity and class experiments encouraged imaginative inquiry and self-reliance.
6. Extension trainers made followup visits to trainees during subsequent seasons. Classes alone made a relatively modest impact on farmers' practices. Farmers diagnosed pest problems better and trusted more in their own decision-making ability when they could practice on their own crop with the extension agent reassuringly accessible to confirm judgments and answer questions.

Philippine Pilot Activities with Master Trainers

A small, dedicated cadre of officers within the Crop Protection Division of the Bureau of Plant Industry, Philippines expanded the training effort begun in Nueva Ecija. The training continued to be conducted according to the guidelines developed by the aforementioned interdisciplinary research team. By 1984, about 200 "master trainers," 4500 extension agents and over 55,000 farmers had been trained in IPM, mostly by staff of Regional Crop Protection Centers (Whitten et al. 1990). Results of two Nueva Ecija surveys comparing the attitudes, knowledge and behavior of trained and untrained farmers illustrate the substantial long-term impact of this training. Farmers in two villages 3 years after training exhibited significant differences in IPM-related understanding, expertise, and level of insecticide use. Trained farmers averaged 3.5 insecticide applications per season as opposed to the 5.7 applications made by untrained farmers (Kenmore et al. 1987). Another survey was carried out in one town over three 1984–1985 cropping seasons 5 years after IPM training. On average, training reduced farmers' pesticide applications from 2.9 to 1.8 per season. Trained farmers' pest control costs dropped by 40% while their rice yields increased by 10%, from 5.03 to 5.52 tons/ha (Kenmore 1987).

The FAO Rice IPM Programme

After 1982, the Philippine training effort received technical and financial assistance from the FAO Inter-Country Programme for Integrated Pest Control in Rice in South and Southeast Asia (hereinafter referred to as the FAO Rice IPM Programme), funded in 1980 by the governments of Australia and the Netherlands and, later, by the Arab Gulf Fund, and implemented by FAO. The FAO Rice IPM Programme, now entering its third 5-year phase, provides technical coordination and training support for national rice IPM research and extension programs in nine countries: Bangladesh, India, Indonesia, Malaysia, the Philippines, Sri Lanka, and Thailand, founding members, and China and Viet Nam, which joined the program in 1989 (Hansen 1987; Hawkins 1989; Whitten et al. 1990).

The FAO Rice IPM Programme emphasizes ongoing evaluation of the implementation and impact of rice IPM extension activities in member countries. Evaluation findings provide guidance for improving extension effectiveness, identify opportunities for innovation, and document the program's value to farmers and its overall return on investment. Because IPM extension aims to change behavior by imparting skills, knowledge, and decision-making principles rather than by promoting a specified technological package, defining "adoption" and deciding who is an "adopter" requires some care. Most evaluators have chosen to gauge IPM extension impact by comparing the knowledge, attitudes, field skills, and practices (particularly the amount of insecticide use) of trained and untrained rice farmers (Kenmore et al. 1987; Adalla et al. 1990; Matteson and Senerath, unpublished manuscript). Knowledge, Attitude, and Practice (KAP) surveys with reference to rice crop protection were conducted in several countries to establish a benchmark for measuring long-term rice IPM extension progress at the national level and to better define farmer training priorities (van de Fliert and Matteson 1989, 1990).

Lessons Learned about Scaling Up

Extension of rice technology in Asia is a mammoth undertaking because of the size of the audience to be reached: almost 200 million rice farmers, mostly smallholders, grow 90% of the world's rice there. With an effective training approach proven on a relatively small scale, after 1984 the FAO Rice IPM Programme began supporting national-level efforts to reach more farmers.

Farmers' awareness of an innovation is greater when they receive the relevant messages through more than one channel and it is common to combine several complementary channels in an overall extension strategy. Besides national extension services, possible channels include other government agricultural or development institutions, schools and universities, nongovernmental development organizations (NGOs), social and religious organizations, and messages conveyed on farms, in homes, and customary gathering places (markets, temples, boutiques, etc.) using traditional or modern media (van de Fliert and Matteson 1989; UNDP 1991).

EXTENSION MEDIA AND STRATEGIC EXTENSION CAMPAIGNS

A variety of media has been used to transmit rice IPM extension messages in Asia. It includes face-to-face contact between the farmer(s) and an extension agent, development officer, or crop protection scout, the mass media (posters, booklets, leaflets, newspapers, magazines, radio/TV), essay contests, community bulletin boards that post information about the status of local rice pest infestations, and field demonstrations. Traditional communications media such as folk songs, folk drama, and puppet shows can be very effective; Indonesian village theater performances about life in the rice paddy are being used to teach important IPM concepts (van de Fliert and Matteson 1989; Stone 1992). Computer-based management information systems and expert systems are still in the research stage but may be of future practical use, particularly as training tools (Teng 1990).

Strategic extension campaigns (SECs) using mass media are an application of commercial marketing and advertising that has been deployed by FAO for agricultural extension. An SEC involves conducting KAP surveys and focus-group interviews for baseline market research, defining objectives, designating target audiences and creating corresponding messages, choosing appropriate extension media and channels, and developing detailed plans for campaign implementation, monitoring, and impact evaluation. A series of practical workshops are held over a year's time to train an appropriate cadre of national personnel via the supervised design and fielding of a pilot SEC (Adhikarya 1989; Swanson 1990).

The FAO Rice IPM Programme and other organizations have fielded mass media campaigns on various IPM themes in Malaysia, Thailand, the Philippines, and Sri Lanka (Escalada and Kenmore 1988; Pfuhl 1988; van de Fliert and Matteson 1989; Whitten et al. 1990). The campaigns sought to create farmer demand for IPM training, motivate extension officers, and convey simple technical and pesticide safety messages. Preliminary results were encouraging. For example, a Malaysian rat control SEC increased farmers' adoption of chronic poison baits from 61 to 98% and of physical control methods from 31 to 60%. The first cropping season's estimated savings to farmers in the form of yield loss prevented was enough to pay for the SEC more than twice over (Swanson 1990).

Recently, disturbing evaluation findings about the impact of certain SECs on farmers' practices have prompted a critical reexamination of the role of SECs in IPM extension strategies. Contrary to expectations, farmers spent more money on herbicides after an SEC on weed control in Malaysia. The Information Recall and Impact Survey showed that impact indicators for messages not related to herbicides showed a small increase (about 5%) after the SEC, but those for messages that could be construed as encouraging increased herbicide use showed much larger increases (over 25%). This was reminiscent of SEC results from the Philippines which showed that even a picture of pesticide application equipment overrode intended messages (Whitten et al. 1990). It appears that 30 years of pesticide promotion may have permeated the Asian information environment so thoroughly that any mention of pesticides, even in an antipesticide statement, tends to obscure the rest of the message and reinforce pesticide use. Experimentation will be necessary to determine whether, and under what conditions, SECs can contribute substantially toward IPM goals. A politically awkward implication of these findings is that IPM projects' customary participation in pesticide safety campaigns should be discontinued because it is a counterproductive use of their funds.

THE ROLE OF NONGOVERNMENTAL ORGANIZATIONS

NGOs are well-suited to play a role in national IPM extension strategies. Their "self-help" approach to social change, by empowering disadvantaged groups through information and organization, is consistent with the training principles and goals of the FAO Rice IPM Programme. They complement national extension systems in that they often focus on farmer groups who are otherwise passed over. Working closely with farmers on a small scale and being problem-oriented, participatory, and flexible, NGO programs can generate innovative ideas. Thus, NGO projects can provide valuable role models for national extension systems, particularly if public extension personnel are trained at project sites (Teng 1990; UNDP 1991).

Some NGOs have already made IPM implementation one of their goals. In 1986, trainers from the FAO Rice IPM Programme and government agencies began teaching IPM to the agricultural staff of several rural community-based NGOs in the Philippines. That effort has grown into a network of 27 NGOs in all major parts of the country, coordinated by a full-time FAO support officer. The participants and trainers are exceptionally motivated and the demand for training is strong (Whitten et al. 1990). Twenty-one Indonesian NGOs network and collaborate with the national rice IPM program, several having contributed group dynamics and nonformal education expertise to that country's present innovative rice IPM field training process (Barfield et al. 1991). The NGO CARE has already designed an IPM field research and training project for Sri Lanka and recently established an organizationwide Pesticide Policy and Implementation Strategy that call for crop protection through IPM in all CARE projects (CARE 1991). Regional workshops are being held to orient field staff and draft national IPM/Pesticide Safety Action Plans.

Regardless of their potential, NGO projects are too small-scale to be the main channel for bringing IPM to Asia's 200 million rice farmers. For instance, all 27 NGOs in the Philippine IPM network together reach fewer than 1600 farmers. Another drawback is that NGO projects depend heavily on donor funding and are hard to institutionalize, which make them relatively short-lived (UNDP 1991). To reach a significant fraction of the extension audience and ensure that promising pilot programs do not deteriorate, the IPM training and extension process must be integrated into national institutions in each participating country.

RICE IPM EXTENSION THROUGH TRAINING AND VISIT SYSTEMS

Philippines Department of Agriculture extension officers were given IPM training in 1984 to expand the scope of the rice IPM extension effort, and in 1984 and 1985, the FAO Rice IPM Programme and Philippine National IPM Program officers presented Intercountry Field IPM Implementation Courses for selected field extension staff from other member countries. The trainees became core instructors for national rice IPM programs that began field-skills training for farmers in the context of Training and Visit (T & V) extension systems (Whitten et al. 1990).

Promoted by the World Bank, which is presently the only source of credit for major extension projects (Swanson 1990), T & V has been implemented in over 40 developing countries and now dominates extension worldwide (Hayward 1990; Swanson et al. 1990). It is a set of management principles which are applied to ministry-based agricultural extension systems to improve management, foster professionalism, provide regular training for extension agents, and preserve a strong field orientation stressing regular visits to farmers (Benor and Harrison 1977; Benor and Baxter 1984). In practice, T & V systems still share some important problems with other large-scale public, multipurpose national agricultural extension systems: a "top-down" orientation with little or no clientele involvement in program development; weak administrative, supervisory, and evaluation skills in mid-level staff; a tendency to train both extension agents and farmers through formal lectures which are seldom problem-oriented or experience-based, with little field practice; an emphasis on periodic message delivery rather than on teaching farmers improved technical and managerial skills; and few or no incentives or rewards for good work by village-level extension agents (Goodell 1983, Byerlee 1987; Axinn 1988, Swanson 1990; UNDP 1991).

In the Philippines, evaluations found that the IPM training process deteriorated after being taken out of the hands of enthusiastic master trainers. There was lack of support for rice IPM extension activities at the provincial and municipal levels. Other programs competed for funds and the attention of multipurpose village-level officers. Much farmer training time was spent in the classroom listening to lectures instead of practicing in the field. The season-long, field-based 60–90 hours of instruction presented in the early 1980s became only three consecutive days with weekly followup visits. In 1990, monitoring found that farmer training sometimes occupied less than 3 days, field time was always less than 25%, lectures were poor, and there were usually no followup visits. Training quality and intensity declined to the point that skills were not being mastered or retained. Though trained farmers consistently did better in evaluations than did untrained ones, the later that farmers were trained during this period, the lower were their rice yields and the greater was their pesticide use (Whitten et al. 1990).

Whereas the Philippine T & V system was laboring under reorganization and civil unrest, IPM extension through T & V was tried in Sri Lanka and Indonesia under favorable conditions. Sri Lanka's system was small and staffed by very well-qualified higher officers who had FAO technical and financial support and were receptive to IPM as a model for farmer training. The Indonesian extension service had an excellent financial record, thousands of well-disciplined staff, good training and extension infrastructure, and a generous budget (Whitten et al. 1990).

With the cooperation of the Sri Lanka Department of Agriculture an attempt was made to modify and augment routine T & V activities to duplicate the rice IPM training approach that had succeeded originally in the Philippines. Sri Lankan extension agents were already interacting with farmer groups. They were to choose a different 20–25 farmer demonstration tract each season and give those farmers intensive instruction in good agricultural practices. Most of those groups consisted of rice farmers, and that activity was harnessed for IPM training. Normally, T & V extension agents' activities follow a biweekly rhythm, so that they visit a given farmer group every 2 weeks. However, they were asked to hold a training class for their rice IPM group each week to ensure enough contact hours and adequate crop surveillance. Every extension officer was given a special 2-week rice IPM train-the-trainers field course and they were supported in the field with handbooks, insect collection and preservation materials, portable chalkboards, and stationery supplies for making visual aids. The FAO Rice IPM Programme ran a seasonal contest for the best farmers' training groups to motivate field-level extension officers and farmers, awarding small prizes and certificates that helped extension agents qualify for promotion.

In spite of these special initiatives, evaluations found that the program did not operate as desired at field level. Neither farmers nor extension officers were motivated for the delivery of enough training to change farmers' pesticide use appreciably. Extension officers kept to their accustomed biweekly visit routine, and even then they did not deliver more than half of the agreed-upon training sessions. The principal reasons given were too many work demands, the difficulty of organizing farmers for weekly classes, and the biweekly activity rhythm of the rest of the T & V system. Individual farmers attended even fewer training classes than were offered. Confident IPM decision making that substantially reduces insecticide use is only possible after adequate supervised field practice; trained farmers showed measurable gains in IPM skills and knowledge but reduced their pesticide use only very little (Matteson and Senerath unpublished manuscript).

In Indonesia, much effort and money were invested in T & V rice IPM training for little benefit. In 1987, a crash program was initiated through the National Agricultural Extension Project. It spent the equivalent of $7 million per year and was given highest (Presidential) priority and extraordinary ministry facilitation—unlimited financial support, good communications, superb training infrastructure, and tested training materials—but training funds and materials arrived late, less than 25% of classes observed actually visited rice fields, and only 10,300 of the targeted 125,000 farmers received training. Trainers used a preemptory approach that did not equip farmers to make pest control decisions by themselves (Whitten et al. 1990).

It would appear that the amount and quality of teaching necessary to extend rice IPM using conventional training methods is difficult to accomplish through big multipurpose "top-down" extension organizations—even with T & V's management improvements plus top-level willingness to be flexible and to initiate special activities. Motivation evaporates when training schemes imposed from above trickle down the bureaucratic hierarchy of large-scale programs. Inertia, indifference, and many conflicting responsibilities interfere with an intense field training effort. Nevertheless, management monitoring like that which discovered the weaknesses in the Philippines and Sri Lanka rice IPM programs provides a basis for constructive change. Because it requires the sort of integrated crop management skills farmers now need to increase their productivity (Byerlee 1987; Röling and van de Fliert 1991), IPM can be an appropriate focus for improving the overall expertise and effectiveness of national extension systems. There is much scope for maximizing the potential of those systems as a component of national rice IPM extension strategies.

"Learning by Doing" Ecosystem Analysis

The Indonesian National Programme for the Development and Training of IPM in Rice-Based Cropping Systems and the regional FAO Rice IPM Programme have been working together to try and overcome the "scale verses quality" motivational problem. They are now fielding a more dynamic, self-replicating IPM training process based on participatory nonformal education (NFE): experience-based learning outside the formal school system that empowers people to find solutions to real-life problems through participation, enhanced self-confidence, dialogue, joint decision making, and self-determination (Freire 1970, 1981; Dilts 1984; Pimbert 1991; Röling and van de Fliert 1991). Long used in small-scale community organizing and development programs in fields such as primary education (Silberman 1973) and health-care worker training (Abbatt and McMahon 1985), participatory NFE is being made central to larger-scale agricultural extension for the first time. Farmers and extension agents are motivated by the excitement of discovering something new and the reinforcement of publicly sharing that discovery with others. This extension approach takes advantage of a more effective training process while following all the guidelines articulated earlier by Philippine rice IPM extension researchers. The robustness of participatory NFE methodology should help maintain the impact of IPM training even if overall training quality declines in the transition to large-scale programs.

APA INI?

Indonesian rice IPM training for extension officers includes new interpersonal, organizational, and evaluation skills they must master to deliver effective training to others. IPM requires creativity and active decision making, so the *process* of training should not be one-way, but rather creative and participatory. The key innovation is that trainers do not allow themselves to be forced into the position of an expert. Trainers facilitate discussion and presentation by trainees by never answering questions directly. Instead, they respond with another question that makes the trainees think for themselves: "What do you think?" "What did you find?" "What did it do?" This is called the "apa ini?" principle, literally "What is this?" "A question answered with a name is a lost opportunity for learning." Participatory NFE methods build on existing knowledge and experience that is common among farmers (Freire 1981; Dilts 1984); the trainer merely provides the structure for the process of learning.

In general, these methods have been warmly accepted by farmers. Hands-on exploration and active participation are a welcome replacement for the boring lecture mode of most extension training. Moreover, the participatory NFE approach circumvents the credibility problem of top-down messages being transmitted through young, usually inexperienced extension workers to much more senior and experienced farmers who have rarely had positive encounters with the extension network. Young persons are able to lead without using insulting directives.

AGRO-ECOSYSTEM ANALYSIS

Research that produces detailed blanket pest control instructions is supposed to provide better decision-making methods to farmers but actually takes the decision-making process away from the intended decision maker. Indeed, one of the paradoxes of modern society is that research designed to give great benefit ends up giving less control to individuals and thus becomes disempowering (Illich 1973). Scientists belonging to the "Farmer First" school of agricultural development (Chambers et al. 1989; Conway and Barbier 1990) reject conventional top-down agricultural research and development as unproductive of truly appropriate innovations. They stress the need for farmers to participate in every step of the process, and thus for agricultural specialists to stimulate farmers to analyze their own situation, to experiment, and to make constructive choices. Placing the farmer at the center of the technology development process is wholly consistent with the IPM objective of empowering **the farmer** as pest management decision maker, independent of constant extension advice.

Farmers' field experience includes a large number of observations which can be brought together by providing a framework with which to analyze the ecosystem of each individual field. Piaget (1977, 1985) used the term "equilibrium" to describe the adequacy of our mental structures to understand the world around us; farmers are learning about an ecological framework for better decision making.

How are Indonesian farmers becoming field ecologists? Training places an emphasis on field skills and basic field ecology. This is not a new approach. Comstock (1911) understood the importance of natural history studies when she wrote a set of field activities for extension workers in New York State. Plant growth and physiology, ecological functions of insects and spiders, ecological interactions, agronomic practices, and pesticide effects on both the ecosystem and farmer health are the core of training.

Each week during the season, the trainees are divided into small groups (usually five people) that observe the state of the field and use a drawing method for analyzing the field situation to learn decision making. This **agro-ecosystem analysis** approach requires trainees to draw the plant in the center of a large piece of newsprint (Fig. 20.1). Tiller number, diseased leaves, water level, rat damage, and weed density are also put into the drawing. Phytophagous insects are drawn to the right of the plant with population density (number per 20 hills) noted, and natural enemies are drawn to the left side of the plant, also with population density noted. With

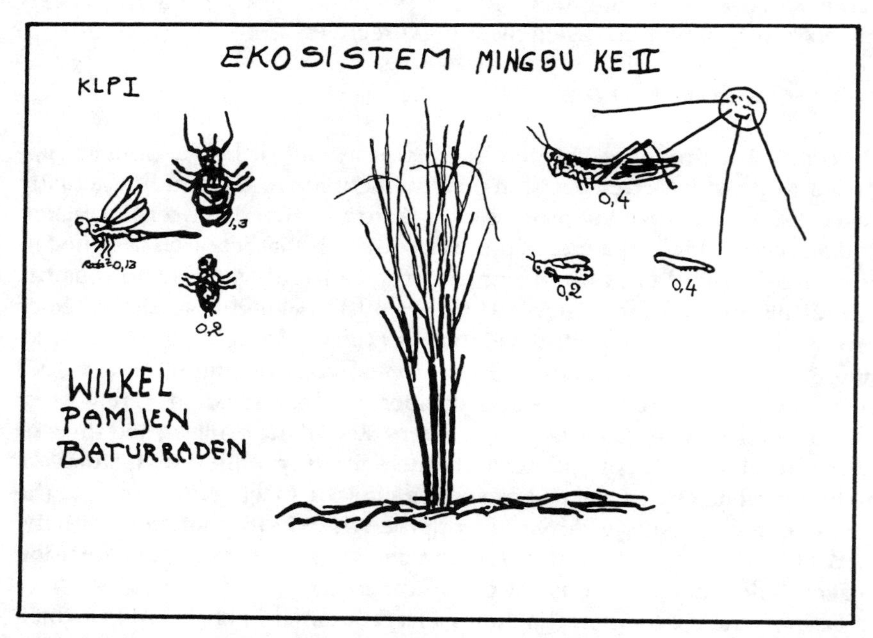

Figure 20.1. Drawing of a rice paddy agroecosystem used by farmer groups to learn management decision making in Indonesia. Title: "Group I, Ecosystem 2nd Week" Wilkel/Pamijen/Baturraden (Farmer's field school location). Phytophagous insects and natural enemies are shown to the right and left of the rice plant respectively along with their densities (No./20 hills).

their drawing as a focus of discussion, group members go through a set of prepared questions which cover agronomic and pest problems. After making decisions on management actions required for the following week (agronomic and pest control), each group must present and defend their summary to the other trainees. The trainer facilitates the process by asking leading questions or adding technical information if necessary. This agro-ecosystem analysis allows trainees to integrate their skills and knowledge, and trainers can immediately evaluate trainees' ability. For example, if a ladybird beetle is drawn as a pest, the trainer will discuss the functional role of predatory beetles in the field.

Agro-ecosystem analysis is complemented by 2- to 4-hour field-oriented activities focusing on special topics concerning plant physiology, insects, rats, population dynamics, economics, diseases, and issues surrounding the use of agricultural poisons. These activities are similar to those employed in modern biology teaching (Novak 1970) or science for young people (Herbert 1959; Cornell 1979). Any part of IPM can be investigated using hands-on exploration: Rat population growth is illustrated by using simple growth rules and counting nails, each representing one rat, to show how many rats are produced from a single pair in a single year; the effect of carbofuran on spiders is observed by setting up a closed experiment with spiders in cups; plant capillary action is demonstrated using dye in a glass with plants; dye is also used to show how much "pesticide" actually ends up on the person using a knapsack sprayer; insect life cycles are discovered by rearing insects in what is called the Insect Zoo. These Zoos are also used to observe predation and parasitism. The philosopher and educator John Dewey championed this "Learn by Doing" approach several decades ago (Dewey 1938).

TRAINING OF TRAINERS

Pest Observers, plant protection agents based at the Directorate of Crop Protection's field laboratories, are the backbone of the Indonesian rice IPM extension program. The program has returned to the master trainer concept by giving Pest Observers the longest, highest-quality IPM training ever offered to developing-country agricultural extensionists. Because trainers tend to teach as they have been taught, the process and content of the 14-month Pest Observer training is similar to that used with farmers, including "apa ini?," group dynamics games, agro-ecosystem analysis, learn-by-doing experiments and a skills test administered pretraining and post-training to measure how much trainees have learned. Pest Observers are responsible, in turn, for mentoring the Directorate of Extension's Field Extension Agents, orienting them and showing them how to establish and conduct IPM Farmer's Field Schools. Training manuals that reflect the Pest Observers' field experience supply written guidance on methodology as well as teaching modules covering special subjects.

For the first 3½ months of their training, Pest Observers become rice farmers, many of them for the first time. They must manage one full crop of rice, including all the field work required (planting, weeding, fertilizing, irrigating, observing, protecting, and harvesting). They learn IPM skills and implement season-long field studies which allow them to see for themselves the effect of fertilizer on yield and pests, the effect of natural enemies on brown planthopper, and the effectiveness of sampling methods, and to compare IPM methods to nationally recommended intensification packages that require prophylactic applications of carbofuran insecticide. These studies illustrate the first three principles of IPM (Table 20.2).

During the following rice season, each Pest Observer is teamed with two Field Extension Agents who are given a week's introductory training and who each then apprentice by helping to conduct a weekly Farmer's Field School for two 25-member farmers' groups. The Pest Observer plans the Field School sessions and teaches them with the assistance of the Field Extension Agents, who meet with the Pest Observer weekly for briefing and class preparation. Each Farmer's Field School has its own fields which are planted and maintained by the trainees. These fields are for testing IPM processes and are not "demonstration plots." Each week for one 10-week growing season, farmers spend about half a day in Field School observing rice paddies, practicing management decisions, and covering special topics (Table 20.4). Activities similar to those used by Pest Observers to learn about specific aspects of IPM are also carried out by farmers. Group dynamics exercises are used to reinforce collaboration within farmer groups and disseminate what is learned. Trainer/facilitators initiate dialogue instead of lecturing (van de Fliert 1991).

Table 20.4. Suggested time schedule for a Farmer's Field School class, Indonesia (modified from van de Fliert 1991).

Hours	Activity
07:30–08:30	**Field observation** in small groups.
08:30–09:45	**Agro-ecosystem analysis.** Groups make drawings of what they found in the field.
09:45–10:00	**Discussion.** Groups present their drawings and discuss the field situation.
10:00–10:15	**Break.**
10:15–12:00	**Special topic.** Lessons, exercises, and discussions on special topics dealing with field problems of the moment.

Pest Observers' next major training activity is planting and tending the dry-season secondary food crop which follows two seasons of wet rice. Soybeans dominate, but groundnuts, mung beans, and vegetables are also typical of the season. The Pest Observers carry out the same sort of IPM-oriented observations and field experiments as in rice, drawing conclusions about the yield impact of various practices. The culmination of Pest Observer training is enrollment in a university for 3½ months of lectures which obtain them a "Diploma 1" degree, a 1-year higher education certificate that qualifies them for a higher salary (Barfield et al. 1991; Röling and van de Fliert 1991).

The first batch of 500 Pest Observers to complete the 14-month course of training returned to the field to follow up the four Farmer's Field Schools they helped initiate, thus contributing to evaluation of the impact of the IPM extension program. During the followup, they also trained promising candidates for farmer-to-farmer training instructors. The program hopes to develop a mixed IPM training team of Pest Observers, Field Extension Agents, and farmers at each rural extension center to carry out the large-scale training planned (Röling and van de Fliert 1991).

IMPACT ON FARMERS

After one season's training, Indonesian farmers are able to analyze field problems and make decisions based on what they have observed in the field. In a followup study of the 50,000 farmers trained in the program's first year (Pincus 1991), insecticide spraying was reduced from an average of 2.8 sprays per farmer to less than one per season, with the majority of farmers not spraying at all. Insecticide costs per hectare decreased dramatically (Fig. 20.2). When farmers did apply pesticides, they could identify a specific target pest. Preventive applications decreased with increasing certainty about pest problems (Fig. 20.3). There was no difference in yields obtained using IPM methods versus those harvested with nationally recommended technical packages.

The current Indonesian rice IPM extension program is young and evaluation of its long-term impact on farmers' practices is not yet possible. So far, the reduction of insecticide use it has brought about appears to approach the potential of IPM practice more closely than that achieved by any other rice IPM training program to date. Although the improved rice varieties currently available harbor significantly damaging insect pest infestations only about once every other season on average, evaluation of the most effective IPM training fielded in other countries shows that, in the long term, trained farmers persist in spraying at least once or twice a season

(Kenmore 1987; Kenmore et al. 1987; Adalla 1990; Matteson and Senerath unpublished manuscript).

There are many other benefits which arise from farmers' understanding of their crop ecosystem. Learning and practicing with IPM for problem solving prepares the farmer to continue learning from the field and from the consequences of each management decision from then on. Farmers develop their own ideas and create IPM programs appropriate to their local conditions. They are less susceptible to the "fear of the unknown" exploited by pesticide salesmen because they can observe, analyze, and understand for themselves the ecological situation in their fields. Increased knowledge is always mentioned by farmers as a benefit of IPM training. As one graduate of an Indonesian IPM farmer's field school put it, "After following the field school I have peace of mind. Because I know now how to investigate, I am not panicked any more into using pesticides as soon as I discover some (pest damage) symptoms" (van de Fliert 1992).

Equipping farmers to take care of their own needs also has logistic value. Government extension and crop protection agencies rarely have enough budget or officers to give adequate advice for all farms, each with its own unique priorities and field situation. Personnel are stretched thin and even for a generic response slow communications often prevent timely action (Matteson et al. 1984).

Ownership of knowledge and freedom from dependence on "messages" liberates farmers from being passive subjects and motivates them to de-

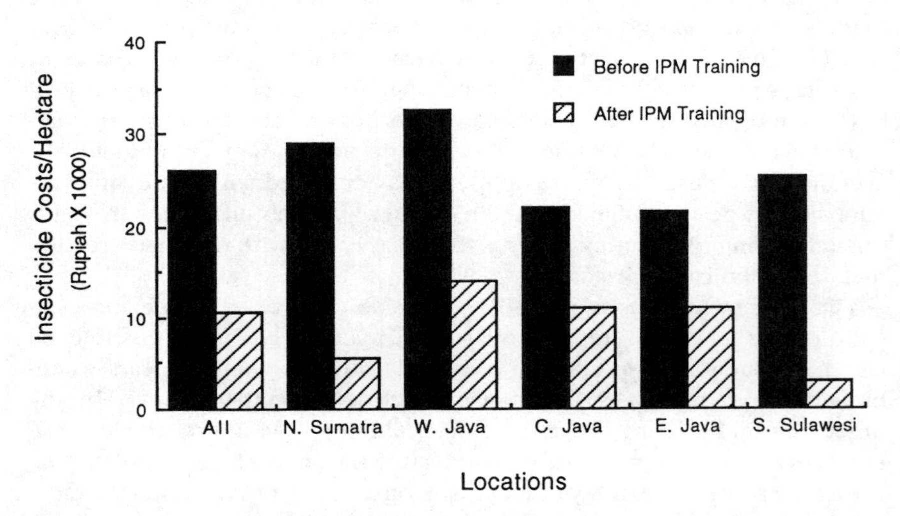

Figure 20.2. Insecticide costs per hectare (rupiah × 1000) before and after rice IPM training. Indonesia, 1990 and 1991 wet seasons.

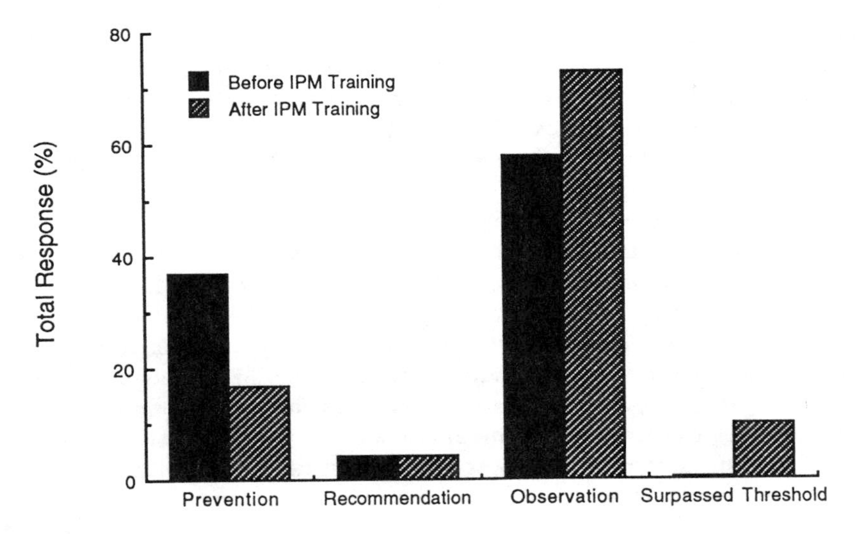

Figure 20.3. Reasons given by farmers for applying pesticides before and after rice IPM training. The percentage of responses in each of four categories (prevention, recommendation, observation of pest/damage, and surpassed economic threshold) is shown. Indonesia, 1990 and 1991 wet seasons.

mand basic agronomic and ecological information that can be used to further refine their judgments. Learners' demand for more education and better answers to harder questions, voiced through linkages with local government, gets stronger. When effective demand from below grows, bureaucracies respond better, leading to further farmer requests for support. As governmental involvement widens from the local level upward, an incremental, self-sustaining spiral of positive feedback is created (Honadle and VanSant 1985). It is this continuing "bottom-up" demand that is expected to keep IPM extension dynamic and extension officers motivated and responsive on the long term.

Topics for Further Investigation

IPM Technology

Because natural enemies of rice pests are such an important part of the IPM curriculum, further research on their life histories, behavior, and interaction with pests and on enhancing their impact will enrich extension training as well as improve the effectiveness of pest control. Another area that needs attention is crop protection decision making. Field surveillance and thresholds need to be reexamined in the light of farmers' attitudes and technology adoption patterns and made more appropriate to farmers' and extension agents' capabilities.

Comparative Studies of Alternative Training Approaches and Organizational Structures and Styles

There is now scope for a potentially useful comparison of the relative effectiveness and cost/benefit ratios of alternative IPM training approaches (duration, frequency, mix of training materials and one-on-one in-field activities, trainer characteristics, methodology, etc.) in different institutional settings with farmers of different characteristics and cultures. Exchange visits and discussion, which are already arranged frequently by the regional FAO Rice IPM Programme, could facilitate comparison both within and among countries (Whitten et al. 1990).

Questions about the quantity of training necessary are raised by the finding of Adalla et al. (1990) that there was not a consistent trend relating the duration or extent of IPM training received by Filipino farmers to their subsequent frequency of pesticide application. Namely, farmers who had attended a one-day seminar and those with three or more days of training reduced their insecticide use at least somewhat (all training beyond 3 days leading to increased reduction), whereas farmers with 2 days of training actually increased their insecticide applications. In Sri Lanka, most farmers attended two or three training classes, and, on average, their pesticide use was almost unchanged although their skills and knowlege had improved (Matteson and Senerath unpublished manuscript). Is there a degree of IPM training that enhances farmers' awareness of pest problems and their ability to recognize them (thus raising their anxiety level) without giving them the skills and/or confidence to do anything but spray more often?

There are many other questions concerning the amount of IPM training necessary to achieve long-term goals at the national level (Röling and van de Fliert 1991). How much refresher training do trained farmers and extension agents need? When new crop protection problems arise how can training-driven demand for applied research best be harnessed to provide solutions, and how much additional farmer training is needed to cope? With a minimum of extra information and perhaps only with the aid of mass media, can farmers trained in rice IPM use those principles in protecting their other crops from pests?

The relative impact of participatory NFE methodology, with special reference to the levels of independence and confidence in decision making it instills among trained farmers as well as the changes in actual behavior it produces, is of particular interest (Whitten et al. 1990). Large-scale sustainability of the participatory NFE approach and the context in which it is used for IPM training in Indonesia is also an issue (Röling and van de Fliert 1991). Can Farmer's Field Schools be replicated for mass IPM extension training? If not, by what means can the integrity of the training process, and therefore its quality, be maintained? If so, given that the Indonesian IPM extension program is the largest and most intensive one ever attempted, what is needed to maintain government political commitment? Because considerable financial support would be necessary, how could an extended program avoid having to absorb too much money at once, which would cause overhasty expansion and dilute quality?

Indonesia's current national IPM extension program is unique in that, instead of being implemented as one of many activities of the national extension system, it is managed and coordinated by the National Development Planning Agency (BAPPENAS) and draws on staff of two sections of the Ministry of Agriculture, Crop Protection and Extension, with Crop Protection officers being pivotal. Does this configuration of technology, knowledge processes, and institutions yield useful insights for overall institutional design and investment for Integrated Agriculture? Integrated Agriculture (sometimes called Alternative Agriculture) consolidates and improves on the Green Revolution introduction of new seed and agrochemical inputs for raising crop yields. Like its component IPM, Integrated Agriculture takes ecological processes in the farming system into account and emphasizes practices and management skills that maximize sustainability and productivity, often through input reduction (National Research Council 1989). Those priorities require mobilization, organization, and training that enable local farmer networks to take locality-specific decisions. IPM is not well served by the conventional linear model of top-down "technology transfer" by which national extension systems have, in the past, promoted packages of blanket recommendations centered on high-yielding crop varieties (Jedlicka 1983). Can conventional extension systems be modified for the large-scale introduction of IPM and other Integrated Agriculture practices, and if so, how? If not, will there be movement toward new extension operations, perhaps accompanied by conflict over funding, staff, and authority? How can such conflicts be resolved (Röling and van de Fliert 1991)?

Targeting Overlooked Groups for Training

The overwhelming majority of participants in village-level IPM field-training groups are male farmers. Certain potentially important audiences have been largely overlooked by rice IPM extension programs. Little or no attention has been paid in most countries to the level of participation by women, and even less to that of landless people. Many farm workers fall into both categories. In Indonesia, the low number of women receiving IPM training is striking, although the official policy of the program is neutral on the subject. When asked why no female farmers were selected to attend IPM field schools, extension officers responded "Oh, we never thought about it" (van de Fliert 1991).

Much more needs to be known about the roles of women and landless laborers in rice production systems, including issues such as their occupational exposure to pesticides. Available evidence does not justify their exclusion from extension training. Large landowners or farmers with an outside job can be virtual absentees on the farm, depending on managers or family members to supervise. Field workers see crops close up at frequent intervals and are well placed to carry out important IPM activities. For instance, rice transplanters can screen out stemborer egg masses and diseased plants, and weeding is an opportunity to scout for pests and their natural enemies. Transplanting and weeding are generally women's jobs in the Philippines, Indonesia, and Sri Lanka. Rice IPM programs are being expanded to include vegetables in rice-based cropping systems, which are often women's responsibility. Women are a natural IPM audience in that they are concerned about the effect of pesticide abuse on family health. In the Philippines they are the main decision makers about buying farm inputs including pesticides (Samonte 1983; Adalla and Hoque 1990).

Extension programs will have to investigate new combinations of messages and extension channels for reaching these additional audience groups most effectively. Landless people are still virtually unstudied as a potential extension audience. The rice pest management KAP survey in Sri Lanka revealed differences in farm women's media habits and sources of agricultural information (particularly their low level of interaction with the extension service) as well as a greater tendency to apply new knowledge (van de Fliert and Matteson 1989). There are similar findings from the Philippines, where women rely mainly on their own formal organizations and informal social communication networks for solutions to farm problems. Philippine men are three times as likely as women to get crop protection information from extension agents, often at organized meetings that women seldom attend (Samonte 1983).

Initial steps were taken to involve women in rice IPM extension in the Philippines. In 1983, the FAO Rice IPM Programme and the National Crop Protection Center held a workshop on the role and potential of the Filipina in rice crop protection (FAO 1983). Farmers, extension workers, training specialists, researchers, and social scientists, mostly female, presented and discussed studies of the role of rural women and possible extension strategies for reaching them. The workshop produced two draft proposals, one on a pilot field project focusing on women as extensionists and one on social research studies of women's roles in crop protection. The field project has verified rice and vegetable IPM technology with farmer participation in five Laguna Province villages and expanded its gender issues focus after finding how large a role farm wives play in pesticide choice, purchase, and use (Adalla 1990; Adalla and Hoque 1990).

Farmer-to-Farmer Transfer of IPM Technology

The spread of simple innovations such as high-yielding crop varieties was often rapid. Because a much greater range and complexity of information and skills are needed in Integrated Agriculture, with IPM perhaps being an extreme case, one would expect farmer-to-farmer flow to be much slower and less effective (Goodell 1984; Byerlee 1987). Nevertheless, IPM practice does appear to spread from trained to untrained farmers. An October 1990 survey of farmers in Central Luzon, Philippines showed that untrained farmers in villages where IPM training has been done spend less on pesticides than do untrained farmers in untrained villages (Table 20.5) (Whitten et al. 1990). Adalla et al. (1990) found that one-third of IPM adopters in Central Luzon and Iloilo were untrained farmers who heard about the new approach via local radio stations, fellow farmers who had attended IPM seminars, or in monthly farmers' meetings.

Formal attempts to promote the lateral spread of rice IPM vary greatly from country to country. For example, in India there are no plans to utilize farmers as change agents for other farmers and there appears to be little spontaneous activity, whereas in China and Indonesia lateral spread is included as a continuous feature of national programs (Whitten et al. 1990). Indonesian Farmer's Field School graduates in West and Central Java are receiving special instruction as "Farmer Facilitators." They train other farmers in pest and natural enemy identification, problem diagnosis, and agroecosystem analysis using funds raised by farmer organizations or from local government budgets. The field-based training process allows farmers to use the rice paddy itself as their primary teaching resource; it presents both the

Table 20.5. Annual expenditures on pesticides (two crops). October 1990 survey, Central Luzon, Philippines (Whitten et al., 1990).

Expenditure	Pesos	US$
IPM trained farmers	725.4	25
Untrained farmers–untrained villages	1654.2	59.1
Untrained farmers–IPM trained villages	774.6	27

curriculum (field problems) and the necessary training materials (plants, diseases, rats, insects, spiders, etc.).

At present, only a tiny fraction of Asia's 200-odd million rice farmers have received rice IPM training (Table 20.6). If there is to be any hope of reaching a critical mass of these potential clients, an understanding of IPM principles and the accompanying knowledge and skills must spread informally from those people reached directly by extension training. Experimentation has just begun on the scope and dynamics of lateral spread and how best to promote it under various circumstances and for different audiences (Röling and van de Fliert 1991).

Table 20.6. Number of IPM trained farmers and trainers in eight Asian countries (modified from Whitten et al., 1990).

Country	Rice Farm Households (millions)	Farmers Trained	Core Trainers Trained	Extension Staff Trained
Bangladesh	10	1,500	25	700
China	129	5,000		440
India	30	50,000	50	
Indonesia	15	60,000	600	1,200
Malaysia	0.4	2,500	25	
Philippines	1.6	175,000	737	11,100
Sri Lanka	0.9	87,100	233	2,600
Vietnam	10	860		
Totals	196.9	381,960	1,670	16,040

Summary

1. Natural enemy studies are the centerpiece of planthopper IPM training because farmers enjoy them and because they underline the need to avoid unnecessary insecticide application.
2. In view of extension experience, the roles of scouting and thresholds and the kind and amount of research effort they deserve need to be reconsidered.
3. Key planthopper control recommendations will not be adopted by farmers unless extension training is extensive and effective enough to change fearful attitudes toward pest attack.
4. The small size of planthoppers requires extension programs to consider the issue of farmers' poor eyesight.
5. Village research in the Philippines from 1978 to 1980 defined principles for effective rice IPM farmer training. Group training classes using

conventional teaching methods were held weekly in the rice paddy for an entire rice-cropping season. Prepared training materials were deemphasized in favor of hands-on field practice using simplified terms and technology, with group discussion. Field demonstrations and class experiments engaged farmers' curiosity and encouraged imaginative inquiry. Trainers made followup visits to help farmers apply what they learned.

6. Philippine master trainers in the Crop Protection Division of the Bureau of Plant Industry pursued that training approach on a pilot scale, reaching about 55,000 farmers by 1984. Training was high quality and had a long-term impact on farmers' pesticide use patterns. After 1982, the rice IPM training program was assisted by the FAO Inter-country Programme for Integrated Pest Control in Rice in South and Southeast Asia, which provides technical coordination and training support to nine Asian countries.
7. From 1985 to 1989, nongovernmental organizations continued small-scale training while attempts were made to reach more farmers with the same methodology through national T & V extension systems in several Asian countries, supported by multimedia strategic extension campaigns (SECs). The quantity and quality of training delivered deteriorated such that it was not changing farmers' practices appreciably. The role of SECs needs further evaluation.
8. The Indonesian national IPM program is addressing the "scale versus quality" problem with a dedicated program that cuts across government agencies. The program follows the same IPM training principles but uses a more robust, self-replicating IPM training process based on participatory nonformal education. Farmers "learn by doing" agro-ecosystem analysis and are empowered as field-level experimenters and decision makers. Training impact is high and farmer demand is expected to maintain the dynamism and quality of the extension program on the long term.
9. More research on natural enemies of pests and appropriate pest control decision making would facilitate the extension process. IPM extension topics meriting further investigation include comparative studies of alternative training approaches and organizational structures and styles, targeting overlooked groups for training, and farmer-to-farmer transfer of rice IPM technology.

ACKNOWLEDGMENTS

The authors are grateful to E. van de Fliert, G. Dively, L. Pedigo, and D. Sammons for their reviews on an earlier draft of this chapter and for contributing useful information.

References

Abbatt, F., and R. McMahon. 1985. Teaching health-care workers, a practical guide. MacMillan Education Inc., London, England. [20]

Abdelrahman, I. 1973. Toxicity of malathion to the natural enemies of California red scale, *Aonidiella aurantii* (Mask.). Australian J. Agric. Res. 24:119–133. [14]

Abdul-Nour, H. 1971. Contribution à l'étude des parasites d'Homoptères Auchenorrhynchues du sud de la France: Dryinidae (Hymenoptères) et Strepsiptères. Thèse, Acad. Montpellier, Univ. Sci. et Tech. Languedoc. [11]

Abraham, C.C., and K.P. Mathew. 1975. The biology and predatory potential of *Coccinella arcuata* Fabricius (Coccinellidae: Coleoptera), a predator of the brown planthopper, *Nilaparvata lugens* Stål. Agric. Res. J. Kerala 13:55–57. [10]

Abraham, C.C., K.P. Mathew, and N.M. Das. 1973. A new record of *Coccinella arcuata* Fabricius (Coccinellidae: Coleoptera) as a predator of the brown planthopper, *Nilaparvata lugens* Stål in Karala. Agric. Res. J. Kerala 11:75. [10]

Abrams, P.A. 1987. On classifying interactions between populations. Oecologia 73:272–281. [7]

Adalla, C. B. 1990. Making modern rice production technology work: the case of participatory IPM Extension and Women Project in the Philippines. Paper presented during the International Research and Development Center (IDRC) Environment Media Seminar, 6–7 March 1990, Toronto, Canada. [20]

Adalla, C. B., and M. M. Hoque. 1990. Gender issues in rice and vegetable production: the case of IPM project in Calamba, Laguna, Philippines. Paper presented at the Workshop on Gender Analysis in Rice Farming Systems Research: Does It Make a Difference? 4–8 June 1990, Bogor, Indonesia. [20]

Adalla, C. B., M. M. Hoque, Z. M. Huelgas, and E. C. Atienza. 1990. Adoption of IPM technology in Iloilo and Central Luzon, Philippines. Pages 183–197. *In* Proceedings of the National Conference and Workshop on Integrated Pest Management in Rice, Corn and Selected Major Crops. National Crop Protection Center, University of the Philippines at Los Baños, College, Laguna, Philippines. [20]

Adam, G. 1984. Plant virus studies in insect cell cultures. Pages 37–62. *In* M. A. Mayo and K. A. Harrap, editors. Vectors in Virus Biology. Academic Press, New York. [12]

Adam, G., and H. T. Hsu. 1984. Comparison of structural proteins from two potato yellow dwarf viruses. J. Gen. Virol. 65:991–994. [12]

Addicott, J.F. 1978. Niche relationships among species of aphids feeding on fireweed. Can. J. Zool. 56:1837–1841. [7]

---

*Numbers in brackets indicate chapters in which reference is cited; i is a reference cited in the Introduction.

Adhikarya, R. 1989. Strategic extension campaigns: a case study of FAO's experiences. Food and Agriculture Organisation of the United Nations, Rome, Italy. [20]

Adkisson, P. A., and V. A. Dyck. 1980. Resistant varieties in pest management systems. Pages 233–251. *In* F. G. Maxwell and P. R. Jennings, editors. Breeding Plants Resistant to Insects. Wiley, New York. [18]

Aguda, R.M., M.C. Rombach, D.J. Im, and B.M. Shepard. 1987. Suppression of populations of the brown planthopper, *Nilaparvata lugens* (Stål) (Hom.; Delphacidae) in field cages by entomogenous fungi (*Deuteromycotina*) on rice in Korea. J. Appl. Ent. 104:167–172. [14]

Akingbohungbe, A. E. 1982. Nomenclatural problems, biology, host plant and possible vector status of Auchenorrhyncha associated with crop plants in Nigeria. Proc. 1st Int. Wksp. Biotax., Classif., Biol. Leafhoppers Planthoppers Econ. Impt. pp. 365–370. [1]

Ali, A. M. H. 1979. Biological investigations on the entomophagous parasites of insect eggs associated with *Juncus* species. Ph.D. dissertation, University of Wales, Cardiff. [11]

Allen, J. C. 1975. Mathematical models of species interactions in time and space. Amer. Natur. 109:319–342. [11]

Altieri, M. A., A. van Schoonhoven, and J. Doll. 1977. The ecological role of weeds in an insect pest management system: A review illustrated by bean (*Phaseolus vulgaris*) cropping systems. PANS 23:195–205. [3]

Altmann, J. 1974. Observational study of behaviour: Sampling methods. Behaviour 49:227–267. [6]

Ammar, E.D. 1973. Factors related to the two wing forms in *Javesella pellucida* (Fab.) (Homoptera:Delphacidae). Zeitschrift fur Angewandte Entomologie. 74:211–216. [4,7]

Ammar, E. D. 1975a. Effect of European wheat striate mosaic, acquired transovarially, on the biology of its planthopper vector *Javesella pellucida*. Ann. Appl. Biol. 79:203–213. [12]

Ammar, E. D. 1975b. Effect of European wheat striate mosaic, acquired by feeding on diseased plants, on the biology of its planthopper vector *Javesella pellucida*. Ann. Appl. Biol. 79: 195–202. [12]

Ammar, E.D. 1977. Biology of the planthopper *Sogatella vibex* (Haupt) in Giza, Egypt. Dtsch. Entomol. Zeit. 24:151–158. [4]

Ammar, E. D. 1985. Internal morphology and ultrastructure of leafhoppers and planthoppers. Pages 127–162. *In* L. R. Nault and J. G. Rodriguez, editors. The Leafhoppers and Planthoppers. Wiley, New York. [12]

Ammar, E. D. 1986. Ultrastructure of the salivary glands of the plant-hopper, *Peregrinus maidis* (Ashmead) (Homoptera: Delphacidae). Int. J. Insect Morphol. Embryol. 15:417–428. [12]

Ammar, E. D. 1987. Ultrastructural studies on the planthopper, *Peregrinus maidis* (Ashmead), vector of maize mosaic and maize stripe viruses. Pages 83–92. *In* M. R. Wilson and L. R. Nault, editors. Proceedings, 2nd Int. Workshop on Leafhoppers and Planthoppers of Economic Importance, July 28–Aug. 1, 1986, Provo, Utah. Commonw. Inst. Entomol., London. [12]

Ammar, E. D., R. E. Gingery, and L. R. Nault. 1985. Two types of inclusions in maize infected with maize stripe virus. Phytopathology 75:84–89. [12]

Ammar, E.D., O. Lamie, and I.A. Khodeir. 1980. Biology of the planthopper *Sogatella furcifera* (= *nigeriensis*) (Horv.) in Egypt. Dtsch. Entomol. Zeit. 27:21–27. [4]

Ammar, E. D., and L. R. Nault. 1985. Assembly and accumulation sites of maize mosaic virus in its planthopper vector. Intervirology 24:33–41. [12]

Anderson, C.E. 1974. A review of structure in several North Carolina salt marsh plants. Pages 307–344. *In* R.J. Reimold and W.H. Queen, editors. Ecology of Halophytes. Academic Press, New York. [2]

Andrewartha, H. G., and L. C. Birch. 1954. The Distribution and Abundance of Animals. University of Chicago Press, Chicago, IL. [11]

Andrzejewska, L. 1962. *Macrosteles laevis* Rib. as an unsettlement index of natural meadow associations of Homoptera. Bull. Acad. Pol. Sci. II. Ser. Sci. Biol. 10:221–226. [3]

Andrzejewska, L. 1965. Stratification and its dynamics in meadow communities of Auchenorrhyncha (Homoptera). Ekologia Polska 13:685–715. [3,13]

Andrzejewska, L. 1976. The influence of mineral fertilization on the meadow phytophagous fauna. Pol. Ecol. Stud. 2:93–109. [2]

Andrzejewska, L. 1978. The effects of treatment and utilization of meadows on Auchenorrhyncha communities. Auchenorrhyncha Newsletter 1:6. [3]

Andrzejewska, L., A. Breymeyer, and E. Olechowicz. 1971. Productivity investigation of two meadows in the Vistula Valley. X. The role of ants as predators in a habitat. Ekol. Pol. 19:213–222. [10]

Anno-Nyako, F. O., H. J. Vetten, D. J. Allen, and G. Thottappilly. 1983. The relation between cowpea golden mosaic and its vector, *Bemisia tabaci*. Ann. Appl. Biol. 102:319–323. [12]

Antolin, M.F., and D.R. Strong. 1987. Long-distance dispersal by a parasitoid (*Anagrus delicatus*, Mymaridae) and its host. Oecologia 73:288–292. [4,9,10,11,14]

Anufriev, G. A. 1968. Planthoppers of the family Derbidae (Homoptera, Auchenorrhyncha) in the fauna of the USSR. Entomol. Rev. 47:75–82. [1]

Anufriev, G. A. 1980. A new species and new genus of leafhopper of the family Delphacidae (Homoptera, Auchenorrhyncha) from Palaearctic. Zool. Zh. 59:208–216. [1]

Argandoña, V.H., J.G. Luza, H.M. Niemeyer, and L.J. Corcuera. 1980. Role of hydroxamic acids in the resistance of cereals to aphids. Phytochemistry 19:1665–1668. [2]

Armstrong, R.W. 1983. Atlas of Hawaii. University of Hawaii Press, Honolulu, HI. [3]

Arnold, S. J., and M. J. Wade. 1984a. On the measurement of natural and sexual selection: theory. Evolution 38:709–719. [6]

Arnold, S. J., and M. J. Wade. 1984b. On the measurement of natural and sexual selection: applications. Evolution 38:720–734. [6]

Arzone, A. 1974. Indagini biologiche sui parassiti oofagi di *Cicadella viridis* (L.) (Hem. Hom. Cicadellidae). II. *Oligosita krygeri* Gir. (Hym. Trichogrammatidae). Ann. Fac. Sci. Agric. Univ. Torino 9:297–318. [11]

Arzone, A. 1977. Reperti biologici ed epidemiologici su *Conomelus dehneli* Nast (Hom. Delphacidae), nuovo per l'Italia come on suo parassita *Tetrastichus mandaris* Walk. (Hym. Eulophidae). Boll. Zool. Agric. Bach. 14:5–16. [13]

Asahina, S., and Y. Tsuruoka. 1968. Records of the insect which visited a weather ship located at the Ocean Weather Station "Tango" on the Pacific. II. Kontyu 36:190–202. [8,9]

Asahina, S., and Y. Tsuruoka. 1969. Records of the insects visited a weather-ship located at the Ocean Weather Station "Tango" on the Pacific, III. Kontyu 37:290–304. [9]

Asahina, S., and Y. Tsuruoka. 1970. Records of the insects visited a weather-ship located at the Ocean Weather Station "Tango" on the Pacific, V. Insects captured during 1968. Kontyu 38:318–330. [9]

Asche, M. 1982a. *Kelisia creticola* nov. spec. und Erganzungen zu den ubrigen Taxa des *Kelisia brucki* Fieber, 1878—Kreises (Homoptera Cicadina Delphacidae). Marburger Entomol. Publ. 1(6):89–116. [1]

Asche, M. 1982b. Beitrage zur Delphaciden-Fauna Jugoslawiens und Bulgariens (Homoptera Cicadina Delphacidae). Marburger Entomol. Publ. 1(7):99–138. [1]

Asche, M. 1982c. Beitrage zur Delphaciden-Fauna der Turkei (Anatolien) (Homoptera Cicadina Delphacidae). Marburger Entomol. Publ. 1(7):71–98. [1]

Asche, M. 1983. Zur Kenntnis der Gattung *Epeurysa* Matsumura, 1900 (Homoptera Auchenorrhyncha Fulgoromorpha Delphacidae). Marburger Entomol. Publ. 1(8):211–226. [1]

Asche, M. 1985. Zur Phylogenie der Delphacidae Leach, 1815 (Homoptera Cicadina Fulgoromorpha). Marburger Entomol. Publ. 2(1):1–910. [1]

Asche, M. 1986. *Kelisia ribocerus* nov. spec. from Greece—an important finding for the phlyogeny of Kelisiinae (Homoptera Fulgoromorpha Delphacidae). Marburger Entomol. Publ. 2(3):179–192. [1]

Asche, M. 1987. Preliminary thoughts in the phylogeny of Fulgoromorpha (Homoptera: Auchenorrhyncha). Proc. 6th Int. Auchen. Meeting. University of Turin Press, Turin, Italy. Pages 47–53. [1]

Asche, M. 1990. Vizcayinae, a new subfamily of Delphacidae with revision of *Vizcaya* Muir (Homoptera: Fulgoroidea)—a significant phylogenetic link. Bishop Mus. Occ. Pap. 30:154–187. [1]

Asche, M., and H. Hoch. 1982. Beitrage zur Delphaciden-Fauna Griechenland II (Homoptera Cicadina Delphacidae). Marburger Entomol. Publ. 1(7):37–70. [1]

Asche, M., and H. Hoch. 1983. *Stenocranus gialovus* nov. spec., eine neue Delphacide aus Sud-griechenland (Homoptera Cicadina Delphacidae). Marburger Entomol. Publ. 1(8):7–24. [1]

Asche, M., and M. R. Wilson. 1989a. The palm-feeding planthopper genus *Ommatissus* (Homoptera: Fulgoroidea: Tropiduchidae). Syst. Entomol. 14:127–147. [1]

Asche, M., and M. R. Wilson. 1989b. The three taro planthoppers: species recognition in *Tarophagus* (Hemiptera: Delphacidae). Bull. Entomol. Res. 79:285–298. [1,10,14]

Asche, M., and M. R. Wilson. 1990. The delphacid genus *Sogatella* and related groups: a revision with special reference to rice-associated species (Homoptera: Fulgoroidea). Syst. Entomol. 15:1–42. [1]

Atsatt, P. R., and D. J. O'Dowd. 1976. Plant defense guilds. Science 193:24–29. [18]

Axinn, G. H. 1988. T & V (Tragic and Vain) extension? Interpaks Exchange 5(3):6–7. [20]

Azwar, R. 1980. Developmental response of brown planthopper *Nilaparvata lugens* (Stål), biotypes on rice varieties with different genes for resistance. MS Thesis, University of the Philippines, Los Baños. [17]

Bach, C. E. 1980. Effects of plant density and diversity in the population dynamics of a specialist herbivore, the striped cucumber beetle, *Acalymma vittata*. Ecology 61:1515–1530. [3]

Bach, C. E. 1984. Plant spatial pattern and herbivore population dynamics: plant factors affecting the movement patterns of a tropical cucurbit specialist (*Acalymma innubum*). Ecology 65:175–190. [3]

Bach, C. E. 1988a. Effects of host plant patch size on herbivore density: patterns. Ecology 69:1090–1102. [3]

Bach, C. E. 1988b. Effects of host plant patch size on herbivore density: underlying mechanisms. Ecology 69:1103–1117. [3]

Bacheller, J. D. 1990. The effect of salinity, soil aeration, and insect predation on the free amino acid, glyclinebetaine, and soluble protein levels of the salt marsh cordgrass, *Spartina alterniflora*. M.S. Thesis. University of South Florida, Tampa, FL [2,6,7]

Backus, E.A. 1985. Anatomical and sensory mechanisms of leafhopper and planthopper feeding behavior. Pages 163–194. *In* The Leafhoppers and Planthoppers. L.R. Nault and J.G. Rodriguez, editors. Wiley, New York. [2,12]

Backus, E. A. 1988. Sensory systems and behaviors which mediate hemipteran plant-feeding: a taxonomic review. J. Insect Physiol. 34:151–165. [12]

Bailey, V. A., A.J. Nicholson, and E.J. Williams. 1962. Interactions between hosts and parasites when some host individuals are more difficult to find than others. J. Theor. Biol. 3:1–18. [11]

Baker, C. F. 1915. Notices of certain Philippine Fulgoridae, one being of economic importance. Philippine J. Sci. 10:137–144. [1]

Baker, H.G. and G.L. Stebbins. 1965. The genetics of colonizing species. Academic Press, New York. [4]

Baker, P.S., R.J. Cooter, P.M. Chang, and H.B. Hashim. 1980. The flight capabilities of laboratory and tropical field populations of the brown planthopper, *Nilaparvata lugens* (Stål) (Hemiptera: Delphacidae). Bull. Entomol. Res. 70:589–600. [9]

Ball, E. D. 1928. Some new genera and species of N. A. Derbidae with notes on others (Fulgoridae). Can. Entomol. 60:196–201. [1]

Ball, E. D. 1930. A new species and variety of *Scolops* with notes on others (Rhynchota Fulgoridae). Pan-Pac. Entomol. 7:9–11. [1]

Ball, E. D. 1933. Notes on the Fulgoridae with some new species. Psyche 40:145–150. [1]

Ball, E. D. 1935a. Some new Issidae, with notes on others (Homoptera-Fulgoridae). Bull. Brooklyn Entomol. Soc. 30:37–41. [1]

Ball, E. D. 1935b. The genus *Bruchomorpha* Newman (Homoptera- Fulgoridae). Bull. Brooklyn Entomol. Soc. 30:197–203. [1]

Ball, E. D. 1937. Some new Fulgoridae from the western United States. Bull. Brooklyn Entomol. Soc. 32:171–183. [1]

Ball, E. D. and A. Hartzell. 1922. A review of the desert leafhopers of the Orgerini (Rhynchota Fulgoridae). Ann. Entomol. Soc. Am. 15:137–153. [1]

Ball, J. C. and D.L. Dahlsten. 1973. Hymenopterous parasites of *Ips paraconfusus* (Coleoptera: Scolytidae) larvae and their contribution to mortality. I. Influence of host tree and tree diameter on parasitization. Can. Entomol. 105:1453–1464. [11]

Ballou, C. H. 1936. Insect notes from Costa Rica in 1935. Insect Pest Surv. Bull. 16 (9, suppl.). [1]

Ballou, J. K., J. H. Tsai, and S. W. Wilson. 1987. Delphacid planthoppers *Sogatella kolophon* and *Delphacodes idonea* (Homoptera: Delphacidae): Descriptions of immature stages and notes on biology. Ann. Entomol. Soc. Am. 80:312–319. [1]

Banttari, E. E., and R. J. Zeyen. 1970. Transmission of oat blue dwarf virus by the aster leafhopper following natural acquisition or inoculation. Phytopathology 60: 399–402. [12]

Barbault, R. 1988. Body size, ecological constraints, and the evolution of life-history strategies. Pages 261–286. *In* M.K. Hecht, B. Wallace, and G.T. Prance, editors. "Evolutionary Ecology 22". Plenum Press, New York. [4]

Barbosa, P. 1988. Natural enemies and herbivore-plant interactions: influence of plant allelochemicals and host specificity. Pages 201–229. *In* P. Barbosa and D. K. Letourneau, editors. Novel Aspects of Insect-Plant Interactions. Wiley, New York. [10,11,18]

Barbosa, P., J.A. Saunders, and M. Waldvogel. 1982. Pages 63–71. *In* J.H. Visser and A.K. Minks, editors. Plant mediated variation in herbivore suitability and parasitoid fitness, Proc. 5th Int. Symp. Plant-Insect Relationships, Wageningen, Pudoc, Wageningen. [14]

Barfield, C. S., J. Mangan, J. Markle, S. Padmanagara, R. Roberts, and Soehardjan. 1991. Report of mid-term review mission. UTF/INS/067/INS, Training and Development of Integrated Pest Management in Rice-based Cropping Systems. July 1991. Available from the National Programme for Development and Training of Integrated Pest Management in Rice-Based Cropping Systems, Jl. Ki Mangunsarkoro No. 5, Jakarta 10310, Indonesia. [20]

Barlow, C.A., and P.A. Randolph. 1978. Quality and quantity of plant sap available to the pea aphid. Ann. Entomol. Soc. Am. 71:46–48. [2]

Barrion, A. A. 1985. Selection of resistant rice varieties and genetics of rice-infesting biotypes on the brown planthopper, *Nilaparvata lugens* (Stål) (Hemiptera: Delphacidae). Ph.D. Thesis, University of the Philippines, Los Baños. [17]

Barrion, A.T., P.C. Pantua, J.P. Bandong, C.G. de la Cruz, F.A. Raymundo, M.D. Lumaban, R.F. Apostol, and J.A. Litsinger. 1981. Food web of the rice brown planthopper in the Philippines. IRRN 6:13–15. [10]

Basilio, R.P., and K.L. Heong. 1990. Brown mirid bug, a new predator of brown planthopper (BPH) in the Philippines. IRRN 15:27–28. [10]

Bateman, A. J. 1948. Intra-sexual selection in *Drosophila*. Heredity 2:349–368. [6]

Batra, S. W. T. 1979. Insects associated with weeds of the northeastern United States: Quickweeds, *Galinsoga ciliata*, and *G. parviflora* (Compositae). Environ. Entomol. 8:1078–1082. [1]

Batra, S. W. T. 1982. Biological control in agroecosystems. Science 215:134–139. [11]

Beamer, R. H. 1929. *Scolops osborni* Ball in Kansas (Homoptera Fulgoridae). J. Kansas Entomol. Soc. 2:70. [1]

Beamer, R. H. 1945. Four new species in the genus *Bakerella* (Homoptera, Delphacidae). J. Kansas Entomol. Soc. 18:149–154. [1]

Beamer, R. H. 1946. The genus *Stenocranus* in America north of Mexico (Homoptera, Delphacidae). J. Kansas Entomol. Soc. 19:1–11. [1]

Beardsley, J. W. 1990. Notes on immigrant delphacid planthoppers in Hawaii (Homoptera: Fulgoroidea). Proc. Hawaiian Entomol. Soc. 30:121–129. [1]

Becker, M. 1975. The biology and population ecology of *Macrosteles sexnotatus* (Fallén) (Cicadellidae, Hemiptera). Ph.D. Dissertation, University of London, London. [11,13]

Beddington, J.R., C.A. Free, and J.H. Lawton. 1978. Characteristics of successful natural enemies in models of biological control of insect pests. Nature 273:513–519. [10,14,18]

Beirne, B. P. 1950. The Canadian Cixiidae (Homoptera: Fulgoroidea). Can. Entomol. 82:93–101. [1]

Beirne, B.P. 1975. Biological control attempts by introductions against pest insects in the field in Canada. Can. Entomol. 197:225–236. [14]

Bell, S.S., E.D. McCoy, and H.R. Mushinsky. 1991. Habitat structure: The physical arrangement of objects in time and space. Chapman and Hall, London. [3]

Benor, D., and M. Baxter. 1984. Training and Visit extension. The World Bank, Washington, DC. [20]

Benor, D., and J. Q. Harrison. 1977. Agricultural extension—the Training and Visit System. The World Bank, Washington, DC. [20]

Benrey, B., and R. F. Denno. In preparation. Density-related patterns of parasitism by *Anagrus* on *Prokelisia* planthopper eggs. [13]

Bentley, J. W. 1989. What farmers don't know can't help them: the strengths and weaknesses of indigenous technical knowledge in Honduras. Agriculture and Human Values 6:25–31. [20]

Bentur, J.S., and M.B. Kalode. 1987. Off-season survival of the predatory mirid bug, *Cyrtorhinus lividipennis* (Reuter). Curr. Sci. 56:956–957. [10]

Berger, P. H., and T. P. Pirone. 1986. The effect of helper component on the uptake and localization of potyviruses in *Myzus persicae*. Virology 153:256–261. [12]

Bergman, J.M., and W.M. Tingey. 1979. Aspects of interaction between plant genotypes and biological control. Bull. Entomol. Soc. Am. 25:275–279. [14,18]

Bernays, E.A. 1986. Diet-induced head allometry among foliage-chewing insects and its importance for graminivores. Science 231:495–497. [2]

Bernays, E., and R.F. Chapman. 1987. Evolution of plant deterrence to insects. Pages 159–173. *In* R.F. Chapman, E.A. Bernays, and J.G. Stoffolano, editors. Perspectives in Chemoreception and Behavior. Springer-Verlag, New York. [1]

Bernays, E., and M. Graham. 1988. On the evolution of host specificity in phytophagous arthropods. Ecology 69:886–892. [1]

Berrigan, D. 1991. The allometry of egg size and number in insects. Oikos 60:313–321. [4]

Besson, E., G. Dellamonica, J. Chopin, K.R. Markham, M. Kim, H. Koh, and H. Fuka. 1985. C-Glycosylflavones from rice plant involved in planthopper feeding. Phytochemistry 24: 1061–1064. [2]

Black, L. M. 1970. Potato yellow dwarf virus. Descriptions of Plant Viruses, No. 35. Commonw. Mycol. Inst./Assoc. Appl. Biol., Kew, Surrey, UK. [12]

Black, L. M. 1979. Vector cell monolayers and plant viruses. Adv. Virus Res. 25:191–270. [12]

Blau, W.S. 1981. Latitudinal variation in the life histories of insects occupying disturbed habitats: A case study. Pages 75–95. *In* R.F. Denno and H. Dingle, editors. Insect Life History Patterns: Habitat and Geographic Variation. Springer-Verlag, New York. [4]

Blum, J. L. 1968. Salt marsh spartinas and associated algae. Ecol. Monogr. 38:199–221. [3]

Blum, M. S., and N. A. Blum, editors. 1979. Sexual Selection and Reproductive Competition in Insects. Academic Press, New York. [6]

Boethel, D.J., and R.D. Eikenberry. 1986. Interactions of Plant Resistance and Parasitoids and Predators of Insects. Ellis Horwood, Chichester. [14]

Bonfils, J. 1982. Description d'une espece nouvelle de Cixiidae nuisible aux plantation de cocotier (Hom. Fulgoromorpha). Bull. Soc. Entomol. France 87:381–384. [1]

Bonman, J.M., G.S. Khush, and R.J. Nelson. 1992. Breeding rice for resistance to pests. Ann. Rev. of Phytopathol. 30:507–528. [15]

Booij, C. J. H. 1981. Biosystematics of the *Muellerianella* complex (Homoptera,Delphacidae), taxonomy, morphology and distribution. Netherlands J. of Zool. 31:572–595. [6]

Booij, C. J. H. 1982a. Biosystematics of the *Muellerianella* complex (Homoptera, Delphacidae): Host-plants, habitats and phenology. Ecol. Entomol. 7:9–18. [i,1,4,15]

Booij, C.J.H. 1982b. Biosystematics of the *Muellerianella* complex (Homoptera, Delphacidae), interspecific and geographic variation in acoustic behavior. Z. Tierpshychol. 58:31–52. [5,6,15]

Booij, C.J.H. 1982c. Biosystematics of the *Muellerianella* complex (Homoptera, Delphacidae), hybridization studies. Genetica 57:161–170. [5]

Booij, C.J.H., and J.A. Guldemond. 1984. Distributional and ecological differentiation between asexual gynogenetic planthoppers and related sexual species of the genus *Muellerianella* (Homoptera, Delphacidae). Evolution 38:163–175. [5,6]

Borden, J.H., and C.E. Slater. 1969. Flight muscle volume change in *Ips confusus* (Coleoptera: Scolytidae). Can. J. of Zoo. 47:29–32. [4]

Borgia, G. 1979. Sexual selection and the evolution of mating systems. Pages 19–90. *In* M.S. Blum and N. A. Blum, editors. Sexual Selection and Reproductive Competition in Insects. Academic Press, New York. [6]

Borgia, G. 1980. Sexual competition in *Scatophaga stercoraria*: size- and density-related changes in male ability to capture females. Behaviour 75:185–206. [6]

Borgia, G. 1981. Mate selection in *Scatophaga stercoraria*: female choice in a male-controlled system. Animal Behaviour 29:71–80. [6]

Borgia, G. 1987. A critical review of sexual selection models. Pages 55–66. *In* J. W. Bradbury and M. B. Andersson, editors. Sexual Selection: Testing the Alternatives. Wiley, New York. [6]

Bourgoin, T. 1985. Morphologie antennaire des Tettigometridae (Hemiptera, Fulgoromorpha). Nouv. Revue Entomol. (N.S.) 2:11–20. [1]

Bourgoin, T. 1986. Valeur morphologique de la lame maxillaire chez les Hemiptera; remarques phylogenetiques. Ann. Soc. Entomol. France. 22:413–422. [1]

Box, G.E.P., and G.M. Jenkins. 1970. Time Series Analysis, Forecasting and Control. Holden-Day, San Francisco. [7]

Bradbury, J. W., and S. L. Vehrencamp. 1977. Social organization and foraging in emballonurid bats. III. Mating systems. Behavioral Ecol. Sociobiol. 2:1–17. [6]

Bradbury, J. W., and M. B. Andersson, editors, 1987. Sexual Selection: Testing the Alternatives. Wiley, New York. [6]

Braun, A.R., G.M. Guerrero, A.C. Belloti, and L.T. Wilson. 1987. Relative toxicity of permethrin to *Mononychellus progresivus* Doreste and *Tetranychus urticae* Koch (Acari:Tetranychidae) and their predators *Amblyseius limonicus* Garman & McGregor (Acari:Phytoseiidae) and *Oligota minuta* Cameron (Coleoptera:Staphylinidae): bioassays and field validation. Environ. Entomol. 16:545–550. [14]

Brĉak, J. 1979. Leafhopper and planthopper vectors of plant disease agents in Central and Southern Europe. Pages 97–154. *In* K. Maramorosch and K.F. Harris, editors. Leafhopper Vectors and Plant Disease Agents. Academic Press, New York. [1]

Breakey, E. P. 1928. The genus *Scolops* (Homoptera, Fulgoridae). Univ. Kansas Sci. Bull. 18:417–455. [1]

Brooks, M.A. 1985. Nutrition, cell culture and symbiosis of leafhoppers and planthoppers. Pages 195–216. *In* The Leafhoppers and Planthoppers. L.R. Nault and J.G. Rodriguez, editors. Wiley, New York. [2,12]

Brown, F. 1989. The classification and nomenclature of viruses: summary of results of meetings of the International Committee on Taxonomy of Viruses in Edmonton, Canada, 1987. Intervirology 30:181–186. [12]

Brown, V. K. 1982. Size and shape as ecological discriminants in successional communities of Heteroptera. Biolog. J. Linnaean Soc. 18:279–290. [3]

Brown, V. K. 1986. Life cycle strategies and plant succession. Pages 105–124. *In* F. Taylor and R. Karban, editors. The Evolution of Insect Life Cycles. Springer-Verlag, New York. [3]

Bruner, S. C., L. C. Scaramuzza, and Y. A. R. Otero. 1945. Catalogo de los insectos que atacan a las plantas economicas de Cuba. Est. Exp. Agron. Bol. 63. [1]

Brunt, A., K. Crabtree, and A. Gibbs. 1990. Viruses of Tropical Crops. Descriptions and Lists from the VIDE Database. C.A.B. International Institute of Biol. Control, Wallingford, Oxon, UK. [12]

Bull, R.M. 1972. A study of the sugar cane leafhopper *Perkinsiella saccharicida* Kirk. (Hom: Delphacidae) in the Bundaberg district of south-eastern Queensland. Proceedings Qd Society Sugar Cane Technology 39th Conf., pp. 173–183. [8]

Bull, R.M. 1981. Population studies on the sugar cane leafhopper (*Perkinsiella saccharicida* Kirk) in the Bundaberg district. Proc. Australian Soc. Sugar Cane Technol. 10:293–303. [8,10]

Bulmer, M.G. 1975. The statistical analysis of density dependence. Biometrics 31:901–911. [7]

Bush, G. L. 1974. The mechanism of sympatric host race formation in the true fruit flies (Tephritidae). Pages 3–23. *In* M. J. White, editor. Genetic Mechanisms of Speciation In Insects. Australia and New Zealand Book Co., Sydney. [18]

Butler, G.W., and R.W. Bailey. 1973. Chemistry and biochemistry of herbage. Academic Press, London. [2]

Butlin, R.K. 1990. Divergence in emergence time of host races due to differential gene flow. Heredity 65:47–50. [15]

Byerlee, D. 1987. Maintaining the momentum in post-Green Revolution Agriculture: a micro-level perspective from Asia. International Development Paper No. 10, Department of Agricultural Economics, Michigan State University, Lansing, MI. [20]

Byerlee, D., and E. H. de Polanco. 1986. Farmers' stepwise adoption of a technological package: Evidence from the Mexican Altiplano. Am. J. Agricultural Econ. 68:519–527. [20]

Byrne, D.N., T.S. Bellows, and M.P. Parrella. 1990. Whiteflies in agricultural systems. Pages 227–261. *In* D. Gerling, editor. Whiteflies: Their bionomics, pest status and management. Intercept Ltd., Andover, UK. [7]

CAB. 1987. Natural Enemy Databank. CAB Intl. Institute of Biol. Control. UK. [14]

Cagampang, G.B., M.D. Pathak, and B.O. Juliano. 1974. Metabolic changes in the rice plant during infestation by the brown planthopper, *Nilaparvata lugens* Stål (Hemiptera: Delphacidae). Appl. Entomol. Zool. 9:174–184. [2,6,7]

Calder, W.A. 1984. Size, Function and Life History. Harvard University Press, Cambridge, MA. [4]

Caldwell, J. S., and D. M. DeLong. 1948. A new species of Issidae from California (Homoptera: Fulgoroidea). Ohio J. Sci. 48:176–177. [1]

Caldwell, J. S., and L. F. Martorell. 1950. Review of the Auchenorynchous Homoptera of Puerto Rico. Part II. The Fulgoroidea except Kinnaridae. J. Agric. Univ. Puerto Rico 34:133–269. [1]

Calisher, C. H., N. Karabatsos, H. Zeller, J-P Digoutte, R. B. Tesh, R. E. Shope, A. P. A. Travassos da Rosa, and T. D. George. 1989. Antigenic relationships among rhabdoviruses from vertebrates and hematophagous arthropods. Intervirology 30: 241–257. [12]

Calvert, P. D., J. H. Tsai, and S. W. Wilson. 1987a. *Delphacodes nigrifacies* (Homoptera: Delphacidae): Field biology, laboratory rearing and descriptions of immature stages. Fla. Entomol. 70:129–134. [1]

Calvert, P. D., and S. W. Wilson. 1986. Life history and description of the immature stages of the planthopper *Stenocranus lautus* (Homoptera: Delphacidae). J. New York Entomol. Soc. 94:118–125. [1,2,4]

Calvert, P. D., S. W. Wilson, and J. H. Tsai. 1987b. *Stobaera concinna* (Homoptera: Delphacidae): Field biology, laboratory rearing and descriptions of immature stages. J. New York Entomol. Soc. 95:91–98. [1,2,4]

Campbell, B.C. 1989. On the role of microbial symbiotes in herbivorous insects. Pages 1–44. *In* E.A. Bernays, editor. Insect-plant interactions. Volume 1. CRC Press, Boca Raton, FL. [2]

Campbell, B.C., and S.S. Duffey. 1979. Tomatine and parasitic wasps: Potential incompatibility of plant antibiosis with biological control. Science 205:700–702. [14]

Caprio, M.A., and B.E. Tabashnik.1992. Gene flow accelerates local adaptation among finite populations: simulating the evolution of insecticide resistance. J. Econ. Entomol. 85:611–620. [15]

CARE. 1991. CARE pest management policy. Annex II: Implementation strategy. Available from CARE, 600 First Avenue, New York, New York 10016, USA. [20]

Cariño, F.O., P.E. Kemmore, and V.A. Dyck. 1979. The FARMCOP sampler for hoppers and predators in flooded rice fields. Int. Rice Res. Newsl. 4(4):21–22. [8]

Carlquist, S. 1970. Hawaii: a natural history. Natural History Press, Garden City, NY. [1]

Carnegie, A. J. M. 1980. Egg mortality of *Numicia viridis* Muir (Homoptera: Tropiduchidae) in sugarcane and in indiginous hosts. J. Entomol. Soc. South Africa 43:215–221. [1]

Caron, R.E. 1981. Interactive effects of soil-applied pesticides, cultivar, and planting date on the soybean-insect complex in North Carolina. Ph.D. Dissertation, North Carolina State University. [14]

Carrow, J.R., and R.E. Betts. 1973. Effects of different foliar-applied nitrogen fertilizers on Balsam Wooly Aphid. Can. J. Forest Res. 3:122–139. [2]

Carson, H. L., and A. R. Templeton. 1984. Genetic revolutions in relation to speciation phenomena: the founding of new populations. Annu. Rev. Ecol. Syst. 15:97–131. [1]

Carter, W. 1941. *Peregrinus maidis* (Ashm.) and the transmission of corn mosaic. Ann. Entomol. Soc. Am. 34: 551–556. [12]

Carver, M., G.F. Gross, and T.E. Westwood. 1991. Hemiptera. Pages 429–509. *In* The Insects of Australia, Vol. I. CSIRO. Cornell University Press, Ithaca, NY. [1]

Casper, R. 1988. Luteoviruses. Pages 235–258. *In* R. Koenig, editor. The Plant Viruses, Vol. 3, Polyhedral Virions with Monopartite RNA Genomes. Plenum Press, New York. [12]

Caswell, H. 1978. Predator-mediated coexistence: a nonequilibrium model. Am. Natur. 112: 127–154. [11]

Cavalieri, A.J., and A.H.C. Huang. 1981. Accumulation of proline and glycinebetaine in *Spartina alterniflora* Loisel. in response to NaCl and nitrogen in the marsh. Oecologia 49:224–228. [2]

Ceiba-Geigy. 1968. Rice Pest Control. Ciba-Geigy. Philippines. [17]

Cendaña, S. M., and F. B. Calora. 1964. Insect pests of rice in the Philippines. Pages 591–616. *In* International Rice Research Institute. The Major Pests of the Rice Plant. Johns Hopkins Press, Baltimore, MD. [3]

Chambers, R., A. Pacey, and L. A. Thrupp, editors. 1989. Farmer First: Farmer Innovation and Agricultural Research. The Bootstrap Press, New York. [20]

Chandra, G. 1980. Taxonomy and biology of the insect parasites of rice leafhoppers and planthoppers in the Philippines and their importance in natural biological control. Philippine Entomol. 4:119–139. [14]

Chang, C.K., and M.E. Whalon. 1987. Substrate specificities and multiple forms of esterases in the brown planthopper, *Nilaparvata lugens* (Stål). Pest. Biochem. Physiol. 27:30–35. [16]

Chang, V.C.S. 1978. Feeding activity of the sugarcane leafhopper: Identification of electronically recorded waveforms. Ann. of the Entomol. Soc. Am. 71:31–36. [2,12]

Chang, V.C.S., and A.K. Ota. 1978. Feeding activities of *Perkinsiella* leafhoppers and Fiji disease resistance of sugarcane. J. Econ. Entomol. 71:297–300. [2]

Chantarasa-ard, S. and Y. Hirashima. 1984. Host range and host suitability of *Anagrus incarnatus* Haliday (Hymenoptera: Mymaridae), an egg parasitoid of delphacid planthoppers. Appl. Entomol. Zool. 19:491–497. [13]

Chantarasa-ard, S., Y. Hirashima, and T. Miura. 1984a. Ecological studies on *Anagrus incarnatus* Haliday (Hymenoptera: Mymaridae), an egg parasitoid of the rice planthoppers. I. Functional response to host density and mutual interference. J. Fac. Agr. Kyushu Univ. 29:59–66. [11]

Chantarasa-ard, S., Y. Hirashima, and T. Miura. 1984b. Ecological studies on *Anagrus incarnatus* Haliday (Hymenoptera: Mymaridae), an egg parasitoid of the rice planthoppers. II. Spatial distribution of parasitism and host eggs in the paddy field. J. Fac. Agr. Kyushu Univ. 29:67–76. [11,14]

Chapman, R.F., E.A. Bernays, and S.J. Simpson. 1981. Attraction and repulsion of the aphid *Cavariella aegopodii*, by plant odors. J. Chem. Ecol. 7:881–888. [2]

Chatterjee, N. C. 1933. Entomological investigation on the spike disease of sandal. The life-history and morphology of *Sarima nigroclypeata*, Mel. (11) Fulgoridae (Homopt.) Indian Forest Rec. Entomol. 18:1–26. [1]

Chatterjee, P.B., and D.K. Choudhuri. 1979. Biology of *Eoeurysa flavocapitata*—a delphacid pest on sugarcane in India. Entomon. 4:263–267. [2,4]

Chelliah, S., L.T. Fabellar, and E.A. Heinrichs. 1980. Effect of sub-lethal doses of three insecticides on the reproductive rate of the brown planthopper, *Nilaparvata lugens*, on rice. Environ. Entomol. 9:778–780. [10,16,18]

Chelliah, S., and E. A. Heinrichs. 1980. Factors affecting insecticide-induced resurgence of the brown planthopper, *Nilaparvata lugens* on Rice. Environ. Entomol. 9:773–777. [10,14, 16,17,18]

Chen, C.C., L.L. Cheng, C.C. Kuan, and R.F. Hou. 1981a. Studies on the intracellular yeast-like symbiote in the brown planthopper, *Nilaparvata lugens* Stål. I. Histological observations and population changes of the symbiote. Zeit. Angew. Entomol. 91:321–327. [2]

Chen, C.C., L.L. Cheng, and R.F. Hou. 1981b. Studies on the intracellular yeast-like symbiote in the brown planthopper, *Nilaparvata lugens* Stål. II. Effects of antibiotics and elevated temperature on the symbiotes and their host. Zeit. Angew. Entomol. 92:440–449. [2]

Chen, C. C., and R. J. Chiu. 1980. Factors affecting transmission of rice transitory yellowing virus by green leafhoppers. Plant Prot. Bull. (Taiwan, R.O.C.) 22:297–306. [12]

Chen, C. L., C. T. Yang, and M. R. Wilson. 1989. Achilidae of Taiwan (Homoptera: Fulgoroidea). Collected Papers on Fulgoroidea of Taiwan. Taiwan Mus. Spec. Publ. Ser. No. 8:1–64. [1]

Chen, R.C. 1983. Studies on lipids as fuel of flight in the brown planthopper (*Nilaparvata lugens* Stål). Acta Entomol. Sinica 26:42–48. [9]

Chen, R.C., and X.N. Cheng. 1980. Take-off behavior of the brown planthopper (*Nilaparvata lugens* Stål) and its synchronous relations to the biological rhythm and environmental factors. J. Nanjing Agric. College 2:1–8. [6,8,9]

Cheng, C.H. 1971. Effect of nitrogen application on the susceptibility in rice to brown planthopper attack. J. Taiwan Agric. Res. 20:21–30. [2]

Cheng, C. H., and W. L. Chang. 1979. Studies on varietal resistance to the brown planthopper in Taiwan. Pages 251–271. *In* Brown Planthopper: Threat to Rice Production in Asia. International Rice Research Institute, Los Baños, Philippines. [17,18]

Cheng, J., and J. Holt. 1990. A systems analysis approach to brown planthopper control on rice in Zhejiang Province, China. I. Simulation of outbreaks. J. Appl. Ecol. 27:85–99. [4,7,8,10,19]

Cheng, J.A., Y. Lou, and Z. Zhu. 1992. Crop variety-related mortality from natural enemies: Interactions in rice. Proc. XIX Intern. Congr. Entomol. Beijing, China, p. 147. [10,14]

Cheng, J.A., G.A. Norton, and J. Holt. 1990. A systems analysis approach to brown planthopper control on rice in Zhejiang Province, China. II. Investigation of control strategies. J. Appl. Ecology 27:100–112. [10,19]

Cheng, J., L. Zhang, S. He, Q. Fan, J. Holt, and G.A. Norton. 1989. A simulation model of population dynamics of rice brown planthopper and its validation. Acta Agric. Univ. Zhejiangensis 15:131–136. [i,7]

Cheng, S.A., J.C. Chen, H. Si, L.M. Yan, T.L. Chu, C.T. Wu, J.K. Chien, and C.S. Yan. 1979. Studies on the migration of brown planthopper *Nilaparvata lugens* Stål. Acta Entomol. Sinica 22:1–21. [7,8,9,19]

Chesson, P. 1978. Predator-prey theory and variability. Annu. Rev. Ecol. Syst. 9:323–347. [7,11]

Chesson, P., and W.W. Murdoch. 1986. Aggregation of risk: relationships among host-parasitoid models. Am. Natur. 127:696–715. [10,11,14]

China, R. E., and R. G. Fennah. 1952. A remarkable new genus and species of Fulgoroidea (Homoptera) representing a new family. Ann. Mag. Nat. Hist. Ser. 12(5):189–199. [1]

Chiu, S. 1979. Biological control of the brown planthopper. Pages 335–355. *In* Brown Planthopper: Threat to Rice Production in Asia. International Rice Research Institute, Los Baños, Phillipines. [8,10,14]

Chiu, S.C., and C.H. Cheng. 1976. Toxicity of some insecticides commonly used for rice insect control to the predators of rice-hoppers (in Chinese). Plant Prot. Bull. (Taiwan) 18:256–267. [10]

Chiu, S.F. 1984. Recent advances in the integrated control of rice insects in China. Bull. Entomol. Soc. Am. 30:41–46. [10]

Chiu, T. H., V. A. Dyck, and N. B. Peña. 1984. Functional response and searching efficiency in *Pseudogonatopus flavifemur* Esaki and Hash. (Hymenoptera: Dryinidae), a parasite of rice planthoppers. Res. Popul. Ecol. 26:74–83. [13,14]

Choi, S.Y., MM. Heu, and J.O. Lee. 1979. Varietal resistance to the brown planthopper in Korea. Pages 219–232. *In* Brown Planthopper: Threat to Rice Production in Asia. International Rice Research Institute, Los Baños, Philippines. [2,10,15]

Choi, S.Y., Y.H. Song, and J.S. Park. 1975. Insecticide resistance to small brown planthopper (*Laodelphax striatellus* Fallen). (1) Local variabilities in susceptibility of small brown planthopper to malathion and NAC. Korean J. Plant Prot. 14:53–58. [16]

Choo, H.Y., H.K. Kaya, and J.B. Kim. 1989. *Agamermis unka* (Mermithidae) parasitism of *Nilaparvata lugens* in rice fields in Korea. J. Nematol. 21:254–259. [14]

Chu, H. J. 1931. Notes on the life -history of *Lycorma delicatula* White in Nanking. Peking Nat. Hist. Bull. 5:33–35. [1]

Chu, Y.I., D.S. Lin, and T. Mu. 1976. Relative toxicity of 9 insecticides against rice insect pests and their predators (in Chinese). Plant Prot. Bull. (Taiwan) 18:369–376. [10]

Chu, Y.I., and P.S. Yang. 1985. Ecology of the brown planthopper (*Nilaparvata lugens* Stål) during the winter season in Taiwan. Chinese J. Entomol. 4:23–34. [2]

Chua, T. H., V.A. Dyck, and N.B. Peña. 1984. Functional response and searching efficiency in *Pseudogonatopus flavifemur* Esaki & Hash (Hymenoptera: Dryinidae), a parasite of rice planthoppers. Res. Popul. Ecol. 26:74–83. [11]

Chua, T.H., and E. Mikil. 1989. Effects of prey number and stage on the biology of *Cyrtorhinus lividipennis* (Hemiptera: Miridae): a predator of *Nilaparvata lugens* (Homoptera: Delphacidae). Environ. Entomol. 18:251–255. [14]

Chung, T.C., and C.N. Sun. 1983. Malathion and MIPC resistance in *Nilaparvata lugens* (Homoptera: Delphacidae). J. Econ. Entomol. 76:1–5. [16]

Chung, T.C., C.N. Sun, and C.T. Hung. 1981. Brown planthopper resistance to several insecticides in Taiwan. Int. Rice Res. Newsl. 6(5):19. [16]

Chung, T.C., C.N. Sun, and C.T. Hung. 1982. Resistance of *Nilaparvata lugens* to six insecticides in Taiwan. J. Econ. Entomol. 75:197–200. [16]

Claridge, M.F. 1985a. Acoustic signals in the Homoptera: behavior, taxonomy and evolution. Annu. Rev. Entomol. 30:297–317. [2,4,5,6,15]

Claridge, M. F. 1985b. Acoustic behavior of leafhoppers and planthoppers: Species problems and speculation. Pages 103–125. *In* L. R. Nault and J. G. Rodriquez, editors. The Leafhoppers and Planthoppers. Wiley, New York. [1,5,6]

Claridge, M.F. 1988. Species concepts and speciation in parasites. Pages 92–111. *In* D.L. Hawksworth, editor. Prospects in Systematics. Clarendon Press, Oxford. [5]

Claridge, M.F. 1990a. Acoustic recognition signals: barriers to hybridization in Homoptera Auchenorrhyncha. Can. J. Zool. 68:1741–1746. [5,6,15]

Claridge, M. F. 1990b. Variation in pest and natural enemy populations—relevance to brown planthopper control strategies. Pages 143–154. *In* B. T. Grayson, M. B. Green, and L. G. Copping, editors. Pest Management in Rice. Elsevier, London. [18]

Claridge, M.F., and J. Den Hollander. 1980. The "biotypes" of the rice brown planthopper, *Nilaparvata lugens*. Entomol. Exp. Appl. 27:23–30. [15,17,18]

Claridge, M.F., and J. Den Hollander. 1982. Virulence to rice cultivars and selection for virulence in populations of the brown planthopper *Nilaparvata lugens*. Entomol. Exp. Appl. 32:213–221. [i,2,10,15,17,18]

Claridge, M.F., and J. Den Hollander. 1983. The biotype concept and its application to insect pests of agriculture. Crop Prot. 2:85–95. [15,17]

Claridge, M.F., J. Den Hollander, and I. Furet. 1982a. Adaptations of brown planthoppers (*Nilaparvata lugens*) populations to rice varieties in Sri Lanka. Entomol. Exp. Appl. 32: 222–226. [8,15,16,17]

Claridge, M.F., J. Den Hollander, and D. Haslam. 1984a. The significance of morphometric and fecundity differences between the "biotypes" of the brown planthopper, *Nilaparvata lugens*. Ent. Exp. Appl. 36:107–114. [15]

Claridge, M. F., J. Den Hollander, and J. C. Morgan. 1982b. Variation within and between populations of the brown planthopper, *Nilaparvata lugens* (Stål). Pages 306–318. *In* W.J. Knight, N.C. Pant, T.S. Robertson, and M. R. Wilson, editors. 1st International Workshop on Leafhoppers and Planthoppers of Economic Importance. Commonwealth Institute of Entomology, London. [17,18]

Claridge, M. F., J. Den Hollander, and J. C. Morgan. 1984b. Specificity of acoustic signals and mate choice in the brown planthopper *Nilaparvata lugens*. Entomol. Exp. Appl. 35:221–226. [5,6,15]

Claridge, M.F., J. Den Hollander, and J.C. Morgan 1985a. Variation in courtship signals and hybridization between geographically definable populations of the rice brown planthopper, *Nilaparvata lugens* (Stål). Biol. J. Linn. Soc. Lond. 24:35–49. [i,5,6,15]

Claridge, M.F., J. Den Hollander, and J.C. Morgan 1985b. The status of weed-associated populations of the brown planthopper, *Nilaparvata lugens* (Stål)—host race or biological species? Zool. J. Linn. Soc. 84:77–90. [2,5,15]

Claridge, M. F., J. Den Hollander, and J. C. Morgan. 1988. Variation in hostplant relations and courtship signals of weed-associated populations of the brown planthopper, *Nilaparvata lugens* (Stål), from Australia and Asia: a test of the recognition species concept. Biol. J. Linn. Soc. 35:79–93. [1,5,6,15]

Claridge, M.F., and J.C. Morgan. 1987. The brown planthopper, *Nilaparvata lugens* (Stål), and some related species: a biotaxonomic approach. Pages 19–32. *In* M.R. Wilson and L.R. Nault, editors. Proc. 2nd Int. Workshop on Leafhoppers and Planthoppers of Economic Importance. CIE, London. [5,6,18]

Claridge, M.F., and J.C. Morgan. in press. Geographical variation in acoustic signals of the planthopper, *Nilaparvata bakeri* (Muir) in Asia: species recognition and sexual selection. Biol. J. Linn. Soc. Lond. [5]

Claridge, M. F., and W. J. Reynolds. 1972. Host plant specificity, oviposition behavior and egg parasitism of some woodland leafhoppers of the genus *Oncopsis* (Hemiptera, Homoptera: Cicadellidae). Trans. Royal. Entomol. Soc. Lond. 124:149–166. [3,11]

Claridge, M. F., and W. J. Reynolds. 1973. Male courtship songs and sibling species in the *Oncopsis flavicollis* species group (Hemiptera:Cicadellidae). J. Entomol. Series B Taxonomy. 42:29–39. [6]

Claridge, M. F., and M. R. Wilson. 1976. Diversity and distribution patterns of some mesophyll-feeding leafhoppers of temperate woodland canopy. Ecol. Entomol. 1:231–250. [3,13]

Claridge, M. F., and M. R. Wilson. 1981. Host plant associations, diversity and species-area relationships of mesophyll-feeding leafhoppers of trees and shrubs in Britain. Ecol. Entomol. 6:217–238. [3,13]

Clausen, C. P. editor. 1978. Introduced parasites and predators of arthropod pests and weeds: a world review. Agricultural Handbook No. 480, USDA. U. S. Government Printing Office, Washington, DC. [13,14]

Clayton, W. D., and S. A. Renvoize. 1986. Genera graminum—grasses of the World. Kew Bull. Additional Ser. 13, Royal Bot. Gardens, Kew, Her Majesties Stationary Office, London. [1]

Cobben, R. H., and G. J. Rozeboom. 1983. De invertebratenfauna van de Zuidlimburgse kalkgraslanden. De Cicaden in bodemvallen (Hemiptera, Homoptera Auchenorrhyncha). Natuurhist. Maandblad 72(6/7):102–110. [1]

Cockbain, A.J. 1961. Fuel utilization and duration of tethered flight in *Aphis fabae* Scop. J. Exp. Biol. 38:163–174. [9]

Comins, H.N., and M.P. Hassell. 1979. The dynamics of optimally foraging predators and parasitoids. J. Anim. Ecol. 48:335–351. [10]

Comstock, A. B. 1911. Handbook of Nature Study. Comstock Publishing Company, Ithaca, NY. [20]

Comstock, J. A. 1942. Notes on *Loxophora dammersi* Van Duzee. Southern Calif. Acad. Sci. 40:160. [1]

Conn, E.E. 1979. Cyanide and cyanogenic glycosides. Pages 387–412. *In* G.A. Rosenthal and D.H. Janzen, editors. Herbivores: their interaction with secondary plant metabolites. Academic Press, New York. [2]

Connell, J. H., and W.P. Sousa. 1983. On the evidence needed to judge ecological stability or persistence. Am. Natur. 121:789–824. [11]

Conner, J. 1989. Density-dependent sexual selection in the fungus beetle, *Bolitotherus cornutus*. Evolution 43:1378–1386. [6]

Connor, E. F., S. H. Faeth, D. Simberloff, and P. A. Opler. 1980. Taxonomic isolation and the accumulation of herbivorous insects: a comparison of introduced and native trees. Ecol. Entomol. 5:205–211. [13]

Conti, M. 1966. Indagini sulla trasmissione del virus del nanismo ruvido del mais (MRDV) per mezzo di *Laodelphax striatellus* Fallen. Ann. Fac. Sci. Agric. Univ. Torino. 3:337–348. [12]

Conti, M. 1980. Vector relationships and other characteristics of barley yellow striate mosaic virus (BYSMV). Ann. Appl. Biol. 95: 83–92. [12]

Conti, M. 1984. Epidemiology and vectors of plant reolike viruses. Pages 111–139. *In* K. F. Harris, editor. Current Topics in Vector Research, Vol. 2. Praeger, New York. [12]

Conti, M. 1985. Transmission of plant viruses by leafhoppers and planthoppers. Pages 289–307. *In* L. R. Nault and J. G. Rodriguez, editors. The Leafhoppers and Planthoppers. Wiley, New York. [6,14]

Conway, G.R. 1984. Pest and Pathogen Control: Strategic, Tactical, and Policy Models. Wiley–IIASA International Series on Applied Systems Analysis. Wiley, New York. [19]

Conway, G. R., and E. B. Barbier. 1990. After the Green Revolution: sustainable agriculture for development. Earthscan Publications Ltd., London. [20]

Cook, A.G., and T.J. Perfect. 1982. Determining the wing-morph of adult *Nilaparvata lugens* and *Sogatella furcifera* from morphometric measurements on the fifth-instar nymphs. Entomol. Exp. Appl. 31:159–164. [8]

Cook, A.G., and T.J. Perfect. 1985a. The influence of immigration on population development of *Nilaparvata lugens* Stål and *Sogatella furcifera* Horvath (Homoptera: Delphacidae) and its interaction with immigration by predators. Crop Prot. 4:423–433. [2,7,8,9,10,14, 15,18,19]

Cook, A.G., and T.J. Perfect. 1985b. Seasonal abundance of macropterous *Nilaparvata lugens* and *Sogatella furcifera* based on presumptive macroptery in fifth-instar nymphs. Ecol. Entomol. 10:249–258. [i,4,7,8]

Cook, A.G., and T.J. Perfect. 1989. The population characteristics of the brown planthopper, *Nilaparvata lugens*, in the Philippines. Ecol. Entomol. 14:1–9. [i,4,8,9,15,18,19]

Cook, A.G., S. Woodhead, V.F. Magalit, and E.A. Heinrichs. 1987. Variation in feeding behavior of *Nilaparvata lugens* on resistant and susceptible rice varieties. Entomol. Exp. Appl. 43: 227–236. [2,18]

Cook, R.M., and S.P. Hubbard. 1977. Adaptive searching strategies in insect parasites. J. Anim. Ecol. 46:115–125. [14]

Coppel, H., and J.W. Mertins. 1977. Biological Insect Pest Suppression. Springer-Verlag, New York. [14]

Corbett, J. B. 1974. The Biochemical Mode of Action of Pesticides. Academic, New York. [16]

Cornell, J. B. 1979. Sharing Nature with Children: A Parents' and Teachers' Nature Awareness Guidebook. Ananda Publications, Nevada City, CA. [20]

Croft, B.A., and A.W.A. Brown. 1975. Responses of arthropod natural enemies to insecticides. Ann. Rev. Entomol. 20:285–336. [14]

Cromartie, W. J. 1975. The effect of stand size and vegetational background on the colonization of cruciferous plants by herbivorous insects. J. Appl. Ecol. 12:517–533. [3]

Cronin, J. T. 1991. Parasitoid foraging behavior and the stabilization of host-parasitoid populations. Ph.D. Dissertation, Florida State University, Tallahassee, FL. [11]

Cronin, J. T., and D.E. Gill. 1989. The influence of host distribution, sex, and size on the level of parasitism by *Itoplectis conquisitor* (Hymenoptera: Ichneumonidae). Ecol. Entomol. 14:163–173. [11]

Cronin, J.T., and D.R. Strong. 1990a. Density-independent parasitism among host patches by *Anagrus delicatus* (Hymenoptera: Mymaridae): Experimental manipulation of hosts. Ecol. Entomol. 59:1019–1026. [i,4,10,11,13]

Cronin, J. T., and D.R. Strong. 1990b. Biology of *Anagrus delicatus* (Hymenoptera: Mymaridae), an egg parasitoid of *Prokelisia marginata* (Homoptera: Delphacidae). Ann. Entomol. Soc. Am. 83:846–854. [11]

Cronquist, A. 1981. An integrated system of classification of flowering plants. Columbia Univ. Press, New York. [1]

Crowley, P. H. 1981. Dispersal and the stability of predator-prey interactions. Am. Natur. 118:673–701. [10,11]

Crowley, P. H., S. Travers, M. Linton, S. Cohn, A. Sih, and R. C. Sargent. 1991. Mate density, predation risk and the seasonal sequence of mate choices: a dynamic game. Am. Natur. 137:567–596. [6]

Culvenor, C.C.J. 1970. Toxic plants—a re-evaluation. Search 1:103–110. [2]

Cumber, R.A. 1952. Studies on *Oliarus akinsoni* Myers (Hem. Cixiidae), vector of "yellow-leaf" disease on *Phormium tenax* Forst. I. Habits and environment, with a note on natural enemies. New Zealand J. Sci. Technol. B 34:92–98. [1]

Cumber, R. A. 1966. Factors influencing population levels of *Scolypopa australis* Walker (Hemiptera-Homoptera: Ricaniidae) in New Zealand. New Zealand J. Sci. 9:336–356. [1]

Dahlgren, R., and K. Bremer. 1985. Major clades of the angiosperms. Cladistics 1:349–368. [1]

Dai, S.M., and C.N. Sun. 1984. Pyrethroid resistance and synergism in *Nilaparvata lugens* Stål (Homoptera: Delphacidae) in Taiwan. J. Econ. Entomol. 77:891–897. [16]

Dale, D. in press. Biology and ecology of rice insects. *In* E.A. Heinrichs, editor. Biology and Management of Rice Insects. Wiley Eastern, New Delhi. [16]

D'Arcy, C. J., P. A. Barnett, A. D. Hewings, and R. M. Goodman. 1981. Purification and characterization of a virus from the aphid *Rhopalosiphum padi*. Virology 112:346–349. [12]

Darwin, C. 1871. The Descent of Man and Selection in Relation to Sex. John Murray, London. [6]

Davies, N. B. 1991. Mating Systems. Pages 263–294. *In* J. R. Krebs and N. B. Davies, editors. Behavioural Ecology: An Evolutionary Approach. Third Edition. The University Press, Cambridge. [6]

Davis, M.A. 1986. Geographic patterns in the flight ability of a monophagous beetle. Oecologia 69:407–412. [4]

Day, M. F., H. Irzkiewicz, and A. Mckinnon. 1951. Observations on the feeding of the virus vector *Orosius argentatus* (Evans) and comparisons with other jassids. Austral. J. Sci. Res. 5:128–142. [12]

DeBach, P. 1974. Biological Control by Natural Enemies. Cambridge University Press, London. [11,14]

DeBach, P., and D. Rosen. 1991. Biological Control by Natural Enemies, 2nd ed. Cambridge Univ. Press, Cambridge. [14]

Demayo, C.G., R.C. Saxena, and A.A. Barrion. 1990. Allozyme variation in local populations of the brown planthopper, *Nilaparvata lugens* (Stål) in the Philippines. Philipp. Ent. 8:737–748. [15]

Dempster, J.P. 1975. Animal population ecology. Academic Press, London, UK. [7]

Dempster, J.P. 1983. The natural control of populations of butterflies and moths. Biol. Rev. 58:461–481. [7,11]

Dempster, J.P., and E. Pollard. 1986. Spatial heterogeneity, stochasticity and the detection of density dependence in animal populations. Oikos 46:413–416. [7]

den Bieman, C. F. M. 1987a. Host plant relations in the planthopper genus *Ribautodelphax* (Homoptera, Delphacidae). Ecol. Entomol. 12:163–172. [1,2,15]

den Bieman, C. F. M. 1987b. Taxonomic evaluation of the *Ribautodelphax collinus* complex (Homoptera, Delphacidae), pages 121–156. *In* Biological and taxonomic differentiation in the *Ribautodelphax collinus* complex (Homoptera, Delphacidae). Meded. Landbouwuniversiteit Wageningen, Netherlands. [1]

den Bieman, C.F.M. 1987c. Biological and taxonomic differentiation in the *Ribautodelphax collinus* complex (Homoptera, Delphacidae). Taxonomic evaluation of the *Ribautodelphax collinus* complex (Homoptera, Delphacidae). Ph.D. Dissertation. Landbouwuniversiteit, Wageningen, Netherlands. [4]

den Bieman, C.F.M. 1987d. Variability in female calling signals in mixed populations of pseudogamous forms and bisexual *Ribautodelphax* species (Homoptera: Delphacidae). Neth. J. Zool. 37:43–58. [i,5]

den Bieman, C.F.M. 1988a. Coexistence of pseudogamous and sexual planthoppers of the genus *Ribautodelphax* (Homoptera, Delphacidae). Ecol. Entomol. 13:383–390. [5,6]

den Bieman, C.F.M. 1988b. Hybridization studies in the planthopper genus *Ribautodelphax* (Homoptera, Delphacidae). Genetica 76:15–26. [5]

den Bieman, C.F.M. 1988c. Karyotypic variation in bisexual species and pseudogamous forms of the planthopper genus *Ribautodelphax* (Homoptera, Delphacidae). Genetica 76:101–110. [5]

den Bieman, C. F. M., and C. J. H. Booij. 1984. New and interesting Dutch Delphacidae (Homoptera, Auchenorrhyncha). Entomol. Berich. 44:117–123. [1]

den Bieman, C.F.M., and P.W.F. de Vrijer. 1987. True parthenogenesis for the first time demonstrated in planthoppers (Homoptera, Delphacidae). Annl. Soc. Entomol. France. (N.S.) 23:3–9. [5]

den Boer, P. J. 1968. Spreading of risk and stabilization of animal numbers. Acta Biotheor. 18:165–194. [10,11]

den Boer, P.J. 1986. Density dependence and the stabilization of animal numbers. 1 The winter moth. Oecologia 69:507–512. [7]

den Boer, P.J. 1988. Density dependence and the stabilization of animal numbers. 3 The winter moth reconsidered. Communication 339 Biological Station, Wijster, Netherlands, pp. 161–168. [7]

Den Hollander, J. 1989. Electrophoretic studies on planthoppers and leafhoppers of agricultural importance. Pages 297–315. *In* H.D. Loxdale and J. den Hollander, editors. Electrophoretic studies on agricultural pests. Clarendon Press, Oxford. [15]

Den Hollander, J., and P. K. Pathak. 1981. The genetics of the "biotypes" of the rice brown planthopper, *Nilaparvata lugens*. Entomol. Exp. Appl. 29:76–86. [15,16,17,18]

Denno, R.F. 1976. Ecological significance of wing-polymorphism in Fulgoroidea which inhabit salt marshes. Ecol. Entomol. 1:257–266. [4,8,13]

Denno, R. F. 1977. Comparisons of the assemblages of sap feeding insects (Homoptera-Hemiptera) inhabiting two structurally different salt marsh grasses of the genus *Spartina*. Environ. Entomol. 6:359–372. [1,3,10,13]

Denno, R. F. 1978. The optimum population strategy for planthoppers (Homoptera: Delphacidae) in stable marsh habitats. Can. Entomol. 110:135–142. [2,3,4,13,15]

Denno, R.F. 1979. The relation between habitat stability and the migration tactics of planthoppers. Misc. Pubs. Entomol. Soc. Am. 11:41–49. [3,4]

Denno, R. F. 1980. Ecotope differentiation in a guild of sap-feeding insects on the salt marsh grass, *Spartina patens*. Ecology 61:702–714. [i,1,2,3,7,13,15]

Denno, R. F. 1983. Tracking variable host plants in space and time. Pages 291–341. *In* R. F. Denno and M. S. McClure, editors. Variable Plants and Herbivores in Natural and Managed Systems. Academic Press, New York. [2,3,4,6,7,8,10,15]

Denno, R.F. 1985a. The role of host plant condition and nutrition in the migration of phytophagous insects. Pages 151–172 *In* D. R. MacKenzie et al., editors. The Movement and Dispersal of Agriculturally Important Biotic Agents. Claitor's Pub. Div., Inc., Baton Rouge, LA. [2]

Denno, R.F. 1985b. Fitness and migration responses of *Prokelisia marginata* to variable host plants. Pages 623–640. *In* M.A. Rankin, editor. Migration: Mechanisms and adaptive significance. Contributions in Marine Science Volume 27. Marine Science Institute, University of Texas at Austin Press, Port Aransas, TX. [2,3,4,6,9]

Denno, R.F., and H. Dingle. 1981. Considerations for the development of a more general life history theory. Pages 1–6. *In* R.F. Denno and H. Dingle, editors. Insect Life History Patterns: Habitat and Geographic Variation. Springer-Verlag, New York. [4]

Denno, R.F., L.W. Douglass, and D. Jacobs. 1985. Crowding and host plant nutrition: Environmental determinants of wing-form in *Prokelisia marginata*. Ecology 66:1588–1596. [2,4,6,7,8]

Denno, R.F., L.W. Douglass, and D. Jacobs. 1986. Effects of crowding and host plant nutrition on a wing-dimorphic planthopper. Ecology 67:116–123. [1,2,4,6,7]

Denno, R.F., and E.E. Grissell. 1979. The adaptiveness of wing-dimorphism in the salt marsh-inhabiting planthopper, *Prokelisia marginata* (Homoptera: Delphacidae). Ecology 60: 221–36. [2,3,4,6,7,8,10]

Denno, R.F., S. Larsson, and K.L. Olmstead. 1990. Host plant selection in willow-feeding leaf beetles (Coleoptera: Chrysomelidae): Role of enemy-free space and plant quality. Ecology 71:124–137. [1,2]

Denno, R.F. and E.S. McCloud. 1985. Predicting fecundity from body size in the planthopper, *Prokelisia marginata* (Homoptera: Delphacidae). Environ. Entomol. 14:846–849. [2,4, 6,7,15]

Denno, R.F., and M.S. McClure, editors. 1983. Variable Plants and Herbivores in Natural and Managed Systems. Academic Press, New York. [2]

Denno, R. F., K. L. Olmstead, and E. S. McCloud. 1989. Reproductive cost of flight capability: A comparison of life history traits in wing dimorphic planthoppers. Ecol. Entomol. 14:31–44. [i,3,4,6,7,8,9,10,15]

Denno, R. F., M. J. Raupp, and D. W. Tallamy. 1981. Organization of a guild of sap-feeding insects: equilibrium versus non-equilibrium coexistence. Pages 151–181. *In* R. F. Denno and H. Dingle, editors. Insect Life History Patterns: Habitat and Geographic Variation. Springer-Verlag, New York. [2,3,4,7,10,13]

Denno, R. F., M. J. Raupp, D. W. Tallamy, and C. F. Reichelderfer. 1980. Migration in heterogeneous environments: Differences in habitat selection between the wing-forms of the dimorphic planthopper, *Prokelisia marginata* (Homoptera: Delphacidae) Ecology 61:859–867. [i,2,3,4,6,8,9,10]

Denno, R. F., and G. K. Roderick. 1990. Population biology of planthoppers. Annu. Rev. Entomol. 35:489–520. [i,1,2,3,4,6,7,8,9,10,14,15,16]

Denno, R.F., and G.K. Roderick. 1991. Influence of patch size, vegetation texture, and host plant architecture on the diversity, abundance, and life history styles of sap-feeding herbivores. Pages 169–196. *In* S. Bell, E.D. McCoy, and H.R. Mushinsky, editors. Habitat structure: The physical arrangement of objects in time and space. Chapman and Hall, London. [1,3,4,7,13]

Denno, R.F., and G.K. Roderick. 1992. Density-related dispersal in planthoppers: Effects of interspecific crowding. Ecology 73:1323–1334. [i,4,6,7,15]

Denno, R.F., G.K. Roderick, K.L. Olmstead, and H.G. Döbel. 1991. Density-related migration in planthoppers (Homoptera: Delphacidae): The role of habitat persistence. Am. Natur. 138: 1513–1541. [i,1,2,3,4,6,7,8,9,10,15]

Denno, R. F., M. E. Schauff, S. W. Wilson, and K. L. Olmstead. 1987. Practical diagnosis and natural history of two sibling salt marsh-inhabiting planthoppers in the genus *Prokelisia* (Homoptera: Delphacidae). Proc. Entomol. Soc. Wash. 89:687–700. [1,4,7,10]

Dent, D. 1991. Insect Pest Management. CABI, Wallingford. [14]

Dethier, V.G. 1954. Evolution of feeding preferences in phytophagous insects. Evolution 8:33–54. [1,2]

Dethier, V.G. 1970. Chemical interactions between plants and insects. Pages 83–102. *In* E. Sodenheimer and J.B. Simeone, editors. Chemical Ecology. Academic Press, New York. [2]

de Vrijer, P. W. F. 1981. Reproductive isolation in the genus *Javesella* Fenn. Acta Entomol. Fenn. 38:50–51. [1,5,14]

de Vrijer, P.W.F. 1984. Variability in calling signals of the planthopper *Javesella pellucida* (F.) (Homoptera: Delphacidae) in relation to temperature, and consequences for species recognition during distant communication. Netherlands. J. Zool. 34:388–406. [i,5]

de Vrijer, P. W. F. 1986. Species distinctiveness and variability of acoustic calling signals in the planthopper genus *Javesella* (Homoptera: Delphacidae). Netherlands J. Zool. 36:162–175. [i,6]

Dewey, J. 1938. Experience and education. Macmillan Publishers, New York. [20]

de Winter, A.J. 1992. The genetic basis and evolution of acoustic mate recognition signals in a *Ribautodelphax* planthopper (Homoptera, Delphacidae): 1. The female call. J. Evol. Biol. 5:249–265. [i,5]

de Winter, A.J., and T. Rollenhagen. 1990. The importance of male and female acoustic behavior for sexual isolation in *Ribautodelphax* planthoppers (Homoptera, Delphacidae). Biol. J. Linn. Soc. Lond. 40:191–206. [5,6]

Dilts, D. R. 1984. Training: reschooling society? Prisma No. 38, Jakarta, Indonesia. [20]

Ding, J. 1982. Two new species of the tribe Tropidocephalini (Homoptera, Delphacidae). J. Nanjing Agric. Coll. 4:42–45. [1]

Ding, J., and G. Hu. 1982. A new species of the genus *Bambusiphaga* from Yunnan (Homoptera: Delphacidae). Acta Entomol. Sinica 25:443–444. [1]

Ding, J., and C. Kuoh. 1981. New species of *Stenocranus* from China (Homoptera: Delphacidae). Acta Zootaxonom. Sinica 6:74–84. [1]

Ding, J., and L. Tian. 1980. A new species of the genus *Opiconsiva* Distant (Homoptera: Delphacidae). J. Nanjing Agric. Coll. 2:1–3. [1]

Ding, J., and L. Tian. 1981. Homoptera: Delphacidae. Insects of Xizang 1:229–232. [1]

Ding, J., L. Yang, C. Hu, and J. Sheng. 1982. A preliminary observation on the green slender planthopper *Saccharosydne procerus* (Matsumura). J. Nanjing Agric. Coll. 2:1–7 [1]

Ding, Z.Z., M.L. Chen, and P.Y. Li. 1981. The reproductive rate and developmental threshold of the brown planthopper, *Nilaparvata lugens* Stål. Acta Entomol. Sinica 24:152–159. [4]

Dingle, H. 1972. Migration strategies of insects. Science 175:1327–1335. [4,8]

Dingle, H. 1974. The experimental analysis of migration and life history strategies in insects. Pages 329–342. *In* L. Barton Browne, editor. The Experimental Analysis of Insect Behaviour. Springer-Verlag, New York. [4]

Dingle, H. 1978. Migration and diapause in tropical, temperate, and island milkweed bugs. Pages 254–276. *In* H. Dingle, editor. The Evolution of Insect Migration and Diapause. Springer-Verlag, New York. [4]

Dingle, H. 1980. Ecology and evolution of migration. Pages 2–101. *In* S.A. Gauthreaux, Jr., editor. Animal Migration, Orientation and Navigation. Springer-Verlag, New York. [9]

Dingle, H. 1982. Function of migration in the seasonal synchronization of insects. Entomol. Exp. Appl. 31:36–48. [9]

Dingle, H. 1985. Migration. Pages 375–415. *In* G.A. Kerkut and L.I. Gilbert, editors. Comprehensive Insect Physiology, Biochemistry and Pharmacology, Vol. 9, Behavior, Pergamon Press, New York. [4]

Dingle, H. 1989. The evolution and significance of migratory flight, Pages 100–114. *In* G.J. Goldsworthy and C.H. Wheeler, editors. Insect Flight. CRC Press, Boca Raton, FL. [9]

Dingle, H., N.R. Blakley, and E.R. Miller. 1980. Variation in body size and flight performance in milkweed bugs (*Oncopeltus*). Evolution 34:371–385. [4]

Dixon, A.F.G. 1970. Quality and availability of food for a sycamore aphid populaiton. Pages 271–287. *In* A. Watson, editor. Animal Populations in Relation to their Food Resources. Blackwell Scientific Publications, Oxford, UK. [2]

Dixon, A.F.G. 1985. Aphid ecology. Blackie and Sons, London. [2,3,4,7]

Dixon, A. F. G., and S. McKay. 1970. Aggregation in the sycamore aphid *Drepanosiphum platanoides* (Schr.) (Hemiptera: Aphididae) and its relevance to the regulation of population growth. J. Animal Ecol. 39:439–454. [3]

Djamin, A., and M.D. Pathak. 1967. The role of silica in resistance to Asiatic rice borer, *Chilo suppressalis* (Walker) in rice varieties. J. Econ. Entomol. 60:347–351. [2]

Dlabola, J. 1970. Beitrag zur Taxonomie und Chorologie einiger palaearktischer Zikadenarten (Homoptera, Auchenorrhyncha). Sonderabdruck Mitt. Muchn. Entomol. Ges. (e. V.) 59:90–107. [1]

Dlabola, J. 1979a. *Tshurtburnella*, *Bubastia* und andere verwandte Taxone (Auchenorrhyncha, Issidae). Acta Entomol. Bohemoslov. 76:266–286. [1]

Dlabola, J. 1979b. Neue Zikaden aus Anatolien, Iran und aus Sudeuropaischen Landern (Homoptera: Auchenorrhyncha). Acta Zool. Acad. Sci. Hungaricae 25(3–4):235–257. [1]

Dlabola, J. 1980. Tribus—einteilung, neue Gattungen und Arten der Subf. Issinae in der eremischen zone (Homoptera, Auchenorrhyncha). Acta Mus. Nat. Pragae 36B:173–247. [1]

Dlabola, J. 1981. Ergebnisse der Tschecholowakisch-Iranischen Entomologischen Expeditionen nach dem Iran (1970 und 1973). Acta Entomol. Mus. Nat. Pragae 40:127–311. [1]

Dlabola, J. 1983. Neue mediterrane, meistens anatolische Issiden (Homoptera, Auchenorrhyncha). Acta Entomol. Bohemoslov. 80:114–136. [1]

Döbel, H. G. 1987. The role of spiders in the regulation of salt marsh planthopper populations. M. S. thesis, Department of Entomology, University of Maryland, College Park, MD. [3,6,8,10,15]

Döbel, H. G., and R. F. Denno. In preparation a. The functional response of wolf spiders to prey density: Influence of prey behavior and size on risk of predation. [10]

Döbel, H. G., and R. F. Denno. In preparation b. The influence of habitat structure on spider/planthopper interactions: Functional and numerical responses, stability and suppression. [10]

Döbel, H. G., and R. F. Denno. In preparation c. The numerical response of wolf spiders to *Prokelisia* planthoppers: An experimental field study. [10]

Döbel, H.G., R.F. Denno, and J.A. Coddington. 1990. Spider (Araneae) community structure in an intertidal salt marsh: Differences along an elevational gradient. Environ. Entomol. 19: 1356–1370. [2,3,10]

Dobzhansky, T. 1937. Genetics and the Origin of Species. Columbia Univ. Press, New York. [5]

Dobzhansky, T., and O. A. Pavlovsky. 1957. Indeterminate outcome of certain experiments on *Drosophila* populations. Evolution 7:198–210. [17]

Doebley, J., M. Durbin, E. Golenberg, M. Clegg, and D. Ma. 1990. Evolutionary analysis of the larage subunit of carboxylase (*rbc*L) nucleotide sequence among the grasses (Gramineae). Evolution 44: 1097–1108. [1]

Doering, K.C. 1932. The genus *Acanalonia* in America north of Mexico (Fulgoridae, Homoptera). Ann. Entomol. Soc. Amer. 25:758–786. [1]

Doering, K. C. 1936. A contribution to the taxonomy of the subfamily Issinae in America north of Mexico (Fulgoridae, Homoptera). Univ. Kansas Sci. Bull. 24:421–467. [1]

Doering, K. C. 1938. A contribution to the taxonomy of the subfamily Issinae in America north of Mexico (Fulgoridae, Homoptera). Part II. Univ. Kansas Sci. Bull. 25:447–575. [1]

Doering, K. C. 1940. A contribution to the taxonomy of the subfamily Issinae in America north of Mexico (Fulgoridae, Homoptera). Part III. Univ. Kansas Sci. Bull. 26:83–167. [1]

Doering, K. C. 1941. A contribution to the taxonomy of the subfamily Issinae in America north of Mexico (Fulgoridae, Homoptera). Part IV. Univ. Kansas Sci. Bull. 27:185–233. [1]

Doering, K. C. 1958. A new species of *Hysteropterum* from grape (Issidae, Fulgoroidea, Homoptera). J. Kansas Entomol. Soc. 31:101–103. [1]

Domingo, I.T., Heinrichs, E.A., and R.C. Saxena. 1983. Occurrence of brown planthopper on *Leersia hexandra* in the Philippines. Int. Rice Res. Newsl. 8:17. [15]

Doutt, R. L., and J. Nakata. 1965. Overwintering refuge of *Anagrus epos* (Hymenoptera: Mymaridae). J. Econ. Entomol. 58:586. [11,14]

Doutt, R. L., and J. Nakata. 1973. The *Rubus* leafhopper and its egg parasitoids: an endemic biotic system useful in grape-pest management. Environ. Entomol. 2:381–386. [11,14]

Downes, W. 1927. A preliminary list of the Heteroptera and Homoptera of British Columbia. Proc. Entomol. Soc. British Columbia 23:3–22. [1]

Dozier, H. L. 1922. A synopsis of the genus *Stenocranus*, and a new species of *Mysidia* (Homoptera). Ohio J. Sci. 22:69–82. [1]

Dozier, H. L. 1926. The Fulgoridae or plant-hoppers of Mississippi, including those of possible occurrence. A taxonomic, biological, ecological, and economic study. Mississippi Agric. Exp. Stn. Tech. Bull. 14:1–152. [1]

Dozier, H. L. 1931. New and interesting West Indian Homoptera. Am. Mus. Nov. 510:1–24. [1]

Draper, N.R., and H. Smith. 1981. Applied Regression Analysis, second edition. Wiley, New York. [7,10]

Drosopoulos, S. 1976. Triploid pseudogamous biotype of the leafhopper *Muellerianella fairmairei*. Nature 263:499–500. [5,6]

Drosopoulos, S. 1977. Biosystematic studies on the *Muellerianella* complex (Delphacidae, Homoptera Auchenorrhyncha). Meded. Landbouwhogeschool Wageningen, Nederland 77–14, 1–134. [1,4,6,7,8,15]

Drosopoulos, S. 1978. Laboratory synthesis of a pseudogamous triploid "species" of the genus *Muellerianella* (Homoptera, Delphacidae). Evolution 32:916–920. [5,6]

Drosopoulos, S. 1982. Hemipterological studies in Greece. Part II. Homoptera–Auchenorrhyncha. On the family Delphacidae. Marburger Entomol. Publ. 1:35–88. [4]

Drosopoulos, S. 1983. Some notes on the genera *Muellerianella* and *Florodelphax* from Greece (Homoptera: Delphacidae) with a description of *Florodelphax mourikisi* n. sp. from Ikaria island. Entomol. Ber. 43:72–75. [1]

Drosopoulos, S. 1985. Acoustic communication and mating behavior in the *Muellerianella* complex (Homoptera—Delphacidae). Behavior 94:183–201. [i,5,6]

Drosopoulos, S., M. Asche, and H. Hoch. 1983. Contribution to the planthopper fauna of Greece (Homoptera, Auchenorrhyncha, Fulgoromorpha, Delphacidae). Ann. Inst. Phytopathol. Benaki (N. S.) 14:19–68. [1,4]

DuBose, W. P. 1960. The genus *Delphacodes* Fieber in North Carolina (Homoptera: Delphacidae). J. Elisha Mitchell Sci. Soc. 76:36–63. [1]

Duffey, S. S., and K. A. Bloem. 1986. Plant defense–herbivore–parasite interactions and biological control. Pages 135–183. *In* M. Kogan, editor. Ecological Theory and Integrated Pest Management Practice. Wiley, New York. [18]

Duffus, J. E. 1987. Whitefly transmission of plant viruses. Pages 73–91. *In* K. F. Harris, editor. Current Topics in Vector Research, Vol. 4, Springer-Verlag, New York. [12]

Dumbleton, L. J. 1934. The apple leafhopper (*Typhlocyba australis* Frogg.). New Zealand J. Sci. Technol. 16:30–38. [13]

Dumbleton, L. J. 1937. Apple leafhopper investigations. New Zealand J. Sci. Technol. 18:866–877. [13]

Dung, W. S. 1981. A general survey on seasonal migrations of *Nilaparvata lugens* (Stål) and *Sogatella furcifera* (Horvath) (Homoptera: Delphacidae) by means of airplane collections. Acta Phytophyl. Sinica 8:74–82. [9]

D'Urso, V., and A. Guglielmino. 1986. Sviluppo postembrionale di *Matutinus putoni* (Costa, A., 1888) (Homoptera, Delphacidae) e note sulla sua biologia. Animalia 13(1/3):77–93. [1]

D'Urso, V., S. Ippolito, and F. Lombardo. 1984. Studio faunistico-ecologico sugli eterotteri terrestri ed omotteri Auchenorrinchi di Monte Manfre (Etna, Sicilia). Animilia (Catania) 11(1–3):155–194. [1]

Dutt, N., and P. Giri. 1978. A dryinid parasite of brown planthopper of paddy, *Nilaparvata lugens* Stål (Homoptera: Delphacidae). Sci. Cult. 144:127–128. [13]

Duvall, M.R., M.T. Clegg, M.W. Chase, W.D. Clark, J.W. Kress, E.A. Zimmer, H.G. Hills, L.E. Eguiarte, J.F. Smith, B.S. Gaut, and G.H. Learn, Jr. In press. Molecular phylogenies for the monocotyledons constructed from *rbc*L sequence data. Ann. Missouri Bot. Gard. [1]

Dyck, V. A., B. C. Misra, S. Alam, C. N. Chen, C. Y. Hsieh, and R. S. Rejesus. 1979. Ecology of the brown planthopper in the tropics. Pages 61–98. *In* Brown Planthopper: Threat to Rice Production in Asia. International Rice Research Institute, Los Baños, Philippines. [2,3,4,7,8,9,10,14,15,17]

Dyck, V. A., and G. C. Orlido. 1977. Control of the brown planthopper (*Nilaparvata lugens*) by natural enemies and timely application of narrow-spectrum insecticides. Pages 58–72. *In* The Rice Brown Planthopper. Food and Fertilizer Technology Centre for the Asian and Pacific Region. Taipei, Taiwan. [10,14]

Dyck, V.A., and B. Thomas. 1979. The brown planthopper problem. Pages 3–17. *In* Brown Planthopper: Threat to Rice Production in Asia. International Rice Research Institute, Los Baños, Philippines. [i,4,7,10,14,16,17,18]

Eberhard, W.G. 1985. Sexual selection and animal genitalia. Harvard University Press, London. [5]

Eberhardt, L.L. 1970. Correlation, regression, and density dependence. Ecology 51:306–310. [7]

Ehler, L.E. 1976. The relationship between theory and practice in biological control. Bull. Entomol. Soc. Am. 22:319–321. [14]

Ehler, L.E. 1977. Enemies of the cabbage looper on cotton in the San Joaquin Valley. Hilgardia 45:73–106. [10]

Ehler, L.E. 1982. Foreign exploration in California. Environ. Entomol. 11:525 [14]

Ehler, L.E. 1990. Introduction strategies in biological control of insects. Pages 111–134. *In* M. Mackauer, L.E. Ehler, and J. Roland, editors. Critical issues in biological control. Intercept, Andover, UK. [10,14]

Ehler, L.E., and R.W. Hall. 1982. Evidence for competitive exclusion of introduced natural enemies in biological control. Environ. Entomol. 11:1–4. [13,14]

Ehler, L.E., and J.C. Miller. 1978. Biological control in temporary agroecosystems. Entomophaga 23:207–212. [10]

Ehlich, P.R., and P.H. Raven. 1964. Butterflies and plants: a study in coevolution. Evolution 18:586–608. [1]

Ekbom, B.S., and X. Rumei. 1990. Sampling and spatial patterns of whiteflies. Pages 107–121. *In* D. Gerling, editor. Whiteflies: Their bionomics, pest status and management. Intercept Ltd., Andover, UK. [7]

El-Shazly, N.Z. 1972. Der Einfluz von Ernahrung und alterbder Muttertieres auf die hamocytare Abwehrreaktion von *Neomyzus circumflexus* (Buck.). Entomophaga 17:203–209. [14]

Emden, H.F. van, and C.H. Wearing. 1965. The role of the aphid host plant in delaying economic damage levels in crops. Ann. Appl. Biol. 56:323–324. [14]

Emeljanov, A. F. 1964. New Cicadina from Kazakstan (Homoptera, Auchenorrhyncha). Trudy Zool. Inst. 34:1–51. [1]

Emeljanov, A. F. 1966. New Palearctic and certain Nearctic leafhoppers (Homoptera, Auchenorrhyncha). Entomol. Rev. 45:53–72. [1]

Emeljanov, A. F. 1977. Leaf-hoppers (Homoptera, Auchenorrhyncha) from the Mongolian People's Republic based mainly on materials of the Soviet-Mongolian zoological expeditions (1967–1969). Insects of Mongolia 5:96–19. [1]

Emeljanov, A. F. 1979. New genera and species of leafhoppers (Homoptera, Auchenorrhyncha) from the USSR and Mongolia. Entomol. Rev. 58:220–233. [1]

Emeljanov, A. F. 1981. Two new species of the genus *Nymphorgerius* (Homoptera, Auchenorrhyncha, Dictyopharidae) from the south of the the Middle Asia. Proc. Zool. Inst. 105: 3–6. [1]

Emeljanov, A. F. 1982. Fulgoroidea (Homoptera) collected in the Mongolian People's Republic by the entomofaunistical group of the Soviet-Mongolian complex biological expedition in 1970–1975. Insects of Mongolia 8:68–123. [1]

Emeljanov, A. F. 1985. A contribution to knowledge of the families Kinnaridae and Meenoplidae (Homoptera, Fulgoroidea). Entomol. Rev. 64:49–65. [1]

Emeljanov, A. F. 1987. The phylogeny of the Cicadina (Homoptera, Cicadina) based on comparative morphological data. Morphological Principles of Insect Phylogeny. Trans. All-Union Entomol. Soc. 69:19–109. [1]

Emeljanov, A. F. 1991. An attempt to construct a phylogenetic tree for planthoppers (Homoptera, Cicadina). Entomol. Rev. 70:24–28. [1]

Emlen, S. T. 1976. Lek organization and mating strategies in the bullfrog. Behavioral Ecol. Sociobiol. 1:283–313. [6]

Emlen, S. T., and L. W. Oring. 1977. Ecology, sexual selection, and the evolution of mating systems. Science 197:215–223. [6]

Endler, J. A. 1986. Natural Selection in the Wild. Princeton University Press. Princeton, NJ. [6,17]

Endo, S., T. Masuda T., and H. Kazano. 1988a. Development and mechanism of insecticide resistance in rice brown planthoppers selected with malathion and MTMC. J. Pestic. Sci. 13:239–245. [i,16]

Endo, S., T. Nagata, S. Kawabe, and H. Kazano. 1988b. Changes of insecticide susceptibility of the white backed planthopper *Sogatella furcifera* Horvath (Homoptera: Delphacidae) and the brown planthopper *Nilaparvata lugens* Stål (Homoptera: Delphacidae). Appl. Entomol. Zool. 23:417–421. [i,16]

Esaki, T., and S. Hashimoto. 1937. Studies on rice leafhoppers. I. Biology and natural enemies. Nojikairyoshiryo No. 127. Ministry of Agriculture and Forestry, Japan. [14]

Escalada, M. M., and P. E. Kenmore. 1988. Communicating integrated pest control to rice farmers at the village level. Pages 221–228. *In* P. S. Teng and K. L. Heong, editors. Pesticide Management and Integrated Pest management in Southeast Asia. Consortium for International Crop Protection, College Park, MD. [20]

Eschrich W. 1970. Biochemistry and fine structure of phloem in relation to transport. Ann. Rev. Plant Physiol. 21:193–214. [2]

Evans, E. W. 1989. Interspecific interactions among phytophagous insects of tallgrass prairie: an experimental test. Ecology 70:435–444. [13]

Evans, J. W. 1963. The phylogeny of the Homoptera. Annu. Rev. Entomol 8:77–94. [1]

Evans, J. W. 1964. The periods of origin and diversification of the superfamilies of the Homoptera-Auchenorhyncha (Insecta) as determined by a study of the wings of Palaeozoic and Mesozoic fossils. Proc. Linn. Soc. Lond. 175:171–181. [1]

Everett, T.R. 1969. Vectors of hoja blanca virus. Pages 111–121. *In* The Virus Diseases of the Rice Plant. Johns Hopkins Press, Baltimore, MD. [2,4,6]

Exsys Inc. 1985. Expert System Development Package: Users Manual. Exsys Inc. Albuquerque, NM. [19]

Fabellar, L.T., and E.A. Heinrichs. 1984. Toxicity of insecticides to predators of rice brown planthopper, *Nilaparvata lugens* Stål (Homoptera: Delphacidae). Environ. Entomol. 13: 832–837. [10,16]

Fairbairn, D.J. 1984. Microgeographic variation in body size and development time in the waterstrider, *Limnoporus notabilis*. Oecologia 61:126–133. [4]

Fairbairn, D.J. 1992. The origins of allometry: Size and shape polymorphism in the common waterstrider, *Gerris remigis* Say (Heteroptera, Gerridae). Biol. J. Linn. Soc. 45:167–186. [4]

Fairbairn, D.J., and L. Desranleau. 1987. Flight threshold, wing muscle histolysis, and alary polymorphism: correlated traits for dispersal tendency in the Gerridae. Ecol. Entomol. 12:13–24. [4]

Fairbairn, D.J., and D.A. Roff. 1990. Genetic correlations among traits determining migratory tendency in the sand cricket, *Gryllus firmus*. Evolution 44:1787–1795. [4]

Falconer, D.S. 1981. Introduction to Quantitative Genetics. 2nd edition. Longman, New York. [4,18]

Falconer, D.S. 1989. Introduction to quantitative genetics, 3rd. ed. Longman, London. [15]

Falk, B. W., K. S. Kim, and J. H. Tsai. 1988. Electron microscopic and physicochemical analysis of a reo-like virus of the planthopper *Peregrinus maidis*. Intervirology 29:195–206. [12]

Falk, B. W., V. A. Klaassen, and J. H. Tsai. 1989. Complementary DNA cloning and hybridization of maize stripe virus RNAs. Virology 173:338–342. [12]

Falk, B. W., and J. H. Tsai. 1985. Serological detection and evidence for multiplication of maize mosaic virus in the planthopper, *Peregrinus maidis*. Phytopathology 75:852–855. [12]

Falk, B. W., J. H. Tsai, and S. A. Lommel. 1987. Differences in levels of detection for the maize stripe virus capsid and major non-capsid proteins in plant and insect hosts. J. Gen. Virol. 68:1801–1811. [12]

FAO. 1983. Proceedings of the FAO/NCPC Workshop on the Role and Potential of the Filipina in Rice Crop Protection, 2–4 February 1983, Los Baños, Philippines. Available from the FAO Inter-country Programme for Integrated Pest Control in Rice in South and Southeast Asia, P. O. Box 1864, Manila, Philippines. [20]

Farrell, B., and C. Mitter. 1990. Phylogenesis of insect/plant interactions: Have *Phyllobrotica* leaf beetles (Chrysomelidae) and the Lamiales diversified in parallel? Evolution 44:1389–1403. [1]

Farrell, B. D., and C. Mitter. In press. Phylogenetic determinants of insect/plant community diversity. *In* R. Ricklefs and D. Schluter, editors. Historical and Geographic Determinants of Community Diversity. University of Chicago Press, Chicago, IL. [1]

Farrell, J. A. K. 1976. Effects of groundnut crop density on the population dynamics of *Aphis craccivora* Koch (Hemiptera, Aphididae) in Malawi. Bull. Entomol. Res. 66:317–329. [3]

Feeny, P. 1976. Plant apparency and chemical defense. *In* J. Wallace and R. Mansell, editors. Biochemical Interactions Between Plants and Insects. Rec. Adv. Phytochem. 10:1–40. [2]

Felt, E. P. 1916. Coquebert's *Otiocerus* (*Otiocerus coquebertii* Kirby). Bull. New York State Mus. 147:68–70. [1]

Fennah, R. G. 1941. Notes on the flatid genus *Ormenis* in the British Lesser Antilles and Trinidad, with descriptions of new species (Homoptera: Fulgoroidea). Proc. Entomol. Soc. Wash. 43:191–210. [1]

Fennah, R. G. 1942a. New or little-known West Indian Kinnaridae (Homoptera: Fulgoroidea). Proc. Entomol. Soc. Wash. 44:99–110. [1]

Fennah, R. G. 1942b. Notes on some West Indian Flatidae. Proc. Entomol. Soc. Wash. 44:155–167. [1]

Fennah, R. G. 1945. The Fulgoroidea, or lanternflies, of Trinidad and adjacent parts of South America. Proc. U.S. Nat. Mus. 95:411–520. [1]

Fennah, R. G. 1948. New pintaliine Cixiidae, Kinnaridae and Tropiduchidae from the Lesser Antilles (Homoptera: Fulgoroidea). Ann. Mag. Nat. Hist. (12). 1:417–437. [1]

Fennah, R. G. 1949a. On a small collection of Fulgoroidea (Homoptera) from the Virgin Islands. Psyche 56:51–65. [1]

Fennah, R. G. 1949b. A new genus of Fulgoroidea (Homoptera) from South Africa. Ann. Mag. Nat. Hist. 12(2):111–120. [1]

Fennah, R. G. 1950. A generic revision of Achilidae (Homoptera: Fulgoroidea) with descriptions of new species. Bull. British Mus. (Nat. Hist.) 1:1–170. [1]

Fennah, R. G. 1952. A revision of *Bladina* Stål (Nogodinidae: Fulgoroidea). Ann. Mag. Nat. Hist. (12) 5:910–928. [1]

Fennah, R. G. 1955. Lanternflies of the family Issidae of the Lesser Antilles (Homoptera: Fulgoroidea). Proc. U.S. Nat. Mus. 105:23–47. [1]

Fennah, R. G. 1956. Homoptera: Fulgoroidea. B. P. Bishop Mus. Ins. Micronesia 6(3)39–211. [1]

Fennah, R. G. 1957. Los insectos de las Islas Juan Fernandez. 29. Fulgoroidea (Homoptera). Rev. Chil. Entomol. 5:375–384. [1]

Fennah, R. G. 1958. Fulgoroidea of south-eastern Polynesia. Trans. Royal Entomol. Soc. Lond. 110:117–220. [1]

Fennah, R. G. 1959a. Delphacidae from the Lesser Antilles (Homoptera: Fulgoroidea). Bull. Brit. Mus. (Nat. Hist.) Entomol. 8:243–265. [1]

Fennah, R. G. 1959b. A new genus of Tettigometridae from southern Rhodesia (Homoptera: Fulgoroidea). Ann. Mag. Nat. Hist. 13:807–809. [1]

Fennah, R. G. 1962a. A new alohine delphacid from San Ambrosio I. (Homoptera: Fulgoroidea). Proc. Biol. Soc. Wash. 75:177–180. [1]

Fennah, R. G. 1962b. A new genus and two new species of Lophopidae from south-east Asia (Homoptera: Fulgoroidea). Ann. Mag. Nat. Hist. 5:725–730. [1]

Fennah, R. G. 1963. The delphacid species-complex known as *Sogata furcifera* (Horvath) (Homoptera: Fulgoroidea). Bull. Entomol. Res. 54:45–79. [1]

Fennah, R. G. 1964. Delphacidae from Madagascar and the Mascarene Islands (Homoptera: Fulgoroidea). Trans. Royal Entomol. Soc. 116:131–150. [1]

Fennah, R. G. 1965a. Delphacidae from Australia and New Zealand (Homoptera: Fulgoroidea). Bull. British Mus. (Nat. Hist.) 17:1–59. [1]

Fennah, R. G. 1965b. New species of Fulgoroidea from the West Indies. Trans. Royal Entomol. Soc. Lond. 117:95–126. [1]

Fennah, R. G. 1967a. A new genus and species of Delphacidae (Homoptera: Fulgoroidea) infesting molasses grass in Kenya. Bull. Entomol. Res. 57:353–356. [1]

Fennah, R. G. 1967b. New species and new records of Fulgoroidea (Homoptera) from Samoa and Tonga. Pac. Ins. 9:29–72. [1]

Fennah, R. G. 1968. A new genus and species of Ricaniidae from Palaeocene deposits in North Dakota. J. Nat. Hist. 2:143–146. [1]

Fennah, R.G. 1969. Damage to sugar cane by Fulgoroidea and related insects in relation to the metabolic state of the host plant. Pages 367–389. *In* J.R. Williams, J.R. Metcalf, R.W. Mungomery, and R. Mathes, editors. Pests of Sugar Cane. Elsevier Publishing Company, New York. [2,4,7,10]

Fennah, R. G. 1974. A new tropiduchid (Homoptera, Fulgoroidea) infesting *Chamaedorea* (Palmaceae). Bull. Entomol. Res. 63:673–675. [1]

Fennah, R. G. 1976. La faune terrestre de l'Ile de Sainte Helene. 3. Fam. Delphacidae. Mus. Royal Afr. Centr., Tervueren Belg. Ann. 215:262–270. [1]

Fennah, R. G. 1978. The higher classification of the Nogodinidae (Homoptera, Fulgoroidea) with the description of a new genus and species. Entomol. Mon. Mag. 113:113–120. [1]

Fennah, R. G. 1980. New and little-known Neotropical Kinnaridae (Homoptera: Fulgoroidea). Proc. Biol. Soc. Wash. 93:674–696. [1]

Fennah, R. G. 1982a. A new species of *Alcestis* (Homoptera: Fulgoroidea: Tropiduchidae) attacking cacao in Brazil. Bull. Entomol. Res. 72:129–131. [1]

Fennah, R. G. 1982b. A tribal classification of the Tropiduchidae (Homoptera: Fulgoroidea), with the description of a new species on tea in Malaysia. Bull. Entomol. Res. 72:631–643. [1]

Fennah, R. G. 1987. A new genus and species of Cixiidae (Homoptera: Fulgoroidea) from Lower Cretaceous amber. J. Nat. Hist. 21:1237–1240. [1]

Fenner, F. 1976. Classification and nomenclature of viruses; 2nd Report of the International Committee on the Taxonomy of Viruses. Intervirology 7: 1–116. [12]

Fernando, H. E. 1975. The brown planthopper problem in Sri Lanka. Rice Entomol. Newsl. 2:34–36. [3]

Fisher, R. A. 1930. The general theory of natural selection. Clarendon Press, Oxford, UK. [15]

Fisk, J. 1980. Effects of HCN, phenolic acids and related compounds in *Sorghum bicolor* on the feeding behavior of the planthopper *Peregrinus maidis*. Entomol. Exp. Appl. 27:211–222. [2,14]

Fisk, J., E.A. Bernays, R.F. Chapman, and S. Woodhead. 1981. Report of studies on the feeding biology of *Peregrinus maidis*. Centre for Overseas Pest Research/ International Crops Research Institute for Semi-Arid Tropics, London (COPR Core Programme Project 27). [2,4,7,10,14]

Fletcher, M.J. 1979. Egg types and oviposition behavior in some fulgoroid leafhoppers (Homoptera, Fulgoroidea). Australian Entomol. Mag. 6:13–18. [1]

Fletcher, M. J. 1985. Revision of the genus *Siphanta* Stål (Homoptera: Fulgoroidea: Flatidae). Australian J. Zool. Suppl. Ser. 110:1–94. [1]

Fletcher, M. J. 1988. The Australian genera of Flatidae (Homoptera, Fulgoroidea). Gen. Appl. Entomol. 20:9–32. [1]

Flynn, J. E., and J. P. Kramer. 1983. Taxonomic study of the planthopper genus *Cedusa* in the Americas (Homoptera: Fulgoroidea: Derbidae). Entomography 2:121–260. [1]

Foelix, R. F. 1982. Biology of Spiders. Harvard University Press, Cambridge, MA. [6]

Force, D.C. 1972. *r*- and *K*-strategists in endemic host-parasitoid communities. Bull. Entomol. Soc. Am. 18: 135–137. [14]

Foster, J. E., H. W. Ohm, F. L. Patterson, and P. L. Taylor. 1991. Effectiveness of deploying single gene resistances in wheat for controlling damage by the Hessian fly (Diptera: Cecidomyiidae). Environ. Entomol. 20:964–969. [18]

Foster, S., L.J. Goodman, and J.G. Duckett. 1983a. Ultrastructure of sensory receptors on the labium of the rice brown planthopper. Cell Tissue Res. 230:353–366. [2]

Foster, S., L.J. Goodman, and J.G. Duckett. 1983b. Sensory receptors associated with the stylets and cibarium of the rice brown planthopper, *Nilaparvata lugens*. Cell Tissue Res. 232: 111–119. [2]

Fowler, S. V. 1987. Field studies on the impact of natural enemies on brown planthopper populations on rice in Sri Lanka. Proc. 6th Auchen. Meeting, Turin, Italy. 6:567–574. [10]

Fowler, S.V., M.F. Claridge, J.C. Morgan, I.D.R. Peries, and L. Nugaliyadde. 1991. Egg mortality of the brown planthopper, *Nilaparvata lugens* (Homoptera: Delphacidae) and green leafhoppers, *Nephotettix* spp. (Homoptera: Cicadellidae), on rice in Sri Lanka. Bull. Entomol. Res. 81:161–169. [8]

Francki, R. I. B., R. G. Milne, and T. Hatta. 1985. Atlas of Plant Viruses, Vol. 1. CRC Press, Boca Raton, FL. [12]

Free, C. A., J.R. Beddington, and J.H. Lawton. 1977. On the inadequacy of simple models of mutual interference for parasitism and predation. J. Anim. Ecol. 46:543–554. [11]

Freire, P. 1970. Pedagogy of the Oppressed. Herder and Herder, New York. [20]

Freire, P. 1981. Education for Critical Consciousness. The Continuum Publishing Corporation, New York. [20]

Freytag, P. H. 1985. The insect parasites of leafhoppers and related groups. Pages 423–467. *In* L. R. Nault and J. G. Rodriguez, editors. The Leafhoppers and Planthoppers. Wiley, New York. [13]

Freytag, P. H. 1986. The parasites of *Graminella nigrifrons* (Forbes). Pages 233–255. *In* M. R. Wilson and L. R. Nault, editors. Proceedings of the second international workshop on leafhoppers and planthoppers of economic importance. Commonwealth Institute of Entomology, London. [13]

Fukuda, H., E. Hama, T. Kuzuya, A. Takahashi, M. Takahashi, H. Tanaka, M. Wakabayashi, and Y. Watanabe. 1984. The Life Histories of Butterflies in Japan. IV. Hoikusha, Osaka, Japan. [2]

Fukuda, K. 1934. Studies on *Liburnia oryzae* Mats. Bull. Dept. Agric. Govt. Res. Inst. Formosa. 99:1–19. [4]

Fukumorita, T., and M. Chino. 1982. Sugar, amino acid and inorganic contents in the rice phloem sap. Plant Cell Physiol. 23:273–283. [2]

Fukumorita, T., Y. Noziri, H. Haraguchi, and M. Chino. 1983. Inorganic content in rice phloem sap. Soil Sci. Plant Nutr. 29:185–192. [2]

Fullaway, D.T. 1918. The corn planthopper (*Peregrinus maidis* Ashm.). Bull. Board Comm. Agric. For. Hawaii Div. Entomol. 4:1–16. [4,10,14]

Fullaway, D.T. 1940. An account of the reduction of the immigrant taro leafhopper (*Megamelus proserpina*) population to insignificant numbers after introduction and establishment of the egg-sucking bug *Cyrtorhinus fulvus*. Proc. Sixth Pacific Sci. Congr. 4:345–346. [10,14]

Fullaway, D.T., and N.L.H. Krauss. 1945. Common Insects of Hawaii. Honolulu: Tongg Publishing Co., Honolulu. [14]

Futuyma, D.J. 1983. Evolutionary interactions among herbivorous insects and plants. Pages 207–231. *In* D.J. Futuyma and M. Slatkin, editors. Coevolution. Sinauer, Sunderland, MA. [1,2]

Futuyma, D.J., and S.J. McCafferty. 1990. Phylogeny and the evolution of host plant associations in the leaf beetle genus *Ophraella* (Coleoptera, Chrysomelidae). Evolution 44:1885–1913. [1]

Futuyma, D.J., and S.C. Peterson. 1985. Genetic variation in the use of resources by insects. Annu. Rev. Entomol. 30:217–238. [1,15]

Futuyma, D.J., and T.E. Philippi. 1987. Genetic variation and covariation in responses to host plants by *Alsophila pometaria* (Lepidoptera: Geometridae). Evolution 41:269–279. [2]

Gaedigk-Nitschko, K., G. Adam, and K.-W. Mundry. 1988. Role of the spike protein from potato yellow dwarf virus during infection of vector cell monolayers, Pages 91–97. *In* J. Mitsuhashi, editor. Invertebrate Cell Systems in Application, Vol. 1. CRC Press, Boca Raton, FL. [12]

Gagne, W. 1972. *Lamenia caliginea* (Stål). Proc. Hawaiian Entomol. Soc. 21:149. [1]

Gallagher, K.D. 1988. Effects of host plant resistance on the microevolution of the rice brown planthopper, *Nilaparvata lugens* (Stål) (Homoptera: Delphacidae). Ph.D. Dissertation. University of California, Berkeley. [15,17]

Gallun, R. L. 1977. Genetic basis of Hessian fly epidemics. Ann. N.Y. Acad. Sci. 287:223–229. [18]

Gallun, R. L., and H. D. Khush. 1980. Genetic factors affecting expression and stability of resistance. Pages 64–85. *In* F. G. Maxwell and P. R. Jennings, editors. Breeding Plants Resistant to Insects. Wiley, New York. [18]

Gallun, R. L., K. J. Starks, and W. D. Guthrie. 1975. Plant resistance to insects attacking cereals. Annu. Rev. Entomol. 20:337–357. [17]

Galvez, G. E. 1968. Transmission studies of the hoja blanca virus with highly active, virus-free colonies of *Sogatodes oryzicola*. Phytopathology 58: 818–821. [12]

Garg, A.K., and G.R. Sethi. 1983. Population build up and the effect of insecticidal treatments on *Brumoides suturalis* (Fabricius)—a predator of paddy pests. Indian J. Entomol. 46:254–256. [10]

Gavarra, M.R., and R.S. Raros. 1975. The predatory wolf spider, *Lycosa pseudoannulata* Boes. et Str. (Araneae: Lycosidae). Philippine Entomol. 2:427–447. [10]

Georghiou G.P., and C.E. Taylor 1977. Pesticide resistance as an evolutionary phenomenon. Pages 759–785. *In* Proc. XV Intl. Cong. Entomol. Entomological Society of America, College Park, MD. [16]

Ghauri, M. S. K. 1964. The male genitalia of *Phalix titan* Fennah (Homoptera: Tettigometridae). Proc. Royal Entomol. Soc. Lond. (B) 33(3–4):53–55. [1]

Ghauri, M. S. K. 1966. *Zophiuma lobulata* sp. n. (Lophopidae: Homoptera), a new pest of coconut in New Guinea. Ann. Mag. Nat. Hist. 13(9):557–561. [1]

Ghauri, M. S. K. 1971 Two new species of the genus *Ketumala* Distant (Hom., Flatidae), one attacking tea in south India. Bull. Entomol. Res. 60:635–638. [1]

Ghauri, M. S. K. 1973. Taxonomic notes on a collection of Fulgoroidea from tea in southern India. Bull. Entomol. Res. 62:541–544. [1]

Gibson, D. O., and G.S. Mani. 1984. An experimental investigation of the effects of selective predation by birds and parasitoid attack on the butterfly *Danaus chrysippus* (L.). Proc. Roy. Soc. Lond. 221:31–51. [11]

Giffard, W.M. 1922. The distribution and island endemism of Hawaiian Delphacidae (Homoptera) with additional lists of their food plants. Proc. Hawaii. Entomol. Soc. 5:103–118. [1,2,3,4]

Gildow, F. E. 1987. Virus-membrane interactions involved in circulative transmission of luteoviruses by aphids, Pages 93–120. *In* K. F. Harris, editor. Current Topics in Vector Research, Vol. 4. Springer-Verlag, New York. [12]

Gildow, F. E., and C. J. D'Arcy. 1988. Barley and oats as reservoirs for an aphid virus and the influence on barley yellow dwarf virus transmission. Phytopathology 78: 811–816. [12]

Gingery, R. E. 1988. The rice stripe virus group, Pages 297–329. *In* R. G. Milne, editor. The Plant Viruses, Vol. 4. Plenum Press, New York. [12]

Gingery, R. E., L. R. Nault, J. H. Tsai, and R. J. Lastra. 1979. Occurrence of maize stripe virus in the United States and Venezuela. Plant Dis. Reptr. 63: 341–343. [12]

Godfray, H. C. J., and M.P. Hassell. 1989. Discrete and continuous insect populations in tropical environments. J. Anim. Ecol. 58:153–174. [11]

Goeden, R. D., and L.T. Kok. 1986. Comments on a proposed "new" approach for selecting agents for the biological control of weeds. Can. Entomol. 118:51–58. [14]

Goeden, R. D., and D. W. Ricker. 1974a. The phytophagous insect fauna of the ragweed, *Ambrosia acanthicarpa*, in southern California. Environ. Entomol. 3:827–834. [1]

Goeden, R. D., and D. W. Ricker. 1974b. The phytophagous insect fauna of the ragweed, *Ambrosia chamissonis*, in southern California. Environ. Entomol. 3:835–839. [1]

Goeden, R. D., and D. W. Ricker. 1975. The phytophagous insect fauna of the ragweed, *Ambrosia confertiflora*, in southern California. Environ. Entomol. 4:301–306. [1]

Goeden, R. D., and D. W. Ricker. 1976a. The phytophagous insect fauna of the ragweed, *Ambrosia dumosa*, in southern California. Environ. Entomol. 5:45–50. [1]

Goeden, R. D., and D. W. Ricker. 1976b. The phytophagous insect faunas of the ragweeds, *Ambrosia chenopodiifolia*, *A. eriocentra*, and *A. ilicifolia*, in southern California. Environ. Entomol. 5:923–930. [1]

Goeden, R. D., and D. W. Ricker. 1986a. Phytophagous insect fauna of the desert shrub *Hymenoclea salsola* in southern California. Ann. Entomol. Soc. Am. 79:39–47. [1]

Goeden, R. D., and D. W. Ricker. 1986b. Phytophagous insect faunas of the two most common native *Cirsium* thistles, *C. californicum* and *C. proteanum*, in southern California. Ann. Entomol. Soc. Am. 79:39–47. [1]

Goeden, R. D., and D. W. Ricker. 1989. Phytophagous insect faunas of the desert shrubs *Bebbia juncea* and *Trixis californica* in southern California. Ann. Entomol. Soc. Am. 82:325–331. [1]

Goodell, G. 1983. Improving administrators' feedback concerning extension, training and research relevance at the local level: new approaches and findings from Southeast Asia. Agric. Admin. 13:39–55. [20]

Goodell, G. 1984. Challenges to international pest management research and extension in the Third World: do we really want IPM to work? Bull. Entomol. Soc. of Am. 30(3):18–26. [20]

Goodell, G. E., P. E. Kenmore, J. A. Litsinger, J. P. Bandong, C. G. Dela Cruz, and M. D. Lumaban. 1982. Rice insect pest management technology and its transfer to small-scale farmers in the Philippines. Pages 25–42. *In* Report of an Exploratory Workshop on the Role of Anthropologists and Other Social Scientists in Interdisciplinary Teams Developing Improved Food Production Technology. International Rice Research Institute, Los Baños, Philippines. [20]

Goodell, G. E., J. A. Litsinger, and P. E. Kenmore. 1981. Evaluating integrated pest management technology through interdisciplinary research at the farmer level. Pages 72–75. *In* Proceedings of the Conference on Future Trends of Integrated Pest Management, 30 May–4 June 1980, Bellagio, Italy. IOBC Special Issue. Centre for Overseas Pest Research, London, England. [20]

Gould, F. 1978. Predicting the future resistance of crop varieties to pest populations: a case study of mites and cucumber. Environ. Entomol. 7:622–626. [15]

Gould, F. 1983. Genetics of plant-herbivore systems: Interactions between applied and basic study. Pages 599–653. *In* R.F. Denno and M.S. McClure, editors. Variable Plants and Herbivores in Natural and Managed Systems. Academic Press, New York. [15]

Gould, F. 1986. Simulation models for predicting durability of insect-resistant germ plasm: A deterministic diploid model. Environ. Entomol. 15:1–10. [18]

Gould, F. 1991a. Arthropod behavior and the efficacy of plant protectants. Annu. Rev. Entomol. 36:305–330. [2]

Gould, F. 1991b. The evolutionary potential of crop pests. Am. Sci. 79:496–507. [18]

Gould, F., G. G. Kennedy, and M. T. Johnson. 1991. Effects of natural enemies on the rate of herbivore adaptation to resistant host plants. Entomol. Exp. Appl. 58:1–14. [10,15,18]

Graham, S. A., and L.G. Baumhofer. 1927. The pine tip moth in Nebraska National Forest. J. Agric. Res. 35:323–333. [11]

Greathead, D. J. 1970. A study of the host relations of *Halictophagus pontifex* Fox (Strepsiptera), a parasite of Cercopidae (Hem., Aphrophorinae), in Uganda. Bull. Entomol. Res. 60:33–42. [11,13]

Greathead, D.J. 1971. A review of biological control in the Ethiopian region. Tech. Commun. of the Commonw. Inst. Biol. Contr. 5. 162 p., London, UK. [14]

Greathead, D. J. 1983. Natural enemies of *Nilaparvata lugens* and other leaf- and planthoppers in tropical agroecosytems and their impact on pest populations. Pages 371–383. *In* W.J. Knight, editor. Proc. 1st International Workshop on Leafhoppers and Planthopper of Economic Importance. Commonwealth Institute of Entomology, London, UK. [10,13]

Greathead, D.J. 1986. Parasitoids in classical biological control. Pages 290–318. *In* J. Waage and D. Greathead, editors. Insect Parasitoids. Academic Press, London. [10,14]

Greathead, D.J., and J.K. Waage. 1983. Opportunities for biological control of biological pests in developing countries. World Bank Technical Paper 11. [14]

Greber, R. S. 1979. Digitaria striate virus—a rhabdovirus of grasses transmitted by *Sogatella kolophon* (Kirk.) Austral. J. Agric. Res. 30: 43–51. [12]

Griffiths, K. J., and C.S. Holling. 1969. A competition submodel for parasites and predators. Can. Entomol. 101:785–818. [11]

Gromko, M. H., D. G. Gilbert, and R. C. Richmond. 1984. Sperm transfer and use in the multiple mating system of *Drosophila*. Pages 371–426. *In* R. L. Smith, editor. Sperm Competition and the Evolution of Animal Mating Systems. Academic Press, New York. [6]

Gross, P. 1991. Influence of target pest feeding niche on success rates in classical biological control. Environ. Entomol. 20:1217–1227. [14]

Guldemond, J.A. 1990. Evolutionary genetics of the aphid *Cryptomyzus*, with a preliminary analysis of the inheritance of host plant preference, reproductive performance, and host-alternation. Entomol. Exp. Appl. 57:65–76. [15]

Gunthart, H. 1984. Zikaden (Hom. Auchenorrhyncha) aus der alpinen Hohenstufe der Schweizer Zentralalpen. Mitt. Schweizer. Entomol. Ges. 57:129–130. [1]

Gunthart, H. 1987. Zikaden (Auchenorrhyncha). Oekolog. Untersuch. Unterengadin 12:203–299. [1]

Gunthart, H. 1990. Hom. Auchenorrhyncha, collected in Istria—Yugoslavia in spring 1974. Scopolia Suppl. 1:97–99. [1]

Gutierrez, A. P. 1986. Analysis of the interaction of host plant resistance, phytophagous and entomophagous species. Pages 198–215. *In* D. J. Boethel and R. D. Eikenbary, editors. Interaction of Host Plant Resistance and Parasitoids and Predators of Insects. Halsted Press, New York. [18]

Gwynne, D. T. 1987. Sex-biased predation and the risky mate-locating behaviour of male tick-tock cicadas (Homoptera:Cicadidae). Animal Behaviour 35:571–576. [6]

Hacker, H. 1925a. The life history of *Oliarus felis* Kirk (Homoptera). Mem. Queensland Mus. 8:113–114 (+2 plates). [1]

Hacker, H. 1925b. *Platybrachys leucostigma* Walk (Homoptera). Mem. Queensland Mus. 8:37–42 (+2 plates). [1]

Hairston, N.G., F.E. Smith, and L.B. Slobodkin. 1960. Community structure, population control and competition. Am. Natur. 44:421–425. [7]

Halkka, O., M. Raatikainen, L. Halkka, and T. Raatikainen. 1977. Coexistence of four species of spittle-producing Homoptera. Ann. Zool. Fennici 14:228–231. [3,13]

Hall, R.W., and L.E. Ehler. 1979. Rate of establishment of natural enemies in classical biological control. Bull. Entomol. Soc. Am. 25:280–282. [14]

Hall, R.W., L.E. Ehler, and B. Bisebri-ershadi. 1980. Rate of success in classical biological control of arthropods. Bull. Entomol. Soc. Am. 26:111–114. [14]

Hall, S.M., and D.A. Baker. 1972. The chemical composition of *Ricinus* phloem exudate. Planta 106:131–140. [2]

Halliday, T. R. 1978. Sexual selection and mate choice. Pages 180–213. *In* J. R. Krebs and N. B. Davies, editors. Behavioural ecology. Blackwell Scientific, Oxford, England. [6]

Halliday, T. R. 1983. The study of mate choice. Pages 3–32 *In* P. Bateson, editor. Mate Choice. Cambridge University Press, Cambridge, England. [6]

Hama, H. 1980. Mechanisms of insecticide resistance in green rice leafhopper and small brown planthopper. Rev. Plant Prot. Res. 13:54–73. [16]

Hama, H. 1983. Changed acetylcholinesterase and resistance in leaf- and planthoppers. Pages 203–208. *In* J. Miyamoto, editor. IUPAC Pesticide Chemistry: Human Welfare and the Environment. Pergamon Press, New York. [16]

Hama, H., and A. Hosoda. 1983. High aliesterase activity and low acetylcholinesterase sensitivity involved in organophosphorus and carbamate resistance of the brown planthopper. Appl. Entomol. Zool. 18:475–485. [16]

Hamilton, K.G.A. 1981. Morphology and evolution of the Rhynchotan head (Insecta: Hemiptera, Homoptera). Can. Entomol. 113:953–974. [1]

Hamilton. K.G.A. 1990. Homoptera: Insects from the Santana Formation, Lower Cretaceous, of Brazil. Bull. Amer. Museum Natural History 195:82–122. [1]

Hanks, L.M., and R.F. Denno. 1993. Natural enemies and plant water relations influence the distribution of an armored scale insect. Ecology 74:1081–1091. [10]

Hansen, M. 1987. Escape from the pesticide treadmill: alternatives to pesticides in developing countries. Institute for Consumer Policy Research, Consumers Union, Mount Vernon, NY. [20]

Harborne, J.B. 1982. Introduction to Ecological Biochemistry. Academic Press, New York. [2]

Harborne, J.B., and C.A. Williams. 1976. Flavonoid patterns in leaves of the Graminae. Biochem. System. Ecol. 4:267–280. [2]

Hare, J.D. 1992. Effects of plant variation on herbivore-natural enemy interactions. Pages 278–298. *In* R.S. Fritz and E.L. Simms, editors. Plant Resistance to Herbivores and Pathogens, Ecology, Evolution and Genetics. University of Chicago Press, Chicago, IL. [10,18]

Harpaz, I. 1972. Morphology and biology of vector species. Pages 154–231. *In* Maize Rough Dwarf, a Planthopper Virus Disease Affecting Maize, Rice, Small Grains and Grasses. Israel University Press, Jerusalem. [4,8,10,14]

Harpaz, I., and M. Klein. 1969. Vector-induced modifications in a plant virus. Entomol. Exp. Appl. 12:99–106. [12]

Harrison, B. D., 1985. Advances in geminivirus research. Annu. Rev. Phytopathol. 23:55–82. [12]

Harrison, B. D., and A. F. Murant. 1984. Involvement of virus-coded proteins in transmission of plant viruses by vectors, Pages 1–36. *In* M. A. Mayo and K. A. Harrap, editors. Vectors in Virus Biology. Academic Press, New York. [12]

Harrison, R.G. 1980. Dispersal polymorphisms in insects. Annu. Rev. Ecol. Syst. 11:95–118. [4]

Hashira, T. 1969. Studies on the mechanism of aphid transmission of stylet-borne virus. (IV) The insertion site of the stylet related to feeding and probing. Toh. J. Agr. Res. 20: 172–187. [12]

Hassan, A.I. 1939. The biology of some British Delphacidae (Homoptera) and their parasites with special reference to Strepsiptera. Trans. Roy. Entomol. Soc. Lond. 89:345–384. [8]

Hassell, M.P. 1978. The Dynamics of Arthropod Predator-prey Systems. Princeton Univ. Press, Princeton, NJ. [10,11,14,18]

Hassell, M. P. 1985a. Insect natural enemies as regulating factors. J. Anim. Ecol. 54:323–334. [7,11]

Hassell, M. P. 1985b. Parasitism in patchy environments: Inverse density dependence can be stabilizing. IMA J. Math. Appl. Med. Biol. 1:123–133. [10,11]

Hassell, M. P. 1986a. Detecting density dependence. Trends Ecol. Evol. 1:90–93. [7,11]

Hassell, M. P. 1986b. Parasitoids and population regulation. Pages 201–224. *In* J. Waage and D. Greathead, editors. Insect Parasitoids. Academic Press, New York. [11]

Hassell, M. P. 1987. Detecting regulation in patchily distributed animal populations. J. Anim. Ecol. 56:705–713. [11]

Hassell, M. P., and R. M. Anderson. 1984. Host susceptibility as a component in host parasitoid systems. J. Anim. Ecol. 53:611–621. [18]

Hassell, M. P., J. Latto, and R.M. May. 1989. Seeing the wood for the trees: detecting density dependence from existing life-table studies. J. Anim. Ecol. 58:883–892. [11]

Hassell, M.P., J.H. Lawton, and J.R. Beddington. 1977. Sigmoid functional responses by invertebrate predators and parasitoids. J. Anim. Ecol. 46:249–262. [10]

Hassell, M.P., and R.M. May. 1973. Stability in insect host-parasite models. J. Anim. Ecol. 42:693–736. [10,11,14]

Hassell, M.P., and R.M. May. 1974. Aggregation in predators and insect parasites and its effects in stability. J. Anim. Ecol. 43:567–594. [14]

Hassell, M. P., R.M. May, S.W. Pacala, and P.L. Chesson. 1991. The persistence of host-parasitoid associations in patchy environments. I. A general criterion. Am. Natur. 138:568–583. [11]

Hatta, T., and R. I. B. Francki. 1982. Similarity in the structure of cytoplasmic polyhedrosis virus, leafhopper A virus and Fiji disease virus particles. Intervirology 18: 203–208. [12]

Hawkins, H. S. 1989. Integrated pest control, a challenge for agricultural extension. J. Ext. Syst. 5:48–53. [20]

Hawkins, J. A., B. H. Wilson, C. L. Mondart, B. D. Nelson, R. A. Farlow, and P. E. Schilling. 1979. Leafhoppers and planthoppers in coastal bermudagrass: Effect on yield and quality and control by harvest frequency. J. Econ. Entomol. 72:101–104. [3]

Hayashi, H., and M. Chino. 1986. Collection of pure phloem sap from wheat and its chemical composition. Plant Cell Physiol. 27:1387–1393. [2]

Hayward, J. A. 1990. Agricultural extension: The World Bank's experience and approaches. Pages 115–134. *In* B. E. Swanson, editor. Report of the Global Consultation on Agricultural Extension, 4–8 December 1989, Rome, Italy. United Nations Food and Agriculture Organization, Rome. [20]

Heads, P.A., and J.H. Lawton. 1983. Studies on the natural enemy complex of the holly leaf miner: the effects of scale on the detection of aggregative responses and the implications for biological control. Oikos 40:267–276. [10,11]

Heady, S.E., and R.F. Denno. 1991. Reproductive isolation in *Prokelisia* planthoppers: Acoustical differentiation and hybridization failure. Insect Behavior 4:367–390. [4,5,6,15]

Heinrichs, E.A. 1979. Chemical control of the brown planthopper. Pages 145–167. *In* Brown Planthopper: Threat to Rice Production in Asia. International Rice Research Institute, Los Baños, Philippines. [7,8,16]

Heinrichs, E. A. 1986. Perspectives and directions for the continued development of insect-resistant rice varieties. Agric. Ecosyst. Environ. 18:9–36. [18]

Heinrichs, E.A. 1988. Variable resistance to homopterans in rice cultivars. ISI Atlas Sci., Plants Anim. 1:213–220. [16]

Heinrichs, E.A., G.B. Aquino, S. Chelliah, S.L. Valencia, and W.H. Reissig. 1982a. Resurgence of *Nilaparvata lugens* (Stål) populations as influenced by methods and timing of insecticide applications in lowland rice. Environ. Entomol. 11:78–84. [i,7,10,16,18]

Heinrichs, E.A., G.B. Aquino, S.L. Valencia, S. de Sagun, and M.B. Arceo. 1986a. Management of the brown planthopper, *Nilaparvata lugens* (Homoptera: Delphacidae), with early maturing rice cultivars. Environ. Entomol. 15:93–95. [15]

Heinrichs, E.A., R.P. Basilio, and S.L. Valencia. 1984a. Buprofezin, a selective insecticide for the management of rice planthoppers and leafhoppers. Environ. Entomol. 77:515–521. [16]

Heinrichs, E.A., L.T. Fabellar, R.P. Basilio, Tu-Cheng Wen, and F. Medrano. 1984b. Susceptibility of rice planthoppers *Nilaparvata lugens* and *Sogatella furcifera* (Homoptera: Delphacidae) to insecticides as influenced by level of resistance in the host plant. Environ. Entomol. 13:455–458. [14,15,16]

Heinrichs, E.A., and F.G. Medrano. 1984. *Leersia hexandra*, a weed host of the brown planthopper *Nilaparvata lugens* (Stål). Crop Prot. 3:77–85. [15]

Heinrichs, E.A., and F.G. Medrano. 1985. Influence of N fertilizer on the population development of brown planthopper (BPH). Int. Rice Res. Newsl. 10:20–21. [2,16]

Heinrichs, E. A., F. G. Medrano, and H. R. Rapusas. 1985. Genetic Evaluation for Insect Resistance in Rice. International Rice Research Institute. Los Baños, Philippines. [17]

Heinrichs, E.A., and O.M. Mochida. 1984. From secondary to major pest status: the case of insecticide-induced rice brown planthopper, *Nilaparvata lugens* resurgence. Prot. Ecol. 7:201–218. [i,10,14,16,17,18]

Heinrichs, E.A., and H.R. Rapusas. 1985. Levels of resistance to the whitebacked planthopper, *Sogatella furcifera* (Homoptera: Delphacidae) in rice varieties with the same major resistance genes. Environ. Entomol. 14:83–86. [2]

Heinrichs, E.A., H.R. Rapusas, G.B. Aquino, and F. Palis. 1986b. Integration of host plant resistance and insecticides in the control of *Nephotettix virescens* (Homoptera: Cicadellidae), a vector of rice tungro virus. J. Econ. Entomol. 79:437–443. [16]

Heinrichs, E. A., W. H. Reissig, S. Valencia, and S. Chelliah. 1982b. Rates and effect of resurgence-inducing insecticides on population of *Nilaparvata lugens* (Homoptera: Delphacidae) and its predators. Environ. Entomol. 11:1269–1273. [10,14,16]

Heinrichs, E.A., and L. Tetangco. 1978. Resistance of the brown planthopper to various insecticides at IRRI. Int. Rice Res. Newsl. 3(3):20. [16]

Heinrichs, E.A., and S.L. Valencia. 1978. Contact toxicity of insecticides to the three biotypes of brown planthopper. Int. Rice Res. Newsl. 3(3):19–20. [16]

Heinrichs, E. A., V. Viajante, and G. Aquino. 1978. Resurgence-inducing insecticides as a tool in field screening of rice against the brown planthopper. Int. Rice Res. Newsl. 3(3):10–11. [17]

Hennig. W. 1981. Insect phylogeny. A.C. Pont, translator. Wiley, New York. [1,12]

Heong, K. L. 1982. Population model of the brown planthopper, *Nilaparvata lugens* (Stål). MARDI Res. Bull. 10:195–209. [10]

Heong, K.L. 1988. Effect of adult density on female longevity and oviposition in the rice brown planthopper, *Nilaparvata lugens* (Stål). J. Plant Prot. Tropics 5:83–86. [4,6,7]

Heong, K.L., G.B. Aquino, and A.T. Barrion. 1991a. Arthropod community structures of rice ecosystems in the Philippines. Bull. Ent. Res. 81:407–416. [15]

Heong, K. L., G. B. Aquino, and A. A. Barrion. 1992. Population dynamics of plant- and leafhoppers and their natural enemies in rice ecosystems in the Philippines. Crop Prot. 11: 371–379. [18]

Heong, K. L., A. T. Barrion, and G. B. Aquino. 1990a. Dynamics of major predator and prey species in ricefields. IRRN 15(6):22–23. [10]

Heong, K. L., S. Bleih, and A. A. Lazaro. 1990b. Predation of *Cyrtorhinus lividipennis* Reuter on eggs of the green leafhopper and brown planthopper in rice. Res. Popul. Ecol. 32:255–262. [10,14]

Heong, K. L., S. Bleih, and E. G. Rubia. 1989. Predation of wolf spider on mirid bug and brown planthopper (BPH). IRRN 14:33. [10]

Heong, K. L., S. Bleih, and E. G. Rubia. 1991b. Prey preferences of the wolf spider, *Pardosa pseudoannulata* (Boesenberg et Strand). Res. Popul. Ecol. 32:179–186. [10,14]

Heong, K. L., and E. G. Rubia. 1989. Functional response of *Lycosa pseudoannulata* on brown planthoppers (BPH) and green leafhoppers (GLH). IRRN 14(6):29–30. [10]

Heong, K. L., and E. G. Rubia. 1990a. Technique for evaluating rice pest predators in the laboratory. IRRN 15:28. [10]

Heong, K. L., and E. G. Rubia. 1990b. Mutual interference among wolf spider adult females. IRRN 15:30–31. [10]

Hepburn, H. R. 1967. Notes on the genus *Epiptera* (Homoptera: Achilidae). J. Georgia Entomol. Soc. 2:78–80. [1]

Herbert, D. 1959. Mr. Wizard's Experiments for Young Scientists. Doubleday Press, New York. [20]

Herdt, R. W. 1987. Equity considerations in setting priorities for third world rice biotechnology research. Development: Seeds of Change 4:19–24. [i,12]

Herzog, D.C., and J.E. Funderburk. 1985. Plant resistance and cultural practices interactions with biological control. Pages 67–88. *In* M.A. Hoy and D.C. Herzog, editors. Biological Control in Agricultural IPM Systems. Academic Press, New York. [14]

Heuttel, M. D. 1976. Monitoring the quality of laboratory-reared insects: a biological and behavioral perspective. Environ. Entomol. 5:807–814. [17]

Hidaka, T. 1989. Report on the project predicting the long-distance migration of insect pests. Res. Rep. Min. Agric. Forest. Fish. Japan. 217. [9]

Hilgendorf, J. H., and R. D. Goeden. 1982. Phytophagous insects reported worldwide from the noxious weeds spiny clotbur, *Xanthium spinosum*, and cocklebur, *X. strumarium*. Bull. Entomol. Soc. Am. 28:147–152. [1]

Hill, M.G. 1976. The population and feeding ecology of five species of leafhopper (Homoptera) on *Holcus mollis* L. Ph.D. Dissertation, University of London, UK. [2]

Hinckley, A.D. 1963. Ecology and control of rice planthoppers in Fiji. Bull. Entomol. Res. 54:467–481. [2,8,10,14]

Hirano, C., and K. Kiritani. 1975. Paddy ecosystem affected by nitrogenous fertilizer and insecticides. Proc. Int. Congr. Hum. Environ., Kyoto, Japan, pp. 197–206. [10]

Hirao, J. 1969. A list of the delphacid planthoppers (Delphacidae) and their seasonal abundance surveyed by a light trap at Fukuyama, Hiroshima. Proc. Chugoku Br. Jap. Soc. Appl. Entomol. Zool. 11:5–8. [9]

Hirao, J. 1979a. Tolerance for starvation by three rice planthoppers (Hemiptera: Delphacidae). Appl. Entomol. Zool. 14:121–122. [9]

Hirao, J. 1979b. Forecasting brown planthopper outbreaks in Japan. Pages 102–112. *In* Brown Planthopper: Threat to Pice Production in Asia. International Rice Research Institute, Los Baños, Philippines. [14,18]

Hirao, J. 1986. Comparisons of bionomics of rice planthoppers between temperate Japan and the tropics. Pages 28–34. *In* International Rice Research Institute (ed.), Seminar on rice insect control, Tsukuba, September 18, 1986, IRRI and NARC. National Agricultural Research Centre, Japan. [9]

Hirashima, Y., K. Aizawa, T. Miura, and T. Wongsiri. 1979. Field studies on the biological control of leafhoppers and planthoppers (Hemiptera: Homoptera) injurious to rice plants in South-east Asia. Progress report of the year 1977. Esakia 13:1–20. [10]

Hoch, H., and F.G. Howarth. 1989a. Reductive evolutionary trends in two new cavernicolus species of a new Australian cixiid genus (Homoptera: Fulgoroidea). System. Entomol. 14: 179–196. [1]

Hoch, H., and F.G. Howarth. 1989b. Six new cavernicolus cixiid planthoppers in the genus *Solanaima* from Australia (Homoptera: Fulgoroidea). System. Entomol. 14:377–402. [1]

Hoch, H., and R. Remane. 1983. Zur Artbildung und Arabgrenzund bie der binsenbesiedelnden Zikaden-Gattung *Conomelus* Fieber, 1866 (Homoptera Auchenorrhyncha Fulgoromorpha Delphacidae). Marburger Entomol. Publ. 1(9):1–114. [1]

Hoch, H., and R. Remane. 1985. Evolution und speciation der Zikaden-Gattung *Hyalesthes* Signoret, 1865 (Homoptera, Auchenorrhyncha, Fulgoroidea, Cixiidae). Marburger Entomol. Publ. 2:1–427. [1]

Hocking, P.J. 1980. The composition of phloem exudate and xylem sap from tree tobacco (*Nicotiana glauca* Grah.). Ann. Botany 45:633–643. [2]

Hoffman, W. E. 1935. Life history notes on a species of *Lawana* (Homoptera, Flatidae). Lingnan Sci. J. 14:521–524. [1]

Hogue, C. L. 1984. Observations on the plant hosts and possible mimicry models of "lantern bugs" (Fulgora spp.) (Homoptera: Fulgoridae). Rev. Biol. Trop. 32:145–150. [1]

Hogue, C. L., T. W. Taylor, A. M. Young, and M. E. Platt. 1989. Egg masses and first instar nymphs of some giant neotropical planthoppers (Homoptera: Fulgoridae). Rev. Biol. Trop. 37:211–226. [1]

Hokkanen, H., and D. Pimentel. 1984. New approach for selecting biological control agents. Can. Entomol. 116:1109–1121. [14]

Holder, M. D. 1990. The sap-feeding guild of prairie cordgrass, *Spartina pectinata*. M.S. Thesis, Department of Biology, Central Missouri State University, Warrensburg, MO. [1,3,4]

Holder, M., and S.W. Wilson. 1992. Life history and descriptions of the immature stages of the planthopper *Prokelisia crocea* (Van Duzee) (Homoptera: Delphacidae). J. New York Entomol. Soc. 100:491–497. [2,4]

Holling, C.S. 1959a. The components of predation as revealed by a study of small mammal predation of the European pine sawfly. Can. Entomol. 91:293–320. [10,11]

Holling, C.S. 1959b. Some characteristics of simple types of predation and parasitism. Can. Entomol. 91:385–398. [10]

Holling, C.S. 1961. Principles of insect predation. Annu. Rev. Entomol. 6:163–182. [10]

Holling, C.S. 1965. The functional response of predators to prey density and its role in mimicry and population regulation. Mem. Entomol. Soc. Can. 45:1–60. [10]

Holling, C.S. 1966. The functional response of invertebrate predators to prey density. Mem. Entomol. Soc. Can. 48:1–86. [10]

Holt, J., J.A. Cheng, and G.A. Norton. 1990. A systems analysis approach to brown planthopper control on rice in Zhejiang Province, China. III. An expert system for making recommendations. J. Appl. Ecol. 27:113–122. [19]

Holt, J., A.G. Cook, T.J. Perfect, and G.A. Norton. 1987. Simulation analysis of brown planthopper (*Nilaparvata lugens*) population dynamics on rice in the Philippines. J. Appl. Ecol. 24:87–102. [8,10,19]

Holt, J., D.R. Wareing, and G.A. Norton. 1992. Strategies of insecticide use to avoid resurgence of *Nilaparvata lugens* (Stål) (Homoptera: Delphacidae), in tropical rice: a simulation analysis. J. Econ. Entomol. 85:1979–1989. [19]

Holt, J., D.R. Wareing, G.A. Norton, and A.G. Cook. 1989. A simulation of the impact of immigration on brown planthopper population dynamics in tropical rice. J. Plant Prot. Tropics 6:173–187. [8,19]

Honadle, G., and J. VanSant. 1985. Implementation for sustainability: lessons from integrated rural development. Kumarian Press, West Hartford, CT. [20]

Hopkins, L., and L. A. Carruth. 1954. Insects associated with salt cedar in southern Arizona. J. Econ. Entomol. 47:1126–1129. [1]

Horn, D. J. 1981. Effect of weedy backgrounds on colonization of collards by green peach aphid, *Myzus persicae*, and its major predators. Environ. Entomol. 10:285–289. [3]

Hosamani, M.M., B.V. Jsyskumsr, and K.M. Sharma. 1986. Sources and levels of nitrogenous fertilizers in relation to incidence of brown plant hopper in Bhadra project. Curr. Res. 15:132–134. [2]

Howard, F. W., J. P. Kramer, and M. Feliz Peralta. 1981. Homopteros Auchenorrhyncha asociados a palmeras en un area de la Republica Dominicana afectada por el amarillamiento letal del cocotero. Folia Entomol. Mex. 47:37–50. [1]

Howard, F. W., and F. W. Mead. 1980. A survey of Auchenorrhyncha (Insecta: Homoptera) associated with palms in southern Florida. Trop. Agric. (Trinidad) 57:145–153. [1]

Howard, L. O., and W.F. Fiske. 1911. The Importation into the United States of the Parasites of the Gipsy-moth and the Brown-tail Moth. U.S. Dept. Agric., Bur. Entomol., Bull. 91. [11]

Howarth, F. G. 1990. Hawaiian terrestrial arthropods: An overview. Bishop Museum Occasional Papers 30:4–26. [1,2,3]

Hoy, M.A., H.E. Van de Bann, J.J.R. Groot, and R.P. Field. 1984. Aerial movement of mites in almonds: implications for pest management. Calif. Agric. 38:21–23. [14]

Hsiao, T.H. 1978. Host plant adaptation among geographic populations of the Colorado potato beetle. Entomol. Exp. Appl. 24:237–247. [1,2,15]

Hsu, H. T., J. H. McBeath, and L. M. Black. 1977. The comparative susceptibilities of cultured vector and nonvector leafhopper cells to three plant viruses. Virology 81:257–262. [12]

Hu, J.Z., Q.H. Li, J.S. Yang, L.P. Yang, and N.Y. Gao. 1986. Influence of N fertilizer level and irrigation on population dynamics of the major insect pests in paddy fields and consequent rice yield. Acta Entomol. Sinica 29:49–55. [2]

Hu, M. 1983. Relationships between the cyclone wave and landing migratory insects. Kunchong Zhishi 3(5):203–209. [9]

Huber, J. T. 1986. Systematics, biology, and hosts of the Mymaridae and Mymarommatidae (Insecta: Hymenoptera): 1758–1984. Entomography 4:185–243. [11,13]

Huffaker, C. B. 1969. Biological Control. Plenum Press, New York. [11]

Huffaker, C.B., Editor. 1980. The New Technology of Pest Control. Wiley, New York. [19]

Huffaker, C.B., R.F. Luck, and P.S. Messenger. 1977. The ecological basis of biological control. Proc. 15th Intern. Congr. Entomol, Washington, DC, pp. 560–586. [10,14]

Huffaker, C.B., and P.S. Messenger, editors. 1976. Theory and Practice of Biological Control. Academic Press, New York. [14]

Huffaker, C. B., P.S. Messenger, and P. DeBach. 1971. The natural enemy component in natural control and the theory of biological control. Pages 16–67. *In* C. B. Huffaker, editor. Biological Control. Plenum Press, New York. [11]

Hulden, I. 1984. Observations on an egg parasite of *Cicadella viridis* (Homoptera, Auchenorrhyncha). Not. Entomol. 64:83–84. [11]

Hunt, R. E., R. E. Gingery, and L. R. Nault. 1988. Evidence for infectivity of maize chlorotic dwarf virus and a helper component in its leafhopper transmission. Phytopathology 78: 499–504. [12]

Hunt, R. E., and L.R. Nault. 1991. Roles of interplant movement, acoustic communication, and phototaxis in mate-location behavior of the leafhopper *Graminella nigrifrons*. Behav. Ecol. Sociobiol. 28:315–320. [4,5,6]

Husain, A. A., A. H. D. Brown, P. B. Hutchinson, and C. A. Wismer. 1967. The testing of sugarcane varieties for resistance to fiji disease in fiji. Pages 1154–1164. *In* Proceedings of the Twelfth I.S.S.C.T. Congress Puerto Rico, 1965. Elsevier, Amsterdam. [3,6]

Hussey, N.W., and C.B. Huffaker. 1976. Spider mites. Pages 178–228. *In* V.L. Delucchi, editor. Studies in Biological Control. Cambridge Univ. Press., Cambridge, UK. [14]

Hutchinson, P. B., and R. I. B. Francki. 1973. Sugarcane Fiji disease virus. Descriptions of Plant Viruses, No. 119. Commonw. Mycol. Inst./Assoc. Appl. Biol., Kew, Surrey, UK. [12]

Hutson, J. C. 1921. Report of the entomologist. Adin. Rep. Ceylon Dep. Agric. 1920:C15–C17. [1]

Ichikawa, T. 1976. Mutual communication by substrate vibration in the mating behavior of planthoppers (Homoptera: Delphacidae). Appl. Entomol. Zool. 11:8–23. [4,5,6]

Ichikawa, T. 1977. Sexual communications in planthoppers. Pages 84–94. *In* The Rice Brown Planthopper. Food and Fertilizer Technology Center for the Asian and Pacific Region, Taipei, Taiwan. [4,6]

Ichikawa, T. 1979. Studies on the mating behavior of four species of auchenorrhynchous Homoptera which attack the rice plant. Mem. Faculty Agric. Kagawa Univ. 34:1–60. [i,5,6,9]

Ichikawa, T. 1982. Density-related changes in male-male competitive behavior in the rice brown planthopper, *Nilaparvata lugens* (Stål) (Homoptera: Delphacidae). App. Entomol. Zool. 17:439–452. [4,5,6]

Ichikawa, T., and S. Ishii. 1974. Mating signal of the brown planthopper, *Nilaparvata lugens* Stål (Homoptera: Delphacidae): Vibration of the substrate. Appl. Entomol. Zool. 9:196–198. [5,6]

Ichikawa, T., M. Sakuma, and S. Ishii. 1975. Substrate vibrations: Mating signal of three species of planthoppers which attack the rice plant. Appl. Entomol. Zool. 10:162–171. [6]

Ikeda, R., and C. Kaneda. 1986. New genes for resistance to the brown planthopper in rice. Japan. J. Breed. 32(Suppl.2):130–131. [17]

Illich, I. 1973. Tools for conviviality. Harper and Row, New York. [20]

Imms, A.D. 1931. Recent Advances in Entomology. Blakistons, Philadelphia. [14]

Inoue, H., and T. Omura. 1982. Transmission of rice gall dwarf virus by the green rice leafhopper. Plant Dis. 66:57–59. [12]

International Rice Research Institute. 1972 Annual Report. Los Baños, Philippines. [17]

International Rice Research Institute. 1976. Annual Report. International Rice Research Institute, Los Baños, Philippines. [17]

International Rice Research Institute. 1979. Brown Planthopper: Threat to Rice Production in Asia. International Rice Research Institute, Los Baños, Philippines. [6]

International Rice Research Institute. 1985. Illustrated guide to integrated pest management in rice in tropical Asia. International Rice Research Institute, Los Baños, Philippines. [20]

International Rice Research Institute. 1990. Work plan for 1990–1994. International Rice Research Institute, Los Baños, Philippines. [17]

International Rice Research Institute. 1991. Report of a workshop on rice pest management in the MUDA scheme in Malaysia. 13–17 August 1990, Alor Setar, Malaysia. International Rice Research Institute, Los Baños, Philippines. [19]

Isman, M.B., S.S. Duffey, and G.G.E. Scudder. 1977. Cardenolide content of some leaf- and stem-feeding insects on temperate North American milkweeds (*Asclepias* spp.). Can. J. Zool. 55:1024–1028. [1]

Israel, P., and P.S. Prakasa Rao. 1968. *Harmonia arcuata* Fabricius (Coccinellidae)—predatory on the rice planthoppers *Sogatella furcifera* Horvath and *Nilaparvata lugens* Stål. Curr. Sci. 37:367–368. [10]

Istock, C.A. 1981. Natural selection and life history variation: Theory plus lessons from a mosquito. Pages 114–127. *In* R.F. Denno and H. Dingle, editors. Insect Life History Patterns: Habitat and Geographic Variation. Springer-Verlag, New York. [4]

Itakura, H. 1973. Relation between planthopper migration and meteorological conditions at the Ocean Weather Station "Tango" during 1973. Plant Prot. (Japan). 27:489–492. [9]

Itô, Y. 1972. On the methods for determining density-dependence by means of regression. Oecologia 10:347–372. [7]

Itô, Y., K. Miyashita, and K. Sekiguchi. 1962. Studies on the predation of the rice crop insect pests using the insecticidal check method. Japan. J. Ecol. 12:1–11 (in Japanese, English summary). [10,14]

Iwanaga, K., F. Nakasuji, and S. Tojo. 1987. Wing polymorphism in Japanese and foreign strains of the brown planthopper, *Nilaparvata lugens*. Entomol. Exp. Appl. 43:3–10. [i,4,7,8,9,15]

Iwanaga, K., and S. Tojo. 1986. Effects of juvenile hormone and rearing density on wing dimorphism and oöcyte development in the brown planthopper, *Nilaparvata lugens*. J. Insect Physiol. 32:585–590. [4,7]

Iwanaga, K., and S. Tojo. 1988. Comparative studies on the sensitivities to nymphal density, photoperiod and rice plant stage in two strains of the brown planthopper, *Nilaparvata lugens* (Stål) (Homoptera: Delphacidae). Japan. J. Appl. Entomol. Zool. 32:68–74. [4]

Iwanaga, K., S. Tojo, and T. Nagata. 1985. Immigration of the brown planthopper, *Nilaparvata lugens*, exhibiting various responses to density in relation to wing-morphism. Entomol. Exp. Appl. 35:101–108. [4,7,9]

Iwao, S. 1968. A new regression method for analyzing the aggregation pattern of animal populations. Res. Popul. Ecol. 10:1–20. [8]

Iwata, T. 1981. Effect of pesticide combinations on the development of resistance in green rice leafhopper, *Nephotettix cincticeps* Uhler. Japan Pesticide Information 39:1–7. [16]

Izadpanah, K., A. A. Ahmadi, S. Parvin, and S. A. Jafari. 1983. Transmission, particle size and additional hosts of the rhabdovirus causing maize mosaic in Shiraz, Iran. Phytopathol. Z. 197:283–288. [12]

Jackson, A. O., R. I. B. Francki, and D. Zuidema. 1987. Biology, structure and replication of plant rhabdoviruses, Pages 427–507. *In* R. R. Wagner, editor. The Rhabdoviruses. Plenum Press, New York. [12]

Jackson, G.V.H. 1980. Diseases and Pests of Taro. South Pacific Commission. Bridge Printery Pty. Ltd. Sydney, Australia. [14]

Jaenike, J. 1986. Genetic complexity of host selection behavior in *Drosophila*. Proc. Nat. Acad. Sci. 83:2148–2151. [15]

Jaenike, J. 1990. Host plant specialization in phytophagous insects. Annu. Rev. Ecol. Syst. 21:243–273. [1,2,15,18]

Janzen, D. H., and C. L. Hogue. 1983. *Fulgora laternaria*. Pages 726–727. *In* D. H. Janzen, editor. Costa Rican Natural History. University of Chicago Press, Chicago, IL. [1]

Jedlicka, A. D. 1983. Technology transfer to subsistence farmers: management process and behavioral techniques. Pages 330–336. *In* D. F. Cusack, editor. Agroclimate Information for Development: Reviving the Green Revolution. Westview Press, Boulder, CO. [20]

Jeffrey, M.H., 1982. A possible correlation between light trap catches of the brown planthopper *Nilaparvata lugens* (Stål) and the stages of development of the rice crop. Centre for Overseas Pest Research Miscellaneous Report No. 57. Centre for Overseas Pest Research, London. [9]

Jeffrey, M.H., and V.A. Dyck. 1983. Effect of moonlight on light-trap catches of brown planthopper. Int. Rice Res. Newsl. 8(6):13. [9]

Jermy, T. 1984. Evolution of insect/host plant relationships. Am. Natur. 124:609–630. [1]

Jervis, M. A. 1980. Ecological studies on the parasite complex associated with typhlocybine leafhoppers (Homoptera: Cicadellidae). Ecol. Entomol. 5:123–136. [11,13]

Jiang, G.H., H.Q. Tan, W.Z. Shen, X.N. Cheng, and R.C. Chen. 1981. The relation between long-distance northward migration of the brown planthopper (*Nilaparvata lugens* Stål) and synoptic weather conditions. Acta Entomol. Sinica 24:251–261. [9]

Jiang, G.H., H.Q. Tan, W.Z. Shen, X.N. Cheng, and R.C. Chen. 1982. The relation between long-distance southward migration of the brown planthopper (*Nilaparvata lugens* Stål) and synoptic weather conditions. Acta Entomol. Sinica 25:147–155. [9]

Johno, S. 1963. Analysis of the density effect as a determining factor of the wing-form in the brown planthopper *Nilaparvata lugens*. Japan. J. Appl. Entomol. Zool. 7:45–48. [4,7,8]

Johnson, B. 1965. Wing polymorphism in aphids. II. Interaction between aphids. Entomol. Exp. Appl. 8:49–64. [7]

Johnson, B., and P.R. Birks. 1960. Studies on wing polymorphism in aphids. I. The developmental process involved in the production of the different forms. Entomol. Exp. Appl. 3:327–339. [7]

Johnson, C.G. 1960. A basis for a general system of insect migration and dispersal by flight. Nature (London) 186:348–350. [8]

Johnson, C. G. 1969. Migration and dispersal of insects by flight. Methuen, London, UK. [3,8,9,13]

Johnson, L. K., and R. B. Foster. 1986. Associations of large Homoptera (Fulgoridae and Cicadidae) and trees in a tropical forest. J. Kansas Entomol. Soc. 59:415–422. [1]

Jones, T.H., J.E.B. Young, G.A. Norton, and J.D. Mumford. 1990. An expert system for management of *Delia coarctata* (Diptera, Anthomyiidae) in the United Kingdom. J. Econ. Entomol. 83:2065–2072. [19]

Julia, J. F. 1982. *Myndus taffini* (Homoptera Cixiidae), vector of foliar decay of coconuts in Vanuatu. Oleagineux 37:413–414. [1,12]

Kabir, M. A. 1984. Genetic analysis of resistance to brown planthopper, *Nilaparvata lugens* (Stål) in some selected rice varieties. MS thesis. University of the Philippines, Los Baños. [17]

Kabir, M. A., and G. S. Khush. 1988. Genetic analysis of resistance to brown planthopper in rice (*Oryza sativa* L.). Plant Breeding 100:54–58. [18]

Kale, H.W. 1965. Ecology and bioenergetics of the long-billed marsh wren *Telmatodytes palustris griseus* (Brewster), in Georgia salt marshes. Pub. Nuttal Ornithol. Club No. 56. [10]

Kamandulu, A. A. N. B., and Bahagiawati A. H.. 1990. Brown planthopper, *Nilaparvata lugens* (Stål), reaction comparison study between Yogyakarta Cisadane population and the IR64 adapted Yogyakarta Cisadane population. Paper prepared for Kongres Biologi Dasar II. February 1990. [17]

Kaneda, C., and R. Kisimoto. 1979. Status of varietal resistance to brown planthopper in Japan. Pages 209–218. *In* The Brown Planthopper: Threat to Rice Production in Asia. International Rice Research Institute, Los Baños, Philippines. [15]

Kanervo, V., O. Heikinheimo, M. Raatikainen, and A. Tinnilä. 1957. The leafhopper *Delphacodes pellucida* (F.) (Hom., Auchenorrhyncha) as the cause and distributor of the damage to oats in Finland. Publ. Finnish State Agric. Res. Board 160:1–56. [2,14]

Kaneshiro, K. Y. 1983. Sexual selection and direction of evolution in the biosystematics of Hawaiian Drosophilidae. Annu. Rev. Entomol. 28:161–178. [3]

Kanno, H., M. Kim, and S. Ishii. 1977. Feeding activity of the brown planthopper, *Nilaparvata lugens* Stål, on rice plants manured with different levels of nitrogen. Japan. J. Appl. Entomol. Zool. 21:110–112. [2]

Karban, R. 1986. Interspecific competition between folivorous insects on *Erigeron glaucus*. Ecology 67:1063–1072. [7]

Karban, R. 1989. Community organization of folivores of *Erigeron glaucus*: effects of competition, predation and host plant. Ecology 70:1028–1039. [13]

Kareiva, P. 1982. Experimental and mathematical analyses of herbivore movement: Quantifying the influence of plant spacing and quality on foraging discrimination. Ecol. Monogr. 52: 261–282. [3]

Kareiva, P. 1983. Influence of vegetation texture on herbivore populations: Resource concentration and herbivore movement. Pages 259–289. *In* R. F. Denno and M. S. McClure, editors.

Variable Plants and Herbivores in Natural and Managed Systems. Academic Press, New York. [3,8]

Kareiva, P. 1987. Habitat fragmentation and the stability of predator-prey interactions. Nature 326:388–391. [3,10]

Kareiva, P. 1990. The spatial dimension in pest-enemy interactions. Pages 213–227. *In* M. Mackauer, L.E. Ehler, and J. Roland, editors. Critical issues in biological control. Intercept, Andover, UK. [10,14]

Kareiva, P., and G. Odell. 1987. Swarms of predators exhibit "preytaxis" if individual predators use area-restricted search. Am. Natur. 130:233–270. [11]

Karim, A. N. M. R., and Q. M. A. Razzaque. 1989. Rice resistance to whitebacked planthopper (WBPH) *Sogatella furcifera* in Bangladesh. Int. Rice Res. Newsl. 14(2):16. [17]

Kartal, V. 1983. Neue Homopteren aus der Turkei II. (Homoptera Auchenorrhyncha). Marburger Entomol. Publ. 1(8):235–248. [1]

Kartohardjono, A., and E.A. Heinrichs. 1984. Populations of the brown planthopper, *Nilaparvata lugens* (Stål) (Homoptera: Delphacidae), and its predators on rice varieties with different levels of resistance. Environ. Entomol. 13:359–365. [10,15,18]

Kaston, B.J. 1948. Spiders of Connecticut. State Geol. Nat. Hist. Survey. Bull. No. 70. Hartford, CT. [10]

Kathirithamby, J. 1978. The effects of stylopisation on the sexual development of *Javesella dubai* Kirschbaum) (Homoptera: Delphacidae). Biol. J. Linn. Soc. 10:163–179. [14]

Kathirithamby, J. 1982. *Elenchus* sp. (Strepsiptera: Elenchiidae), a parasitoid of *Nilaparvata lugens* (Stål) (Homoptera: Delphacidae) in peninsular Malaysia. Proc. Int. Conf. Plant Prot. Tropics, pp. 349–361. [14]

Kathirithamby, J. 1985. Parasitism of *Nilaparvata lugens* (Stål) (Homoptera: Delphacidae) by a strepsipteran parasitoid in Tanjung Karang, West Malaysia. J. Plant Prot. Tropics 2:41–44. [11]

Kaufmann, W. G., and R. V. Flanders. 1985. Effects of variably resistant soybean and lima bean cultivars on *Pediobius foveolatus* (Hymenoptera: Eulophidae), a parasitoid of the Mexican bean beetle, *Epilachna varivestis* (Coleoptera: Coccinellidae). Environ. Entomol. 14: 678–682. [18]

Kellogg, E.A., and C.S. Campbell. 1987. Phylogenetic analysis of the Gramineae. Pages 310–322. *In* T.R. Soderstrom, K.H. Hilu, C.S. Campbell, and M.E. Barkworth, editors. Grass Systematics and Evolution. Smithsonian Institution Press, Washington, DC. [1]

Kenmore, P.E. 1980. Ecology and outbreaks of a tropical insect pest of the green revolution, the rice brown planthopper, *Nilaparvata lugens* (Stål). Ph.D. Dissertation. University of California, Berkeley. [2,4,6,7,8,10,14,17]

Kenmore, P. E. 1987. Crop loss assessment in a practical integrated pest control program for tropical Asian rice. Pages 225–241. *In* P. S. Teng, editor. Crop Loss Assessment and Pest Management. American Phytopathological Society, St. Paul, MN. [i,20]

Kenmore, P.E. 1991. Indonesia's integrated pest management—A model of Asia. FAO Rice IPC Programme, FAO, Manila, Philippines. [i,10,17]

Kenmore, P. E., F.O. Cariño, C. A. Perez, V.A. Dyck, and A. P. Gutierrez. 1984. Population regulation of the rice brown planthopper (*Nilaparvata lugens* Stål) within rice fields in the Philippines. J. Plant Prot. Tropics 1:19–37. [i,6,7,8,10,14,16,17,18,19,20]

Kenmore, P. E., J. A. Litsinger, J. P. Bandong, A. C. Santiago, and M. M. Salac. 1987. Philippine rice farmers and insecticides: thirty years of growing dependency and new options for change. Pages 98–108. *In* J. Tait and B. Napompeth, editors. Management of Pests and Pesticides: Farmers' Perceptions and Practices. Westview Press, Boulder, CO. [i,20]

Kennedy, J. S. 1961. A turning point in the study of insect migration. Nature 189:785–791. [6]

Kennedy, J.S., C.O. Booth, and W.J.S. Kershaw. 1959. Host finding by aphids in the field. I. Gynoparae of *Myzus persicae* (Sulzer). Ann. Appl. Biol. 47:410–423. [2]

Kennedy, J.S., C.O. Booth, and W.J.S. Kershaw. 1961. Host finding by aphids in the field. III. Visual attraction. Ann. Appl. Biol. 49:1–21. [2]

Kershaw, J. C. W., and G. W. Kirkaldy. 1910. A memoir on the anatomy and life-history of the homopterous insect Pyrops candelaria (or "candle-fly"). Zool. Jahrb. Abt. Syst. Geogr. Biol. Thiere 29:105–124. [1]

Khafagi, R. M. 1986. The biological relationships of *Macrosteles viridigriseus* (Homoptera) and its parasite *Anteon pubicorne* (Hymenoptera). Ph.D. Dissertation, University of Newcastle upon Tyne, UK. [13]

Khan, M. Q., and A.S. Rao. 1956. The influence of the black ant (*Camponotus compressus* F.) on the incidence of two homopteran crop pests. Indian J. Entomol. 18:199–200. [14]

Khan, Z.R., and R.C. Saxena. 1984. Electronic device to record feeding behavior of whitebacked planthopper on susceptible and resistant rice varieties. Int. Rice Res. Newsl. 9:8–9. [2]

Khan, Z.R., and R.C. Saxena. 1985. Behavioural and physiological responses of *Sogatella furcifera* (Homoptera: Delphacidae) to selected resistant and susceptible rice cultivars. J. Econ. Entomol. 78:1280–1286. [2]

Khush, G. S. 1979. Genetics of and breeding for resistance to the brown planthopper. Pages 321–332. *In* Brown Planthopper: Threat to Rice Production in Asia. International Rice Research Institute, Los Baños, Philippines. [2,15,17,18]

Khush, G. S. 1984. Breeding rice for resistance to insects. Prot. Ecol. 7:147–165. [i,18]

Khush, G. S., A. N. M. R. Karim, and E. R. Angeles. 1986. Genetics of resistance of rice cultivar ARC 10550 to Bangladesh brown planthopper biotype. J. Genet. 64(2/3):121–125. [17]

Kidd, N.A.C. 1976. Aggregation in the lime aphid (*Eucallipterus tiliae* L.). Oecologia 25:175–185. [7]

Kifune, T. 1986. A new host of *Elenchus japonicus* (Strepsiptera, Elenchidae) in Kyushu, Japan. Kontyu 54:525. [13]

Kilin, D., T. Nagata, and T. Masuda. 1981. Development of carbamate resistance in the brown planthopper (*Nilaparvata lugens* Stål) (Homoptera: Delphacidae). Appl. Entomol. Zool. 16:1–6. [16]

Kim, J.B., J.S. Hyun, K.B. Uhm, D.J. Cho, W.K. Shin, and Y.S. Lee. 1987. Development of small brown planthopper *Laodelphax striatellus* Fallen population in the fields at the Southern Korea. Research Reports, Rural Dev. Adm., Korea Republic, Plant Environment, Mycology & Farm Products Utilization 29:282–289. [8]

Kim, M., H. Koh, and H. Fukami. 1985. Isolation of C-Glycosylflavones as probing stimulant of planthoppers in rice plant. J. Chem. Ecol. 11:441–452. [2]

Kimmins, F. M. 1986. Ultrastructure of the stylet pathway of *Brevicoryne brassicae* in host plant tissue, *Brassicae oleracea*. Entomol. Exp. Appl. 41:283–290. [12]

Kimmins, F.M. 1989. Electrical penetration graphs from *Nilaparvata lugens* on resistant and susceptible rice varieties. Entomol. Exp. Appl. 50:69–79. [2,18]

King, A.B.S., and J.L. Saunders. 1984. The invertebrate pests of annual food crops in Central America. Overseas Development Administration, London. [2,4,7,10]

Kiritani, K. 1977. Recent progress in the pest management for rice in Japan. JARQ 11:40–49. [10]

Kiritani, K. 1979. Pest management in rice. Annu. Rev. Entomol. 24:279–312. [4,10,14]

Kiritani, K., N. Hokyo, T. Sasaba, and F. Nakasuji. 1970. Studies on population dynamics of the green rice leafhopper, *Nephotettix cincticeps* Uhler: regulatory mechanisms of the population density. Res. Popul. Ecol. 12:137–153. [11]

Kiritani, K., and N. Kakiya. 1975. An analysis of the predator-prey system in the paddy field. Res. Popul. Ecology 17:29–38. [10]

Kiritani, K., and S. Kawahara. 1973. Food-chain toxicity of granular formulations of insecticides to a predator, *Lycosa pseudoannulata* of *Nephotettix cincticeps*. Botyu-Kagaku 38:69–75. [10]

Kiritani, K., S. Kawahara, T. Sasaba, and F. Nakasuji. 1971. An attempt of rice pest control by integration of pesticides and natural enemies. Gensei 22:19–23 (in Japanese, English summary). [10]

Kiritani, K., S. Kawahara, T. Sasaba, and F. Nakasuji. 1972. Quantitative evaluation of predation by spiders on the green rice leafhopper, *Nephotettix cincticeps* Uhler, by a sight-count method. Res. Popul. Ecol. 13:187–200. [10,14]

Kirkaldy, G.W. 1906. Leaf-hoppers and their natural enemies (Part IX Leafhoppers—Hemiptera). Report of Work of the Experiment Station of the Hawaiian Sugar Planters' Association. Bulletin No. 1, Part 9, pp. 1–479. [3,4]

Kirkpatrick, M. 1987. The evolutionary forces acting on female mating preferences in polygynous animals. Pages 67–82. *In* J.W. Bradbury and M.B. Andersson, editors. Sexual selection: Testing the alternatives. John Wiley and Sons, NY. [6]

Kisimoto, R. 1956a. Effect of crowding during the larval period on the determination of the wing-form of an adult plant-hopper. Nature 178:641–642. [7]

Kisimoto, R. 1956b. Studies on the polymorphism in the planthopper (Araeopidae, Homoptera). Preliminary Report. Oyo-Kontyu 12:56–61. [2,7]

Kisimoto, R. 1956c. Factors determining the wing-form of adult, with special reference to the effect of crowding during the larval period of the brown planthopper, *Nilaparvata lugens* Stål. Studies of the polymorphism in the planthoppers (Homoptera, Araeopidae). I. Oyo-Kontyu 12:105–111. [4]

Kisimoto, R. 1958. Studies on the diapause in the planthoppers. Effect of photoperiod on the induction and the completion of diapause in the fourth larval stage of the smaller brown planthopper, *Delphacodes striatella* Fallen. Japan. J. Appl. Entomol. Zool. 2:128–134. [4]

Kisimoto, R. 1959. Studies on the polymorphism in the planthoppers (Homoptera: Araeopidae). IV. On the stages susceptible to the effect of crowding determining the wing-forms in the brown planthopper, *Nilaparvata lugens* Stål. Japan. J. Ecol. 9:94–97. [8]

Kisimoto, R. 1965. Studies on polymorphism and its role playing in the population growth of the brown planthopper, *Nilaparvata lugens* Stål. Bull. Shikoku Agric. Exp. Sta. 13:1–106. [i,4,6,7,8,9,10]

Kisimoto, R. 1969. Net trap, a survey technique for the flight activity and the aerial density of planthoppers and leafhoppers. Plant Prot. (Japan) 23:245–248. [9]

Kisimoto, R. 1971. Long distance migration of planthoppers *Sogatella furcifera* and *Nilaparvata lugens*. Pages 201–216. *In* Proc. Symp. Rice Insects. Trop. Agric. Res. Cent., Minist. Agric. For., Tokyo. [4,8,9,16]

Kisimoto, R. 1973. Leafhoppers and planthoppers. Pages 137–156. *In* A. J. Gibbs, editor. Viruses and Invertebrates. Elsevier, New York. [12]

Kisimoto, R. 1976. Synoptic weather conditions inducing long distance immigration of planthoppers, *Sogatella furcifera* Horvath and *Nilaparvata lugens* Stål. Ecol. Entomol. 1:95–109. [i,3,4,6,8,9,10,15]

Kisimoto, R. 1977. Bionomics, forecasting of outbreaks and injury caused by the rice brown planthopper. Pages 27–41. *In* Food and Fertilizer Technology Center for the Asian and Pacific Region, The Rice Brown Planthopper. Food and Fertilizer Technology Center for the Asian and Pacific Region, Taipei, Taiwan. [4,9,14,19]

Kisimoto, R. 1979. Brown planthopper migration. Pages 113–124. *In* Brown Planthopper: Threat to Rice Production in Asia. International Rice Research Institute, Los Baños, Philippines. [3,4,8,9]

Kisimoto, R. 1981. Development, behaviour, population dynamics and control of the brown planthopper, *Nilaparvata lugens* Stål. Rev. Plant Prot. Res. 14:26–58. [1,4,7,8,9,10,14]

Kisimoto, R. 1987. Ecology of planthopper migration. Pages 41–54. *In* M.R. Wilson and L.R. Nault, editors. Proc. 2nd Int. Workshop on Leafhoppers and Planthoppers of Economic Importance. Common. Inst. Entomol., London. [4,7,9]

Kisimoto, R. 1989. Geographical variation of short-wingform productivity of the brown planthopper, *Nilaparvata lugens* Stål. 1st Asia-Pacific Conf. Entomol., Abstr. [9]

Kisimoto, R. 1991. Long-distance migration of rice insects, Pages 167–195. *In* E.A. Heinrichs and T.A. Miller, editors. Rice Insects: Management Strategies. Springer-Verlag, New York. [9]

Kisimoto, R., and V.A. Dyck. 1976. Climate and rice insects, Pages 367–390. *In* International Rice Research Institute (ed.), Proceedings of the Symposium on Climate and Rice. International Rice Research Institute, Manila. [9]

Kisimoto, R., J. Hirao, Y. Hirahara, and A. Tanaka. 1982. Synchronization in migratory flight of planthoppers, *Nilaparvata lugens* Stål and *Sogatella furcifera* Horvath (Hemiptera:Delphacidae), in south-western Japan. Japan. J. Appl. Entomol. Zool. 26:112–118. [9]

Kisimoto, R., and M. A. Watson. 1965. Abnormal development of embryos induced by inbreeding *Delphacodes pellucida* Fab. and *Delphacodes dubia* Kirschbaum (Araepedae, Homoptera), vectors of European wheat striate mosaic virus. J. Invertebr. Pathol. 7: 297. [12]

Kiyosawa, S. 1982. Genetics and epidemiological modeling of breakdown of plant disease resistance. Annu. Rev. Phytopathol. 20:93–117. [18]

Knight, J.D. 1991. Flypast: A database of aphid suction trap catches for the UK. Silwood Centre for Pest Management, Ascot, U.K. [19]

Kobayashi, S. 1975. The effect of *Drosophila* release on the spider population in a paddy field. Appl. Entomol. Zool. 10:268–274. [6,10]

Kobayashi, S., and T. Hiwada. 1968. A significance of the early seasonal and the earlier cultures of rice plants in the multiplication of the smaller brown planthopper. Japan. J. Ecol. 18: 179–185. [10]

Kobayashi, S., and H. Shibata. 1973. Seasonal changes in population density of spiders in paddy fields, with reference to the ecological control of the rice insect pests. Japanese J. Appl. Entomol. Zool. 17:193–202. [14]

Kogan, M. 1977. The role of chemical factors in insect/plant relationships. Proc. Int. Cong. Entomol. 15:211–227. [2]

Kogan, M. 1982. Plant resistance in pest management. Pages 93–134. *In* R. L. Metcalf and W. H. Luckman, editors. Introduction to Insect Pest Management. Second edition, Wiley, New York. [18]

Kramer, J. P. 1973. Revision of the American planthoppers of the genus *Stobaera* (Homoptera: Delphacidae) with new distributional data and host plant records. Proc. Entomol. Soc. Wash. 75:379–402. [1,2,4]

Kramer, J. P. 1976. Revision of the Neotropical planthoppers of the genus *Bladina* (Homoptera: Fulgoroidea: Nogodinidae). Trans. Am. Entomol. Soc. 102:1–40. [1]

Kramer, J.P. 1977. Taxonomic study of the planthopper genus *Oecleus* in the United States (Homoptera: Fulgoroidea: Cixiidae). Trans. Am. Entomol. Soc. 103:379–449. [1]

Kramer, J. P. 1978. Taxonomic study of the American planthopper genus *Cyrpoptus* (Homoptera: Fulgoroidea: Fulgoridae). Proc. Biol. Soc. Wash. 91:305–335. [1]

Kramer, J.P. 1979. Taxonomic study of the planthopper genus *Myndus* in the Americas (Homoptera: Fulgoroidea: Cixiidae). Trans. Am. Entomol. Soc. 105:301–389. [1]

Kramer, J.P. 1981. Taxonomic study of the planthopper genus *Cixiius* in the United States and Mexico (Homoptera: Fulgoroidea: Cixiidae). Trans. Am. Entomol. Soc. 107:1–68. [1]

Kramer, J.P. 1983. Taxonomic study of the planthopper family Cixiidae in the United States (Homoptera: Fulgoroidea). Trans. Am. Entomol. Soc. 109:1–58. [1]

Krampl, F., and J. Dlabola. 1983. A new genus and species of epipyropid moth from Iran ectoparasitic on a new *Mesophantia* species, with revision of the host genus (Lepidoptera, Epipyropidae; Homoptera, Flatidae). Acta Entomol. Bohemoslov. 80:451–472. [1]

Krauss, N. L. H. 1953. Notes on insects associated with *Lantana* in Cuba. Proc. Hawaiian Entomol. Soc. 15:123–125. [1]

Krebs, J. R., and N. B. Davies, editors. 1978. Behavioural Ecology: An Evolutionary Approach. First edition. The University Press, Cambridge, England. [6]

Krebs, J. R., and N. B. Davies, editors. 1984. Behavioural Ecology: An Evolutionary Approach. Second edition. The University Press, Cambridge, England. [6]

Krebs, J. R., and N. B. Davies, editors. 1991. Behavioural Ecology: An Evolutionary Approach. Third edition. The University Press, Cambridge, England. [6]

Krishna, T. S., and D. V. Seshu. 1984. Genetics of resistance to brown planthopper in rice and association of certain plant characters with resistance. Indian J. Genet. Plant Breeding. 44(2):336–342. [17]

Kulshreshtha, J. P., A. Anjaneyulu, and S. Y. Padmanabhan. 1974. The disastrous brown planthopper attack in Kerala. Indian Farming 24:5–7. [3]

Kuno, E. 1968. Studies on the population dynamics of the rice leafhoppers in a paddy field. Bull. Kyushu Agric. Exp. Sta. 14:131–246. [8,9,10]

Kuno, E. 1973a. Statistical characteristics of the density-dependent population fluctuation and the evaluation of density-dependence and regulation in animal populations. Res. Popul. Ecol. 15:99–120. [7]

Kuno, E. 1973b. Population ecology of rice leafhoppers in Japan. Rev. Plant Prot. Res. 6:1–16. [7,8]

Kuno, E. 1977. Distribution pattern of the rice brown planthopper and field sampling techniques. Pages 135–147. *In* The Rice Rice Brown planthopper. Food and Fertilizer Technology Center for the Asian and Pacific Region, Taipei, Taiwan. [8]

Kuno, E. 1979. Ecology of the brown planthopper in temperate regions. Pages 45–60. *In* Brown Planthopper: Threat to Rice Production in Asia. International Rice Research Institute, Los Baños, Philippines. [i,4,6,7,8,9,10,14,15,18,19]

Kuno, E. 1984. Pest status, dynamics and control of the rice leaf- and planthopper populations in Japan. Prot. Ecol. 7:129–145. [8]

Kuno, E., and V.A. Dyck. 1985. Dynamics of Philippine and Japanese populations of the brown planthopper: Comparison of basic characteristics. Proc. ROC–JAPAN Seminar on the ecology and control of the Brown Planthopper. Nat. Sci. Council ROC. 4:1–9. [7,8,10,14]

Kuno, E., and N. Hokyo. 1970. Comparative analysis of the population dynamics of rice leafhoppers, *Nephotettix cincticeps* Uhler and *Nilaparvata lugens* Stål, with special reference to natural regulation of their numbers. Res. Popul. Ecol. 12:154–184. [i,7,8,10,13,14]

Kushmayadi, A., E. Kuno, and H. Sawada. 1990. The spatial distribution of the brown planthopper *Nilaparvata lugens* Stål (Homoptera Delphacidae) in West Java, Indonesia. Res. Popul. Ecol. 32:67–83. [8]

Kuwahara, M., K. Hiraoka, and K. Sakano. 1956. Experimental studies on the biology and hibernation in the brown planthopper, *Nilaparvata lugens* (Stål) (Hom., Auchenorrhyncha). Norinsyo-Byogaityu-Hasseiyosatu-Siryo 56:54–65. [4]

Kwon, Y. J. 1982. New and little known planthoppers of the Family Delphacidae (Homoptera: Auchenorrhyncha). Korean J. Entomol. 12:1–11. [1]

Lallemand, V. 1938. Homopteres recueillis a Eala (Congo Belge) par J. Ghesquire. Bull. Mus. Royal Hist. Nat. Belgique 14(58):1–4. [1]

Lamp, W. O., R. J. Barney, E. J. Armbrust, and G. Kapusta. 1984. Selective weed control in spring-planted alfalfa: Effect on leafhoppers and planthoppers (Homoptera: Auchenorrhyncha), with emphasis on potato leafhopper. Environ. Entomol. 13:207–213. [3,13,15]

Lande, R. 1982. A quantitative theory of life history evolution. Ecology 63:607–615. [4]

Lande, R., and S. J. Arnold 1983. The measurement of selection on correlated characters. Evolution 37:1210–1226. [6]

Langer, R.H.M. 1979. How grasses grow. Edward Arnold Limited, London, UK. [2,7]

Lastra, R., and O. Carballo. 1983. Maize virus disease problems in Venezuela, Pages 83–86. *In* D. T. Gordon, J. K. Knoke, L. R. Nault, and R. M. Ritter, editors. Proceedings International Maize Virus Disease Colloquium and Workshop Aug. 2–6, 1982, Wooster, Ohio. The Ohio State Univ., Ohio Agric. Res. Devel. Ctr., Wooster, OH. [12]

Lastra, J. R., and J. Esparza. 1976. Multiplication of vesicular stomatitis virus in the leafhopper *Peregrinus maidis* (Ashm.), a vector of a plant rhabdovirus. J. Gen. Virol. 32: 139–142. [12]

Latto, J., and M.P. Hassell. 1987. Do pupal predators regulate the winter moth? Oecologia 74:153–155. [7]

Lauterer, P. 1983. *Fagocyba cerricola* sp. n. and new an interesting records of leafhoppers from Czechoslovakia (Homoptera, Auchenorrhyncha). Acta Mus. Moraviae 68:139–152. [1]

Lawrence, W. S. 1986. Mate choice and competition in *Tetraopes tetraophthalamus*: Effects of local sex ratio variation. Behavioral Ecol. Sociobiol. 18:289–296. [6]

Lawton, J. H. 1978. Host plant influences on insect diversity: the effects of space and time. Symp. Roy. Entomol. Soc. Lond. 9:105–125. [3]

Lawton, J.H. 1983. Plant architecture and the diversity of phytophagous insects. Ann. Rev. Entomol. 28:23–29. [3]

Lawton, J. H. 1986. The effect of parasitoids on phytophagous communities. Pages 265–287. *In* J. Waage and D. Greathead, editors. Insect Parasitoids. Academic Press, New York. [11]

Lawton, J.H., and M.P. Hassell. 1983. Interspecific competition in insects. *In* C.B. Huffaker and R.L. Rabb, editors. Ecological Entomology. Wiley, New York. [7]

Lawton, J. H., and S. McNeill. 1979. Between the devil and the deep blue sea: On the problem of being a herbivore. Symp. Brit. Ecol. Soc. 20:223–244. [18]

Lawton, J.H., and D.R. Strong. 1981. Community patterns and competition in folivorous insects. Am. Natur. 118:317–338. [1,7]

Lee, C. E., and Y. J. Kwon. 1977. Studies on the spittlebugs, leafhoppers and planthoppers (Auchenorrhyncha, Homoptera, Hemiptera). Nature and Life (Kyungpook J. Biol. Sci.) 7:55–111. [1]

Lee, C. E., and Y. J. Kwon. 1980. A study on the local distribution of planthoppers in Korea (Auchenorrhyncha, Delphacidae). Nature and Life (Kyungpook J. Biol. Sci.) 10:23–42. [1]

Lee, J. O., and J.S. Park. 1977. Biology and control of the brown planthopper (*Nilaparvata lugens*) in Korea. Pages 199–213. *In* The Rice Brown Planthopper. Food and Fertilizer Technology Centre for the Asian and Pacific Region, Taipei, Taiwan. [10]

Lees, A.D. 1966. The control of polymorphism in aphids. Adv. Insect Physiol. 3:207–277. [7]

Lees, A.D. 1967. The production of the apterous and alate forms in the aphid, *Megoura viciae* Buckton, with special reference to the role of crowding. J. Insect Physiol. 13:289–318. [7]

Lefevre, G. Jr., and V. B. Johnson. 1962. Sperm transfer, storage, displacement, and utilization in *Drosophila melanogaster*. Genetics 47:1719–1736. [6]

Le Quesne, W. J. 1960. Hemiptera (Fulgoromorpha). Handbks. Ident. British Ins. 2(3):1–68. [1]

Le Quesne, W. J. 1972. Studies on the coexistence of three species of *Eupteryx* (Hemiptera: Cicadellidae) on nettle. J. Entomol. 47:37–44. [13]

Le Quesne, W. J. 1983. The leafhoppers and allied insects (Auchenorhyncha: Hemiptera) of Jersey. Ann. Bull. Soc. Jersiaise 23:363–368. [1]

Levins, R. 1968. Evolution in Changing Environments. Princeton Univ. Press, Princeton, NJ. [6]

Levins, R., and R. MacArthur. 1969. An hypothesis to explain the incidence of monophagy. Ecology 50:910–911. [1,2]

Lim, G.S., and K.L. Heong. 1984. The role of insecticides in rice integrated pest management. Pages 19–39. *In* Judicious and Efficient Use of Insecticides in Rice. Proceedings of the FAO/IRRI Workshop, 21–23 February 1983. International Rice Research Institute, Los Baños, Philippines. [i,19]

Lim, G.S., K.L. Heong, and P.A.C. Ooi. 1978. The brown planthopper of rice in Malaysia. Technical Leaflet No. 15, Ministry of Agriculture, Kuala Lumpur, Malaysia. [10]

Lin, K.S. 1974. Notes on some natural enemies of *Nephotettix cincticeps* (Uhler) and *Nilaparvata lugens* (Stål) in Taiwan [in Chinese with English summary]. J. Taiwan Agric. Res. 23:91–115. [14]

Lindroth, C.H., H. Andersson, H. Bödvarsson, and S.H. Richter. 1973. Surtsey, Iceland. The development of a new fauna, 1963–1970. Terrestrial Invertebrates Entomol. Scand. Suppl. 5:1–280. [4]

Linnavuori, R. 1952. Studies on the ecology and phenology of the leafhoppers (Homoptera) of Raisio (S.W. Finland). Ann. Zool. Soc. Vanamo 14:1–32. [4]

Linnavuori, R. 1964. Hemiptera of Egypt, with remarks on some species of the adjacent Eremian region. Ann. Zool. Fenn. 1:338–355. [1]

Liquido, N. J. 1978. Morphology, cytology and breeding behavior of *Nilaparvata lugens* (Stål) (Delphacidae: Homoptera). MS thesis. University of the Philippines, Los Baños. [17]

Litsinger, J. A. 1984. Assessment of need-based insecticide application for rice. Paper presented at MA–IRRI Technology Transfer Workshop, 19 March 1984. [20]

Litsinger, J. A., J. P. Bandong, and C. G. Dela Cruz. 1984. Verifying and extending integrated pest control technology to small-scale farmers. Pages 326–354. *In* P. C. Matteson, editor. Proceedings of the International Workshop on Integrated Pest Control for Grain Legumes, 3–9 April 1983, Goiania, Goias, Brazil. EMBRAPA, Brasilia, Brazil. [20]

Liu, C.-H. 1985. Study on the long-distance migration of the brown planthopper in Taiwan. Chin. J. Entomol. 4(2):49–54. [9]

Liu, H.Q., Z.J. Liu and W.H. Zhu. 1983. Results of net-trapping of brown planthoppers over the China Seas. Acta Entomol. Sinica 26: 109–113. [9]

Lloyd, J. E. 1975. Aggressive mimicry in photuris fireflies: signal repertoires by femme fatales. Science 187:452–453. [6]

Loevinsohn, M.E. 1984. The ecology and control of rice pests in relation to the intensity and asynchrony of cultivation. Ph.D. Dissertation. University of London, London, UK. [8,19]

Loevinsohn, M. E. 1991. Brown planthopper (BPH) dispersal range under natural conditions in the Philippines. Int. Rice Res. Newsl. 16:27. [6]

Lopez-Abella, D., R. H. E. Bradley, and K. F. Harris. 1988. Correlation between stylet paths made during superficial probing and the ability of aphids to transmit viruses, Pages 251–285. *In* K. F. Harris, editor. Advances in Disease Vector Research, Vol. 5, Springer-Verlag, New York. [12]

Luck, R. F., R. Van den Bosch, and R. Garcia. 1977. Chemical insect control—a troubled strategy. BioScience 27:606–611. [10]

Luff, M.L. 1983. The potential of predators for pest control. Agric. Ecosyst. Environ. 10:159–181. [10]

Luong, M. C. 1990. Development of a brown planthopper (BPH) biotype and change in varietal resistance in Mekong Delta. Int. Rice Res. Newsl. 15(5):12. [17]

Luong, M. C., and R. C. Saxena. 1989. Reaction of brown planthopper (BPH) of varieties originating from *Oryza officinalis*. Int. Rice Res. Newsl. 14(6):9–10 [17]

Mabry, T.J., and J.E. Gill. 1979. Sesquiterpine lactones and other terpenoids. Pages 502–537. *In* G.A. Rosenthal and D.H. Janzen, editors. Herbivores: Their Interaction with Secondary Plant Metabolites. Academic Press, New York. [2]

MacArthur, R.H., and E.D. Wilson. 1967. The Theory of Island Biogeography. Princeton Univ. Press, Princeton, NJ. [4]

MacGarvin, M. 1982. Species-area relationships of insects on host plants: Herbivores on rosebay willowherb. J. Anim. Ecol. 51:207–223. [3]

MacNally, R. C., and J. M. Doolan. 1986. An empirical approach to guild structure: habitat relationships in nine species of eastern-Australian cicadas. Oikos 47:33–46. [13]

Madden, L. V., and L. R. Nault. 1983. Differential pathogenicity of corn stunting mollicutes to leafhopper vectors in *Dalbulus* and *Baldulus* species. Phytopathology 73:1608–1614. [12]

Madden, L. V., L. R. Nault, S. E. Heady, and W. E. Styer. 1984. Effect of maize stunting mollicutes on survival and fecundity of *Dalbulus* leafhopper vectors. Ann. Appl. Biol. 105:431–441. [12]

Maes, J. M., and L. B. O'Brien. 1988. Catalogo de los Fulgoroidea (Homoptera) de Nicaragua. Rev. Nicaragua Entomol. 2:27–42. [1]

Magyarosy, A. C., and E. S. Sylvester. 1979. The latent period of beet curly top virus in the beet leafhopper, *Circulifer tennellus*, mechanically injected with infectious phloem exudate. Phytopathology 69:736–738. [12]

Mahmud, F.S. 1980. Alary polymorphism in the small brown planthopper *Laodelphax striatellus* (Homoptera:Delphacidae). Entomol. Exp. Appl. 28:47–53. [4,9]

Manjunath, T.M. 1977. A note on oviposition in the macropterous and brachypterous forms of the rice brown planthopper, *Nilaparvata lugens*, Homoptera, Delphacidae. Proc. Indian Acad. Sci. Section B. 86:405–408. [4]

Manti, I. 1989. The role of *Cyrtorhinus lividipennis* Reuter (Hemiptera, Miridae) as a major predator of the brown planthopper *Nilaparvata lugens* Stål (Homoptera, Delphacidae). Ph.D. Dissertation, University of the Philippines at Los Baños, Philippines. [10,14]

Manti, I., and B. M. Shepard. 1990. Predation of brown planthopper (BPH) eggs by *Cyrtorhinus lividipennis* Reuter. IRRN 15(6):25. [10]

Maramorosch, K. 1950. Influence of temperature on incubation and transmission of the wound-tumor virus. Phytopathology 40:1071–1093. [12]

Marciano, V. P., A.M. L. Mandac, and J. C. Flynn. 1981. Insect management practices of rice farmers in Laguna, Philippines. IRRI Agricultural Economics Department Paper No. 81–03. International Rice Research Institute, Los Baños, Philippines. [20]

Markham, P. G., M. S. Pinner, B. Raccah, and R. Hull. 1987. The acquisition of a caulimovirus by different aphid species: comparison with a potyvirus. Ann. Appl. Biol. 111:571–587. [12]

Markow, T. A., M. Quaid, and S. Kerr. 1978. Male mating experience and competitive courtship success in *Drosophila melanogaster*. Nature 276:821–822. [6]

Marshall, G.A.K. 1916. The fauna of British India. Coleoptera, Rhynchophora, Curculionidae. Taylor and Francis, London, England. [4]

Martinez, C. R., and G. S. Khush. 1974. Sources and inheritance of resistance to brown planthopper in some breeding lines of rice *Oryza sativa*. Crop Sci. 14(2):264–267. [17]

Martorell, L. F., and J. Adsuar. 1952. Insects associated with papaya virus diseases in the Antilles and Florida. J. Econ. Entomol. 45:863–869. [1]

Matsumoto, S., K. Ninomiya, and S. Yoshizimi. 1971. Characteristic features of "Baiu" front associated with heavy rainfall. J. Met. Soc. Japan. 49:267–281. [9]

Matsumoto, B.M., and T. Nishida. 1966. Predator-prey investigations on the taro leafhopper and its egg predator. Univ. Hawaii, Hawaii Agric. Exp. Sta. Technical Bull. 64:3–32. [i,2,10,14]

Matteson, P. C. 1986. First rice IPC demonstration results from Sri Lanka: guidance for the national implementation program. Paper presented at the 2nd International Conference on Plant Protection in the Tropics, 17–20 March 1986, Genting Highlands, Malaysia. [20]

Matteson, P. C., M. A. Altieri, and W. C. Gagné. 1984. Modification of small farmer practices for better pest management. Annu. Rev. Entomol. 29:383–402. [20]

Matthews, R. E. F. 1991. Plant Virology. Third edition. Academic Press, New York. [12]

Mattson, W.J. 1980. Herbivory on relation to plant nitrogen content. Ann. Rev. Ecol. Syst. 11:119–161. [2,7]

Mattson, W.J., and N.D. Addy. 1975. Phytophagous insects as regulators of forest primary production. Science 190:515–522. [2]

May, R.M., and M.P. Hassell. 1988. Population dynamics and biological control. Phil. Trans. Roy. Soc. Lond. 318:129–169. [14]

May, Y. Y. 1971. The biology and population ecology of *Stenocranus minutus* (Fabricius) (Delphacidae, Hemiptera), Ph.D. Dissertation, University of London, London, UK. [11,13]

May, Y.Y. 1975. Study of two forms of the adult *Stenocranus minutus*. Trans. Roy. Entomol. Soc. Lond. 127:241–254. [4]

May, Y. Y. 1978. A population study of *Stenocranus minutus* (Fab.) (Hemiptera: Delphacidae). Res. Popul. Ecol. 20:61–78. [13]

Maynard Smith, J. 1987. Sexual selection—A classification of models. Pages 9–21. *In* J. W. Bradbury and M. B. Andersson, editors. Sexual Selection: Testing the Alternatives. Wiley, New York. [6]

Mayr, E. 1942. Genetics and the Origin of Species from the viewpoint of a Zoologist. Columbia University Press, New York. [5]

Mayr, E. 1963. Animal species and evolution. Harvard University Press, Cambridge, MA. [1]

Mayse, M. 1978. Effects of spacing between rows on soybean arthropod populations. J. Anim. Ecol. 15:439–450. [3,13]

McCauley, D. E., and M. J. Wade 1978. Female choice and the mating structure of a natural population of the soldier beetle, *Chauliognathus pennsylvanicus*. Evolution 32:771–775. [6]

McClay, A. S. 1983. Biology and host-specificity of *Stobaera concinna* (Stål) (Homoptera: Delphacidae), a potential biocontrol agent for *Parthenium hysterophorus* L. (Compositae). Folia Entomol. Mex. 56:21–300. [1,2]

McClure, M. S. 1974. Biology of *Erythroneura lawsoni* (Homoptera: Cicadellidae) and coexistence in the sycamore leaf-feeding guild. Environ. Entomol. 3:59–68. [13]

McClure, M.S. 1979. Self regulation in hemlock scale populations: Role of food quantity and quality. Misc. Pubs. Entomol. Soc. Am. 11:33–40. [2]

McClure, M.S. 1980a. Foliar nitrogen: A basis for host suitability for elongate hemlock scale, *Fiorinia externa* (Homoptera: Diaspididae). Ecology 61:72–79. [2]

McClure, M.S. 1980b. Competition between exotic species: Scale insects on hemlock. Ecology 61:1391–1401. [7]

McClure, M.S. 1983. Competition between herbivores and increased resource heterogeneity. Pages 125–153. *In* R.F. Denno and M.S. McClure, editors. Variable Plants and Herbivores in Natural and Managed Systems. Academic Press, New York. [7]

McClure, M. S., and P. W. Price. 1975. Competition and coexistence among sympatric *Erythroneura* leafhoppers (Homoptera: Cicadellidae) on American sycamore. Ecology 56:1388–1397. [3,7,13]

McClure, M. S., and P. W. Price. 1976. Ecotope characteristics of coexisting *Erythroneura* leafhoppers (Homoptera: Cicadellidae) on sycamore. Ecology 57:928–940. [3,13]

McCoy, E.D., and J.R. Rey. 1981. Patterns of abundance, distribution, and alary polymorphism among salt marsh Delphacidae (Homoptera: Fulgoroidea) of northwest Florida. Ecol. Entomol. 6:285–291. [4,9]

McDaniel, L. L., E. D. Ammar, and D. T. Gordon. 1985. Assembly, morphology and accumulation of a Hawaiian isolate of maize mosaic virus. Phytopathology 75:1167–1172. [12]

McDermott, B. T. 1952. A revision of the genus *Megamelanus* and its allies (Homoptera, Fulgoridae, Delphacinae). J. Kansas Entomol. Soc. 25:41–59. [1]

McEvoy, P. B. 1986. Niche partitioning in spittlebugs (Homoptera: Cercopidae) sharing shelters on host plants. Ecology 67:465–478. [3,13]

McGuire, M.R. 1989. Control of the potato leafhopper: Past practices, current research, and future considerations. Misc. Publ. Entomol. Soc. Am. 72:50–57. [14]

McKinion, J.M., and Lemmon, H.E. 1985. Expert systems for agriculture. Comput. Electron. Agric. 1:31–40. [19]

McLachlan, A. 1983. Life-history tactics of rain-pool dwellers. J. Anim. Ecol. 52:545–561. [4]

McLain, D. K. 1982. Density dependent sexual selection and positive phenotypic assortative mating in natural populations of the soldier beetle, *Chauliognathus pennsylvanicus*. Evolution 36:1227–1235. [6]

McLean, D. L., and M. G. Kinsey. 1967. Probing behavior of the pea aphid, *Acyrthosiphon pisum*. I. Definitive correlation of electronically recorded waveforms with aphid probing activities. Ann. Entomol. Soc. Am. 60:400–416. [2]

McLean, D. L., and M. G. Kinsey. 1984. The precibarial valve and its role in the feeding behavior of the pea aphid, *Acyrthosiphon pisum*. Bull. Entomol. Soc. Am. 30:26–31. [12]

McMillian, W.W. 1963. Reproductive system and mating behavior of *Sogata orizicola* (Homoptera: Delphacidae). Ann. Entomol. Soc. Am. 56:330–334. [4,5,6]

McNair, J. N. 1986. The effects of refuges on predator-prey interactions: a reconsideration. Theor. Pop. Biol. 29:38–63. [11]

McNaughton, S.J., and J.L. Tarrants. 1983. Grass leaf silification: Natural selection for an inducible defense against herbivores. Proc. Nat. Acad. Sci. 80:790–791. [2]

McNeill, S., and R. A. Prestidge. 1982. Plant nutritional strategies and insect herbivore community dynamics. Pages 225–235. *In* J. H. Visser and A. K. Minks, editors. Proc. 5th Int. Symp. Insect-Plant Relationships, Wageningen, Pudoc, Netherlands. [1,3]

McNeill, S., and T. R. E. Southwood. 1978. The role of nitrogen in the development of insect/plant relationships. Pages 77–98. *In* J. B. Harborne, editor. Biomedical aspects of plant and animal coevolution. Academic Press, London. [2,3]

Mead, F.W., and J.P. Kramer. 1982. Taxonomic study of the planthopper genus *Oliarus* in the United States (Homoptera: Fulgoroidea: Cixiidae). Trans. Am. Entomol. Soc. 107:381–569. [1]

Medler, J. T. 1990. *Sosephena* and *Trisephena*, two new genera from New Guinea with tricarinate frons (Homoptera: Flatidae). Bishop Mus. Occ. Papers 30:204–218. [1]

Medrano F. G., and E. A. Heinrichs. 1984. A method for purifying brown planthopper (BPH) *Nilaparvata lugens* biotypes. Int. Rice Res. Newsl. 9(4):16–17. [17]

Medrano, F. G., and E. A. Heinrichs. 1985. Response of resistant rices to brown planthopper (BPH) collected in Mindanao, Philippines. Int. Rice Res. Newsl. 10(6):14–15. [17]

Metcalf, Z. P. 1946a. A new species of *Delphacodes* from Alberta (Fulgoroidea). Can. Entomol. 78:63–65. [1]

Metcalf, Z. P. 1946b. Homoptera. Fulgoroidea and Jassoidea of Guam. Insects of Guam–II. B. P. Bishop Mus. Bull. 189:105–148. [1]

Metcalf, Z. P. 1947. A new genus of Lophopidae from Brazil. Proc. Entomol. Soc. Wash. 49: 283–240. [1]

Metcalf, Z. P., and S. C. Bruner. 1930. Cuban Fulgorina. 1. The families Tropiducidae and Acanaloniidae. Psyche 37:395–424. [1]

Metcalf, Z. P., and S. C. Bruner. 1948. Cuban Flatidae with new species from adjacent regions. Ann. Entomol. Soc. Am. 51:63–118. [1]

Metcalf, Z.P., and V. Wade. 1966. A catalogue of the fossil Homoptera (Homoptera: Auchenorrhyncha). A general catalogue to the Homoptera, Supplement to Fasc. I—Membracidae. Waverly Press, Baltimore, MD. [1]

Metcalfe, J. R. 1969. Studies on the biology of the sugar-cane pest *Saccharosydne saccharivora* (Westw.) (Hom., Delphacidae). Bull. Entomol. Res. 59:393–408. [1,2,4,7,10]

Metcalfe, J. R. 1970. Studies on the effect of the nutrient status of sugar-cane on the fecundity of *Saccharosydne saccharivora* (Westw.) (Hom., Delphacidae). Bull. Entomol. Res. 60:309–325. [1,2,6,7]

Metcalfe, J.R. 1971. Observations on the ecology of *Saccharosydne saccharivora* (Westw.) (Hom., Delphacidae. Bull. Entomol. Res. 60:565–597. [8,10]

Metcalfe, J. R. 1972. An analysis of the population dynamics of the Jamaican sugar- cane pest *Saccharosydne saccharivora* (Westw.) (Hom., Delphacidae). Bull. Entomol. Res. 62:73–85. [i,11,13]

Meyerdirk, D. E., and W. G. Hart. 1982. Survey of Auchenorrhyncha (Insecta: Homoptera) assiciated with the Canary Island date palm in southern Texas. Florida Entomol. 65:327–334. [1]

Meyerdirk, D. E., and M. A. Hussein. 1985. Population dynamics of the beet leafhopper, *Circulifer tenellus* (Baker) and associated *Empoasca* spp. (Homoptera: Cicadellidae) and their egg parasitoids on sugar beets in southern California. J. Econ. Entomol. 78:346–353. [13]

Michelsen, A., F. Fink, M. Gogala, and D. Traue. 1982. Plants as transmission channels for insect vibrational songs. Behav. Ecol. Sociobiol. 11:269–281. [5]

Miller, J.C., and L.E. Ehler. 1990. The concept of parasitoid guild and its relevance to biological control. Pages 159–170. *In* M. Mackauer, L.E. Ehler, and J. Roland, editors. Critical Issues in Biological Control. Intercept, Andover, UK. [14]

Miller, W. R., and F. E. Egler. 1950. Vegetation of the Wequetequock-Pawcatuck tide-marshes, Connecticut. Ecol. Monogr. 20:143–172. [3]

Milne, R. G. 1988. Taxonomy of the rod-shaped filamentous viruses, Pages 3–50. *In* R. G. Milne, editor. The Plant Viruses, Vol. 4, The Filamentous Plant Viruses. Plenum Press, New York. [12]

Minks, A.K., and P. Harrewijn. 1987. World Crop Pests: Aphids, Their Biology, Natural Enemies, and Control. Vol. A. Elsevier, New York. [7]

Mitchell, W.C., and P.A. Maddison. 1983. Pests of Taro. Pages 180–268. In J.A. Wang, editor. Taro. A Review of *Colocasia esculenta* and Its Potentials. University of Hawaii Press, Honolulu. [1,2]

Mitomi, M., T. Ichikawa, and H. Okamoto. 1984. Morphology of the vibration-producing organ in adult rice brown planthopper, *Nilaparvata lugens* (Stål) (Homoptera, Delphacidae). Appl. Entomol. Zool. 19:407–417. [5]

Mitsuhashi, J. 1979. Artificial rearing and aseptic rearing of leafhopper vectors: Application in virus and MLO research. Pages 369–412. *In* K. Maramorosch and K. Harris, editors. Leafhopper Vectors and Plant Disease Agents. Academic Press, New York. [1]

Mitsuhashi, J., and K. Koyama. 1974. Folic acid as a dietary factor affecting the wing morph of the planthopper, *Laodelphax striatellus* (Hemiptera, Delphacidae). Entomol. Exp. Appl. 17:77–82. [4]

Mitter, C., and D. R. Brooks. 1983. Phylogenetic aspects of coevolution. Pages 65–98. *In* D. J. Futuyma and M. Slatkin, editors. Coevolution. Sinauer Press, Sunderland, MA. [1,12]

Mitter, C., and B. Farrell. 1991. Macroevolutionary espects of insect-plant relationships. Pages 36–78. *In* E. A. Bernays, editor. Insect-plant interactions, Vol. III. CRC Press, Boca Raton, FL. [1]

Mitter, C., and D.J. Futuyma. 1983. An evolutionary-genetic view of host-plant utilization by insects. Pages 427–458. *In* R.F. Denno and M.S. McClure, editors. Variable Plants and Herbivores in Natural and Managed Systems. Academic Press, New York. [15]

Miura, T. 1976a. Parasitism of *Gonatocerus* sp. (Hymenoptera: Mymaridae), an egg parasite of the green rice leafhopper, *Nephotettix cincticeps* Uhler in the paddy field. Bull. Faculty Agric. Shimane Univ. 10:43–48. [13]

Miura, T. 1976b. Parasitic activity of *Paracentrobia andoi* (Ishii) and *Gonatocerus* sp. (Hymenoptera: Mymaridae), two egg parasites of the green rice leafhopper, *Nephotettix cincticeps* Uhler in the paddy field. Bull. Faculty Agric. Shimane Univ. 10:49–55. [13]

Miura, T., Y. Hirashima, M. T. Chûjô, and Y. Chu. 1981. Egg and nymphal parasites of rice leafhoppers and planthoppers. A result of field studies in Taiwan in 1979 (Part 1). Esakia 16:39–50. [11,13]

Miura, T., Y. Hirashima, and T. Wongsiri. 1979. Egg and nymphal parasites of rice leafhoppers. A result of field studies in Thailand in 1977. Esakia 13:21–44. [13,14]

Miyai, S., K. Kiritani, and T. Sasaba. 1978. An empirical model of *Lycosa*-hoppers interaction system in the paddy field. Prot. Ecol. 1:9–21. [10]

Miyake, T., and A. Fujiwara. 1962. Studies on the hibernation and diapause of the white-backed planthopper, *Sogatella furcifera* Horvath and the brown planthopper, *Nilaparvata lugens* Stål. Bull. Hiroshima Agric. Exp. Stat. 13:4–73. [8]

Miyashita, K. 1963. Outbreaks and population fluctuation of insects, with special reference to agricultural insect pests in Japan, Bull. Nat. Inst. Agric. Sci., Series C. 15:99–170. [10]

Miyata, J. 1989. Problems in the control of insecticide-resistant rice plant- and leafhoppers. Pestic. Sci. 26:261–269. [16]

Mobberley, D. G. 1956. Taxonomy and distribution of the genus *Spartina*. J. Sci. Iowa St. Coll. 30:471–754. [3]

Mochida, O. 1964. On oviposition in the brown planthopper, *Nilaparvata lugens* (Stål) (Hom., Auchenorrhyncha). II. The number of eggs in an egg group, especially in relation to the fecundity. Japan. J. Appl. Entomol. Zool. 8:141–148. [4]

Mochida, O. 1970. A red-eyed form of the brown planthopper, *Nilaparvata lugens* (Stål) (Hom., Auchenorrhyncha). Bull. Kyushu Agric. Exp. Sta. 15:141–273. [4,9]

Mochida, O. 1973. The characters of the two wing-forms of *Javesella pellucida* (F.) (Homoptera:Delphacidae), with special reference to reproduction. Trans. Roy. Entomol. Soc. Lond. 125:177–225. [4,7,8]

Mochida, O. 1974. Long-distance movement of *Sogatella furcifera* and *Nilaparvata lugens* (Homoptera: Delphacidae) across the East China Sea. Rice Entomol. Newsl. 1:18–22. [9]

Mochida, O., and V.A. Dyck. 1977. Bionomics of the brown planthopper, *Nilaparvata lugens*. Pages 192–198. *In* The Rice Planthopper. Tokyo, 1976. Asian and Pacific Council, Food and Fertilizer Technology Center, Taipei. [14]

Mochida, O., and R. Kisimoto. 1971. A review of the studies on *Javesella pellucida* (F.) (Hom., Delphacidae) and associated subjects. Rev. Plant Prot. Res. 4:1–57. [1]

Mochida, O., and T. Okada. 1971. A list of the Delphacidae (Homoptera) in Japan with special reference to host plants, transmission of plant diseases, and natural enemies. Bull. Kyushu Agric. Exp. Sta. 15:737–843. [1]

Mochida, O., and T. Okada. 1973. Supplementary notes to "A list of the Delphacidae (Homoptera) in Japan with special reference to host plants, transmission of plant diseases, and natural enemies." Kontyu 41:166–169. [1]

Mochida, O., and T. Okada. 1979. Taxonomy and biology of *Nilaparvata lugens* (Hom., Delphacidae). Pages 21–43. *In* Brown Planthopper: Threat to Rice Production in Asia. International Rice Research Institute, Los Baños, Philippines. [1,4,8,10]

Montgomery, D. C., and E.A. Peck. 1982. Introduction to Linear Regression Analysis. Wiley, New York. [11]

Moran, V. C. 1980. Interactions between phytophagous insects and their *Opuntia* hosts. Ecol. Entomol. 5:153–164. [3]

Moratorio, M. S. 1977. Aspects of the biology of *Anagrus* spp. (Hymenoptera: Mymaridae) with special reference to host-parasitoid relationships. Ph.D. Dissertation, University of London, London. [11,14]

Morgan, F.D. 1984. Psylloidea of South Australia. Handbook of the flora and fauna of South Australia. Woolman, Government Printer, South Australia. [7]

Morgan, L. W., and R. H. Beamer. 1949. A revision of three genera of delphacine fulgorids from America north of Mexico (Homoptera—Fulgoridae -Delphacinae). J. Kansas Entomol. Soc. 22:97–120, 121–142. [1]

Mori, A., and K. Kiritani. 1971. Accumulative effect of larval density on the determination of wing-form in the brown planthopper (*Nilaparvata lugens* Stål). Japan. J. Ecol. 21:146–152. [4,7,8]

Mori, K., and F. Nakasuji. 1990. Genetic analysis of the wing-form determination of the small brown planthopper, *Laodelphax striatellus* (Hemiptera: Delphacidae. Res. Popul. Ecol. 32:279–287. [4]

Morris, M. G. 1971. The management of grassland for the conservation of invertebrate animals. Pages 527–552. *In* E. Duffey and A. S. Watt, editors. The scientific management of animal and plant communities for conservation. Blackwell, Oxford, UK. [3,13]

Morris, M. G. 1973. The effects of seasonal grazing on the Heteroptera and Auchenorrhyncha (Hemiptera) of chalk grassland. J. Appl. Ecol. 10:761–780. [3,13]

Morris, M.G. 1974. Auchenorrhyncha (Hemiptera) of the Burren with special reference to species associations of the grassland. Proc. Roy. Irish Acad. 74:7–30. [2,3,13]

Morris, M. G. 1981. Responses of grassland invertebrates to management by cutting. III. Adverse effects on Auchenorrhyncha. J. Appl. Ecol. 18:107–123. [3]

Morris, M. G., and K. H. Lakhani. 1979. Responses of grassland invertebrates to management by cutting. I. Species diversity of Hemiptera. J. Appl. Ecol. 16:77–98. [3,13]

Morris, R.F. 1959. Single-factor analysis in population dynamics. Ecology 40:580–588. [7]

Morrison, G., and P. Barbosa. 1987. Spatial heterogeneity, population "regulation" and local extinction in simulated host-parasitoid interactions. Oecologia 73:609–614. [11]

Morrison, G., and D.R. Strong. 1980. Spatial variations in host density and the intensity of parasitism: Empirical examples. Environ. Entomol. 9:149–152. [10]

Mound, L.A., B.S. Heming, and J.M. Palmer. 1980. Phylogenetic relationships between the families of recent Thysanoptera (Insecta). Zool. J. Linn. Soc. 69:111–141. [1]

Mountford, M. D. 1988. Population regulation, density dependence, and heterogeneity. J. Anim. Ecol. 57:845–858. [7,11]

Muir, F. 1917. The Derbidae of the Philippine Islands. Philippine J. Sci. 12:49–106. [1]

Muir, F. 1923. On the classification of the Fulgoroidea (Homoptera). Proc. Hawaiian Entomol. Soc. 5:205–247. [1]

Muir, F. 1924a. On a new cixiid attacking coconut palms (Homopt.). Bull. Entomol. Res. 14:456. [1]

Muir, F. 1924b. New and little known fulgorids from the West Indies (Homoptera). Proc. Hawaiian Entomol. Soc. 5:461–471. [1]

Muir, F. 1926. Contributions to our knowledge of South American Fulgoroidea (Homoptera). Part I. The family Delphacidae. Bull. Exp. Sta. Hawaiian Sugar Planters' Assoc. Entomol. Ser. Bull. 18:2–38. [1]

Muir, F. 1928. Notes on some African Derbidae (Homoptera).—II*. Ann. Mag. Nat. Hist. 10(1): 498–525. [1]

Muir, F. 1930a. On the classification of the Fulgoroidea. Ann. Mag. Nat. Hist. 10(6):461–478. [1]

Muir, F. 1930b. New Derbidae from Sierra Leone (Homoptera Fulgoroidea. Ann. Mag. Nat. Hist. 10(5):81–84. [1]

Müller, H.J. 1957. Über die Diapause von *Stenocranus minutus* Fabr. (Hom. Auchenorrhyncha). Beitr. Entomol. Berlin 7:203–226. [4]

Müller, H.J. 1958. Über den Einfluss der Photoperiode auf Diapause und Körpergrässe der Delphacidae *Stenocranus minutus* Fabr. (Hom. Auchenorrhyncha). Zool. Anz. 160:294–312. [4]

Mumford, J.D., and G.A. Norton. 1985. Economics of decision making in pest management. Annual Review of Entomology 29:157–174. [19]

Mumford, J.D. and G.A. Norton. 1989. Expert systems in pest management: implementation on an international basis. AI Appl. 3:67–69. [19]

Munson, M.A., P. Baumann, M.A. Clark, L. Baumann, N.A. Moran, D.J. Voegtlin, and B.C. Campbell. 1991. Aphid-eubacterial endosymbiosis: Evidence for its establishment in an ancestor of four aphid families. J. Bacteriol. 173:6321–6324. [1]

Munz, P.A., and D.D. Keck. 1965. A California flora. University of California Press, Berkeley. [4]

Murant, A. F., B. Raccah, and T. P. Pirone. 1988. Transmission Vectors. Pages 237–273. *In* R. G. Milne, editor. The Plant Viruses, Vol. 4, The Filamentous Plant Viruses. Plenum Press, New York. [12]

Murdoch, W.W. 1969. Switching in general predators: Experiments on predator and stability of prey populations. Ecol. Monogr. 39:335–354. [10,11]

Murdoch, W.W. 1970. Population regulation and population inertia. Ecology 51:497–502. [7]

Murdoch, W.W. 1975. Diversity, complexity, stability and pest control. J. Appl. Ecol. 12:795–807. [10]

Murdoch, W. W. 1979. Predation and the dynamics of prey populations. Fortschr. Zool. 25: 295–310. [11]

Murdoch, W.W. 1990. The relevance of pest-enemy models to biological control. Pages 1–24. *In* M. Mackauer, L.E. Ehler, and J. Roland, editors. Critical issues in biological control. Intercept, Andover, UK. [10,14]

Murdoch, W. W. 1992. Ecological theory and biological control. Pages 197–221. *In* L. Botsford and S. Jain, editors. Applied Population Biology. Kluwer Academic Publishers, Boston, MA. [11]

Murdoch, W.W., J. Chesson, and P.L. Chesson. 1985. Biological control in theory and practice. Am. Natur. 125:344–366. [10]

Murdoch, W. W., F. C. Evans, and C. H. Peterson. 1972. Diversity and pattern in plants and insects. Ecology 53:819–828. [3,13]

Murdoch, W. W., R.F. Luck, S.J. Walde, J.D. Reeve, and D.S. Yu. 1989. A refuge for red scale under control by *Aphytis*: structural aspects. Ecology 70:1707–1714. [10,11]

Murdoch, W. W., and A. Oaten. 1975. Predation and population stability. Adv. Ecol. Res. 9:2–131. [3,10,11,14]

Murdoch, W.W., J.D. Reeve, C.B. Huffaker, and C.E. Kennett. 1984. Biological control of the olive scale and its relevance to ecological theory. Am. Natur. 123:371–392. [10,14]

Murdoch, W.W., and A. Stuart-Oaten. 1989. Aggregation by parasitoids and predators: Effects on equilibrium and stability. Am. Natur. 134:288–310. [10]

Murphy, F., and A. Harrison. 1980. Electron microscopy of the rhabdoviruses on animals. Pages 65–107. *In* D. H. Bishop, editor. Rhabdoviruses, Vol. 1. CRC Press, Boca Raton, FL. [12]

Murphy, G.I. 1968. Pattern in life history and the environment. Am. Natur. 102:391–403. [4]

Myers, J.G. 1922. Life history of *Siphanta acuta* (Walk.), the large green planthopper. New Zealand J. Sci. Technol. 5:256–263. [1]

Myers, J. G. 1923. A contribution to the study of New Zealand leaf-hoppers and plant-hoppers (Cicadellidae and Fulgoroidea). Trans. New Zealand Inst. 54:407–429. [1]

Myers, J. G. 1929. Observations on the biology of two remarkable cixiid planthoppers (Homoptera) from Cuba. Psyche 36:283–292. [1]

Nachmann, G. 1987. Systems analysis of acarine predator-prey interactions. II. The role of spatial processes in system stability. J. Anim. Ecol. 56:267–281. [11]

Nagata, T. 1982. Insecticide resistance and chemical control of the brown planthopper, *Nilaparvata lugens* Stål (Homoptera: Delphacidae). Bull. Kyushu Nat. Agric. Exp. Sta. 22: 49–164. [9,16]

Nagata, T. 1983. Insecticide resistance in rice pests, with special emphasis on the brown planthopper (*Nilaparvata lugens* Stål). Pages 599–606. *In* Proc. 10th Int. Congress of Plant Protection, Vol. 2, Brighton, England. [16]

Nagata, T., and T. Masuda. 1980. Insecticide susceptibility and wing-form ratio of the brown planthopper, *Nilaparvata lugens* (Stål) (Hemiptera: Delphacidae) and the white backed planthopper, *Sogatella furcifera* (Horvath) (Hemiptera: Delphacidae) of southeast Asia. Appl. Entomol. Zool. 15:10–19. [4,9,16]

Nagata, T., and O. Mochida 1984. Development of insecticide resistance and tactics for prevention. Pages 93–106. *In* Proceedings of the FAO Workshop on Judicious and Efficient Use of Insecticides on Rice. International Rice Research Institute, Los Baños, Philippines. [i,16]

Nagata, T., T. Masuda, and S. Moriya. 1979. Development of insecticide resistance in the brown planthopper, *Nilaparvata lugens* Stål (Hemiptera: Delphacidae). Appl. Entomol. Zool. 14: 264–269. [16]

Nagata, T., and S. Moriya. 1969. Resistance to gamma BHC in the brown planthopper, *Nilaparvata lugens* Stål. Proc. Assoc. Plant Prot. Kyushu 15:113–115 (in Japanese with English summary). [16]

Nagata, T., and S. Moriya. 1974. Resistance in the brown planthopper, *Nilaparvata lugens* Stål, to lindane. Japan. J. Appl. Entomol. Zool. 18:73–80 (in Japanese with English summary). [16]

Naito, A. 1977. Feeding habits of leafhoppers. Japan. Agric. Res. Quart. 11:115–119. [12]

Nakamura, K. 1974. A model of the functional response of a predator to prey density involving the hunger effect. Oecologia 16:265–278. [10]

Nakamura, K. 1977. A model for the functional response of a predator to varying prey densities; based on the feeding ecology of wolf spiders. Bull. Nat. Inst. Agric. Sci., Ser. C 31:29–69. [10]

Nakasuji, F. 1987. Egg size in skippers (Lepidoptera: Hesperiidae) in relation to their host specificity and to leaf toughness of host plants. Ecol. Res. 2:175–183. [2]

Nakasuji, F., and V. A. Dyck. 1984. Evaluation of the role of *Microvelia douglasi atrolineata* (Bergroth) (Heteroptera: Veliidae) as predator of the brown planthopper *Nilaparvata lugens* (Stål) (Homoptera: Delphacidae). Res. Popul. Ecol. 26:134–149. [i,8,10,14]

Napompeth, B. 1973. Ecology and population dynamics of the corn planthopper, *Peregrinus maidis* (Ashmead) (Homoptera: Delphacidae), in Hawaii. Ph.D. Dissertation. University of Hawaii, Manoa. [i,2,4,7,8,10,14]

Nasu, S. 1963. Studies on some leafhoppers and planthoppers which trasmit virus diseases of rice plant in Japan. Bull. Kyushu Agric. Exp. Sta. 8:153–349. [12]

Nasu, S. 1965. Electron microscopic studies on transovarial passage of rice dwarf virus. Japan. J. Appl. Entomol. Zool. 9:225–237. [12]

Nasu, S. 1969. The Virus Diseases of the Rice Plant. Johns Hopkins Press, Baltimore, MD. [4]

National Academy of Sciences. 1986. Pesticide Resistance: Strategies and Tactics for Management. National Academy Press, Washington, DC. [16]

National Research Council, National Academy of Sciences. 1989. Alternative Agriculture. National Academy Press, Washington, DC. [20]

Nault, L. R. 1985. Evolutionary relationships between maize leafhoppers and their host plants. Pages 309–330. *In* L. R. Nault and J. G. Rodriguez, editors. The Leafhoppers and Planthoppers. Wiley, New York. [i,12]

Nault, L. R. 1987. Origin and evolution of Auchenorrhyncha-transmitted, plant infecting viruses. Pages 131–149. *In* M. R. Wilson and L. R. Nault, editors. Proceedings, 2nd Int. Workshop on Leafhoppers and Planthoppers of Economic Importance, July 28–Aug. 1, 1986, Provo, Utah. Commonw. Inst. Entomol., London. [12]

Nault, L. R. 1990. Evolution of an insect pest: maize and the corn leafhopper, a case study. Maydica 35:165–175. [12]

Nault, L. R., and E. D. Ammar. 1989. Leafhopper and planthopper transmission of plant viruses. Annu. Rev. Entomol. 34:503–529. [i,6,12]

Nault, L. R., and R. H. E. Bradley. 1969. Acquisition of maize dwarf mosaic virus by the green bug, *Schizaphis graminum*. Ann. Entomol. Soc. Am. 62:403–406. [12]

Nault, L. R., R. E. Gingery, and D. T. Gordon. 1980. Leafhopper transmission and host range of maize rayado fino virus. Phytopathology 70:709–712. [12]

Nault, L. R., and D. T. Gordon. 1988. Multiplication of maize stripe virus in *Peregrinus maidis*. Phytopathology 78:991–995. [12]

Nault, L. R., and J. G. Rodriguez, editors. 1985. The Leafhoppers and Planthoppers. Wiley, New York. [6,14]

Nelson, G., and N. Platnick. 1981. Systematics and Biogeography: Cladistics and Vicariance. Columbia Univ. Press. New York. [1]

Ng, E.K. 1978. Pest survey and legislation. Pages 193–210. *In* Proc. Plant Prot. Conf., Kuala Lumpur, Malaysia. [10]

Nielson, M.W., and H. Don. 1974. Probing behavior of biotypes of the spotted alfalfa aphid on resistant and susceptible clones. Entomol. Exp. Appl. 17:477–486. [1,2]

Niemela, P., J. Tuomi, and E. Haukioja. 1980. Age-specific resistance in trees: Defoliation of tamaracks (*Larix laricinia* by larch bud moth (*Zeiraphera improbana*). Report Kevo Subarctic Res. Sta. 16:49–57. [3]

Nijhout, H.F., and D.E. Wheeler. 1982. Juvenile hormone and the physiological basis of insect polymorphism. Quart. Rev. Biol. 57:109–133. [4]

Nishida, T. 1975. Causes of brown planthopper outbreaks. Rice Entomol. Newsl. 2:38. [2,3]

Noda, H. 1977. Histological and histochemical observation of intracellular yeastlike symbiotes in the fat body of the smaller brown planthopper, *Laodelphax striatellus* (Homoptera: Delphacidae). Appl. Entomol. Zool. 12:134–141. [2]

Noda, H. 1986a. Pre-mating flight of rice planthopper migrants (Homoptera: Delphacidae) collected on the East China Sea. Appl. Entomol. Zool. 21:175–176. [9]

Noda, H. 1986b. Damage to ears of rice plants caused by the white-backed planthopper, *Sogatella furcifera* (Homoptera: Delphacidae). Appl. Entomol. Zool. 21:474–476. [2]

Noda, H., and T. Saito. 1979a. The role of intracellular yeastlike symbiotes in the development of *Laodelphax striatellus* (Homoptera: Delphacidae). Appl. Entomol. Zool. 14:453–458. [2]

Noda, H., and T. Saito. 1979b. The effects of high temperature on the development of *Laodelphax striatellus* (Homoptera: Delphacidae) and on its intracellular yeastlike symbiotes. Appl. Entomol. Zool. 14:64–75. [2]

Noda, H., K. Wada, and T. Saito. 1979. Sterols in *Laodelphax striatellus* with special reference to the intracellular yeastlike symbiotes as a sterol source. J. Insect Physiol. 25:443–447. [2]

Noda, T., and K. Kiritani. 1989. Landing places of migratory planthoppers, *Nilaparvata lugens* (Stål) and *Sogatella furcifera* (Horvath) (Homoptera: Delphacidae) in Japan. Appl. Entomol. Zool. 24:59–65. [9]

Norton G.A., and C.S. Holling, editors. 1979. Pest Management: Proceedings of an International Conference. IIASA Proceedings Series, Volume 4. Pergamon Press, Oxford, UK. [19]

Norton, G.A., J. Holt, K.L. Heong, J.A. Cheng, and D.R. Wareing. 1991. Systems analysis and rice pest management. Pages 287–321. *In* E.A. Heinrichs and T.A. Miller, editors. Rice Insects: Management Strategies. Springer-Verlag, New York. [19]

Norton, G.A., and J.D. Mumford. 1982. Information gaps in pest management. Pages 589–597. *In* K.L. Heong, B.S. Lee, T.M. Lim, C.H. Tesh, and Y. Ibrahim, editors. Proceedings of the

International Conference on Plant Protection in the Tropics. Malaysia Plant Protection Society, Kuala Lumpur, Malaysia. [19]

Norton, G.A., and M.J. Way. 1990. Rice pest management systems—past and future. Pages 1–14. *In* B.T. Grayson, M.B. Green, and L.G. Copping, editors. Pest Management in Rice. Published for the Society of Chemical Industry by Elsevier Applied Science, London. [19]

Novak, J. D. 1970. The Improvement of Biology Teaching. Bobbs-Merrill Company, Inc., New York. [20]

Nuss, D. L. 1984. Molecular Biology of Wound Tumor Virus. Adv. Virus Res. 29:57–93. [12]

Oaten, A., and W.W. Murdoch. 1975a. Functional response and stability in predator-prey systems. Am. Natur. 109:299–318. [10,11]

Oaten, A., and W.W. Murdoch. 1975b. Switching, functional response, and stability in predator-prey systems. Am. Natur. 109:299–318. [10]

Obata, T., M. Kim, H. Koh, and H. Fukami. 1981. Planthopper attractants in the rice plant. Japan. J. Appl. Entomol. Zool. 25:47–51. [2]

O'Brian, P.J., G.W. Elzen, and S.B. Vinson. 1985. Toxicity of azinphosmethyl and chlordimeform to parasitoid *Bracon mellitus*: lethal and reproductive effects. Environ. Entomol. 14:891–894. [14]

O'Brien, L. B. 1967. *Caliscelis bonellii* (Latreille), a European genus of Issidae new to the United States (Homoptera: Fulgoroidea). Pan-Pacific Entomol. 43:130–133. [1]

O'Brien, L. B. 1971. The systematics of the tribe Plectoderini in America north of Mexico (Homoptera: Fulgoroidea, Achilidae). Univ. California Publ. Entomol. 64:1–79. [1]

O'Brien, L. B. 1986a. A new species of *Dictyssa* (Homoptera: Fulgoroidea: Issidae) from Baja California. Ann. Inst. Biol. Univ. Nal. Auton. Mexico, Ser. Zool. 56(1):137–140. [1]

O'Brien, L. B. 1986b. Five new species of Fulgoroidea (Homoptera) from the western United States and Mexico. Southwestern Entomol. 11:67–74. [1]

O'Brien, L. B. 1988. Taxonomic changes in North American Issidae (Homptera: Fulgoroidea). Ann. Entomol. Soc. Am. 81:865–869. [1]

O'Brien, L. B. 1992. Tropiduchidae (Homptera: Fulgoroidea) of the United States. Ann. Entomol. Soc. Am. 85:121–126. [1]

O'Brien, L. B., and S. W. Wilson. 1985. Planthopper systematics and external morphology. Pages 61–102. *In* L. R. Nault and J. G. Rodriguez, editors. The Leafhoppers and Planthoppers. Wiley, New York. [1,8,12,15]

O'Brien, R.D. 1976. Acetylcholinesterase and its inhibition. Pages 271–296. *In* C.F. Williamson, editor. Insecticide Biochemistry and Physiology. Plenum, New York. [16]

O'Connell, C. 1991. Acoustic communication and sexual selection in a group of Hawaiian planthoppers (Homoptera: Delphacidae). Masters Thesis. University of Hawaii, Manoa. [15]

Ofori, F. A., and R. I. B. Francki. 1985. Transmission of leafhopper A virus, vertically through eggs and horizontally through maize in which it does not multiply. Virology 144:152–157. [12]

Oh, R.J. 1979. Repeated copulation in the brown planthopper, *Nilaparvata lugens* Stål. (Homoptera:Delphacidae). Ecol. Entomol. 4:345–353. [4,5,6]

Ohkubo, N. 1967. Study on the density effect of the adult of the brown planthopper, *Nilaparvata lugens* Stål. Japan. J. Ecol. 17:230–233. [8,9]

Ohkubo, N. 1973. Experimental studies on the flight of planthoppers by the tethered flight technique. I. Characteristics of flight of the brown planthopper *Nilaparvata lugens* (Stål) and effects of some physical factors. Japan. J. Appl. Entomol. Zool. 17:10–18. [9]

Ohkubo, N. 1981. Behavioral and ecological studies on the migratory flight of rice planthoppers. Ph. D. dissertation, Kyoto University, Japan. [9]

Ohkubo, N., and R. Kisimoto. 1971. Diurnal periodicity of the flight behavior of the brown planthopper, *Nilaparvata lugens* Stål, in the 4th and 5th emergence periods. Japan. J. Appl. Entomol. Zool. 15:8–16. [8,9]

Oka, I. N. 1979. Cultural control of the brown planthopper. Pages 357–369. *In* Brown Planthopper: Threat to Rice Production in Asia. International Rice Research Institute, Los Baños, Philippines. [3,10,14,18]

Okada, T. 1977. Taxonomic characters of identification of the rice brown planthopper (*Nilaparvata lugens*) and its related species in the asian and Pacific region. Proc. Intern. Seminar, Food and Fertilizer Technology Center for the Asian and Pacific Region, Taipei, Taiwan. 1:1–26. [4]

Okada, T., and J. Hirao. 1981. Delphacid planthoppers caught by a light trap at Chikugo, Fukuoka, during 1966–1975. Proc. Assoc. Plant Prot. Kyushu 27:71–73. [9]

Okumura, T. 1963. Induction of diapause in eggs of the planthoppers, *Sogatella furcifera* and *Nilaparvata lugens*, by changing rearing conditions of their adult life. Japan. J. Appl. Entomol. Zool. 7:285–290. [8]

Okuyama, S., K. Yora, and H. Asuyama. 1968. Multiplication of the rice stripe virus in its vector, *Laodelphax striatellus* Fallen. Ann. Phytopathol. Soc. Japan. 34:255–264. [12]

Olmstead, R.G., H.J. Michaels, K.M. Scott, and J.D. Palmer. 1992. Monophyly of the Asteridae and identification of their major lineages inferred from DNA sequences of *rbc*L. Ann. Missouri Bot. Garden 70:249–265. [1]

Oloumi-Sadeghi, H., L. R. Zavaleta, W. O. Lamp, E. J. Armbrust, and G. Kapusta. 1987. Interactions of the potato leafhopper (Homoptera: Cicadellidae) with weeds in an alfalfa ecosystem. Environ. Entomol. 16:1175–1180. [3,13]

Ooi, P. A. C., 1980. Seasonal abundance of the whitebacked planthopper and brown planthopper and predators in insecticide free rice fields in Malaysia. Int. Rice Res. Newsl. 5(1): 13–14. [17]

Ooi, P.A.C. 1982. Attempts at forecasting rice planthopper populations in Malaysia. Entomophaga 27:89–96. [10]

Ooi, P. A. C. 1984. Insecticides disrupt natural control of *Nilaparvata lugens* in Sekinchan, Malaysia. Pages 109–120. *In* M. Y. Hussein and A. G. Ibrahim, editors. Biological Control in the Tropics. Universiti Pertanian Malaysia, Serdang, Malaysia. [20]

Ooi, P.A.C. 1988. Ecology and surveillance of *Nilaparvata lugens* (Stål)—implications for its management in Malaysia. Ph. D. Dissertation. University of Malayia, Kuala Lumpur. [i,10, 14,17]

Ooi, P. A. C., and K. L. Heong. 1988. Operation of a brown planthopper surveillance system in the Tanjung Karang Irrigation Scheme in Malaysia. Crop Prot. 7:273–278. [20]

Ooi, P.A.C., and B.M. Shepard. In press. Predators and parasitoids of rice insect pests. *In* E.A. Heinrichs, editors. Management of Rice Insects. Wiley-Eastern, New Delhi. [10,18]

Orita, S. 1969. *Gonatocerus* sp. (Hymenoptera: Mymaridae), a natural enemy of green rice leafhopper, *Nephotettix cincticeps* Uhler. Proc. Assoc. Plant Prot. Hokusiku 17:67–69. [13]

Orita, S. 1972. Some notes on *Lymaenon* sp. (Hymenoptera: Mymaridae), an egg parasite of green rice leafhopper, *Nephotettix cincticeps* Uhler (Homoptera: Cicadellidae) and its distribution in Hokuriku District. Bull. Hokuriku Agric. Exp. Sta. 14:122–127. [13]

Osborn, A.W. 1974. Population dynamics of the Australian sugarcane leafhopper, *Perkinsiella saccharicida* Kirkaldy. Proc. Int. Soc. Sugar Cane Technol. 15:463–467. [2,10]

Osborn, H. 1903. A subterranean root-infesting fulgorid (*Myndus radicis* n. sp.). Ohio Natur. 4:42–44. [1]

Osborn, H. 1926. Faunistic and ecological notes on Cuban Homoptera. Ann. Entomol. Soc. Am. 19:335–366. [1]

Ossiannilsson, F. 1949. Insect drummers. A study on the morphology and function of the sound-producing organ of Swedish Homoptera, Auchenorrhyncha, with notes on their sound-production. Opuscula Entomol. Suppl. 10:1–145. [5]

Ossiannilsson, F. 1953. On the music of some European leafhoppers (Homoptera, Auchenorrhyncha) and its relation to courtship. Trans. IXth Int. Congr. Entomol. Amsterdam, 1953. 2:139–141. [5]

Ossiannilsson, F. 1978. The Auchenorrhyncha (Homoptera) of Fennoscandia and Denmark. Part 1: Introduction, infraorder Fulgoromorpha. Fauna Entomol. Scand. 7(1):3–222. [1,4]

Ôtake, A. 1967. Studies on the egg parasites of the smaller brown planthopper, *Laodelphax striatellus* (Fallén) (Hemiptera: Delphacidae). I. A device for assessing the parasitic activity, and the results obtained in 1966. Bull. Shikoku Agric. Exp. Sta. 17:91–103. [11]

Ôtake, A. 1968. Studies on the egg parasites of the smaller brown planthopper *Laodelphax striatellus* (Fallén) (Hemiptera: Delphacidae). II. Development of *Anagrus* nr. *flaveolus* Waterhouse (Hymenoptera: Mymaridae) within its host. Bull. Shikoku Agric. Exp. Sta. 18: 161–169. [11]

Ôtake, A. 1970. Studies on the egg parasites of the smaller brown planthopper, *Laodelphax striatellus* (Fallén) (Hemiptera: Delphacidae). IV. Seasonal trends in parasitic and dispersal activities, with special reference to *Anagrus* nr. *flaveolus* Waterhouse (Hymenoptera: Mymaridae). Appl. Entomol. Zool. 5:95–104. [11]

Ôtake, A. 1976. Trapping of *Anagrus* nr. *flaveolus* Waterhouse (Hymenoptera: Mymaridae) by the eggs of *Laodelphax striatellus* (Fallén) (Hemiptera: Delphacidae). Physiol. Ecol. Japan 17:473–475. [11]

Ôtake, A. 1977. Natural enemies of the brown planthopper. Pages 42–57. *In* The Rice Brown Planthopper. Food and Fertilizer Technology Center for the Asian and Pacific Region (ASPAC), Taipei, Taiwan. [8,10]

Ôtake, A., and N. Hokyo. 1976. Rice plant- and leafhopper incidence in Malaysia and Indonesia. Report of a research tour January to March. Shiryo N. 33. Trop. Agric. Res. Centre, Tokyo, Japan. [10,14]

Ôtake, A., P.H. Somasundaram, and M.B. Abeykoon. 1976. Studies on populations of *Sogatella furcifera* Horvath and *Nilaparvata lugens* Stål (Hemiptera: Delphacidae) and their parasites in Sri Lanka. Appl. Entomol. Zool. 11:284–294. [8,11,14]

Oya, S. and J. Hirao. 1982. Catches of migrating rice planthoppers on the East China Sea and northern Kyushu, Japan in late June 1981. Proc. Assoc. Pl. Prot. Kyushu 28:117–121. [9]

Oyama, M. 1972. Observation of mating behaviour of planthoppers and leafhoppers. Presented paper 32nd Annual Meeting of the Entomological Society of Japan, Matsuyama. [6]

Ozaki, K. 1969. The resistance to organophosphorus insecticides of the green rice leafhopper, *Nephotettix cincticeps* Uhler and the smaller brown planthopper, *Laodelphax striatellus* Fallen. Rev. Plant Prot. Res. 2:1–15. [16]

Ozaki, K. 1983. Suppression of resistance through synergistic combinations with emphasis on planthoppers and leafhoppers infesting rice in Japan. Pages 599–613. *In* T. Saito and G.P. Georghiou, editors. Pest Resistance to Pesticides. Plenum Press, New York. [16]

Ozaki, K., and T. Kassai. 1984. The insecticidal activity of pyrethroids against insecticide-resistant-strains of planthoppers, leafhoppers and the housefly. J. Pestic. Sci. 9(1):61–66. [16]

Ozaki, K., Y. Sasaki, M. Veda, and T. Kassai. 1973. Results of the alternate selection with mixtures with two or three ones on *Laodelphax striatellus* Fallen. Botyu-Kagaku 38:222–230. [16]

Pacala, S. W., and M.P. Hassell 1991. The persistence of host-parasitoid associations in patchy environments. II. Evaluation of field data. Am. Natur. 138:584–605. [11]

Pacala, S. W., M.P. Hassell, and R.M. May. 1990. Host-parastoid associations in patchy environments. Nature (London) 344:150–153. [11]

Packard, A. S. 1890. Insects injurious to forest and shade trees. Rept. U.S. Entomol. Comm. 5:1–955. [1]

Padgham, D.E. 1983a. The influence of the host-plant on the development of the adult brown planthopper, *Nilaparvata lugens* (Stål) (Hemiptera: Delphacidae), and its significance in migration. Bull. Entomol. Res. 73:117–128. [2,9,15]

Padgham, D.E. 1983b. Flight fuels in the brown planthopper *Nilaparvata lugens*. J. Insect Physiol. 29:95–99. [2,9]

Padgham, D. E., T. J. Perfect, and A. G. Cook. 1987. Flight behaviour in the brown planthopper, *Nilaparvata lugens* (Stål) (Homoptera:Delphacidae). Insect Sci. Appl. 8:71–75. [6,9]

Padgham, D. E., and S. Woodhead. 1986. Resistance mechanisms in rice to the brown planthopper, *Nilaparvata lugens*. TDRI/IRRI Collaborative Project A1107/C0453. Report for 1986, Jan-Oct. Tropical Development and Research Institute, London. [17]

Padgham, D.E., S. Woodhead, and H.R. Rapusas. 1989. Feeding responses of the brown planthopper, *Nilaparvata lugens* (Stål) (Hemiptera: Delphacidae), to resistant and susceptible host-plants. Bull. Entomol Res. 79:309–318. [15]

Page, R. D. M. 1990. Component analysis: A valiant failure? Cladistics 6:119–136. [1]

Paguia, P., M. D. Pathak, and E. A. Heinrichs. 1980. Honeydew excretion measurement techniques for determining differential feeding activity of biotypes of *Nilaparvata lugens* on rice varieties. J. Econ. Entomol. 73:35–40. [17]

Paik, W. H. 1977. Historical review of the occurrence of the brown planthopper in Korea. Pages 230–247. *In* The Rice Brown Planthopper. Asian and Pacific Region Food and Fertilizer Technology Centre, Taipei, Taiwan. [16]

Panda, N., and E.A. Heinrichs. 1983. Levels of tolerance and antibiosis in rice varieties having moderate resistance to the brown planthopper, *Nilaparvata lugens* (Stål), (Homoptera: Delphacidae). Environ. Entomol. 12:1204–1214. [16]

Park, J.S. 1973. Studies on the recent occurrence tendency of major insect pests of the rice plant. Symp. Plant Environ. Res., Research Rep. Off. Rural Development, Ministry of Agric., Korea. Pages 91–102. [9]

Parker, G. A. 1970. The reproductive behaviour and the nature of sexual selection in *Scatophaga stercoraria* L. (Diptera:Scatophagidae). II. The fertilization rate and the spatial and temporal relationships of each sex around the site of mating and oviposition. J. Anim. Ecol. 39:205–228. [6]

Partridge, L., and J. A. Endler. 1987. Life history constraints on sexual selection. Pages 265–277. *In* J. W. Bradbury and M. B. Andersson, editors. Sexual Selection: Testing the Alternatives. Wiley, New York. [6]

Partridge, L., A. Hoffmann, and J. S. Jones. 1987. Male size and mating success in *Drosophila melanogaster* and *D. pseudoobscura* under field conditions. Anim. Behaviour 35:468–476. [6]

Paterson, H.E.H. 1978. More evidence against speciation by reinforcement. S. Afr. J. Sci. 74: 369–371. [5]

Paterson, H.E.H. 1985. The recognition concept of species. Pages 21–29. *In* E.S. Vrba, editor. Species and Speciation. Transvaal Museum Monograph No. 4. Transvaal Museum, Pretoria, South Africa. [5]

Pathak, M. D. 1968. Ecology of common insect pests of rice. Annu. Rev. Entomol. 13:257–294. [3]

Pathak, M.D., F. Andres, N. Galacgac, and R. Raros. 1971. Resistance of rice varieties to striped rice borers. Int. Rice Res. Inst. Tech. Bull. 11. [2]

Pathak, M. D., and G. S. Khush. 1979. Studies of varietal resistance in rice to the brown planthopper at the international rice research institute. Pages 285–301. *In* Brown Planthopper: Threat to Rice Production in Asia. International Rice Research Institute, Los Baños, Philippines. [18]

Pathak, M. D., and R. C. Saxena. 1980. Breeding approaches in rice. Pages 421–455. *In* F. G. Maxwell and P. E. Jennings, editors. Breeding Plants Resistant to Insects. Wiley, New York. [17,18]

Pathak, P.K., and E.A. Heinrichs. 1982. Selection of biotype populations 2 and 3 of *Nilaparvata lugens* by exposure to resistant rice varieties. Environ. Entomol. 11:85–90. [2,10,15,16, 17,18]

Pathak, P. K., R. C. Saxena, and E. A. Heinrichs. 1982. Parafilm sachet for measuring honeydew excretion of *Nilaparvata lugens* on rice. J. Econ Entomol. 75(2):194–195. [17]

Pedgley, D.E. 1982. Windborne pests and diseases: meteorology of airborne organisms. Ellis Horwood, Chichester, UK. [9]

Pekkarinen, A., and M. Raatikainen. 1973. The Strepsiptera of eastern Fennoscandia. Notulae Entomol. 53:1–10. [13]

Peña, N., and M. Shepard. 1986. Seasonal incidence of parasitism of brown planthoppers, *Nilaparvata lugens* (Homoptera: Delphacidae), green leafhoppers, *Nephotettix* spp., and whitebacked planthoppers, *Sogatella furcifera* (Homoptera: Cicadellidae) in Laguna Province, Philippines. Environ. Entomol. 15:263–267. [11,13,14]

Perfect, T.J. and A.G. Cook. 1982. Diurnal periodicity of flight in some Delphacidae and Cicadellidae associated with rice. Ecol. Entomol. 7:317–326. [9]

Perfect, T.J., and A.G. Cook. 1987. Dispersal patterns of the rice brown planthopper, *Nilaparvata lugens* (Stål), in a tropical rice-growing system and their implications for crop protection. J. Plant. Prot. Trop. 4:121–127. [7,8]

Perfect, T.J., A.G. Cook, D.E. Padgham, and J.M. Crisostomo. 1985. Interpretation of the flight activity of *Nilaparvata lugens* (Stål) and *Sogatella furcifera* (Horváth) (Hemiptera: Delphacidae) based on comparative trap catches and marking with rubidium. Bull. Entomol. Res. 75:93–106. [6,8]

Perkins, J.H. 1982. Insects, experts and the insecticide crisis—the quest for new pest management strategies. Plenum Press, New York. [19]

Peterson, C. H. 1984. Does a rigorous criterion for environmental identity preclude the existence of multiple stable points? Am. Natur. 124:127–13. [11]

Petterson, J. 1970. Studies on *Rhopalosiphum padi* (L). I. Laboratory studies on olfactometric responses to the winter host, *Prunus padus*. Lantbrukshogskolans Annaler 36:381–399. [2]

Pfeiffer, J. W., and R. G. Wiegert. 1981. Grazers on *Spartina* and their predators. Pages 87–112. *In* L. R. Pomeroy and R. G. Wiegert, editors. The Ecology of a Salt Marsh. Springer-Verlag, New York. [6,10]

Pfuhl, E. H. 1988. Radio-based communication campaigns: a strategy for training farmers in IPM in the Philippines. Pages 251–255. *In* P. S. Teng and K. L. Heong, editors. Pesticide Management and Integrated Pest Management in Southeast Asia. Consortium for International Crop Protection, College Park, MD. [20]

Piaget, J. 1977. The Development of Thought: Equilibration of Cognitive Structures. Viking Press, New York. [20]

Piaget, J. 1985. The Equilibration of Cognitive Structures: The Central Problem of Educational Development. University of Chicago Press, Chicago, IL. [20]

Picard, F. 1921. La faune entomolique du figuier. Rev. Appl. Entomol. 9:23–24. [1]

Pielou, E. C. 1974. Population and Community Ecology. Principles and Methods. Gordon and Breach, New York. [11]

Pilson, D. 1992. Insect distribution patterns and the evolution of host use. Pages 120–139. *In* R. S. Fritz and E. L. Simms, editors. Plant Resistance to Herbivores and Pathogens, Ecology, Evolution and Genetics. University of Chicago Press, Chicago, IL. [18]

Pimbert, M. P. 1991. Designing integrated pest management for sustainable and productive futures. Gatekeeper Series No. 29. International Institute for Environment and Development, London. [20]

Pimentel, D. 1963. Introducing parasites and predators to control native pests. Can. Ent. 95:785–792. [14]

Pincus, J. 1991. Impact study of Farmer Field Schools. Available from the National Programme for Development and Training of Integrated Pest Management in Rice-Based Cropping Systems, Jl. Ki Mangunsarkoro No. 5, Jakarta, Indonesia. [20]

Pinto, J. D., and S. I. Frommer. 1980. A survey of the arthropods on jojoba (*Simmondsia chinensis*). Environ. Entomol. 9:137–143. [1]

Pitcairn, M. J., W.M. Getz, and D.W. Williams. 1990. Resource availability and parasitoid abundance in the analysis of host-parasitoid data. Ecology 71:2372–2374. [11]

Plapp, F.W., and D.L. Bull. 1978. Toxicity and selectivity of some insecticides to *Chrysopa carnea*, a predator of the tobacco budworm. Environ. Entomol. 7:431–434. [14]

Poethke, H.J., and M. Kirchberg. 1987. On the stabilizing effect of density-dependent mortality factors. Oecologia 74:156–158. [7]

Polis, G.A., C.A. Myers, and R.D. Holt. 1989. The ecology and evolution of intraguild predation: Potential competitors eat each other. Annu. Rev. Syst. Ecol. 20:297–330. [10]

Pollard, D. G. 1973. Plant penetration by feeding aphids (Hemiptera, Aphidoidea): a review. Bull. Entomol. Res. 62:631–714. [12]

Pongprasert, S., and P. Weerapat. 1979. Varietal resistance to the brown planthopper in Thailand. Pages 273–283. *In* Brown Planthopper: Threat to Rice Production in Asia. International Rice Research Institute, Los Baños, Philippines. [10]

Port, G. 1978. Studies on the feeding biology of Auchenorrhyncha (Homoptera). Ph.D. Dissertation, University of London. [2]

Prestidge, R.A. 1982a. Instar duration, adult consumption, oviposition and nitrogen utilization efficiencies of leafhoppers feeding on different quality food (Auchenorrhyncha: Homoptera). Ecol. Entomol. 7:91–101. [2,3,4,6,15]

Prestidge, R.A. 1982b. The influence of nitrogenous fertilizer on the grassland Auchenorrhyncha Homoptera. J. Appl. Ecol. 19:735–750 [2,3]

Prestidge, R.A. and S. McNeill. 1982. The role of nitrogen in the ecology of grassland Auchenorrhynch. Pages 257–281. *In* J.A. Lee, I.H. Rorison, and S. McNeill, editors. Nitrogen as an Ecological Factor. Symp. British Ecol. Soc. 22. Blackwell Scientific Publications., Oxford, UK. [2,3,4,6,8]

Prestidge, R.A., and S. McNeill. 1983. Auchenorrhyncha-host plant interactions: Leafhoppers and grasses. Ecol. Entomol. 8:331–339. [i,2,4,15]

Price, P.W. 1973. Reproductive strategies in parasitoid wasps. Am. Natur. 107:684–693. [4]

Price, P.W. 1974. Strategies for egg production. Evolution 28:76–84. [4]

Price, P.W. 1975. Reproductive strategies of parasitoids. Pages 87–111. *In* P.W. Price, editor. Evolutionary Strategies of Parasitic Insects and Mites. Plenum Press, New York. [14]

Price, P.W. 1983. Hypotheses on organization and evolution in herbivorous insect communities. Pages 559–596. *In* R.F. Denno and M.S. McClure, editors. Variable Plants and Herbivores in Natural and Managed Systems. Academic Press, New York. [3,7]

Price, P.W. 1984. Insect Ecology, Second edition. Wiley, New York. [14]

Price, P. W. 1988. Inversely density-dependent parasitism: The role of plant refuges for hosts. J. Anim. Ecol. 57:89–96. [11]

Price, P.W., C.E. Bouton, P. Gross, B.A. McPherson, J.N. Thompson, and A.E. Weis. 1980. Interactions among three trophic levels: Influence of plants on interactions between herbivores and natural enemies. Annu. Rev. Ecol. Syst. 11:41–65. [1,2,11]

Price, P. W., and Clancy, K. M. 1986. Interactions among three trophic levels: Gall size and parasitoid attack. Ecology 67:1593–1600. [11]

Price, T., and D. Schluter. 1991. On the low heritability of life-history traits. Evolution 45: 853–861. [15]

Procter, W. 1938. Biological survey of the Mount Desert Region. Part IV. The Insect Fauna with References to Methods of Capture, Food Plants, the Flora and other Biological Features. Wistar Institute of Anatomy Biology, Philadelphia. [1]

Pschorn-Walcher, H. 1977. Biological control of forest insects. Annu. Rev. Entomol. 22:1–22. [14]

Purcell, A. H. 1982. Evolution of the insect vector relationship, Pages 121–156. *In* M.S. Mount and G. H. Lacey, editors. Phytopathogenic Prokaryotes, Vol. 1. Academic Press, New York. [12]

Purcell, A. H., and L. R. Nault. 1991. Interactions among plant pathogenic prokaryotes, plants and insect vectors. Pages 383–405. *In* P. Barbosa, V. A. Krischik, and C. Jones, editors. Microorganisms, Plants, and Herbivores. Wiley, New York. [12]

Qingcai, W., and M. A. Jervis. 1988. Foraging for patchily distributed preys by *Cyrtorhinus lividipennis* (Reuter). J. Southwest Agric. Univ., China 10:245–252. (in Chinese, English summary). [10]

Quartau, J. A. 1975. On the origin of two flatids (Homoptera Auchenorrhyncha) of the Salvage Islands. Bocagiana Mus. Mun. Funchal 37:1–8. [1]

Quartau, J. A. 1981. Ecological notes on *Batrachomorphus* Lewis (Insecta, Homoptera, Cicadellidae) in Africa. Bol. Soc. Portug. Entomol. 14:1–10. [13]

Raatikainen, M. 1960. The biology of *Calligypona sordidula* (Stål) (Hom., Auchenorrhyncha). Ann. Entomol. Fennici 26:229–242. [4]

Raatikainen, M. 1967. Bionomics, enemies and population dynamics of *Javesella pellucida* (F.) (Homoptera: Delphacidae). Ann. Agric. Fennae 6:1–149. [i,3,4,6,8,9,10,11,14]

Raatikainen, M. 1972. Dispersal of leafhoppers and their enemies to oat fields. Ann. Agric. Fennae 11:146–153. [4,9,11]

Raatikainen, M., and A. Vasarainen. 1964. Biology of *Dicranotropis hamata* (Boh.) (Hom., Araeopidae). Ann. Agric. Fenniae 3:311–323. [4]

Raatikainen, M., and A. Vasarainen. 1976. Composition, zonation and the origin of the leafhopper fauna of oat fields in Finland. Ann. Zool. Fennici. 13:1–24. [2,4]

Raatikainen, M., and A. Vasarainen. 1990. Biology of *Metadelphax propinqua* (Fieber) (Homoptera, Delphacidae). Entomol. Fennica 3:145–149. [2,4]

Rabb, R.L., and J.R. Bradley. 1968. The influence of host plants on parasitism of eggs of the tobacco hornworm. J. Econ. Entomol. 61:1249–1252. [14]

Rahman, K. A., and R. Nath. 1940. Bionomics and control of the Indian sugar-cane leafhopper, *Pyrilla perpusilla*, Wlk. (Rhynchota, Fulg.) in the Punjab. Bull. Entomol. Res. 31:179–190. [1]

Raman, K. 1981. Studies on the foliar application of insecticides on the resurgence of the brown planthopper, *Nilaparvata lugens* (Stål) in rice. M.S. Thesis, Tamil Nadu Agric. University, Coimbatore, India. [16]

Raman, K., and S. Uthamasamy. 1983. Influence of foliar application of insecticides on the resurgence of brown planthopper, *Nilaparvata lugens* (Stål) in rice. Entomon 8 (1):41–45. [16,18]

Ramaraju, K., P. C. Sundara Babu, and K. Gunathilagaraj. 1989. Whitebacked planthopper (WBPH) *Sogatella furcifera* (Horvath) survival and nymph emergence on some rice varieties. Int. Rice Res. Newsl. 14(6):9 [17]

Ramos, J. A. 1957. A review of the Auchonrhynchous Homoptera of Puerto Rico. J. Agric. Univ. Puerto Rico 41:38–117. [1]

Randles, J. W., and D. Hanold. 1989. Coconut foliar decay virus particles are 20 nm icosahedra. Intervirology 30:177–180. [12]

Raupp, M. J., and R. F. Denno. 1979. The influence of patch size on a guild of sap-feeding insects that inhabit the salt marsh grass *Spartina patens*. Environ. Entomol. 8:412–417. [1,3,10,15]

Rausher, M. 1981. The effect of native vegetation on the susceptibility of *Aristolochia reticulata* (Aristolochiaceae) to herbivore attack. Ecology 62:1187–1195. [3]

Reddingius, J. 1971. Gambling for existence. A discussion of some theoretical problems in animal population ecology. Acta Biotheor. (Leiden) 20(Suppl.):1–208. [11]

Reddingius, J., and P.J. den Boer. 1970. Simulation experiments illustrating stabilization of animal numbers by spreading of risk. Oecologia 15:240–284. [10,11]

Reddy, D. V. R., and L. M. Black. 1974. Deletion mutations of the genome segments of wound tumor virus. Virology 61:458–473. [12]

Reddy, P. S., and K. L. Heong. 1991. Co-variation between insects in a ricefield and important spider species. IRRN 16:24. [10]

Redfield, A. C. 1972. Development of a New England salt marsh. Ecol. Monogr. 42:201–237. [3]

Reeve, J. D. 1988. Environmental variability, migration and persistence in host-parasitoid systems. Am. Natur. 132:810–836. [10,11]

Reeve, J.D., and W.W. Murdoch. 1986. Biological control by the parasitoid *Aphytis melinus*, and population stability of the California red scale. J. Anim. Ecol. 55:1069–1082. [10]

Reichert, S.E., and T. Lockley. 1984. Spiders as biological control agents. Annu. Rev. Entomol. 29:299–320. [10]

Reimer, N.J., and R.D. Goeden. 1982. Life history of the delphacid planthopper *Stobaera tricarinata* (Say) on western ragweed, *Ambrosia psilostachya* Decandolle, in southern California (Hemiptera-Homoptera:Delphacidae). Pan-Pacific Entomol. 58:105–108. [2,4]

Reinert, J.A. 1980. Phenology and density of *Haplaxius crudus* (Homoptera: Cixiidae) on three southern turfgrasses. Environ. Entomol. 9:13–15. [1]

Reissig, W.H., E.A., Heinrichs, and S.L. Valencia. 1982a. Insecticide induced resurgence of the brown planthopper, *Nilaparvata lugens* on rice varieties with different levels of resistance. Environ. Entomol. 11:165–168. [i,7,14,16]

Reissig, W. H., E. A. Heinrichs, and S. L. Valencia. 1982b. Effects of insecticides on *Nilaparvata lugens*, and its predators: spiders, *Microvelia atrolineata*, and *Cyrtorhinus lividipennis*. Environ. Entomol. 11:193–199. [8,10,16]

Reissig, W. H., E. A. Heinrichs, J. A. Litsinger, K. Moody, L. Fiedler, T. W. Mew, and A. T. Barrion. 1985. Illustrated guide to integrated pest management in rice in tropical Asia. International Rice Research Institute, Manila, Philippines. [20]

Remane, R. 1985a. Vorlaufige Anmerkungen zur Evolution und Speziation der Gattung Issus F. auf den Mittelatlantishen Inseln (Kanaren, Madeira) (Homoptera Auchnorrhyncha Fulgoromorpha Issidae). Marburger Entomol. Publ. 1(10):1–168. [1]

Remane, R. 1985b. Kinnaridae in der SW—Palaarktis: Zwei neue Taxa von den Kanaren (Homoptera Fulgoromorpha). Marburger Entomol. Publ. 1(10):241–264. [1]

Renault, T.R., and C.A. Miller. 1972. Spiders in a fir-spruce biotype: Abundance, diversity, and influence on spruce budworm densities. Can. J. Zool. 50:1039–1046. [10]

Reuter, O. M. 1875 Remarques sur le polymorphisme des hemipteres. Ann. Soc. Entomol. France 5:225–236. [3,4]

Rey, J. R. 1981. Ecological biogeography of arthropods on *Spartina* islands in northwest Florida. Ecol. Monogr. 51:237–265. [3]

Reynolds, D.R., and M.R. Wilson. 1989. Aerial samples of micro-insects migrating at night over central India. J. Plant Prot. Trop. 6:89–101. [9]

Rhoades, D.F. 1983. Herbivore population dynamics and plant chemistry. Pages 155–220. *In* R.F. Denno and M.S. McClure, editors. Variable Plants and Herbivores in Natural and Managed Systems. Academic Press, New York. [2]

Riley, J. R., X. N. Cheng, X.X. Zhang, D. R. Reynolds, G.M. Xu, A. D. Smith, J.Y. Cheng, A.D. Bao, and B.P. Zhai. 1991. The long-distance migration of *Nilaparvata lugens* (Stål) (Delphacidae) in China: Radar observations of mass return flight in the autumn. Ecol. Entomol. 16:471–489. [6,9]

Riley, J. R., D. R. Reynolds, and R. A. Farrow. 1987. The migration of *Nilaparvata lugens* (Stål) (Delphacidae) and other Hemiptera associated with rice during the dry season in the Philippines: A study using radar, visual observations, aerial netting and ground trapping. Bull. Entomol. Res. 77:145–169. [6,8,9]

Ripper, W.E. 1956. Effect of pesticides on balance of arthropod populations. Annu. Rev. Entomol. 1:403–438. [14]

Risch, S. 1981. Insect herbivore abundance in tropical monocultures and polycultures: An experimental test of two hypotheses. Ecology 62:1325–1340. [3]

Rivera, C. T., S. H. Ou, and T. T. Iida. 1966. Grassy stunt disease of rice and its transmission by the planthopper *Nilaparvata lugens* Stål. Plant Dis. Rep. 50:453–456. [12]

Robinson, J. V. 1981. The effect of architectural variation in habitat on a spider community: an experimental field study. Ecology 62:73–80. [10]

Robinson, R. A. 1976. Plant Pathosystems. Springer-Verlag, Berlin. [17]

Robinson, T. 1979. The evolutionary ecology of alkaloids. Pages 413–448. *In* G.A. Rosenthal and D.H. Janzen, editors. Herbivores: Their Interaction with Secondary Plant Metabolites. Academic Press, New York. [2]

Roderick, G. K. 1987. Ecology and evolution of dispersal in California populations of a salt marsh insect, *Prokelisia marginata*. Ph.D. Dissertation. University of California, Berkeley. [3,4,6,7,10,11,13,14,15]

Roderick, G.K. 1992. Post-colonization evolution of natural enemies. Thomas Say Publ. Entomol. Soc. Am. 1:71–86. [15]

Roderick, G.K., and R.L. Caldwell. 1992. An entomological perspective on animal dispersal. Pages 274–290. *In* N.C. Stenseth and W.Z. Lidicker, editors. Animal Dispersal. Chapman and Hall, London. [15]

Roff, D. A. 1974a. Spatial heterogeneity and the persistence of populations. Oecologia 15:245–258. [11]

Roff, D. A. 1974b. The analysis of a population model demonstrating the importance of dispersal in a heterogeneous environment. Oecologia 15:259–275. [11]

Roff, D.A. 1977. Dispersal in dipterans: Its costs and consequences. J. Anim. Ecol. 46:443–456. [4]

Roff, D.A. 1978. Size and survival in a stochastic environment. Oecologia 36:163–172. [4]

Roff, D.A. 1981. On being the right size. Am. Natur. 118:405–422. [4]

Roff, D.A. 1984. The cost of being able to fly: A study of wing polymorphism in two species of crickets. Oecologia 63:30–37. [4]

Roff, D.A. 1986a. The evolution of wing dimorphism in insects. Evolution 40:1009–1020. [4,7]

Roff, D.A. 1986b. Evolution of wing polymorphism and its impact on life cycle adaptation in insects. Pages 204–221. *In* F. Taylor and R. Karban, editors. The Evolution of Insect Life Cycles, Springer-Verlag, New York. [4,8]

Roff, D.A. 1986c. The genetic basis of wing dimorphism in the sand cricket, *Gryllus firmus* and its relevance to the evolution of wing dimorphisms in insects. Heredity 57:221–231. [4]

Roff, D.A. 1990. Antagonistic pleiotropy and the evolution of wing dimorphism in the sand cricket, *Gryllus firmus*. Heredity 65:169–177. [4]

Roff, D.A. 1991. Life history consequences of bioenergetic and biomechanical constraints on migration. Am. Zool. 31:205–215. [4]

Roff, D.A. 1992. The Evolution of Life Histories: Theory and Analysis. Chapman and Hall, New York. [4]

Roland, J. 1990. Parasitoid Aggregation: Chemical Ecology and Population Dynamics. Pages 185–211. *In* M. Mackauer, L.E. Ehler, and J. Roland, editors. Critical issues in biological control. Intercept, Andover, UK. [10]

Röling, N., and E. van de Fliert. 1991. Beyond the green revolution: capturing efficiencies from integrated agriculture. Indonesia's IPM programme provides a generic case. Informal report. Available from the National Programme for Development and Training of Integrated Pest Management in Rice-Based Cropping Systems, Jl. Ki Mangunsarkoro No. 5, Jakarta 10310, Indonesia. [20]

Romena, A.M., and E.A. Heinrichs. 1989. Wild rices *Oryza* spp. as sources of resistance to rice insects. J. Plant Prot. Trop. 6:13–21. [16]

Romena, A. M., H. R. Rapusas, and E. A. Heinrichs. 1986. Evaluation of rice accessions for resistance to the whitebacked planthopper *Sogatella furcifera* (Horv th) Homoptera: Delphacidae). Crop Prot. 5:334–340. [i,18]

Romstöck-Völkl, M. 1990. Host refuges and spatial patterns of parasitism in an endophytic host-parasitoid system. Ecol. Entomol. 15:321–331. [11]

Root, R. B. 1973. Organization of a plant arthropod association in simple and diverse habitats. The fauna of collards (*Brassica oleracea*). Ecol. Monogr. 43:95–124. [3]

Rose, D.J.W. 1972. Dispersal and quality in populations of *Cicadulina* species (Cicadellidae). J. Anim. Ecol. 41:589–609. [4]

Rosenberg, L.J., and J.I. Magor. 1983a. Flight duration of the brown planthopper, *Nilaparvata lugens* (Homoptera: Delphacidae). Ecol. Entomol. 8:341–350. [4,6,9]

Rosenberg, L.J., and J.I. Magor. 1983b. A technique for examining the long-distance spread of plant virus diseases transmitted by the brown planthopper, *Nilaparvata lugens* (Homoptera: Delphacidae), and other wind-borne insect vectors, Pages 229–238. *In* R.T. Plumb and J.M. Thresh, editors. Plant virus epidemiology, the spread and control of insect-borne diseases. Blackwell Scientific Publications, Oxford, UK. [9]

Rosenberg, L.J., and J.I. Magor. 1984. Flight duration of the brown planthopper. Int. Rice Res. Newsl. 8(2):30. [9]

Rosenberg, L.J., and J.I. Magor. 1986. Modelling the effects of changing windfields on migratory flights of the brown planthopper, *Nilaparvata lugens* Stål. Pages 345–356. *In* G.D. McLean, R.G. Garrett, and W.G. Ruesink, editors. Plant virus epidemics: Monitoring, modelling and predicting outbreaks. Academic Press, Sydney. [9]

Rosenberg, L.J., and J.I. Magor. 1987. Predicting windborne displacements of the brown planthopper, *Nilaparvata lugens* from synoptic weather data. 1. Long-distance displacements in the north-east monsoon. J. Anim. Ecol. 56:39–51. [8,9]

Rosenthal, G.A., and D.H. Janzen. 1979. Herbivores: Their Interaction with Secondary Plant Metabolites. Academic Press, New York. [2]

Ross, H. H. 1957. Principles of natural coexistence indicated by leafhopper populations. Evolution 11:113–129. [13]

Ross, H. H. 1958. Further comments on niches and natural coexistence. Evolution 12:112–113. [13]

Rothschild, G. H. L. 1966. A study of a natural population of *Conomelus anceps* Germar) (Homoptera: Delphacidae) including observations on predation using the precipitin test. J. Anim. Ecol. 35:413–434. [i,10]

Rothschild, G. H. L. 1967. Notes on two hymenopterous egg parasites of Delphacidae (Hem.). Entomol. Mon. Mag. 103:5–9. [13]

Rothschild, M.J. 1973. Secondary plant substances and warning coloration in insects. Pages 59–83. *In* H.G. van Emden, editor. Insect/plant relationships. Symp. Roy. Entomol. Soc. London. Blackwell Scientific Publications, Oxford. [2]

Rothschild, M.J., J. von Euw, and T. Reichstein. 1970. Cardiac glycosides in the Oleander aphid *Aphis nerii*. J. Insect Physiol. 16:1141–1145. [2]

Roush, R.T. 1989. Designing resistance management programs: how can you chose? Pestic. Sci. 26:423–441. [16]

Roush, R.T., and B.E. Tabashnik, editors. 1990. Pesticide Resistance in Arthropods. Chapman and Hall, New York. [16]

Royama, T. 1977. Population persistence and density dependence. Ecol. Monogr. 47:1–35. [11]

Royama, T. 1981. Fundamental concepts and methodology for the analysis of animal population dynamics, with particular reference to univoltine species. Ecol. Monogr. 51:473–493. [11]

Ruesink, W.G. 1976. Status of the systems approach to pest management. Ann. Rev. Entomol. 21:27–44. [19]

Russell, F.C. 1947. The chemical composition and digestibility of fodder shrubs and trees. Pubs. Common. Agric. Bureau 10:185–231. [2]

Safriel, U.N., and U. Ritte. 1980. Criteria for the identification of potential colonizers. Biol. J. Linn. Soc. 13:287–297. [4]

Sahad, K. A., and Y. Hirashima. 1984. Taxonomic studies on the genera *Gonatocerus* Nees and *Anagrus* Haliday of Japan and adjacent regions, with notes on their biology (Hymenoptera, Mymaridae). Bull. Inst. Trop. Agric. Kyushu Univ. 7:1–78. [13]

Sahlberg, J. 1871. Öfversigt af Finlands och den Sksndinaviska halföns Cicadariae. Not. Sallsk. Fauna Flora Fennica Förhandl. 12:1–506. [4]

Sakai, T., and K. Sogawa. 1976. Effects of nutrient compounds on sucking response of the brown planthopper, *Nilaparvata lugens* (Homoptera: Delphacidae). Appl. Entomol. Zool. 11:82–88. [2]

Sakaluk, S. K., and J. J. Belwood. 1984. Gecko phonotaxis to cricket calling song: a case of satellite predation. Anim. Behaviour 32:659–662. [6]

Salim, M., and E.A. Heinrichs. 1986. Impact of varietal resistance in rice and predation on the mortality of *Sogatella furcifera* (Horvath) (Homoptera: Delphacidae). Crop Prot. 5:395–399. [10,18]

Salim, M., and E.A. Heinrichs. 1987. Insecticide-induced changes in the levels of resistance of rice cultivars to the whitebacked planthopper *Sogatella furcifera* (Horvath) (Homoptera: Delphacidae). Crop Prot. 6:28–32. [16]

Salyk, R.P., and D.J. Sullivan. 1982. Comparative feeding behavior of two aphid species: Bean aphid (*Aphis fabae* Scopoli) and pea aphid (*Acyrthosiphon pisum* Harris) (Homoptera: Aphididae) J. New York Entomol. Soc. 90:87–93. [7]

Samal, P., and B.C. Misra. 1975. Spiders, the most effective natural enemies of the brown planthopper in rice. Rice Entomol. Newsl. 3:31. [10]

Samonte, V. 1983. The role and potential of the Filipina in rice pest control. Pages 83–90. *In* Proceedings of the FAO/NCPC Workshop on the Role and Potential of the Filipina in Rice Crop Protection, 2–4 February 1983, Los Baños, Philippines. Available from the FAO Intercountry Programme for Integrated Pest Control in Rice in South and Southeast Asia, Manila, Philippines. [20]

Sasaba, T., and K. Kiritani. 1972. Evaluation of mortality factors with special reference to parasitism of the green rice leafhopper, *Nephotettix cincticeps* Uhler (Hemiptera: Deltocephalidae). Appl. Entomol. Zool. 7:83–93. [11,13]

Sasaba, T., K. Kiritani, and S. Kawahara. 1973. Food preference of *Lycosa* in paddy fields. (in Japanese). Bull. Kochi Inst. Agric. Sci. 5:61–64. [10,14]

Sasamoto, K. 1961. Resistance of the rice plant applied with silicate and nitrogenous fertilizers to the rice stem borer, *Chilo suppressalis* (Walker). Pages 1–73. *In* Proc. Fac. Liberal Arts and Educ. No. 3, Yamanasaki Univ., Japan. [2]

Satomi, H., and H. Itakura. 1970. Synoptic weather conditions at flight of planthoppers at the Ocean Weather Station "Tango" on the Pacific. Abst. Ann. Meet. Jap. Soc. Appl. Entomol. Zool. 3. [9]

Sawicki, R.M., and I. Denholm. 1987. Management of resistance in cotton pests. Trop. Pest Manag. 33:262–272. [16]

Saxena, R.C., and A.A. Barrion. 1985. Biotypes of the brown planthopper *Nilaparvata lugens* (Stål) and strategies in deployment of host plant resistance. Insect Sci. Appl. 6:271–289. [15,16,17,18]

Saxena, R.C., C.G. Demayo, and A.A. Barrion. 1991. Allozyme variation among biotypes of the brown planthopper *Nilaparvata lugens* in the Philippines. Biochem. Gen. 314:115–123. [15]

Saxena, R.C., and H.D. Justo. 1984. Trapping air borne insects aboard inter island ships in the Philippine archipelago with emphasis on the brown planthopper. Int. Rice Res. Newsl. 9(5):16. [9]

Saxena, R. C., and Z. R. Khan. 1989. Factors affecting resistance of rice varieties to planthopper and leafhopper pests. Agric. Zool. Rev. 3:97–132. [18]

Saxena, R.C., S.H. Okech, and N.J. Liquido. 1981. Wing morphism in the brown planthopper, *Nilaparvata lugens*. Insect Sci. Appl. 1:343–348. [4,6,8]

Saxena, R.C., and M.D. Pathak. 1979. Factors governing susceptibility and resistance of certain rice varieties to the brown planthopper. Pages 303–317. *In* Brown Planthopper: Threat to Rice Production in Asia. International Rice Research Institute, Los Baños, Philippines. [2,10]

Saxena, R. C., and L. M. Rueda. 1982. Morphological variations among three biotypes of the brown planthopper *Nilaparvata lugens* in the Philippines. Insect Sci. Appl. 3:193–210. [17,18]

Schaffer, W.M. 1974. Selection for optimal life histories: the effects of age structure. Ecology 55:291–303. [4]

Schlinger, E. I. 1958. Notes on the biology of a mud egg-case making fulgorid, *Hysteropterum beameri* Doering (Homoptera: Fulgoridae). J. Kansas Entomol. Soc. 31:104–106. [1]

Schluter, D. 1988. Character displacement and the adaptive divergence of finches on islands and continents. Am. Natur. 131:799–824. [1]

Schoener, T.W. 1983. Field experiments on interspecific competition. Am. Natur. 122:240–285. [7]

Schoener, T.W. 1988. Ecological interactions. Pages 255–297. *In* A.A. Myers and P.S. Giller, editors. Analytical Biogeography. Chapman and Hall, New York. [7]

Schoener, T., and D.H. Janzen. 1968. Notes on environmental determinants of tropical versus temperate insect size patterns. Am. Natur. 102:207–224. [4]

Schultz, C.A., and J. Meijer. 1978. Migration of leafhoppers (Homoptera: Auchenorrhyncha) into a new polder. Holarctic Ecol. 1:73–78. [4]

Schumacher, F. 1920. Der genwartige Stand unserer Kenntnis von der Homopteren-Fauna der Insel Formosa. Rev. Appl. Biol. 8:85–86. [1]

Scudder, G. G. E. 1963. Studies on the Canadian and Alaska Fulgoromorpha (Hemiptera). I. The genera *Achrotile* (sic) Fieber and *Laccocera* Van Duzee (Delphacidae). Can. Entomol. 95: 167–177. [1]

Sein, F. 1932. Artificial transmission and other studies on sugar cane mosaic. Int. Soc. Sugar Cane Technol. Bull. 84. [1]

Sein, F. 1933. Soil animals and root disease in Puerto Rico. Int. Soc. Sugar Cane Technol. Bull. 91. [1]

Seino, H., Y. Shiotsuki, S. Oya, and Y. Hirai. 1987. Prediction of long distance migration of rice planthoppers to Northern Kyushu considering low-level jet stream. J. Agric. Met. Tokyo. 43:203–208. [9]

Sekido, S., and K. Sogawa. 1976. Effects of salicylic acid on probing and oviposition of the rice plant- and leafhoppers (Homoptera: Delphacidae and Deltocephalinae). Appl. Entomol. Zool. 11:75–81. [2]

Service, P.M, and M. Rose. 1985. Genetic covariance among life-history components: the effects of novel environments. Evolution. 39:943–945. [15]

Settle, W.H., and L.P. Wilson. 1990. Invasion by the variegated leafhopper and biotic interactions: Parasitism, competition and apparent competition. Ecology 71:1461–1470. [11]

Seyedodeslami, H., and B.A. Croft. 1980. Spatial distributions of overwintering eggs of the white apple leafhopper, *Typhlocybga pomaria*, and parasitism by *Anagrus epos*. Environ. Entomol. 9:624–628. [11,13]

Shepard, B. M., E. R. Ferrer, P. E. Kenmore, and J. P.. Sumangil. 1986. Sequential sampling: planthoppers in rice. Crop Prot. 5: 319–322. [18]

Shepard, B. M., E. R. Ferrer, J. Soriano, and P. E. Kenmore. 1989. Presence/absence sampling of planthoppers and major predators in rice. J. Plant Prot. Trop. 6:113–118. [20]

Shepard, B.M., Z.R. Khan, M.D. Pathak, and E.A. Heinrichs. 1990. Management of insect pests of rice in Asia. Pages 255–278. *In* D. Pimentel, editor. CRC Handbook of Pest Management in Agriculture. Second edition. CRC Press, Boca Raton, FL. [10,16,18]

Sheppard, C., P. G. Martin, and F. W. Mead. 1979. A planthopper (Homoptera: Cixiidae) associated with red imported fire ant (Hymenoptera: Formicidae) mounds. J. Georgia Entomol. Soc. 14:140–144. [1]

Shikata, E. 1979. Rice viruses and MLO's and leafhopper vectors, Pages 515–527. *In* K. Maramorosch and K. F. Harris, editors. Leafhopper Vectors and Plant Disease Agents. Academic Press, New York. [12]

Shikata E., and G.E. Galvez. 1969. Fine flexuous threadlike particles in cells of plants and insect hosts infected with rice hoya blanca virus. Virology 39:635–651. [12]

Shimazu, M. 1979. Resting spore formation of *Entomophthora sphaerosperma* Fresenius (Entomophthorales: Entomophthoracea) in the brown planthopper, *Nilaparvata lugens* (Stål) (Hemiptera: Delphacidae). Appl. Entomol. Zool. 14:383–388. [14]

Shinkai, A. 1962. Studies on insect transmissions of rice virus diseases in Japan. Bull. Nat. Inst. Agric. Sci. Ser. C. 14:1. [12]

Silberman, C. E., editor. 1973. The Open Classroom Reader. Vintage Books, New York. [20]

Silvanima, J.V.C., and D.R. Strong. 1991. Is host-plant quality responsible for the population pulses of salt-marsh planthoppers (Homoptera: Delphacidae) in northwestern Florida? Ecol. Entomol. 16:221–232. [2,15]

Simberloff, D. 1981. What makes a good island colonist? Pages 193–205. *In* R. F. Denno and H. Dingle, editors. Insect Life History Patterns: Habitat and Geographic Variation. Springer-Verlag, New York. [3]

Simms, E.L., and M.D. Rausher. 1992. Uses of quantitative genetics for studying the evolution of plant resistance. Pages 42–68. *In* R.S. Fritz and E.L. Simms, editors. Plant Resistance to Herbivores and Pathogens. University of Chicago Press, Chicago. [15,18]

Simons, J. N. 1962. The pseudo-curly top disease in South Florida. J. Econ. Entomol. 55:358–363. [12]

Sinha, R. C. 1960. Comparison of the ability of nymph and adult *Delphacodes pellucida* Fabricius, to transmit European wheat striate mosaic virus. Virology 10:344–352. [12]

Sinha, R. C. 1967. Response of wound-tumor virus infection in insects to vector age and temperature. Virology 31:746–748. [12]

Sinha, R. C. 1973. Viruses and leafhoppers, Pages 491–511. *In* A. J. Gibbs, editor. Viruses and Invertebrates. Elsevier, New York. [12]

Sinha, R. C. 1981. Vertical transmission of plant pathogens, Pages 109–121. *In* J. J. McKelvey, B. F. Eldridge and K. Maramorosch, editors. Vectors of Disease Agents. Praeger, New York. [12]

Sirrine, F. A., and B. B. Fulton. 1914. The cranberry toad-bug. New York Agric. Exp. Sta. Bull. 377:91–112. [1]

Sivapragasam, A., and A. Asma. 1985. Development and reproduction of the mirid bug, *Cyrtorhinus lividipennis* (Heteroptera: Miridae) and its functional response to the brown planthopper. Appl. Entomol. Zool. 20:373–379. [10,14]

Slade, N.A. 1977. Statistical detection of density dependence from a series of sequential censuses. Ecology 58:1094–1102. [7]

Slatkin, M. 1987. Gene flow and the geographic population structure of natural populations. Science 236:787–792. [15]

Slykhuis, J.T., and M.A. Watson. 1958. Striate mosaic of cereals in Europe and its transmission by *Delphacodes pellucida* (Fab.). Ann. Appl. Biol. 46:542–553. [12]

Smith, A.D.M., and D.A. Maelzer. 1986. Aggregation of parasitoids and density independence of parasitism in field populations of the wasp *Aphytis melinus* and its host, the red scale Aonidiella aurantii. Ecol. Entomol. 11:425–434. [10]

Smith, C.M. 1990. Adaptation of biochemical and genetic techniques to the study of plant resistance to insects. Am. Entomol. 36:141–146. [15]

Smith, E. S. C. 1980. *Zophiuma lobulata* Ghauri (Homoptera: Lophopidae) and its relation to the Finschhafen coconut disorder in Papua New Guinea. Papua New Guinea Agric. J. 31: 37–45. [1]

Smith, H.S. 1919. On some phases of insect control by the biological method. J. Econ. Entomol. 12:288–292. [14]

Smith, J. 1976. Influence of crop background on aphids and other phytophagous insects on Brussel sprouts. Ann. Appl. Biol. 83:1–13. [3]

Smith, J., J. A. Litsinger, J. P. Bandong, M. D. Lumaban, and C. G. Dela Cruz. 1989. Economic thresholds for insecticide application to rice: profitability and risk analysis to Filipino farmers. J. Plant Prot. Trop. 6:19–24. [20]

Smith, K. M. 1925. A comparative study of the feeding methods of certain hemiptera and of the resulting effects upon the plant tissue, with special reference to the potato plant. Ann. Biol. Sci. 13:109–139. [12]

Smith, L. M., W. O. Lamp, and E. J. Armbrust. 1992. Potato leafhopper (Homoptera: Cicadellidae) acceptance of alfalfa as a host: The role of non-host stimuli. J. Entomol. Sci. 27:56–64. [3]

Smith, R. F. 1978. Phases in the development of integrated pest control. *In* Proceedings of the Third Pest Control Conference, 2–5 October 1976. Ain Shams University Press, Cairo, Egypt. [20]

Sogawa, K. 1970a. Studies on the feeding habits of the brown planthopper. I. Effects of nitrogen-deficiency of host plant on insect feeding. Japan. J. Appl. Entomol. Zool. 14:101–106. [2]

Sogawa, K. 1970b. Studies on the feeding habits of the brown planthopper. II. Honeydew excretion. Japan. J. Appl. Entomol. Zool. 14:134–139. [2]

Sogawa, K. 1971a. Feeding behavior of the brown leafhopper and varietal resistance of rice to this insect. Pages 195–200. *In* Symposium on Rice Insects. Proceedings of the Symposium on Tropical Agriculture Researches. Serial No. 5. TARC, Tokyo. [2,7]

Sogawa, K. 1971b. Effects of feeding of the brown planthopper on the components in the leaf blade of rice plants. Japan J. Appl. Entomol. Zool. 15:175–179. [2]

Sogawa, K. 1972. Studies on the feeding habits of the brown planthopper. III. Effects of amino acids and other compounds on the sucking response. Japan J. Appl. Entomol. Zool. 16:1–7. [2]

Sogawa, K. 1973. Feeding of the rice plant- and leafhoppers. Rev. Plant Prot. Res. 6:31–41. [2]

Sogawa, K. 1976. Studies on the feeding habits of the brown planthopper, *Nilaparvata lugens* (Stål) (Hemiptera: Delphacidae). V. Probing stimulatory effect of rice flavonoid. Appl. Entomol. Zool. 9:204–213. [2]

Sogawa, K. 1977. Feeding physiology of the brown planthopper. Pages 95–116. *In* The Rice Brown Planthopper. Food and Fertilizer Technology Center for the Asian and Pacific Region. Taipei, Taiwan. [2,15]

Sogawa, K. 1978a. Electrophoretic variations in esterase among biotypes of the brown planthopper. Int. Rice Res. Newsl. 3(5):8–9. [17]

Sogawa, K. 1978b. Quantitative morphological variations among biotypes of the brown planthopper. Int. Rice Res. Newsl. 3(5):9. [17]

Sogawa, K. 1981a. Biotypic variations in the brown planthopper, *Nilaparvata lugens* (Homoptera: Delphacidae) at IRRI, the Philippines. Appl. Ent. Zool. 16:129–137. [15]

Sogawa, K. 1981b. Hybridization experiments on three biotypes of the brown planthopper, *Nilaparvata lugens* (Homoptera: Delphacidae) at IRRI, Philippines. Appl. Entomol. Zool. 16:193–199. [15,16,17,18]

Sogawa, K. 1982. The rice brown planthopper: Feeding physiology and host plant interactions. Annu. Rev. Entomol. 27:49–73. [i,1,2,3,6,7,10,12,15,17]

Sogawa, K., D. Kilin, and A. H. Bhagiawati. 1984a. Characterization of the brown planthopper population on IR42 in North Sumatra, Indonesia. Int. Rice Res. Newsl. 9(1):25. [17]

Sogawa, K., A. Kusumayadi, and J. S. Sitio. 1984b. Monitoring brown planthopper (BPH) biotypes by rice garden in North Sumatra. Int. Rice Res. Newsl. 9(3):15–16. [17]

Sogawa, K., and M.D. Pathak. 1970. Mechanism of brown planthopper resistance in Mudgo variety of rice. Appl. Entomol. Zool. 5:145–158. [2,18]

Sokal, R.R., and R.J. Rohlf. 1981. Biometry. Freeman, San Francisco, CA. [4]

Solbreck, C. 1978. Migration, diapause, and direct development as alternative life histories in a seed bug, *Neacoryphus bicrucis*. Pages 195–217. *In* H. Dingle, editor. The Evolution of Insect Migration and Diapause. Springer-Verlag, New York. [4]

Solow, A. R., and J. H. Steele. 1990. On sample size, statistical power, and the detection of density dependence. J. Anim. Ecol. 59:1073–1076. [11]

Song, Y.H., J.S. Park, J.O. Lee, and S.Y. Choi. 1974. Studies on the resistance of rice to the leaf- and planthoppers. Res. Rep. Off. Rural Development, Korea. 16:11–20. [2]

Sonku, Y., and Y. Sakurai. 1973. Transmission of rice stripe virus by *Laodelphax striatellus* Fallen. 1. Mode of stylet insertion and virus secretion. Ann. Phytopathol. Soc. Japan 39:53–60. [2]

Sonoda, S., M. Muraji, and F. Nakasuji. 1992. The effects of diet combination on the development and fecundity of the semi-aquatic insect, *Microvelia douglasi* (Heteroptera: Veliidae). Appl. Entomol. Zool. 27:162–164. [10]

Soper, R.S. 1985. Pathogens of leafhoppers and planthoppers. Pages 469–488. *In* L.R. Nault and J.G. Rodriguez, editors. The Leafhoppers and Planthoppers. Wiley, New York. [14]

Soper, R. S., G. E. Shewell, and D. Tyrell. 1976. *Cokondamyia auditrix* Nov. sp. (Diptera: Sarcophagidae), a parasite which is attracted by the mating song of its host, *Okanagana rimosa* (Homoptera:Cicadidae). Can. Entomol. 108:61–68. [6]

Southwood, T. R. E. 1960. The abundance of the Hawaiian trees and the number of their associated insect species. Proc. Hawaii Entomol. Soc. 17:299–303. [13]

Southwood, T. R. E. 1961. The number of species of insect associated with various trees. J. Anim. Ecol. 30:1–8. [13]

Southwood, T. R. E. 1962. Migration of terrestrial arthropods in relation to habitat. Biol. Rev. 27:171–214. [3,4,8]

Southwood, T.R.E. 1973. The insect/plant relationship—an evolutionary perspective. Pages 3–30. *In* H.F. van Emden, editor. Insect/plant relationships. Blackwell Scientific Publications, Oxford, UK. [2]

Southwood, T. R. E. 1977a. Habitat, the templet for ecological strategies. J. Anim. Ecol. 46: 337–365. [3,4,8,9]

Southwood, T.R.E. 1977b. The relevance of population dynamics theory to pest status. Pages 35–54. *In* J.M. Cherrett and G.R. Sagar, editors. Origins of Pest, Parasite, Disease and Weed Problems. Blackwell Scientific Publications, Oxford, UK. [14]

Southwood, T. R. E. 1978. The components of diversity. Symp. Roy. Entomol. Soc. Lond. 9:19–40. [3]

Southwood, T. R. E., V. K. Brown, and P. M. Reader. 1979. The relationships of plant and insect diversities in succession. Biol. J. Linn. Soc. 12:327–348. [3]

Southwood, T.R.E., and H.N. Comins. 1976. A synoptic population model. J. Anim. Ecol. 45: 949–965. [19]

Spangler, S. M., and J. A. MacMahon. 1990. Arthropod faunas of monocultures and polycultures in reseeded rangelands. Environ. Entomol. 19:244–250. [1]

Sparks, A.N., R.D. Jackson, J.E. Carpenter, and R.A. Muller. 1986. Insects captured in light traps in the Gulf of Mexico. Ann. Entomol. Soc. Am. 79:132–139. [9]

Speight, M.R. 1983. The potential of ecosystem management for pest control. Agric. Ecosys. Environ. 10:183–199. [14]

Speight, M., and J. Lawton. 1976. The influence of weed-cover on the mortality imposed on artificial prey by predatory ground beetles in cereal fields. Oecologia 23:211–223. [3]

Spiller, N.J. 1990. An ultrastructural study of the stylet pathway of the brown planthopper, *Nilaparvata lugens* Stål. Entomol. Exp. Appl. 54:191–193. [2]

Spiller, N. J., F. M. Kimmins, and M. Llewellyn. 1985. Fine structure of aphid stylet pathways and its use in host plant resistance studies. Entomol. Exp. Appl. 38:293–295. [12]

Srivastava, K. M., B. P. Singh, V. C. Dwadash Shreni, and B. N. Srivastava. 1977. Zinnia yellow net disease-transmission, host range, and agent-vector relationship. Plant Dis. Rep. 61: 550–554. [12]

Stapley, J.H. 1975. The problem of the brown planthopper (*Nilaparvata lugens*) on rice in the Solomon Islands. Rice Entomol. Newsl. 2:37. [10,14]

Stapley, J.H. 1976. The brown planthopper and *Cyrtorhinus* spp. predators in the Solomon Islands. Rice Entomol. Newsl. 4:17. [8,10]

Stapley, J. H., Y. Y. May, and W. G. Golden. 1977. Observation in the Solomon Islands. Rice Entomol. Newsl. 4:15–16. [10]

Stapley, J. H., Y. Y. May-Jackson, and W. G. Golden. 1979. Varietal resistance to the brown planthopper in the Solomon Islands. Pages 233–239. *In* Brown Planthopper: Threat to Rice Production in Asia. International Rice Research Institute, Los Baños, Philippines. [10]

Starks, K. J., R. Muniappan, and R. D. Eikenbary. 1972. Interaction between plant resistance and parasitism against the greenbug on barley and sorghum. Ann. Entomol. Soc. Am. 65:650–655. [18]

Stearns, S.C. 1976. Life-history tactics: A review of the ideas. Quart. Rev. Biol. 51:3–47. [4]

Stearns, S.C. 1977. The evolution of life history traits: A critique of the theory and a review of the data. Ann. Rev. Ecol. Syst. 8:145–171. [4]

Stearns, S.C. 1980. A new view of life-history evolution. Oikos 35:266–281. [4]

Stenseth, N.C., L.R. Kirkendall, and N. Moran. 1985. On the evolution of pseudogamy. Evolution 39:294–307. [5]

Stephan, D. L. 1975. Comparative morphology of the fulgoroid families comprising the Flataria (Homoptera: Auchenorrhyncha). M.S. thesis, North Carolina State University, Raleigh. [1]

Stern, V.M., R.F. Smith, R. van den Bosch, and K.S. Hagen. 1959. The integrated control concept. Hilgardia 29:81–101. [14]

Stickney, F. S., D. F. Barnes, and P. Simmons. 1950. Date palm insects in the United States. U.S. Dept. Agric. Circ. 846. [1]

Stiling, P. D. 1979. Ecological studies on leafhoppers occurring on stinging nettles (*Urtica dioica* L.). Ph.D. dissertation, University College, Cardiff. [13]

Stiling, P. D. 1980a. Host plant specificity, oviposition behaviour and egg parasitism in some leafhoppers of the genus *Eupteryx* (Hemiptera: Cicadellidae). Ecol. Entomol. 5:79–85. [13]

Stiling, P. D. 1980b. Competition and coexistence among *Eupteryx* leafhoppers (Hemiptera: Cicadellidae) occurring on stinging nettles (*Urtica dioica* L.). J. Anim. Ecol. 49:793–805. [3,11,13]

Stiling, P. D. 1987. The frequency of density dependence in insect host-parasitoid systems. Ecology 68:844–856. [10,11,13,14]

Stiling, P. D. 1988. Density-dependent processes and key factors in insect populations. J. Anim. Ecol. 57:581–593. [11,13]

Stiling, P. 1990. Calculating the establishment rates of parasitoids in classical biological control. Am. Entomol. 36:225–230. [14]

Stiling, P. D., B. Brodbeck, and D. R. Strong. 1982. Foliar nitrogen and larval parasitism as determinants of leafminer distribution patterns on *Spartina alterniflora*. Ecol. Entomol. 7:447–452. [13]

Stiling, P. D., B. Brodbeck, and D. R. Strong. 1984. Intraspecific competition on *Hydrellia valida* (Diptera: Ephydridae), a leaf miner of *Spartina alterniflora*. Ecology 65:660–662. [13]

Stiling, P. D., B. Brodbeck, and D. R. Strong. 1991a. Population increases of planthoppers on fertilized salt-marsh cord grass may be prevented by grasshopper feeding. Florida Entomol. 74:88–97. [13]

Stiling, P. D., and D. R. Strong. 1982a. Egg density and the intensity of parasitism in *Prokelisia marginata* (Homoptera: Delphacidae). Ecology 63:1630–1635. [i,10,11,13,14,15]

Stiling, P. D., and D. R. Strong. 1982b. The parasitoids of the planthopper *Prokelisia marginata* (Homoptera: Delphacidae). Florida Entomol. 65:191–192. [8,13]

Stiling, P. D., A. Throckmorton, J. Silvanima, and D. R. Strong. 1991b. Does scale affect the incidence of density dependence? A field test with insect parasitoids. Ecology 72:2143–2154. [i,11,13]

Stiling, P. D., A. Throckmorton, J. Silvanima, and D. R. Strong. 1991c. Biology of and rates of parasitism by nymphal and adult parasites of the salt-marsh-inhabiting planthoppers *Prokelisia marginata* and *P. dolus*. Florida Entomol. 74:81–87. [13]

Stone, R. 1992. Researchers score victory over pesticides—and pests—in Asia. Science 256: 1272–1273. [20]

Storey, H. H. 1928. Transmission studies of maize streak disease. Ann. Appl. Biol. 15:1–25. [12]

Stratton, G. E., and G. W. Uetz. 1981. Acoustic communication and reproductive isolation in two species of wolf spiders. Science 214:575–577. [6]

Stratton, G. E., and G. W. Uetz. 1983. Communication via substratum-coupled stridulation and reproductive isolation in wolf spiders (Araneae: Lycosidae). Anim. Behaviour 31:164–172. [6]

Strickland, E. H. 1940. Additional Fulgoridae taken in Alberta. Can. Entomol. 72:87–88. [1]

Strong, D. R. 1979. Biogeographical dynamics of insect-host plant communities. Annu. Rev. Entomol. 24:89–119. [3,13]

Strong, D.R. 1984. Density-vague ecology and liberal population regulation in insects. Pages 313–327. *In* P.W. Price, C.N. Slobodchikoff, and W.S. Gaud, editors. A New Ecology: Novel Approaches to Interactive Systems. Wiley, New York. [7]

Strong, D.R. 1986. Density-vague population change. TREE 1:39–42. [7]

Strong, D. R. 1988. Parasitoid theory: from aggregation to dispersal. Trends Ecol. Evol. 3:277–280. [11]

Strong, D. R. 1989. Density independence in space and inconsistent temporal relationships for host mortality caused by a fairyfly parasitoid. J. Anim. Ecol. 58:1065–1076. [11,13,15]

Strong, D. R. 1992. Nonequilibrium themes for ecological theory: implications for fungal communities: Its organization and role in the ecosystem. Pages 1–15. *In* G. C. Carroll and D. T. Wicklow, editors. The Fungal Community. Marcel Decker, New York. [11]

Strong, D. R., M. F. Antolin, and S. Rathbun. 1990. Variance and patchiness in rates of population change: A planthopper's case history. Pages 75–90. *In* B. Shorrocks and I. Swingland, editors. Living in a Patchy Environment. Oxford University Press, London. [i,6,7,11]

Strong, D. R., J. H. Lawton, and T. R. E. Southwood. 1984. Insects on plants: Community patterns and mechanisms. Harvard Univ. Press, Cambridge, MA. [1,2,3,7]

Strong, D. R., and P. D. Stiling. 1983. Wing dimorphism changed by experimental density manipulation in a planthopper, *Prokelisia marginata* (Homoptera: Delphacidae). Ecology 64:206–209. [3,11,15]

Stroud, C. P. 1950. A survey of the insects of White Sands National Monument, Tularosa Basin, New Mexico. Am. Midl. Nat. 44:659–677. [1]

Strübing, H. 1958a. Lautausserung—der entscheidende Faktor fur das Zusammenfinden der Geschlechter bei Kleinzikaden (Homoptera-Auchenorrhyncha). Zool. Beitr. N. F. 4:15–21. [5]

Strübing, H. 1958b. Paarungsverhalten und Lautausserung von Kleinzikaden, demonstriert an Beispielen aus der Familie der Delphacidae. Zool. Beitr. IV. 1:8–14. [5]

Strübing, H. 1959. Lautgebung und Paarungsverhalten von Kleinzikaden. Verh. D. Zool. Gesellschaft Munster/Westf. 1959:118–120. [5]

Strübing, H. 1977. Lauterzengung oder Substratvibration als Kommunikationsmittel bei Kleinzikaden. Zool. Beitr. N. F. 24:155–164. [5]

Strübing, H., and A. Hasse. 1975. Ein Beitrag zur Neuen Systematik—demonstriert am Beispiel zweier *Javesella*—Arten (Homoptera—Cicadina: Delphacidae). Zool. Beitr. N.F. 21:517–543. [5,15]

Strübing, H., and T. Rollenhagen. 1988. A new recording system for vibratory signals and its application to different species of the family Delphacidae (Homoptera-Cicadina). Zool. Jb. Physiol. 92:245–268. [5]

Suenaga, H. 1963. Analytical studies on the ecology of the two species of planthoppers, the white back planthopper (*Sogatella furcifera* Horvath) and the brown planthopper (*Nilaparvata lugens* Stål) with special reference to their outbreaks. Bull. Kyushu Agric. Exp. Sta. 8:1–152. [3,4,10]

Suenaga, H., and K. Nakatsuka. 1958. Review on the forecasting of leaf and planthoppers infesting rice. Spec. Rep. Forecasting Dis. Insects. Japanese Ministry of Agriculture Forestry No 1. [16]

Sun, C.N., T.C. Chung, and S.M. Dai. 1984. Insecticide resistance in the brown planthopper, *Nilaparvata lugens* Stål (Homoptera: Delphacidae). Prot. Ecol. 7:167–181. [16]

Sun, C.N., and S.M. Dai. 1984. Brown planthopper (BPH) resistance to a synthetic pyrethroid. Int. Rice Res. Newsl. 9(4):12. [16]

Suzuki, Y., and K. Kiritani. 1974. Reproduction of *Lycosa pseudoannulata* (Boesenberg et Strand) (Araneae: Lycosidae) under different feeding conditions. Japan. J. Entomol. Zool. 18:166–170. [10]

Swain, T. 1979. Tannins and lignins. Pages 657–682. *In* G.A. Rosenthal and D.H. Janzen, editors. Herbivores: Their Interaction with Secondary Plant Metabolites. 1979. Academic Press, New York. [2]

Swaine, G. 1971. Agricultural Zoology in Fiji, Foreign and Commonwealth Office, Overseas Development Administration, Overseas Research Publ. 18. Her Majesty's Stationary Office, London. [14]

Swaminanthan, S., and T.N. Ananthakrishnan. 1984. Population trends of some monophagous and polyphagous fulgoroids in relation to biotic and abiotic factors (Insecta: Homoptera). Proc. Indian Acad. Sci. (Anim. Sci.) 93:1–8. [11,13]

Swanson, B. E., editor. 1990. Report of the Global Consultation on Agricultural Extension, 4–8 December 1989, Rome, Italy. United Nations Food and Agriculture Organization, Rome. [20]

Swanson, B. E., B. J. Farner, and R. Bahal. 1990. The current status of agricultural extension worldwide. Pages 43–75. *In* B. E. Swanson, editor. Report of the Global Consultation on Agricultural Extension, 4–8 December 1989, Rome, Italy. United Nations Food and Agriculture Organization, Rome. [20]

Swezey, O. H. 1904. A preliminary catalogue of the described species of the Family Fulgoridae of North America, north of Mexico. Ohio Dept. Agric. Div. Nurs. Orch. Insp. Bull. 3. [1]

Swezey, O. H. 1906. Observations on the life history of *Oliarus koanoa* Kirkaldy. Proc. Hawaiian Entomol. Soc. 1(2):83–84. [1]

Swezey, O.H. 1936. Biological control of the sugar cane leafhopper in Hawaii. Bull. Exp. Sta. Hawaii. Sugar Plant. Assoc. 21:57–101. [i,4,10,14]

Sylvester, E. S. 1958. Aphid transmission of plant viruses. Proc. 10th Int. Cong. Entomol. 3:195–204. [12]

Sylvester, E. S. 1965. The latent period of pea-enation mosaic virus in the pea aphid, *Acyrthosiphon pisum* (Harris)—an approach to its estimation. Virology 25:62–67. [12]

Sylvester, E. S. 1980. Circulative and propagative virus transmission by aphids. Annu. Rev. Entomol. 25:257–286. [12]

Sylvester, E. S., and R. Osler. 1977. Further studies on the transmission of filaree red-leaf virus by the aphid *Acyrthosiphon pelargonii zerozalphum*. Environ. Entomol. 6:39–42. [12]

Sylvester, E. S., and J. Richardson. 1963. Short-term probing behavior of *Myzus persicae* (Sulzer) as affected by fasting, anesthesia, and labial manipulation. Virology 20:302–309. [12]

Sylvester, E. S., J. Richardson, and N. W. Frazier. 1974. Serial passage of strawberry crinkle virus in the aphid *Chaetosiphon jacobi*. Virology 59:301–306. [12]

Synave, H. 1951. Catalogue des Fulgoroidea de Belgique (Hemiptera-Homoptera). Bull. Ann. Soc. Entomol. Belgique 87:136–140. [1]

Synave, H. 1961. Meenoplidae (Homoptera—Fulgoroidea). Parc Nat. Garamba Fasc. 20:3–37. [1]

Synave, H. 1962a. Lophopidae (Homoptera—Fulgoroidea). Parc Nat. Garamba Fasc. 28(1):3–36. [1]

Synave, H. 1962b. Achilidae (Homoptera—Fulgoroidea). Parc Nat. Garamba Fasc. 33(1):3–37. [1]

Synave, H. 1964. Description de deux nouveaux Hemipteres Homopteres. Bull. I. F. A. N. 26(A):1313–1316. [1]

Synave, H. 1973. Monographie des Derbidae Africains (Homoptera—Fulgoroidea). Etud. Cont. Africain. Fasc. 2:3–223. [1]

Synave, H. 1978. Fulgoroides Africains nouveaux (Homopter). Bull. I. F. A. N. 40(A):587–604. [1]

Tabashnik, B.E. 1989. Managing resistance with multiple pesticide tactics: theory, evidence and recommendations. J. Econ. Entomol. 82:1263–1269. [16]

Tahvanainen, J. O., and R. B. Root. 1972. The influence of vegetation diversity on the population ecology of a specialized herbivore, *Phyllotreta cruciferae* (Coleoptera: Chrysomelidae). Oecologia 10:321–346. [3]

Takagi, M. 1972. Effects of the wing forms of females and their conditions under which adult females were reared on the progeny generation in the brown planthopper. Japan. J. Ecol. 23:118–124. [4,7,8]

Takahashi, S. 1919. Notes on insects injurious to hemp and on *Thyestes gebleri*, Fald. Rev. Appl. Entomol. 7:155–156. [1]

Takara, J., and T. Nishida. 1983. Spatial distribution of the migrants of the corn delphacid *Peregrinus maidis* (Homoptera: Delphacidae) in corn fields. Proc. Hawaii Entomol. Soc. 24:327–334. [8]

Takeda, M. 1974. Mating behavior of the Brown Planthopper, *Nilaparvata lugens* Stål. Japan. J. Appl. Entomol. Zool. 18:43–51. [6]

Tallamy, D. W., and R. F. Denno. 1979. Responses of sap-feeding insects (Homoptera: Hemiptera) to simplification of host plant structure. Environ. Entomol. 8:1021–1028. [1,2,3,10,13,15]

Tallamy, D.W., and R.F. Denno. 1981. Alternative life history strategies in risky environments: An example from lacebugs. Pages 130–147. *In* R.F. Denno and H. Dingle, editors. Insect Life History Patterns: Habitat and Geographic Variation. Springer-Verlag, New York. [4]

Tamaki, G., and W.W. Allen. 1969. Competition and other factors influencing the population dynamics of *Aphis gossypii* and *Macrosiphoniella sanborni* on greenhouse chrysanthemums. Hilgardia 39:447–505. [7]

Tan, H., M. Ruizeng, J. Cheng, and H. Yao. 1984. Relation between landing of brown planthoppers (*Nilaparvata lugens* Stål) in long distance migration and vertical air stream and rain fall. J. Nanjing Agric. Coll. 2:18–25. [9]

Tanaka, S., and H. Shiota. 1970. Latent period of potato leaf roll virus in the green peach aphid (*Myzus persicae* Sulzer). Ann. Phytopathol. Soc. Japan 36:106–111. [12]

Tay, E. B. 1972. Population ecology of *Cicadella viridis* (L.) and bionomics of *Graphocephala coccinea* (Forster) (Homoptera: Cicadellidae). Ph.D. Dissertation, University of London. [11,13]

Taylor, F., and R. Karban. 1986. The Evolution of Insect Life Cycles. Springer-Verlag, New York. [4]

Taylor, L.R. 1974. Insect migration, flight periodicity and the boundary layer. J. Anim. Ecol. 43: 225–238. [9]

Taylor, R.A.J. 1985. Migratory behavior in the Auchenorrhyncha. Pages 259–288. *In* L.R. Nault and J.G. Rodriguez, editors. The Leafhoppers and Planthoppers. Wiley, New York. [4,7,9]

Taylor, R.A.J., and D. Reling. 1986. Preferred wind direction of long-distance leafhopper (*Empoasca fabae*) migrants and its relevance to the return migration of small insects. J. Anim. Ecol. 55:1103–1114. [9]

Taylor, T.H.C. 1955. Biological control of insect pests. Ann. Appl. Biol. 42:190–196. [14]

Teng, P. S. 1990. Integrated pest management in rice: an analysis of the status quo with recommendations for action. Dept. of Plant Pathology, IRRI, Manila, Philippines. [20]

Terry, F.W. 1905. Leaf-hoppers and their natural enemies. (Forficulidae, Syrphidae and Hemerobiidae). Exp. Sta. Hawaiian. Sug. Plant. Assoc. Entomol. Ser. Bull. 1, Pt. 5, pp. 163–181. [10]

Teulon, D. A. J., and D.R. Penman. 1986. Temporal distribution of Froggatt's apple leafhopper (*Typhlocyba froggatti* Baker) and the parasite *Anagrus armatus* (Ashmead) in an abandoned orchard. New Zealand J. Zool. 13:93–100. [11,13]

Thomas, M.J., K. Balakrishna Pilai, K.V. Mammen, and N.R. Nair. 1979. Spiders check planthopper population. IRRN 4:18–19. [10]

Thompson, C.R., J.C. Nickerson, and F.W. Mead. 1979. Nymphal habit of *Oliarus vicarius* (Homoptera: Cixiidae), and possible association with *Aphaenogaster* and *Paratrechina* (Hymenoptera: Formicidae). Psyche 86:321–325. [1]

Thompson, J.N. 1988a. Evolutionary ecology of the relationship between oviposition preference and performance of offspring in phytophagous insects. Entomol. Exp. Appl. 47:3–14. [1,2,15]

Thompson, J.N. 1988b. Coevolution and alternative hypotheses on insect/plant interactions. Ecology 69:893–895. [1]

Thompson, J.N., W. Wehling, and R. Podolski. 1990. Evolutionary genetics of host use in swallowtail butterflies. Nature 344:148–150. [15]

Thorarinsson, K. 1990. Biological control of the cottony-cushion scale: Experimental tests of the spatial density-dependence hypothesis. Ecology 71:635–644. [10]

Thornbury, D. W., G. M. Hellmann, R. E. Rhoads, and T. P. Pirone. 1985. Purification and characterization of potyvirus helper component. Virology 144:260–267. [12]

Thornhill, R. 1978. Some arthropod predators and parasites of adult scorpionflies (Mecoptera). Environ. Entomol. 7:714–716. [6]

Thornhill, R. 1979. Male and female sexual selection and the evolution of mating strategies in insects. Pages 81–121. *In* M.S. Blum and N. A. Blum, editors. Sexual Selection and Reproductive Competition in Insects. Academic Press, New York. [6]

Thornhill, R., and Alcock, J. 1983. The Evolution of Insect Mating Systems. Harvard Univ. Press. Cambridge, MA. [6]

Throckmorton, A. E. 1989. The effects of tidal inundation, spider predation, and dispersal on the population dynamics of *Prokelisia marginata*. Ph. D. dissertation, The Florida State University, Tallahassee. [10]

Tingey, W.M. 1985. Plant defensive mechanisms against leafhoppers. Pages 217–234. *In* L.R. Nault and J.G. Rodriguez, editors. The Leafhoppers and Planthoppers. Wiley, New York. [2]

Tjallingii, W.F. 1978. Electronic recording of penetration behavior by aphids. Entomol. Exp. Appl. 24:521–530. [2]

Tonkyn, D.W., and R.F. Whitcomb. 1987. Feeding strategies and the guild concept among vascular feeding insects and microorganisms. Pages 179–199. *In* K.F. Harris, editor. Current Topics in Vector Research, Vol. 4, Springer-Verlag, New York. [2]

Travers, S. E., and A. Sih. 1991. The influence of starvation and predators on the mating behavior of a semiaquatic insect. Ecology 72:2123–2134. [6]

Trexler, J. C., C.E. McCulloch, and J. Travis. 1988. How can the functional response best be determined? Oecologia 76:206–214. [11]

Trivers, R. L. 1972. Parental investment and sexual selection. Pages 136–179. *In* B. G. Campbell, editor. Sexual Selection and the Descent of Man. Aldine, Chicago. [6]

Triwidodo, H. 1992. Personal communication on research results in West Java. Bogor Institute of Agriculture. College of Agriculture. Department of Entomology. Bogor. West Java. Indonesia. [17]

Trujillo, G. E., J. M. Acosta, and A. Pinero. 1974. A new corn virus disease found in Venezuela. Plant Dis. Rep. 58:122–126. [12]

Tsai, J.H., and O.H. Kirsch. 1978. Bionomics of *Haplaxius crudus* (Homoptera: Cixiidae). Environ. Entomol. 7:305–308. [1]

Tsai, J.H., and S.W. Wilson. 1986. Biology of *Peregrinus maidis* with descriptions of immature stages (Homoptera: Delphacidae). Ann. Entomol. Soc. Am. 79:395–401. [4,10]

Tsai, J.H. N.L. Woodiel, and O.H. Kirsch. 1976. Rearing techniques for *Haplaxius crudus* (Homoptera: Cixiidae). Florida Entomol. 59:41–43. [1]

Tsai, J. H., and T. A. Zitter. 1982. Characteristics of maize stripe virus transmission by the corn Delphacid. J. Econ. Entomol. 75:397–400. [12]

Tsai, P., F. Hwang, W. Feng, Y. Fu, and Q. Dong. 1964. Study on *Delphacodes striatella* Fallen (Homoptera: Delphacidae) in north China. Acta Entomol. Sinica 13:552–571. [4]

Tsaur, S-C. 1989a. Additions to the fanua [*sic*] of Taiwanese Meenoplidae. J. Taiwan Mus. 42:25–29. [1]

Tsaur, S-C. 1989b. Two flatid nymphs from Taiwan (Homoptera: Fulgoroidea). J. Taiwan Mus. 42:31–35. [1]

Tsaur, S-C., T-C. Hsu, and J. Van Stalle. 1988a. Cixiidae of Taiwan, Part (I). Pentastirini. J. Taiwan Mus. 41:35–74. [1]

Tsaur, S-C., T-C. Hsu, and J. Van Stalle. 1988b. Cixiidae of Taiwan, Part (V). Cixiini except *Cixius*. J. Taiwan Mus. 44:1–78. [1]

Tsaur, S-C., C-T. Yang, and M. R. Wilson. 1986. Meenoplidae of Taiwan (Homoptera: Fulgoroidea). Monogr. Taiwan Mus. No. 6:81–118. [1]

Tsurumachi, M., and K. Yasuda. 1989. Long-distance immigration and infestation of the rice planthoppers and leaffolder in Okinawa. Research Reports on Tropical Agriculture, Tropical Agricultural Research Centre, Ministry of Agriculture, Forestry and Fisheries, Japan. 66:50–54. [9]

Turnbull, A.L. 1962. Quantitative studies of the food of *Linyphia triangularis* Clerck (Araneae: Linyphiidae). Can. Entomol. 94:1233–1249. [10]

Uetz, G. W. 1991. Habitat structure and spider foraging. Pages 325–348. *In* E. D. McCoy, S. S. Bell, and H. R. Mushinsky, editors. Habitat Structure and Diversity. Chapman and Hall, London. [10]

UNDP. 1991. Programme Advisory Note. Agricultural Extension. United Nations Development Programme Technical Advisory Division, Bureau for Programme Policy and Evaluation, New York. [20]

Usinger, R.L. 1939. Distribution and host relationships of *Cyrtorhinus* (Hemiptera: Miridae). Proc. Hawaiian Entomol. Soc. 10:271–273. [10]

van de Fliert, E. 1991. Two IPM Farmers Field Schools, a case study in kabupaten Grobogan. Internal report. National Programme for Development and Training of Integrated Pest Management in Rice-Based Cropping Systems, Indonesia. [20]

van de Fliert, E. 1992. Risk and reward of IPM implementation and good farmers and IPM farmers. Internal report. National Programme for Development and Training of Integrated Pest Management in Rice-Based Cropping Systems, Indonesia. [20]

van de Fliert, E., and P. C. Matteson. 1989. Integrated pest control channels for extension in Sri Lanka. J. Ext. Syst. 5:33–47. [20]

van de Fliert, E., and P. C. Matteson. 1990. Rice integrated pest control training needs identified through a farmer survey in Sri Lanka. J. Plant Prot. Trop. 7:15–26. [20]

van der Broek, L. J., and C. C. Gill. 1980. The median latent periods for three isolates of barley yellow dwarf virus in aphid vectors. Phytopathology 70:644–646. [12]

van der Plank, J. E. 1978. Genetic and Molecular Basis of Plant Pathogenesis. Springer-Verlag, Berlin. [17]

Van Driesche, R. G. 1983. The meaning of "percent parasitism" in studies of insect parasitoids. Environ. Entomol. 12:1611–1622. [11,13]

Van Driesche, R. G., T.S. Bellows, J. S. Elkington, J. R. Gould, and D.N. Ferro. 1991. The meaning of percentage parasitism revisited: solutions to the problem of accurately estimating total losses from parasitism. Environ. Entomol. 20:1–7. [11]

Van Duzee, E. P. 1889. Hemiptera from Muskoka Lake District. Can. Entomol. 21:1–11. [1]

Van Duzee, E. P. 1894. A list of the Hemiptera of Buffalo and vicinity. Bull. Buffalo Soc. Nat. Sci. 5:167–204. [1]

Van Duzee, E. P. 1905. List of Hemiptera taken in the Adirondack Mountains. New York State Mus. Sci. Serv. Bull. 97:546–556. [1]

Van Duzee, E. P. 1907. Notes on Jamaican Hemiptera: A report on a collection of Hemiptera made on the island of Jamaica in the spring of 1906. Bull. Buffalo Soc. Nat. Sci. 8:3–79. [1]

Van Duzee, E. P. 1914. A preliminary list of the Hemiptera of San Diego County, California. Trans. San Diego Soc. Nat. Hist. 2:1–57. [1]

Van Duzee, E. P. 1916. Notes on some Hemiptera taken near Lake Tahoe, California. Univ. California Publ. Tech. Bull. Entomol. 1:229–249. [1]

Van Duzee, E. P. 1923. Expedition of the California Academy of Sciences to the Gulf of California in 1921. The Hemiptera (True bugs, etc.). Proc. California Acad. Sci. 4th Ser. 12(11):123–200. [1]

van Emden, H. F. 1966. Plant insect relationships and pest control. World Rev. Pest Contr. 5:115–123. [18]

Van Etten, C.H., and H.L. Tookey. 1979. Chemistry and biological effects of glucosinolates. Pages 471–500. *In* G.A. Rosenthal and D.H. Janzen, editors. Herbivores: Their Interaction with Secondary Plant Metabolites. Academic Press, New York. [2]

van Lenteren, J.C. 1986. Parasitoids in the greenhouse: Success with seasonal inoculative release systems. Pages 341–374. *In* J. Waage and D. Greathead, editors. Insect parasitoids. Academic Press, London. [10]

Van Stalle, J. 1986a. Revision of Afrotropical Pentastirini (Homoptera, Cixiidae) II: The genus *Pentastiridius* Kirschbaum, 1868. Bull. Ann. Soc. Roy. Belge Entomol. 122:81–105. [1]

Van Stalle, J. 1986b. Revision of Afrotropical Pentastirini (Homoptera, Cixiidae) III: *Norialsus* gen. nov. J. Entomol. Soc. South Africa 49:197–230. [1]

Van Stalle, J. 1986c. *Helenolius*, a cixiid bug genus endemic to St Helena (Insecta: Homoptera). J. Nat. Hist. 20:273–278. [1]

Van Stalle, J. 1986d. A review of the genus *Proutista* Kirkaldy (Homoptera, Derbidae) in New Guinea. Indo-Malayan Zool. 3:87–96. [1]

Van Stalle, J. 1987a. Revision of Afrotropical Pentastirini (Homoptera, Cixiidae) V: The genus Oliarus Stål, 1862. Mus. Roy. Afrique Cent., Tervuren, Belgique, Sci. Zool. 252:1–173. [1]

Van Stalle, J. 1987b. A revision of the Neotropical species of the genus *Mnemosyne* Stål, 1866 (Homoptera, Cixiidae). Bull. Inst. Roy. Sci. Nat. Belgique Entomol. 57:121–139. [1]

Van Stalle, J. 1988. A revision of the Afrotropical species of the genus *Cixius* Latreille, 1804 (Homoptera, Cixiidae). J. African Zool. 102:223–252. [1]

Van Zwaluwenburg, R. H. 1917. Insects affecting coffee in Porto Rico. J. Econ. Entomol. 10:515–517. [1]

Varley, G.C., and G.R. Gradwell. 1963. Predatory insects as density dependent mortality factors. Proc. XVI Int. Cong. Zool. Washington DC 1:240. [7]

Varley, G. C., and G.R. Gradwell. 1968. Population models for the winter moth. Pages 132–142. *In* T. R. E. Southwood, editor. Insect Abundance. Blackwell Scientific Publications, Oxford, UK. [11]

Varley, G.C., G.R. Gradwell, and M.P. Hassell. 1973. Insect population ecology, an analytical approach. Blackwell Scientific Publications, Oxford, UK. [7]

Varma, P. M. 1963. Transmission of plant viruses by whiteflies. Bull. Nat. Inst. Sci. India 24:11–33. [12]

Vehrencamp, S. L., and J. W. Bradbury 1984. Mating systems and ecology. Pages 279–298. *In* J. R. Krebs and N. B. Davies, editors. Behavioural ecology. Blackwell Scientific Publications, Oxford, UK. [6]

Vepsäläinen, K. 1978. Wing dimorphism and diapause in *Gerris*: Determination and adaptive significance. Pages 218–253. *In* H. Dingle, editor. Evolution of Insect Migration and Diapause. Springer-Verlag, New York. [4]

Verma, J.S. 1955. Biological studies to explain the failure of *Cyrtorhinus mundulus* (Breddin) as an egg-predator of *Peregrinus maidis* (Ashmead) in Hawaii. Proc. Hawaiian Entomol. Soc. 15:623–634. [10,14]

Verma, J.S. 1956. Effects of demeton and schradan on *Peregrinus maidis* (Ashm.) and its egg-predator, *Cyrtorhinus mundulus* (Breddin). J. Econ. Entomol. 49:58–63. [10]

Via, S. 1984a. The quantitative genetics of polyphagy in an insect herbivore. I. Genotype-environment interaction in larval performance on different host plant species. Evolution 38:881–895. [15]

Via, S. 1984b. The quantitative genetics of polyphagy in an insect herbivore. II. Genetic correlations in larval performance within and among host plants. Evolution 38:986–905. [15]

Via, S. 1990. Ecological genetic and host adaptation in herbivorous insects: The experimental study of evolution in natural and agricultural systems. Annu. Rev. Entomol. 35:421–446. [15,17,18]

Vickery, W.L., and T.D. Nudds. 1984. Detection of density-dependent effects in annual duck censuses. Ecology 65:96–104. [7]

Viktorov, G. A. 1968. The influence of the population density upon the sex ratio of *Trissolcus grandis* Thoms. (Hymenoptera: Scelionidae). Zool. Zh. 47:1035–1039. [11]

Vilbaste, J. 1965. On the Altaic Cicadine fauna. Inst. Zool. Bot., Acad. Sci. Estonian SSR, Tartu, 156 pp. [1]

Vilbaste, J. 1968. On the Cicadine fauna of the Primosk region. Inst. Zool. Bot., Acad. Sci. Estonian SSR, Valgus Publ. Tallin, 195 pp. [1]

Vilbaste, J. 1971. Eesti tirdid (Homoptera: Cicadinea). Eesti Nsv. Tead. Akad., Zool. and Bot. Institute, Kirjastus Valgus, Tallinn, 284 pp. [1]

Vilbaste, J. 1974. Preliminary list of Homoptera—Cicadinea of Latvia and Lithuania. Eesti Nsv Tead. Akad. Toimetised 23:131–162. [1]

Villiers, A. 1977. Atlas des Hemipteres. Societe Nouvelle des Ed. Boubee & Co., Paris. [1]

Vince, S.W., I. Valiela, and J.M. Teal. 1981. An experimental study of the structure of herbivorous insect communities in a salt marsh. Ecology 62:1662–1678. [2,6,10]

Vinson, S. B. 1976. Host selection by insect parasitoids. Annu. Rev. Entomol. 21:109–133. [11]

Vinson, S. B. 1981. Habitat location. Pages 51–77. *In* D. A. Nordlund, R. L. Jones, and W. J. Lewis, editors. Semiochemicals: Their Role in Pest Control. Wiley, New York. [11]

Vinson, S. B., and P. Barbosa. 1987. Interrelationships of nutritional ecology of parasitoids. Pages 673–695. *In* F. Slansky, Jr. and J. G. Rodriguez, editors. Nutritional Ecology of Insects, Mites, Spiders, and Related Invertebrates. Wiley, New York. [18]

Voss, G. 1983. The biochemistry of insecticide resistant hopper races. Pages 351–357. *In* Proc. Int'l. Workshop on Biotaxonomy, Classification and Biology of Leafhoppers and Planthoppers (Auchenorryncha) of economic importance. Commonwealth Institute of Entomology, London. [16]

Waage, J. 1990. Ecological theory and the selection of biological control agents. Pages 135–157. *In* M. Mackauer, L.E. Ehler, and J. Roland, editors. Critical issues in biological control. Intercept, Andover, UK. [10,14]

Waage, J., and M.P. Hassell. 1982. Parasitoids as biological control agents—a fundamental approach. Parasitology 84:241–268. [10]

Wada, T., H. Seino, Y. Ogawa, and T. Nakasuga. 1987. Evidence of autumn overseas migration in the rice planthoppers, *Nilaparvata lugens* and *Sogatella furcifera*: analysis of light trap catches and associated weather patterns. Ecol. Entomol. 12:321–330. [9]

Wade, M. J. 1979. Sexual selection and variance in reproductive success. Am. Natur. 114:742–747. [6]

Wade, M. J. 1980. Effective population size: the effects of sex, genotype, and density on the mean and variance of offspring numbers in the flour beetle, *Tribolium castaneum*. Genet. Res. 36:1–10. [6]

Wade, M., and S. J. Arnold. 1980. The intensity of sexual selection in relation to male sexual behavior, female choice, and sperm precedence. Anim. Behaviour 28:446–461. [6]

Wagner, W.L., D.R. Herbst, and S.H. Sohmer. 1990. Manual of the flowering plants of Hawaii. Vols 1 and 2. Bishop Museum Special Publication 83. University of Hawaii Press and Bishop Museum Press, Honolulu. [3,4]

Waibel, H. 1986. The Economics of Integrated Pest Control in Irrigated Rice. Springer-Verlag, Berlin. [20]

Waibel, H. 1987. Farmers' practices and recommended economic threshold levels in irrigated rice in the Philippines. Pages 191–197. *In* J. Tait and B. Napompeth, editors. Management of Pests and Pesticides: Farmers' Perceptions and Practices. Westview Press, Boulder, CO. [20]

Walde, S. J., R.F. Luck, D.S. Yu, and W.W. Murdoch. 1989. A refuge for red scale: the role of size-selectivity by a parasitoid wasp. Ecology 70:1700–1706. [11]

Walde, S. J., and W.W. Murdoch. 1988. Spatial density dependence in parasitoids. Annu. Rev. Entomol. 33:441–466. [10,11,13,14]

Walker, I. 1979. Some British species of *Anagrus* (Hymenoptera: Mymaridae). Zool. J. Linn. Soc. 67:181–202. [11]

Walker, G. P. 1988. The role of leaf cuticle in leaf age preference by bayberry whitefly (Homoptera: Aleyrodidae) on lemon. Ann. Entomol. Soc. Am. 81:365–369. [12]

Waloff, N. 1973. Dispersal by flight of leafhoppers (Auchenorrhyncha: Homoptera). J. Appl. Ecol. 10:705–730. [4,7,9,10]

Waloff, N. 1975. The parasitoids of the nymphal and adult stages of leafhoppers (Auchenorrhyncha: Homoptera). Trans. Roy. Entomol. Soc. Lond. 126:637–686. [11,13]

Waloff, N. 1979. Partitioning of resources by grassland leafhoppers (Auchenorrhyncha: Homoptera). Ecol. Entomol. 4:134–140. [i,3,7,13]

Waloff, N. 1980. Studies on grassland leafhoppers (Auchenorrhyncha, Homoptera) and their natural enemies. Adv. Ecol. Res. 11:81–215. [i,1,2,3,4,6,7,8,9,10,14]

Waloff, N. 1981. The life history and descriptions of *Halictophagus silwoodensis* sp. n. (Strepsiptera) and its host *Ulopa reticulata* (Cicadellidae) in Britain. Syst. Entomol. 6:103–113. [13]

Waloff, N. 1983. Absence of wing polymorphism in the arboreal, phytophagous species of some taxa of temperate Hemiptera: An hypothesis. Ecol. Entomol. 8:229–232. [3,4]

Waloff, N,. and M.A. Jervis. 1987. Communities of parasitoids associated with leafhoppers and planthoppers in Europe. Adv. Ecol. Res. 17:281–402. [i,11,13]

Waloff, N., and M. G. Solomon. 1973. Leafhoppers (Auchenorrhyncha: Homoptera) of acidic grassland. J. Appl. Ecol. 10:189–212. [1,2]

Waloff, N., and P. Thompson. 1980. Census data and analysis of populations of some leafhoppers (Auchenorrhyncha, Homoptera) of acidic grassland. J. Anim. Ecol. 49:395–416. [4,10,11,13,14]

Wang, H., J. Zhou, and G. Liu. 1982. Studies on the biology of the pseudoannulate wolf spider, *Lycosa pseudoannulata* Boes. et Str. Acta Entomol. Sinica 28:69–79. [10]

Wang, S.C., T.Y. Ku, and Y.I. Chu. 1988. Resistance patterns in brown planthopper *Nilaparvata lugens* (Homoptera: Delphacidae) after selection with four insecticides and their combinations. Plant Prot. Bull. Taiwan 30(1):59–67. [16]

Ward, C. R., C. W. O'Brien, L. B. O'Brien, D. E. Foster, and E. W. Huddleston. 1977. Annotated checklist of New World insects associated with *Prosopis* (mesquite). U.S. Dept. Agric. Tech. Bull. 1557. [1]

Wasserman, S.D., and D.J. Futuyma. 1981. Evolution of host plant utilization in laboratory populations of the southern cowpea weevil, *Callosobruchus maculatus* Fabricius (Coleoptera: Bruchidae). Evolution 35:606–617. [1,2]

Watanabe, N. 1967. The density effect on the appearance of two wing forms in the brown planthopper, *Nilaparvata lugens* and smaller brown planthopper, *Laodelphax striatellus*. Japan. J. Appl. Entomol. Zool. 11:57–61. [8]

Waterhouse, D.F., and K.R. Norris. 1987. Biological Control: Pacific Prospects. Inkata Press, Melbourne. [14]

Watson, M. A., and R. C. Sinha. 1959. Studies on the transmission of European wheat striate mosaic virus by *Delphacodes pellucida* Fabricius. Virology 8:139–163. [12]

Way, M.J., and M. Cammell. 1970. Aggregation behavior in relation to food utilization by aphids. Pages 229–247. *In* A. Watson, editor. Animal populations in relation to their food resources. Blackwell Scientific Publications, Oxford, UK. [2,7]

Weaving, A. J. S. 1980. Observations on *Hilda patruelis* Stål (Homoptera: Tettigometridae) and its infestation of the groundnut crop in Rhodesia. J. Entomol. Soc. South Africa 43:151–167. [1]

Weis, A. E. 1983. Patterns of parasitism by *Torymus capite* on hosts distributed in small patches. J. Anim. Ecol. 52:867–877. [11]

Weis, A. E., and W. G. Abrahamson. 1985. Potential selective pressures by parasitoids on a plant-herbivore interaction. Ecology 66:1261–1269. [11]

Weis-Fogh, T. 1952. Fat combustion and metabolic rate of flying locusts (*Schistocerca gregaria* Forskal). Phil. Trans. Roy. Soc. 237B:1–36. [9]

Wellhouse, W. H. 1920. Wild hawthorns as hosts of apple, pear and quince pests. J. Econ. Entomol. 13:388–391. [1]

Wellings, P.W., and A.F.G. Dixon. 1987. The role of weather and natural enemies in determining aphid outbreaks. Pages 313–346. *In* P. Barbosa and J.C. Schultz, editors. Insect Outbreaks. Academic Press, New York. [10]

West-Eberhardt, M.J. 1983. Sexual selection, social competition, and speciation. Quart. Rev. Biol. 58:155–183. [5,6]

Whalley, P. E. S. 1956. On the identity of species of *Anagrus* (Hym., Mymaridae) bred from leaf-hopper eggs. Entomol. Mon. Mag. 92:147–149. [13]

Whalon, M.E., H.E. van de Baan, and K. Untung. 1990. Resistance management of brown planthopper, *Nilaparvata lugens*, in Indonesia. Pages 155–165. *In* B.T. Grayson et al., editors. Pest Management in Rice. Elsevier Applied Science, London. [16,18]

Wheeler, A. G., Jr., and E. R. Hoebeke. 1982. Host plants and nymphal descriptions of *Acanalonia pumila* and *Cyarda* sp. near *acutissima* (Homoptera, Fulgoroidea: Acanaloniidae and Flatidae). Florida Entomol. 65:340–349. [1]

Wheeler, A. G. Jr., and S. W. Wilson. 1987. Life history of the issid planthopper *Thionia elliptica* (Homoptera: Fulgoroidea) with description of a new *Thionia* species from Texas. J. New York Entomol. Soc. 95:440–451. [1]

Wheeler, A. G., Jr., and S. W. Wilson. 1988. Notes on the biology and immatures of the issid planthoppers *Thionia bulltata* and *T. simplex* (Homoptera: Fulgoroidea). J. New York Entomol. Soc. 96:266–273. [1]

Whitcomb, R. F. 1972. Transmission of viruses and mycoplasmas by auchenorrhynchous Homoptera, Pages 168–203. *In* C. I. Kado and H. O. Agrawal, editors. Principals and Techniques in Plant Virology. Van Nostrand Reinhold, New York. [12]

Whitcomb, R. F., and L. M. Black 1969. Demonstration of exvectorial wound tumor virus. Annu. Rev. Phytopathol. 7: 86–87. [12]

Whitham, T.G. 1978. Habitat selection by *Pemphigus* aphids in response to resource limitation and competition. Ecology 59:1164–1176. [7]

Whitham, T.G. 1979. Territorial behavior of *Pemphigus* gall aphids. Nature 279:324–325. [7]

Whitmore, R.W., K.P. Pruess, and J.T. Nichols. 1981. Leafhopper and planthopper populations on eight irrigated grasses grown for livestock forage. Envrion. Entomol. 10:114–118. [15]

Whittaker, J.B. 1969a. Quantitative and habitat studies of the froghoppers and leafhoppers (Homoptera, Auchenorrhyncha) of Wytham Woods. Berkshire. Entomol. Mon. Mag. 105: 27–37. [2]

Whittaker, J. B. 1969b. The biology of Pipunculidae (Diptera) parasitizing some British Cercopidae (Homoptera). Proc. Roy. Entomol. Soc. Lond. 44:17–24. [13]

Whittaker, J. B. 1971. Population changes in *Neophilaenus lineatus* (L.) (Homoptera: Cercopidae) in different parts of its range. J. Anim. Ecol. 40:425–433. [11]

Whittaker, J. B. 1973. Density regulation in a population of *Philaenus spumarius* (L.) (Homoptera: Cercopidae). J. Anim. Ecol. 42:163–172. [11]

Whitten, M. J., L. R. Brownhall, K. G. Eveleens, W. Heneveld, M. A. R. Khan, S. Li, Z. Mihyu, K. Sogawa, and S. M. Yassin. 1990. Mid-term review of FAO Intercountry Programme for the Development and Application of Integrated Pest Control in Rice in South and Southeast Asia, Phase II. Mission report, November 1990. Available from the FAO Inter-country Programme for Integrated Pest Control in Rice in South and Southeast Asia, Manila, Philippines. [20]

Wicklund, C. 1975. The evolutionary relationship between adult oviposition preferences and larval host plant range in *Papilio machaon* L. Oecologia 18:185–197. [2]

Wickremasinghe, N., and Y. Elikawela. 1982. Pesticide usage and induction of resistance in pests and vectors. Paper presented at the International Workshop on Resistance to Insecticides in Public Health and Agriculture, Colombo, Sri Lanka. 22–26 February 1982. [16]

Wilhoit, L. R. 1992. Evolution of Herbivore Virulence to Plant Resistance: Influence of Variety Mixtures. Pages 91–119. *In* R. S. Fritz and E. L. Simms, editors. Plant Resistance to Herbivores and Pathogens, Ecology, Evolution and Genetics. University of Chicago Press, Chicago. [18]

Williams, C.B. 1964. Patterns in the Balance of Nature. Academic Press, New York. [3]

Williams, D. W. 1984. Ecology of the blackberry-leafhopper-parasite system and its relevance to California grape agroecosystems. Hilgardia 52:1–33. [11,14]

Williams, F.X. 1931. The insects and other invertebrates of Hawaiian sugarcane fields. Hawaiian Sugar Planters, Assoc. Expt. Sta. Honolulu, Hawaii. [10]

Williams, F.X. 1932. Handbook of the Insects and other Invertebrates in Hawaiian Sugar Cane Fields. Hawaiian Sugar Planters' Assoc. Exp. Sta., Honolulu. [14]

Williams, J.R. 1957. The sugar-cane Delphacidae and their natural enemies in Mauritius. Trans. Roy. Entomol. Soc. Lond. 109:65–110. [2,4,7,10]

Williams, J. R. 1975. The Cixiidae (Fulgoroidea: Homoptera) of Mauritius. J. Nat. Hist. 9:169–204. [1]

Williams, J. R., and R. G. Fennah. 1980. Ricaniidae (Hemiptera: Fulgoroidea) from Mauritius, with a description of *Trysanor cicatricosus* spec. nov., gen. nov. J. Entomol. South Africa 43:7–22. [1]

Williams, J. R., and D. J. Williams. 1988. Homoptera of the Mascarene Islands—an annotated catalogue. Entomol. Mem. Dept. Agric. Wat. Suppl. Rep. S. Africa No. 72. [1]

Willis, H. L. 1982. Collection of *Euklastus harti* Metcalf in Wisconsin (Homoptera: Derbidae). Entomol. News 93:51–53. [1]

Wilson, M. R. 1984. *Numicia gauhati* sp. n. (Homoptera: Fulgoroidea: Tropiduchidae) from rice in India, with taxonomic notes on other Indian *Numicia* species. Bull. Entomol. Res. 74:469–472. [1]

Wilson, M. R. 1986. An Indian tropiduchid planthopper *Tambinia verticalis* Distant (Hemiptera: Fulgoroidea) breeding on coconut in Zanzibar. Bull. Entomol. Res. 76:385–388. [1]

Wilson, M. R. 1987. The Auchenorrhyncha (Homoptera) associated with palms. Proc. 2nd Inter. Workshop on Leafhoppers and Planthoppers of Economic Importance, pp. 327–342. [1]

Wilson, M. R. 1989. The planthopper family Achilixiidae (Homoptera, Fulgoroidea): A synopsis with a revision of the genus *Achilixius*. Syst. Entomol. 14:487–506. [1]

Wilson, M.R., and M.F. Claridge. 1985. The leafhopper and planthopper fauna of rice fields. Pages 381–404. *In* L.R. Nault and J.G. Rodriguez, editors. The Leafhoppers and Planthoppers. Wiley, New York. [2,3,7,13]

Wilson, M. R., and M. F. Claridge. 1991. Handbook for the identification of leafhoppers and planthoppers of rice. CAB International, Wallingford, Oxford, UK. [1,18]

Wilson, M. R., and D. J. Hilburn. 1991. Annotated list of the Auchenorrhynchous Homoptera (Insecta) of Bermuda. Ann. Entomol. Soc. Am. 84:412–419. [1]

Wilson, S. W. 1982a. The planthopper genus *Prokelisia* in the United States (Homoptera: Fulgoroidea: Delphacidae). J. Kansas Entomol. Soc. 55:532–546. [1,4,10]

Wilson, S. W. 1982b. Description of the fifth instar of *Apache degeerii* (Homoptera: Fulgoroidea: Derbidae). Great Lakes Entomol. 15:35–36. [1]

Wilson, S. W. 1985. Description of the immature stages of *Delphacodes bellicosa* (Homoptera: Fulgoroidea: Delphacidae). Pan-Pacific. Entomol. 61:72–78. [1]

Wilson, S. W., and J. E. McPherson. 1979. The first record of *Megamelus palaetus* in Illinois (Homoptera: Fulgoroidea: Delphacidae). Great Lakes Entomol. 12:227. [1]

Wilson, S. W., and J. E. McPherson. 1980. A list of the host plants of the Illinois Acanaloniidae and Flatidae (Homoptera: Fulgoroidea). Trans. Illinois Acad. Sci. 73(4):21–29. [1]

Wilson, S. W., and J. E. McPherson. 1981a. Life history of *Megamelus davisi* with descriptions of immature stages. Ann. Entomol. Soc. Am. 74:345–350. [1,4]

Wilson, S. W., and J. E. McPherson. 1981b. Notes on the biology of *Nersia florens* (Homoptera: Fulgoroidea: Dictyopharidae) with descriptions of eggs, and first, second, and fifth instars. Great Lakes Entomol. 14:45–48. [1]

Wilson, S. W., and L. B. O'Brien. 1986. Descriptions of nymphs of *Itzalana submaculata* Schmidt (Homoptera: Fulgoridae), a species new to the United States. Great Lakes Entomol.19:101–105. [1]

Wilson, S. W., and L. B. O'Brien. 1987. A survey of planthopper pests of economically important plants (Homoptera: Fulgoroidea). Pages 343–360. *In* M.R. Wilson and L.R. Nault, editors. Proc. 2nd Inter. Workshop Biotax., Class., Biol., Leafhoppers and Planthoppers (Auchenorrhyncha) of economic importance. Commonwealth Institute of Entomology, London. [1,3,15]

Wilson, S. W., and J. H. Tsai. 1984. *Ormenaria rufifascia* (Homoptera: Fulgoroidea: Flatidae): Descriptions of nymphal instars and notes on field biology. J. New York Entomol. Soc. 92:307–315. [1]

Wilson, S. W., and A. G. Wheeler, Jr. 1984. *Pelitropis rotulata* (Homoptera: Tropiduchidae): Host plants and descriptions of nymphs. Florida Entomol. 67:164–168. [1]

Wilson, S. W., and A. J. Wheeler, Jr. 1986. *Pentagramma longistylata* (Homoptera: Delphacidae): Descriptions of immature stages. J. New York Entomol. Soc. 94:126–133. [1]

Wilson, S. W., and A. G. Wheeler, Jr. 1992. Host plant and descriptions of nymphs of the planthopper *Rhabdocephala brunnea* (Homoptera: Fulgoridae). Ann. Entomol. Soc. Amer. 85:258–264. [1]

Wink, M., T. Hartman, L. Witte, and J. Rheinheimer. 1982. Interrelationship between quinolizidine alkaloid producing legumes and infesting insects: Exploitation of the alkaloid-containing phloem sap of *Cytisus scoparius* by the broom aphid, *Aphis cytisorum*. Z. Naturforsch. 37:1081–1086. [2]

Wirtner, P. M. 1905. A preliminary list of the Hemiptera of western Pennsylvania. Ann. Carnegie Mus. 3:183–228. [1]

Withcombe, C. L. 1926. Studies on the aetiology of sugar-cane froghopper blight in Trinidad. Ann. Biol. Sci. 13:64–108. [12]

Witsack, W. 1971. Experimentell-ökologische Untersuchungen über Dormanz- Formen von Zikaden (Homoptera-Auchenorrhyncha). I. Zur Form und Induktion der Embryonaldormanz von *Muellerianella brevipennis* (Boheman) (Delphacidae). Zool. Jb. Syst. 98:316–340. [4]

Wood, T.K. 1984. Life history of tropical membracids (Homoptera: Membracidae. Sociobiology 8:299–344. [7]

Wood, T.K., and S.I. Guttman. 1982. The ecological and behavioral basis for reproductive isolation in the sympatric *Enchenopa binotata* complex (Homoptera: Membracidae). Evolution 36:233–242. [15]

Woodhead, S., and E.A. Bernays. 1977. Changes in the release rates of cyanide in relation to palatability of *Sorghum* to insects. Nature 270:235–236. [2]

Woodhead, S., and D. E. Padgham. 1988. The effect of plant surface characteristics on resistance of rice to the brown planthopper, *Nilaparvata lugens*. Entomol. Exp. Appl. 47:15–22. [1,2,18]

Woodward, T. E., J. W. Evans, and V. F. Eastopo. 1970. Hemiptera. *In* C. S. I. R. O. The insects of Australia. Melbourne Univ. Press, Carlton, Australia. [1]

Wootton, R. J. 1981. Palaeozoic insects. Annu. Rev. Entomol. 26:319–344. [1]

Wray, D. L. 1950. Insects of North Carolina. Second Supplement. North Carolina Dept. Agric. [1]

Wright, S. 1932. Evolution in Mendelian population. Genetics. 16(2):97–159. [17]

Wright, S. 1968. Evolution and the genetics of populations, Vol. 1: Genetic and biometric foundations. Chicago Univ. Press, Chicago. [15]

Wu, J.T., E.A. Heinrichs, and F.G. Medrano. 1986. Resistance of wild rices, *Oryza* spp., to the brown planthopper, *Nilaparvata lugens* (Homoptera: Delphacidae). Environ. Entomol. 15: 648–653. [10]

Wu, L., and H. Wang. 1987. Functional response of a predator, *Pirata subpiraticus* (Araneae: Lycosidae) to its prey. Chinese J. Biol. Contr. 3:7–10. [10]

Wu, Z.W., Y.L. Xia , Y. Li, Q.P. Yang, and Q.Z. Wang. 1987. Preliminary studies on field distribution pattern of *Sogatella furcifera* and sequential sampling techniques. Insect Knowledge 24(4):196–199. [8]

Wylie, H. G. 1965. Some factors that reduce the reproductive rate of *Nasonia vitripennis* (Walk.) at high adult population densities. Can. Entomol. 97:970–977. [11]

Yadav, R. P., and J. P. Chaudhary. 1984. Laboratory studies on the biology of *Ooencyrtus papilionis* Ashmead (Hymenoptera: Encyrtidae), an egg-parasitoid of the sugarcane leafhopper (*Pyrilla perpusilla* Walker). J. Entomol. Res. 8:162–166. [13]

Yamashina, T. 1974. Rice insect pest resistance to insecticides in Japan [in Japanese]. Pestic. Sci. 2:91–106. [16]

Yang, C. T. 1989. Delphacidae of Taiwan (II) (Homoptera: Fulgoroidea). Nat. Sci. Council (Rep. China) Spec. Publ. 6:1–334. [1]

Yang, J., and C. Yang. 1986. Delphacidae of Taiwan (I). Asiracinae and the Tribe Tropidocephalini (Homoptera: Fulgoroidea). Taiwan Mus. Spec. Publ. Ser. 6:1–79. [1]

Yang, J. T., C. T. Yang, and M. R. Wilson. 1989. Tropiduchidae of Taiwan (Homoptera: Fulgoroidea). Taiwan Mus. Spec. Publ. Ser. No. 8:65–115. [1]

Yasumatsu, K., and T. Torii. 1968. Impact of parasites, predators, and diseases on rice pests. Annu. Rev. Entomol. 13:295–324. [10]

Yasumatsu, K., T. Wongsiri, S, Navavichit, and C. Tirawat. 1975. Approaches toward an integrated control of rice pests. Part I. Survey of natural enemies of important rice pests in Thailand. Plant Prot. Serv. Tech. Bull. No. 24, Dept. Agric., Minist. Agric. Thai and UNDP 9/FAO THA 68/526. [10,14]

Ye, Z.Z., S.M. He, L.Q. Lu, H.G. Zheng, B.C. Fen, and J. Yao. 1981. Observations on migration in *Nilaparvata lugens* Stål. Kunchong Zhishi 18:97–100. [9]

Yoshihara, T., K. Sogawa, M.D. Pathak, B. Juliano, and S. Sakamura. 1979a. Soluble silicic acid as a sucking inhibitory substance on rice against the brown planthopper (Delphacidae, Homoptera). Entomol. Exp. Appl. 26:314–322. [2]

Yoshihara, T., K. Sogawa, M.D. Pathak, and R. Villareal. 1979b. Comparison of oxalic acid concentration in rice varieties resistant and susceptible to the brown planthopper. Int. Rice Res. Newsl. 4:10–11. [2]

Yoshihara, T., K. Sogawa, M.D. Pathak, B. Juliano, and S. Sakamura. 1980. Oxalic acid as a sucking inhibitor of the brown planthopper in rice (Delphacidae, Homoptera). Entomol. Exp. Appl. 27:49–155. [2]

Zeigler, H. 1975. Nature of transported substances. Pages 59–100. *In* M.H. Zimmerman and J.A. Milburn, editors. Transport in Plants. I. Phloem Transport. Springer-Verlag, Berlin. [2]

Zera, A.Z., and K.C. Tiebel. 1989. Differences in juvenile hormone esterase activity between presumptive longwinged and shortwinged *Gryllus rubens*: implications for the hormonal control of winglength polymorphism. J. Insect Physiol. 35:7–17. [4]

Zettler, F. W. 1967. A comparison of species of Aphididae with species of three other aphid families regarding virus transmission and probe behavior. Phytopathology 57:398–400. [12]

Zhang, S. X., G. Z. Shou, and X. H. Shu. 1982. Study on the efficiency of virus transmission by the vector *Laodelphax striatellus* for wheat rosette stunt. Sci. Agric. Sinica 1982 (5):78–82. [12]

Zhang, Z. 1983. A study on the development of wing dimorphism in the rice brown plant-hopper *Nilaparvata lugens* Stål. Acta Entomol. Sinica 26:260–267. [4]

Zhang, Z.G., J.S. Wu, Y.F. Jin, and Q.M. Xia. 1983. Studies on the feeding behavior of *Nilaparvata lugens* (Stål). Insect Knowledge 20:49–52. (In Chinese). [19]

Zhang, Z.T., and L.Y. Chen. 1987. Analysis of courtship signals of planthoppers (*Nilaparvata lugens* and *Sogatella furcifera*) and leafhopper (*Nephotettix cincticeps*). Kexue Tongbao (Sci. Bull.) 20:1583–1586. [i,5]

Zhang, Z.T., W.Z. Korg, J.L. Gao, and T.H. Shao. 1988. Acoustic signal-producing organ of brown planthopper. Int. Rice Res. Newsl. 13:38–39. [5]

Zhu, S.Y., C.Z. Wu, J.Y. Du, X.Q. Huang, Y.H. Shua, and F.L. Hong. 1982. A summary report on the long-distance migration of the brown planthopper. Guangdong Agric. Sci. 4:22–24. [9]

Zimmerman, E. C. 1948. Homoptera: Auchenorhyncha. Insects of Hawaii 4:1–268. [1,2,3,4, 10,14]

Zwölfer, H. 1971. The structure and effect of parasite complexes attacking phytophagous host insects. Pages 405–418. *In* P.J. den Boer and G.R. Gradwell, editors. Dynamics of Populations. Center for Agricultural Publishing and Documentation, Wageningen. [1]

Zwölfer, H. 1978. Mechanismen and Ergebnisse der Co-evolution von phytophagen und entomophagen Insekten und hoheren Pflanzen. Sonderbd. Naturwiss. Ver. Hambg. 2:7–50. [1]

Taxonomic Index

Subject Index